# Encyclopedic Dictionary
# of Hydrogeology

# Encyclopedic Dictionary of Hydrogeology

**D.J. Poehls** and **Gregory J. Smith**

ELSEVIER

Amsterdam • Boston • Heidelberg • London • New York • Oxford
Paris • San Diego • San Francisco • Singapore • Sydney • Tokyo
Academic press is an imprint of Elsevier

Academic Press is an imprint of Elsevier
30 Corporate Drive, Suite 400, Burlington, MA 01803, USA
525 B Street, Suite 1900, San Diego, CA 92101-4495, USA
32 Jamestown Road, London NW1 7BY, UK
Radarweg 29, PO Box 211, 1000 AE Amsterdam, The Netherlands

First edition 2009

Notice
No responsibility is assumed by the publisher for any injury and/or damage to persons
or property as a matter of products liability, negligence or otherwise, or from any use
or operation of any methods, products, instructions or ideas contained in the material
herein. Because of rapid advances in the medical sciences, in particular, independent
verification of diagnoses and drug dosages should be made

**Library of Congress Cataloging-in-Publication Data**
A catalog record for this book is available from the Library of Congress

**British Library Cataloguing in Publication Data**
A catalogue record for this book is available from the British Library

ISBN: 978-0-12-558690-0

For information on all Academic Press publications
visit our web site at books.elsevier.com

Printed and bound in the USA

09  10     10  9  8  7  6  5  4  3  2  1

Working together to grow
libraries in developing countries

www.elsevier.com | www.bookaid.org | www.sabre.org

ELSEVIER    BOOK AID
            International    Sabre Foundation

# Contents

# Preface

The proliferation of environmental regulations and subsequent concerns for soil and groundwater pollution has helped the field of hydrogeology evolve from its geologic roots and move from its early hydraulic affiliations with water supply and geotechnical concerns into a more complex discipline. The study of water as an environmental science has drawn from a number of disciplines, including advanced mathematics and computer modeling to water well drilling.

As practicing hydrogeologists, we have found ourselves referring to a myriad of sources during our professional activities for consistent definitions. Further, in this search, we have also discovered an evolution in the definition of terms. In response, we have compiled an extensive list of terms typically encountered in the field of hydrogeology.

We have endeavored to be as complete as possible without becoming "a series." This necessitates excluding various subject words that, although related, are not directly needed by the practicing hydrogeologist or is better covered in a separate book.

Often a word will have more than one meaning, depending on usage, or may have a shaded meaning. We have defined the word only in its relation to the practice of hydrogeology. Users are referred to the American Geological Institute *Dictionary of Geological Terms* and *Glossary of Hydrology* or the *Dictionary of Science and Technology* for further nuances of meaning and for geological words not covered in this text.

Instead of adhering to simple definitions, we expanded them to include content pertinent to the understanding of the term or phrase; hence the title *Encyclopedic Dictionary of Hydrogeology*. All entries open with the term's basic definition and occasionally a brief history relating to the term which may enhance understanding or provide a frame of reference in which the concept will have greater meaning. Multiple or differing interpretations of a term are subordinated as additional sub-entries. Contradictory meanings are stated explicitly, and the preferred usage is so indicated.

This reference is designed to complement textbooks on hydrogeology and serve as a desk reference for the practicing geologist and hydrogeologist or any person endeavoring to understand aspects of hydrogeology. Measurements are usually in Système Internationale (S.I.) but occasionally English units have been used or included if in common use. Detailed conversion tables are incorporated as an appendix.

Words that are italicized within the text are also defined and should be reviewed for a clearer understanding of a concept. At the end of an entry, there may be entries following Cf: cross reference. When an entry is cross referenced, it would either be a related word or possibly its antonym that could help clarify or expand the understanding of the entry. The cross referenced word is defined in the text only if it is italicized. If not italicized, we suggest using another reference book if our definition is still unclear or if the reader needs more advanced understanding.

We have tried to include all terms in common use and some that have been "shelved" to make way for modern terminologies. Unavoidably, there are terms that should have been included and were not and the best explanation is that we just did not think of it. If you find errors, omissions, or other suggestions that would enhance or complete this work in future revisions, we would like to hear from you.

We have relied extensively on other works to ensure accuracy, and as these are definitions it can be expected that phraseologies have been duplicated. With the exception of figures that have been directly copied with permission from other works, we have not referenced each work within the text as this would have led to a multi-volume set.

# Aa

**Abandoned well** A production or *monitoring well* that is either in bad repair or no longer in use and is therefore permanently removed from service. Well abandonment usually fulfills a procedure set by the regulating authority. Cf. *Sealing abandoned wells.*

**Abiotic degradation** The breakdown of a chemical in groundwater that is a result of non-biological action. Abiotic reactions may be in the form of, for example *hydrolysis* or oxidation. An example of hydrolysis reaction where 1,1,1-trichloroethane is hydrolyzed to form vinylidene chloride then vinyl chloride and finally ethene is illustrated below:

$$CH_3CCl_3 \rightarrow CH_2{:}CCl_2 \rightarrow CH_2{:}CHCl \rightarrow CH_2{:}CH_2$$

**Abnormal pore pressure** A subsurface *fluid pressure* that is significantly different from the *hydrostatic pressure* for a given depth. Cf. *Pore pressure.*

**ABS casing** See *Acrylonitrile Butadiene Styrene casing (ABS).*

**Absolute humidity** The ratio of the mass of water vapor in a sample of air to the volume of the sample. Cf. *Humidity; Relative humidity.*

**Absolute ownership rule** See *rule of capture.*

**Absolute permeability** See *Permeability; Permeability, absolute.*

**Absolute viscosity/dynamic viscosity ($\mu$)** The measure of the resistance of a fluid to the shearing that is necessary for the fluid to *flow* and is independent of the medium through which it flows. Water that is resistant to relative motion, which is a *Newtonian fluid*, is proportional to the fluid property of *viscosity.* Absolute viscosity and density can be combined into a physical parameter called *kinematic viscosity, v:*

$$\nu = \frac{\mu}{\rho}$$

where:
$\mu =$ absolute viscosity $[M{\cdot}T^{-1}{\cdot}L^2]$
$\rho =$ density $[M{\cdot}L^{-3}]$

Absolute (dynamic) and kinematic viscosity decrease as molecular motion increases with increasing temperature.

**Absorbed water** The water that enters the ground surface and is mechanically held within the soil. Cf. *Adsorbed water.*

**Absorbing well** See *Drainage well; Relief well.*

**Absorption** The natural assimilation or incorporation of fluids into *interstices*, i.e. liquids in solids. In absorption, the dissolved molecules are incorporated within the structure of the solid (such as soil or a rock mass) and the fluid is held mechanically. The absorbed water includes, but is not limited to, gravity flow of water from *streams* or other earth openings, and movement of *atmospheric water.* Cf. *Absorbed water; Adsorption; Infiltration; Percolation.*

**Absorption loss** The volume of water lost through mechanical incorporation of water into surface and subsurface materials, e.g. rock and soil. Absorption loss is typically an important parameter during the initial filling of a reservoir or other means of impounding water, such as a dam.

**Abstraction** The merging of two or more subparallel *streams* into a single stream course as a result of competition between adjacent, *consequent stream* paths, e.g. gullies and ravines. Abstraction is the simplest type of *stream capture*, in which the stream having the more rapid erosive action drains water from the competing stream. Water abstraction is that part of *precipitation* that does not become *direct runoff*, but is *transpiration*, stored, *evaporation*, or *absorbed.*

**Accident** An interruption in a watercourse, e.g. river, that interferes with, or sometimes stops, the normal development of the *river system.*

**Accordant junction** In watercourses, having surface elevations at the same level at their place of junction, as with two accordant streams. Cf. *Discordant junction.*

**Acid mine water/acid mine drainage** Water containing free sulfuric acid ($H_2SO_4$) due to the weathering of iron pyrites exposed to oxygen during mining operations. The oxidation of sulfides produces $H_2SO_4$ and sulfate salts, causing the water that drains from mine and the remaining solid waste to have an acidic pH typically between 2.0 and 4.5, as demonstrated by the equation:

$$FeS_{2(s)} + \frac{7}{2}O_{2(s)} + H_2O \rightarrow Fe^{2+} + 2H^+ + 2SO_4^{2-}$$

Surface and underground mines are generally extended below the *water table*, requiring *dewatering* of the mining operation; water drained (low-quality drainage) or pumped from the mine excavation may be highly mineralized. Characteristics of acid mine water include high iron, aluminum, and sulfate content. *Leaching* of old mine tailings and settling ponds can also lead to groundwater impact.

**Acid rain** *Precipitation* that becomes a weak *carbonic acid* ($H_2CO_3$) by dissolving atmospheric carbon dioxide ($CO_2$) as it falls. The burning of coal, oil, wood, and other fossil fuels releases additional $CO_2$ into the atmosphere while consuming oxygen. As the precipitation falls toward the earth, the water droplets adsorb excess $CO_2$ and become more acidic, with a pH of 5.0 or less. Nitric and sulfuric acids ($HNO_3$ and $H_2SO_4$) from the addition of $SO_x$ and $NO_x$ gases also lower the pH of precipitation. Biological activity on land surfaces and in water bodies can be seriously affected in areas where the bedrock or soils do not naturally buffer or neutralize acid rain. Cf. *Adsorbtion.*

**Acoustic Doppler systems** A downhole measuring device used to measure *groundwater* flow and direction. Acoustic waves are induced into the groundwater in a *well*, and the instrument utilizes the "Doppler Effect" to determine direction and velocity of groundwater movement.

**Acoustic log** Often called a sonic tool. A geophysical instrument designed to measure the time a pulsed compressional sound wave takes to travel 0.3 m (1 ft; the interval transit time). The tool consists of a transmitter emitting sound pulses at a frequency of about 23,000 Hz and a receiver located several feet away to time the arrival of the bursts. **Figure A-1** shows a typical acoustic log tool.

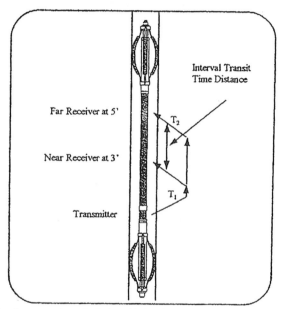

**Figure A-1** A schematic of an acoustic sonic device with borehole compensators to maintain an equal distance of transmitter and receiver in the borehole. (Welenco, 1996. Reprinted with permission from Welenco Inc.)

Cf. *Acoustic logging; Acoustic well logging.*

**Acoustic streaming** Unidirectional *flow* currents in a fluid that are due to the presence of sound waves. Cf. *Acoustic well logging.*

**Acoustic televiewer (ATV) logging** The use of a device that projects and receives sound waves to provide high-resolution information on the location and character of *secondary porosity* traits such as fractures and solution openings. ATV logging has applications in determining the strike and dip of planar features. The ATV probe, also called a borehole televiewer or variable density log, is a rotating 1.3-MHz transducer that is both the transmitter and the receiver of high-frequency acoustic energy, which is reflected from the borehole wall and does not penetrate the formation. Each time the probe rotates past magnetic north, it is signaled by a trigger pulse activating the sweep on an oscilloscope that represents a 360° scan of the borehole wall. The brightness of the oscilloscope trace is proportional to the amplitude of the reflected acoustic signal, so openings in the wall appear as dark areas. **Figure A-2** is an ATV log of a fracture zone in a producing geothermal well at Roosevelt Hot Springs, Utah.

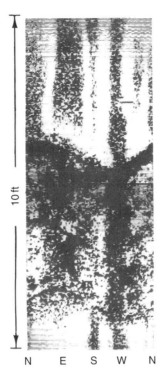

10 ft

N  E  S  W  N

**Figure A-2** The acoustic televiewer printout depicting a producing zone in a geothermal well. (Keys, 1989. Reprinted with permission from the National Water Well Association.)

Cf. *Acoustic well logging.*

**Acoustic velocity logging** Commonly called sonic logging or transit-time logging. A method for estimating the location where *groundwater flow* may be concentrated in semi-consolidated or consolidated rocks by determining the formation's relative *porosity.* Acoustic velocity logging can also be used to locate the top of the *static water level* in deep holes and to detect *perched water* tables. The interval transit time for a formation depends on the elastic properties of the formation, which are related to lithology and porosity. In general, waves travel faster through denser formations. Therefore, an increasing travel time or a slower traveling wave for a given material indicates increased porosity. Travel time, $\Delta t$, measured by acoustic velocity logging, is the interval transit time measured in microseconds ($\mu$s). The time-average equation, also called the Wylie equation, relates sonic travel time to formation porosity, $\eta$:

$$\eta = \frac{\Delta t - \Delta t_{ma}}{\Delta t_f - \Delta t_{ma}}$$

where:
$\Delta t$ = change in formation travel time from log [T]
$\Delta t_{ma}$ = change in matrix travel time [T]
$\Delta t_f$ = change in fluid travel time [T]

An acoustic velocity probe is centralized within the borehole so that the travel path to and from the rock is consistent. The borehole must contain fluid to transmit the pulse to the borehole wall. The interval from transmission to reception of the acoustic pulse is the time the pulse takes to travel laterally through the drilling fluid to the borehole wall, through the rock formation, and back to the receiver. The radius of investigation of the probe is approximately three times the wavelength, or the velocity divided by the transmitter frequency. A lower transmitter frequency increases the area of the investigation but decreases the resolution of small features such as fractures. Velocities and transit times for some common types of rocks and fluids are presented in **Table A-1.**

**Table A-1**  Rock and fluid compression-wave velocity and transit time (single values are averages). (Keys, 1989. Reprinted with permission from National Water Well Association.)

| Rock or fluid type | Velocity (feet per second) | Transit time (microseconds per foot) |
|---|---|---|
| Sandstone | | |
| Slightly consolidated | 15,000–17,000 | 58.8–66.7 |
| Consolidated | 19,000 | 52.6 |
| Shale | 6,000–16,000 | 62.5–167.0 |
| Limestone | 19,000–21,000+ | 47.6–52.6 |
| Dolomite | 21,000–24,000 | 42.0–47.6 |
| Anhydrite | 20,000 | 50 |
| Granite | 19,000–20,000 | 50.0–52.5 |
| Gabbro | 23,600 | 42.4 |
| Freshwater | 5,000 | 200 |
| Brine | 5,300 | 189 |

Cf. *Acoustic log.*

**Acoustic waveform logging**  A method of obtaining information on the lithology and structure of a formation by analyzing the amplitude changes of acoustic signals. Acoustic waveform logs have been used to estimate the vertical compressibility of artesian *aquifers*. These compressibility values are used to plot the effects that changes in net stress have on the *storage coefficient* of the aquifer. Acoustic waveform data are also needed to accurately interpret *cement bond logs*, used extensively during well installation. The acoustic waveforms are recorded digitally and displayed on an oscilloscope, or a variable–density log or *acoustic televiewer log (ATV)* can be made. In photographs, troughs in the waveform produce dark bands on the log and peaks produce light bands. Analysis of the various components of acoustic signals yields considerable information on lithology and structure, depending on interpretation. Acoustic waveform logs have not been used extensively in groundwater hydrology; however, potential hydrologic applications include prediction of the subsidence and fracturing characteristics of rocks. Cf. *Acoustic well logging.*

**Acoustic well logging**  The determination of the physical properties of a *borehole* by the emission and analysis of sound waves. Acoustic logging uses a transducer to transmit an acoustic wave through the fluid in the borehole and into the surrounding medium. Four popular types of acoustic logging are as follows: *acoustic velocity* (sonic), *acoustic waveform*, *cement bond*, and *acoustic televiewer*. The log types are differentiated by the frequencies transmitted, the signal recorded, and the purpose of the log. All acoustic logs require fluid in the borehole to couple the signal to the surrounding wall. Acoustic logs provide data on *porosity*, lithology, cement, and the location and character of fractures. Cf. *Acoustic log.*

**Acre-foot**  The volume of water required to cover 1 acre to a depth of 1 ft; hence, 1 acre-foot (acre-ft) is equivalent to 43,560 cubic feet (ft$^3$), or $3.259 \times 10^5$ gal. (1233.5 m$^3$). The acre-foot is a convenient unit for measuring irrigation water, *runoff volume*, and reservoir capacity. For other conversions, see: **Appendix A-1**.

**Acre-foot per day**  The unit volume rate of water *flow* or volume per unit time flow. For conversions, see: **Appendix A-1**.

**Acre-inch**  The volume of water required to cover 1 acre to a depth of 1 in.; hence, 1 acre-inch (acre-in.) is equivalent to 3630 ft$^3$ (102.8 m$^3$). The acre-inch is a unit-volume measurement for water flow. For other conversions, see: **Appendix A-1**.

**Acrylonitrile butadiene styrene (ABS) casing**  A thermoplastic casing that is resistant to *corrosion* and acid treatment, light weight, relatively inexpensive, and easy to install. ABS casing is desirable for many installations where high strength is not required, as in *monitoring wells*, which are typically not in service for an extended period of time. Standardization of thermoplastic *well casing* is described in ASTM (American Society for Testing and Materials) Standard F-480, "Thermoplastic Water Well Casing, Pipe and Couplings Made in Standard Dimension Ratios (SDR)." Cf. *Polyvinyl-chloride casing (PVC)*.

**Activated carbon**  Also called activated charcoal. A powdered, granular, or pelleted form of amorphous carbon used in water conditioning as an adsorbent for organic matter and certain dissolved gasses. In the environmental industry, activated carbon is used to adsorb organic contaminants. Granular-activated carbon (GAC), usually produced by roasting cellulose-based substances such as wood or coconut shells in the absence of air, is most often used in treatment systems for contaminated groundwater extracted from the subsurface. Various organic contaminants are physically and/or chemically affixed to the GAC when water is passed through the carbon bed. The water is usually in contact with the GAC for about 10 min. In time, the GAC is depleted and the contaminant breaks through. Just before *breakthrough*, the GAC is replaced. Alternatively, the GAC can be regenerated by heating the material unit until the contaminants are driven off. The ability of GAC to remove organic contaminants results from its enormous available surface area. A vast network of pores is generated inside each carbon granule when the coal is crushed and heated. One pound (0.5 kg) of carbon granules has an effective surface area equal to 100–135 acres (40.5–54.6 hectares). The ability of activated carbon to effectively remove a particular organic material, the approximate capacity of the carbon for the application, and the estimated carbon dosage required can be determined from the appropriate *adsorption isotherm* test. The *Freundlich* *isotherm* and the *Langmuir isotherm* are used to represent the adsorption equilibrium. When the adsorbed material in GAC is in equilibrium with the influent concentration, the GAC is loaded to capacity, and that portion of the bed is exhausted. Reactivation of the carbon restores its ability to adsorb contaminants, although 5–10% of the GAC may be lost during regeneration. Cf. *Air stripping; Aeration.*

**Activation** The process of treating *bentonitic clay* with acid to improve its adsorptive properties or to enhance its bleaching action.

**Activation log** See *neutron-activation log.*

**Activity coefficient ($\gamma_i$)** A correction factor used when applying the law of mass action to chemical reactions occurring in natural waters. The driving force of a chemical reaction is related to the concentration of the reactants and the concentration of the products; the law of mass action expresses this relationship when the reaction is at equilibrium and, therefore, is useful in the interpretation of chemical interactions between *groundwater,* and activity coefficient, $\gamma_i$, is used as a correction factor to convert measured ion concentrations to effective concentrations or activities, and the calculation of activities makes it possible to apply the law of mass action to natural waters. To obtain values for the activity coefficient of common inorganic constituents, the graphical relation to ionic strength, $I$, can be used, as shown in **Figure A-3**. Mathematical corrections for *dissolved solids* effects are based on the *Debye–Huckel equation* or Davies equation.

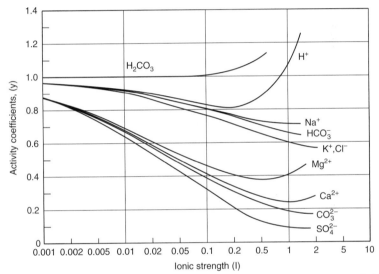

**Figure A-3** The activity coefficient plotted versus the ionic strength. (Freeze and Cherry, 1979. Reprinted with permission from Prentice-Hall, Inc.)

**Actual evapotranspiration (AE)** The volume of water lost as a result of *evaporation* from the surface of plants and from soil. The steady-state *hydrologic budget* equations provide an approximation of the hydrologic regime in a *watershed* basin. The equations are based on average annual parameters, including the time-dependent parameter *evapotranspiration.* Most budgets are calculated using *potential evapotranspiration (PE)*, and although AE closely approaches the value of PE, AE may in fact be considerably less. There is an upper limit to the amount of water an ecosystem loses by evapotranspiration. Often there is not enough water available from *soil moisture,* and therefore the term "actual evapotranspiration" is used to describe the amount of evapotranspiration that occurs under field conditions. The summer months incur the majority of water loss from evapotranspiration, with little or no water loss occurring during the colder winter months. **Figure A-4a and b** shows the AE and PE in a region with dry summers, moist winters, and limited *soil moisture storage capacity,* and a region with little variation in *precipitation* and ample soil moisture storage capacity. In **Figure A-4a**, the AE is less than the PE, especially if the soil moisture storage capacity is limited. When the PE is less than the precipitation, the soil moisture will be tapped; when available soil moisture is depleted, new precipitation will be partitioned to ET, storage, runoff, etc. When precipitation is evenly distributed through the year, the AE will be closer to the PE value (**Figure A-4b**).

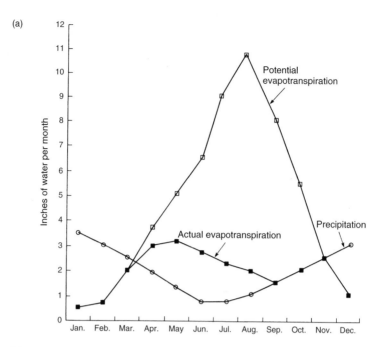

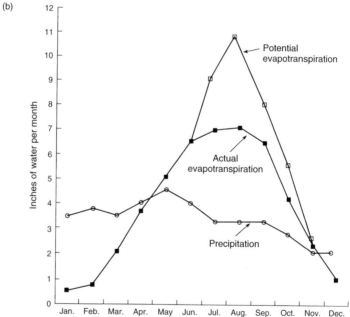

**Figure A-4** (a) Plots of potential and actual evapotranspiration for location of warm summers and cool winters. Limited soil moisture storage capacity for dry summers and moist winters. (b) Plot of PE and AE with little seasonal variation in precipitation and plenty of soil moisture storage capacity. (Thornthwaite, 1957.)

**Adhesion water** See *adhesive water*.

**Adhesive water** The molecular attraction between the walls of *interstices* and the adjacent molecules of water forming a continuous surface. Cf. *Cohesion; Attached water*.

**Adjusted stream** A watercourse that *flows* essentially parallel to the strike of the underlying beds, typically in the least-resistant bedrock surface over which it flows.

**Admixture** A material other than water, *aggregate* or *cement*, used as a *grout* ingredient or as an additive to cement-based grouts to produce some desired change in properties. Cf. *Well bore*.

**Adsorbed water** Water attracted by physicochemical forces to the particle surfaces in a soil or rock mass, and having properties that may differ from those of *pore water* at the same temperature and pressure, due to its altered molecular arrangement. Adsorbed water adheres to the surface of soil or minerals in ionic or molecular layers. Adsorbed water does not include water that is chemically combined within clay minerals. Cf. *Absorbed water; Absorption; Adsorption*.

**Adsorption** The surface retention of solid, liquid, or gas molecules, atoms, or *ions* by a solid or liquid. Adsorption differs from *absorption*, which is the penetration of substances into the bulk of the solid or liquid. When a solution contacts a solid, such as a soil particle, a portion of the *solute* transfers from the solution to the solid, i.e. adsorbs, until the concentration of solute in solution is in equilibrium with the concentration of the adsorbate attached to the solid. The chemical composition of natural waters can be affected by the adsorptive capacity of soils. Adsorption can be an important mechanism that reduces the apparent rate at which the solute in *groundwater* is moving and may make it more difficult to remove solute from the subsurface. The adsorption process entails the removal of chemicals or ions from solution, and their subsequent retention on the surface of soil particles by physical or chemical bonding. Adsorption, a physical process caused by van der Waals forces, may be relatively weak; if the bonds are physical, the chemicals are easily removed or desorbed by a change in solution concentration of the adsorbate. If the bonds formed between the adsorbate and soil are chemical, the process is almost always irreversible. A specific soil's adsorptive capacity is a function of its mineralogy, particle size, ambient temperature, *soil moisture*, organic carbon content, tension, *pH, Eh*, and activity of the ion. Clays tend to be strong adsorbers because they have both a large surface area per unit volume and an electrically charged surface that attracts many solutes. Cf. *Absorption; Desorption*.

**Adsorption isotherm** The relationship of the mass of *solute* adsorbed per unit mass of soil as a function of the concentration of the solute, which is plotted on a log–log graph. They are called isotherms because the adsorption experiments are conducted at a constant temperature. The concentration of solute remaining in solution, $C$, is a function of the amount adsorbed onto the solid surface, $C^*$. When plotted, the resulting curve is described by the equation:

$$\log C^* = b \log C + \log K_d$$

where:
$b$ = slope of the line
$K_d$ = intercept of the line with the axis, or *distribution coefficient* $[L^2 \cdot M^{-1}]$

Many trace-level solutes in contact with geologic media plot as a straight line on the log–log plot. Adsorption of a single-solute species by a solid has been mathematically calculated by several different types of equations such as the *Langmuir* and *Freundlich isotherms*.

**Advancing-slope grouting** A method of applying *grout* in which the front of a mass of grout is caused to move horizontally by use of a suitable grout-injection sequence.

**Advection** Also called advective transport. A process by which solutes (dissolved constituents) and/or heat are transported by the bulk motion of *groundwater flow*. Groundwater components (e.g. non-reactive contaminants and *dissolved solids*) that are traveling by advection move at the same rate and direction as the *average linear velocity* of the groundwater, which can be the most important way of transporting solute away from the source. The rate of groundwater flow is calculated from *Darcy's law* with this equation:

$$v_x = \frac{K}{n_e} \frac{dh}{dl}$$

where:
$v$ = average linear velocity $[L \cdot T^{-1}]$
$K$ = *hydraulic conductivity* $[L \cdot T^{-1}]$
$n_e$ = *effective porosity* [dimensionless]
$dh/dl$ = hydraulic gradient $[L \cdot L^{-1}]$

**Advection–dispersion equation** The mathematical expression that describes the transport of dissolved reactive constituents in moving water. The terms in this differential equation are *advection, dispersion, diffusion*, and decay reaction. When determining the advective transport of solutes, both the dispersion and the chemical reactions taking place need to be considered. Contaminant movement is strongly influenced by *heterogeneities* in the medium that change the *average linear velocity* of flow. *Fick's law* of diffusion is used to approximate the dispersive process. The chemical reactions that occur are also represented in a mathematical formula incorporating *adsorption* (defined by a retardation factor) and *hydrolysis* and decay (described by a first-order rate constant). The advection–dispersion equation is derived by writing a mass balance equation using the mathematical equations for dispersion (Fick's law) and the mathematical equation representing the chemical reactions. The equation in one dimension is:

$$D_x \frac{\partial^2 C}{\partial x^2} - \bar{v}_x \frac{\partial C}{\partial x} = \frac{\partial C}{\partial t}$$

where:
$C$ = solute concentration [M·L$^{-3}$]
$D$ = longitudinal dispersion coefficient [L]
$\bar{v}_x$ = average linear velocity [L·T$^{-1}$]
$t$ = time since release [T]
$x$ = length of the flow path [L]

**Advective transport** See *advection; advection–dispersion equation.*

**AE** See *actual evapotranspiration.*

**Aeration** More commonly called *air stripping.* A process utilized in *groundwater remediation* that brings air into contact with water, usually by bubbling air through the water, to remove dissolved gases such as carbon dioxide ($CO_2$), volatile organic compounds (VOCs), and hydrogen sulfide ($H_2S$), or to oxidize dissolved materials like iron compounds. Aeration typically promotes biological degradation of organic matter. Aeration of *well screen* increased pumping rates may cause *drawdown* to extend into the screen from time to time. Aeration of the screen for extended periods can lead to higher incrustation rates, cascading water, and reduced *well efficiency.*

**Aeration zone** See *zone of aeration; unsaturated zone.*

**Aerobic** Organisms (especially bacteria), activities, and conditions that can exist only in the presence of dissolved oxygen ($O_2$) in *pore waters.* In aerobic chemical reactions, $O_2$ behaves as an electron acceptor (and thus is reduced), ultimately forming water. Cf. *Anaerobic.*

**Aerobic decomposition** The breakdown of organic material by oxygen-utilizing organisms that occurs in the presence of atmospheric oxygen ($O_2$). In secondary sewage treatment, microbes degrade effluent into harmless components by aerobic decomposition. Cf. *Anaerobic decomposition.*

**Aerobic process** A process requiring the presence of oxygen ($O_2$). Cf. *Aerobic.*

**Affluent feeder** See *tributary.*

**Afflux** A rise in water levels in the *upstream* direction resulting from a restriction or obstruction of the waterway or the difference between flood levels in the upstream and *downstream* direction of a *weir* (manmade restriction).

**Aggrading stream** A watercourse that is actively building up its channel or *floodplain* or is up-building approximately at grade by being supplied with more *load* than it is capable of transporting. Cf. *Stream load; Load; Degrading stream.*

**Aggregate** A mixture of relatively inert granular mineral material, such as natural sand, gravel, slag, and/or crushed stone, used with cementing material for making mortars and concretes. In a *seive analysis* the fine aggregate passes through a No. 4 (6.4 mm) screen, whereas coarse aggregate does not. Aggregate can also be mixed with a cementing agent (such as Portland cement and water) to form a grout material (ASTM). Cf. *Grout; Admixture; Alkali aggregate reaction.*

**Aggregation** The uneven consolidation of particles in suspension, leading to the formation of *aggregates.* Drilling fluids are made with clay additives that support suspended particles when the fluid is at rest. The actual strength of the fluid is dependent on how well the clay particles have mixed. If mixing is incomplete and the clay platelets are not evenly dispersed throughout the liquid, the platelets may aggregate (stack face to face), thus limiting the *viscosity* of the fluid and decreasing *gel strength.* Cf. *Flocculation; Deflocculation.*

**Air development** A *well-development* method using compressed air to surge the well and settle the *filter pack* upon completion. An air compressor is used to inject air into the *well casing,* lifting the water to the surface, as shown in **Figure A-5**. The air supply is then shut off as the *well bore* water reaches the top of the casing, thereby allowing the aerated water column to fall. Two commonly used methods of air development are backwashing and *surging.* In backwashing, air is delivered down the *air line* inside a drop pipe in the well casing. The compressed air pumps water out of the well through a discharge pipe. The discharge is then closed off and the air supply is directed down a bypass air line into the well above the *static water level.* The *casing water* is forced back into the formation, agitating and breaking down *bridges* of sand grains. When the water level reaches the bottom of the drop pipe, air begins to escape from the discharge pipe; the air supply will be cut off; and the *aquifer* is allowed to return to static level. This is repeated until the well is developed or sediment free. Air development by surging requires that 35–60% of the drop pipe be submerged beneath the water surface when the well is pumping.

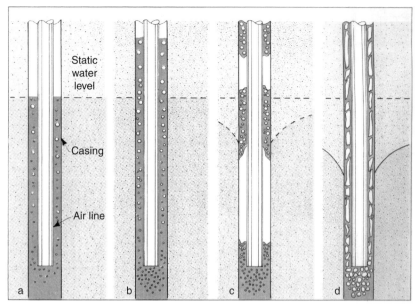

**Figure A-5** Well development technique using air. (a) introduction of small quantity of air. (b) air volume increases and water column rises as it is displaced by air. (c) aerated slugs of water are lifted out of the well. (d) continued increase of air volume will cause aerated water to continue to flow out the top of the well. (Driscoll, 1986. Reprinted by permission of Johnson Screens/a Weatherford Company.)

Cf. *Air-lift pumping; Backwash in well development.*
**Air entrainment** The incorporation of air into a surface water body as a result of surface turbulence.
**Air line** An installation within a well for the purpose of determining the depth to water. An air line is most often utilized during *pumping tests* when many measurements are needed during the critical first 10 min of the test when the greatest change in *drawdown* occurs, and hence more frequent data collection is required. Air lines are less accurate than acoustical or *electrical sounders* and have generally been replaced by data loggers. **Figure A-6** presents a typical air-line apparatus. The air line, a rigid material such as *Poly Vinyl Chloride (PVC)* or copper tubing, is extended from the top of the well to a depth several feet below the lowest anticipated water level drop, and the exact total length of the tubing is recorded. The air line must be completely airtight and is pressurized until all the water has been forced out of the line, when the pressure in the tube ceases to build and balances the water pressure. A stabilized gauge reading indicates the pressure necessary to support a column of water equal to the distance from the bottom of the air line to the water level. If the gauge measures feet of *head*, $h$, this indicates the submerged length of the air line in feet, which is subtracted from the total length of the line, $L$, to give the depth to water below the measuring point. If the gauge reading is other than feet of head, a simple conversion must be applied (**Appendix A-1**). The depth to water, $d$, is thus determined by the equation:

$$d = L - h$$

where:
$L$ = total length of air line [L]
$h$ = gauge reading of head [L]

For measurements during a pumping test, the well water level drops or the depth to water increases, the submerged length of the air line decreases, and the pressure reading on the gauge decreases.

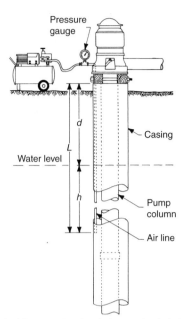

**Figure A-6** The use of the air-line method for measuring downhole water levels. (Driscoll, 1986. Reprinted by permission of Johnson Screens/a Weatherford Company.)

**Air sparging** A remedial method to promote biodegradation and to strip volatile compounds from *groundwater* by injecting air below the *water table*. Cf. *Air stripping*.

**Air stripping** Also called *aeration*. A mass-transfer process in which a *solute* in water is transferred to solution in a gas, usually air. Air stripping is a common cost-effective treatment technology for removing or reducing the concentration of organic chemicals in *groundwater*, especially *volatile organic compounds (VOCs)*. The water is mixed with air in a chamber or tower filled with packing material that disperses the water to enhance air contact. The tower contains beds of materials such as 1/4- to 2-in. (6.4–51 mm) metal, ceramic, or plastic spheres, tellerettes, saddles, or rings (**Figure A-7**). The water containing solutes is introduced to the top of the bed and trickles downward through the bed

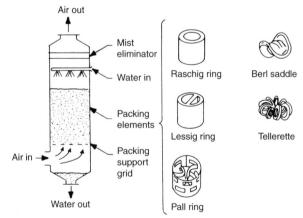

**Figure A-7** An example of a packed air-stripping tower and the types of packing used. (Driscoll, 1986. Reprinted by permission of Johnson Screens/a Weatherford Company.)

material. At the same time, air is passed upward through the bed and brought into contact with the water, thus removing contaminants from the water. Removal of the chemical depends on the length of contact time, air/water ratio, temperature, *vapor pressure*, and *solubility* of the contaminant. Best results can be achieved if the contact time (determined by the height of the tower) is long, the rates of liquid and air flow are adjusted properly, and the surface area of the packing material in the tower is optimal. If the solubility of the contaminant and the vapor pressure of the pure material are known, it is possible to predict the minimum air-to-water ratio required to achieve complete removal of the contaminant. **Figure A-8** shows the ratios of air and water for four trihalomethanes.

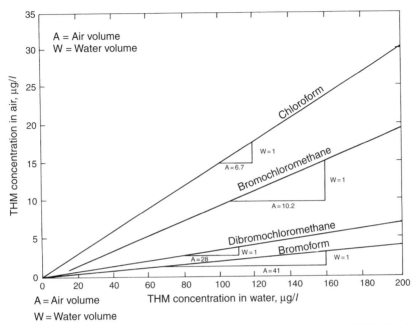

**Figure A-8** Curves showing the minimum air-to-water ratio that produce complete removal of VOCs, the theoretical equilibrium lines for trihalomethanes. (Driscoll, 1986. Reprinted by permission of Johnson Screens/a Weatherford Company.)

The slope of the curve indicates the ease with which the contaminant can be removed. The steeper the curve, the less air is required. In practice, however, complete removal may not be possible, even with large volumes of air. Cf. *Counter-current-packed tower.*

**Air well** A method of water collection utilized in desert climates, by which rocks are piled in a tower configuration allowing for the collection of water from *condensation* of atmospheric moisture on the cooler rock pile. Cf. *Atmospheric water.*

**Air-entry pressure** Also called the bubble point or bubbling pressure. The *capillary pressure* at which air, gases, or vapors enter into the *interstices* within a saturated *porous medium*, displacing water. Liquid saturation is 100% above the air-entry pressure.

**Air-lift pumping** See: **Appendix C – Drilling Methods.** Cf. *Air development.*

**Air–space ratio (Ga)** The volume of water that can be drained from a saturated soil or rock under the action of force of gravity divided by the total volume of voids. American Society for Testing and Materials (ASTM) standard terminology D653-90. Cf. *Specific yield; Zone of saturation.*

**Air–water partition coefficient** See *Henry's law constant.*

**Alignment test** A means of determining the alignment of a *well casing*, by setting a tripod over the *well bore* and lowering a plumb bob in the casing for shallow wells, or using an *inclinometer* inside a specially grooved plastic or aluminum casing. The grooves keep the instrument centered and aligned in the casing, and the instrument can sense a deviation from vertical up to 12°. During installation, a well bore can become skewed, and subsequently, the well casing itself will be off center. If a well is out of alignment beyond a certain limit, the pump cannot be properly set. A straight well is one in which all casing sections can be joined in alignment, and a plumb well is one whose center does not deviate from vertical. Cf. *Plumbness.*

**Alkali aggregate reaction** A chemical interaction between $Na_2O$ and $K_2O$ in cement, certain silicate minerals in cement, and certain silicate minerals in *aggregate*, which causes expansion. The alkali aggregate reaction results in weakening and cracking of Portland *cement grout* used in well completions. Cf. *Reactive aggregate*; *Aggregate*.

**Alkali lake** Also called an alkaline lake, or salt lake. An interior water body containing a significant amount of sodium carbonate ($Na_2CO_3$), potassium carbonate ($K_2CO_3$), sodium chloride (NaCl), and other alkaline compounds that make the waters salty. Alkali lakes commonly occur in arid regions. Cf. *Potash lake*; *Soda lake*.

**Alkalinity** The quantitative capacity of water to neutralize an acid. That is, the measure of how much acid can be added to a liquid without causing a significant change in pH. Alkalinity is not the same as pH because water does not have to be strongly basic (high pH) to have high alkalinity. In the water industry, alkalinity is expressed in mg $L^{-1}$ of equivalent calcium carbonate. The following chemical equilibrium equations show the relationships among the three kinds of alkalinity: carbonate ($CO_3^{2-}$), bicarbonate ($HCO_3^-$), and hydroxide alkalinity ($OH^-$). Total alkalinity is the sum of all three kinds of alkalinity.

$$CO_2 + H_2O \leftrightarrow H_2CO_3 \leftrightarrow H^+ + HCO^{3-} \, (pH \, 4.5 \, to \, pH \, 8.3) 2H^+ + CO_3^{2-}$$

Above pH 9.5 (usually well above pH 10), $OH^-$ alkalinity can exist, or $CO_3^{2-}$ and $OH^-$ alkalinities can coexist together. Individual kinds of alkalinity can be differentiated through chemical analyses. At neutral pH, bicarbonate alkalinity is the most common form of alkalinity. Bicarbonate alkalinity is also formed by the oxidation of organic matter in water. Oxidation of organic matter occurs in the biodegradation of natural or manmade organic chemicals in water (surface or groundwater), and the detection of bicarbonate is diagnostic of biodegradation reactions occurring. Cf. *Electrochemical sequence*.

**Allowable sediment concentration** Guideline for the amount of sediment acceptable in a given *water supply*. The US Environmental Protection Agency (EPA) and the National Water Well Association (NWWA) (1975) have recommended the following limits on sediment concentration based on how the water is to be used:

- 1 mg $L^{-1}$ – water to be used directly in contact with, or in the processing of, food and beverage.
- 5 mg $L^{-1}$ – water for homes, institutions, municipalities, and industries.
- 10 mg $L^{-1}$ – water for sprinkler irrigation systems, industrial evaporative cooling systems, and any other use in which a moderate amount of sediment is not especially harmful.
- 15 mg $L^{-1}$ – water for flood-type irrigation.

Sediment concentration is determined by averaging five samples collected during a *pumping test*. Water that contains less than 8 mg $L^{-1}$ of sand, silt, or clay is considered to be "sand free." Sediments in water supplies are destructive to *pumps*, and although *well development* reduces sediment concentrations, it may not effectively eliminate them.

**Alluvial river** A watercourse located within a *floodplain* in which the depth of the alluvium deposited by the *river* is equal to or greater than the depth to which scour occurs during *flood*.

**Alluvial-dam lake** An interior water body formed as a result of the collection of surface *runoff* that is blocked or dammed by an alluvial deposit, preventing the water from reaching mean sea level.

**AMC** See *antecedent moisture conditions*.

**American doctrine/American rule** See *Rule of reasonable use*.

**Anabranch** A diverging offshoot of a *stream* or *river* that loses itself in sandy soil or rejoins the main flow *downstream*. A *braided stream* consists of many intertwined channels or anabranches separated by islands. Individual anabranches of a braided stream continually shift and change. This shifting is a means of energy dissipation and occurs when *bed material* is coarse and *heterogeneous* and *banks* are easily erodible. Cf. *Anastomosing stream*.

**Anaerobic** A condition in which dissolved oxygen (DO) in water is depleted or there is an absence of free oxygen ($O_2$). Under anaerobic conditions, soil minerals and minerals dissolved in groundwater such as nitrate, pyrolusite ($MgO_2$) and ferric hydroxide ($Fe(OH)_3$), and sulfate act as electron acceptors. Cf. *Aerobic*.

**Anaerobic decomposition** The microbial breakdown of organic matter that occurs in the absence of oxygen. Methane is sometimes a by-product of anaerobic decomposition. Cf. *Anaerobic degradation*.

**Anaerobic degradation** The breakdown of organic compounds that occurs in oxygen-deficient *groundwater* systems. A complete degradation process converts hydrocarbons to carbon dioxide ($CO_2$) and water. Environmental factors such as dissolved oxygen (DO) concentration, *pH*, temperature, oxidation–reduction potential (*Eh*), *salinity*, and compound concentration all affect the rate and extent of degradation. Organic compounds can also be degraded by anaerobic microbes. Cf. *Anaerobic decomposition*.

**Anastomosing stream** A watercourse with a branching and recombining pattern, as in a *braided stream*. Cf. *Anabranch*.

**Anisotropic** Varying in a physical property with direction. An anisotropic medium displays directional differences in *hydraulic conductivity*. A *porous medium* is anisotropic if the geometry of the voids is not uniform, e.g. the *permeability* is greater in the $x$ and $y$ directions than the $z$ direction. **Figure A-9** displays the effects of grain shape and orientation of sediments on flow conditions.

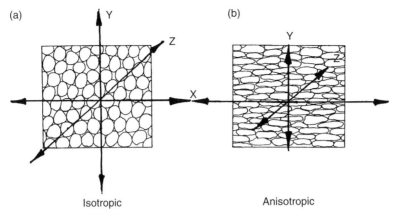

(a) Isotropic  (b) Anisotropic

**Figure A-9** An aquifers grain size, shape, and orientation determines the isotropic or anisotropic nature. The hydraulic conductivity in an isotropic environment (a) will be equal in the X, Y and Z directions, while in an anisotropic environment (b) the hydraulic conductivity will be greater in the X direction. (After Fetter, 1994.)

Cf. *Isotropic.*

**Anisotropy** Variation in a physical property with direction. Cf. *Anisotropic; Isotropic.*

**Annual flood series** Maximum annual peak flows of *streams* or *rivers* for a series of years constitute a data set that is analyzed for the likelihood and magnitude of the occurrence of *floods.* The set of data, called the annual series, is useful for flood prevention, control, and protection, that is, populations in flood plains. Cf. *Flood frequency analysis; Recurrence intervals.*

**Annual series** A data set of yearly event measurements used in probability analysis for a *flood peak.* To provide reliable predictions, probability analysis must start with data that is relevant, adequate, and accurate. When the analysis is to address *flood probabilities* less than 0.5, the annual series or annual maximum series data set comprises the largest event occurring each year. Flood, *rainfall,* and other hydrologic events can also be modeled using a *partial-duration series* (PDS) or *peaks-over-threshold* (POT) approach, in which the data set includes all peaks above a threshold value.

**Annular space** See *Annulus.*

**Annular velocity** Also called annular uphole velocity, or uphole velocity. The speed at which *drilling fluid* must move in order to remove cuttings from the *borehole.* The successful mixture of a drilling fluid is dependent on its fluid *viscosity,* annular velocity, and additive concentrations.

**Annulus** Also called the annular space. The space between the *well casing* and the *borehole* wall, or between two concentric strings of casing, or between the casing and tubing.

**Antecedent moisture conditions (AMCs)** The amount of water in *storage* at the beginning of an *precipitation* event that is relevant to that event. The three classes of AMCs are dry, average, and wet. The dimensionless *runoff curve number* (CN) is determined by the antecedent wetness of the *drainage basin;* standard values for CN are calculated for each AMC group in a runoff analysis. Because no single observation can define the AMC, storm characteristics are determined from data collected by a storm *gauge network. Groundwater discharge* in some basins has been used as an index of basin moisture conditions, but this is not applicable to all environments. The AMC index most commonly used is the *antecedent precipitation index* (API).

**Antecedent precipitation index (API)** Used in the estimation of *runoff* to determine the AMCs for a *catchment basin.* The rate at which moisture is depleted from a particular basin under specified meteorological conditions is approximately proportional to the amount of moisture in *storage.* During periods of no *precipitation,* the *soil moisture* should decrease logarithmically with time. The mathematical representation of this decrease is:

$$I_t = I_0 k^t$$

where:
$I_0$ = initial API value [L]
$I_t$ = reduced API value $t$ days later [L]
$k$ = recession factor ranging normally between 0.85 and 0.9

Setting $t$ equal to 1 day, the index for any day is equal to that of the previous day multiplied by the recession factor, $k$. **Table A-2** lists the values of $k^t$ for various combinations of $k$ and $t$. If precipitation occurs, the amount of precipitation is added to the index.

**Table A-2** Values of $k^t$ for various values of $k$, the recession factor, and $t$, time. (Linsley et al., 1982. Reprinted with permission from McGraw-Hill, Inc.)

| | | | | | $k$ | | | | | |
|---|---|---|---|---|---|---|---|---|---|---|
| $t$ | 0.80 | 0.82 | 0.84 | 0.86 | 0.88 | 0.90 | 0.92 | 0.94 | 0.96 | 0.98 |
| 1 | 0.800 | 0.820 | 0.840 | 0.860 | 0.880 | 0.900 | 0.920 | 0.940 | 0.960 | 0.980 |
| 2 | 0.640 | 0.672 | 0.706 | 0.740 | 0.774 | 0.810 | 0.846 | 0.884 | 0.922 | 0.960 |
| 3 | 0.512 | 0.551 | 0.593 | 0.636 | 0.681 | 0.729 | 0.779 | 0.831 | 0.885 | 0.941 |
| 4 | 0.410 | 0.452 | 0.498 | 0.547 | 0.600 | 0.656 | 0.716 | 0.781 | 0.849 | 0.922 |
| 5 | 0.328 | 0.371 | 0.418 | 0.470 | 0.528 | 0.590 | 0.659 | 0.734 | 0.815 | 0.904 |
| 6 | 0.262 | 0.304 | 0.351 | 0.405 | 0.464 | 0.531 | 0.606 | 0.690 | 0.783 | 0.886 |
| 7 | 0.210 | 0.249 | 0.295 | 0.348 | 0.409 | 0.478 | 0.558 | 0.648 | 0.751 | 0.868 |
| 8 | 0.168 | 0.204 | 0.248 | 0.299 | 0.360 | 0.430 | 0.513 | 0.610 | 0.721 | 0.851 |
| 9 | 0.134 | 0.168 | 0.208 | 0.257 | 0.316 | 0.387 | 0.472 | 0.573 | 0.693 | 0.834 |
| 10 | 0.107 | 0.137 | 0.175 | 0.221 | 0.279 | 0.349 | 0.434 | 0.539 | 0.665 | 0.817 |
| 11 | 0.086 | 0.113 | 0.147 | 0.190 | 0.245 | 0.314 | 0.400 | 0.506 | 0.638 | 0.801 |
| 12 | 0.069 | 0.092 | 0.123 | 0.164 | 0.216 | 0.282 | 0.368 | 0.476 | 0.613 | 0.785 |
| 13 | 0.055 | 0.076 | 0.104 | 0.141 | 0.190 | 0.254 | 0.338 | 0.447 | 0.588 | 0.769 |
| 14 | 0.044 | 0.062 | 0.087 | 0.121 | 0.167 | 0.229 | 0.311 | 0.421 | 0.565 | 0.754 |
| 15 | 0.035 | 0.051 | 0.073 | 0.104 | 0.147 | 0.206 | 0.286 | 0.395 | 0.542 | 0.739 |
| 16 | 0.028 | 0.042 | 0.061 | 0.090 | 0.129 | 0.185 | 0.263 | 0.372 | 0.520 | 0.724 |
| 17 | 0.023 | 0.034 | 0.052 | 0.077 | 0.114 | 0.167 | 0.242 | 0.349 | 0.500 | 0.709 |
| 18 | 0.018 | 0.028 | 0.043 | 0.066 | 0.100 | 0.150 | 0.223 | 0.328 | 0.480 | 0.695 |
| 19 | 0.014 | 0.023 | 0.036 | 0.057 | 0.088 | 0.135 | 0.205 | 0.309 | 0.460 | 0.681 |
| 20 | 0.012 | 0.019 | 0.031 | 0.049 | 0.078 | 0.122 | 0.189 | 0.290 | 0.442 | 0.668 |
| 21 | 0.009 | 0.015 | 0.026 | 0.042 | 0.068 | 0.109 | 0.174 | 0.273 | 0.424 | 0.654 |
| 22 | 0.007 | 0.013 | 0.022 | 0.036 | 0.060 | 0.098 | 0.160 | 0.256 | 0.407 | 0.641 |
| 23 | 0.006 | 0.010 | 0.018 | 0.031 | 0.053 | 0.089 | 0.147 | 0.241 | 0.391 | 0.628 |
| 24 | 0.005 | 0.009 | 0.015 | 0.027 | 0.047 | 0.080 | 0.135 | 0.227 | 0.375 | 0.616 |
| 25 | 0.004 | 0.007 | 0.013 | 0.023 | 0.041 | 0.072 | 0.124 | 0.213 | 0.360 | 0.603 |
| 26 | 0.003 | 0.006 | 0.011 | 0.020 | 0.036 | 0.065 | 0.114 | 0.200 | 0.346 | 0.591 |
| 27 | 0.002 | 0.005 | 0.009 | 0.017 | 0.032 | 0.058 | 0.105 | 0.188 | 0.332 | 0.579 |
| 28 | 0.002 | 0.004 | 0.008 | 0.015 | 0.028 | 0.052 | 0.097 | 0.177 | 0.319 | 0.568 |
| 29 | 0.002 | 0.003 | 0.006 | 0.013 | 0.025 | 0.047 | 0.089 | 0.166 | 0.306 | 0.557 |
| 30 | 0.001 | 0.003 | 0.005 | 0.011 | 0.022 | 0.042 | 0.082 | 0.156 | 0.294 | 0.545 |
| 40 | | | | 0.002 | 0.006 | 0.015 | 0.036 | 0.084 | 0.195 | 0.446 |
| 50 | | | | | | 0.005 | 0.015 | 0.045 | 0.130 | 0.364 |
| 60 | | | | | | | 0.007 | 0.024 | 0.086 | 0.298 |

**Antecedent runoff conditions (ARCs)** An index of *runoff* potential for a storm event, which attempts to account for the variation of the USDA Soil Conservation Service (SCS) *runoff curve number* at a given location using storm rainfall and runoff data. ARCs are used primarily for design applications.

**Antecedent stream** A watercourse that existed prior to the present topography or was established before local uplift began and incised its channel at the same rate that the land rose. Cf. *Stream*.

**API** See *antecedent precipitation index*.

**Appropriation doctrine** See *Prior appropriation right*.

**Aquiclude** A relatively impervious formation, e.g. a tight clay or intact shale formation, capable of adsorbing water slowly but not transmitting it fast enough to furnish an appreciable water supply for a *well* or *spring*. To transmit water, a formation must have many open spaces (pores or *interstices*) that are relatively large. A clay or fine-grained sediment has a large pore volume but the open space is typically very small. Therefore, the clay has a large water-holding capacity but cannot transmit the water. An aquiclude can be a confining layer of low *permeability* located so as to form an upper and/or lower boundary to a groundwater flow system. Cf. *Confining layer; Aquifuge; Aquitard; Adsorption; Confining bed; Aquifer confined*.

**Aquifer** A water-bearing or saturated formation that is capable of serving as a *groundwater reservoir* supplying enough
water to satisfy a particular demand, as in a body of rock that is sufficiently *permeable* to conduct *groundwater* and to
yield economically significant quantities of water to wells and *springs*. Aquifers occur in unconsolidated sands, silts,
gravels, or sediment mixtures and also in consolidated formations such as sandstones, limestones, dolomites, basalts,
and fractured plutonic and metamorphic rock. **Figure A-10a–f** shows the many different aquifers and their associated
boundaries.

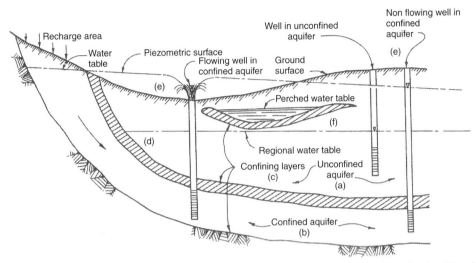

**Figure A-10** (a) Unconfined or water-table aquifer. (b) Confined aquifer. (c) Confining aquifer – alternating deposition of
confining layers. (d) Confining aquifer – upwarping. (e) Confined aquifer with artesian and flowing wells. (f) Perched aquifer.
(U. S. Dept. of Interior, 1985.)

Cf. *Water horizon.*

**Aquifer, artesian** A common misnomer referring to artesian conditions as only water that discharges aboveground
under natural pressure conditions. The *hydraulic head* at the point of measurement is above the ground surface elevation. The *hydrostatic head* can be determined by capping the well near the ground surface and measuring the shut-in
head with a pressure gauge. The water rises approximately 2.31 ft (0.7 m) above the gauge for every 1 psi (6.9 kPa) of
recorded pressure. See **Figure A-10e.** Cf. *Aquifer, confined.*

**Aquifer, confined** Also called an artesian aquifer, or confined *groundwater.* A formation that contains water bounded
above and below by impermeable beds or by beds of distinctly lower *permeability* than that of the water-bearing formation itself. The *piezometric pressure* or *head* is sufficient to cause the water within the formation to rise above the confining layer, or in the case of an artesian aquifer, flow above the ground surface. When a well is installed through an
impervious layer into a confined aquifer, water rises in the well above the confining unit. The water level in such a well
is determined by the *confining pressure* at the top of the aquifer or just below the confining layer. The elevation to which
this water rises is called its *potentiometric level,* which has an imaginary potentiometric surface representing the
confined pressure (*hydrostatic head*) throughout all or part of the confined aquifer. See **Figure A-10.** Cf. *Confining bed;
Aquifer; Aquifer, artesian; Confining pressure.*

**Aquifer development** Also called *aquifer stimulation.* Methods used to increase *well yield* beyond that obtained
through typical *well development.* Aquifer development may employ acids, explosives, and *hydrofracturing.* Acid is
used in limestone or dolomite aquifers or in any aquifer that is cemented by calcium carbonate ($CaCO_3$). Acid added
to the borehole dissolves *carbonate* minerals naturally occurring within the aquifer and opens up fractures and cre-
vices, thereby increasing the overall *hydraulic conductivity.* Explosives can be used in rock wells to increase *specific
capacity* within an aquifer. Explosive charges up to 2000 lb (454–907 kg) are detonated in igneous rock terrains.
This aquifer development process has increased yields on an average of 3–60 gpm (16–327 m$^3$ day$^{-1}$) at 60% of the
maximum *drawdown* in igneous and metamorphic rock aquifers. **Figure A-11** shows the placement of the explosive

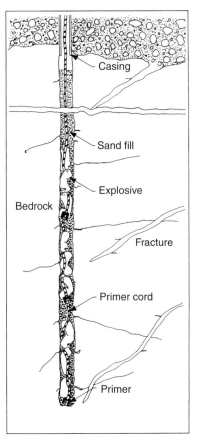

**Figure A-11** The safe and effective way of setting the explosive agent, sand, and igniting equipment in a borehole to be blasted during aquifer development. (Driscoll, 1986. Reprinted by permission of Johnson Screens/a Weatherford Company.)

agent, sand, and igniting equipment in a well to be blasted. In aquifer development by hydrofracturing, the pressure of overlying rock is overcome by using high-pressure pumps to create new fractures. The overburden pressure is approximately equal to 1 psi (6.9 kPa) for every foot (0.3 m) of depth below ground; therefore, a pump capable of generating pressures greater than 200 psi (1390 kPa) can hydrofracture the pressures existing at 200 ft (61 m) below ground. The pressure created in the production zone causes breaks in the rock that spread radially, providing additional interconnections between the water-bearing fractures and the well bore. Hydrofracturing has been known to increase yields from 50 to 130%. Generally, *well development* techniques are employed before aquifer development is initiated.

**Aquifer, leaky** See *Aquifer, leaky confined.*

**Aquifer, leaky confined** Also called a semi-confined aquifer. A water-bearing formation with an underlying or overlying confining layer of low *permeability* that stores *groundwater* and also transmits it slowly from the aquifer. Under pumping conditions, water will drain into the aquifer. The term "leaky confined" describes parameters of the *confining layer* and not the aquifer itself: a leaky confined aquifer leaks as a result of a leaky confining layer. See **Figure A-10**. Cf. *Aquitard; Aquiclude.*

**Aquifer loss** The *laminar flow* portion of the *step-drawdown test* for a pumping well. Jacob referred to the laminar term as the aquifer loss and the turbulent term as the well loss or *head loss* attributable to inefficiency. *Drawdown* in a pumping well is a function of the *hydraulic head* in the aquifer immediately adjacent to the *well screen* or laminar flow from the aquifer and losses in head due to turbulent flow through the well screen. The *Jacob step-drawdown test* was developed to determine the relative proportion of laminar and turbulent flow occurring at individual pumping rates to determine the optimum pumping rate and pump-setting depth for a given well. The *Jacob equation* for turbulent flow expresses the drawdown in a well as the sum of a first-order (laminar) component and a second-order (turbulent) component:

$$s = BQ + CQ^2$$

where:
$s$ = drawdown [L]
$Q$ = discharge [$L^3 \cdot T^{-1}$]
$BQ$ = laminar term
$CQ^2$ = turbulent term

In reality, it has been shown that $BQ$ includes portions of the well loss and that $CQ^2$ includes some aquifer loss. Although the step-drawdown test is still used to determine optimum pumping rates, the *Bierschenk method* is most often used to transform data from a step-drawdown test into the ratio of laminar to total head loss in a well.

**Aquifer, perched** A body of *groundwater* (a saturated zone) that is unconfined and open to the overlying *unsaturated zone* but is separated from an underlying main body of groundwater by an unsaturated zone. Also known as *perched groundwater*, as shown in **Figure A-10f**. A perched aquifer often occurs when a layer of lower *permeability* occurs as a lens within more permeable material. Perched aquifers typically are not large enough and usually too transient to be a reliable water supply source. Cf. *Aquifer, unconfined; Zone of saturation.*

**Aquifer recharge** See *recharge.*

**Aquifer, semi-confined** See *aquifer, leaky confined.*

**Aquifer stimulation** The development of a *groundwater reservoir* in a semi-consolidated or completely consolidated formation by physically altering the formation to improve its hydraulic properties. Cf. *Aquifer development.*

**Aquifer test** An evaluation that generally involves withdrawing or adding a measured amount of water to a well and recording the resulting change in *head* in the *aquifer*. The change in head in the aquifer is used to determine the hydraulic properties of the aquifer (e.g. *hydraulic conductivity, transmissivity,* and *storativity*). Cf. *Pumping test; Slug test; Jacob distance-drawdown straight-line method; Jacob time-drawdown straight-line method; Theis non-equilibrium equation; Theis non-equilibrium type curve, W(u).*

**Aquifer, unconfined** Also called a *water-table* aquifer, free *groundwater*, unconfined groundwater, and occasionally *phreatic water*. Groundwater that is not contained under pressure beneath relatively impermeable sediments or rocks and is exposed to the atmosphere through openings in the overlying soils or in the *vadose zone*. The *water table* is the upper boundary of the *zone of saturation* in which the absolute pressure equals the atmospheric pressure and the pressure due to the water equals zero. Under unconfined conditions, the water table is free to rise and fall. During periods of *drought*, the water table may drop as outflow to *springs, streams,* and wells reducing the volume of water in storage. When *precipitation* commences, aquifer *recharge* is generally rapid and the water table rises in response. Recharge of an unconfined aquifer can also occur through lateral groundwater movement or through seepage from an underlying leaky confined aquifer.

**Aquifer yield** The maximum rate of withdrawal that can be sustained by an aquifer without causing an unacceptable decline in the *hydraulic head* in the *aquifer*. An unacceptable decline would be based on management goals when *groundwater* is considered a mineable resource. The optimum development of a water resource system would be incomplete without reference, and therefore, research of the aquifer yield. Cf. *Safe yield; Potential yield; Basin recharge; Basin yield.*

**Aquitard** Also called a leaky confining layer, or leaky *confining bed*. A confining unit that retards but does not prevent the *flow* of water to or from an adjacent aquifer. An aquitard is the less-permeable bed in a stratigraphic sequence. An aquitard does not readily yield water to wells or *springs* but may serve as a storage unit for *groundwater* and can transmit water slowly from one aquifer to another. Most water-bearing formations yield some water and therefore are classified as either aquifers or aquitards. Cf. *Aquiclude.*

**ARC** See *antecedent runoff conditions.*

**Archimedes' principle** The rule that the apparent weight of a solid that is completely immersed in a fluid is reduced by an amount equal to the weight of the fluid it displaces.

**Area of influence of a well** Commonly called the *radius of influence*. The area surrounding a pumping or *recharge well* within which the *potentiometric surface* is altered by the action of that well at its maximum steady *discharge* or recharge rate. Cf. *Radial flow.*

**Area relations of catchments** One of many equations relating the form and hydrologic performance of a *catchment area*. A *catchment basin* tends to elongate as it increases in size. Large rivers appear to conform to the equation:

$$L = 1.27A^{0.6}$$

17

where:
$L$ = main-channel length [L]
$A$ = *drainage area* [L$^2$]

argument of the well function (u): In the *Theis-type curve*, the curve is derived from the plot of the *well function* of $u$, $W(u)$, and $1/u$. The value of $u$ is equal to:

$$u = \frac{r^2 S}{4Tt}$$

where:
$r$ = radial distance from a pumping well to the point of interest [L]
$S$ = storativity/*storage coefficient* [dimensionless]
$T$ = aquifer *transmissivity* [L$^2$·T$^{-1}$]
$t$ = time since commencement of constant pumping rate [T]

**Arid-zone hydrology** The properties, circulation, and distribution of water in a climatic zone where the *evaporation* rate is higher than the *precipitation rate*. Annual rainfall in an arid zone is generally less than 25 cm.

**Armored mud ball** A spherical mass of silt or clay generally between 5 and 10 cm in diameter, which forms within a *stream* channel and becomes coated or studded with coarse sand and fine gravel as it rolls along *downstream*.

**Artesian** For *groundwater*, the state of being under sufficient fluid pressure to rise above the ground surface in a well that penetrates the aquifer. In this sense, "artesian aquifer" is synonymous with confined aquifer; therefore, *artesian discharge, artesian head, artesian leakage, artesian pressure,* and *artesian spring* also describe water and occurrences in a confined condition under *hydrostatic pressure* (see **Figure A-10e**). In the legal field, the term "artesian" is applied to those aquifers that are confined above and below by impermeable layers, but the water within an artesian aquifer is classified as *percolation*. Cf. *Aquifer, artesian; Aquifer, confined; Confining bed.*

**Artesian aquifer** See *aquifer, artesian; aquifer, confined; artesian.*

**Artesian basin** A topographical concavity containing a defined *watershed* that includes a confined aquifer with a *potentiometric surface* above the land surface in the lower parts of the basin. The *fluid pressure* within the aquifer of an artesian basin is sufficient to cause water to *discharge* on the surface in low-lying areas. Cf. *Aquifer, artesian; Aquifer, confined.*

**Artesian discharge** Groundwater emitted from a well, spring, or *aquifer* under *confining pressure*. Cf. *Aquifer, artesian; Aquifer, confined.*

**Artesian flow** The movement of water from a confined aquifer in which the *pressure* is great enough to cause the water to *discharge* above the ground surface. Although not all *flow* from wells penetrating a confined aquifer is emitted above the ground surface, the term "artesian flow" has classically been used for only those aboveground discharges.

**Artesian head** The *hydraulic head* of the *artesian* or confined aquifer. Cf. *Aquifer, confined; Confining bed.*

**Artesian leakage** The portion of the *groundwater* flowing from a confined aquifer into the overlying or underlying *confining beds*. Cf. *Aquifer, confined; Leakage factor.*

**Artesian pressure** The *fluid pressure* of *groundwater* in an *aquifer* that is under pressure greater than atmospheric, where the fluid pressure at any point in the aquifer will rise above the ground surface. Cf. *Aquifer, artesian; Artesian.*

**Artesian spring** See *Spring, artesian*

**Artesian water** Water discharged or pumped from a confined *aquifer*. Cf. *Artesian; Aquifer, artesian.*

**Artesian well** Also called an overflow well. A well in which the water rises above the top of the *aquifer* under *artesian pressure*, whether or not it flows out at the land surface. The term "artesian well" is most often interpreted to mean a flowing artesian well.

**Artificial brine** *Groundwater* that has an increased ionic concentration as the result of solution mining of an underground deposit of salt or other soluble rock material. Cf. *Brine; Brackish water; Chebotarev's succession.*

**Artificial recharge** A *groundwater* management technique to recharge an *aquifer*, whereby surface water is purposely transferred into the groundwater system at rates greater than natural recharge. Artificial recharge commonly takes place during the wet season in order to restore depleted supplies or to increase and expand the amount of available water for use during the dry season. *Recharge basins*, or *spreading zones*, are used to recharge unconfined aquifers. A mound of water forms below the recharge basin and disperses through the unconfined aquifer when recharge ceases. *Recharge wells* must be used to artificially recharge confined aquifers. A recharge well discharges water into the surrounding aquifer either under a gravity *head* or, more commonly, under a head maintained by an injection pump. Cf. *Aquifer, confined; Aquifer, unconfined.*

**Atmometer** An instrument used in experimental *evaporation* and *transpiration* studies for estimating temporal and spatial variations in *potential evapotranspiration*. The atmometer automatically supplies water from a reservoir to an exposed, wetted surface. The loss of reservoir fluid is an indicator of transpiration. Cf. *Livingston atmometer.*

**Atmospheric water** The water in the atmosphere in the gaseous, liquid, or solid state. Constituents of the earths atmosphere is listed in **Table A-3**.

**Table A-3** Composition of the Earth's atmosphere by constituent and percent volume (Plummer et al., 2001. Reprinted with permission from McGraw-Hill, Inc.)

| Gas | Volume (%) |
|---|---|
| Nitrogen ($N_2$) | 78.1 |
| Oxygen ($O_2$) | 20.9 |
| Argon (A) | 0.934 |
| Water ($H_2O$) | Up to 1.0 |
| Carbon dioxide ($CO_2$) | 0.031 (variable) |
| Neon (Ne) | 0.0018 |
| Helium (He) | 0.00052 |

**Attached groundwater** The water that is held on the *interstice* walls within the *unsaturated zone* and measured as the *specific retention* of the soil. Cf. *Adhesive water; Pellicular water.*

**Attenuated flood wave** A pattern of water movement generated by lateral inflow along all the channels of a stream system. *Flood wave* attenuation is calculated mathematically from theoretical study in hydraulic routing methods such as surges in canals, impulse waves in still water (including *seiches* and *tides*), and waves released from dams.

**Attipulgite clay** Also called palygorskite. A chain-lattice clay mineral with the formula $(Mg, Al)_2Si_4O_{10}(OH)\cdot 4H_2O$, which has valuable bleaching and adsorbent properties. Attipulgite clay is also used as an additive to increase the *viscosity* of *drilling fluid.* Attipulgite can be used in both *freshwater* and saltwater conditions, unlike the primary drilling fluid additive *montmorillonite,* which only hydrates in freshwater. Cf. *Montmorillonite.*

**ATV logging** See *acoustic televiewer (ATV) logging.*

**Augmented water supply** A water supply that is increased during a prolonged *drought* through proper planning. Augmentation methods include fully utilizing the storage capacity of *groundwater reservoirs,* reusing *groundwater,* ensuring multiplicity of groundwater use, treating *brackish water,* and conserving water.

**Available moisture** Also called available water. The maximum portion of *soil water* that is accessible to plants. The available *soil moisture* moves through root membranes as a result of osmotic pressure caused by concentration differences between the sap in the root cells and the water in the soil. Available moisture is the difference between the moisture content at *field capacity* and the wilting point of plants and represents the useful storage capacity of the soil. The available moisture in soil also limits the rate of *evaporation* from soil surfaces.

**Available water** See *available moisture.*

**Average annual flood damage** See *flood damage analysis.*

**Average discharge** The arithmetic average of discharge for a given basin or watercourse for all complete water years on record, whether consecutive or not, as calculated by the United States Geological Society (USGS).

**Average linear velocity ($V_x$ or v)** The average rate at which water moves through all microscopic pathways through a porous medium. On the macroscopic scale, *specific discharge* is the volumetric flux divided by the cross-sectional area for *flow,* but because flow is limited to that portion of the cross-sectional area occupied by voids (*effective porosity*), the average linear groundwater velocity is defined by the equation:

$$\bar{v} = \frac{Q}{n_e A} = -\frac{Kdh}{n_e dl}$$

where:
$\bar{v}$ = average linear velocity $[L \cdot T^{-1}]$
$Q$ = outflow rate $[L^3 \cdot T^{-1}]$
$n_e$ = areal or effective porosity [dimensionless]
$A$ = cross-sectional area $[L^2]$
$K$ = hydraulic conductivity $[L \cdot T^{-1}]$
$dh/dl$ = gradient $[L \cdot L^{-1}]$
Cf. *Advection.*

**Average pore water velocity in the vadose zone** In the *vadose zone,* soil pores contain water, air, and potentially other gases (e.g. methane). Water is not the wetting fluid for all of the soil pores. As a result, equations for saturated porous media are not applicable. The pore water velocity is governed by gravity *flow* and *precipitation,* as presented below:

$$v = \frac{w}{\theta}$$

where:

$v$ = average pore water velocity [$L \cdot T^{-1}$]

$w$ = net groundwater recharge rate [$L \cdot T^{-1}$]

$\theta$ = average volumetric water content [dimensionless]

Cf. *Effective velocity*.

**Average velocity** *Stream* discharge divided by the area of a cross-section normal to the streamflow, or as the volume of *groundwater* discharging through a cross-sectional area divided by the aquifer's *porosity.* Cf. *Manning's equation; Chezy's formula.*

**Axial jet** A *hypopycnal inflow* pattern in which the inflowing body of water spreads into the main water body in the shape of a cone with an apical angle of about 20°. Cf. *Plane jet.*

**Axial stream** A watercourse having a *flow path* along the axis of anticline, syncline, or the longest and deepest dimension of a valley.

**Background water quality** Surface or *ground water* chemistry that has not been impacted by the natural or the man-made event under investigation. In environmental studies, background water quality is determined by sampling a well(s) upgradient from the point of release of the chemical(s) impacting groundwater. Natural phenomena can also impact water quality such as earthquakes and landslides.

**Backhand drainage** A condition in which the general course of a *tributary* within a *drainage basin* is opposite that of the main stream channel on both sides in that basin.

**Backswamp deposit** A flood deposit, consisting of extensive layers of silts and clays, which forms behind a natural levee within a poorly drained *flood basin*. Recent backswamp deposits, or older ones that occur at the surface, are prime agricultural lands and subject to periodic flooding. Buried backswamp deposits are poor water-storage basins and often act as *aquitards* above or below more permeable deposits. Within a buried valley, the deposition of coarse materials into erosional cuts in a backswamp deposit forms *aquifers* that resemble buried rivers, as in parts of the present-day Mojave River Basin and Santa Clara River Basin in southern California.

**Backwash in water treatment** A method of cleaning and rejuvenating *filters* and *well screens*, typically used in a water-treatment facility, in which the normal direction of water flow is reversed. This flow reversal loosens the filter material and flushes away solids that may have accumulated on the *upstream* surface of the filter beds. If this backwash operation is not regularly or effectively carried out, the normal flow through the beds becomes restricted and may be cut off completely. In addition, pressure buildup due to clogging of the beds may damage upstream facilities and equipment, and *breakthrough* of constituents in the water being treated will be evident *downstream*.

**Backwash in well development** Also called *rawhiding*. A reversal of the *flow* of water through the *well screen* and *filter pack* by *surging* the well. Backwash loosens fine-grained and/or *bridged* material in the screen, filter pack, or formation material so that it may be removed by pumping or bailing. An example of bridging is shown in **Figure B-19**. The goal is to improve well production by minimizing the effects of drilling that interfere with production and by removing fine material around the well to improve communication with the surrounding *aquifer*. **Figure B-1** illustrates backwash in *well development*.

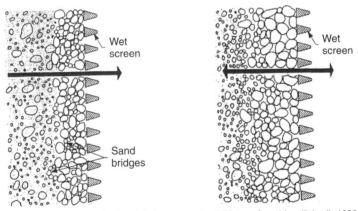

Wet screen

Wet screen

Sand bridges

**Figure B-1** The reverse flow action in backwash helps prevent the bridging of particles. (Driscoll, 1986. Reprinted by permission of Johnson Screens/a Weatherford Company.)

**Backwater** A relatively stagnant body of water, typically joined to a main *stream* or other water body that is retarded, backed up, or turned back upon itself by an obstruction or opposing current such as a *tide*, or water in a back area, side channel, or depression that is separated from another body of water by a manmade or natural obstruction.

**Backwater curve/dropdown curve** The *longitudinal profile* of the surface, or the flow form of a *stream*, at the point where the stream depth exceeds the normal depth as a result of an obstruction or constriction of the stream channel. In uniform channels, a backwater curve forms an upward concave feature. The term "backwater curve" is also used to describe all *surface profiles* in streams.

**Backwater effect** An increase in the height of a *stream* surface in the *upstream* direction caused by the retardation of flow, either from an obstruction or an overflow of the main stream onto low-lying land, which leads to a backup in its *tributaries*. The backwater effect can also cause an increase in stream width or a decrease in *stream velocity*. Cf. *Backwater*.

**Bacterial degradation** The chemical alteration of compounds in soil, surface water, and *groundwater* that occurs as a consequence of redox reactions catalyzed by microbial enzymes. Microbial enzymes increase the rate of redox reactions by decreasing the activation energies of these reactions. Bacteria can also degrade some harmful groundwater contaminants, such as gasoline constituents to less toxic compounds that can be cleaned up using cheaper and less time-consuming remediation techniques. Naturally occurring bacterial degradation, or *natural attenuation*, may proceed at an unacceptably slow rate due to high concentrations of chemical contaminants, imbalances among the chemicals to be remediated, or nutrient or terminal acceptor depletion. By supplying the bacteria with nutrients, or amending the environment through artificial means such as venting the soil or adding appropriate chemicals or terminal electron acceptors, the degradation process can be accelerated. An alternative approach is to introduce bacteria that are not native to the soil or water of the area requiring remedial action but which are known to metabolize and degrade the chemicals of concern (COC). Nutrient supplements may also be necessary when utilizing non-native bacteria. In groundwater, bacteria are the most important microorganisms involved in degrading chemical compounds, while in other aqueous environments, algae, fungi, yeasts, and protozoa can also contribute. All of these microorganisms have been utilized in soil and water remediation. It should be noted, however, that enhanced biodegradation can be limited by the ability to deliver the nutrients/terminal electron acceptors/primary substrate, etc. to the desired zone.

**Bactericide** A strong oxidizing chemical employed to rehabilitate a well by destroying bacteria (particularly *iron bacteria*) but having no other undesirable effect on well installation or water production. Bacteria in wells can be controlled by chemical treatments or by physical methods. Bactericides are typically cheaper and more effective, but for maximum effectiveness, their application must be followed by physical agitation of the well. Common chemicals used in the control of iron bacteria are chlorine and hypochlorites such as calcium hypochlorite ($Ca(OCl)_2$) and sodium hypochlorite ($NaOCl$), chlorine dioxide ($ClO_2$), and potassium permanganate ($KMnO_4$). These chemicals oxidize or "burn" organic compounds, kill bacteria, and also dissolve and loosen the organic sludge that clogs *well screens* and formation material within the *production zone*.

**Bacteriological water quality standards** Guidelines, based on bacterial content, for appraising the suitability of water sources for various intended uses. In the United States, the bacteriological *drinking water standard* is based on total *coliform bacteria*, with a recommended concentration limit of 1 coliform cell per 100 mL. Coliform concentrations somewhat above this recommended limit are generally not harmful but serve as a generalized measure that can be used to compare water sources. Cf. *Water quality; Septic system.*

**Bacteriostatic** Preventing the growth of bacteria without killing them. As in a bacteriostatic agent or chemical added in the attempt to keep bacteria levels constant.

**Bagnold dispersive stress** The forces resulting from the collective mutual collisions of current *flow* or the *shear stress* between layers in a fluid, quantified by British geographer Ralph A. Bagnold. Shear stress is caused by the impact between cohesionless particles that collide during current flow. Because stress increases with the square of particle diameter, larger particles are subjected to the greatest stress and are forced to the bed surface, where the stress is zero.

**Bail bottom/bailer bottom** A method of placing a *well screen* at a desired depth within the well *borehole* by using a connecting ring (bail) and hook to lower the screen down the *well casing*, as shown in **Figure B-2**.

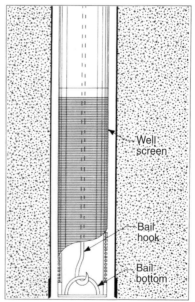

Well
screen

Bail
hook

Bail
bottom

**Figure B-2** Placing a well screen with the use of the bail or bailer bottom. (Driscoll, 1986. Reprinted by permission of Johnson Screens/a Weatherford Company.)

**Bail-down procedure** A method of *well screen* installation in which sediment from below the screen is removed, allowing the screen to settle into position below the *well casing*. After drilling a *borehole* for well installation, it is not always possible or desirable to pull back the casing to expose the well screen. **Figure B-3** demonstrates the bail-down procedure and the tools used. The bottom of the well screen is fitted with a *bail-down shoe*, and the screen is lowered through the casing. Often, special connection fittings are used to attach the screen to a length of pipe called the bailing pipe. The screen is then worked down below the casing into the formation by operating the *bailer* or drilling tools through the bailing pipe. The added weight of the bailing pipe aids in lowering the screen into place. Once the bailing operation is complete, the native formation materials settle around the screen.

**Bail-down shoe** A fitting attached to the bottom of a *well screen*, as shown in **Figure B-3**, during a well installation using the *bail-down procedure*.

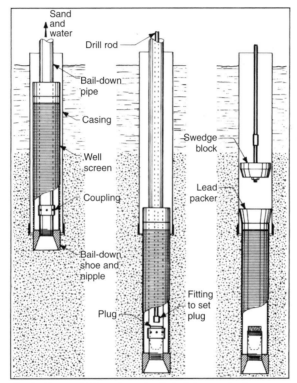

**Figure B-3** The screen is fitted with a special shoe when using the bail-down method to remove sediment from below the screen causing the screen to settle. (Driscoll, 1986. Reprinted by permission of Johnson Screens/a Weatherford Company.)

**Bail-down test/bail test** A method of estimating *hydraulic conductivity* and potentially *specific storage* by rapidly removing a known volume of water from a small-diameter well in a single removal stroke and then recording the water level in the well as it recovers. The same effect can be obtained by rapidly introducing or removing a solid cylinder (i.e. *bailer*) of a known volume. Care must be taken to ensure that sufficient water is displaced during the test to flow into or out of the geologic formation of interest. An alternate method is to bail the well, remove water from the well using the bailer, and record the rate of water removal and the *drawdown* obtained; the drawdown will stabilize after bailing at a relatively constant rate. The data obtained from the bail-down test (water level versus time) can be interpreted to determine hydraulic conductivity for a point *piezometer*. For confined aquifers, the *Hvorslev method* or the *Cooper–Brederhoeft–Papadopulos method* (also appropriate for semi-confined conditions) is recommended. To analyze for hydraulic conductivity in unconfined conditions were the well completed just below the water table, the *Bouwer and Rice method* is recommended. By determining the *specific capacity*, the well's *potential yield* for a specified drawdown is estimated. A bail-down test is similar to a *slug test*; the procedure is called a bail-down test when water is removed and a slug test when water is added. The slug test is used to calculate hydraulic conductivity, *transmissivity*, and the storage coefficient.

**Bailer** A device used to collect a liquid sample from a well or *piezometer* while causing minimal disturbance to any water stratification within the *well casing*. Typically, a bailer is a rigid tube with a *check valve* in the bottom that allows liquid to flow into the tube as it is lowered but prevents backflow, or loss of water, as it is raised. Bailers designed to collect samples below the surface of the *water table* may also have a check valve at the top to minimize mixing with water in the well above the desired sample depth. A bailer is also used for collecting formation samples and bailing or cleaning out mud and slurried material during drilling. Several types of bailer are shown in **Figure B-4a**, and a common bailer construction with check valve is shown in **Figure B-4b**.

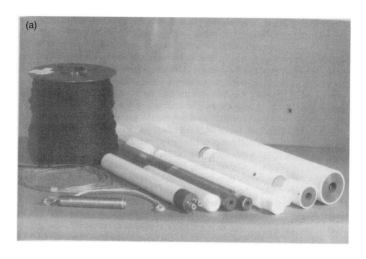

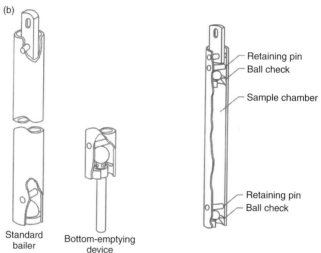

**Figure B-4** (a) Typical bailers in use, PVC, stainless steel, or Teflon. (Sanders, 1998.) (b) Construction of a bailer with either single or double check valves. (Sanders, 1998.)

**Bailer test** See *bail-down test/bail test; slug test.*

**Bank** Also called a *ripa*. The margin or land area that borders a water body and confines the *flow* of water to a natural channel during normal flow periods. The *left bank* and *right bank* of a watercourse channel are designated as they would be viewed by an observer facing the *downstream* direction. Cf. *Riparian; Riverain.*

**Bank stability** The resistance to change in the contours of a *stream* or other watercourse margin. Bank stability can be attained naturally or artificially. Stream maturity, natural benching, and vegetation increase bank stability, as do man-made protections such as retaining walls, *riparian* structures, and modified *drainage*.

**Bank storage** Also called lateral storage. Water that is temporarily *absorbed* into the *porous media* along the margins of a watercourse during high water stages in the channel. At lower water stages, this stored water drains from the *bank* until it is depleted; the excess water may return to the channel as *effluent seepage* when the stage falls below the *water table*. Effluent seepage from bank storage can affect the stream *hydrograph* and must be taken into account during *hydrograph separation* or hydrograph analysis. **Figure B-5a** shows a typical change in bank storage resulting from a flood. The subsurface water contributions to streamflow in the upper reaches of a *watershed* will increase the buildup of a *flood wave* in a stream. The *groundwater* streamflow interaction in the lower reaches is the bank storage and mediates the flood wave. **Figure B-5b** shows how flow from an increased water stage, influenced by a flood wave, can be induced into the stream banks. As the stage recedes, the flow is reversed. **Figures B-5c–f** shows a graph of the effect of bank storage on the stream hydrograph, bank storage volume, and associated rates of inflow and outflow. A flush of cool bank storage water can also decrease the overall stream temperature.

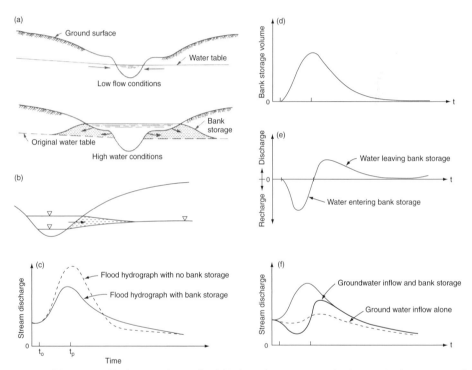

**Figure B-5** (a) Variations of bank storage during a flood. (Linsley et al., 1982. Reprinted with permission from McGraw-Hill, Inc.) (b) A river shows an increase in river stage by the arrival of a flood wave. (Freeze and Cherry, 1979. Reprinted with permission from Prentice-Hall, Inc.) (c–f) The modifications of the flood wave due to the effects of back storage. (Freeze and Cherry, 1979. Reprinted with permission from Prentice-Hall, Inc.)

Cf. *Influent stream; Effluent stream.*

**Bankfull discharge/bankfull flow** The maximum flow, in dimensions of $L^3 \cdot T^{-1}$, that a watercourse channel can transmit without overflowing its *banks*. Occurrence of bankfull discharge varies between watercourses, but *recurrence intervals* are in the order of once in 1.5–10 years. During these intervals, the stream is most effective in transporting sediment along the channel and in shaping its bed and banks. The height of the water at bankfull discharge is the *bankfull stage*. Cf. *Loaded stream*.

**Bankfull stage** The time period during which a watercourse first overflows its natural *banks*. Bankfull stage is the height of the water at *bankfull discharge*. It has been observed that, on average, a stream reaches bankfull stage, or height, once a year and overflows its banks once every 2.33 years or that the *discharge* in a watercourse will equal or exceed bankfull in 2 out of 3 years. Cf. *Channel capacity*.

25

**Barbed tributary** A smaller class stream that joins the main stream channel in an *upstream* direction. The junction is a sharp bend, with an acute angle between the barbed tributary and the main channel flowing *downstream*.

**Barometric efficiency (B)** The ratio of the water level fluctuation in a *well* to changes in the atmospheric pressure. Atmospheric (barometric) pressure has an inverse relationship with well water level: an increase in *pressure* creates a decrease in the water level observed. The barometric efficiency equation relates how water levels in a confined aquifer respond to changes in atmospheric pressure:

$$B = \frac{\gamma dh}{dp_A}$$

where:

$B$ = barometric efficiency [$M \cdot L^{-1} \cdot T^2$]
$\gamma$ = amplitude [L]
$dh$ = change in *head* [L]
$dp_A$ = change in atmospheric pressure [$M \cdot L^{-1} \cdot T^2$]

When $dh$ and $dp_A$ are plotted on arithmetic graph paper, the result is a straight line, the slope of which is the barometric efficiency, typically in the range of 0.20–0.75 in dimensions of $M \cdot L^{-1} \cdot T^{-2}$. Increases in atmospheric pressure acting on a water column are added to, and decreases are subtracted from, the pressure of the water in the well. Cf. *Barometric pressure effects on groundwater levels; Aquifer, confined.*

**Barometric pressure effects on groundwater levels** Variations in atmospheric pressure (as measured by a barometer) that can produce water level fluctuation in a *well*, particularly with a confined *aquifer* that is not in equilibrium with atmospheric pressure. For example, the water level in the well drops with an increase in the atmospheric (barometric) pressure. Because this effect can directly influence water level measurements, it is imperative when trying to determine the *groundwater gradient* in a particular well field that the measurements be collected within a short time period to ensure that they reflect the same atmospheric pressure. The effects of barometric pressure are also significant in the interpretation of water level data during *aquifer tests*. *Barometric efficiency* measurements are taken during the aquifer test and corrections are applied to the water level data to nullify the barometric pressure effect. Atmospheric pressure at sea level is 760 mm or 29.92 in. of mercury, 14.66 lb in$^{-2}$ (psi), 1 bar, 1,000,000 dyn cm$^{-2}$, and 1,013.25 millibars. Cf. *Capillarity.*

**Barrier, aquifer effects** A groundwater reservoir's limit or terminus formed by a relatively impermeable formation (such as bedrock or fine-grained sediments that are not a source of water), by a thinning of the saturated formation, or by *erosion* or other discontinuity. The barrier affects groundwater movement and becomes apparent during *aquifer tests* when the *cone of depression* from pumping intersects a discontinuity, which in turn alters the application of standard aquifer solutions, i.e. the *Theis method* or *Jacob equation*. Physically, the observed *drawdown* appears to be accelerated when it encounters a hydrologic barrier. To determine the distance and location of barriers, *image well theory* is used in the calculation of time-drawdown data. An average or effective line of zero flow is also established for boundary calculations in aquifer simulations, as shown in **Figure B-6**.

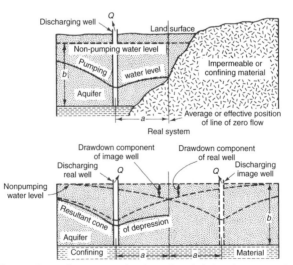

**Figure B-6** Boundary representations in aquifer calculations. (USGS, 1962.)

Cf. *Boundary of an aquifer; Boundary conditions; Impermeable boundary.*

**Barrier boundaries** See *boundary conditions*.

**Barrier lake** A water body that is impounded or retained by a naturally occurring obstruction such as a landslide, alluvial delta, glacial moraine, ice, or lava flow. A *lagoon* formed by a shore dune or sandbank is a type of barrier lake.

**Barrier spring** See *spring, barrier*.

**Barrier well** A pumping or *recharge well* used to intercept, or induce sufficient *head* to prevent, the spread of a contaminant or constituent of concern (COC), such as saltwater. Cf. *Salinity*.

**Basal groundwater** A freshwater layer that floats above seawater in the *groundwater* of oceanic islands, typically, basalt or coral atolls. The freshwater present on these islands results from *rainfall*, which *percolates* into the island soils and rocks. This water either collects in small shallow *aquifers* or eventually reaches the saltwater level, where it forms a *freshwater lens* on top of the saltwater in hydrodynamic equilibrium. Basal groundwater is often the principal source of *potable water* for island communities.

**Basal till** Unconsolidated material deposited and compacted beneath a glacier and as a result has relatively high bulk density.

**Basal tunnel** Also called a Maui-type well. An excavation along the *water table* in basaltic regions to supply *freshwater*.

**Base discharge** A US Geological Survey (USGS) measurement of *discharge* for a given *gauging station* above which *peak discharge* data are determined. Base discharge represents the groundwater discharge contribution to streamflow.

**Base exchange** Displacement of a cation from the surface of a solid to which it is bound by a cation in solution. Certain types of base exchange in clays can make them more flocculent. A clay particle with its anions may act as a salt, with the colloidal clay particle taking the part of the anion. Some of the cations may replace others, making the clay more flocculent. Cf. *Flocculation; Ion exchange*.

**Base runoff** The natural flow of a *stream* consisting of the *effluent* groundwater and the *delayed runoff* from slow passage of fluid through lakes or swamps. Base runoff is the sustained or fair-weather runoff of a *drainage basin*. Cf. *Baseflow; Direct runoff; Fair-weather runoff*.

**Baseflow** That portion of a watercourse's total *flow* that is not attributed to *direct runoff* from *precipitation* or snowmelt but rather is due to *groundwater* seeping into the watercourse below its banks as *groundwater discharge*. As shown in **Figure B-7**, stream discharge receives a flow contribution from groundwater when the *groundwater gradient* is toward

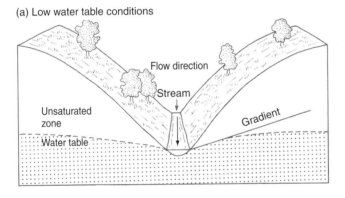

(a) Low water table conditions

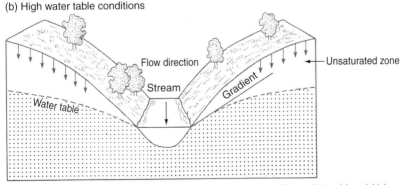

(b) High water table conditions

**Figure B-7** How baseflow affects the groundwater gradient in low water table conditions (a) and high water table conditions (b). (USGS, 1998.)

27

the stream, or when the *water table* in the vicinity is higher than the *free surface* of the stream. This sustained or "fair-weather" flow may represent 100% of streamflow in periods of *drought*. If the water table is lower than the stream's free surface, loss from the stream may occur through seepage back into the streambed. Two main components of a stream hydrograph (a plot of *discharge* versus time) are baseflow and surface runoff (such as the volume of water derived from a storm event). **Figure B-8** is a *hydrograph* showing the time relation between the occurrence of surface runoff and the delayed increase in the baseflow contributed from groundwater.

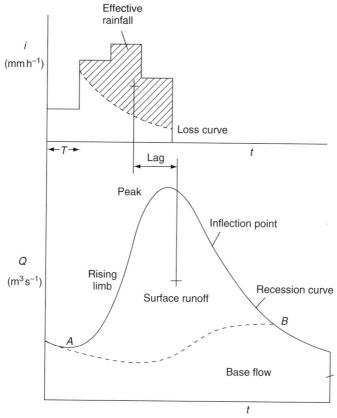

**Figure B-8** A hydrograph displaying the time relation between precipitation and the increase of baseflow contributed by groundwater. (Shaw, 1983. Reprinted with permission from Van Nostrand Reinhold Co. Ltd.)

Cf. *Influent; Effluent.*

**Baseflow recession** The decreasing rate of stream *discharge* that occurs when a stream is supplied only by *groundwater discharge* over an extended period of time. Baseflow recession usually takes the form of a decaying exponential curve, because the stream is *dewatering* the *aquifer* in the vicinity as the stream's *flow* continues over time. The baseflow–recession portion of the *hydrograph* shown in **Figure B-9** results from *groundwater storage* in the basin. The characteristics or shape of the baseflow–recession curve for a basin is a function of the basin's geomorphology, i.e. the nature of the *drainage system*. When most of the *drainage* takes place in the subsurface, as in limestone regions, the baseflow–recession curve is relatively flat, whereas in an area of low permeability, as in granitic regions, the curve is typically very steep. As the *water table* declines, so does the seepage from the groundwater into the streambed. The baseflow equation shows that flow varies logarithmically with time:

$$Q = Q_0 e^{-at}$$

where:
  $Q$ = flow at some time $t$ after the recession started [$L^3$]
  $Q_0$ = flow at the start of the recession [$L^3$]
  $a$ = recession constant for the basin
  $t$ = time since the recession began [T]

Baseflow–recession curves are used in the estimation of groundwater flow to evaluate water budgets and to evaluate single-storm water precipitation and storm flow. Also, the time at which the hydrograph takes up the shape of a baseflow–recession curve corresponds to the time at which surface runoff ends.

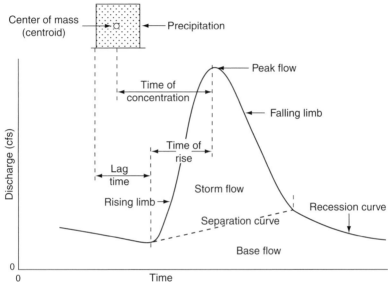

**Figure B-9** A baseflow recession hydrograph showing the results of groundwater storage in basin. (Black, 1996. Reprinted with permission from Ann Arbor Press, Inc.)

**Baseflow–recession hydrograph** A graph with a decaying exponential form that depicts a watercourse's reduced *discharge* as a result of the decreasing *groundwater* contribution over time. *Baseflow* decreases because the *water table* declines as groundwater seeps into the stream. *Baseflow recession* for a particular site is similar from year to year. **Figure B-10** shows six consecutive annual baseflow–recession curves for a stream in a climate with a dry summer season and stream hydrographs of successive years. Hydrographs exhibiting this behavior, however, must not automatically be assumed to be controlled by baseflow considerations; other causes, such as declining snowmelt contribution to a stream, can mimic a decaying exponential form. A simple method of determining the total potential *groundwater discharge,* $V_{tp}$, is to use the equation:

$$V_{tp} = \frac{Q_0 t_1}{2.3}$$

where:
  $V_{tp}$ = total potential groundwater discharged by a complete groundwater recession [$L^3$]
  $Q_0$ = baseflow at the start of the recession [$L^3 \cdot T^{-1}$]
  $t_1$ = time required for baseflow to go from $Q_0$ to $0.1Q_0$ [T]

*Groundwater recharge* between baseflow recessions can be determined by calculating the difference between the remaining potential groundwater discharge at the end of one recession and at the beginning of the next recession.

29

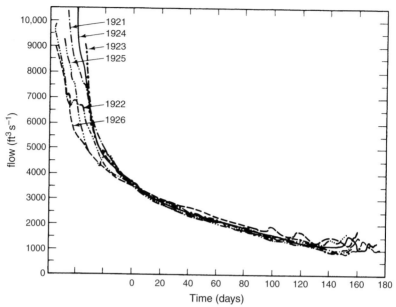

**Figure B-10** For six consecutive years, the annual baseflow recession for summer. (Fetter, 1994. Reprinted with permission from Prentice-Hall, Inc.)

Cf. *Groundwater recession curve.*

**Basic data stations** Also known as *gauging stations.* Locations that are included in streamflow network to obtain hydrologic data for future use, including land and water features that vary in location and records of processes that vary with location and time, e.g. *precipitation, evapotranspiration,* streamflow, and *groundwater* elevations. Basic data station are set up to develop synthetic records for major streams in the area of investigation or *drainage basin,* as well as for a number of *tributaries.* Once the data are sufficient for synthetic records to be derived, the station use may be discontinued.

**Basic-stage flood** See *partial-duration flood.*

**Basin accounting** Calculating the *water balance* for a *drainage basin,* considering all the factors that can add, remove, or store water in the basin. The factors capable of adding water are commonly forms of *precipitation,* including snowmelt, and factors capable of removing water, including *runoff, evapotranspiration,* and *infiltration.* Basin accounting can be performed by measuring as many of these factors as possible and estimating the rest, but a numerical model is more commonly used once the basic parameters of the basin have been measured or estimated from similar situations. The Stanford Watershed Model (SWM), shown as a flow diagram in **Figure B-11,** is one of the earliest such models and a good example of conceptual basin accounting. The outputs of most interest in basin accounting are runoff (for flooding considerations) and infiltration (for aquifer *recharge* considerations).

**Basin area** For a watercourse segment of a given *stream order,* the projection on a horizontal plane of the total area, $A_u$, of its *drainage basin* as bounded by the basin perimeter and the area contributing *overland flow,* including all *tributaries* of lower order. Cf. *Law of basin areas.*

**Basin, closed** A region with no surface outlet for *drainage.* The *runoff* from a closed basin ends in a depression or *lake* within its area and only escapes through *evaporation* or *infiltration* into the ground.

**Basin, drainage** A surface area that contributes *runoff* water to a *stream* channel, *river, lake, reservoir,* or other water body. A *drainage basin* is bounded by a drainage *divide* that separates it from other basins. See *drainage basin.*

**Basin elongation ratio ($R_e$)** The diameter of a circle having the same area as a specified *drainage basin* divided by the maximum length of that basin.

**Basin, flood** The expanse covered by water during the highest known or predicted flood in a given basin, such as the flat area between the sloping sides of a *river* valley and the natural levee built up by the river. A *floodplain* commonly contains enriched soils and swampy vegetation.

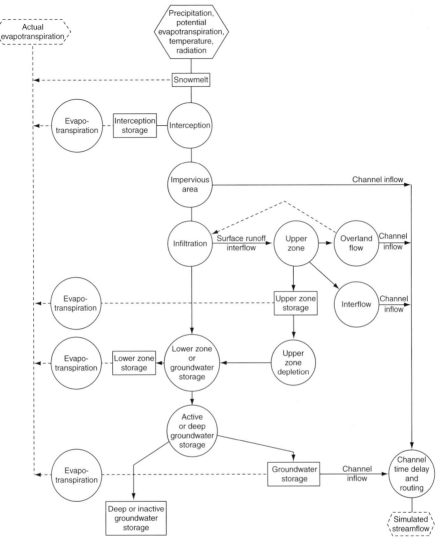

**Figure B-11** One of the earliest models for conceptual basin accounting – The Stanford Watershed Model IV. (Linsley et al., 1982. Reprinted with permission from McGraw-Hill, Inc.)

**Basin, groundwater** A system of aquifers or a single *aquifer*, of any shape, that has reasonably well-defined boundaries and definite areas of *recharge* and *discharge*. Cf. *Boundary, aquifer*.

**Basin lag** The time elapse between the centroid of a *rainfall* event over a given basin and the centroid of *runoff* at the mouth of the basin, or the time difference between the centroid of the rainfall and the peak of the *hydrograph*, occurrence of *peak discharge*. Basin lag, measured in hours, is a function of the basin's size and shape and is an important parameter in developing synthetic *hydrographs* for predicting future runoff events.

**Basin length ($L_b$)** The horizontal distance of a straight line extending from the mouth of a *stream*, parallel to the principal drainage line of the stream, to the farthest point on the drainage *divide* of its basin or *watershed* area.

**Basin order** A designation applied to the entire basin contributing to a *stream* segment of a given order. The basin order corresponds to the designated order of the stream it feeds. A first-order basin, therefore, consists of all of the *drainage area* contributing to a first-order stream. Cf. *Stream order*.

**Basin perimeter (P)** The length of the boundary that encloses the entire area of a *drainage basin*. Cf. *Boundary conditions*.

**Basin recharge** Water gained by an subsurface regime from the difference between water additions to a basin (*precipitation* including snowmelt) and surface losses from the basin (*runoff* and *evapotranspiration*), with consideration of the *storage* capacity of the *vadose zone* above the *aquifer*. When the entire vadose zone has enough moisture, such that additional water simply passes through to the *water table*, then no losses occur to the vadose zone. Any other condition, however, results in moisture to be stored in the vadose zone until passing through to the aquifer or is lost to the atmosphere. In arid climates and areas with large depths to *groundwater*, basin *recharge* may only occur under standing bodies of water.

**Basin relief (H)** The difference in elevation between the mouth of a *stream* and the highest point within, or on the perimeter of, its *drainage basin*. Basin relief is the maximum relief in a drainage basin.

**Basin yield** The maximum rate of *groundwater* withdrawal sustainable within the *hydrogeologic system* of a basin while ensuring that the *hydraulic head* throughout that basin can be maintained at satisfactory levels without causing unacceptable changes to any other portion of the basin. The concept of yield can be applied at several scales relating to the development of *groundwater resources*, i.e. *well yield*, *aquifer yield*, and basin yield. These yields must be viewed in terms of a balance between the benefits received from groundwater usage and the undesirable changes induced by groundwater pumping. Basin yield depends on how the effect of pumping is transmitted through the basin and on the changes in rates of *recharge* and *discharge* induced by withdrawals. An equation for the transient *hydrologic budget* for the saturated portion of a groundwater basin is:

$$Q(t) = R(t) - D(t) + \frac{dS}{dt}$$

where:
$Q(t)$ = total rate of groundwater withdrawal [$L^3 \cdot T^{-1}$]
$R(t)$ = total rate of groundwater recharge to the basin [$L^3 \cdot T^{-1}$]
$D(t)$ = total rate of groundwater discharge for the basin [$L^3 \cdot T^{-1}$]
$dS/dt$ = rate of change of storage in the saturated zone of the basin [dimensionless]
Cf. *Safe yield; Optimal yield; Basin accounting*.

**Basin-area ratio ($R_a$)** Within a designated *drainage basin*, the mean basin area of a watercourse with a given *stream order* divided by the mean basin area of the watercourse with the next-lower stream order. Cf. *Basin area*.

**Basin-circularity ratio ($R_c$)** The area of a *drainage basin* divided by the area of a circle with the same perimeter as the basin.

**Baumé degrees** The *specific gravity* of a solution determined by the acid concentration. As the degrees Baumé increase, the strength of the solution increases.

**Baumé gravity** A scale for relating the *specific gravity* of crude oil to that of water, devised by Antoine Baumé. This scale is based on the equation:

$$°(Baume') = \frac{140}{\text{specific gravity at } 60°F} - 130$$

As an example, water at 60°F would result in a Baumé scale reading of 100°Baumé. The Baumé gravity scale was replaced in 1921 by the American Petroleum Institute (API) gravity scale; however, it is still in use today to define acid concentration.

**Bay/embayment** A body of water egressing into land and forming a broad curve. The area of a bay is larger than that of a cove but typically smaller than a gulf. The international designation of a bay, for the purposes of territorial rights, is a body of water consisting of a baymouth less than 24 nautical miles wide and an area that is equal to or greater than the area of a semi-circle with a diameter equal to the width of the baymouth. Cf. *Dead storage*.

**Bayou** A term used in the Gulf Coast region of the United States for a small *tributary*, creek, or secondary watercourse, typically containing sluggish or *stagnant water*, off a larger, higher order body of water. A bayou typically follows a sinuous path through the alluvial lowlands of a large river valley and is a channel that carries flood water or allows access of tidal water through swamps or marshland. Cf. *Tides*.

**Bazin's average velocity equation** An equation for calculating the speed of water flowing in an open channel, proposed by H. E. Bazin in 1897. Bazin's average velocity equation has been modified and a more complex accurate equation has been proposed in *Chezy's formula*.

**Beaded drainage** Also called a beaded *stream*. A pattern resulting from thawing of ground ice in a series of pools 1–4 m deep that are connected by small watercourses or stream segments. Cf. *Drainage pattern*.

**Bed material** The soil and rock particles constituting the foundation of a water body (or former water body). The bed material may be the remains of the watercourse's *suspended load*, *bed load*, or even residual material, depending on the combination of flow characteristics, formation characteristics, and properties of the *drainage area* supplying eroded materials. The dynamics of river mechanics are integrally related to the bed materials and their behavior. The effect of discrete particles of the bed material on water flow is a function of the *grain size* and shape and the aggregate of material forms definitive structures on the bed itself.

**Bed roughness** The relief at the base of a flowing fluid, which comprises the bedforms (or form roughness) and the projections from the sediment bottom (or grain roughness) and which represents the frictional effect that the bed exerts on fluid flow. Bed roughness is considered to be smooth or rough depending on whether elements project through the viscous sublayer at the base of the flow. Cf. *Grain angularity*.

**Bed-material load** A subset of the *sediment load* of a *stream*, that includes those particles of a *grain size* normally found in the stream bed. The bed-material load of a high-velocity flow stream consists predominantly of coarser material, while the bed-material load of a low-velocity stream consists of finer material. Cf. *Bed load*.

**Bed load** Also called bottom load, or traction load. The coarse soil particles, pebbles, cobbles, and boulders transported on or immediately above the bottom of a stream by sliding, rolling, *traction*, or *saltation*, as opposed to those materials moving in suspension, are considered to be *suspended load*. The bed load is often defined as that portion of the stream load in which the predominant transfer of energy is by grain-to-grain contact rather than fluid-to-grain transfer. For rounded and spherical particles, the dividing point between grain-to-grain and fluid-to-grain energy transfer occurs at a volume concentration of solids of about 0.09. In an area with no fine material sources, streams flowing over a bed of pebbles and larger material are usually clear (having no suspended load). Even under strong flow conditions, these streams predominantly carry bed load. With a fine-grained source or bed material, suspended load is likely to dominate under all flow conditions. If the conditions of flow change, particles considered to be bed load may become suspended load and vice versa. Cf. *Flow layers, vertical*; *Suspended load*; *Traction carpet*.

**Bed-load function** The rate for a given channel at which various streamflows transport the different particle sizes of the *bed-load* material.

**Beer's law** A relationship between the intensity of light at the surface of a column of pure water and the intensity of light at any depth within that column, as defined by the equation:

$$q_z = q_0 e^{-nz}$$

where:
$q_z$ = intensity of light at depth $z$
$q_0$ = intensity of light at the water surface, $n$
$e^{-nz}$ = extinction coefficient for a specific wavelength of light

The standard unit of measurement for light intensity is the candela. Light is very quickly transformed to heat or scattered in water and approximately 50% of incident light is lost in the first meter (3.3 ft) of depth. By a depth of 10 m (33 ft), light intensity is only 0.1% of the incident light at the surface. If the water is not pure – and most water bodies have some dissolved or suspended material in them – the effective extinction coefficient is higher.

**Beheaded stream** The remaining lower part of a watercourse that has suffered diminished flow as a result of the capture of its *headwaters* by another stream or from the loss of the upper part of its *drainage area*. Cf. *Stream capture*.

**Beheading** The diversion of a stream's *headwaters* via *stream capture*, which cuts off the upper part of the stream, or via the removal of the upper part of a stream's *drainage area*. Beheading causes the stream's *flow* to diminish. Cf. *Betrunking*.

**Belanger's critical velocity** The flow rate ($L \cdot T^{-1}$) at which the fluid's minimum energy value is attained. Belanger's critical velocity is that condition of fluid flow in open channels for which the velocity head equals one-half the mean depth of the channel. See *critical velocity*; *Reynold's number*; *Kennedy's critical velocity*; *Unwin's critical velocity*; *laminar flow*; *turbulent flow*.

**Belt of soil moisture** See *belt of soil water*.

**Belt of soil water** Subdivision of the *unsaturated zone*. The upper limit of the belt of soil water is land surface and below by the *intermediate belt*, within the unsaturated zone. Plants and roots are supported in this zone and the *soil water* is available for growth. See *zones of water*; *soil water*; *Zone of saturation*.

**Beneficial use** Water law referring to any use of water which results in gain or benefit to the user. Cf. *Prior appropriation*; *Rule of reasonable use*.

**Bentonite/bentonitic clay** A colloidal material of hydrated aluminum silicate, predominantly composed of the mineral sodium *montmorillonite*, and typically formed by the breakdown and alteration of volcanic ash and tuffs. Clays like bentonite have a plate-like structure in which the edges of the plates are positively charged. As a consequence, clay particles swell when exposed to water because the electrically charged water molecules are attracted to the plate surfaces, thereby forcing them apart (**Figure B-12**). These hydrated clay particles occupy more space, increasing the *viscosity* of the fluid. Because of its relatively high density and viscosity, bentonite is used as a constituent of *drilling fluid* to "float" or carry cuttings out of the *borehole*. Bentonite is also added above the installed *filter pack* of a well to seal off the filter pack from surface materials or from the cement grout which may be added to seal the *well casing*. In addition, bentonite is mixed with *grout* or *cement* to counteract the tendency to shrink as these materials dry and set. The proper ratio of bentonite to grout must be maintained to achieve the optimum size characteristics without degrading the structural properties of the grout. A *cement–bentonite* mixture tends to make a better annular well seal than cement alone because as cement ages, surface water leaches through it and may cause high *pH* in wells.

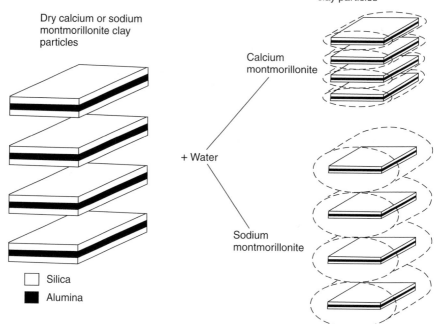

Hydrated montmorillonite clay particles

Dry calcium or sodium montmorillonite clay particles

Calcium montmorillonite

+ Water

Sodium montmorillonite

☐ Silica

■ Alumina

**Figure B-12** The hydration of calcium and sodium montmorillonite in bentonite clays. (Driscoll, 1986. Reprinted by permission of Johnson Screens/a Weatherford Company.)

Cf. *Activation.*

**Bentonite barrier** A clay formation that can act as a formidable obstacle to groundwater migration because of the clay's low *hydraulic conductivity,* ability to swell, and attract water molecules. The volume of the void spaces in a *bentonitic clay* deposit is generally high, thereby giving it a high *porosity,* but the actual void size is very small. Because water is strongly attracted to the positively charged surface of clay particles, the influence of the water gradient is reduced, and thus, water does not easily move through a clay deposit. Bentonite barriers have been used to inhibit migration of contaminants.

**Bernoulli effect** The observed reduction of *pressure* in a fluid as the velocity of *flow* increases. Cf. *Bernoulli equation.*

**Bernoulli equation** Swiss engineer Daniel Bernoulli's mathematical description of the conservation of energy expressing the total energy potential of flowing water due to *pressure energy,* velocity energy, and elevation energy in the steady flow of an ideal, frictionless, incompressible fluid. The *Bernoulli effect* or theorem shows that in fluid flow, pressure is reduced as flow velocity increases, or that fluid velocity increases as the area available for flow decreases. In 1738, Bernoulli theorized that in uniform steady flow in a conduit or stream channel, the sum of the velocity head, *pressure head,* and *elevation head* at any given point is equal to the sum of these heads, or sum of the energy potentials, at any other point, plus (if upstream) or minus (if *downstream*) the losses in head due to friction between the two points. Under irrotational conditions, this energy total remains constant along *equipotential lines,* as shown in the Bernoulli equation:

$$\phi = \frac{p}{\rho} + \frac{V^2}{2g} + z$$

where:

$\phi$ = energy potential, or total energy per unit mass, $E_{\text{tm}}$ [L]

$p$ = pressure [M·L$^{-2}$]

$\rho$ = density of the liquid [M·L$^{-3}$]

$V$ = liquid velocity [L·T$^{-1}$]

$g$ = gravitational constant [L·T$^{-2}$]

$z$ = elevation above some datum [L]

The first term, $p/\rho$, is the pressure head attributed to the forces containing the fluid; $V^2/2g$ is the velocity head resulting from the movement of the water; and $z$ is the elevation energy, which can be converted to velocity of pressure energy due to the position above a datum.

**Betrunking** The loss of the lower part of a *stream* as a result of the submergence of the valley in which it flows, e.g. a stream delta, or of coastal recession causing the upper branches of a *drainage basin* to *discharge* into the sea as independent streams. Cf. *Beheading*.

**Bierschenk method** A straight-line step-drawdown test analysis of specific *drawdown* and *discharge* plotted on arithmetic graph paper and used to identify the parameters of *aquifer loss* and the well loss coefficient. The Bierschenk method is the calculated equivalent of a drawdown–discharge relationship established in the field through a set of *pumping tests*, each with different discharges yielding different drawdown values. To simplify the field method, a continuous pumping test is conducted with discharge increments of a specified duration, resulting in each successive step containing residual drawdown from the preceding step. The plot of the specific drawdown, $s_w/Q$ (the inverse of the *specific capacity*, $Q/s_w$), versus pumping rate, $Q$, should be a straight line. The intercept of this line and the $s_w/Q$ axis equals the aquifer loss factor, $A$, and the slope of the line equals the well loss coefficient, $C$.

**Bifurcation** The separation or branching of a higher order *stream* into two or more separate streams. Cf. *Bifurcation ratio; Stream order*.

**Bifurcation ratio** The number of streams of a given order within a basin divided by the number of streams in the next-lower order. The bifurcation ratio was defined by R. E. Horton in 1945 as part of his attempt to classify basins by the amount of stream branching. The bifurcation ratio generally is similar within a basin and has been found to be between 2 and 4; the mean value for a wide variety of basins is estimated to be about 3.5. A bifurcation ratio of exactly 2.0 is shown in **Figure B-13**. Bifurcation ratio has been used to determine the likelihood of flooding; the higher the ratio, the greater the probability of flooding.

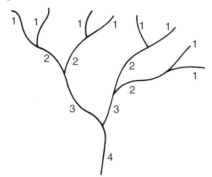

**Figure B-13** A sketch of a stream order with a bifurcation ratio of 2.0. (Linsley et al., 1982. Reprinted with permission from McGraw-Hill, Inc.)

Cf. *Stream basin; Stream order*.

**Bioavailability** An organism's capacity to take up and concentrate a contaminant in water.

**Biochemical oxygen demand (BOD)** See *biological oxygen demand*.

**Biohydrology** The study of the interactions of water, plants, and other organisms by investigating the effects of water on the biota and also the physical and chemical changes caused by the biota in the aquatic environment.

**Biointermediate elements** Chemical elements (barium, calcium, carbon, and radon) that are partially depleted in surface waters as the result of biological activity. Cf. *Biolimiting elements; Biounlimiting elements*.

**Biolimiting elements** Chemical elements (nitrogen, phosphorus, and silicon) that are required by living organisms and are depleted in surface waters, relative to their deep-water concentration, by biological activity, thereby limiting further biological production until the elements are replaced. Replacement of biolimiting elements usually occurs by upwelling. Cf. *Biointermediate elements; Biounlimiting elements*.

**Biological organisms in water** The forms of life found in aquatic conditions. Biological organisms are much more limited in *groundwater* than in surface waters. The two major groups of organisms in surface waters are those utilizing light and carbon dioxide ($CO_2$) to synthesize carbon compounds (autotrophs) and those utilizing carbon compounds for energy and as components to form other carbon compounds (heterotrophs). Because light is required for photosynthesis, autotrophs are rarely present in groundwater. Among the heterotrophs in surface waters (zooplankton, macroinvertebrates, fish, and decomposing bacteria), only the smallest (the bacteria) are found in groundwater unless the pore spaces are large, as in karst regimes and possibly some *fracture porosity* situations. Two types of decomposing bacteria are found in groundwater: aerobic bacteria that use oxygen in the process of converting organic material into $CO_2$, water, and inorganic oxides and anaerobic bacteria that use the reduction process instead of oxidation for energy to break down organic materials. These bacteria can be important in the cleanup of contamination in groundwater. The relationship between water quality parameters and water body materials is shown in a flow diagram in **Figure B-14**. Cf. *Bacterial degradation*.

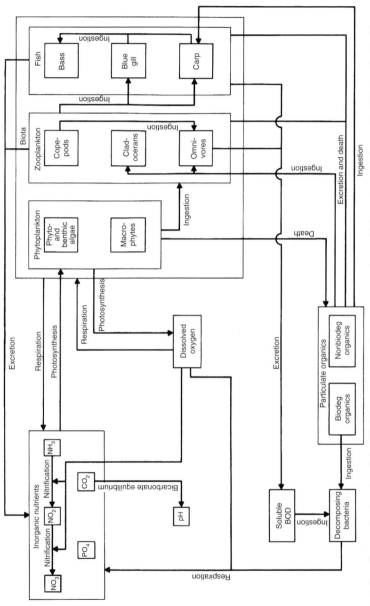

**Figure B-14** A flow diagram showing the relationship between water quality parameters and material interactions. (Linsley et al., 1982. Reprinted with permission from McGraw-Hill, Inc.)

**Biological oxygen demand (BOD)** Also called biochemical oxygen demand. A measure of the amount of oxygen ($O_2$) that is removed from aquatic environments by the metabolic reactions of *aerobic* microorganisms, which indicates the concentration of organic matter in water. Oxygen is consumed by microorganisms, e.g. bacteria, during metabolism of carbon- and nitrogen-containing compounds under aerobic conditions. The consumption of oxygen by bacteria results from the oxidation of organic matter; therefore, a relatively high BOD of $1^{-1}$ mg occurs in the presence of a high concentration of organic matter. In the BOD test, only biologically reactive carbon is oxidized and is measured as the weight in milligrams of oxygen used by 1 L of effluent kept in dark at 20°C for 5 days. The chemical oxygen demand (COD) test is an approximate substitute for the BOD test. The COD test measures the total quantity of $O_2$ required for oxidation of a waste to $CO_2$ and water and can be performed in 3 h. The range of possible readings can vary considerably: Water from an exceptionally clear lake might show a BOD of less than 2 mL $L^{-1}$ of water. Raw sewage may give readings in the hundreds and food processing wastes may be in the thousands. Cf. *Chemical oxygen demand.*

**Bioremediation** The controlled use of naturally occurring or manufactured microorganisms to modify, control, and/or reduce contaminant levels. An organic compound is oxidized (made to lose electrons) by an electron acceptor, which then becomes reduced (gains electrons). $O_2$, nitrate ($NO_3^-$), sulfate ($SO_4^{-2}$), and $CO_2$ are electron acceptors that can convert organic wastes into biomass, $CO_2$, methane ($CH_4$), and inorganic salts. Bioremediation occurs naturally and can also be enhanced by an engineered delivery of nutrients to the microorganisms or an addition of available electron acceptors to the contaminated zone within the groundwater system.

**Biounlimiting elements** Chemical elements (such as boron, magnesium, strontium, and sulfur) that show no measurable decrease in concentration as a result of biological activity in surface waters as compared to deeper waters. Cf. *Biolimiting elements; Biointermediate elements.*

**Bitter lake** A confined body of water with a high sodium sulfate ($Na_2SO_4$) content in addition to the typical carbonates and chlorides contained in salt lakes although the concentration of carbonates and chlorides is generally less than that typically found in salt lakes. The water is said to have a bitter taste. Carson Lake in Nevada and the Great Bitter Lake in Egypt are examples of bitter lakes. Cf. *Salinity.*

**Black smoker** See *hydrothermal vent.*

**Blaney–Criddle method** An indirect estimation of *potential evapotranspiration* (PE), used to predict the seasonal consumptive use of water by a particular agricultural crop. The Blaney–Criddle method is similar to the *Thornthwaite equation* but adds a crop factor that varies with the growing season, thereby restricting its use to agricultural crops. The Blaney–Criddle method does not account, however, for the effects of wind and relative humidity. The equation is:

$$E_t = K^* \left( \sum{}^m \right) PT$$

where:
$E_t$ = estimated evapotranspiration (in.)
$K$ = annual or seasonal consumptive use coefficient (particular to a crop and empirically determined)
$m$ = monthly summation of the percentage of daylight hours in a given period [T]
$P$ = given period [T]
$T$ = mean monthly temperature (°F)
Cf. *Hargreave's equation.*

**Blank sample** A sample prepared by the laboratory or sometimes in the field by filling a sample bottle with distilled or deionized water. A quality assurance/quality control (QAQC) technique to determine whether the samples have become contaminated by the sampling process or through transit.

**Blasting** See *explosives, use of.*

**Blind creek** A little used term in hydrology (but may be found in literature) for a small watercourse that is dry until *precipitation* exceeds *infiltration* rates. Cf. *Intermittent stream.*

**Blind drainage** A *drainage basin* that has no surface outlet, or an individual *depression storage* of appreciable area relative to the *drainage basin*. In karstic topography, blind drainage may take the form of a *stream* draining or disappearing into a tunnel at the lowest point in a valley. In arid or semi-arid regions, blind drainage may lead into a valley with a sill, where the *flow* into the valley is removed by *evaporation* before the water level in the *lake* or pond can reach the elevation of the sill. Blind drainage also results when water flows into basins below sea level, such as in Death Valley in the western United States or the Dead Sea between Israel and Jordan.

**Blowing well** More commonly called a *breathing well* or *blow well*. A water well in which air moves through the point of entry. Cf. *Flowing well.*

**Blowout** An uncontrolled or even violent and usually unexpected escape of water, oil, gas, or *drilling fluid* from a *borehole* or well that occurs when formation pressure exceeds the *hydrostatic head* of the fluid, or a violent release of fluids as a result of the failure of equipment regulating the *flow* in a well.

**Blow well** See *artesian well; aquifer, artesian; Blowing well; Flowing artesian well.*

**Boathook bend** The area where a *tributary* joins the main stream in an *upstream* direction, forming a sharp curvature in the shape of a boathook. Cf. *Barbed tributary.*

**Boca** A Spanish term, commonly used in the United States, for the area where the mouth of a *stream* flows out of a channel, canyon, gorge, or precipitous valley onto a plain.

**BOD** See *biological oxygen demand.*

**Bog** An area of wet, spongy ground with soil composed mainly of decayed vegetable matter. Sphagnum moss grows out from the edge of the bog in tangled mats strong enough to support trees, shrubs, and even people. In northern Europe, bogs (also called "heaths") were once the gloomy retreats of society's outcasts. They became known as "heathens" and "bogeymen". Bog waters are characteristically acidic and contain little to no oxygen. Cf. *Fen*; *Marsh*; *Swamp*.

**Boiling point variation with elevation** Reduction of the temperature at which a liquid boils with increasing elevation above sea level, due to the decrease of atmospheric pressure. Water boils when its *vapor pressure* at that temperature exceeds the atmospheric pressure at that location. At sea level, with atmospheric pressure at 760 mm of mercury, the boiling point of water is 100°C (212°F). At the top of Mount Everest in Tibet approximately 30,000 ft above sea level, atmospheric pressure is approximately 215 mm of mercury and the boiling point of water drops to about 68°C. The variations of pressure, temperature, density, and boiling point with elevation are shown in **Table B-1**.

**Table B-1** The effect of elevation change to pressure, temperature, density, and boiling point of water. ("U.S. Standard Atmosphere, 1962," National Aeronautics and Space Administration, U.S. Air Force, and U.S. Weather Bureau.)

| Elevation from mean sea level (m) | Pressure | | | Air temp. (°C) | Air density (kg/m³) | Boiling point (°C) |
|---|---|---|---|---|---|---|
| | mm Hg | Millibars | cm $H_2O$ | | | |
| −500 | 806.15 | 1074.78 | 1096.0 | 18.2 | 1.285 | 101.7 |
| 0 | 760.00 | 1013.25 | 1033.2 | 15.0 | 1.225 | 100.0 |
| 500 | 716.02 | 954.61 | 973.4 | 11.8 | 1.167 | 98.3 |
| 1000 | 674.13 | 898.76 | 916.5 | 8.5 | 1.112 | 96.7 |
| 1500 | 634.25 | 845.60 | 862.3 | 5.3 | 1.058 | 95.0 |
| 2000 | 596.31 | 795.01 | 810.7 | 2.0 | 1.007 | 93.4 |
| 2500 | 560.23 | 746.92 | 761.6 | −1.2 | 0.957 | 91.7 |
| 3000 | 525.95 | 701.21 | 715.0 | −4.5 | 0.909 | 90.0 |
| 3500 | 493.39 | 657.80 | 670.8 | −7.7 | 0.863 | 88.3 |
| 4000 | 462.49 | 616.60 | 628.8 | −11.0 | 0.819 | 86.7 |
| 4500 | 433.18 | 577.52 | 588.9 | −14.2 | 0.777 | 85.0 |
| 5000 | 405.40 | 540.48 | 551.1 | −17.5 | 0.736 | 83.3 |

**Boiling spring** A *discharge* of water that is agitated by heat and appears to "boil" as it emerges but is actually at a temperature much less than boiling. The appearance of boiling water may also be caused by strong, turbulent, vertical *eddy* flow from the *spring*.

**Bond strength** The adhesion, friction, and longitudinal *shear stresses* that resist separation from the materials in contact with set *grout*.

**Border spring** See *spring, boundary*.

**Bored well** A hole dug into an *aquifer* using a hand-operated and/or power-driven auger system. The term "bored well" is usually restricted to wells between 3 and 30 m in unconsolidated formations. In areas where these formation materials are competent or consolidated enough to maintain an open hole without casing, bucket augers or solid-stem augers are often used to drill the hole and withdraw sediments, and then the *well casing*, *well screen*, and pumps are placed. See: **Appendix C – Drilling Methods**. Cf. *Bore*; *Borehole*.

**Borehole** Also called a boring or a well bore. A relatively small-diameter hole (e.g. 8–16 cm) advanced below the ground surface, typically by various drilling rigs, (See: **Appendix C – Drilling Methods**) selected on the basis of the geology, size, and depth of the installation required, as well as the availability of rig type and personnel expertise. The advancement of a borehole is the most direct method of soil or *groundwater* sample collection, geologic interpretation, well installation, and measurement or recording of geophysical properties. Boreholes are typically advanced by a drill rig but historically or in remote locations have also been advanced by hand or with the use of animal labor. Geologic details obtained from a borehole are representative of the immediate area around it. Boreholes advanced in other selected sites can provide general spatial information on the *hydrogeology* of an area, but caution is advised in extrapolation significantly beyond the immediate area of borehole placement and borehole-to-borehole geology relations. During the installation of a well, the hole is called a boring until the well is installed, and then it is called a well. Cf. *Well bore*.

**Borehole dilution test** A method to obtain estimates of groundwater velocity in the vicinity of tested wells as an alternative to the large-scale pump test to evaluate area *hydraulic conductivity*. This method is based on the rate of dilution of a tracer that is initially placed within the *well screen* zone below a *packer*. The rate of dilution is proportional to the ambient velocity of groundwater moving through the well screen.

**Borehole geochemical probe** A water-quality measuring device lowered on a cable into a well to take direct readings of chemical parameters such as *dissolved oxygen*, *Eh*, *pH*, *specific conductivity*, temperature, and *turbidity*. Vertical profiles of geochemical variations in an *aquifer* can be logged by slowly lowering the probe into a *borehole* and recording the parameters measured. Cf. *Borehole geophysics*.

**Borehole geophysics** A field of geophysics emphasizing the use of wells or *boreholes* in which physical data are collected as a function of depth. Instruments and equipment are lowered into open boreholes or wells to measure and record naturally occurring parameters within a formation. Many different borehole geophysical techniques borrowed from the petroleum industry are applicable to water wells. **Table B-2** lists some of the available logging techniques with associated parameters useful for *hydrogeology* and hydrologic investigation, and **Figure B-15** shows an example of their interpretation.

**Table B-2** Summary of various geophysical tools and potential uses. (Welenco, 1996. Reprinted with permission from Welenco Inc.)

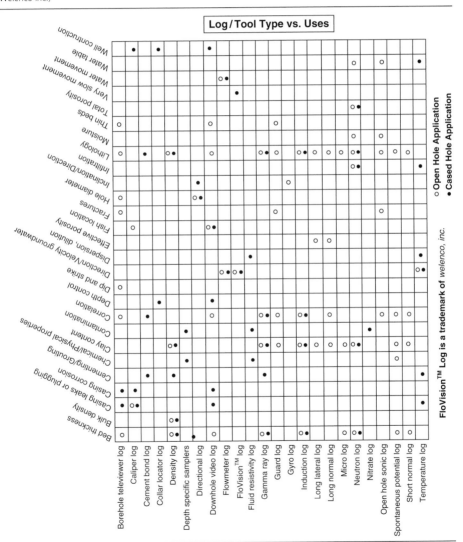

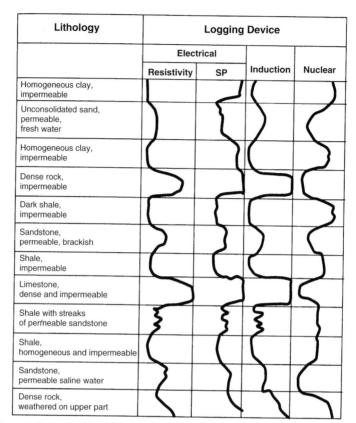

| Lithology | Logging Device | | | |
|---|---|---|---|---|
| | Electrical | | Induction | Nuclear |
| | Resistivity | SP | | |
| Homogeneous clay, impermeable | | | | |
| Unconsolidated sand, permeable, fresh water | | | | |
| Homogeneous clay, impermeable | | | | |
| Dense rock, impermeable | | | | |
| Dark shale, impermeable | | | | |
| Sandstone, permeable, brackish | | | | |
| Shale, impermeable | | | | |
| Limestone, dense and impermeable | | | | |
| Shale with streaks of permeable sandstone | | | | |
| Shale, homogeneous and impermeable | | | | |
| Sandstone, permeable saline water | | | | |
| Dense rock, weathered on upper part | | | | |

**Figure B-15** Conceptual example of using borehole geophysics information to interpret subsurface lithology. (Maidment, 1993. Reprinted with permission from McGraw-Hill, Inc.)

**Borehole infiltration test** A method of assessing *hydraulic conductivity* in unsaturated soils. A *borehole* is advanced in the soil of the area to be tested, filled with water, and the rate of inflow is then measured while a constant *head* is maintained in the borehole. Hydraulic conductivity can be determined by the following relationships (valid for $H > 7h_m$ and $h > d/4$):

$$\text{For constant head tests: } k = \frac{Q}{7\pi dh_m}$$

$$\text{For variable head tests: } Q = \frac{\Delta h}{\Delta t}\pi\frac{d^2}{4}$$

$$k = \frac{\Delta hd}{\Delta t28h_m}$$

where:
$k$ = hydraulic conductivity [L·T$^{-1}$]
$Q$ = volumetric flow rate [L$^3$·T$^{-1}$]
$d$ = diameter of borehole [L]
$h$ = depth of borehole [L]
$h_m$ = mean height of water in borehole [L]
$H$ = height between bottom of borehole and water table [L]
Cf. *Infiltration test.*

**Borehole log** See *well logging.*
**Borehole permeability test/hydraulic conductivity test** See *slug test; bail-down test/bail test; packer test; pumping test.*
**Borehole resistivity** See *electric logging.*
**Borehole storage effect** The influence on water level that is due to water removed from the *well casing* and surrounding *gravel pack* when well pumping begins. Removal of this water from *borehole* storage, or *dead storage*, affects the water level in the well (i.e. time response to pumping) but is not indicative of *aquifer* water.
**Borehole survey** See *well logging.*
**Borehole televiewer** See *acoustic televiewer (ATV) logging.*
**Boring** See *borehole.*
**Bottom load** See *bed load.*
**Bottom water** Also called *deep water.* A water mass that cools and becomes denser than the surrounding water and subsequently moves as a mass to the deepest portion or bottom of the water body. The bottom water mass is relatively dense and cold (e.g. North Atlantic bottom water, 1–2°C).
**Bouguer anomaly value** The measured value of gravity at a specific location subtracted from the theoretical value of gravity. The resulting value is used when making a *Bouguer correction* to gravity survey data; the correction adjusts for the gravitational variation resulting from a difference in the mass of rock between the gravity station and a common datum (usually sea level). Gravity and magnetic surveys have been used to delineate potential *aquifers* contained in basin fill areas and buried stream channels. After the gravity data are corrected with the Bouguer anomaly value, along with adjustments for tidal effects, terrain, and latitude, a contour map of the corrected data can be used to delineate *groundwater basins* and depths to bedrock. **Figure B-16** shows the use of a corrected gravity profile to delineate basin fill and a potential aquifer.

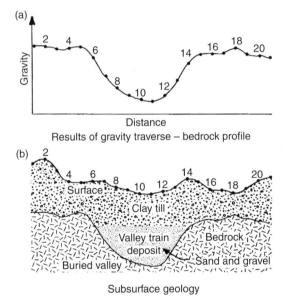

**Figure B-16** Gravity instruments are used to locate buried bedrock valleys lying below unconsolidated deposits. These valleys often contain sand and gravel deposits that can yield high volumes of groundwater. (Driscoll, 1986. Reprinted by permission of Johnson Screens/a Weatherford Company.)

**Bouguer correction** An adjustment to gravity data for the difference in the mass of material between the measurement point (the gravity station) and a common datum such as *sea level.* The Bouguer correction accounts for the distance and additional mass between sea level and a measurement point above sea level, or for the distance and the mass that is missing in the interval between sea level and a measurement point below sea level. In either case, the correction is applied the same way, and a density must be assumed for the mass that is removed or added. Cf. *Bouguer anomaly value.*
**Bound water** Water that cannot be removed or used without changing the structure or composition of the material associated with it, e.g. plant, animal, or soil, and is therefore not interactive as *free water.*

41

**Boundary, aquifer** The horizontal or vertical limit of a *groundwater reservoir* with relatively *homogeneous* properties. Many analytical techniques for evaluating the properties and associated production potential of an *aquifer* assume an infinite areal extent; if this assumption is not valid, the conclusions reached may be erroneous. Aquifer boundaries can be caused by changes in aquifer properties and can often be anticipated on the basis of known geological conditions. If the detailed geology of the area is not known, *pumping tests* may be needed to reveal the boundaries of an aquifer. Cf. *Boundary conditions; Boundary, specified-head.*

**Boundary conditions** The change in *hydraulic head* or recharge/discharge conditions that are defined at the vertical and horizontal limits of the area to be modeled. Three basic features of a region must be known to solve *flow equations* for its aquifer system: the size and shape of the system, the boundary and initial conditions, and the physical and chemical properties of the system. The most commonly encountered boundary condition is reduced hydraulic conductivity or an *impervious boundary* potentially resulting from fine-grained layers occurring vertically or bedrock occurring either vertically or horizontally. *Recharge boundaries* or hydraulic-head controls occur at the intersection of the aquifer with a body of water. Boundaries may be defined as no-flow (impermeable boundaries; $\partial h/\partial n = 0$, or flow = 0); flux boundary, where the flux is a specified value; or specified-head boundary, where the head is constant or stipulated as a time varying function. Cf. *Boundary aquifer; Boundary, flux; Specified-head boundary; Barrier, aquifer effects.*

**Boundary, flux** A *aquifer* condition simulated by varying the flow at the limits of an aquifer, or boundary node, or changing value, to solve a set of equations that forms a mathematical model of the physical and/or chemical processes occurring. Cf. *Boundary, aquifer; Boundary conditions.*

**Boundary, head-dependent** In computer modeling, this condition establishes a flux based on a stipulated head. See *Cauchy condition.*

**Boundary, no-flow** In numerical groundwater modeling, a *boundary condition* across which *flow/flowlines* cannot occur.

**Boundary, specified-head** A *aquifer* condition simulated by setting the *head* at the limits of an aquifer, or boundary node, to a known head value or time varying head value to solve a set of equations that forms a mathematical model of the physical and/or chemical processes occurring. Cf. *Boundary, aquifer; Boundary conditions.*

**Boundary spring** See *spring, boundary.*

**Boundary value problem** Where an ordinary differential equation or a partial differential equation is used to define a domain ($\Omega$), which has values assigned to the physical boundaries of the domain. In numerical modeling, the physical boundaries may be described as no flow boundaries, constant head boundaries, and constant flux boundaries.

**Bounding formation** A geologic formation that acts as a limit, either vertical or horizontal, to an *aquifer* and is usually less permeable than the aquifer. Bedrock formations are one type of bounding formation and often form the bottom or a horizontal limit of an aquifer. Fine-grained sedimentary formations can form lower and upper bounding formations if their *hydraulic conductivity* is three orders of magnitude difference from that of the aquifer formation itself. In some depositional environments such as river valleys, erosional effects on fine-grained deposits and subsequent deposition of coarser material can produce horizontal aquifer boundaries. Cf. *Permeability/permeable.*

**Boussinesq equation** A differential flow equation describing two-dimensional, unconfined, transient flow. In the unconfined water-table *aquifer*, water is derived from storage by vertical drainage, resulting in a decline in the saturated thickness, *drawdown*, near the pumping well over time. The Boussinesq equation is linearized by assuming that the drawdown in the aquifer is very small compared to the saturated thickness; the variable thickness in the differential equation is then replaced by an average thickness, $b$. The equation becomes a partial differential equation in which the head, $h$, is described in terms of directions, $x$, $y$, and $z$, and time, $t$:

$$\frac{\partial^2 h}{\partial x^2} + \frac{\partial^2 h}{\partial y^2} = \frac{S_y}{Kb}\frac{\partial h}{\partial t}$$

where:
$S_y$ = specific yield [dimensionless]
$K$ = vertical conductivity [L·T$^{-1}$]
$b$ = average saturated thickness [L]
$h$ = head [L]
$t$ = time [T]

**Bowen's ratio (R)** The mathematical relationship between heat lost by *conduction* and energy utilized by *evaporation* (i.e. between sensible heat and *latent heat*). Bowen's ratio may be used to evaluate *evapotranspiration*, $E_t$, and calculate evaporative losses from water bodies, thus avoiding the difficulty of measuring conductive heat loss. Through manipulation of the energy budget equation for a surface water body or reservoir surface, Bowen's ratio, $R$, can be reduced to a function of the air and water temperatures, the atmospheric pressure, and the measured saturation vapor pressure of the air over the water:

$$R = \left(0.61\frac{P}{1000}\right) \times \frac{(T_w - T_a)}{(e_w - e_a)}$$

where:

$P$ = atmospheric pressure (millibars)

$T$ = temperature of water, w, and air, a (°C)

$e$ = vapor pressure gradient of water, w, and air, a (millibars)

Values for $R$ vary between 1 and −1, with a negative sign indicating dew or frost. Combined with the evaporative heat loss, the conductive heat loss can then be calculated instead of measured.

**Bouwer and Rice method** One of many methods of *slug test* data analysis, in which initial water levels are measured and then the well is either *bailed down*, or a known volume of water is added, or a slug is lowered or raised within the well. The changes in water level during the recovery period are then measured. Use of this method is specific to a well completed near the water table. Cf. *Hvorslev method*.

**Brackish water** Water having saline or *total dissolved solids* (TDS) content that is intermediate between the *freshwater* of streams and the saltwater of oceans. The lower limit of the brackish water category has been defined in relation to *drinking water quality* and is generally accepted as one part per thousand of TDS. The upper limit has no particular functional relationship to health or other characteristics and has been somewhat arbitrarily accepted in *hydrogeology* as 10 parts per thousand of TDS. Limits set by the US Department of Agriculture are more than 1500 mg $L^{-1}$ and less than 15,000 mg $L^{-1}$ TDS. The *salinity* range for oceans (euhaline seas) is 30–35‰, brackish seas or waters in the range of 0–29‰, and metahaline seas from 36 to 40‰. These waters are all grouped as homoiohaline because their salinity is derived from the ocean (thalassic) and essentially invariant, in contrast to poikilohaline environments in which the salinity variation is biologically significant. The Thalassic series classifies brackish waters and freshwater by their percent salinity into the following zones:

| Thalassic series | |
| --- | --- |
| >300‰ | |
| | Hyperhaline |
| 60–80‰ | |
| | Metahaline |
| 40‰ | |
| | Mixoeuhaline |
| 30‰ | |
| | Polyhaline |
| 18‰ | |
| | Mesohaline |
| 5‰ | |
| | Oligohaline |
| 0.5‰ | |

Cf. *Chebotarev's succession*.

**Braided stream** A watercourse that has divided and reunited in several channels, or *anabranches*, resembling in plan view the strands of a complex braid. Braided streams tend to be wide and shallow. The causes of the channel deviation are usually obstructions, which can be materials deposited by the stream as a result of the stream's inability to maintain its load-carrying capacity or in some cases vegetation or other materials not related to the streamflow. The path changes produce deposits with an irregular areal distribution. An overloaded and *aggrading* stream flowing in a wide channel or *floodplain* leads to braiding. The reuniting of the divergent streams is generally due to the tendency of flowing water to take the steepest *gradient*. **Figure B-17** shows a graph of the relationship between a river's *discharge* and a channel slope, demonstrating that braided channels occur on steeper slopes than meanders. Sediment transport and *bank* erosion are associated with steeper slopes and also with coarse *heterogeneous* materials, which are all conditions contributing to braiding.

**Branch** A small stream or *tributary* of a main *stream*. Cf. *Distributary*.

**Branching bay** A *bay* displaying a dendritic pattern as a result of the flooding of a river valley by seawater. Cf. *Estuary*.

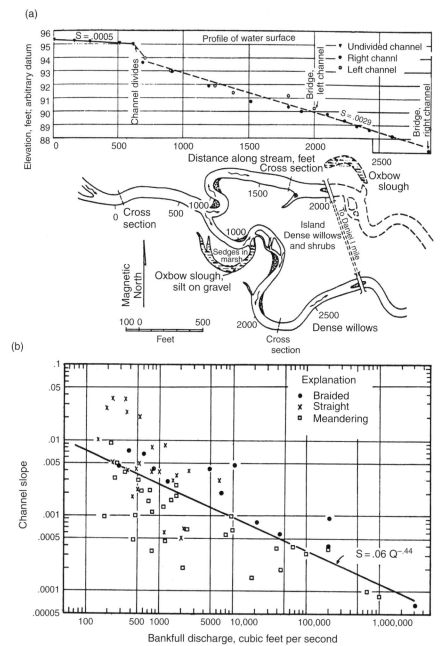

**Figure B-17** The discharge to slope relationship between braided and non-braided streams. (a) Charts the profile of the water surface in the map view. (b) Graphs the channel slope vs. the bankfull discharge for braided, straight and meandering streams. (Leopold et al., 1992. Reprinted with permission from Dover Publication, Inc.)

**Breakthrough/breakthrough curve** In contaminant hydrogeology, it is the concentration-versus-time plot at a monitoring point downgradient from a source. **Figure B-18** illustrates the *longitudinal dispersion* of a *tracer (solute)* with steady flow and continuous supply of the solute at a given concentration, $C$, at time 0, $t_0$, at the solute's first appearance, $t_1$, and at breakthrough, $t_2$. The tracer/solute front spreads along the flow path, which increases with the distance traveled. The behavior of the tracers or solutes shown in **Figure B-18** migrating through porous materials reflects *mechanical dispersion* and *molecular diffusion*, which cause some of the molecules to move faster than the *average linear velocity* of the water and cause some to move slower.

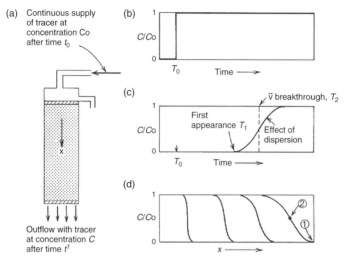

**Figure B-18** A tracer in a porous medium displaying the longitudinal dispersion and eventual breakthrough. (a) The supply of tracer and (b)–(d) demonstrates the time for the appearance of the tracer as a result of longitudinal dispersion. (Freeze and Cherry, 1979. Reprinted with permission from Prentice-Hall, Inc.)

**Breathing well** A fluid-production unit set below the ground surface with a portion of its *well screen* within the *vadose* or *unsaturated zone*, which is a *porous medium* and permeable enough to freely transmit air but is capped above by an impermeable unit restricting direct access to the atmosphere. A breathing well alternately "sucks and blows" air in response to changes in atmospheric pressure. Cf. *Blowing well*; *Flowing artesian well*.

**Breccia porosity** The interparticle *porosity* found in brecciated sedimentary rock.

**Bridge/bridging** Particles of rock or sediment that are locked together and form an obstruction of the well bore or well annulus. Bridging can result from caving of the formation in the well bore, intrusion of gravel or boulders, improperly set *filter pack* material, or *overpumping* during *well development*. **Figure B-19** shows the formation of a bridge caused by

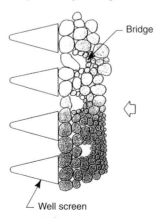

**Figure B-19** Bridging caused by overpumping and flow occurring in only one direction. (Driscoll, 1986. Reprinted by permission of Johnson Screens/a Weatherford Company.)

pumping the well in excess of its capacity, when water is flowing in only one direction. The *surging* method breaks down such bridges and allows proper development of the formation around the well, as shown previously in **Figure B-1**. Improper placement of the filter pack during well installation causes bridging of sand grains in the *annulus*, leaving void spaces that tend to collapse at the start of pumping, thereby leading to the collapse of the well-completion material. Bridging in the annular seal can be avoided by moving or tapping the casing from side to side as the sand is placed downhole or by tamping the sand as it is placed with a small-diameter pipe.

**Bridge seal** A type of seal used to properly abandon a producing well. The bridge seal is the bottom-most *cement* or weighted wood plug that is placed beneath a major *aquifer*. During well abandonment, it is critical to seal off the aquifers that were intercepted both above and below by the well. Cement may be placed on top of the bridge seal to complete the bottom seal. Well seals are illustrated in **Figure B-20**.

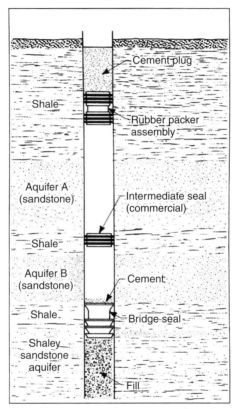

**Figure B-20** A bridge seal commonly used to abandon a producing well. (Driscoll, 1986. Reprinted by permission of Johnson Screens/a Weatherford Company.)

Cf. *Sealing abandoned wells*.

**Bridging of sand grains** See *bridge/bridging*.

**Brine** Water with high levels of salt or *dissolved solids* (*DO*), typically containing elevated levels of $Ca^{2+}$, $Na^+$, $K^+$, and $Cl^-$. Briny pore fluid often occurs in deep sedimentary basins such as oil fields, geothermal areas encouraging mineralized fluids, and restricted basins such as the Salton Sea in California. Chemical characteristics of waters can be very different at depth, where temperature and pressure affect mineral solubilities and ion complexing. Therefore, deeper *groundwater* can become more saline or briny than water at shallower depths. In 1955, according to Chebotarev, groundwater evolves chemically toward the composition of seawater in an anion evolution sequence based on mineral availability and mineral solubility. In *Chebotarev's succession*, deep groundwater flow in a sedimentary basin evolves to a rich brine, at which point $Cl^-$ is the dominant anion species. Cf. *Brackish water; Freshwater; Artificial brine; Salinity; Thalassic series*.

**Brine front** In both petroleum production and *groundwater* supply, the intrusion of *brine* into the petroleum or *freshwater*. In petroleum formations, the brine is usually beneath the petroleum in a stratigraphic or structural trap. In groundwater supply, saltwater can intrude into coastal *aquifers* if production exceeds *recharge*. In this case, the position of the brine front may determine which production wells must be shut down as their water becomes too saline for use. Cf. *Thalassic series*.

**Broken stream** A *stream* that disappears and reappears along its watercourse. Broken streams often occur in arid regions.

**Brooks and Corey model** A model to describe the unsaturated behavior of soils. Suction properties constitute a key element for the functional representation of unsaturated flow conditions. *Water retention curves* developed by Brooks and Corey in 1964 are governed by the following equations:

For $h < h_b$:

$$S_e = \frac{h}{h_b}^{(-\lambda)}$$

$$K_r = \frac{h}{h_b}^{(-2-3*\lambda)}$$

For $h \geq h_b$:

$$S_e = 1$$
$$K_r = 1$$

where:
  $S_e$ = effective saturation = $(\text{VMC} - \text{RMC})/(\theta - \text{RMC})$
  VMC = volumetric moisture content $[\text{L}^3 \cdot \text{L}^{-3}]$
  RMC = residual moisture content $[\text{L}^3 \cdot \text{L}^{-3}]$
  $K_r$ = relative hydraulic conductivity $[\text{L} \cdot \text{T}^{-1}]$
  $h$ = pressure head [L]
  $h_b$ = bubbling or air entry pressure head [L]
  $\lambda$ = pore size distribution index [dimensionless]
  $\theta$ = porosity $[\text{L}^3 \cdot \text{L}^{-3}]$
  Cf. *Buckingham–Darcy equation; van Genuchten equations.*

**Bubbler stage gauge** A device used to record the water *flow* stage at a *gauging station* by recording the pressure required to maintain a small flow of gas from an orifice submerged in the *stream*. The bubbler stage gauge is based on the variable height of water above the gauge producing a variable pressure at the gauge level. The automated device measures the water pressure near the bottom of the stream and equates the pressure with the water depth at which the gauge is placed.

**Bubbling pressure** See *air-entry pressure.*

**Buckingham–Darcy equation** In 1907 Buckingham modified Darcy's equation or *Darcy's law* for steady flow to describe flow in unsaturated soils. The Buckingham–Darcy equation assumes that the air in voids is interconnected, continuous, and able to easily escape as water moves in, thereby offering negligible resistance to water flow. Buckingham also assumed that unsaturated flow is isosmotic and isothermal so that the effects of salt and temperature variations on liquid movement are negligible. In the modified equation for unsaturated soil, $K(\theta) = K$ in that the *hydraulic conductivity*, $K$, is a function of the volumetric *soil water* content, $\theta$, the soil water potential, $h$, is negative because of the capillary suction forces, a function of $\theta$; and thereby $h = h(\theta)$ and becomes the soil water matric potential head. The vertical flow, $q$, is defined as:

$$q = -K(\theta)\left[\frac{\partial h(\theta)}{\partial z} - 1\right]$$

where:
$z$ = soil depth [L]
$\theta$ = volumetric soil water content

The Darcy's equation for steady flow then becomes an equation of soil water *diffusivity*, $D$:

$$q = -D(\theta)\frac{\partial \theta}{\partial z} - K(\theta)$$

where:

$$D(\theta) = K(\theta)\frac{dh}{d\theta}$$

When $\theta$ is assumed to be a unique function of $h$, the term $K(\theta)$ can be written as $K(h)$, the matric potential head. Cf. *Head; Flow, steady.*

**Buffering capacity** The ability of water to maintain the localized *pH* by neutralizing added acids or bases. A low-alkaline body of water (*soft water*) has a low buffering capacity and is vulnerable to acid input, e.g. *acid rain.* Acidic water bodies are likely higher in *hardness*, alkalinity, and buffering capacity. Phosphates, carbonates, and bicarbonates provide buffering capacity against pH changes in natural waters and are important in *water quality.* This buffering capacity is provided when dissolved $CO_2$ is consumed by reaction faster than it can be replaced by atmospheric gas exchange or produced by bacterial decomposition. Cf. *Buffering intensity; Hardness.*

**Buffering intensity** The measure in natural water of how fast a system can react to maintain its normal *pH* value. The speed of buffering reactions is important in the equilibrium of water, which involves *precipitation* and dissolution of solid phases. Rapid interactions aid in maintaining the natural pH, but if the reactions are slow or limited, then the *buffering capacity* is less.

**Buildup test** See *recovery test.*

**Bulk concentration (m)** The mass of a constituent per bulk volume of soil, $m$, is expressed by the following equation:

$$m = \theta_w c_w + \theta_a c_a + \theta_o c_o + \rho_b c_s$$

where:
$\theta_w$ = volumetric water content $[L^3]$
$\theta_a$ = volumetric air content $[L^3]$
$\theta_o$ = volumetric content of a *non-aqueous-phase liquid* (NAPL) or oil $[L^3]$
$c_w$ = concentration of a constituent in the water phase $[M \cdot L^{-3}]$
$c_a$ = concentration of a constituent in the air phase $[M \cdot L^{-3}]$
$c_o$ = concentration of a constituent in the NAPL phase $[M \cdot L^{-3}]$
$\rho_b$ = soil bulk density $[M_{bulk} \cdot L^{-3}]$
$c_s$ = soil phase concentration, mass of constituent sorbed per unit mass of soil $[M \cdot M^{-1}]$

According to thermodynamic equilibrium principles and mass-transfer kinetic factors, individual chemical constituents partition themselves among the various phases. **Figure B-21** is a schematic of the presence of four phases: water, air, NAPL, and soil. Compounds can migrate across the physical interface existing between each of the phases. In multi-phase systems, investigations of the fate and transport of chemical constituents are based on the assumption that the *pore spaces* in the medium are filled by the sum of the fluids present, leading to the calculation of the bulk concentration to facilitate an understanding of solute concentrations.

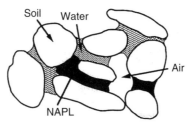

**Figure B-21** Compounds are capable of migrating across the physical interface existing between four phases. (Maidment, 1993. Reprinted with permission from McGraw-Hill, Inc.)

**Bulk mass density** The oven-dried mass of a sample of a soil or rock divided by its field volume, which can be used to calculate the *porosity* of the material. Porosity can be measured in a laboratory by saturating a sample, measuring its volume, weighing it, oven-drying it, and then converting the weight of the lost water to a volume. In practice, however, it can be difficult to completely saturate a sample; therefore, the following equation is useful:

$$n = 1 - \frac{\rho_b}{\rho_s}$$

where:

$n$ = porosity [dimensionless]
$\rho_b$ = bulk mass density of the sample [$M \cdot L^{-3}$]
$\rho_s$ = particle mass density [$M \cdot L^{-3}$]

B

The particle mass density is the oven-dried mass divided by the volume of the solid particles, as determined by a water displacement test or simply approximated as $\rho_s = 2.65$ g cm$^{-3}$ for most soils.

**Bulk water partition coefficient ($B_w$)** In a multiphase system, the determination of solute transport is simplified by comparing concentrations in one phase, e.g. air, oil, or soil, to concentrations in another phase or to the bulk or total concentration. **Figure B-21** illustrates the concept of multiphase equilibrium. In *soil leaching*, the logical comparison is the concentration in other phases to the concentration in water, whereas for volatilization, comparison to the air phase is more appropriate. *Bulk concentration, m*, is the mass of a constituent per bulk volume of soil and is expressed in terms of the concentration of a single phase, with water as the reference phase, by the equation:

$$m = (\theta_w + \theta_a K_h + \theta_o K_o + \rho_b K_d)c_w = B_w c_w$$

where:

$K_h$ = *Henry's law* air–water partitioning constant [dimensionless]
$K_o$ = Henry's law NAPL–water partitioning constant [dimensionless]
$K_d$ = Henry's law soil–water partitioning constant [$L^3 \cdot M^{-1}$]
$\theta_w$ = concentration in water [$M \cdot L^{-3}$]
$\rho_b$ = soil bulk density [$M \cdot L^{-3}$]
$\theta_w$ = volumetric water content
$\theta_a$ = volumetric air content
$\theta_o$ = volumetric content of NAPL/organic immiscible liquid/oil.

**Buoyant density** The dry density, $\rho_d$, minus the mass density of water, $\rho$, as expressed by the equation:

$$\rho = \rho_d - \rho$$

In *groundwater*, the downward-directed gravity force is due to the buoyant weight of the saturated *porous medium*, and if the dry density of the medium, $\rho_d$, equals 2.0 g cm$^{-3}$, it has a buoyant density equal to the density of water, $\rho = 1.0$ g cm$^{-3}$. See **Appendix A** for unit conversions.

**Buried channel** The remains of a former *stream* channel or *tributary*, now concealed beneath surficial deposits. Cf. *Paleochannel.*

**Buried river** The bed of a former *river* that has been concealed beneath sedimentary deposits, lava, pyroclastic material, or till. Cf. *Paleochannel.*

**By-channel** A small natural watercourse, such as a *stream* or *branch*, that diverts or carries excess water along the main stream. Cf. *Distributary; Tributary.*

**Calcium hypochlorite** The chemical formula $(Ca(OCl)_2)$ used as a dry mixture in place of handling chlorine gas for the purpose of removing *iron bacteria* in *well rehabilitation*. Calcium hypochlorite contains about 65% available chlorine. When hydrated, the chlorine is dissolved in water, killing the bacteria. Cf. *Polyphosphates; Chlorination, shock.*

**Calcium rich deposits, hydraulic properties of** Formations mainly composed of limestone and dolomite having characteristic water-transmission qualities. When first deposited, calcium-rich deposits are quite massive, with little *void space* for *fluid storage*. When these formations become lithified and stressed, fractures and faults are formed. *Carbonic acid* $(H_2CO_3)$, formed in the atmosphere from carbon dioxide and water, is a mild acid that creates solutioning in the rock. Through secondary solutioning, many deposits of limestone and dolomite can become large capacity reservoirs for *groundwater storage*. The transmission of water largely occurs through these secondary features. Cf. *Karst spring.*

**Caliche** A massive form of calcite $(CaCO_3)$ found in semi-arid regions precipitated from *groundwater*. Usually, this is found as a very hard layer in sedimentary basins that retards and confines *groundwater flow*.

**Caliper log** A downhole geotechnical logging tool that measures the diameter of the *borehole*. The caliper log is probably the simplest form of a downhole logging tool, and also the most useful one. Changes in borehole diameter provide indications of rock *competence*. More massive formations provide diameters consistent with the drill bit diameter, whereas increases (washouts) or decreases (squeezing in) in borehole diameter would be indicative of less competent rock. Cf. *Geophysical exploration methods.*

**Camera, down hole** A television camera and lighting equipment that is raised or lowered within a *borehole* to log the hole. This geotechnical downhole logging technique has seen significant improvement from its earliest development. Washouts and casing integrity can be observed with the downhole camera. *Suspended solids* in the water can obscure the picture, reducing its utility. Cf. *Geophysical exploration methods.*

**Canister float** See *float measurement of discharge.*

**Cap formation** A term that has its origin in the petroleum industry to describe a stratum or mass that provides a trap for oil or natural gas accumulation. "Cap formation" is sometimes used to describe a confining layer in an aquifer system that may result in *artesian* conditions. Cf. *Confining bed.*

**Capillarity** The forces resulting from the surface tension of the fluid and the *pore space* in which it resides.

**Capillary barrier** The *capillary fringe*, as a result of the water pressure being less than atmospheric, may act as a barrier to the vertical migration of water or of liquid pollutants, such as gasoline or trichloroethene. A capillary barrier may exist where a silt layer exists over coarser sand or gravel. Under these conditions, water infiltrating through the silt layer is held up above the interface with the higher *hydraulic conductivity* coarse-grained layer until the water entry pressure for the coarse-grained layer is reached. The capillary fringe's ability to act as a barrier is transitory, depending on the quantity of liquid pollutant present, the *specific gravity* of the pollutant (and hence the pressure exerted), and natural fluctuations in the *water table.*

**Capillary fringe** Also called the tension-saturated zone, or zone of capillarity. The saturated zone of soil moisture above the *water table* in which the water is at less than atmospheric pressure. In contrast to the capillary fringe, water at the water table is, by definition, at atmospheric pressure. Because its *fluid pressure* is less than atmospheric, there can be no natural outflow to the atmosphere form the capillary fringe. This zone has varying thickness, depending on the pore size of the matrix material. Generally, the smaller the pore space, the thicker the capillary fringe (**Figure C-1**).

**Capillary pressure** The water pressure in saturated *pore spaces* that is less than atmospheric. Cf. *Air-entry pressure.*

**Capillary water** Water that resides under negative pressure in the capillary zone.

**Capture zone** A term used in environmental applications to designate the region of an *aquifer* that contributes water to an *extraction well*. All the *flow lines* originating within the capture zone terminate at the extraction well; groundwater outside the capture zone will bypass the extraction well. **Figure C-2** illustrates the interaction between pumping wells.

**Carbon adsorption** A water treatment method that utilizes granular-*activated carbon* (GAC) for the removal of pollutants and impurities in water. This represents one of the oldest forms of water treatment, where water is passed through beds of carbon to remove impurities. The effectiveness of removal is governed by the compound's *octanol–carbon partition coefficient*. The higher the octanol–carbon partition coefficient, the greater the affinity of the compound for adsorption. The adsorption follows an elution process, whereby there is actually adsorption/desorption occurring as the compound and water passes through the carbon beds. Cf. *Activated carbon.*

**Carbon-14 dating** A method for dating the age of a particular material using the ratio of carbon 14 ($^{14}C$) to carbon 12 ($^{12}C$) based on the half-life of $^{14}C$. Carbon 14, the radioactive isotope of carbon 12 (the stable form), has a half-life of 5730 years. The amount of carbon 14 in the environment has remained relatively stable. When organisms die, they stop taking in carbon 14, which allows the measurements of relative concentrations in various organic materials to determine their age. This technique is useful for dating organic objects less than 50,000 years old. Cf. *Dating of groundwater; Radiocarbon dating of groundwater; Environmental isotopes.*

**Carbonate** A sediment formed by the organic or inorganic precipitation of carbonate minerals, such as limestone and dolomite. Cf. *Hardness; Softening of water.*

**Carbonic acid** $H_2CO_3$, formed in the atmosphere from carbon dioxide and water. Cf. *Acid rain.*

**Casing** See *well casing.*

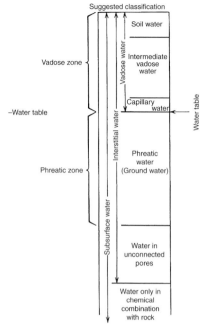

Figure C-1 Classification of subsurface water. (Davis and DeWiest, 1966. Reprinted by permission of John Wiley and Sons, Inc.)

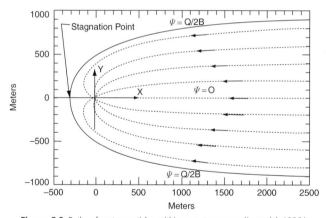

Figure C-2 Paths of water particles within a capture zone. (Javandel, 1986.)

**Casing storage** The water present in a well within the casing. This water influences early *drawdown* in pump tests, for the *well casing* is being evacuated at the start of the test, and little drawdown response in the *observation wells* until the storage is removed to an elevation corresponding to the pumping level within the well.

**Casing water** In a *monitoring well*, the water that occupies the *well casing*. The casing water is purged prior to obtaining a sample of the formation water.

**Catchment Area** A surface water hydrologic term used to denote the area that is being drained by *rivers*, storm sewer, or any other natural or manmade surface water collection system. Cf. *Catchment basin; Drainage basin; Drainage system.*

**Catchment Basin** The term is used interchangeably with the more common term *watershed* or *drainage basin*. Outside the United States, the term catchment is often used. Cf. *Groundwater basin; Recharge basin; Antecedent precipitation index; Drainage basin*.

**Cathodic protection** A means of protecting against *corrosion*, whereby an anode having higher conductance is provided to corrode instead of the object to be protected. Sacrificial anodes are used for protection of underground storage tanks or wells in situations where the soils are corrosive, or where there are stray electrical currents in the ground leaking from underground power lines.

**Cation exchange** An ion exchange process in which cations in solution are exchanged for other cations from a matrix. Cf. *Ion exchange; Cation exchange capacity*.

**Cation exchange capacity** The excess of counter ions in the zone adjacent to the charged surface or layer which can be exchanged for other cations. This is determined by the following:

1. adjusting the *pH* of the pore water to 7.0,
2. saturation of the exchange sites with ammonia ($NH_4^-$) by mixing the soil in a solution of ammonium acetate,
3. removal of the adsorbed $NH_4^-$ on the exchange sites by leaching with a strong solution of NaCl, where Na replaces the $NH_4^-$ on the exchange sites, and
4. determination of the $NH_4^-$ content of the leaching solution after equilibrium is attained.

The *cation exchange* is expressed as the number of milliequivalents of cations that can be exchanged in a sample with dry mass of 100 g. The cation exchange capacity determined by the test is for conditions simulated by the test. For many materials, adsorption is strongly pH dependent.

**Cauchy condition** On the boundary of a numerical model domain, the Cauchy conditions enforce the continuity of the mass flux on each side of the boundary. In numerical or analytical computer *groundwater* modeling, this is a general term boundary value problem establishing continuity of mass flux from one side of the boundary to another to model groundwater flow conditions. Also known as a constant head boundary where the *piezometric head* is defined to simulate groundwater flow conditions. Cf. *Boundary, specified-head*.

**Cave** An inlet with a curved form, the smallest form of a *bay*.

**Cavitation** A phenomena of cavity formation within a moving fluid. Typically encountered in pumps, resulting when the absolute pressure reaches the *vapor pressure* of the fluid, causing the formation of vapor pockets. The vapor pockets result in corrosion of the impellers and metal parts. It has also been observed that hydroxyl radicals are formed under cavitation conditions. This observation has resulted in the use of cavitation to create hydroxyl radicals for water treatment. Cf. *Point-of-use water treatment systems*.

**Cement** A mixture used in *grouting* well casings that consists of a dry powder made from silica, alumina, lime, iron oxide, and magnesia which hardens when water is added. An ingredient in concrete. Cf. *Cement grout; Portland cement; Slurry; Montmorillonite; Flocculation*.

**Cement–bentonite** A mixture of typically 95% *Portland cement* to 5% sodium bentonite (by weight). The addition of *bentonite* provides for a consistent *well seal* with lesser cracks than bentonite or cement alone. Cf. *Montmorillonite*.

**Cement bond log** Also known as *acoustic log*. A downhole geophysical logging tool used to determine the integrity of the cement bonding between the *well casing* and the *borehole walls*. The logging measures the travel time and attenuation of sonic waves. The average velocity of a sonic wave through a formation is affected by the matrix and the *pore fluid*. The matrix velocity is directly affected by the lithology. As such, not only do these logs determine how well the casing is bonded to the formation, but also it is used to determine lithology, fracture zones, and the top of the *static water level*. Cf. *Geophysical exploration method*.

**Cement grout** Grouting or cementing a *well casing* involves the placement of a suitable *slurry* of *Portland cement* or clay. Cement is the most common grouting material and the term cementing is used interchangeably with grouting. The classifications of cement used in water *wells* are listed in **Appendix B**. Cf. *Aggregate; Alkali aggregate; Reactive aggregate; Flocculation*.

**Centering guides/centralizers** Devices placed on the *well casing* to ensure consistent annular spacing between the casing and the *borehole* walls. Typically positioned at 40 ft intervals, centering guides are used on wells deeper than 40 ft to keep the casing centered in the borehole. As the length of casing increases, even steel casing can flex. Further, the borehole itself can be out of alignment. Centralizers are used to assure the *filter pack* thickness around the screened interval, and the grout thickness around the casing is consistent. This will ensure consistent flow into the screened interval and seals around the casing.

**Centrifugal sand sampler** A device used during the development of water wells to determine the maximum allowable sediment in the water. Pumped water passes through a cyclone device that causes the sediment to be separated and collected in a cone for measurement.

**CERCLA** See *Comprehensive Environmental Response, Compensation, and Liability Act*.

**Channel capacity** A measurement of a manmade or natural channel's ability to convey water. Channel capacity is influenced by cross-sectional area, *stream* slope, and bank height above normal flow stage. Cf. *Bankfull stage*.

**Channel efficiency** A measure of the ability of a *river* channel to move water and sediment. Channel efficiency is determined by calculating the channel's *hydraulic radius* (cross-sectional area/length of bed and *bank*). The most efficient channels are generally semicircular in cross section.

**Channel improvements for flood mitigation** See *flood mitigation, channel improvements for*.

**Channel roughness** See *roughness*.

**Channel shape** The shape of the channel or its cross section, affects the wetted perimeter. The wetted perimeter is a measure of the extent to which water is in contact with its channel. The greater the wetted perimeter, the greater the friction between the water and the *banks* and the bed of the channel, and the slower the flow of river.

**Channel storage** The *hydrologic cycle* is based on the movement and storage of water. Several types of storage are named by their location. Channel storage exists in the open portion of a stream or river that is visible, in the nonvisible porous banks, as *bank storage*, and in the bed of the stream or river. Channel storage may, therefore, be larger than the stream or river's visual water limits.

**Chebotarev's succession/Chebotarev's sequence** The chemical evolution of *groundwater* along its *flow* path as it evolves toward the composition of seawater. The chemical evolution is illustrated below:

Travel along flow path →
$$HCO_3^- \rightarrow HCO_3^- + SO_4^{2-} \rightarrow SO_4^{2-} + HCO_3^- \rightarrow SO_4^{2-} + Cl^- \rightarrow Cl^- + SO_4^{2-} \rightarrow Cl^-$$
Increasing age →

For large sedimentary basins, the Chebotarev sequence can be described as having three main zones:

- The upper zone – where there is active groundwater flushing through well-leached rocks. The dominant anion is $HCO_3^-$, and the water is low in *total dissolved solids* (TDS).
- The intermediate zone – where there is increasing total dissolved solids and less active groundwater flushing. Sulfate is the dominant anion.
- The lower zone – characterized by high total dissolved solids and very sluggish groundwater flow. Chloride is the dominant anion.

The Chebotarev sequence can be modified by bacterial activity. Cf. *Electrochemical sequence; Brine; Brackish water; Artificial brine.*

**Check valve** A mechanism used to prevent backflow of the pumped fluid when the pump is turned off. Check valves are also found in items such as *bailers* and sump pump *discharge* piping.

**Chelating agent (sequestering agent)** A substance having the ability to "surround" metals ions (such as iron, calcium, and magnesium) in solution and keep them from combining chemically with other *ions*. Chelating agents are used in *groundwater* treatment to prevent fouling on treatment systems.

**Chezy's formula** A mathematical relationship under *flow* conditions in which the shear stress is proportional to velocity squared. Chezy's formula, used to describe steady uniform flow in prismatic open channels, is:

$$V = CR_h^{1/2}S^{1/2}$$

where:
$V$ = velocity $[L \cdot T^{-1}]$
$R_h$ = hydraulic radius, defined as $A/P$, where $A$ is the cross-sectional area perpendicular to flow and $P$ is the wetted perimeter of the channel $[L]$
$C$ = Chezy constant
$S$ = slope of the channel $[L \cdot L^{-1}]$

Manning found that the Chezy constant, $C$, can be determined by:

$$C = \frac{C_m}{n}R^{1/6}$$

where:
$C_m$ = 1.49 in English units and 1.0 in SI units
$n$ = porosity
Cf. *Manning's equation/Manning's formula; Bazin's average velocity equation; Steady-state flow.*

**Chlorination, shock** The use of a chlorine solution having a concentration of 1000 mg $L^{-1}$ chlorine or greater to remove biological fouling in wells, piping, and treatment systems. Cf. *Well rehabilitation; Calcium hypochlorite; Poly phosphates.*

**Clastic** Pertaining to a rock or sediment composed primarily of broken fragments that are derived from pre-existing rocks or minerals that have been transported some distance from their place of origin.

**Clay barriers** Usually refers to manmade features to control the movement of water. It can also be natural. The low *hydraulic conductivity* associated with clay tends to impede *groundwater* movement and for most engineering applications is considered impervious. Cf. *Clay cap; Clay liner.*

**Clay caps** Manmade *clay barriers* to minimize *infiltration* of water for such applications as landfill covers. May also be referred to as a *Resource Conservation and Recovery Act* (RCRA) cap in engineered landfill covers in the United States.

**Clay liners** Similar to *clay caps*, however, in the instance of landfills, used as the base or lateral barriers for water *infiltration* or leachate control.

**Clay slurry wall** Engineered water control measures usually constructed in trenches. The term "clay slurry wall" may be used interchangeably with "*grout curtain*," but a grout curtain usually refers to measures constructed using a series of pressure-grouted boreholes.

**Clean Water Act (CWA)** The law that authorizes establishment of the regulatory program to restore and maintain the physical and biological integrity of the waters of the United States. As part of the CWA, the *National Pollutant Discharge Elimination System* (NPDES) was established to regulate industrial and municipal point-source discharges.

**Coagulation** Chemical coagulation occurs when floc-forming compounds are added to destabilize or suspend colloidal material. Colloidal material, being negatively charged, repels other particles. Coagulants neutralize the negative particles, allowing them to combine with other particles and settle out of suspension. Cf. *Flocculation; Colloidal dispersion.*

**Coefficient of permeability** In older textbooks of geology, it is a synonym for *hydraulic conductivity.* In contemporary hydrogeology, this term is rarely used, and typically the term *permeability* is interchanged with hydraulic conductivity. However, strictly speaking, permeability is a characteristic of the matrix in which flow is occurring given in units of $L^2$. Hydraulic conductivity on the other hand is a characteristic of the saturated media, expressed in terms of $L \cdot T^{-1}$.

**Coefficient of storage** See *storage coefficient.*

**Coefficient of transmissivity** See *transmissivity.*

**Cohesion** The molecular tendency of parts of a body of like composition to be held together as a result of intermolecular attractive forces. Cf. *Adhesive water; Interstices.*

**Coliform bacteria** A genus of bacteria that is usually associated with animal or human waste. Coliform bacteria find their way into *groundwater* through barnyard runoff, or through *septic systems.*

**Collapse strength** The strength of the *well screen* against the forces induced by *well development* or pumping (**Figure C-3**). The main factor influencing collapse strength is the massiveness of the materials used to construct the screen. For continuous wire-wound screens, the "height" of the screen provides the principal factor in determining the strength. The height of the screen is the dimension measured from the outside edge of the screen to the inside edge. For slotted pipe, the height would be the thickness of the pipe.

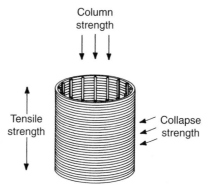

**Figure C-3** The structural components of a continuous slot screen are selected to withstand three major stresses that are placed on a well during construction and use of the well. The column and tensile strength required during construction are provided by longitudinal rods. The shape and massiveness of the wrapping wire provide the necessary collapse resistance needed during development and the long-term use of the well. The actual tensile, column, and collapse strengths are dependent on the construction material of the screen. (Driscoll, 1986. Reprinted by permission of Johnson Screens/a Weatherford Company.)

**Collection lysimeter** See *lysimeter.*

**Collector well** Also called a Ranney well or Ranney collector. A type of *well* constructed from a central caisson with one or more horizontal screens radiating from the caisson (**Figure C-4**). The horizontal-screened sections produce water, whereas the central caisson serves as a large collection or storage tank. Although expensive to construct, these wells can have large yields, with minimal *drawdown* when completed in permeable deposits.

**Colloid** Extremely small soil particles, 0.0001–1 μm in size, which will not settle out of a solution; intermediate between a true *dissolved solid* and a suspended solid which will settle out of suspension.

**Colloidal dispersion** An intimate mixture in which small particles are permanently dispersed throughout a solvent. The degree of colloidal dispersion is important to the *gel strength* of a *drilling fluid.* The suspended particles are intermediate in size between visible particles and individual molecules. Cf. *Gel; Flocculation; Deflocculation.*

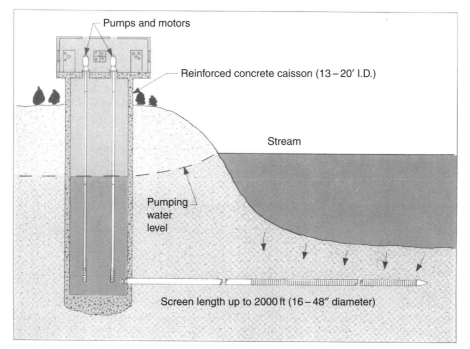

**Figure C-4** Collector well with screen jacked out from a large caisson. (Driscoll, 1986. Reprinted by permission of Johnson Screens/a Weatherford Company.)

**Column strength** Refers to the strength of a *well screen* to vertical loading (**Figure C-3**). For a wire-wound screen, the column strength is directly proportional to the *yield strength* of the vertical rods used in the construction of the screen.

**Combined aeration–granular activated carbon (GAC) systems** A water treatment system that combines *air stripping* and GAC. A combined aeration–GAC treatment system is usually employed where there is a mixture of chemicals to be removed from the *groundwater* having different volatilities. The air stripping removes the more volatile components, whereas the GAC removes the less volatile in a step, also referred to as carbon polishing.

**Combining weight** See *equivalents per liter (epi)*.

**Common-ion effect** A phenomenon by which the addition of ions of one mineral can influence the *solubility* of another mineral to a greater degree than the effect exerted by the change in *activity coefficients*. For instance, in an aqueous solution saturated with calcite ($CaCO_3$), the addition of an electrolyte that does not contain an ion in common with $CaCO_3$ will cause the solubility of $CaCO_3$ to increase because of its ionic strength. However, if an electrolyte containing an ion in common with $CaCO_3$ such as calcium or carbonate is added, then calcite may precipitate.

**Common law** Legislation set at the state or provincial level governing the development of *groundwater resources* in North America. Common law therefore varies from state to state and from province to province. In general, the groundwater resource is the property of the people, held in trust by the state. However in the western United States, the property owner who first developed the resource has established water rights that can be transferred with the sale of the property. Generally, this has resulted with the long-time residents having greater water rights than later residents.

**Competence** A geotechnical term, generally used in relation to slope stability. A competent slope is considered a stable slope.

**Complementary error function** The complimentary error function has its origins in probability and statistics. This function appears in a number of analytical solutions to *boundary value problems* describing solute transport in *porous media*, including cases involving diffusive transport and *advective–dispersive transport*. The form of the function is as follows:

$$\text{erfc}(x) = 1 - \text{erfc}(x) = \frac{2}{\sqrt{\pi}} \int_x^\infty e^{-u^2} \, du$$

The complementary error function has the following properties:

$$\text{erfc}(0) = 1$$

$$\text{erfc}(\infty) = 0$$

Cf. *Error function*; *Diffusion equation*.

**Complexation** Many metal ions form complexes with ligands or chelants that can donate electrons to empty orbitals on the metal. Cf. *Chelating agent*.

**Compliance monitoring network** In the United States, under the Resource Conservation Recovery Act (RCRA), monitoring of groundwater quality in the vicinity of a closed hazardous waste storage facility is required. To monitor groundwater quality, a compliance *monitoring well* network is designed and installed to ensure that the closed facility does not experience releases to groundwater that impact groundwater quality. Refer to **Appendix B** for *maximum contaminant levels* (MCLs).

**Composite pumping cones** An effect on water levels that results from pumping of multiple wells with overlapping *radii of influence*. A composite pumping cone is a combination of the *cone of depression* for each of these wells (**Figure C-5**). The *drawdown* at any point within the composite pumping cones is the sum of the *distance–drawdown relationships* from the individual wells. *Dewatering* of excavations and other features is typically engineered using composite pumping cones.

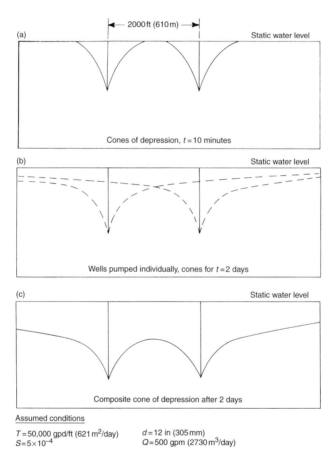

Figure C-5 Interference of cones of depression during pumping between adjacent wells tapping the same confined aquifer. Composite cone is for both wells pumping simultaneously under the assumed conditions. (Driscoll, 1986. Reprinted by permission of Johnson Screens/a Weatherford Company.)

Cf. *Image well theory.*

**Comprehensive Environmental Response, Compensation, and Liability Act (CERCLA)** Also called the *Superfund Act*. This is the Act that established the framework for the cleanup of uncontrolled hazardous waste sites in the United States. Originally promulgated in 1984, it was re-authorized as the *Superfund Amendments and Reauthorization Act* (SARA) in 1989.

**Compressibility** *Porous media* will compress when subjected to an applied load or when the effective stress at grain-to-grain contacts increases because of a reduction in *fluid pressure*. The compressibility measures the reduction in pore volume as a function of the increase in effective stress; solid rock has a small compressibility, whereas clays can have compressibility orders of magnitude greater than that of rock.

**Conceptual model** In performing *hydrogeologic* investigations, it is helpful to develop a framework that explains the *groundwater* flow, contaminant transport, location of sources of pollution, and physical, chemical, and biological processes taking place in the subsurface, called a conceptual model. When performing mathematical modeling or additional investigation, data are reviewed in the context of the conceptual model, where the model may be updated.

**C**

**Condensation** The change in state from a gas or *vapor* to a liquid. Cf. *Latent heat of condensation.*

**Condensation nuclei** The solid or liquid particle onto which a gas or *vapor* condenses.

**Conductance, specific** See *specific electrical conductance.*

**Conduction, heat transfer by** The process whereby heat is transferred by the physical contact between substances.

**Conductivity, electric** The transmission of electrical current through a substance, measured in mhos. Cf. *Specific electrical conductance.*

**Conductivity, hydraulic** See *hydraulic conductivity.*

**Conductivity, thermal** See *thermal conductivity.*

**Conduit flow** The phenomena where flow of a fluid or gas is transmitted predominantly through distinct areas of high *permeability.* An example of conduit flow is water movement through a gravel channel within a *braided stream* deposit, or *karst spring.*

**Cone of depression** Also called a drawdown cone, or a pumping cone. A cone-shaped depression in water levels in the immediate vicinity of the well under pumping conditions. Under homogeneous *isotropic* conditions this forms a cone, within which the *hydraulic gradient* is toward the well. However, under field conditions, the cone may be elongated in any direction due to *anisotropy, heterogeneities* in the *permeability* of the aquifer material. The interference of cones of depression is illustrated in **Figure C-5** composite pumping cones. Cf. *Capture zone.*

**Confined aquifer** See *aquifer, confined.*

**Confining bed** The geologic strata that are of a lower *hydraulic conductivity* than the underlying strata under saturated conditions. This forms a layer that provides for confining the *piezometric head* within an aquifer above atmospheric pressure. Cf. *Artesian conditions; Aquiclude; Zone of saturation.*

**Confining layer** See *confining bed.*

**Confining pressure** From the *Bernoulli equation,* the energy in any water mass consists of three components: *pressure,* velocity, and *elevation head* (energy derived from the elevation of the water body). Under confined conditions, the *pressure head* can exceed the elevation head, such that *artesian* conditions exist.

**Confluent feeder** See *tributary.*

**Conjunctive use of water supplies** The optimal allocation of water available from a water supply basin, taking into account both the *groundwater* and the surface *runoff.* Cf. *Watershed.*

**Connate water** Also called fossilized *brine* or original formation water. Water present in the pore spaces of a rock when the rock was formed. Cf. *Formation water; Natural water.*

**Consequent stream** Also called original *stream.* A watercourse having a route that was determined by the original slope of the land. Cf. *Obsequent stream; Longitudinal consequent stream; Resequent stream.*

**Conservation of groundwater resources** Generally, developing water holdings and limiting their use to maintain them within a *water balance* that accounts for *basin recharge* and the *conjunctive use of the water supplies.*

**Conservation of mass** Also called the continuity principle. A natural law stating that matter cannot be created or destroyed but can be changed to a different state (solid, liquid, or gas). For steady-state conditions, mass in = mass out of a unit volume. Under transient conditions, mass in − mass out = temporal rate of change of mass storage in unit volume.

**Constant flux boundary** See *second type boundary.*

**Constant head boundary** See *boundary, head dependent; head dependent boundary.*

**Constant head permeameter** An apparatus used to determine the hydraulic conductivity of sediments. With a constant-head *permeameter,* the *hydraulic conductivity* is found by measuring the area of the sample, rate of flow, and *hydraulic gradient.* Cf. *Guelph permeameter.*

**Constant rate pumping test** See *pumping test*

**Consumptive use** The development of a water resource for a specific need, such as for industry, animal watering, irrigation, or drinking water supplies.

**Contact spring** See *spring, contact.*

**Contact time** The amount of time that water is in contact with a specific media. For treatment through *adsorption,* contact time is a treatment design parameter.

**Contaminant** See *pollutant.*

**Contamination** The degradation of natural water and/or soil qualities typically resulting from human intervention. The degree of perceived impact to quality depends on the intended end use or uses of the impacted media. Cf. *Pollutant; Pollution.*

**Continental divide** See *divide; groundwater divide.*

**Continuity principle** See *conservation of mass.*

**Continuous slot screen** A widely used variety of the *well screen* that serves as a filtering device at the intake portion of wells installed in unconsolidated or semi-consolidated *aquifers.* Also known as wire-wound screen, **Figure C-3** depicts the structural components. Cf. *Gauze number; Louver screen; Open area; Slot opening; Well design.*

**Convection** The transference of heat or gas or liquid resulting from differences in temperature and associated differences in density. A heated fluid will rise to the top of a column, radiate heat away, and then fall to be reheated again. A fluid can be trapped in this cycle, becoming part of a convection cell. Convection cells can form at all scales, from millimeters across to a global scale.

**Convective transport** This refers to the convective transport of heat in groundwater *flow* systems, which only occurs in moving *groundwater*.

**Convergence criterion** Mathematical models involve the simultaneous solution of many differential equations describing groundwater *flow* and/or transport and/or chemical/biological transformations. The solution of these equations is iterative and tends to converge on a unique solution. However, the time required to compute this unique solution may be impractically long. Therefore, a convergence criterion is established, which defines the conditions under which the solution is stable, which will then define the solution of the simultaneous differential equations.

**Cooper–Bredehoeft–Papadopulos method** A technique for interpreting *slug test* data based on the Theis analysis for *pumping tests*. As with the *Theis method*, the Cooper–Bredehoeft–Papadopulos method involves curve matching to determine *transmissivity* and coefficient of storage. As shown in **Figure C-6**, *drawdown* data from a slug test is matched against type curves for different values of $\alpha$. A match point or vertical axis is matched, and values of $t$ and $W$ are read off the horizontal scales of the matched axis of the types plot. For ease of calculation, it is convenient to select a match point of $W = 1$. The transmissivity $T$ is given by:

$$T = \frac{Wr^2}{t}$$

where:

$W$ = match value on types curve
$t$ = time [T]
$r$ = radius of well or piezometer [L]

Units of measurement (SI or English) must be consistent with the system one is using. Refer to **Appendix A** for unit conversions.

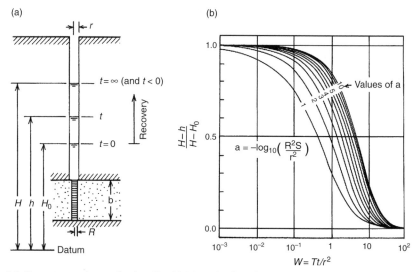

**Figure C-6** Piezometer test in a confined aquifer. (a) Schematic of test layout; (b) type curves with which to analyze the data. (Freeze and Cherry, 1979. Reprinted with permission by Prentice-Hall, Inc.)

Cf. *Storage coefficient*.

**Core sample** A geologic specimen obtained from soil or rock drilling techniques that show a continuous sequence of the geologic material.

**Coriolis force** A force resulting from the rotation of the earth that causes a clockwise circulation of water and air in the northern hemisphere and a counter-clockwise rotation in the southern hemisphere. Cf. *Ferrel's law*.

**Correlation length** The length of a curve over which a correlation or regression is performed, comparing sets of data.

**Correlation method** Also called a regression technique. The identification of the function that most appropriately describes the data trends and performance of statistical calculations to determine the parameters of the equation that describe the data trends.

**Correlative rights** The *groundwater* law in the United States according to which competing landowners share the water resource during *drought* based on relative areal extent of the land owned by the competitors. Cf. *Beneficial use; Prior appropriation; Rule of capture; Rule of reasonable use; Water law.*

**Corrosion** The act or process of dissolving or wearing away metals, typically relating to well installations. Cf. *Chelating agent; Chlorination shock; Well rehabilitation.*

**Counter-current-packed tower** Also called as an air stripper. A cylindrical tower filled with a media designed to increase the exposed surface area of the water by decreasing droplet size. The water to be treated is introduced at the top of the tower and allowed to cascade through the media. Air is introduced from the bottom of the tower and passes counter-current to the water flow to promote the vaporization of volatile compounds from the water (**Figure C-7**).

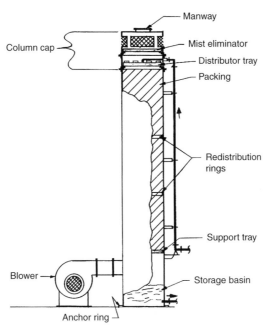

**Figure C-7** Components of a counter-current-packed tower, more commonly known as an air stripper. Water enters through distributor tray and flows under gravity through packing media in the tower. Air is blown from the bottom of the tower to remove volatile organic compounds. The packing media is designed to provide for maximum residence time within the tower and to breakup the water droplets maximizing the surface area of the water droplets for treatment. (Nyer, 1985. Reprinted with permission from Van Nostrand & Reinhold, Inc.)

**Coupled flow** See *groundwater flow.*

**Critical hydraulic gradient** The upward *hydraulic gradient* at which soil will become cohesionless and its bearing capacity is reduced to zero and quick conditions result. This is defined as:

$$i_{critical} = \frac{h_{critical}}{L} = \frac{\gamma - \gamma_w}{\gamma_w} = \frac{\gamma_b}{\gamma_w}$$

where:
$h_{critical}$ = groundwater head causing flow [L]
$L$ = length of vertical flow path [L]
$\gamma$ = soil bulk unit weight [M·L$^{-3}$]
$\gamma_w$ = specific gravity of water [M·L$^{-3}$]
$\gamma_b$ = bulk density of soil [M·L$^{-3}$]

**Critical velocity** Also known as critical flow. The rate (L·T$^{-1}$) at which the fluid flow changes from *laminar flow* to *turbulent flow*, where the friction becomes proportional to a power of the velocity higher than the first power. Cf. *Reynold's number; Beleanger's critical velocity; Kennedy's critical velocity; Unwin's critical velocity; Supercritical flow.*

**Cross-hole tomography** Tomography is a technique whereby three-dimensional images can be generated from two-dimensional cross-sectional slices through a media. Cross-hole tomography refers to generating three-dimensional pictures of subsurface conditions through the use of downhole geophysical methods. These geophysical techniques may include bore-hole radar, acoustic, and resistivity to generate the three-dimensional images. Cf. *Geophysical exploration methods.*

**Cubic law** This mathematical formulation can be used to describe *laminar flow* between two plates and as such has applications to fracture flow. The general form of the equation is:

$$Q = \frac{\rho g}{\mu} \frac{Wb^3}{12} i$$

where:

$Q$ = flow [$L^3 \cdot T^{-1}$]
$\rho$ = density of the fluid [$M \cdot L^{-3}$]
$\mu$ = viscosity of the fluid [$M \cdot L^{-1} \cdot T$]
$g$ = acceleration of gravity [$L \cdot T^{-2}$]
$i$ = head gradient [$L \cdot L^{-1}$]

Cf. *Fracture velocity.*

**Cumulative frequency curve** Graphs and histograms of water *discharge* versus percent of total load quantity, dissolved and suspended, show that the extremely high flows (approximately 10,000 cubic feet per second, cfs), though effective in erosion and transportation, are so infrequent that their contribution to the total work of a *stream* is less than that of more modest discharges which occur more often. Cf. *Annual flood series; Flood frequency analysis; Recurrence interval.*

**Cutthroat flume** See *flume, cutthroat.*

**CWA** See *Clean Water Act.*

**DAD** See *depth–area–duration relations*.

**Dalton's law** The natural law of partial pressures. Dalton's law states that in a mixture of gases, the total *pressure* equals the sum of the partial pressures (the partial pressure of a *vapor* is also called the *vapor pressure*). As a result of the equilibrium established between gas and water by the exchange of molecules across the liquid–gas interface, *groundwater* can contain a mixture of dissolved gases from exposure to the earth's atmosphere prior to subsurface *infiltration*, contact with soil gases during infiltration through the *unsaturated zone*, or chemical or biochemical reactions within the water itself. If sufficient energy and water are available, the vapor pressure gradient affects *evaporation*. The law of partial pressures may be written as:

$$P \text{ total} = P_1 + P_2 + P_3 + \ldots P_n$$

Cf. *Thornthwaite equation*.

**Darcian flow** The movement of *groundwater* or gas according to *Darcy's law*, which posits a linear relationship between *discharge* and *gradient*. Cf. *Groundwater gradient*.

**Darcy** A unit of *permeability* expressed in the dimension of area, $L^2$. The darcy, abbreviated as $D$, is often used in the petroleum industry as a unit of intrinsic or *absolute permeability*. A darcy is equivalent to 1 cm$^3$ of fluid of 1 centipoise *viscosity*, flowing in 1 s under a pressure differential of 1 atmosphere, through a *porous medium* having a cross-sectional area of 1 cm$^2$ and a length of 1 cm:

$$1 \text{ darcy} = \frac{(1 \text{ centipoise} \times 1 \text{ cm}^3 \text{ s}^{-1})/1 \text{ cm}^2}{1 \text{ atmosphere}/1 \text{ cm}}$$

This expression can be converted to square centimeters, because 1 centipoise $= 0.01$ dyn-s cm$^{-2}$ and 1 atmosphere $= 1.0132 \times 10^6$ dyn cm$^{-2}$; therefore, 1 darcy $= 9.87 \times 10^{-13}$ m$^2$, or $1.062 \times 10^{-11}$ ft$^2$, or $0.987 \times 10^{-8}$ cm$^2$. Refer to **Appendix A** for unit conversions.

**Darcy flux/Darcian velocity/Darcy's flux velocity/bulk velocity** More commonly called *specific discharge*. A empirical relationship posited by the French engineer Henri Darcy in 1856, that the *flow* of water through a *porous medium* is analogous to pipeflow. Darcy (or Darcian, Darcinian) flux or velocity is more commonly called the specific discharge, an easily measured macroscopic concept that provides a way to quantify the energy (*head*) required to move water through an *aquifer*. As shown in **Figure D-1**, this macroscopic concept is clearly different from the microscopic velocities associated with actual paths of individual water particles as they move through grains of sand. The real microscopic velocities, however, are impossible to measure. The Darcian approach replaces the actual *porous* medium through which water flows with a more consistent continuum for which macroscopic parameters such as *hydraulic conductivity* can be defined.

**Figure D-1** The concept of groundwater flow in the macroscope and macroscopic view. (Freeze, 1979. Reprinted with permission from Prentice-Hall, Inc.)

Cf. *Darcy's law*.

**Darcy's equation** See *Darcy's law*.

**Darcy's law** A law that relates the rate of fluid *flow* to the *flow path* and *hydraulic head* gradient, assuming that the flow is *laminar* and that inertia can be neglected. Darcy's law is expressed as the basic equation describing the flow of *groundwater* through *porous media* and is equally important in many other applications of flow such as studies of gas and oil. Through laboratory experimentation, the French engineer Henri Darcy determined that the rate of flow through a column of saturated sand is proportional to the difference in the *hydraulic head* between the two ends of the column and inversely proportional to the length of the column. **Figure D-2** shows an example of such an experimental apparatus. As groundwater *discharge* often cannot be measured directly, equations of groundwater flow have been derived from the laws of *conservation of mass* and energy and Darcy's law. Darcy's law is used to define the relation between discharge and *head*, $h$, a variable that can be measured directly. Formulas expressing Darcy's law have been written in several formats, and derivations of the law are used in all aspects of fluid flow through porous media. The many equations based on Darcy's law state that the flow of water through a porous medium with a given cross-sectional area is related to the product of the *hydraulic gradient* and a constant of proportionality termed *hydraulic conductivity*, $K$. Some of the most widely used variations of the Darcy flow equation are:

$$Q = -KA\left(\frac{(h_1 - h_2)}{L}\right)$$

$$Q = -KA\left(\frac{dh}{dL}\right)$$

where:

    $K$ = hydraulic conductivity of the aquifer material [L·T$^{-1}$]
    $Q$ = rate of flow [L$^3$·T$^{-1}$]
    $A$ = area normal to the direction of flow [L$^2$]
    $h_1 - h_2$ = difference in hydraulic head between two points [L]
    $L$ = distance along flow path [L]
    $dh/dl$ = hydraulic gradient [L·L$^{-1}$]

The quantity $dh$ represents the difference in head between two points, and $dl$ is the distance between these two points. The negative sign in the equations indicates that flow is in the direction of decreasing hydraulic head. The use of the negative sign requires careful determination of the sign of the gradient. If the value of the head at point $X_2$, $h_2$, is greater than that at point $X_1$, $h_1$, then flow is from point $X_2$ to $X_1$. If $h_1 > h_2$, then flow is from $X_1$ to $X_2$. Energy is lost through friction between the moving water and the confining walls of the pores. Darcy's law states that this energy loss is proportional to the velocity of laminar flow; the faster the flow, the higher the energy loss. Darcy's law is valid for groundwater flow in any direction in space and can be written in three dimensions as:

$$q_x = -K_x \frac{dh}{dx}$$

$$q_y = -K_y \frac{dh}{dy}$$

$$q_z = -K_z \frac{dh}{dz}$$

where:

    $q_{x,y,z}$ = *specific discharge* or Darcy velocity in the $x,y$, and $z$ directions [L·T$^{-1}$]
    $K$ = hydraulic conductivity with components in the $x$, $y$, and $z$ directions [L·T$^{-1}$]

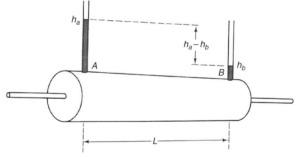

**Figure D-2** Demonstration of Darcy's experiment. (Freeze and Cherry, 1979. Reprinted with permission from Prentice-Hall, Inc.)

Cf. *Advection; Non-Darcian flow.*

**Darcy–Weisbach resistance coefficient** Also known as the Darcy–Weisbach friction coefficient, or Darcy $f$ which represents the effect of *roughness of* a channel bank and bed particles as well as losses resulting from the dynamic channel character in the form of vegetation along the banks and floodplain. The resistance to flow can also be altered by change of seasons, thereby affecting the extent of vegetation, change in water temperature, and manmade channel alterations. The coefficient varies with the magnitude of flow and is defined for each channel reach based on observation. Cf. *Roughness coefficient; Manning's equation.*

**Dating of groundwater** The assessment of relative concentrations of radioactive environmental isotopes, such as tritium ($^3$H), chlorofluorocarbons (CFCs), and carbon-14 ($^{14}$C), have been used to determine the age of *groundwater*. Prior to the use of aboveground thermonuclear tests in 1953, $^3$H had occurred naturally at very low levels and atmospheric $^{14}$C had been derived entirely from the natural nitrogen transmutation caused by bombardment of cosmic rays. Tritium concentration in *meteoric water* is measured in tritium units (TU): 1 TU is equal to 1 atom of $^3$H per $10^{18}$ atoms of hydrogen, which is equivalent to 7.1 disintegrations of $^3$H per min per l of water. The $^3$H content of meteoric precipitation in the northern hemisphere has fluctuated widely from less than 25 TU prior to 1953 up to more than 2200 TU in 1964 following the extensive atmospheric testing of nuclear devices. In the southern hemisphere, bomb-produced $^3$H levels are lower now and approaching preatmospheric testing levels. $^3$H in groundwater has also decayed to levels approximating preatmospheric testing levels in both northern and southern hemispheres, such that tritium cannot be effectively used as a dating tool for groundwater. The phase out of CFC use in aerosols beginning in 1996 (China and India had until 2006 to comply with the Montreal protocol of 1987), the relative abundance of CFCs in groundwater (compared to atmospheric conditions has been used to date young groundwater. Dissolved carbon in groundwater may be derived from several sources having different $^{14}$C concentrations and carbon-isotope compositions, and it is necessary to determine the source of the carbon to gain meaningful hydrologic information from this data. Despite the potential multiplicity of sources, useful groundwater dating results have been obtained under favorable geologic and hydrologic conditions. When water moves below the *water table* and becomes isolated from the earth's carbon dioxide ($CO_2$) reservoir, radioactive decay causes the $^{14}$C content in the dissolved carbon to gradually decline. The specific activity of $^{14}$C in carbon that was in equilibrium with the earth's atmosphere prior to thermonuclear testing is approximately 10 disintegrations per minute per gram. Substitution of this specific activity in the equation for the radioactive decay of carbon yields a maximum apparent age of the dissolved inorganic carbon. Non-radioactive (stable) environmental isotopes such as oxygen-18 ($^{18}$O) and deuterium ($^2$H) can also provide valuable hydrologic information. The distributions of $^{18}$O and $^2$H have been used as indicators of groundwater source areas and as *evaporation* indicators in surface water bodies. Cf. *Carbon-14 dating; Radiocarbon dating of groundwater; Tracer.*

**Dead sea** An archaic term for a body of water from which *evaporites* have been or are being precipitated and which no longer supports typical aquatic organisms. The term had described the type locality as the Dead Sea in the Middle East, which actually displays a wide variety of aquatic life, counter to the original definition.

**Dead storage** Also called off-channel storage. The areas of inactivity in the plan view of a *river* or *stream*. When shown in cross section, dead storage is where the *flow velocity* in the *downstream* direction is negligible relative to the active stream velocity. **Figure D-3** shows both the plan and the cross-sectional views of dead storage. These areas only serve to cache a portion of the passing flow. Examples of dead storage are seen in embayments, ravines, and *tributaries* that are connected to the flow channel but do not convey flow.

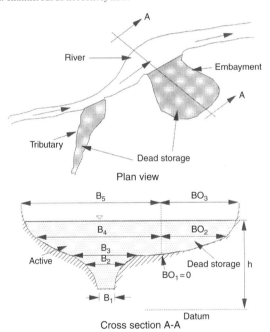

**Figure D-3** Plan and cross-sectional view of dead storage. (Maidment, 1993. Reprinted with permission from McGraw-Hill, Inc.)

Cf. *Braided stream.*

**Dead zone** In contaminant transport models or investigations, the area where a *tracer* or *influent* is trapped along the *bank* or bottom of a *river* or *stream*. The tracers are used to determine the areas of little or no water movement along a water course.

**Debouchment** The mouth of a *river* or channel, or the point of emergence of a *tributary* into the next-order stream. Cf. *Stream order.*

**Debris flood** A flood classification between a *turbid* flood, existing in *streams* at high elevations, and a mudflow, in which the *flow* has incorporated earth material in excess of the stream's *bedload* capacity yet continues to move downgradient. The waters of a debris flood incorporate rock fragments, soil, and mud. Cf. *Debris flow.*

**Debris flow** A variety of rapid mass movement involving the downslope movement of high-density coarse clast-bearing mudflows, usually on alluvial fans. More than half of the particles of the moving mass of rock fragments, soil, and mud are larger than sand size. Slow debris flows may move less than 1 m year$^{-1}$, while rapid ones may reach 160 km h$^{-1}$.

**Debye–Huckel equation** A method of calculating the value of the *activity coefficient* ($\gamma_i$) for an individual *ion* at a dilute concentration. The Debye–Huckel equation is valid for solutions with an ionic strength of 0.1 or less, which is approximately 5000 mg L$^{-1}$. The law of mass action expresses *solute* concentrations as activities or in terms of activity coefficients, which are valid for solutions with any ionic strength. The chemical activity of an ionic species is equal to its *molality* times the activity coefficient of that species, written as:

$$a_i = m_i \gamma_i$$

where:

$a_i$ = activity of the solute species $I$ [dimensionless]
$m_i$ = molality
$\gamma_i$ = activity coefficient of the ion [kg ·mol$^{-1}$]

Except for waters with extremely high salt concentration, $\gamma_i$ is less than 1 for ionic species. The activity coefficient is therefore an adjustment factor used to convert concentrations into a form suitable for most thermodynamically based equations. The value of $\gamma_i$ at dilute concentrations can be obtained using the Debye–Huckel equation for individual ion activities:

$$-\log \gamma_i = \frac{-Az_i^2 \sqrt{I}}{1 + Ba_i \sqrt{I}}$$

where:

$\gamma_i$ = activity coefficient of the ion [kg ·mol$^{-1}$]
$A$ = constant relating to the solvent [for water at 25°C, 0.5085]
$z_i$ = ionic charge
$B$ = constant relating to the solvent [for water at 25°C, 0.3281]
$a_i$ = constant relating to the effective diameter of the ion in solution
$I$ = ionic strength of the solution.

**Table D-1** lists the values of $a_i$ for various ions and the parameters $A$ and $B$ at 1 bar [Freeze and Cherry, 1979, adapted from Kielland (1937) and Butler (1964)].

**Table D-1** Kielland table for ion-activity coefficients for the Debye–Huckel equation: $log\gamma = -Az^2 \sqrt{I}/1+ åB\sqrt{I}$. (Freeze and Cherry, 1979. Reprinted with permission from Prentice-Hall, Inc.)

### Values of the Ion-Size Parameter å for Common Ions Encountered in Natural Water:

| $a \times 10^8$ | Ion |
|---|---|
| 2.5 | $NH_4^+$ |
| 3.0 | $K^+, Cl^-, NO_3^-$ |
| 3.5 | $OH^-, HS^-, MnO_4^-, F^-$ |
| 4.0 | $SO_4^{2-}, PO_4^{3-}, HPO_4^{2-}$ |
| 4.0–4.5 | $Na^+, HCO_3^-, H_2PO_4^-, HSO_3^-$ |
| 4.5 | $CO_3^{2-}, SO_3^{2-}$ |
| 5 | $Sr^{2+}, Ba^{2+}, S^{2-}$ |
| 6 | $Ca^{2+}, Fe^{2+}, Mn^{2+}$ |
| 8 | $Mg^{2+}$ |
| 9 | $H^+, Al^{3+}, Fe^{3+}$ |

### Parameters A and B at 1 Bar

| Temperature (°C) | A | B ($\times 10^{-8}$) |
|---|---|---|
| 0 | 0.4883 | 0.3241 |
| 5 | 0.4921 | 0.3249 |
| 10 | 0.4960 | 0.3258 |
| 15 | 0.5000 | 0.3262 |
| 20 | 0.5042 | 0.3273 |
| 25 | 0.5085 | 0.3281 |
| 30 | 0.5130 | 0.3290 |
| 35 | 0.5175 | 0.3297 |
| 40 | 0.5221 | 0.3305 |
| 50 | 0.5319 | 0.3321 |
| 60 | 0.5425 | 0.3338 |

**Deep percolation** Also called *groundwater* percolation. The gravity-driven *drainage* of soil water below the maximum effective depth of the root zone. Deep percolation represents that portion of the *infiltration* in excess of the soil's *field capacity*.

**Deep water** Also called *bottom water*. The lowest of the lateral zones within a body of water that result from density differences caused by the cooling of the water mass. Cf. *Intermediate water*.

**Deep-well disposal** See *deep-well injection*.

**Deep-well injection** Also called deep-well disposal. The disposal of liquid waste through *wells* constructed especially for that purpose. These wells ideally penetrate deep, *porous*, and *permeable* formations and are confined vertically by relatively impermeable beds. Deep-well injection can be used to dispose of materials such as the saline water brought to the surface in oil wells or a variety of wastes from industrial processes. *Injection wells* typically range in depth between 200 and 4000 m below ground surface and are located in sandstones, carbonate rocks, and basalt.

**Deeper pool test** See *discovery well*.

**Defeated stream** A watercourse that is unable to erode the land surface in response to rapid uplift or increase in topographic elevation and thereby becomes ponded or diverted away from its original, more direct course. The flow of a defeated *stream* resumes as a *consequent stream*. Cf. *Obsequent stream; Beheaded stream; Stream capture*.

**Deferred junction** Also called *yazoo*. An area where the main *stream* is joined by a *tributary* whose course is elongated in the *downstream* direction by a barrier along the main stream. This barrier continues for a considerable distance before the two water courses are connected. Cf. *False stream*.

**Deferred tributary** A watercourse flowing parallel to, and eventually into, a main *stream*. Cf. *Yazoo stream; Tributary*.

**Deflation lake** A water body in a basin formed by wind *erosion* in an arid or a semi-arid region. A deflation lake is typically very shallow and may exist only intermittently during the rainy season. Cf. *Dune lake*.

**Deflocculation** The *dispersion* of suspended particles to become more evenly distributed throughout a liquid, as opposed to *aggregation*. A dispersing agent, or deflocculant, prevents fine soil particles in suspension from coalescing to form flocs. In a *drilling fluid* at rest, the suspended particles have a tendency to orient themselves, or flocculate, to balance the surface electrical charge that is the source of the drilling fluid's *gel strength*. The addition of an appropriate thinner deflocculates the solids by reducing the attractive forces and separating the clay platelets. Clay particles mixed with water can be either flocculated (in an aggregated condition) or deflocculated, as shown in **Figure D-4**. In soils where

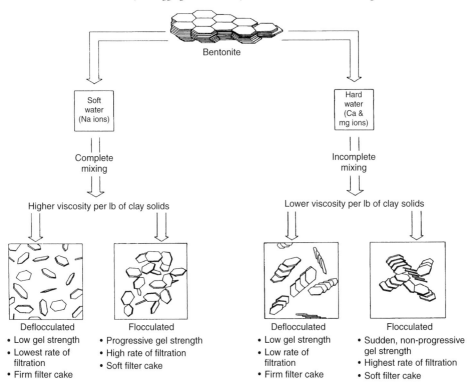

**Figure D-4** Clay particles, bentonite, mixed with water either become dispersed or aggregate as shown. (Driscoll, 1986. Reprinted by permission of Johnson Screens/a Weatherford Company.)

silt predominates and swelling clays are less than 25% of the sediment, the clay aggregates the silt grains, but as a consequence of wetting, deflocculation disintegrates the aggregates and disperses the silt. Deflocculation of clays can be responsible for structural failures of earth dams. Cf. *Flocculation*.

**Defluent** A watercourse or *stream* that flows from a glacier or a *lake*.

**Degrading stream** A confined watercourse that is actively eroding or down-cutting its channel or valley. Parameters such as the velocity and *gradient* allow a degrading *stream* to transport a greater *bed load* than that supplied to it either by its own *erosion* or *tributary* additions, thereby enabling the stream to continue its erosional activity. Cf. *Aggrading stream*; *Stream load*; *Load*.

**Degree** An expression of a water's *hardness*. Cf. *Grain*; *Salinity*; *Softening of water, processes for*.

**Degree of saturation** The extent to which the voids in a rock or soil contain water, oil, or gas. Degree of saturation is usually expressed as a percentage of the medium's total void or pore space. Cf. *Zone of saturation*; *Void ratio*; *Porosity*.

**Degrees Baumé** See *Baumé, degrees*.

**Delay index** A scale reflecting the observation that water-table *drawdown* in a *piezometer* adjacent to a pumping well in an unconfined *aquifer* declines at a slower rate than that predicted by the Theis solution or *Theis method*. In 1963 Boulton proposed a semi-empirical mathematical solution that presented three segments of the time-drawdown curve in an unconfined aquifer. Boulton assumed that the empirically defined delay index required to solve his equation was an aquifer constant, but it has since been demonstrated that the delay is related to the vertical components of *flow* induced in the flow system and is a function of the radius, *r*, and perhaps the time, *t*. The solution of *Neuman* in 1975 reproduces the three segments of the time-drawdown curve and does not require the use of the delay index. Cf. *Jacob time-drawdown straight-line method*.

**Delayed runoff/delayed flow** Water from *precipitation* that *percolates* into the ground and is temporarily inhibited from reaching an *aquifer* or discharging into a *stream* through a seep or *spring*. The delay may be by any means, including temporary storage in the form of snow and ice. Delayed runoff is a portion of the *groundwater runoff*.

**Delta** A low, nearly flat, alluvial land mass at or near the mouth of a *stream* or *river*. The area is typically fan-shaped and extends beyond the general associated coastline. The accumulation of sediments result from erosion *upstream* that is not removed by *tides*, waves, or currents. Most deltas are partially below water and sub-aerial. Cf. *Drowned coast*.

**Delta lake** A body of water formed within a *delta* or along its margin. The water of a delta lake is isolated by sediment bars formed across a shallow embayment by *stream* deposition. A briny delta lake results when part of the sea is enclosed along the delta margin by the accretion of deltaic deposits. Cf. *Bay*; *Brine*.

**Dendritic drainage pattern** See *drainage pattern*.

**Dendrohydrology** The study of the periodicity of *river* flow and flooding using dated tree-ring series.

**Denitrification** A redox process by which nitrate ($NO_3^-$) in *anaerobic* systems can be converted by bacteria to gaseous nitrogen species such as elemental nitrogen gas ($N_2$), nitrous oxide ($N_2O$), or nitric acid ($HNO_3$). Denitrification consumes organic matter and reduces inorganic compounds in *groundwater*, as shown in the equation:

$$CH_2O + \frac{4}{5}NO_3^- = \frac{2}{5}N_2(g) + HCO_3^- + \frac{1}{5}H^+ + \frac{2}{5}H_2O$$

A relatively large group of bacteria accomplishes denitrification by using $NO_3^-$ as an oxygen source in respiration. Biological denitrification can cause complete reduction in total nitrogen in percolating water.

**Dense non-aqueous-phase liquid (DNAPL)** A fluid that is denser than water with measurable aqueous solubility and can therefore displace water. If present in sufficient volume, DNAPL may continue gravitational migration (displacing water in soil pores) in *groundwater* until it encounters a layer of low permeability inhibiting further migration. **Figure D-5** shows a schematic representation of DNAPL *infiltration*. At a low-permeable *boundary*, pressure and gravity forces spread a DNAPL along the *confining bed*. DNAPL migration as a separate phase is effectively controlled by the geologic properties of the site, largely ignoring hydrologic influences. As in the overlying *vadose zone*, some of the DNAPL is held in the pore space within the saturated zone (residual saturation) and can serve as a source of contaminants for an extended period of time to a dissolved plume, depending on the *aqueous solubility* of the compounds. In the vadose zone, rainwater infiltration may dissolve organic vapors or the residual *non-aqueous-phase liquid* (NAPL) and transport these organic components to the underlying saturated zone. Cf. *Light non-aqueous phase liquid (LNAPL)*.

**Density current** Also known as density gradient. A gravity-driven flow resulting from density differences in surface water or in a *groundwater basin*, attributed to differences in temperature, *salinity*, total dissolved solids (TDS), or concentration of the *suspended load*. For example, the inflowing water to a standing water body, e.g. a *lake*, may be denser than the lake's surface water; therefore, the inflowing water plunges below the lake surface as a density current and carries its load to the bottom. Cf. *Salinity current*; *Turbidity current*; *Thalassic series*.

**Density log** See *gamma–gamma log*.

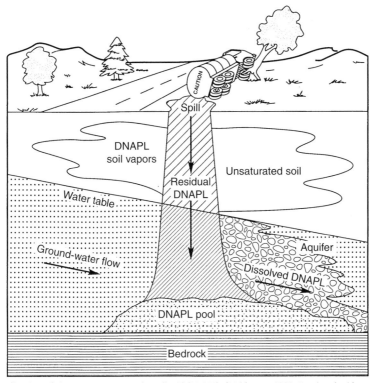

Labels in figure: Spill; DNAPL soil vapors; Residual DNAPL; Unsaturated soil; Water table; Ground-water flow; Aquifer; Dissolved DNAPL; DNAPL pool; Bedrock; CAUTION

**Figure D-5** Infiltration of dense non-aqueous-phase liquid (DNAPL). (Maidment, 1993. Reprinted with permission from McGraw-Hill, Inc.)

**Density of water** The mass per unit volume of water, expressed in SI units of $kg \cdot m^{-3}$ and English units of $lb_m \cdot ft^{-3}$ (See: **Appendix A-1**). The density of a water solution is a function of temperature, and of the concentration and type of *solutes*. To a lesser degree, it is also influenced by pressure, but because water is essentially incompressible, pressure effects on water density are normally ignored. The density of water decreases as temperature increases, except between 0 and 4°C, for which density increases with increasing temperature. Pure water has a maximum density of approximately $1000 \, kg \, m^{-3}$ at 4°C, which decreases to $994 \, kg \, m^{-3}$ at 35°C. Density, $\rho$ is the index used to determine a fluid's *specific gravity* (i.e. specific gravity of the fluid = density of the fluid). Density values are also required to convert concentrations in weight-per-weight units (ppm) to weight-per-volume units ($mg \, L^{-1}$) or the reverse, especially for highly mineralized solutions. Concentrations expressed in milligrams per liter are applicable only at the temperatures at which the determination was made. The density value is a useful indicator of *salinity* for *brines* in which sodium ($Na^+$) and chloride ($Cl^-$) are predominant. Density effects related to temperature, solute concentrations, or suspended sediment concentration can be observed in water movements within reservoirs and *lakes* and may sometimes be observed in groundwater circulation patterns. For example, vertical density differences in a water body result in *density stratification*, which may control the *flow* of water and constituents through the water body. The density of water also controls buoyancy. The properties of water are presented in **Appendix B**. Cf. *Thalassic series*.

**Density stratification** The vertical layering produced within a water body by density differences. The lightest or least-dense water layer occurs at the top and the heaviest at the bottom. Density stratification is caused by temperature changes, and also by differences in the amount of dissolved material, as when a surface *freshwater lens* overlies saltwater. Cf. *Thermal stratification; Temperature logging.*

**Dependable yield** The minimum water supply available on demand from a basin, with the possibility of a decrease on the average of once every *n* number of years. Dependable yield is most often used in describing *groundwater* availability but has also been applied, to a lesser degree, to entire water basins that include surface waters.

**Depletion** Withdrawal of water from a *stream*, surface water body, or *groundwater basin* at a rate greater than that of water *replenishment*, bringing about the ultimate exhaustion of the resource.

**Depletion curve** The portion of a *hydrograph* extending from the termination of the *recession curve* to the point of inflow due to water becoming available for streamflow. A depletion curve graphically represents the decrease of water in a stream channel, surface soil, or *groundwater basin*. Curves are also drawn for *baseflow, direct runoff,* and *runoff*.

**Depression-focused recharge** Locations, natural or manmade, that allow for ponding of rainwater to localize areas of aquifer recharge. Cf. *Recharge*.

**Depression head** See *drawdown*.

**Depression spring** See *spring; depression*.

**Depression storage** Ground surface depressions that begin to fill with water when *rainfall intensity* exceeds the local *infiltration capacity*. An example of depression storage is water trapped in puddles. Water in depressions at the end of a *precipitation* event either evaporates or contributes to *soil moisture* and/or subsurface flow following *infiltration*. The volume of water in depression storage, $V_s$, can be expressed by the equation:

$$V_s = S_d \left(1 - e^{-kP_l}\right)$$

where:
$S_d$ = depression storage capacity of the basin [$L^3$]
$e$ = base of natural logarithms (2.718281828...)
$k = 1/S_d$
$P_l$ = volume of precipitation in excess of interception and infiltration [$L^3$]

The value of $S_d$ for most basins lies between 10 and 50 mm (0.5 and 2.0 in.). The above equation ignores *evaporation* from depression storage during the storm, as that factor is usually negligible. Cf. *Detention basin*.

**Depth of circulation** An understanding of the maximum depth of circulating water is critical to understanding *flow systems*. In rock formations and thus in their resident water, a *geothermal gradient* is observed, i.e. a temperature increase with depth. Therefore, temperature measurements can aid in determining the depth of circulation. The deeper water is stored, the warmer it is when it issues from a *spring* or well. The value of the geothermal gradient varies with depth; an average value is 3°C/100 m. Using a simplified approach, the depth of circulation can be approximated using its relationship to the geothermal gradient, expressed by the equation:

$$depth(m) = \frac{T_{measured} - T_{surface}}{\Delta T/100}$$

where:
$T_{measured}$ = spring or well temperature (°C)
$T_{surface}$ = local average annual surface temperature (°C)
$\Delta T$ = local geothermal gradient from geophysical studies (°C)

If water varies in a temperature profile, it may vary in chemical parameters as well, or the chemical parameters may change as a result of the geothermal gradient. Temperature variations resulting from seasonal changes may significantly affect chemical data collected at the same location but during different times of the year. Cf. *Temperature logging*.

**Depth–area–duration (DAD) curves or relations** Storm analyses conducted for planning considerations, to determine the maximum amount of *precipitation* occurring for various durations over areas of various sizes. For a storm with a single major storm center, the isohyets are interpreted as boundaries of individual areas. The average storm precipitation is computed within each isohyet, and the storm total is distributed through successive increments of time (usually 6 h) in accordance with the distribution recorded at nearby stations, yielding the time distribution of average rainfall over areas of various sizes. From these data, the maximum rainfall for various durations (6, 12, 18 h, etc.) can be selected for each area size. These maxima are plotted as shown in **Figure D-6**, and an enveloping depth–area curve is drawn for each duration. Storms with multiple centers are divided into different zones for analysis. Cf. *Gauging station*.

**Deranged drainage** Within a *watershed* basin, flow that has been disturbed from its original *drainage pattern* by forces such as tectonic activity, glaciation, or mass movement. Cf. *Defeated stream; Beheaded stream*.

**Dereliction** The drying of a region or depression as a result of the recession or withdrawal of water. Cf. *Relict lake*.

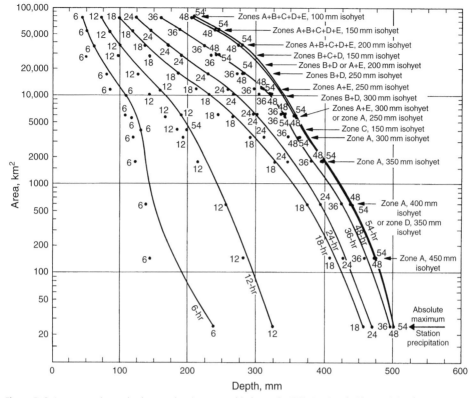

**Figure D-6** A storm maximum depth–area–duration curves. (Linsley et al., 1982. Reprinted with permission from McGraw-Hill, Inc.)

**Desalination** The removal of dissolved salts from seawater, saline water, or *brackish water* in order to make it potable. Desalination can be used to augment *groundwater* supplies, especially in times of *drought*. Regulations limit *total dissolved solids (TDS)* levels in groundwater to 500 mg L$^{-1}$ or less for drinking water use (See: **Appendix B– Drinking Water Standards)**. *Distillation* is a common method to remove salts and other chemicals, but desalination technologies such as *reverse-osmosis* systems are capable of treating seawater with a TDS level of 35,000 mg L$^{-1}$. In coastal areas where groundwater is of poor quality or in very short supply, desalination may offer a more cost-effective water supply than obtaining higher quality surface water from great distances. Cf. *Thalassic series*.

**Desiccation** A complete or nearly complete loss of water from the *pore spaces* within soils or sediments, usually attributed to regional climatic changes. *Evaporites* often form as a result of the desiccation of water bodies in arid regions.

**Design flood** The probabilistic estimate of a *flood* having a specific magnitude that will be equaled or exceeded with a given frequency. For example, an estimate of 0.1 means the design flood has the probability of being exceeded 10 times in a 100-year cycle. The purpose of a design flood is to estimate a selected exceedance probability for a given area from *rainfall intensity, rainfall duration,* and *rainfall frequency* data so that protective measures can be taken. Estimation of actual floods from actual storms is accomplished using parameters derived from individual observed events, whereas when estimating a design flood, the values of such parameters are derived statistically through probability analyses of data from historically recorded events. Cf. *Flood mitigation; Channel improvements for; Flood frequency analysis; Gringorten formula.*

**Design storm** The *rainfall* estimate corresponding to an enveloping depth–duration curve for a selected frequency. Cf. *Design flood.*

**Desorption** A geochemical process that entails the removal of an adsorbed or absorbed substance, which is partially responsible for controlling the concentration of solutes in natural waters. Sorption of a dissolved ionic species onto a soil particle is generally reversible as desorption. Cf. *Sorption; Absorption; Adsorption.*

**Desynchronization of flood peaks** A process by which *runoff* is spread out over a longer time period to lessen potential destruction. In forest management, the process entails cutting wide-open north-south strips above the snow line to produce seasonally early runoff on south-facing slopes; snowmelt runoff is thereby desynchronized in different parts of the *catchment area.* Desynchronization of flood peaks can also be achieved through a land *drainage* scheme near the catchment outflow, accelerating the arrival of flood peaks from this area ahead of the peak flow from the rest of the catchment, which arrives later.

**Detention** See *detention basin.*

**Detention basin** In *floodplain* management, a method of confining excess water to minimize property damage and reduce the *flood* threat to the population. When properly designed and located, storm-water detention basins can lessen the impact of area development by implementing controls to accommodate increases in *runoff* from the development. Detention surfaces are also a method of recharging groundwater reserves. Cf. *Recharge basin; Retention basin.*

**Detention storage** The volume of water from a precipitation event existing as *overland flow* but not including the *depression storage* volume. Detention storage in the subsurface is that water bound to soil particles by electrochemical forces in non-capillary pores, or water held back from further migration. The combination of detention storage and *retention storage* (water in the *capillary pores*) is collectively called the *soil water.* Cf. *Surface detention; Hygroscopic storage; Recharge; Retention basin; Field capacity.*

**Development** See *well development; aquifer development; aquifer stimulation.*

**Dewatering** The removal of water from a formation for construction, remediation, or mining practices. Dewatering improves the compaction characteristics of soils in the bottoms of excavations for basements, freeways, and other structures. *Drainage* below the *water table* also serves to reduce *uplift* pressures and uplift gradients at the bottom of an excavation, thus providing protection against bottom heave and *piping.* Excavation slope stability is also improved by dewatering to reduce *pore pressure* on the slopes, thereby preventing sloughing or slope failures. **Figure D-7** illustrates three common dewatering methods. Generally, the pumping rate necessary to establish a predrained condition may be greater than the rate necessary to sustain the *drawdown.* In equilibrium conditions, the total volume of water that a dewatering system is required to pump from an unconfined aquifer to produce a certain drawdown in the designated area is calculated using the *Thiem equation:*

$$Q = \frac{K(H^2 - h^2)}{0.733 \log R/r}$$

where:
$Q = discharge$ $[L^3 \cdot T^{-1}]$
$K = hydraulic\ conductivity$ $[L \cdot T^{-1}]$
$H =$ saturated thickness of the aquifer before pumping [L]
$h =$ elevation of water in the well above the bottom of the aquifer while pumping [L]
$R =$ radius of the *cone of depression* [L]
$r =$ radius of the well [L]

The depth, $h$, of the water table at a distance, $r$, from the well, when $r > 1.5H$, is:

$$h = \sqrt{H^2 - \frac{0.733Q \log R/r}{K}}$$

The equation for the volume of water to be removed from a confined condition is:

$$Q = \frac{Kb(H - h)}{0.366 \log R/r}$$

where:
$b =$ thickness of the aquifer [L]
$H =$ distance from the static water level to the bottom of the aquifer [L]

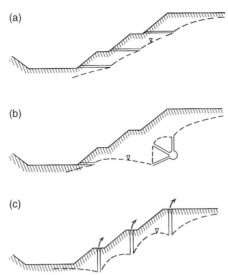

**Figure D-7** Methods of dewartering: (a) installation of horizontal drains; (b) radially drain holes in a drainage gallery. (c) system of well points. (Freeze and Cherry, 1979. Reprinted with permission from Prentice-Hall, Inc.)

Cf. *Dewatering well*; *Specific yield*; *Basin yield.*

**Dewatering well** A well that is part of a system intended to remove water from a formation or excavation and is generally in place prior to excavating so that the excavation can proceed "dry." The design of a dewatering system is an array of pumped wells or a *well point system* distributed according to the hydraulics of the area. The desired cone of draw-down or *cone of depression* in the *water table* at the excavation is created by planned mutual *interference* between the individual drawdown cones of each well or *well point*, as shown in **Figure D-8**. The *drawdown*, $s$, within the cone is expressed by the equation:

$$s = \frac{0.366Q \log R/r}{Kb}$$

where:
$Q = discharge\ [\mathrm{L^3 \cdot T^{-1}}]$
$R = $ radius of the cone of depression [L]
$r = $ radius of the well [L]
$K = hydraulic\ conductivity\ [\mathrm{L \cdot T^{-1}}]$

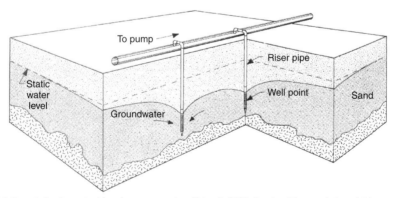

**Figure D-8** Dewatering by mutual interference pumping. (Driscoll, 1986. Reprinted by permission of Johnson Screens/a Weatherford Company.)

*Transmissivities* and storativities of the formation to be dewatered are determined during well installation, and the design of the remainder of the dewatering system is based on those values. **Figure D-9** shows examples of dewatering wells.

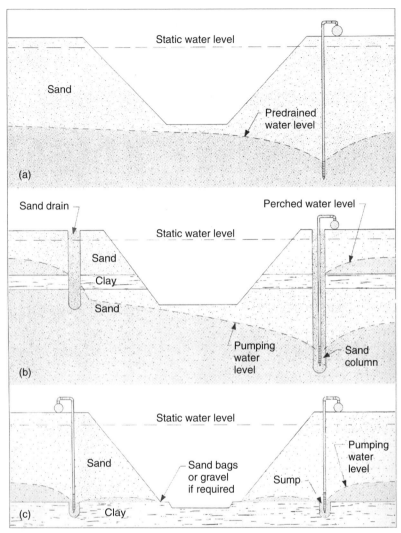

**Figure D-9** Dewatering with well points and sand drains in a trench. (a) Uniform sediments can be dewatered with well points on one side. (b) A clay layer above the subgrade may require sand drains on the opposite side to handle perched water. (c) Clay at and below subgrade may require well points set partially into the clay on both sides of trench. (Driscoll, 1986. Reprinted by permission of Johnson Screens/a Weatherford Company.)

**Dewpoint** The temperature at which a space becomes saturated with water *vapor* when the air is cooled under constant *pressure* and constant water vapor content. At the dewpoint, the saturation *vapor pressure*, $e_s$, is equal to the existing vapor pressure, $e$. The dewpoint determines whether the condensate will be water, which occurs when the ambient temperature is above 32°F, or ice, which occurs below 32°F. Cf. *Relative humidity*; *Absolute humidity*.

**Dialysis** See *osmosis*.

**Dialysis membrane sampler** See *peeper*.

**Diamond drilling** See: **Appendix C – Drilling Methods**.

**Diatomaceous earth filter** A thin membrane precoated with diatomaceous earth, which serves as a water-filtering surface. Diatomaceous earth is a light-colored, soft, siliceous material composed of the shells of diatoms. The largest deposits of diatomaceous earth are marine, but some are of lake origin. To meet federal standards for *maximum* contaminant levels (MCLs) in drinking water, a combination of *coagulation*, *flocculation*, and *filtration* is employed to reduce or eliminate *turbidity*. If the concentration of turbid matter is low, a diatomaceous earth filter is used for the filtration process. When the filter becomes clogged, the obstruction is removed by *backwashing*. See the list of maximum contaminant levels (MCL) in **Appendix B – Properties of Water**.

**Differential pressure** The difference in *pressure* between two sides or locations, e.g. the pressure at the bottom of a well and at the wellhead, or between flowing pressure at the wellhead and the pressure in the gathering line. In fluvial processes, the difference in pressure that occurs between *upstream* and *downstream* when there is a restriction of streamflow is a pressure differential. Cf. *Osmosis*; *Reverse osmosis*.

**Differential temperature log** A record of the rate of change in temperature versus depth. Cf. *Temperature log*.

**Diffluent** A watercourse, such as a *stream*, that splits or branches into two or more segment such as a tributary. Cf. *Confluent*; *Effluent stream*; *Influent stream*.

**Diffuse recharge** A condition in which the amount of water infiltrated per unit area of an *aquifer* outcrop is approximately equal. Diffuse *recharge* most commonly occurs when recharge to the aquifer occurs through openings in the bedrock.

**Diffused aeration** One of several methods of *air stripping*, using an equipment configuration of a basin in which water flows through from top to bottom while air is dispersed through diffusers at the bottom of the basin. Compared to other air-stripping systems, one drawback of diffused aeration is the lower air-to-water ratio.

**Diffusion** Often called self-diffusion, *molecular diffusion*, or ionic diffusion. A mixing process that occurs in response to a concentration *gradient*. Diffusion results in mass mixing or mass transport of ionic and molecular species and occurs regardless of the bulk hydraulic movement of the solution. Diffusive movement is in the direction of decreasing concentration and ceases only when the concentration gradient no longer exists, i.e. when masses are in equilibrium. Although diffusion is accounted for in *groundwater* transport problems, its overall contribution to the spread of contaminants or *solutes* may be negligible and is important as a *dispersion* process only at low velocities or for mass transport in fractured rock and clay, heterogenous sands with clay interbeds, and low-permeability units. Cf. *Advection–dispersion equation*; *Advection*; *Osmosis*; *Reverse osmosis*; *Differential pressure*.

**Diffusion cell** A testing cell used to determine *diffusion coefficents* for solutes passing through porous media.

**Diffusion coefficient** A *diffusivity* term used in expressions of *Fick's first law* relating concentration gradient to the solute flux that is due to *diffusion*. The diffusion coefficients for *electrolytes* in aqueous solutions are known parameters. The major ions in *groundwater* ($Na^+$, $K^+$, $Mg^{2+}$, $Ca^{2+}$, $Cl^-$, $HCO_3^-$, and $SO_4^{2-}$) have diffusion coefficients ranging from $1 \times 10^{-9}$ to $2 \times 10^{-9}$ $m^2 s^{-1}$ at 25°C. Diffusion coefficients are temperature dependent, becoming smaller with decreasing temperature; the effect of ionic strength then becomes very small. Cf. *Fick's first law*.

**Diffusion equation** Defined for homogeneous *isotropic* media as:

$$\frac{\partial^2 h}{\partial x^2} + \frac{\partial^2 h}{\partial y^2} + \frac{\partial^2 h}{\partial z^2} = \frac{\rho g(\alpha + n\beta)}{K} \frac{\partial h}{\partial t}$$

where:

$h$ = change in *head* in the $x$, $y$, and $z$ directions [L]
$\rho$ = *density of water* [M·L$^{-3}$]
$g$ = gravitational constant [L·T$^{-2}$]
$n$ = *porosity* [L$^3$·L$^{-3}$]
$\beta$ = compressibility of the fluid [L$^3$·M$^{-1}$]
$\alpha$ = compressibility of the *aquifer* [L$^3$·M$^{-1}$]
$K$ = *hydraulic conductivity* [L·T$^{-1}$]
$t$ = time [T]

The solution of $h(x, y, z, t)$ is the value of the *hydraulic head* at any time and at any point in a *flow* field and requires the input of the hydrogeologic parameters $K$, $\alpha$, and $n$ and the fluid parameters $\beta$ and $\rho$. Boundary conditions for the domain are also required for modeling purposes. Cf. *Fick's first law*; *Equations of groundwater flow*; *Error function*.

**Diffusion porosity** The pore space volume through which mass is transferred by *diffusion*, divided by the total volume of the *porous medium*. Cf. *Peclet number*.

**Diffusive model of dispersion** A concept of the transport of a given constituent in the direction of its decreasing concentration. In the diffusive model, dispersive mass flux is assumed to be driven by a concentration *gradient*, as described by *Fick's law* and expressed as the equation:

$$F_x = \bar{v}_x nC - nD_x \frac{\partial C}{\partial x}$$

where:

$\bar{v}_x = v/n$
$F_x$ = dissolved mass flux in the $x$ direction [M·L$^{-2}$·T$^{-1}$]
$v_x$ = velocity in $x$ direction [L·T$^{-1}$]
$n$ = *porosity* [L$^3$·L$^{-3}$]
$D_x$ = *dispersion coefficient* in the $x$ direction [L]
$C$ = concentration [M·L$^{-3}$]

The dispersion coefficient includes both the contributions from true *molecular diffusion* and hydrodynamic mixing. For flux in the $x$ direction, the equation can also be written as:

$$F_x = -E_x \frac{\partial C}{\partial x}$$

where:
$E_x = diffusivity\ [L^2 \cdot T^{-1}]$

This equation states that the flux is proportional to the gradient of concentration. Cf. *Dispersion; Diffusion; Peclet number.*

**Diffusivity (D)** A single parameter that describes the transmission properties of *transmissivity, T,* or *hydraulic conductivity, K,* and the storage properties of storativity, *S,* or *specific storage, $S_s$.* The *hydraulic diffusivity, D,* of a formation or *aquifer* is defined by the equation:

$$D = \frac{T}{S} = \frac{K}{S_s}$$

where:
$T = transmissivity\ [L^2 \cdot T^{-1}]$
$S = storativity\ [dimensionless]$
$K = hydraulic\ conductivity\ [L \cdot T^{-1}]$
$S_s = specific\ storage\ [L^{-1}]$

The diffusivity has dimensions of $L^2 T^{-1}$. The term "diffusivity" is not commonly used today.

**Dike spring** See *spring, dike.*

**Dimictic** Overturning or undergoing a period of deep circulation twice per year, as in an interior water body. In a temperate climate, a *freshwater* lake typically overturns in the spring and fall. *Thermal stratification* occurs during the summer season when surface water layers warm and no longer mix with the dense, cold *bottom water.* During the winter season, the surface layers cool, thereby becoming denser than the water below, resulting in reverse stratification and causing an overturn of the water column. Cf. *Monomictic;Thalassic series; Density stratifications.*

**Dimple spring** See *spring, depression.*

**Dip stream** A unique type of *consequent stream* having its watercourse and *flow* in the dip direction of the strata that it transverses. Cf. *Cataclinal stream.*

**Dipmeter logging** A *geophysical exploration method* that utilizes resistivity measurements to determine the strike and dip of strata or fractures. Common dipmeters use four pads located 90° apart, oriented with respect to magnetic north by a magnetometer within the probe, to detect resistivity anomalies. A computer correlates the anomalies and calculates their true depth. Dipmeter logging is often used in oil exploration to determine the strike and dip of bedding planes. The technique has also been used in investigations of fracture media to predict the direction of *groundwater* flow from measurements in *boreholes*, but with limited success. Fracture flow is more irregular and generally has several related intersections, thereby having a wider range of dip angle within a short depth interval. It has been demonstrated that *acoustic televiewer (ATV) logging* provides more information on the location, orientation, and character of fractures than dipmeter logging and can be substantially less expensive.

**Direct air rotary method** See: **Appendix C – Drilling Methods.**

**Direct intake** Aquifer *recharge* that occurs from, and directly through, the *zone of saturation.*

**Direct precipitation** *Rainfall* that enters an inland water body, e.g. a *lake* or *stream*, without passing through or being a part of any phase of land *runoff.*

**Direct rotary drilling** See: **Appendix C – Drilling Methods.**

**Direct runoff** Also called immediate runoff or direct surface runoff. The excess *rainfall* or *precipitation* that enters a body of water immediately after an event. The excess may eventually become *flood waters*, which are generated at the surface as *Horton overland flow* or from a subsurface route as *throughflow*. Cf. *Base runoff; Storm runoff; Surface runoff.*

**Direction of groundwater flow** The regional or local *groundwater flow* direction and the *hydraulic gradient* can be determined from measurements of water elevation relative to the ground surface, in three wells situated in a triangular array, called a three-point problem as demonstrated in **Figure D-10**. Necessary information includes the water level elevation or *total head* at each well, the relative geographic position of each well, and the distance between the wells.

**Disappearing stream** A surface watercourse that vanishes underground in a sink. Cf. *Sink hole.*

**Discharge** The removal of water (outflow) from the *zone of saturation* across the *water table* surface, along with associated *flow* within the saturated zone toward the water table. Examples of *groundwater* discharge at the surface are water from *springs*, water-fed swamps and *lakes*, and water pumped from wells. Discharge is also a term for the flow rate at a given instant expressed as volume per unit of time, $L^3 \cdot T^{-1}$, e.g. cubic feet per second (ft$^3$ s$^{-1}$) or cubic meters per hour (m$^3$ h$^{-1}$). When identifying *discharge areas* within a hydrologic regime, several characteristic features of discharge are noted, such as

- elevation in the *water table*, displayed on water contour maps
- water temperatures a few degrees above the average annual ambient temperature, reflecting warming while circulating at depth

- water *salinity* or sulfate above that of surface water, if the discharge has passed through rocks containing soluble sodium chloride (NaCl) or gypsum (CaSO$_4$·2H$_2$O).

When discharge water shows considerable variability in parameters such as temperature and chloride (Cl$^-$) content, it may indicate that more than one type of water is contributing to discharge. Cf. *Specific discharge*; *Flow rate*; *Streamflow*.

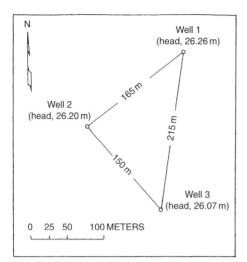 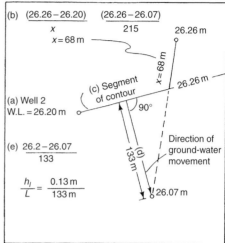

**Figure D-10** Calculation of hydraulic gradient and determination of direction of groundwater flow. (USGS Water Supply Paper.)

**Discharge area** The portion of the *drainage basin* in which the net saturated *flow* of *groundwater* is directed toward the *water table*. In a discharge area, a component of the groundwater flow direction near the surface is also upward toward the surface, and the water table is usually at or very near the surface; whereas in a *recharge area*, the water table usually lies at some depth. **Figure D-11** shows, in cross section, a discharge area (region AE), a recharge area (region ED), and the hinge line (point E) that separates them, and the impermeable boundary at the bottom of the isotropic system (BC).

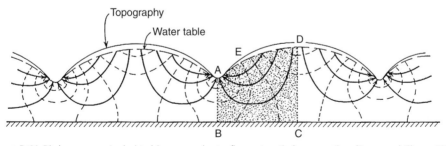

**Figure D-11** Discharge area, A, depicted in a groundwater flow net vertical cross section. (Freeze and Cherry, 1979. Reprinted with permission from Prentice-Hall, Inc.)

**Discharge, mean annual ($Q_m$)** An average measurement of the *runoff* or water emitted from a *drainage basin* over a year. In most areas, runoff is not uniform throughout the study area. Mean annual *discharge* represents the integrated discharge/runoff for the entire basin above a *gauging station*. These data are used to create a map showing the geographic variations in the discharge/runoff. Channel width, channel depth, and meander wavelength are directly related to discharge, whereas *gradient* is inversely related to discharge. These generalized relationships with mean annual discharge, $Q_m$, are expressed by the equation:

$$Q_m = \frac{bd\lambda}{S}$$

where:

$Q_m$ = mean annual discharge (where *T* is implicitly defined at one year) [$L^3 \cdot T^{-1}$]
$b$ = channel width [L]
$d$ = depth [L]
$\lambda$ = meander wavelength [L]
$S$ = gradient [$L \cdot L^{-1}$]

**Discharge velocity/discharge rate** The speed at which a given volume of water flows through a given unit of the *porous medium's* total area perpendicular to the direction of *flow*. Discharge velocity can also represent the volume of water per unit time discharged from a well (the pumping rate) or the volume of water per unit time flowing past a designated measuring point in a stream channel. These measurements of discharge velocity are expressed in dimensions of $L^3 \cdot T^{-1}$. Refer to **Appendix A** for conversion values. Cf. *Specific discharge; Darcy flux.*

**Discharge–drawdown relationship** The directly proportional association, when all subsurface factors are constant, between the amount of water removed from a well and the reduction in water level at any point in the corresponding *cone of depression*. In theory, if a well's *discharge* is doubled, the amount of *drawdown* is also doubled in accordance with the discharge–drawdown relationship. Cf. *Thiem equation; Theis equation.*

**Discordant junction** The merger of two or more watercourses with different surface elevations at the point of joining. Typically, a *tributary* emanating from a higher region cascades down into the main or next-order stream at a discordant junction. Cf. *Accordant junction.*

**Discovery well** Also called a successful wildcat, outpost well, deeper pool test, or shallower pool test. In the *water well* industry, the first well in an area to yield sufficient amounts of water for production purposes, or in the petroleum industry, the first well to encounter gas or oil in a previously unproven area or at a previously unproductive depth.

**Discrete-film zone** See *belt of soil water.*

**Disk-tension infiltrometer** Also called a constant-head permeameter. A measuring device that estimates hydraulic parameters of the *unsaturated zone* of soil near the surface by allowing *infiltration* under positive moisture suction, or at a *pressure head* that is less than zero. Cf. *Disk-tension permeameter.*

**Dismembered river system** A watercourse pattern consisting of a trunk or main river and its associated *tributaries* or branches, in which the lower portion at its *discharge* point (*sea level*) has been flooded by the sea. As a result, the *streams* that were formerly tributaries of the river are separated from the dismembered river system and enter the sea by separate mouths. Cf. *Beheaded stream.*

**Dispersal function** An expression of the movement of water out of the diverse storage sites within a *watershed*. The dispersal function describes the movement of mobilized chemicals within a watershed. A stream *hydrograph* is a record of the stream's character and the ultimate fate of *runoff*, including mobilized chemicals described by the dispersal function.

**Dispersant/dispersing agent** A compound such as a polyphosphate that is intentionally added to surface water or *groundwater* to cause or aid the spread of contaminants or the breakup of chemicals or coagulants. Chemically or physically altering a contaminant to make it more degradable by microbes can enhance microbial activity and benefit aquifer remediation techniques. Dispersants are also used before or during *well development* to aid in removing clays that occur naturally in the aquifer or those introduced by *drilling fluids* during borehole advancement. Polyphosphates are *surged* through the *well screen* to force them through the formation and separate the clay aggregates for easy removal. In *grouting*, a dispersing agent is added to separate particulate grout ingredients by reducing their interparticle attraction. Cf. *Deflocculation.*

**Dispersed phase** Solid material in the form of a *colloid* that is suspended in a fluid (the dispersion medium).

**Dispersion** Also called hydrodynamic dispersion. The tendency of a fluid to spread out from the path that it would be expected to follow according to the advective hydraulics of a *flow system*. Mechanical or hydraulic dispersion is caused entirely by the movement of the fluid, resulting in mixing. When comparing the relative magnitudes of mixing caused by hydrodynamic dispersion and *diffusion*, dispersion is far more significant. In groundwater flow, dispersion has an effect similar to that of *turbulent flow* in surface water and will expand a contaminant plume to a size beyond that expected from *advection* alone, as shown in **Figure D-12**. Dispersion in a direction perpendicular to the mean direction of groundwater flow is termed lateral or *transverse dispersion*, and dispersion parallel to the mean direction of flow is termed *longitudinal dispersion*. Longitudinal dispersion is normally more dominant and varies with the distance traveled and the media through which flow occurs. **Figure D-13** shows the variations of dispersivity with distance observed in two types of media. Dispersion is closely related to advection and changes in velocity. In a field settings, larger scale heterogeneity results in larger spatial variations in velocity than are observed at the pore scale, resulting in greater plume spreading.

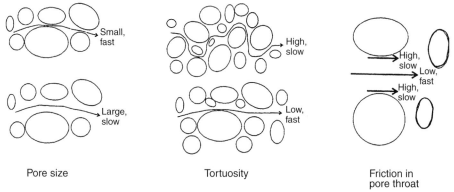

**Figure D-12** Hydrologic factors that are capable of spreading a contaminant in addition to adjective forces. (Air Force Center for Environmental Excellence, 1995.)

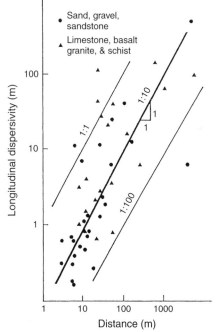

**Figure D-13** Dispersivity variation with distance. (Driscoll, 1986. Reprinted by permission of Johnson Screens/a Weatherford Company.)

Three mechanisms are known to cause dispersion (**Figure D-14**). First, the drag exerted on a fluid by the *roughness* of the pore surfaces forces molecules to travel at different velocities at different points across the individual *pore channel*. The second dispersive mechanism is differing pore sizes along flow paths. Because of differences in surface area and roughness relative to the volume of water in individual pores, different pore channels have different bulk fluid velocities. The third dispersive mechanism is related to the tortuosity, branching, and interfingering of pore channels, which cause variation in the travel path in the transverse direction.

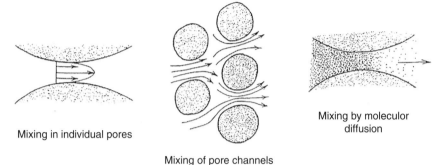

Mixing in individual pores

Mixing of pore channels

Mixing by moleculor diffusion

**Figure D-14** Schematic of dispersion on a microscopic scale. (Freeze and Cherry, 1979. Reprinted with permission from Prentice-Hall, Inc.)

Cf. *Advection–dispersion equation*; *Mechanical dispersion.*
**Dispersion coefficient ($D_x$)** Also called the hydrodynamic dispersion coefficient. A function of fluid *flow* or velocity and certain intrinsic properties of an *aquifer*: *heterogeneity, hydraulic conductivity,* and *porosity.* The coefficient is used in calculations of *dispersion,* which is one of two mechanisms of *solute* transport. Dispersion and *advection* are the physical processes that control the *flux* into and out of a volume of a non-reactive constituent in a *homogeneous* medium. The process of hydrodynamic dispersion is a result of the mechanical mixing due to differential advection and the *molecular diffusion* dependent on concentration. Transport by dispersion, *D,* is expressed by the equation:

$$D = nD_x \frac{\partial C}{\partial x} dA$$

where:
$n$ = porosity [$L^3 L^{-3}$]
$D_x$ = dispersion coefficient in the $x$ direction [$L^2 \cdot T^{-1}$]
$C$ = concentration of solute [$M \cdot L^{-3}$]
$dA$ = cross-sectional area [$L^2$]

The dispersion coefficient, $D_x$, is related to the *dispersivity* and *diffusion coefficient* of a *porous medium* by the equation:

$$D_x = \alpha_x \bar{v}_x + D^*$$

where:
$\alpha_x$ = dispersivity [L]
$\bar{v}_x$ = average linear velocity [$L \cdot T^{-1}$]
$D^*$ = coefficient of molecular diffusion for the solute in a porous medium [$L^2 \cdot T^{-1}$].

At a low velocity, diffusion is the important contributor to dispersion, and therefore the dispersion coefficient equals the diffusion coefficient: $D_x = D^*$. At higher velocities, mechanical mixing is the dominant dispersive process, in which case $D_x = \alpha_x v$.
**Dispersive stress** See *Bagnold dispersive stress.*
**Disphotic zone** In a body of water, the depth or area in which there is only dim light and therefore a decrease in photosynthesis. Cf. *Eutrophic.*
**Displacement entry pressure** The *fluid pressure* that must be exerted to displace a wetting fluid from the void spaces of a *porous medium* by an invading non-wetting fluid. For example, the wetting fluid may be water and the non-wetting fluid a non-aqueous phase liquid (*NAPL*) such as trichloroethene. Mean pore aperture size is the major control on displacement entry pressure.

**Displacement grouting** The injection of a *grout* mixture into a formation in such a manner that it forces a movement or shift of the natural formation material. When intentional and controlled, displacement grouting may aid in well production. Cf. *Penetration grouting*.

**Disposal well** A *well* drilled specifically for discarding *brines* or other fluids below the surface of the ground in order to prevent contamination of the surface by such wastes.

**Dissolution** The process by which minerals dissolve into water. Dissolution controls the mixture of cations (*ions* of positive electrical charge) and anions (ions of negative electrical charge). The form and concentration of the dissolved constituents is dependent on the water's contact with the atmosphere during *precipitation* and with the soil during either *infiltration* or *groundwater flow*. The range and typical concentrations of dissolved constituents for *water quality* parameters in *streams* and *rivers* are presented in **Appendix B**.

**Dissolved gases in water** When water is exposed to a gas phase, an equilibrium exists between the liquid and the gas through the exchange of molecules across the liquid–gas interface. Surface waters, for example, experience a continual and free exchange of molecules between the water surface and the earth's atmosphere. *Groundwater* also contains dissolved gases from exposure to the atmosphere during *infiltration* through the *unsaturated zone*, or from gas production below the *water table* by chemical or biochemical reactions involving the groundwater, minerals, organic matter, and bacterial activity. The most abundant dissolved gases in groundwater are nitrogen ($N_2$), oxygen ($O_2$), carbon dioxide ($CO_2$), methane ($CH_4$), hydrogen sulfide ($H_2S$), and nitrous oxide ($N_2O$). As $N_2$, $O_2$, and $CO_2$ are components of the atmosphere, it is not surprising that they are found in subsurface water. $CH_4$, $H_2S$, and $N_2O$ often exist in significant concentrations in groundwater as the product of biogeochemical processes that occur in non-aerated subsurface zones. The concentrations of these gases are indicators of geochemical conditions. Other species of dissolved gases that occur in groundwater in minute amounts can provide information on water source, age, or other factors of hydrologic or geochemical interest. *Well casings* and *well screens* are affected by dissolved gases, causing *corrosion* and *incrustation*. Cf. *Dalton's law*.

**Dissolved load** The part of the total *stream load* that is carried in solution. Approximately 90% of the dissolved load normally consists of five *ions*: chloride ($Cl^-$), sulfate ($SO_4^{2-}$), bicarbonate ($HCO_3^-$), sodium ($Na^+$), and calcium ($Ca^{2+}$). Cf. *Solution load*; *Dissolved solids*.

**Dissolved oxygen (DO)** The amount of oxygen gas ($O_2$) in water, expressed by weight as parts per million (ppm) or by weight per volume as milligrams per liter ($mg\,L^{-1}$). Review **Appendix A** for unit conversions. The primary source of DO, either directly or indirectly, is the atmosphere. Rain and surface waters in equilibrium with the atmosphere become saturated with DO. When this water infiltrates the soil column and reaches the *zone of saturation*, it is isolated from further contact with air. The concentration of DO in air-saturated water depends on *pressure*, which is controlled by altitude and temperature. Increasing the temperature decreases the oxygen's *solubility*, increasing the *salinity*. **Figure D-15** shows a plot of the concentration of DO in freshwater as a function of the ambient temperature. Understanding the DO in water can be useful in characterizing the following:

- the degree of *aerobic/anaerobic* conditions
- biological activity (an increase in biological activity may be indicated by a decrease in the DO)
- confined systems (older confined groundwater is likely to have decreased oxygen concentrations)
- variations or water stratification in depth profiles.

The levels of DO in *groundwater* can also be important in considerations of well design and in remediation design for contaminated *aquifers*. DO levels above $2\,mg\,L^{-1}$ suggest corrosive conditions. In *groundwater remediation*, levels of

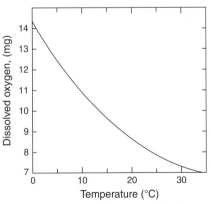

**Figure D-15** The variation of dissolved oxygen content in fresh water as related to temperature. (Mazor, 1991. Reprinted with permission from Halsted Press, a division of John Wiley & Sons, Inc.)

DO may directly affect remedial processes or indicate aquifer conditions that influence contaminant behavior and fate. DO is removed from groundwater in many ways. Organic carbon, a common component of sedimentary formations, tends to react with DO fairly rapidly, so high levels of DO may indicate low levels of organic carbon. This information is particularly important in considerations of contaminant transport and cleanup, because many contaminant chemicals are absorbed and held in place by organic carbon. DO levels are usually highest in the upper region of an unconfined shallow aquifer. Cf. *Well rehabilitation; Well maintenance.*

**Dissolved solids** The total amount of organic and inorganic material in solution, or the *dissolved load,* contained in a sample of water. A general groundwater classification system based on *total dissolved solids* (TDS) is as follows:

| Classification | TDS (mg L$^{-1}$) |
| --- | --- |
| Freshwater | 0 to 1000 |
| Brackish water | 1000 to 10,000 |
| Saline water | 10,000 to 100,000 |
| Brine | Greater than 100,000 |

Cf. *Activity coefficient; Debye–Huckel equation; Advection.*

**Dissolved solute transport** See *dispersion; dispersion coefficient; dissolution; distribution coefficient; dense non-aqueous-phase liquid; light non-aqueous-phase liquid.*

**Distance–drawdown method** See *Hantush inflection point method; Hantush Jacob formula.*

**Distance–drawdown relationship** A correlation between the rate of water removal from a well and the water level in an *unconfined aquifer,* or depressurization in a *confined aquifer.* At a given *discharge* rate, the *drawdown* at any point on the *cone of depression* is inversely proportional to the log of the distance from the pumping well. Simultaneous measurement of drawdown in a minimum of three *observation wells* is required to construct a distance–drawdown graph with a fair degree of confidence. When the drawdown is plotted on standard graph paper, the curve represents the observed cone of depression, as shown in **Figure D-16**. The drawdown curve for the same well configuration, discharge, and time period becomes a straight line when plotted on a semi-logarithmic

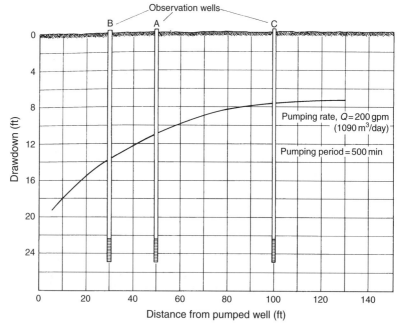

**Figure D-16** Defining the cone of depression by plotting the drawdown in three observation wells. (Driscoll, 1986. Reprinted by permission of Johnson Screens/a Weatherford Company.)

paper, as shown in **Figure D-17**. Observation wells located further from the pumped well fall on the straight-line graph, and therefore the effect of pumping at any distance from the pumped well can be determined. The trace of the cone of depression on the straight-line curve allows for the calculation of *transmissivity, T,* and the *coefficient of storativity, S,* from the equations for *T* and *S.*

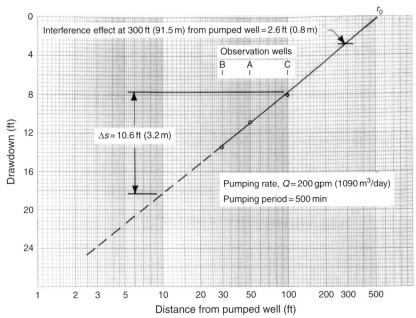

**Figure D-17** The resulting straight line when cone of depression traces are plotted on semi-log paper. (Driscoll, 1986. Reprinted by permission of Johnson Screens/a Weatherford Company.)

Cf. *Jacob distance–drawdown straight line method; Jacob step–drawdown test.*

**Distillation** The removal of impurities from a liquid by heating it and condensing the *vapor.* A distillation unit boils water, collects the steam, and then condenses it to almost pure liquid, leaving the unwanted minerals or *ions* in the concentrated residue. Distillation is the most common method of removing salts from seawater. Cf. *Salinity.*

**Distributary** A *branch* of a *river* or *stream* flowing away from the main channel as a divergent stream and not returning. Cf. *Tributary.*

**Distribution coefficient (Kd)** Also referred to as a partition coefficient. A mathematical expression that describes *solute* partitioning between liquid and solid or gas and solid. The expression is valid if the reactions that cause the partitioning are fast and reversible and if the *adsorption isotherm* is linear. Many known contaminants in water meet these criteria. The distribution coefficient, $K_d$, is a measure of the tendency for a solute to be sorbed to the soil and can be expressed as the mass of solute on the solid phase per unit mass of solid phase divided by the concentration of solute in solution. The test to determine the distribution coefficient for a solute involves mixing the soil and fluid, either liquid or gas, and measuring the solute concentration in the fluid before and after mixing, which gives the mass of the solute sorbed by the soil. The mass of the solute sorbed per unit mass of soil is plotted versus the solute concentration in the fluid (only valid for metals at constant *pH*). If the graph, or isotherm, is linear, the slope is the distribution coefficient of the solute. The *Freundlich isotherm,* a non-linear model, is used if appropriate. The dimensions for the distribution coefficient are $L^2 \cdot M^{-1}$ and measured $K_d$ values are normally reported as milliliters per gram (mL g$^{-1}$). When the $K_d$ value is orders of magnitude larger than 1, the solute is essentially immobile. Refer to **Appendix A** for unit conversions.

**Distribution graph** A graph having the same scale as the *unit hydrograph* and ordinates of the percentage of the total surface *runoff* that occurs during successive, but arbitrarily chosen, uniform time increments. The unit hydrograph theory states that all unit storms produce nearly identical distribution graphs, regardless of their magnitude.

**Diversion** The process by which one *stream* changes the course of another stream, as by aggradation or *stream capture.* Cf. *Diverted stream.*

**Diverted stream** A stream for which the course or *drainage* has been changed, e.g. by piracy or *stream capture.* Cf. *Beheaded stream.*

**Divide** In the case of a continental divide, an aboveground topographic separation or high ground that forms a physical boundary between two adjacent *drainage basins,* dividing the surface waters that *flow* in one direction from those that flow in the opposite or different direction; in the case of a *groundwater divide*, a belowground surface boundary in a *water table* or *potentiometric surface* from which the *groundwater* moves away in both directions. Water above or below, in adjacent *aquifers*, or even in the same aquifer but at greater depths may not be affected and can therefore flow across the surface divide. A groundwater divide may also separate adjacent *groundwater basins* where an elevation in the water table causes segregation of *groundwater reservoirs*. Cf. *Discharge area*.

**Diviner** See *dowser.*

**Divining rod** A forked wooden stick, metallic wire, or similar handmade object used in *dowsing*. Used by a *dowser*, a divining rod is said to dip downward sharply or, in the case of a wire, crosses perpendicularly when held over a body of *groundwater* or a mineral deposit. No scientific evidence supports the use of a divining rod, but the practice is still active today in some areas.

**DNAPL** See *dense non-aqueous-phase liquid.*

**DO** See *dissolved oxygen.*

**Domestic water** Water for residential use, supplied to homes and the general population. Cf. *Supplied water.*

**Dominant discharge** The typical volume of flowing water that determines the character of a natural channel. Dominant discharge is a relationship between the channel's maximum discharge and mean discharge, which in turn determines the channel's sediment composition, *flow*, and *flood frequency*.

**Double porosity** See *dual porosity.*

**Downgradient** A groundwater term synonymous with the surface water term *downstream*, indicating the direction in which *groundwater* will *flow*, but typically including a mathematical degree of elevation change. Cf. *Drainage basin*; *Groundwater gradient*; *Hydraulic gradient*.

**Downhole logging** See *geophysical exploration methods*; *electric logging.*

**Downstream** The direction of *stream* or channel flow in response to gravity-driven movement from higher to lower elevations. The lower elevation is the downstream direction. The elevated ground bordering the stream, i.e. its banks, are designated right or *left bank* as to an observer facing down stream. Cf. *Anabranch.*

**Down-the-hole air hammer** See: **Appendix C – Drilling Methods.**

**Dowser** Also called a diviner, or water witch. One who practices *dowsing.*

**Dowsing** The practice of locating *groundwater*, mineral deposits, or other objects by means of a *divining rod* or a pendulum. No scientific evidence supports the use of a divining rod but the practice is still active today in some areas.

**Drainage** The movement by which the waters of an area *flow* off, or drain, into surface *streams* or subsurface conduits, either natural or manmade. Drainage is a natural or artificial means of discharging water from an area by a system of surface and subsurface passages, e.g. streamflow, sheet flow, and *groundwater flow*. Manmade drainage systems are often engineered to lower *water table* elevations and intercept seepages. Cf. *Dewatering*; *Overland flow*; *Discharge area*.

**Drainage basin** An area where surface *runoff* collects and from which it is carried by a *drainage system*, such as a *river* or *stream* and its *tributaries*. The term has also been substituted for catchment area, *drainage* area, feeding ground, gathering ground, or hydrographic basin depending on whether its use is in geology, hydrology, or other arena. Cf. *Watershed*; *Catchment basin.*

**Drainage basin/drainage area** A region defined by a drainage *divide* and containing a system that collects the water within the region and then transports it to a particular *stream* channel, network of channels, *lake, reservoir*, or other body of water. In a *fluvial* system, a drainage basin is divided into three parts called Zones 1, 2, and 3, as shown in **Figure D-18**. Zone 1 is uppermost and consists of the *watershed* and sediment-source area. Progressing in the

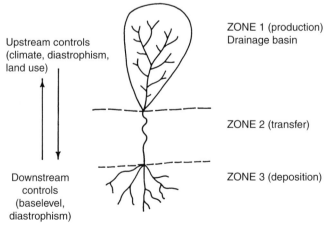

**Figure D-18** Drainage basin zones in a fluvial system. (Schumm, 1977. Reprinted with permission from John Wiley & Sons, Inc.)

*downstream* direction, Zone 2 is the transfer or predominant *transport* zone, and Zone 3 is the sediment sink or zone of deposition. A drainage basin underlain by relatively impermeable materials is capable of handling a large *discharge*, inhibits the contribution to *groundwater*, and has a small *low-flow* discharge in dry periods. High sediment yields are anticipated from drainage basins underlain by shales and lower sediment yields from equally erodible but permeable underlying sandstones. Cf. *Drainage basin relief ratio (Rh); River basin; Watershed; Hydrographic basin; Area relations of catchments; Catchment basins; Law of basin areas; Flow system.*

**Drainage basin relief ratio (*Rh*)** A relief characteristic that correlates with the rate of sediment loss for the basin being studied. The relief ratio, *Rh*, of a *drainage basin* is determined by the equation:

$$Rh = \frac{H}{L}$$

where:

$H$ = elevation difference between the highest and the lowest points within the basin [L]

$L$ = horizontal distance along the longest dimension of the basin parallel to the main stream line [L].

Cf. *Law of basin areas; Area relations of catchments; Flow system.*

**Drainage coefficient** The depth of water or amount of *runoff* drained from an area within a 24-h period.

**Drainage density** The ratio of the total length of all *streams* within a *drainage basin* (expressed in miles or kilometers of channel lengths) to the area of that basin (expressed in square miles or square kilometers). Refer to **Appendix A** for conversion values. Drainage density integrates both soil and rock characteristics and is a measure of the topographic texture of the area, in which variations are dependent on climate, vegetation, relief, and lithology. Drainage density is directly related to *flood runoff*, inversely related to *baseflow*, and also related to the number of channels per unit area, or channel frequency; as the topographic texture becomes finer, existing streams are lengthened and new low-order streams are added to the drainage network. Drainage density varies with climate. In an area with little to no *precipitation*, the drainage density is near zero, but when precipitation increases in an arid region, *erosion* begins, and the drainage density invariably increases. Drainage density also parallels sediment yield. Maximum drainage densities and maximum sediment yields in the same climatic area suggest that the high sediment yields reflect increased channel development and, therefore, a more efficiently drained system. Cf. *Stream order; Basin order; Law of basin areas.*

**Drainage divide** See *divide; discharge area.*

**Drainage gallery** In *grouting*, an opening or passageway from which grout holes or drainage curtain holes, or both, are drilled. Cf. *Grout curtain; Infiltration gallies.*

**Drainage lake** An interior body of water, such as an *open lake*, from which water is removed by a surface outlet and whose water level is controlled by the degree of its *effluent.*

**Drainage network** See *drainage pattern.*

**Drainage network analysis** The study of the organization of the pattern of streams in a *drainage basin*. Classical work in drainage network analysis focused on the relations between the importance, or order, of a stream segment and its frequency, and certain "laws" of drainage network composition were derived. The modern approach emphasizes the importance of random processes in the explanation of these laws and is more concerned with the density of the drainage network. Cf. *Basin accounting; Law of basin areas.*

**Drainage pattern** Also called drainage network. The configuration, arrangement, or spatial relationship of the watercourses in a given area. The areal drainage pattern that develops is related to

- initial slope of the surface
- surface and subsurface lithology, and variations therein
- structure and geologic events (e.g. tectonics)
- geologic and geomorphic history
- climate and climatic history.

Drainage patterns can be classified into four major divisions, as illustrated in **Figure D-19**:

1. Dendritic: the most common pattern, a characteristic irregular arrangement of *tributary* streams diverging randomly at almost any angle, which in plan view resembles the branching of tree limbs. A dendritic surface drainage pattern indicates that the underlying rocks offer uniform resistance to *erosion*.
2. Trellis, fault-trellis, or rectangular: a structurally controlled pattern of parallel or subparallel streams that developed along the bedrock strike-and-dip features.
3. Radial: a pattern resulting from a domed or centrally high elevation feature, such as a volcano, and from local geologic (bedrock) and geomorphic features and history.
4. Rectangular: develops on regularly fractured bedrock with tributaries commonly having 90° bends joining other streams at right angles.

The growth of a drainage pattern is described in stages. The first stage is initiation, or the beginning and early development of a shallow, skeletal drainage pattern on the undissected area. This is followed by elongation, or the headward growth of the main streams, and elaboration, or the filling in of previously undissected areas by small tributaries. Maximum extension is the maximum development of the pattern. The final stage is abstraction, or the loss of tributaries as the basin's relief is reduced over time.

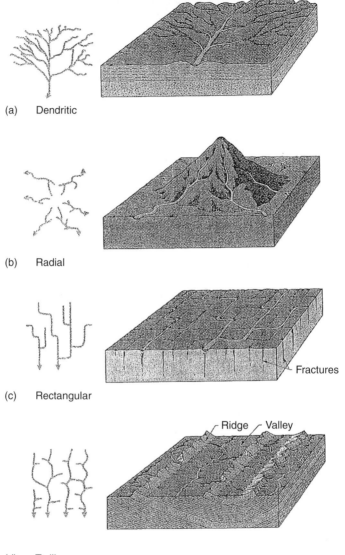

(a)  Dendritic

(b)  Radial

(c)  Rectangular

(d)  Trellis

**Figure D-19** Drainage pattern often reveal characteristics of the underlying material as shown by four major pattern classifications. (McGeary, 2001. Reprinted with permission from McGraw Hill, Inc.)

Cf. *Law of basin areas.*

**Drainage ratio** See *drainage basin relief ratio; drainage density; drainage network analysis.*

**Drainage system** A grouping of all of the *streams,* water bodies, *tributaries,* or modes of water conveyance contributing to a given surface water body and by which a region is drained. Cf. *Drainage pattern; Drainage network; Stream order; Drainage basin.*

**Drainage texture** See *drainage pattern; stream order.*

**Drainage well** A vertical shaft typically drilled in a masonry dam to intercept seepage before it reaches the *downstream* side. Cf. *Capture zone; Cone of depression; Drawdown; Well design; Seepage face.*

**Draw** A small ravine or shallow gulch, usually dry, but which can contain water after a *rainfall*, or a sag or depression leading from a valley to a gap between two topographically higher areas.

**Drawdown** Also called the depression head. In an unconfined *aquifer*, the measurement of the *water table* that results from the removal of water by a pumping well, or in a confined aquifer, the reduction of the *pressure head* that results from the removal of water. Drawdown is the difference between the static and the dynamic *potentiometric* water levels. A drawdown can be marked by the difference between the height of the water table and that of the water in a well, or by the difference between the water table or *potentiometric surface* and the *pumping water level*. Drawdown measurements are given in length, L. **Figure D-20** depicts the effect of aquifer *transmissivity* on the drawdown and associated *radius of influence* of a pumping well, and on the resulting *drawdown curve*. In surface water, drawdown is also the vertical distance by which the level of a *reservoir* is lowered when water is withdrawn.

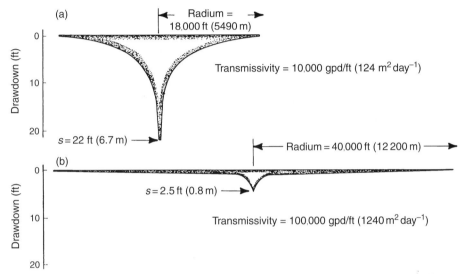

**Figure D-20** The shape, depth, and extent of a cone of depression as effected by differences in the coefficients of transmissivity when pumping rate is constant. (Driscoll, 1986. Reprinted by permission of Johnson Screens/a Weatherford Company.)

**Drawdown cone** See *drawdown*; *cone of depression*.

**Drawdown curve** The profile line defined by the intersection of the water level surface around a *pumping well* and a vertical plane passing through the well, which graphically demonstrates the relationship between *drawdown* and the radial distance from the pumping well(s). **Figure D-21** shows a drawdown curve where the water level is determined from *observation wells*. In an unconfined *aquifer*, the drawdown curve depicts the level to which the formation remains saturated. In a confined *aquifer*, the curve represents the *hydrostatic pressure* in the aquifer. At any given point on the drawdown curve, the difference between the water level indicated by the curve and the *static water level* is the drawdown in the aquifer. Drawdown in the vicinity of a pumping well is directly proportional to the pumping rate and the time since pumping began and is inversely proportional to the *transmissivity*, the storativity, and the square of the distance between the pumping well and any point on the *cone of depression*. Cf. *Aquifer, unconfined*; *Distance–drawdown relationship*.

**Drift** (1) The noted difference between the velocity of fluid *flow* and the velocity of an object moving within the fluid relative to a fixed external measuring point. (2) Drift is also a drilling term relating the degree of variance of the actual drilling angle from the intended drilling angle. (3) Drift is an abbreviation of glacial drift (till). (4) Applied to the movement of continents, as in continental drift. Cf. *Plumbness and alignment*; *Alignment test*; *Inclinometer*.

**Drill cuttings** See *driller's log*.

**Drill mud** See *drilling fluid/drilling mud*.

**Driller's log** The brief, often vernacular descriptions of the gross characteristics of well cuttings, noted by the drilling crew as a well is drilled. In groundwater investigations, an analysis of the geologic materials is critical in determining well location and, therefore, the capacity of the materials to yield sufficient, good quality water. The most direct method of learning of the character of the formations below ground surface is by drilling through them. The driller's description of the geologic character, depth, and intervals of the formations drilled through are recorded. The driller's log is included as part of a driller's report. The driller's log may also be referred to as the lithologic log as it does record the lithology encountered while drilling. Lithologic logs when completed by a geologist are typically more detail oriented. Other important parameters that can be recorded include the drilling action and penetration rate, i.e. the noise and motion of the drilling rig; a drilling-time account because the character of the material largely determines the rate at which penetration proceeds; water level fluctuations indicating whether an impervious material has been penetrated, i.e. a *confining layer*, and heaving sands, if encountered.

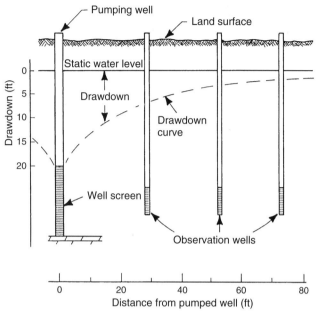

**Figure D-21** The variation of drawdown with distance from a pumped well. (Driscoll, 1986. Reprinted by permission of Johnson Screens/a Weatherford Company.)

**Drilling fluid/drilling mud** Also called drill mud, or foam. A dense fluid or *slurry* used in rotary drilling (See: **Appendix C**) to prevent caving of the borehole walls, as a circulation medium to carry cuttings away from the bit and out of the hole and to seal fractures or permeable formations or both, preventing loss of circulation fluid. The most common drilling fluid is a water–*bentonite* mixture; however, many other materials may be added or substituted to increase density or decrease velocity. Cf. *Attipulgite clay*; *Montmorillonite*; *Cement grout*; *Flocculation*.

**Drilling methods** See: **Appendix C – Drilling Methods**.

**Drilling rig** See: **Appendix C – Drilling Methods**.

**Drilling-time log** An incremental record of the time, in minutes per foot or meter, required to advance a predetermined depth while drilling. A drilling-time log can help to determine the characteristics of the formation; an increase in drilling rate or a sudden drop in the drill line potentially indicates a more *porous medium*, or in the case of consolidated or karst topography, a break in the structure or a conduit for *flow*. Cf. *Spontaneous potential (SP) logging*.

**Drill-through casing driver** See: **Appendix C – Drilling Methods**.

**Drinking water standards** The criteria that serve as a basis for appraisal of the results of chemical analyses of water in terms of suitability of the water for the intended uses. The most important water quality standards are those established for drinking water. Recent United States drinking water standards as issued by the US Environmental Protection Agency (EPA) are presented in **Appendix B – Properties of water**. Cf. *Potable water*; *Maximum contaminant levels (MCLs)*.

**Driven well** Also called a tube well. Typically, a shallow well of small diameter (3–10 cm), installed by driving a series of connected lengths of pipe into unconsolidated material without drilling, boring, or jetting. *Well points* driven by an aboveground hammer weighing up to 1000 lb can reach depths to 50 ft below ground if subsurface conditions are favorable. **Figure D-22** shows a typical drive point used for a driven well. The advantage of a driven well is that the *well screen*, or screened interval, is placed without the use of a *filter pack* and, therefore, is set into natural material with minimal disturbance. Cf. *Dug well*; *Kanat*; *Jetted well*.

**Drive pipe** A thick type of casing that is driven or forced into a *borehole* to shut off water or prevent caving. A drive pipe may be fitted at its lower end with a sharp steel or diamond shoe, which is employed when difficulty is encountered in inserting the casing. Cf. *Driven well*.

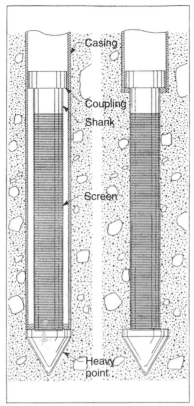

**Figure D-22** Setting a well, in rocky terrain, by using a drive point pushed by the casing which is protecting the screen above the drive point. (Driscoll, 1986. Reprinted by permission of Johnson Screens/a Weatherford Company.)

**Drop** The surface water elevation change measured from a location *upstream* and a location *downstream* of a natural or manmade constriction in the *stream* channel.

**Drop pipe** See *eductor pipe*.

**Drought** A climatological term associated with a sustained period of lower-than-normal moisture levels and water supply. A drought is a period of deficient *precipitation* relative to the normal levels in the environment under study. As drought in one region may represent normal conditions in another, the definition of drought is directly dependent on the location and the topic of discussion such as a meteorological, climatological or hydrologic condition, agricultural perspective, and economic consideration. Cf. *Aquifer, unconfined*; *Baseflow*; *Evaporation*; *Hydrologic cycle*.

**Drowned coast/drowned river mouth/drowned valley** A shoreline with long, narrow channels suggesting that subsidence of the coast has transformed the lower portions of *river* valleys into tidal estuaries; or a river terminus that has been invaded, submerged, or widened by seawater intrusion; or a valley that is partly submerged by the intrusion of a sea or a *lake*. Cf. *Dismembered river system*; *Delta*.

**Dry basin** An interior area in a climate so arid that the *drainage* is negligible, and therefore the area contains no perennial lake. Cf. *Drought*.

**Dry detention basin** Also called an extended detention basin. A system for stormwater collection and treatment that is designed for peak flow attenuation and is an effective method of reducing *pollutants* in *stormwater runoff*. The dry detention basin is the most common type of detention basin in the United States, Canada, and Australia. Cf. *Peak discharge*.

**Dry lake** A basin that formerly contained a standing surface water body, which disappeared when *evaporation* processes exceeded *recharge*. A dry lake can be a *playa*, or a salt- and mineral-encrusted land in an arid or semi-arid region, occasionally covered by an *intermittent lake*.

**Dual porosity** Also called double porosity. The two main types of voids, or pores, exhibited in a water-bearing formation. Dual *porosity* comprises *primary porosity*, or matrix porosity, and *secondary porosity*, i.e. joints and fractures resulting from stresses applied to the formation. Cf. *Pore spaces*.

**Dual-wall reverse-circulation rotary drilling** See: **Appendix C – Drilling Methods.**

**du Boys formula or equation** The computation of the volume of *bed load* material. The separation of the bed load and *suspended load* for water and sediments in a channel is often arbitrary. According to the du Boys formula, the transport rate depends on an empirical coefficient related to the size and shape of the *load*, the bed shear, and the magnitude of *shear stress* at which transport begins (the critical shear). The du Boys formula is:

$$G_i = \gamma \frac{\tau_0}{w} (\tau_0 - \tau_c)$$

where:
$G_i$ = rate of bed-load transport per unit width of stream [$L \cdot T^{-1} \cdot L^{-1}$]
$\gamma$ = empirical coefficient [$L^6 \cdot M^{-2} \cdot T$]
$\tau_0$ = bed shear [$M \cdot L^{-2}$]
$w$ = specific weight of water [$M \cdot L \cdot T^2 \cdot L^{-3}$]
$\tau_c$ = critical shear [$M \cdot L^{-2}$]

The coefficient and values of critical shear for bed load movement are listed in **Table D-2**.

**Table D-2**   Values for use in du Boys formula. (Linsley et al., 1982. Reprinted with permission from McGraw-Hill.)

| Particle diameter, mm | $\gamma$ | | $\tau_c$ | |
|---|---|---|---|---|
| | $ft^6/lb^2 \cdot S$ | $m^6/kg^2 \cdot S$ | $lb/ft^2$ | $kg/m^2$ |
| $\frac{1}{8}$ | 0.81 | 0.0032 | 0.016 | 0.078 |
| $\frac{1}{4}$ | 0.48 | 0.0019 | 0.017 | 0.083 |
| $\frac{1}{2}$ | 0.29 | 0.0011 | 0.022 | 0.107 |
| 1 | 0.17 | 0.0007 | 0.032 | 0.156 |
| 2 | 0.10 | 0.0004 | 0.051 | 0.249 |
| 4 | 0.06 | 0.0002 | 0.090 | 0.439 |

**Dug well** A shallow, usually large-diameter well excavated with hand tools or power machinery instead of by drilling or driving. A dug well is commonly used for individual *domestic water* supply, as extraction from it may be limited. Although dug wells are typically an ancient method of water extraction, they are still in use in developing countries. Cf. *Driven well; Kanat.*

**Dune lake** A basin-formed body of water resulting from the blockage of the mouth of a stream by the migration of sand dunes along a shore. A dune lake occupies a deflation basin among dunes. Cf. *Deflation lake.*

**Dupuit–Forchheimer assumptions** See *Dupuit–Forchheimer theory.*

**Dupuit–Forchheimer equation** An analytical expression of groundwater problems in unconfined aquifers. The existence of a seepage surface creates complex *boundary conditions*, as shown in **Figures D-23 and D-24**. Accepting the assumptions of the *Dupuit–Forchheimer theory* enables the derivation of an equation expressing *aquifer* discharge, Q, at a radial distance, r, as:

$$Q(r) = 2\pi r h(r) q(r)$$

where:
$h(r)$ = saturation thickness at distance r [L]
$q(r)$ = time-dependent *specific discharge* at distance r [$L^3 \cdot T^{-1}$]

Substituting this equation into that of *Darcy's law* for steady *radial flow* in a unconfined aquifer yields:

$$h^2(R) - h^2(r) = \frac{Q}{\pi K} \ln\left(\frac{R}{r}\right)$$

where:
$h^2(r)$ = saturated thickness at the *radius of influence* r [L]
$K$ = *hydraulic conductivity* [$L \cdot T^{-1}$]

When R>r, or in terms of *drawdown*, s(r) = H−h(r), where H is the saturation thickness before the water is extracted, the equation becomes:

$$s^2(r) - s^2(R) = \frac{Q}{\pi K} \ln\left(\frac{R}{r}\right)$$

At distances of $r < 1.5H$, however, the Dupuit–Forchheimer equation for drawdown measurement does not yield satisfactory results. Unlike the results for confined aquifers, the relationship between the distance and the drawdown is not linear.

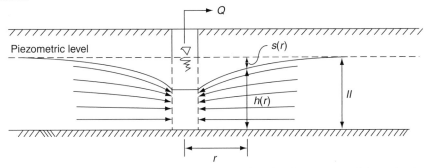

**Figure D-23** Complex boundary conditions exist in unconfined aquifers. Flow lines are not parallel to each other and a seepage surface exists. (Sen, 1995. Reprinted with permission from CRC Press, Inc.)

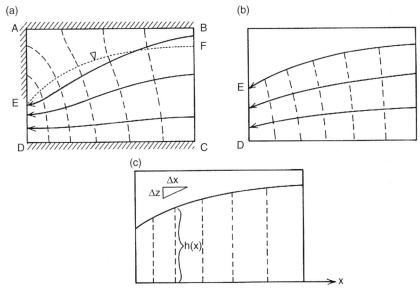

**Figure D-24** An outflow boundary of a seepage face in (a) saturated–unsaturated flow net, (b) free-surface flow net, (c) Dupuit–Forchheimer flow net. (Freeze and Cherry, 1979. Reprinted with permission from Prentice-Hall, Inc.)

Cf. *Seepage face*; *Aquifer, confined.*

**Dupuit–Forchheimer theory** The concept underlying a mathematical method of establishing an empirical approximation of an actual *flow* field in an unconfined system bounded by a *free surface.* This approach, pioneered by Dupuit in 1863 and advanced by Forchheimer in 1930, is often used in the development of a *seepage face* on a free-outflow boundary, such as a stream *bank* or the downstream face of an earth dam, for modeling or construction of a *flow net,* shown in **Figure D-24**. The analytical treatment of flow in an unconfined aquifer is complex because of the reduction in saturation thickness and decrease in *transmissivity* as the flow approaches a well. The Dupuit–Forchheimer theory is based on the following assumptions: (1) flow lines are horizontal and *equipotential lines* are vertical and (2) the *hydraulic gradient* is equal to the slope of the free surface and is invariant with depth. Cf. *Dupuit–Forchheimer equation*; *Aquifer, unconfined.*

**Duration curve** A graph of the frequency distribution of mean daily *flows* at a particular location on a watercourse, which indicates how often a given quantity is equaled or exceeded in a given period. The position of the duration curve indicates the magnitude of the flow, and the slope of the curve is a measure of flow variability. In **Figure D-25**, which shows the duration curve for two rivers with equal *drainage areas*, the daily flows form straight lines on the log-normal graph (a log-normal distribution). The duration curve, or flow frequency distribution curve, is useful in the analysis of the flow values associated with principal *transport* of sediment and *dissolved loads*.

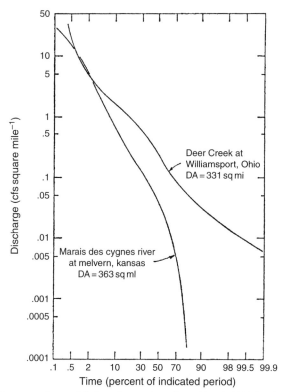

**Figure D-25** Flow duration curves of mean daily flows for two rivers. (Leopold et al., 1992. Reprinted with permission from Dover Publication, Inc.)

Cf. *Flow-duration curve/flood-duration curve*; *Flood-frequency curve*; *Flood-frequency analysis*.

**Duration of rainfall/duration of precipitation** The length of time (minutes, hours, days; refer to **Appendix A** for conversion values) characterizing a particular *rainfall* or *precipitation* event. Other characteristics of a rainfall event depend on the relationships between the duration, amount (inches or millimeters), intensity (in. $h^{-1}$, cm s$^{-1}$), and frequency of occurrence. These relationships are summarized as follows:

- the greater the duration, the greater the amount
- the greater the duration, the lower the intensity
- the more frequent the storm, the shorter the duration
- the more frequent the storm, the lower the intensity.

Determining the mean rainfall for a storm event depends on the pattern of the rainfall.

**Durov–Zaprorozec diagram** An illustration, originated by Durov and described by Zaprorozec in 1972, visually describing the differences in major-ion chemistry in groundwater samples. As shown in **Figure D-26**, the Durov–Zaprorozec diagram is based on a percentage plot of cations and anions in separate triangles, in a manner similar to that of a *Piper trilinear diagram*. From a sample plotted on both triangles, lines are extended that intersect at a point in the central rectangle. That point represents the major-ion concentration on a percentage basis. From that point, lines are extended to adjacent scaled rectangles, representing analysis of the hydrochemical data in terms such as total major-ion concentrations, *total dissolved solids* (TDS), ionic strength, specific conduction, *hardness*, or *pH*.

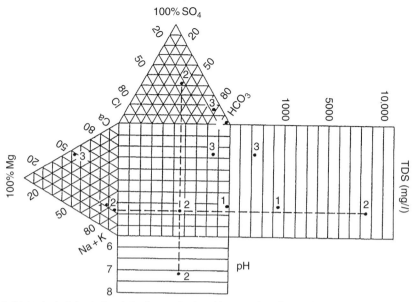

**Figure D-26** Method of chemical analysis of water developed by Durov by milliequivalents per liter. (Freeze and Cherry, 1979. Reprinted with permission from Prentice-Hall, Inc.)

Cf. *Fingerprint diagram; Stiff diagram.*

**Dye tracer** A method used in investigations to determine *groundwater flow* paths, primarily in karst or fractured terrains, by which a colored or radioactive compound is injected into a source well, and the surrounding points of *discharge* or in-place *monitoring wells* are observed for the appearance of the injected compound. Other methods of *flow path* determination, including injection of a physical entity such as ground peanut shells or micrspheres, have been used in areas where regulatory agencies do not allow the injection of chemicals into groundwater. In grouting, a dye *tracer* is an additive used primarily to change the color of the *grout* or water for visualization. Cf. *Radioactive tracer.*

**Dynamic equation** A formula associating *storage* with *flow rates*, used to solve problems of *flood routing* in *rivers* by considering the hydraulics of *open channel flow*. Hydrologic methods of solving such problems are typically based on the principle of continuity, or conservation of mass, and the dynamic equation. These methods are based in turn on the solution of the two basic differential equations (the Saint Vernant equations) governing gradually varying non-steady flow in open channels. The dynamic equation is derived by considering either the energy or the momentum equation for the short length of the channel. As shown in **Figure D-27**, the loss in *head* over the length of the *reach*, d$x$, is attributable to two main components: friction and acceleration. The equation for the head loss due to friction, $h_f$, is:

$$h_f = S_f dx$$

where:

$S_f$ = total storage [dimensionless]

The equation for the head loss due to acceleration, $h_a$, is:

$$h_a = S_a dx = \frac{1}{g}\frac{\partial v}{\partial t}$$

where:

$S_a$ = total storage [dimensionless]
$v$ = mean velocity [L·T$^{-1}$]
$g$ = gravitational constant [L·T$^{-2}$]
$t$ = time period [T].

91

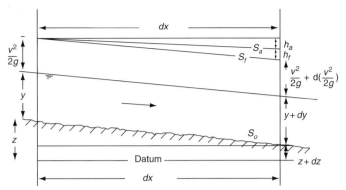

**Figure D-27** Graph of the dynamic equation. (Shaw, 1983. Reprinted with permission from Van Nostrand Reinhold Co. Ltd.)

Cf. *Equations of groundwater flow.*

**Dynamic equilibrium aquifer** An *aquifer* for which, in natural conditions, the volume of water *recharge* is equal to the volume of water *discharge*, causing the *potentiometric surface* to be steady and the volume of water in *storage* to be constant.

**Dynamic head** The *total head* against which a pump, within a well, functions, which is equal to the *lifting head* plus the *friction head.*

**Dynamic pressure** The force exerted by a fluid against a surface such as a *stream* channel or soil particles. The reaction to the dynamic pressure can change the direction and velocity of fluid *flow.*

**Dynamic viscosity coefficient/viscosity coefficient ($\mu$)** The measure of the resistance of a fluid to the shearing that is necessary for fluid *flow.* The dynamic viscosity coefficient has dimensions of $M \cdot T^{-1} \cdot L^{-2}$ and is expressed in units of centipoise ($N\,s\,m^{-2} \cdot 10^{-3}$). See: **Appendix A** for conversion units. Cf. *Absolute viscosity.*

**Dynamic wave** A wave of a certain speed within a watercourse, for which the inertia of the water is significant in determining its motion. Dynamic waves result from precipitation *runoff*, landslides, or reservoir releases. Mathematical procedures can be used to predict a change in magnitude, speed, or shape of a *flood wave* along a watercourse such as a *river, stream, reservoir, estuary,* canal, *drainage* ditch, or storm sewer.

**Dystrophic lake** A body of water, usually shallow, characterized by a lack of nutrient matter and by a high oxygen consumption in the bottom layers, or *hypolimnion.* Typically, a dystrophic lake is brownish or yellowish, containing unhumified or dissolved humic matter and minimal bottom fauna, often resulting from the depletion or near-depletion of oxygen in the water. The term "dystrophic lake" is often associated with acidic peat bogs and is an old categorization of standing water. Cf. *Oligotrophic lake; Eutrophic; Mesotrophic lake.*

**Early stiffening** See *false set.*

**Earth resistivity** See *surface resistivity.*

**Earth tide** Deformation of the solid earth as it rotates within the gravitational field of the sun and the moon causes water level fluctuations in *wells* competed in confined *aquifers.* Due to these effects, water levels peak near moonrise and moonset. Fluctuations in water levels due to earth tides are considerably smaller than those caused by barometric fluctuations. Cf. *Tides; Aquifers, confined; Ebbing well.*

**Ebbing spring** See *spring, periodic.*

**Ebbing well** A *well* screened in a coastal area subject to tidal influences. Cf. *Earth tide; Tides.*

**Economic yield** The maximum estimated rate at which water may be withdrawn from an *aquifer* without depleting the *reservoir* or affecting the quality of the water. Cf. *Safe yield; Aquifer yield; Potential yield.*

**Eddy** A circular movement of water giving the appearance of a whirlpool that may *flow* in a direction different than the main current. An eddy is usually the result of an obstruction in a *stream* during conditions of fast or *turbulent flow* in a *river.*

**Eddy viscosity** The turbulent transfer of momentum by fluid moving contrary to the main current (an *eddy*) giving rise to an internal fluid friction, similar to the action of molecular viscosity in *laminar flow,* but on a much larger scale.

**Eductor pipe** Also called a drop pipe. A casing used for air-lift pumping and *well development,* as depicted in **Figure E-1.** Eductor pipes are used for large-diameter wells or when limited volumes of air are available for development or when the *static water level* is low in relation to well depth. **Table E-1** presents recommended sizes of eductor pipe and *air line* for air-lift pumping.

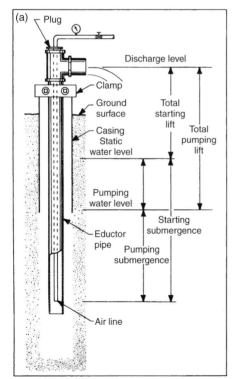
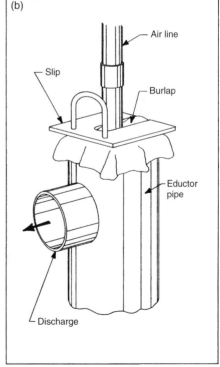

**Figure E-1** The use of eductor pipe in air-lift pumping (a) shows a cross-section through a well, and (b) are the details of a well head. (Driscoll, 1986. Reprinted by permission of Johnson Screens/a Weatherford Company.)

**Table E-1** Recommended pipe sizes for air-lift pumping. (Driscoll, 1986. Reprinted by permission of Johnson Screens/a Weatherford Company.)

| Pumping rate[a] | | | | Size of well casing if eductor pipe is used | | Size of eductor pipe (or casing if no eductor pipe is used) | | Minimum size of air line | |
|---|---|---|---|---|---|---|---|---|---|
| gpm | | m$^3$ day$^{-1}$ | | in. | mm | in. | mm | in. | mm |
| 30 to | 60 | 164 to | 327 | 4 | 102, or larger | 2 | 51 | ½ | 13 |
| 60 to | 80 | 327 to | 436 | 5 | 127, or larger | 3 | 76 | 1 | 25 |
| 80 to | 100 | 436 to | 545 | 6 | 152, or larger | 3½ | 89 | 1 | 25 |
| 100 to | 150 | 545 to | 818 | 6 | 152, or larger | 4 | 102 | 1¼ | 32 |
| 150 to | 250 | 818 to | 1360 | 8 | 203, or larger | 5 | 127 | 1½ | 38 |
| 250 to | 400 | 1360 to | 2180 | 8 | 203, or larger | 6 | 152 | 2 | 51 |
| 400 to | 700 | 2180 to | 3820 | 10 | 254, or larger | 8 | 203 | 3 | 64 |
| 700 to | 1000 | 3820 to | 5450 | 12 | 305, or larger | 10 | 254 | 3 | 64 |
| 1000 to | 1500 | 5450 to | 8180 | 16 | 406, or larger | 12 | 305 | 4 | 102 |

[a] *Actual pumping rate is dependent on percent submergence.*

**Effective grain size** In a clast or sediment, the *grain size* for which 90% of the sediment is coarser and 10% is finer. Effective grain size is the 90%-retained size in a grain size analysis. Hazen, in 1893, determined from a series of experiments that the *hydraulic conductivity* varies in proportion to the square of the effective grain size. Other sediment characteristics influence hydraulic conductivity. For example, as the *uniformity coefficient* gets larger, it reduces the impact of the effective grain size as illustrated in **Table E-2**.

**Table E-2** Comparative data for filter sands. (Driscoll, 1986. Reprinted by permission of Johnson Screens/a Weatherford Company.)

| | Fine sand 0.008″ to 0.012″ (0.2 mm to 0.3 mm) | Coarse sand 0.033″ to 0.046″ (0.84 mm to 1.17 mm) | Fine gravel 0.046″ to 0.093″ (1.17 mm to 2.36 mm) |
|---|---|---|---|
| Effective size in inches (mm) | 0.008 (0.2) | 0.034 (0.86) | 0.048 (1.22) |
| Uniformity coefficient | 1.2 | 1.2 | 1.4 |
| Hydraulic conductivity, in gpd ft$^{-2}$ (m day$^{-1}$) | 540 (22) | 9600 (391) | 13,000 (529) |
| Porosity, in percent | 37 | 37 | 37 |

Cf. *Grain size distribution; Grain size gradation curve and conductivity; Sieve analysis.*
**Effective particle size** See *effective grain size.*
**Effective permeability** The ability of geologic material to conduct one fluid, such as a gas in the presence of another, such as oil or water. Cf. *Relative permeability; Permeability, absolute; Permeability /permeable.*
**Effective porosity** For a given mass of rock or soil, the percentage that consists of interconnected *pore spaces.* Effective porosity can be calculated from the *specific discharge* divided by the mean velocity of a conservative *tracer.* Cf. *Porosity; average linear velocity; Effective velocity.*
**Effective rainfall** See *rainfall excess.*
**Effective stress** The difference between the total stress and the *pore pressure.* This is also known as the stress at an arbitrary point in the subsurface that is not borne by the fluid.

$$\text{effective stress} = \text{total stress} - \text{pore pressure}$$

With the total stress constant, if the fluid pressure changes, the effective stress changes by an equal amount.

**Effective uniform depth** The average depth of water that would be present if a given amount of *precipitation* was evenly distributed in a given *drainage basin*.

**Effective velocity** A representation of the actual velocity of water moving through interconnected *pore spaces*. Effective velocity is calculated as the *Darcy flux* (which represents the bulk flow through a volume of geologic material) divided by the *effective porosity* of the material. Cf. *Average linear velocity; Average pore water velocity*.

**Efficiency, well** See *step drawdown test*.

**Effluent** Outflow, as in *discharge* from a body of water or treatment system. Cf. *Effluent seepage; Effluent stream*.

**E-logging** See *electric logging*.

**Effluent seepage** *Seepage* that is *effluent* from one zone to another. Examples of effluent seepage include seepage from the subsurface to a surface water body and from a dam to a *stream*. Cf. *Bank storage; Seepage velocity; Seepage face*.

**Effluent stream** A watercourse that is receiving *groundwater discharge* as its *baseflow*. Cf. *Influent stream; Bank storage; Gaining stream*.

**Eh** The oxidation potential of an aqueous solution is the *Eh*. The energy gained in the transfer of 1 mol of electrons from an oxidant to $H_2$. The $E$ symbolizes the electromotive force, while the $h$ indicates that the potential is on the hydrogen scale. *Eh* is mathematically defined by a relation known as the *Nernst equation*:

$$Eh \text{ (volts)} = Eh^\circ + \frac{2.3RT}{nF} \log\left(\frac{\text{oxidant}}{\text{reducant}}\right)$$

where:
$Eh^\circ$ = a standard or reference condition at which all substances involved are at unit activity
$F$ = Faraday constant ($9.65 \times 10^4 \, C \cdot mol^{-1}$)
$R$ = universal gas constant ($E \cdot mol^{-1} \cdot {}^\circ K^{-1}$)
$T$ = temperature ($^\circ K$)
$n$ = number of electrons in the half reaction

*Eh* has been used in many investigations because of the ease in measuring the field electrode potentials as voltage. *Eh* was widely used prior to the 1970s; however, *pE* is becoming more commonly used in redox studies because its formulation follows from half-cell representations of redox reactions in combination with the law of mass action. *pE* and *Eh* are related by:

$$pE = \frac{nF}{2.3RT} Eh$$

For reactions at 25°C, for the transfer of a single electron, the equation becomes:

$$pE = 16.9Eh$$

Cf. *Eh–pH diagram; Oxidation potential*.

**Eh–pH diagram** A graph that shows the equilibrium occurrence of ions or minerals as domains relative to *Eh* (or *pE*) and *pH*. *Eh–pH* diagrams are an important aid in understanding the processes that control the occurrence and mobility of minor and trace elements. In constructing *Eh–pH* diagrams, the equilibrium conditions or stability field of dissolved species and minerals in aqueous environments are of interest. Therefore, it is appropriate to establish the *Eh–pH* conditions under which water is stable:

$$O_2 + 4H^+ + 4e = 2H_2O$$

$$2H^+ + 2e = H_{2 (g)}$$

for conditions at 25°C, the corresponding redox conditions are:

$$pE = 20.8 - pH + \frac{1}{4}\log P_{O_2}$$

$$pE = -pH - \frac{1}{2}\log P_{H_2}$$

These relations plot as straight lines on the *Eh–pH* diagram shown in **Figure E-2a**. **Figure E-2b** presents the *Eh–pH* diagram for the Fe–$H_2O$ system (only two oxidation states are shown, as iron in solution in groundwater is normally present as $Fe^{3+}$ and $Fe^{2+}$), and the completed diagram showing stability fields is shown in **Figure E-2c**. (Note: the boundaries are defined at specific concentration values on these plots.) The use of *Eh–pH* diagrams has become widespread in geology, limnology, oceanography, and petrology.

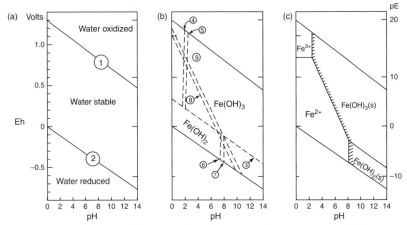

**Figure E-2** *Eh–pH* diagram for 25°C and 1 bar. (a) Stability field for water; (b) construction lines for the Fe–H₂O system; (c) completed diagram showing stability fields for major dissolved species and solid phases. (Freeze and Cherry, 1979. Reprinted with permission by Prentice-Hall, Inc.)

Cf. *Eh; Nernst equation; Oxidation potential.*

**Elastic storage coefficient** See *specific storage.*

**Electric log** A record of the electrical resistance providing a "picture" of drilled rock formations. See *electric logging; geophysical exploration methods.*

**Electric logging** Commonly referred as E-logging or resistivity logging. This is a downhole logging technique to measure the resistance of the rock formations that have been drilled through. Usually combined with measurements of *spontaneous potential* (SP), it provides a relatively inexpensive means of logging a hole. A good log provides a detailed picture of the thicknesses of the various strata and an indication of the water quality by measuring the apparent resistivity of the strata. Variations in resistivity are caused mainly by differences in the character of the geologic material and the mineral content of the water contained within these strata. This permits the well driller to place the screens in the most desirable downhole location without having to rely on logging the cuttings. Electric logging can only be done in uncased holes filled with water or *drilling fluid*. The fluid will affect the measured resistivities. Additionally, "filter cake" from the drilling fluid on the *borehole* walls may also mask the resistivity of the formation, potentially hiding good producing zones behind the filter cake. Polymeric drilling fluids reduce this problem and provide more accurate readings of the formation resistance. **Figure E-3** illustrates three different electrode arrangements. Although a single electrode has a limited depth of investigation, it has good vertical resolution, uses a single conductor cable, and is inexpensive. A two-electrode arrangement is called a normal log, arranged in either a short normal separation (16 in or 406 mm) or a long normal separation (64 in or 1630 mm). The spacing of the current and potential electrodes determines the penetration distance into the formation. However, larger spacing also reduces the vertical resolution. A three electrode arrangement consists of two potential electrodes and one current electrode. This provides an improvement over the one- or two- electrode arrangement, giving greater distance penetration into the formation while maintaining the vertical resolution. **Figure E-4** provides a simplified log of apparent resistivity for various geologic media. Cf. *Geophysical exploration methods.*

**Electrical analog models** Also called an electrical resistance model. This representation of a water *flow* system by electrical current. Electric analog models can be used to illustrate *groundwater flow* because the basic equation governing electric circuits ($V = IR$, where $V$ = voltage or potential, $I$ = current, and $R$ = resistance) is analogous to Darcy's law ($v = Ki$), electrical circuits can be used to model groundwater flow. Electrical analog models were most commonly used in the early 1960s and have been replaced by digital computer models since the mid-1970s as they are expensive to set up and only model one flow system. Cf. *Geophysical exploration methods.*

**Electrical conductance** A measure of the ease with which a conducting electric current can flow through a material under the influence of an applied potential. This is the reciprocal of *electrical resistivity* and is measured in mhos per unit length.

**Electrical conductance of water** Commonly referred to as *specific electrical conductance* when measured in the field and is the conductance of a cubic centimeter of any substance compared with the same volume of water. Pure or deionized water has a very low *electrical conductance*. A minute amount of dissolved mineral matter will increase the conductance of the water because the ions are electrically charged and will move toward a current source that will neutralize them. Anions (negative charge) move toward a positive electrode (anode), while cations (positive charge) are attracted to the negative electrode (cathode). Because of this relationship, there is a strong correlation between electrical conductance and *total dissolved solids* (TDS) in water. For most *groundwater*, a factor of 0.55–0.75 times the specific conductance yields a reasonable approximation of the concentration of TDS in mg L⁻¹. Refer to **Appendix A** for unit conversions.

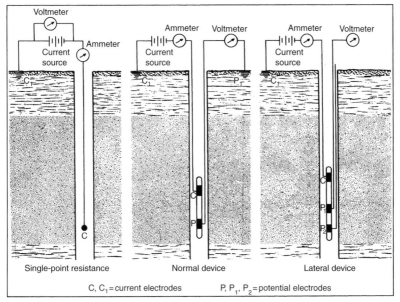

Single-point resistance     Normal device     Lateral device

C, C$_1$ = current electrodes     P, P$_1$, P$_2$ = potential electrodes

**Figure E-3** Electrode arrangements and circuits for three electric logging procedures. Each produces a resistivity curve that differs in some details from the others; these differences assist in interpreting the logs. (Driscoll, 1986. Reprinted by permission of Johnson Screens/a Weatherford Company.)

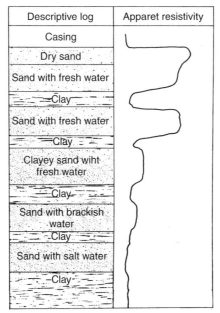

| Descriptive log | Apparet resistivity |
|---|---|
| Casing | |
| Dry sand | |
| Sand with fresh water | |
| Clay | |
| Sand with fresh water | |
| Clay | |
| Clayey sand wiht fresh water | |
| Clay | |
| Sand with brackish water | |
| Clay | |
| Sand with salt water | |
| Clay | |

**Figure E-4** Electric log of a series of sand and clay beds. In this particular example, brackish and saltwater in the bottom sand formations causes lower electrical resistivities, resulting in reduced contrast between the sand and the clay. (Driscoll, 1986. Reprinted by permission of Johnson Screens/a Weatherford Company.)

**Electrical profiling** A surface geophysical technique for variations in subsurface resistivity in the subsurface at various depth along a line to produce a profile view of the resistivity features and hence a geologic interpretation. Electrodes are laid out on the surface in an array. The distance between the electrodes dictates the depth of the measurement. Typically, the depth of penetration of the electrical current is equal to one-half of the distance between the potential electrode and the current electrode. Varying this distance produces the depth measurements, while moving the electrode array produces the profile. Two different electrode arrays are utilized: a *Wenner array* and a *Schlumberger array*, as illustrated in **Figure E-5**.

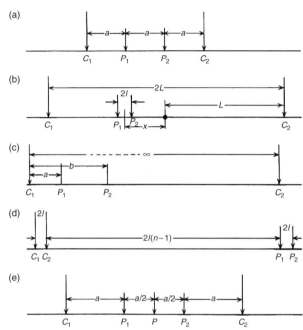

**Figure E-5** Common electrode spread for resistivity surveys. (a) Wenner array; (b) Schlumberger array; (c) three-point array; (d) double-dipole array; (e) Lee-partition array. (Telford et al., 1976. Reprinted with permission of Cambridge University Press, Inc.)

Cf. *Electric logging.*
**Electrical resistance model** See *electrical analog model.*
**Electrical resistivity** The opposite of *electrical conductance*. Electrical resistivity is measured in ohms. Refer to **Appendix A** for units and properties.
**Electrical resistivity log** See *electric logging.*
**Electrical sounder** A device for measuring the depth to water within a well, involving the use of a weighted electrical cable with gradations and a probe at the end cable with a light, ammeter, or annunciator that indicates the depth at which the sounder makes contact with the water. **Figure E-6** illustrates the use of an electrical sounder for measuring water levels.
**Electrochemical sequence** The changes in subsurface redox (oxidation–reduction) conditions that result from the presence of organic substances in *groundwater flow* systems. Under natural conditions, infiltrating meteoric waters interact with soil organics and minerals. After the oxygen in the groundwater is consumed, oxidation of organic matter can still occur, but the oxidizing agents are $NO_3^-$, $MnO_2$, $Fe(OH)_3$, $SO_4^{2-}$. As these oxidizing agents are consumed, the groundwater environment becomes more and more reduced to the point where methanogenic conditions may result if there are sufficient oxidizable organic material, sufficient nutrients, and temperature conditions conducive to bacterial growth. The oxidation of organic matter results in lowering pH. When organic *pollutants* are released to the groundwater, redox conditions that would normally take place in the soil zone persist to the saturated zone. Such releases of organic matter into groundwater flow systems can create localized changes in the groundwater chemistry and can modify the *Chebotarev sequence*. Half-reactions in **Table E-3** represent the oxidation of a simple hydrocarbon and the resultant reduction of minerals in the subsurface. Cf. *Alkalinity; Brine; Total dissolved solids (TDS); Zone of saturation.*

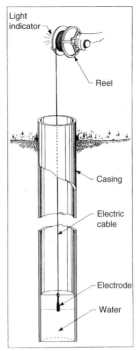

Light indicator

Reel

Casing

Electric cable

Electrode

Water

**Figure E-6** Electric sounder for measuring depth to water consists of an electrode, two-wire cable, and a light or annunciator that indicates a closed circuit when the electrode touches the water. (Driscoll, 1986. Reprinted by permission of Johnson Screens/a Weatherford Company.)

**Table E-3**   The processes resulting from electrolyte reactions. (Freeze and Cherry, 1979. Reprinted with permission from Prentice Hall, Inc.)

| Process | reaction |
|---|---|
| Denitrification: | $CH_2O + {}^4/_5NO_3^- \rightarrow {}^2/_5N_{2(g)} + HCO_3^- + {}^1/_5H^+ + {}^2/_5H_2O$ |
| Manganese reduction: | $CH_2O + 2MnO_{2(s)} + 3H^+ \rightarrow 2Mn^{2+} + HCO_3^- + 2H_2O$ |
| Iron reduction: | $CH_2O + 4Fe(OH)_3 + 7H^+ \rightarrow 4Fe^{2+} + HCO_3^- + 10H_2O$ |
| Sulfate reduction: | $CH_2O + {}^1/_2SO_4^{2-} \rightarrow {}^1/_2HS^- + HCO_3^- + {}^1/_2H^+$ |
| Methane reduction: | $CH_2O + {}^1/_2H_2O \rightarrow {}^1/_2CH_4 + {}^1/_2HCO_3^- + {}^1/_2H^+$ |

**Electrolyte**   A substance that dissociates into positive ions (cations) and negative ions (anions) when dissolved in water, resulting in the increased electrical conductivity of the solution. Cf. *Electrical conductance.*

**Electromagnetic resistivity mapping**   The performance of *electrical resistivity* measurements over a gridded area to map conductivity or resistivity anomalies. Any electrode array can be used, but the most common are the *Wenner* and the *Schlumberger arrays*. In all cases, the resistivity is plotted on a map at the mid-point of the potential electrodes as they are moved around the gridded area being mapped. Cf. *Geophysical exploration methods.*

**Electromagnetic surveys**   Any of several techniques involving the propagation of time-varying, low-frequency electromagnetic fields in and over the earth. Electromagnetic surveys are best suited for locating conductors at relatively shallow depths (i.e. generally below 3 m or 10 ft). Measurements can be done quickly, making it a relatively inexpensive exploration technique. Electromagnetic methods (EMs) represent the second most common geophysical technique, with magnetics being the most common in mineral exploration. Generally, inductive coupling (i.e. the source of the electromagnetic energy need not be in indirect contact with the ground) is used to transmit the signal to the detector. Almost all EM field sets include a portable power source. Some methods make use of low-frequency radio waves in the range of 5–25 kHz. Another method, known as AFMAG (audio frequency magnetic fields), makes use of atmospheric energy resulting from world-wide thunderstorm activity. Because of the use of inductive coupling, AFMAG permits the use of EM systems in aircraft. For a more detailed treatment of this subject matter, the reader is directed to Telford et al. (1976). Electromagnetic surveys are used in the search for *groundwater* and in the investigation of groundwater contamination. At frequencies above 100 mHz (micro-wave frequencies), electromagnetic waves propagate into the ground in straight lines to depths of a few feet to tens of feet (but with considerable less resolution with depth) depending on the electrical conductivity of the terrain. These microwave frequency instruments are called *ground penetrating radar*. At lower frequencies, the electromagnetic waves diffuse more slowly into the earth rather than traveling a straight line. This results in sampling a much larger volume of earth, reducing the resolution, while increasing the penetration distance. Ground penetrating radar records the time for the wave to travel to an interface between two formations. The interface is defined by changes in magnetic permeability, electrical conductivity, and dielectric constant, which constitute an electromagnetic impedance. Electromagnetic waves at these high frequencies provide excellent vertical and lateral resolution, but the depth penetration is limited from a few feet to nominally 50 ft (15.2 m) depending on the frequency of the input signal and the ground conductivity. The acquisition of data is quick and easy making this a very cost- effective technique to map out investigation areas. Terrain conductivity is another EM. This can be used in either the frequency or the time domain mode. Depths of up to 200 ft (61 m) can be mapped using a two-person frequency domain configuration. Shallower depths (20 ft or 6.1 m) can be measured using a one-person instrument. In this method, a transmitter coil is energized with an alternating current at an audio frequency, with a receiver coil a short distance away. The coils can be placed on the ground, perpendicular to the ground or suspended above the ground on a harness worn by the operators. The magnetic field produced by the alternating current in the primary or transmitting coil induces small electric currents in the earth which are sensed by the receiver coil. For groundwater exploration and environmental work, terrain conductivity surveys are useful in locating gravel bodies, saline water, cavities in carbonate rock, contaminant plumes in groundwater, and bedrock topography. Time domain techniques are similar to ground penetrating radar methods; however, at lower frequencies, the transmitters operate at a few tens of hertz. The current is abruptly terminated and the decay of the signal with depth is measured. The penetration depth is governed by the size of the loop that is laid out on the ground [varying from 150 (45.7 m) to 1500 ft (457 m) on a side]. In general, the depth of penetration is 1.5 times the length of one side of the loop or 1.5 times the diameter, if laid out in a circle. Time domain techniques are used to penetrate to great depths and for detecting *saltwater intrusion*.

**Electro-osmosis** The osmotic movement of a liquid under the application of an electrical field. Electro-osmosis is used in *dewatering* clay soils. The application of electricity charges the clay platelets, resulting in the migration of the polar molecules of water under the influence of the electrical field. Cf. *Flocculation*.

**Electroviscosity** The *viscosity* of a fluid is influenced by electric properties. Fluids with lower *electrical conductance* tend to have greater viscosities than fluids with higher electrical conductance.

**Elevation head** The energy required to raise the water from a datum to its elevation. If the datum is *sea level*, then the elevation head is equal to the elevation of the measuring point above the sea level. Cf. *Darcy's law; Static head; Head, total; Total head; Pressure head; Groundwater energy.*

**E-logging** See *electric logging*.

**Embayment** See *bay/embayment.*

**Energy budget** See *groundwater budget; water balance; hydrologic budget; basin accounting.*

**English Rule of Capture** See *Rule of capture.*

**Entrance velocity** The rate at which water enters the screened interval of a *well*. Entrance velocities that are too high can result in chemical *incrustation* and *corrosion* of the *well screen*. A general rule of thumb that has developed is that entrance velocities should have an average of 0.1 ft s$^{-1}$ (0.3 m s$^{-1}$). Well screens designed to have an average entrance velocity of 0.1 ft s$^{-1}$ or less require less maintenance over the long term, limit sand entering the well screen, and create minimal *head loss* as the water enters the screen. Refer to **Appendix A** for conversion units. **Table E-4** presents maximum screen entrance velocities.

**Table E-4** Maximum entrance velocity versus hydraulic conductivity. (Driscoll, 1986. Reprinted by permission of Johnson Screens/a Weatherford Company.)

| Hydraulic conductivity of aquifer | | Maximum screen entrance velocity | |
|---|---|---|---|
| gpd ft$^{-2}$ | m$^2$ day$^{-1}$ | ft s$^{-1}$ | cm s$^{-1}$ |
| More than 6000 | More than 245 | More than 0.10 | More than 3.05 |
| 6000 | 245 | 0.10 | 3.05 |
| 5000 | 204 | 0.10 | 3.05 |
| 4000 | 163 | 0.10 | 3.05 |
| 3000 | 122 | 0.10 | 3.05 |
| 2500 | 102 | 0.08 | 2.54 |
| 2000 | 82 | 0.08 | 2.54 |
| 1500 | 61 | 0.07 | 2.03 |
| 1000 | 41 | 0.07 | 2.03 |
| 500 | 20 | 0.05 | 1.52 |
| Less than 500 | Less than 20 | Less than 0.03 | Less than 1.02 |

Cf. *Well maintenance*.

**Environmental isotopes** The use of stable and radioactive isotopes for studying the behavior of water. These can be used in three ways:

1. As *tracers* to track the movement of water. For example, rain water during a heavy storm is often depleted in the heavy isotopes of $^2$H, deuterium, or stable $^{18}$O, with respect to the most abundant isotopes of $^1$H and $^{16}$O, respectively.
2. Using isotope fractionation where water changes from one phase to another (e.g. liquid to vapor), the ratio of isotopes of an element also changes. Or stable isotope concentrations change when certain geochemical or hydrological process takes place. For example, the isotopic composition of carbon and oxygen forming calcium carbonate is different for marine and freshwater origins.
3. Age dating by radioactive decay.

Cf. *Carbon-14 dating; Radiocarbon dating of groundwater.*

**Environmental Protection Agency (EPA)** An independent agency of the United States government established in 1970 responsible for protecting the environment and maintaining it for future generations.

**EPA** See *environmental protection agency.*

**Ephemeral rivers** A watercourse that only flows in response to *precipitation* in the immediate locality and remains dry otherwise. An ephemeral *river* or *stream* is located above the *water table.*

**Epilmnion** The oxygen-rich uppermost layer of water in a lake. The epilmnion overlies the metalimnion in a thermally stratified lake. In summer months, this layer is warmer than the underlying layers and is relatively uniformly mixed as a result of wind and wave action. Cf. *Hypolimnion; Eutrophic; Disphotic; Thermal stratification; Monomictic; Thermocline.*

**Episodic erosion** Also called discontinuous *erosion.* The transport of earth materials that may be related to storms or catastrophic events such as landslides or earthquakes.

**epl** See *equivalents per liter.*

**epm** See *equivalents per million.*

**Equations of groundwater flow** *Groundwater flow,* just as any field of science and engineering, uses mathematics to describe the physical processes. The basic law of flow is *Darcy's law.* This combined with an equation of continuity that describes the conservation of mass through *porous media* results in a partial differential equation. Equations of groundwater flow are developed for (1) steady-state saturated flow, (2) transient saturated flow, and (3) transient unsaturated flow.

Steady-State Saturated Flow: The law of conservation of mass requires that for *steady-state flow* through a porous media, the rate of fluid flow into any elemental control volume must be equal to the mass flow, leaving the same volume. Referring to **Figure E-7**,

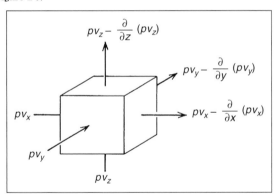

**Figure E-7** Elemental control volume for flow through porous media. (Freeze and Cherry, 1979. Reprinted with permission by Prentice-Hall, Inc.)

the law of continuity translates this into mathematical form as:

$$\frac{\partial(\rho v_x)}{\partial x} + \frac{\partial(\rho v_y)}{\partial y} + \frac{\partial(\rho v_z)}{\partial z} = 0$$

For most engineering calculations, water is considered an incompressible fluid, such that $\rho(x,y,z) = \text{constant}$. Therefore, the $\rho$ term, or fluid density, can be eliminated. Further, substituting for Darcy's law for the specific discharge ($v$), $v_x$, $v_y$, and $v_z$, yields the equation for steady-state flow through an anisotropic saturated porous media.

$$\frac{\partial}{\partial x}\left(K_x \frac{\partial h}{\partial x}\right) + \frac{\partial}{\partial y}\left(K_y \frac{\partial h}{\partial y}\right) + \frac{\partial}{\partial z}\left(K_z \frac{\partial h}{\partial z}\right) = 0$$

where:

$K$ = the *hydraulic conductivity* in each of the $x$, $y$, and $z$ directions [L·T$^{-1}$]

$h$ = the *hydraulic head* [L]

In isotropic medium, $K_x = K_y = K_z$, and if the medium is also homogenous, then $K(x, y, z)$ = constant, the equation for steady-state flow can be reduced to:

$$\frac{\partial^2 h}{\partial x^2} + \frac{\partial^2 h}{\partial y^2} + \frac{\partial^2 h}{\partial z^2} = 0$$

This equation is one of the most basic differential equation in mathematics, know as the *Laplace equation*. The solution of this equation describes the hydraulic head at any point in a three-dimensional flow field.

Transient Saturated Flow: Under transient conditions, the law of conservation of mass stipulates that the net rate of fluid mass flow into an elemental volume be equal to the time rate of change of fluid mass storage within that same elemental volume. In this instance, the equation of continuity takes the form of:

$$-\left[\frac{\partial(\rho v_x)}{\partial x} + \frac{\partial(\rho v_y)}{\partial y} + \frac{\partial(\rho v_z)}{\partial z}\right] = \frac{\partial(\rho n)}{\partial t}$$

The right-hand side of the equation can be expanded, as:

$$\frac{\partial(\rho n)}{\partial t} = n\frac{\partial \rho}{\partial t} + \rho\frac{\partial n}{\partial t}$$

The first term on the right-hand side of the equation is the mass-flow rate of water produced by the expansion of water under a change in its density, $\rho$. The second term on the right-hand side of the equation is the mass-flow rate of water produced by the compaction of a porous medium as reflected by the change in its *porosity*, $n$. Changes in porosity and density are governed by changes in head, such that the volume of water produced by these two processes for a unit decline in head is defined as the *specific storage*. The mass rate of water produced is $\rho S_s \partial h/\partial t$ where $S_s$ is the specific storage. Therefore, the equation of continuity takes the form of:

$$-\left[\frac{\partial(\rho v_x)}{\partial x} + \frac{\partial(\rho v_y)}{\partial y} + \frac{\partial(\rho v_z)}{\partial z}\right] = \rho S_s \frac{\partial h}{\partial t}$$

We can eliminate $\rho$ from both sides of the equation, for water is typically considered an incompressible fluid under most conditions and therefore a constant. Further, inserting Darcy's law:

$$\frac{\partial}{\partial x}\left(K_x \frac{\partial h}{\partial x}\right) + \frac{\partial}{\partial y}\left(K_y \frac{\partial h}{\partial y}\right) + \frac{\partial}{\partial z}\left(K_z \frac{\partial h}{\partial z}\right) = S_s \frac{\partial h}{\partial t}$$

This is the equation for transient flow in saturated anisotropic media. If the media is homogenous and isotropic, the equation can be simplified to:

$$\frac{\partial^2 h}{\partial x^2} + \frac{\partial^2 h}{\partial y^2} + \frac{\partial^2 h}{\partial z^2} = \frac{S_s}{K}\frac{\partial h}{\partial t}$$

This equation is known as the *diffusion equation*.

Transient Unsaturated Flow: Flow through the elemental volume that may be only partially saturated, the equation of continuity must incorporate a change in moisture content as well. The *degree of saturation* $\theta'$ is defined as $\theta' = \theta/n$, where $\theta$ is the moisture content and $n$ is the porosity. Therefore, the equation that was presented under transient saturated flow becomes:

$$-\left[\frac{\partial \rho v_x}{\partial x} + \frac{\partial \rho v_y}{\partial y} + \frac{\partial \rho v_z}{\partial z}\right] = n\theta'\frac{\partial \rho}{\partial t} + \rho\theta'\frac{\partial n}{\partial t} + n\rho\frac{\partial \theta'}{\partial t}$$

The first two terms on the right side of the equation represent changes in fluid density and porous space, which can be negligible under most considerations. Inserting the following,

$$v_x = -k(\psi)\frac{\partial h}{\partial x} \quad \text{and} \quad n d\theta' = d\theta$$

where $\psi$ is the pressure head, results in:

$$\frac{\partial}{\partial x}\left[K(\psi)\frac{\partial h}{\partial x}\right] + \frac{\partial}{\partial y}\left[K(\psi)\frac{\partial h}{\partial y}\right] + \frac{\partial}{\partial z}\left[K(\psi)\frac{\partial h}{\partial z}\right] = \frac{\partial\theta}{\partial t}$$

Specific moisture is defined as $C = d\theta/d\psi$, and $h = \psi + z$, this can be rewritten as the pressure head-based equation for transient flow for unsaturated porous media, also known as "Richards' equation"

$$\frac{\partial}{\partial x}\left[K(\psi)\frac{\partial\psi}{\partial x}\right] + \frac{\partial}{\partial y}\left[K(\psi)\frac{\partial\psi}{\partial y}\right] + \frac{\partial}{\partial z}\left[K(\psi)\left(\frac{\partial\psi}{\partial z} + 1\right)\right] = C(\psi)\frac{\partial\psi}{\partial t}$$

**E**

Cf. *Governing equation for groundwater flow; Dynamic equation.*

**Equilibrium, chemical** From a thermodynamic perspective, the equilibrium state is a state of maximum stability toward which a closed physicochemical system proceeds by irreversible processes as reported by Stumm and Morgan in 1981. **Figure E-8** presents this concept from a mechanical perspective. The conditions of stable, metastable, and unstable are represented by peaks and troughs depicted by the energy or entropy function. The energy of a closed system can be described in terms of a particular function known as the Gibbs free energy. The energy of the system is at its lowest at point c, which is the most stable state. State c is the most stable state as it has the lowest absolute Gibb's free energy under closed system conditions at constant temperature and pressure. Natural processes proceed toward equilibrium states but never away from them. The driving force in a chemical reaction is commonly represented by the Gibb's free energy of the reaction, commonly denoted as $G_r$. For closed systems at constant temperature and pressure, $G_o$ represents the change in internal energy of that closed system as a measure of the reaction's ability to do mechanical work. The condition of mechanical equilibrium can be defined as:

$$\Sigma\text{free energy : products} - \Sigma\text{free energy : reactants} = 0$$

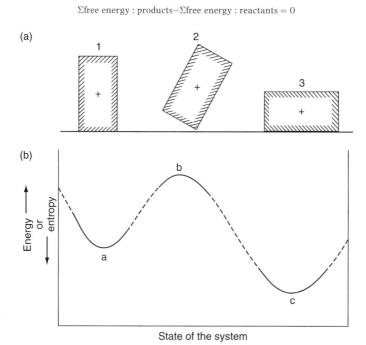

Figure E-8 Concepts in mechanical and chemical equilibrium. (a) Meta-stable, unstable, and stable equilibrium in a mechanical system. (b) Meta-stability, instability, and stability for different energetic states of a thermodynamic system. (Stumm and Morgan, 1981. Reprinted with permission by John Wiley & Sons, Inc.)

**Equilibrium concentration** The concentration of a given chemical in water at a point where a reaction is at equilibrium conditions.

**Equilibrium well equations** Mathematical expressions adapting *Darcy's law* to water *flow* to a *well* have been derived by various investigators for relating well discharge to drawdown. There are two basic equations: one for unconfined flow and one for confined flow as shown below:

In SI units for unconfined flow (See: **Appendix A** for conversion units)

$$Q = \frac{1.366K(H^2 - h^2)}{\log(R/r)}$$

E

where:
$Q =$ well yield or pumping rate $[L^3 \cdot T^{-1}]$
$K = $ *hydraulic conductivity* of the water-bearing formation $[L \cdot T^{-1}]$
$H = $ *static head* measured from the bottom of the aquifer [L]
$h = $ level of water in the well while pumping, measured from the bottom of the aquifer [L]
$R = $ radius of the *cone of depression* [L]
$r = $ radius of the well [L]

For unconfined flow in English engineering units (See: **Appendix A** for conversion units)

$$Q = \frac{K(H^2 - h^2)}{0.141 \log(R/r)}$$

where:
$Q = $ *well yield* or pumping rate (gpm)
$K = $ hydraulic conductivity for the water-bearing formation (ft day$^{-1}$)
$H = $ static head measured from the bottom of the aquifer (ft)
$h = $ level of the water in the well while pumping, measured from the bottom of the aquifer (ft)
$R = $ radius of the cone of depression (ft)
$r = $ radius of the well (ft)

In SI units, the equation for confined aquifers becomes (See: **Appendix A** for conversion units)

$$Q = \frac{2.73Kb(H - h)}{\log(R/r)}$$

where:
$b = $ aquifer thickness (m).

For confined aquifers, the equation in English units becomes

$$Q = \frac{Kb(H - h)}{70.75 \log(R/r)}$$

where:
$b = $ aquifer thickness (ft).

The following simplifying assumptions were used to derive the equilibrium equations:

- The water-bearing formation materials have a uniform hydraulic conductivity within the *radius of influence* of the well.
- The aquifer thickness is assumed to be constant at the initiation of pumping for unconfined aquifers.
- The well is 100% efficient, meaning the drawdown within the well is equal to the water level just outside the well bore.
- The well is *fully penetrating* the aquifer.
- The water table or piezometric surface is not inclined.

- Flow within the well is laminar.
- The cone of depression has reached equilibrium.
- Flow to the well is radial.
- The vertical component of flow is ignored.

Cf. *Image well theory*; *Equations for groundwater flow*.

**Equipotential line** A contour line or isopleth joining points of equal *hydraulic head*. Equipotential lines are drawn in plan view or cross section to depict *groundwater flow*. Groundwater flow is perpendicular to the equipotential lines in an *isotropic* porous medium. Drawings showing both equipotential lines and *flow lines* are called *flow nets* as shown in **Figure E-9**.

E

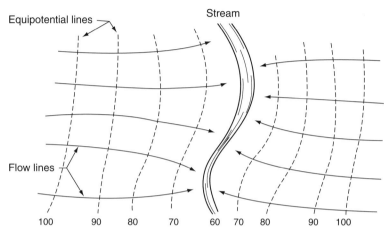

**Figure E-9** Plan view of a flow net depicting groundwater flow to a stream. The dashed lines represent points of equal groundwater elevation or equi-potentials. (Driscoll, 1986. Reprinted by permission of Johnson Screens/a Weatherford Company.)

**Equipotential surface** An imaginary surface defining the *fluid potential* or *total head* in an aquifer. Cf. *Groundwater energy*.

**Equivalent homogeneous porous medium** An analogy used in the depiction of *groundwater flow* to simplify problems of *heterogeneity* or *anisotropy*. Geologic media are heterogeneous, and as such have preferential paths of groundwater flow or anisotropic flow. These anisotropies are the result of depositional sedimentary environments or fracturing and faulting of rock. Anisotropic flow can be difficult to model or requires considerable computer memory, such that often an assumption is made that the flow is *homogenous* and *isotropic*, or it is stated that the groundwater flow in an equivalent homogeneous *porous medium* can be represented by a certain model. This simplification can provide a general depiction of groundwater flow but can be invalid in such media as *braided stream* deposits, fractured bedrock, till sequences, and karst terrain.

**Equivalent porous medium model** A concept where a geologic unit such as a fractured rock mass with a limited number of interconnected, open fractures at the scale of analysis is treated as it were a conventional *porous medium*, without having to identify explicit properties of each individual fracture within the unit. Cf. *Pore spaces*.

**Equivalents per liter (ep/)** In evaluating chemical equilibrium in *groundwater*, the concentrations must be expressed on a molar basis. Equivalents per liter (ep$\ell$) are the number of moles of solute, multiplied by the valence of the solute species, in 1 L of solution:

$$\mathrm{ep}\ell = \frac{\text{moles of solute} \times \text{valence}}{\text{litre of solution}}$$

Typically, chemical analyses are received from the laboratory in mg L$^{-1}$ or µg L$^{-1}$ (See: **Appendix A** for conversion units). Presented in **Table E-5** are factors to convert from mg L$^{-1}$ to ep$\ell$.

**Table E-5** Conversion factors to convert from mg $L^{-1}$ to meq $L^{-1}$. (Walton, 1970. Reprinted with permission from McGraw-Hill, Inc.)

| To convert from mg $L^{-1}$ | To meq I, multiply by | To convert from mg $L^{-1}$ | To meq $l^{-1}$ multiply by |
|---|---|---|---|
| **Cations** | | **Anions** | |
| Al | 0.1112 | Br | 0.0125 |
| Ba | 0.0146 | Cl | 0.0282 |
| Ca | 0.0499 | $CO_3$ | 0.0333 |
| $Cr^{+3}$ | 0.0577 | F | 0.0526 |
| $Cu^{+2}$ | 0.0315 | $HCO_3$ | 0.0164 |
| $Fe^{+3}$ | 0.0537 | $HSO_4$ | 0.0103 |
| H | 0.9921 | I | 0.0079 |
| K | 0.0256 | $NO_2$ | 0.0217 |
| Mg | 0.0823 | $NO_3$ | 0.0161 |
| $Mn^{+2}$ | 0.0364 | $PO_4$ | 0.0316 |
| Na | 0.0435 | $SiO_3$ | 0.0263 |
| $NH_4$ | 0.0554 | $SO_4$ | 0.0206 |

*Cf. Equivalents per million (epm).*
**Equivalents per million (epm)** The number of moles of solute multiplied by the *valence* of the solute species in $10^6$ g of solution. Or this can be stated as the number of milligram equivalents of solute per kilogram of solution:

$$\text{epm} = \frac{\text{moles of solute} \times \text{valence}}{10^6 \text{ g of solution}}$$

**Figure E-10** is useful when determining equivalents per million. Equivalents per million (epm) can also be stated as the number of milligram equivalents of solute per kilogram of solution. (See: **Appendix A** for conversions)

a. To convert $63 \text{ mg/}\ell \text{ Mg}^{+2}$ to epm:

Atomic weight Mg $= 24.32$
Valence $= 2$

Equivalent weight $= \dfrac{24.32}{2} = 12.16$

$63 \text{ mg/}\ell \text{ Mg}^{+2} = \dfrac{63}{12.16} = 5.19 \text{ epm}$

b. To convert $5 \text{ mg/}\ell \text{ NO}_3^-$ to epm:

Atomic weight N $= 14.0$
Atomic weight O $= 16.0$
Molecular weight $NO_3^- = 62.0$
Valence $= 1$

Equivalent weight $= \dfrac{62.0}{1} = 62$

$5 \text{ mg/}\ell \text{ NO}_3^- = \dfrac{5}{62} = 0.08 \text{ epm}$

c. To check the following analysis, tabulate the epm of anions and cations and then add each column.

| Ion | mg/$\ell$ | epm cations | epm anions |
|---|---|---|---|
| $Ca^{+2}$ | 42 | 2.10 | |
| $Mg^{+2}$ | 27 | 2.22 | |
| $HCO_3^-$ | 196 | | 3.21 |
| $SO_4^{-2}$ | 15 | | 0.31 |
| $Cl^-$ | 72 | | 2.03 |
| $NO_3^-$ | 5 | — | 0.08 |
| Total | | 4.32 | 5.63 |

**Figure E-10** Method of calculating the equivalent parts per million (epm) of a water sample. (Driscoll, 1986. Reprinted by permission of Johnson Screens/a Weatherford Company.)

*Cf. Equivalents per liter.*
**Erosion** The natural process or processes that *transport* earth materials from one location to another via water (e.g. *rainfall*, waves, currents, and glacial ice), wind, and gravity. Cf. *Episodic erosion.*

**Error function** The error function has its origins in probability and statistics. This function appears in a number of analytical solutions to *boundary value problems* describing solute transport in a *porous medium*, including cases involving diffusive transport and advective–dispersive transport. The error function is defined as:

$$\text{erf}(x) = \frac{2}{\sqrt{\pi}} \int_0^x e^{-u^2} \, du$$

The error function has the following special properties:

$\text{erf}(0) = 0$
$\text{erf}(\infty) = 1$
$\text{erf}(-x) = -\text{erf}(x)$

Cf. *Complimentary error function; Advection-dispersion equation; Diffusion equation.*

**Estuary** Also known as branching bay, drowned river mouth, and firth. A semi-enclosed coastal body of water which has a connection with the open sea. In an estuary, the *sea water* is significantly diluted with *freshwater*. Cf. *Bay/embayment; Tidal estuaries; Delta; Beheaded stream.*

**Eutrophic** A characterization of surface water bodies containing high levels of plant nutrients. As a result, the *hypolimnion* of a eutrophic lake is deficient in oxygen. Cf. *Thermal stratification; Epilmnion; Disphotic zone.*

**Eutrophication** The process whereby water becomes devoid of oxygen in stagnant conditions through the action of bacteria. Cf. *Chebarotev succession.*

**Evanescent lake** A short-lived lake formed after a heavy rain.

**Evaporation** The conversion of a liquid to the vapor state by the addition of latent heat below the boiling point of the liquid. The rate of evaporation is expressed as the depth of liquid water removed from a specified surface per unit time, generally in inches or centimeters per day, month, or year. Dissolved solids are concentrated as an evaporate by driving off water through heating. Cf. *Arid-zone hydrology; Atmometer; Playa lake; Dry lake; Livingston atmometer.*

**Evaporation pan** See *Pan coefficients, evaporation.*

**Evaporite** A sediment produced from *seawater* or salt lakes in tidal areas under arid conditions as a result of extensive *evaporation.* Cf. *Tides.*

**Evapotranspiration** The process whereby water is evaporated from plants, soils, and surface water bodies. Cf. *Actual evapotranspiration (AE); Potential evapotranspiration (PE).*

**Exceedance probability** See *recurrence interval.*

**Excess pore pressure** *Pore pressure* in excess of *hydrostatic pressure.*

**Exchange capacity** See *cation exchange capacity.*

**Exotic stream** A *stream* whose waters were derived from a different region, climate, or physiographic area and flows through another physiographic region (e.g., the Nile river).

**Explosives, use of** Also called blasting, or shooting. The application of explosive materials to break up large boulders that are impeding drilling progress or to improve the *specific capacity* of the *well.* Given the appropriate rock type and the size and depth of the well, blasting can be used with good success to improve specific capacity. However, because of the many unknowns inherent in dealing with subsurface conditions, it is difficult to predict success, especially in sedimentary rocks such as sandstone.

**Exsurgence** The rising of a *stream* from a cave where the driving head is the result of downward *percolation* of water from overlying strata. Cf. *Spring, karst.*

**Extended detention basin** See *dry detention basin.*

**Extraction wells** Also known as *pumping wells, recovery wells,* or *production wells.* Wells used to extract or produce water from an *aquifer.*

**Facet** A collective term for both the aspect and the slope of a *watershed*.

**Fair–Hatch equation** A formula used to estimate the saturated *hydraulic conductivity, K*, of a granular *porous medium* as related to the *grain size distribution* of the medium. The Fair–Hatch equation for *K*, is:

$$K = \left(\frac{\rho g}{\mu}\right)\left[\frac{n^3}{(1-n)^2}\right]\left[\frac{1}{m\left(\frac{\theta}{100}\sum\frac{P}{d_{\mathrm{m}}}\right)^2}\right]$$

where:
$\rho = $ *fluid density* $[\mathrm{M \cdot L^{-3}}]$
$g = $ gravitational constant $[\mathrm{L \cdot T^{-2}}]$
$\mu = $ *viscosity* $[\mathrm{L^2 \cdot T}]$
$n = $ *porosity* $[\mathrm{L^3 \cdot L^{-3}}]$
$m = $ packing factor ($\sim$5)
$\theta = $ sand shape factor (between 6.0 for spherical grains and 7.7 for angular)
$P = $ percentage of sand held between adjacent sieves (%)
$d_{\mathrm{m}} = $ geometric mean of the rated sizes of adjacent sieves $[\mathrm{L}]$.

For non-uniform soils, the Fair–Hatch equation is based on the expectation that *dm* is a certain representative *grain size* and that determining the relation between hydraulic conductivity and soil texture requires the choice of a representative grain diameter. Cf. *Hazen method; Kozeny–Carmen equation.*

**Fair-weather runoff** See *base runoff.*

**Falling limb of a hydrograph** See *recession curve; hydrograph.*

**Falling velocity** See *Settling velocity.*

**Falling-head permeameter** See *permeameter.*

**Falling-head test** An in situ method of estimating the *hydraulic conductivity* of an *aquifer* or *confining bed* by adding a slug of water to a *well casing* to cause an instantaneous rise in the water level. The rate of the *recovery* to the initial or static *water level* is correlated with hydraulic conductivity, i.e. a faster return to the initial level indicates higher hydraulic conductivity. The falling-head test is one of two types of *slug test*, the other being a *rising-head test*. Cf. *Aquifer test; Hvorslev method/Hvorslev piezometer test.*

**False set** Also called early or premature stiffening, or a hesitation or rubber set. The rapid hardening of a grouting mixture while *grouting*, without the generation of much heat (*heat of hydration*). A false set typically occurs in a well bore. If the *grout* is freshly mixed and becomes rigid, its plasticity can usually be regained by adding more water to the mixture while mixing. Cf. *Flash set.*

**False stream** Water residing in an elongated basin or ditch along the edge of a *floodplain*. The basin containing the false stream slopes toward the side of the valley, away from the main stream. Cf. *Tributary; Deferred junction; Yazoo stream.*

**Fault-dam spring** See *spring, fault.*

**Fault spring** See *spring, fault.*

**Fault-trellis drainage pattern** See *drainage pattern.*

**Fault-trough lake** See *sag pond.*

**Fault zone permeability** Fault zones may behave either as important conduits for fluid *flow* in the earth's crust or may act as barriers to large scale flow, depending on their *permeability* characteristics. The quantification of crustal fluid fluxes is important for modeling and understanding a number of linked hydromechanical and fluid–rock interaction processes. Different fault rocks have been determined to have widely differing permeabilities. As a result, the fine detail of internal structure in fault zones can strongly influence fluid flow patterns. Laboratory-determined permeability data show wide variation with fault rock microstructure (e.g. gouge microclast size), controlled by structural position in the fault zone and slip zone intersections. Cf. *Drainage pattern; Sag pond; Secondary porosity.*

**Fecal coliform bacteria** A type of bacteria typically found in the intestinal tracts of mammals but which is occasionally detected in *groundwater*, indicating possible contamination with pathogens. The criteria for swimming is fewer than 200 colonies/100 mL, for fishing and boating fewer than 1000 colonies/100 mL, and for *domestic water* supply fewer than 2000 colonies/100 mL. Cf. *Supplied water; Maximum contaminant levels (MCLs).*

**Feeder** See *tributary.*

**Feeding ground** See *drainage basin/drainage area.*

**Fen** A *wetland* area typically deriving its water from *groundwater* that is rich in calcium (Ca) and magnesium (Mg). As a result of its groundwater source, a fen is less acidic than a *bog*, and its vegetation is more alkaline. The decaying alkaline vegetation contained in a fen may ultimately be the foundation material of a peat deposit. Cf. *Marsh.*

**Ferrel's law** The natural law stating that the centrifugal force produced by the rotation of the earth causes rotational deflection of currents of water and air. Ferrel's law explains the *coriolis force* by which currents in the northern hemisphere are deflected to the right and currents in the southern hemisphere are deflected to the left.

**Fick's first law** The law stating that the *flux* of a *solute* through a liquid under steady-state conditions is proportional to the concentration *gradient* of the solute in the liquid. Fick's first law is expressed by the equation:

$$F_x = -D_x \frac{dC}{dx}$$

where:

$F_x$ = mass flux of solute per unit area per unit time $[M \cdot L^{-2} \cdot T^{-1}]$
$D_x$ = *diffusion coefficient* $[L^2 \cdot T^{-1}]$
$C$ = solute concentration $[M \cdot L^{-3}]$
$dC/dx$ = concentration gradient $[M \cdot L^{-3} \cdot L^{-1}]$

The negative sign indicates a positive flux in the direction of negative gradient, or the direction of decreasing concentration. The *diffusion* coefficients for the major cations/anions in water range from $1 \times 10^{-9}$ to $2 \times 10^{-9}$ m$^2$ s$^{-1}$ and are temperature dependent, e.g. the coefficients are approximately 50% less at 5°C. The transfer of biologically important gases such as carbon dioxide ($CO_2$) and oxygen ($O_2$) between the atmosphere and the water is described by Fick's first law as a net flux of gas through a *stagnant water* layer of variable thickness at the surface. The diffusive *transport* of volatiles, however, occurs partly in the liquid phase and partly in the gaseous phase. Fick's first law for diffusion in a *porous medium* process is stated as:

$$F_x = -D^* \frac{\partial \theta C}{\partial z}$$

where:

$\theta$ = porosity [dimensionless]
$D^*$ = apparent diffusion coefficient $[L^2 \cdot T^{-1}]$
$C$ = solute concentration $[M \cdot L^{-3}]$

Cf. *Fick's laws; Fick's second law; Advection-dispersion equation.*

**Fick's laws** Laws governing the *diffusion* of a *solute* within a liquid. In *hydrogeology*, the liquid of interest is usually water, but in environmental investigations, it may be a *non-aqueous-phase liquid* (NAPL) such as a petroleum hydrocarbon, chlorinated solvent, or toxic oil. Cf. *Fick's first law; Fick's second law; Advection-dispersion equation; Diffusion coefficient.*

**Fick's second law** The law that relates the concentration of a diffusing substance to space and time. A one-dimensional *diffusion equation* derived from *Fick's first law* and the equation of continuity expresses Fick's second law as:

$$\frac{\partial C}{\partial t} = D_x \frac{\partial^2 C}{\partial x^2}$$

where:

$\partial C/\partial t$ = change in concentration with time $[M \cdot L^{-3} T^{-1}]$
$D_x$ = *diffusion coefficient* $[L^2 T^{-1}]$,
$C$ = solute concentration $[ML^{-3}]$

This equation, with appropriate conditions, can be solved to predict the concentration as a function of distance and time away from a source. Cf. *Advection-dispersion equation; Fick's laws.*

**Field blank** An aliquot of purified water that is handled exactly the same as the *water quality* samples being taken in the field, which is used to detect any contamination introduced by sampling in the ambient conditions that may affect sample chemistry. A sample container is opened at the sampling area and filled with distilled or deionized water, and then the field blank is sent to the laboratory along with the field samples for analysis of the constituents being tested. Cf. *Trip blank.*

**Field capacity** Also called field-moisture capacity. The *moisture content* of a soil after *gravity drainage* is complete, or at the point when the force of gravity acting on the soil water equals the surface tension at which gravity drainage ceases. Field capacity is the amount of water held in *pore spaces* by surface tension on the soil particles as determined after a certain time period versus the *specific retention* of a soil, which is not based on time. Soils may have a moisture content less than field capacity as a result of evaporation and transpiration. Field capacity is the water left in soil that has been saturated with a moisture content of $\theta_s$ and allowed to drain for 24 h to 2 days, or an unspecified time, until the remaining water reaches equilibrium with the gravitational forces causing *drainage*, at a moisture content of $\theta_f$. Field capacity is the dividing point between *detention storage* and *retention storage*. Loosely defined, however, it is the maximum volume of water a soil zone can hold. Cf. *Normal moisture capacity; Available moisture; Free water.*

**Field-moisture capacity** See *field capacity.*

**Field-moisture deficiency** The volume of water needed to restore a given *soil water* to *field capacity.*

**Field-moisture equivalent** The minimum water content of a soil at which additional water is not absorbed when placed on a smooth surface of the soil mass but rather beads on the soil surface or spreads out and gives the soil a shiny appearance. The field-moisture equivalent is expressed as a percentage of the soil mass' dry weight.

**Filler** A small watercourse, e.g. a *stream*, that discharges into and fills a *lake*. Cf. *Tributary*.

**Film water** See *adhesive water; attached groundwater; fixed groundwater; pellicular water*.

**Filter** A designed pervious material in single or multiple layers that is installed or utilized to provide selective *drainage*, or a *porous medium* used to separate suspended materials from liquid. A filter membrane or other medium allows certain liquids to move through it while restricting the movement of soil particles or dissolved constituents of larger-diameter molecules. In surface water and *groundwater* filtration, the main objective is to reduce or eliminate *turbidity* caused by solids and organic matter that do not settle out of water. The filters used in the past were mainly rapid (coarse) and slow (fine) sand beds. Mixed-media beds, however, have since been shown to be more effective and less susceptible to clogging and *head loss*. Mixed-media beds consist of layers of materials having different sizes and densities. This arrangement keeps the filter layers from mixing during *flow* and backwashing and maintains the maximum volume and area for the *filtration* process. If only two media are used (i.e. a dual-media filter), the coarser material is placed over the finer one to utilize more than the surface for filtration. A typical three-media filter might consist of anthracite (coal) particles with a density of 1.4 and a size of about 1 mm, sand with a density of 2.7 and a size of 0.45 mm, plus garnet or illmenite mineral with a density of about 4 and a size of 0.3 mm. Cf. *Backwash in well development*.

**Filter cake** Also called mud cake. A low-*permeability* film created by the *drilling fluid* on the porous face of a *borehole*, which is intended to prevent fluid loss into the surrounding formation. Two types of constituents are used with drilling fluids to enhance the formation of filter cake: clays and polymeric *colloids*. These materials can be used alone or in combination, depending on the formation characteristics. *Hydraulic pressure* in the borehole forces the drilling fluid into the formation as the hole is drilled. During penetration of the fluid, added clay particles collect on the wall of the borehole, forming a cake. Fluid loss continues until the filter cake becomes thick enough to stop penetration. When a polymer is used, it forms a film on the borehole wall rather than penetrating the formation; the film is made up of insoluble particles surrounded by a viscous layer of fluid. Polymer filter cakes are generally thinner than clay filter cakes. Cf. *Montmorillonite; Flocculation*.

**Filter pack** Material that is placed in the *annulus* in the soil zone between the *well screen* and the *borehole* wall to prevent formation material or *fines* from entering or clogging the screen and thereby reducing well productivity. The material used as a filter pack is usually well-sorted, rounded, siliceous sand or gravel. The *grain size distribution* of the filter pack is selected by multiplying the 70%-retained size of the finest formation sample by a factor between 4 and 10; therefore, the filter pack does not restrict the *flow* from the layers of coarsest material. A factor of 4 is considerably more conservative than would be chosen for a production well, but in a *monitoring well* it is typically more important to keep out fine material than to maximize pumping rates. **Figure F-1** presents the curve for selecting the proper filter pack material. The screen *slot opening* is then selected based on the size that retains 90% or more of the filter pack material.

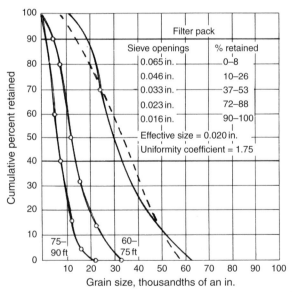

**Figure F-1** To properly select the filter pack material, the grain size is plotted versus the cumulative percent retained. (Driscoll, 1986. Reprinted by permission of Johnson Screens/a Weatherford Company.)

Cf. *Air development; Bride/bridging*.

**Filter velocity** See *specific discharge; Darcy flux/Darcian velocity/Darcy's flux velocity/bulk velocity.*

**Filtering of water samples** The removal of the *suspended load* and *colloids* from water samples collected for chemical analysis. Water analysis may require *filtration*, acidification, or preservation of the sample, depending on the parameters of interest. In filtration, the water is sampled and is immediately passed through a 0.45-μm filter An in-line *filter* can be attached to the sampling discharge line, thereby avoiding exposure of the sample to ambient air. A sample to be analyzed for metals, dissolved phosphorus, phosphate, or radioactive constituents may be filtered first and then acidified, because the addition of acid may dissolve the sediment or colloidal material in the water. Cf. *Maximum contaminant levels.*

**Filtration** The separation of *suspended solids* and/or colloidal material from a liquid by passing the liquid through a *porous medium.* Cf. *Filter; Diatomaceous earth filter.*

**Filtration ratio** The concentration of ionic species in an input solution divided by the concentration in the *effluent*, to determine the efficacy of a clay membrane in retarding the *flow* of *ions.*

**Filtration spring** See *spring; filtration.*

**Final set** The time taken by a *grout* mixture to achieve its ultimate degree of stiffness. The final set is stiffer than the initial set. Final set is typically stated as an empirical value of the time that the grout takes to set sufficiently to resist penetration of a weighted test needle.

**Fineness modulus** The total percentage of a sample of soil retained in each of a specified series of sieves, and dividing the sum by 100. In the United States, the US Standard sieve sizes are as follows:

| Sieve No | Particle size |
|----------|---------------|
| No. 200 | 75 μm |
| No. 100 | 149 μm |
| No. 50 | 297 μm |
| No. 30 | 590 μm |
| No. 16 | 1190 μm |
| No. 8 | 2380 μm |
| No. 4 | 4760 μm |
| 3/8 in. | 9.5 mm |
| 3/4 in. | 19 mm |
| 1½ in. | 38 mm |
| 3 in. | 76 mm |
| 6 in. | 150 mm |

Cf. *Grain size analysis; Grain size distribution.*

**Fines** The portion of a soil sample that is fine enough to pass through a No. 200 (75 μm) US standard sieve. Cf. *Grain size analysis; Fineness modulus.*

**Fingering** The type of *dispersion* resulting from lens-type *heterogeneities*, in which a fluid and/or constituent is transported more rapidly in the areas of higher *hydraulic conductivity*, causing a distinct pattern. Also caused by fluids of different density or *viscosity*, and in infiltration through the unsaturated zone with unstable fronts. Fingering is shown in **Figure F-2**.

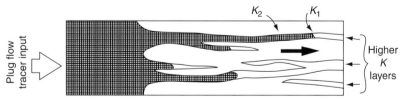

**Figure F-2** The fingering advancement of contaminants resulting from layered beds and lenses. (Freeze, R. A. 1979. Reprinted with permission from Prentice-Hall, Inc.)

**Finger lake** A body of water restricted to a long, narrow basin. Typically, the basin of a finger *lake* is formed by glacial processes. Cf. *Yazoo stream; Deferred tributary; False stream.*

**Fingerprint diagram** An illustration of the pattern and relative abundance of dissolved *ions* versus the relative *salinity* in a water sample. **Figure F-3** is an example of a fingerprint diagram in which the shape of each line gives the pattern of ions and the position of the line indicates relative salinity. Before constructing a fingerprint diagram and interpreting the data, several parameters should be considered:

- The choice of graph paper (scale; millimeter, semi-logarithmic, and log) can change the observed patterns.
- The choice in the number of cycles in semi-logarithmic graphs depends on the difference between the lowest concentration and the highest concentration in the data set.
- The order in which the ions are displayed on the horizontal axis must be selected, i.e. by geochemical importance, increasing or decreasing concentration, anions versus cations, or a combination of the above.

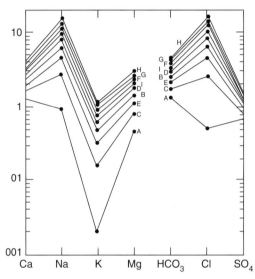

**Figure F-3** A fingerprint diagram displaying cations on the left and anions on the right. (Mazor, 1991. Reprinted with permission from Halsted Press, a division of John Wiley & Sons, Inc.)

Cf. *Stiff diagram; Piper trilinear diagram; Durov-zaprorozec diagram.*

**Fingertip channel** A small, unbranched stream course that typically occurs at the head or beginning of a *drainage pattern*. Cf. *Headwater;Tributary.*

**Finite difference technique** The numerical solution of differential equations describing hydrologic processes can be performed by discetizing the domain (*aquifer(s)* or portions thereof) over which the simulation is to be performed. Discretization is the process in which an ordinary differential equation, or a partial differential equation is approximated by an algebraic formula. There are three popular approaches to discretizing a given equation: (1) finite difference technique, (2) *finite volume technique*, and (3) *finite element technique*. These three techniques are roughly of comparable complexity when it comes to their implementation. The finite difference and finite element techniques are the most common approaches in hydrologic modeling. The finite difference technique involves the construction of a grid over the domain. As a first step toward solving this differential equation, the numerical solution only at a set of N points (called nodes) is to be determined. To ensure that these equations are computed as reliably as possible, these N nodes are spaced equally, at a distance equal to x (**Figure F-4**). Where steeper *hydraulic gradients* are encountered (e.g. near a pumping well), the nodes need to be placed more closely together. This process of placing the nodes in a given region is called grid generation and is a highly developed technique. It is important to note that this process of grid generation must be carried out first and the grid plotted to ensure that the nodes are equally placed. The differential equations are approximated by a Taylor's series about the node i. If there are n nodes at most, a (n − 1) order polynomial can be fit. The Taylor series is typically truncated for computational efficiency, which can also result in truncation errors. The simulation proceeds by solving the matrix form of the set of difference algebraic equations (the Taylor series approximations) for a series of discrete steps in time. Given *transmissivity, storativity,* and pumping rate for each node and boundary and initial conditions, the solution of the set of equations in the mathematical model generates *hydraulic head* for each particular time step.

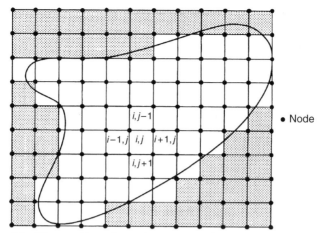

**Figure F-4** A square finite difference grid superimposed on a map area. (From Chorley, D. W., F. W. Schwartz, and A. S. Crowe. 1982. Inventory and potential applications of groundwater flow and chemistry models. Prepared for the Research Management Division, Alberta Environment, by the University of Alberta, Department of Geology. RMD 82/7. 73 pp.)

Cf. *Boundary conditions.*

**Finite element technique** This method for hydrologic modeling involves applying an integral expression from a differential equation valid over the entire region of interest. The finite element method is based on integration techniques, whereas the *finite difference technique* is based on differencing techniques to solve equations of flow. A complex region is divided into smaller subregions, called elements (**Figure F-5**). Commercially available modeling packages typically have integrated finite element generation algorithms The following steps are involved in applying the finite element method. The first step involves developing an integral representation of the differential equations governing flow. Second, dependent variables (e.g. *hydraulic head*) at the nodes are represented as interpolation functions (also called shape functions) referenced to the coordinates of the nodes. These interpolation functions are then substituted into the integral solutions, yielding a set of algebraic equations expressed in nodal coordinates. For transient analysis, the time derivative is approximated by a finite difference equation. Finally, the algebraic equations are derived for each node and are combined along with *boundary conditions* and time into a matrix equation, which may be solved using various matrix solution techniques.

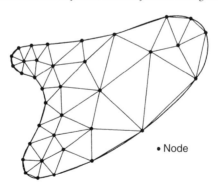

**Figure F-5** A linear triangular finite element representation of a map area. (From Chorley, D. W., F. W. Schwartz, and A. S. Crowe. 1982. Inventory and potential applications of groundwater flow and chemistry models. Prepared for the Research Management Division, Alberta Environment, by the University of Alberta, Department of Geology. RMD 82/7. 73 pp.)

**Finite resource** An asset that is formed or concentrated at a much slower rate than its rate of consumption and that eventually becomes non-renewable.

**Finite volume techniques** In using hyperbolic functions in computer modeling, discontinuities in the domain such as river bank flooding lead to computational difficulties. Classical finite difference methods, in which derivatives are approximated by finite differences, can be expected to break down near discontinuities in the domain where the differential equation does not hold. Finite volume methods, are based on the integral form instead of the differential

113

form of a function. Rather than pointwise approximations at grid points, the domain is broken into grid cells and approximate the total integral of the function over each grid cell, or actually the cell average of the function, which is this integral divided by the volume of the cell. These values are modified in each time step by the flux through the edges of the grid cells, and the primary problem is to determine good numerical flux functions that approximate the correct fluxes reasonably well, based on the approximate cell averages. This method is used for two-dimensional unsteady flow models such as RBFVM-2D (University of California, Berkeley), which is based on the finite-volume method with a combination of unstructured triangular and quadrilateral grids in a river basin system. This model deals with the wetting and drying processes for *flood-plain* and *wetland* studies, dam breaking phenomena involving discontinuous flows, *subcritical* and *supercritical flows*, and other cases where there is a discontinuity in the domain. The computations of *tributary* inflows and regulated flows through gates, *weirs*, and culverts or bridges are also included. Cf. *Unsteady state groundwater flow.*

**FIRM** See *flood insurance rate map.*

**First flush** The occurrence of significantly higher concentrations of water-borne *pollutants* at the beginning of a storm because of the greater *rainfall intensity* that results in higher *runoff* and greater boundary *shear stress* and erosive potential. First flush is also affected by the availability of pollutant constituents that accumulated during previous dry weather.

**First-order reaction** A reaction whose rate of change over time can be described by the following relationship:

$$\frac{dC}{dt} = -\lambda C$$

where:

$C$ = concentration [M·L$^{-3}$]

$t$ = time [T]

$\lambda$ = the decay constant [dimensionless]

The above equation can be rewritten in the form of:

$$\frac{dC}{C} = -\lambda dt$$

Integrating with respect to time yields:

$$\ln\frac{C}{C_o} = -\lambda t$$

Substituting $C = {}^1/_2 C_o$ to determine half life yields the following relationship:

$$t_{1/2} = \frac{0.693}{\lambda}$$

**First type boundary** A *boundary conditions* for a modeled domain of contaminant transport where $C = C_o$.

**Fishing tool** An implement used in the recovery of lost equipment, the "fish," from a *borehole*. A fishing tool is particularly helpful during drilling or *well installation* when equipment, tools, and broken gear are most likely to fall into, or lodge and break off in, the borehole, thus impeding drilling or installation progress, *well development*, or *water sampling*. Specialized tools manufactured for recovery or "fishing" are shown in **Figure F-6**.

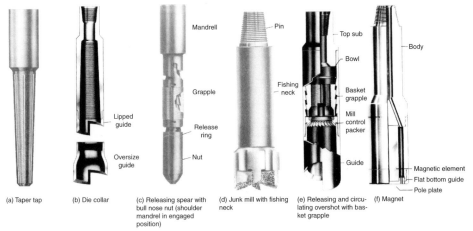

(a) Taper tap   (b) Die collar   (c) Releasing spear with bull nose nut (shoulder mandrel in engaged position)   (d) Junk mill with fishing neck   (e) Releasing and circulating overshot with basket grapple   (f) Magnet

**Figure F-6** Common fishing tools used in rotary drilling methods. (Driscoll, 1986. Reprinted by permission of Johnson Screens/a Weatherford Company.)

Cf. *Well maintenance.*

**Fissure spring** See *spring, fissure*.

**Fissure water** Water contained in open fractures or joints, usually only found in sufficient quantities near the earth's surface often resulting in *artesian* flow.

**Fixed groundwater** *Groundwater* that is held in soil or rock pore *interstices* as a result of the small diameter of the void spaces. Fixed groundwater is typically not economical for use because it moves so slowly through the *pore spaces* and/or is not available for withdrawal. Cf. *Adhesive water; Pellicular water*.

**Fixed moisture** The water retained in a soil that is less than the water represented by the *hydroscopic coefficient*. Cf. *Gravity water; Gravimetric moisture content; Soil water*.

**Fixed ring infiltrometer** A device used to measure in situ soil *infiltration*. The assembly consists of two concentric cylinders of height = 400 mm and radius = 150 mm (inner cylinder) and 300 mm (outer cylinder). A barrel (208 mm) is used as a mariotte to supply water to each cylinder, while the flux of the inner cylinder is measured with a calibrated flow tube-type flow meter (**Figure F-7**). See: **Appendix A** for conversion units.

**F**

**Figure F-7** Double ring infiltrometer used for measuring infiltration rate and hydraulic conductivity in soils

**Flash** A sudden increase in the water level of a watercourse, typically caused by damming or another restriction to streamflow and resulting in a pool of water or a *marsh*.

**Flash flood** A sudden increase in the velocity and/or volume of a watercourse or water body, typically of short duration and caused by excessive *precipitation* over a short time span, and which often results in *overbank flow*. In arid regions, a flash flood can occur in a usually dry ravine experiencing a sudden and dramatic precipitation event. A flash flood can also result from *river* blockage or dam failure.

**Flash set** Also called a quick set or grab set. The very rapid development of rigidity in a freshly mixed *grout*. A considerable amount of heat (*heat of hydration*) is usually generated when a grout mixture flash sets, which can cause PVC *well casing* to warp, depending on the percentage of *cement* in the mixture. Cf. *False set*.

**Flashy hydrologic behavior** Behavior attributed to small *watersheds* that exhibit higher high flows and lower *low flows* than other watersheds in terms of *runoff* per unit area, or higher ratios of maximum to minimum flows.

**Flexible wall permeameter** A device for measuring the *permeability* undisturbed or compacted fine-grained soils such as clay, clay and sand tills, silt, and peat muds. These materials typically have saturated permeabilities ranging from $1 \times 10^{-3}$ to $1 \times 10^{-9}$ cm s$^{-1}$. Refer to **Appendix A** for units and property conversions. Cf. *Permeameter; Guelph permeameter; Constant head permeameter*.

**Float measurement of discharge** A direct method of assessing *discharge* by placing a buoyant object in a *stream* or other watercourse and timing the object as it travels a given distance (**Figure F-8**). Float measurement is considered a rough estimate of discharge because the velocity measured only represents the layer or layers of the stream that contact the float body, or the stream's surface velocity. As a stream velocity profile usually varies from zero at the bottom to a maximum value at the surface, a float typically overestimates the total stream discharge. A surface float usually travels at about 1.2 times the mean velocity of the stream, and a float extending to mid-depth travels at about 1.1 times the mean stream velocity (**Table F-1**). Surface floats can be modified, as shown in **Figure F-8**, to force them to travel at the mean velocity of the watercourse. If a modified float is not employed, a correction factor of 0.7 is used for a watercourse 1 m in depth, and a factor of 0.8 is used for a watercourse 6 m deep or more.

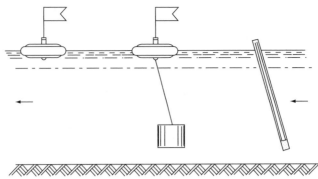

**Figure F-8** A surface float, canister float, and rod float, respectively. (Shaw, 1983. Reprinted with permission from Van Nostrand Reinhold Co. Ltd.)

**Table F-1** Velocity calculations from the float method. (Sanders, 1998.)

| Section of stream | Width[a] (ft) | Trial | Distance[a] (ft) | Time[a] (s) | Surface velocity (ft s⁻¹) | Average surface velocity in this section (ft s⁻¹) | Weighted surface velocity (ft s⁻¹) | Notes |
|---|---|---|---|---|---|---|---|---|
| 1 | 11 | 1 | 100 | 23.5 | 4.26 | 4.20 | 1.32 | |
| | | 2 | 100 | 25.0 | 4.00 | | | |
| | | 3 | 100 | 23.0 | 4.34 | | | |
| 2 | 10 | 1 | 100 | 22.0 | 4.55 | 4.61 | 1.32 | |
| | | 2 | 100 | 21.3 | 4.69 | | | |
| | | 3 | 100 | 21.7 | 4.61 | | | |
| | | 4 | 100 | 21.7 | 4.61 | | | |
| 3 | 14 | 1 | 100 | 25.1 | 3.98 | 3.98 | 1.59 | |
| | | 2 | 100 | – | – | | | Float caught in branch |
| | | 3 | 100 | 25.2 | 3.97 | | | |
| | | 4 | 100 | 25.1 | 3.98 | | | |

a = per section.
Surface velocity = sum of weighted velocities = 4.23 ft s⁻¹.
Average velocity = multiply uncorrected velocity by 0.85 to account for variation of velocity with depth in stream = 3.59 ft s⁻¹.
Weighted surface velocity is calculated by weighting sectional values according to the width of the section:

$$\text{Weighted velocity} = \frac{\text{width of section}}{\text{total stream width}} \times \text{velocity measured in that section.}$$

Cf. *Chezy's formula; Flumes; Flow meter.*
**Float shoe** Part of a *borehole* packing system used to facilitate *grout* placement in a borehole that cannot be backfilled before grouting. At the bottom of the interval to be grouted, an external packer is installed in the casing string to support the float shoe, which keeps the grout from penetrating past the packer, as shown in **Figure F-9**. In this way, an area in the formation that might be damaged by grouting or might absorb unacceptable amounts of grout can be isolated from the rest of the borehole.

**Figure F-9** An external packer equipped with a float shoe to aid in the placement of grout. (Driscoll, 1986. Reprinted by permission of Johnson Screens/a Weatherford Company.)

**Flocculation** The formation of loose open-structured particles, or flocs, that are suspended in solution. Flocs are an *aggregation* of minute particles tightly held together. Flocculation and *deflocculation* are an important physical condition of the *gel strength* of a *drilling fluid*, as shown in **Figure F-10**. Flocculation is also observed in the settling-out process of clay particles in saltwater. The *coagulation* of the finer clay particles usually forms a gelatinous mass.

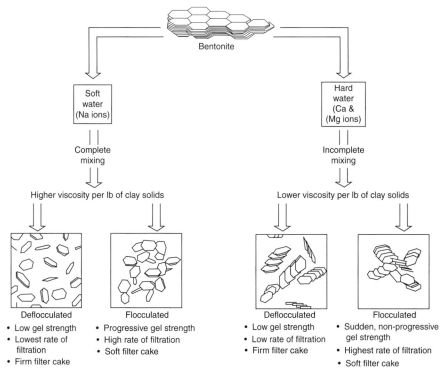

**Figure F-10** The flocculation results of either completely mixing or the incomplete mixing of clay particles. (Driscoll, 1986. Reprinted by permission of Johnson Screens/a Weatherford Company.)

Cf. *Dispersion.*

**Flood** A sudden increase in the volume and/or velocity of a body of water, causing *overflow* or breaching of natural or manmade confines, which drowns or covers land areas that are usually dry. A flood is literally any *peak flow*, but the term "flood" has been accepted for a relatively high *stream* or channel flow that exceeds the capacity of its *banks* at any given location or *reach*, as measured by *gauge height* or *discharge*. Therefore, a watercourse can be flooded in one location and not another. Cf. *Design flood*.

**Flood above a base** See *partial-duration flood series*.

**Flood absorption** A reduction in a *flood stage* discharge by the storage of water in a *reservoir, lake,* channel, or other water body. Cf. *Absorption*.

**Flood basin** An area of land that was historically inundated by water during the region's highest known *flood*. A flood basin usually contains heavy soils and is either lacking vegetation or limited to swampy vegetation. Cf. *Backswamp deposit; Marsh*.

**Flood control** Engineered preventative measures implemented to reduce damage caused by flooding. Flood control measures include but are not limited to diversion to *floodways*, or, around populated or critical areas to be protected, containment in constructed *reservoirs*. A large percentage of funding for flood control research is directed at predicting the magnitude of *flood peaks* from storm or snowmelt events. Cf. *Design flood*.

**Flood crest** The highest stage or the top of the *flood wave*. Cf. *Flood peak*.

**Flood damage analysis** Associating the available data relating peak flow and damage with predicted probabilities of various levels of peak flow to produce a damage-probability relation for a given *stream*, and the points of protection interest, in an area susceptible to flooding. A *flood* damage probability curve is constructed by combining the flow-damage curve and flow-probability curve, as shown in **Figure F-11**. The average annual flood damage is equal to the area under the damage-probability curve. A *reservoir* can be placed in the model of the stream in order to assess the resulting reduction of peak flow relations (and damage relations), and alterations in reservoir design can be evaluated for comparative damage reduction.

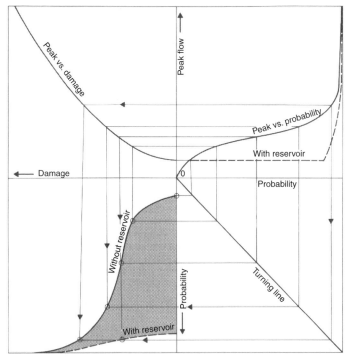

**Figure F-11** A flood damage analysis utilizing the flood-damage probability curve. (Linsley, 1982. Reprinted with permission from McGraw-Hill, Inc.)

Cf. *Peak discharge; Flood probability; Gringorten formula*.

**Flood damage curve** See *flood damage analysis*.

**Flood damage probability curve** See *flood damage analysis*.

**Flood discharge** A watercourse's *discharge* during the *flood stage*. Cf. *Flood; Flood flow; Flood damage analysis; Flood peak*.

**Flood forecasting** A method using the characteristics of a *river basin* and the measurements of a historical *precipitation* event to predict the timing, *discharge*, and height of future *flood peaks*. Flood forecasting is used in a warning system for areas at *flood* risk. Cf. *Flood prediction*.

**Flood frequency** An average time interval between *floods* that equal or exceed a given magnitude. Cf. *Flood-frequency analysis; Gringorten formula; Recurrence interval*.

**Flood-frequency analysis** A calculation of flooding likelihood that is based on the average time interval between past *peak flow* events. Flood-frequency analysis focuses on the maximum instantaneous *discharge*. Many governmental agencies keep peak flow records that include annual maximum instantaneous *flood-peak* discharges and stages, dates of occurrence, and associated *partial-duration series (PDS)* for many locations within a district. In the United States, the *flood-frequency curve* is computed using an annual *flood series* with at least 10 years of data, if available. A flood with a *recurrence interval* of 10 years would have a 10% chance of occurring in any year; therefore, flood-frequency analysis is not a forecast of flooding but of probability. Cf. *Flow-duration curve; Gringorten formula; Design flood; Peak discharge; Gringorten formula*.

**Flood-frequency curve** A graph of the number of times in a given interval that a *flood* of a given magnitude is equaled or exceeded. **Figure F-12a** is a flood–frequency curve plotting *peak discharge* versus the percentage of time that discharge is equaled or exceeded. The regional flood–frequency curve shown in **Figure F-12b** plots the ratio of the flood *flow* to the annual flood on the ordinate, and the *recurrence interval* for a particular flood flow can be traced to the abscissa. Cf. *Unit hydrograph*.

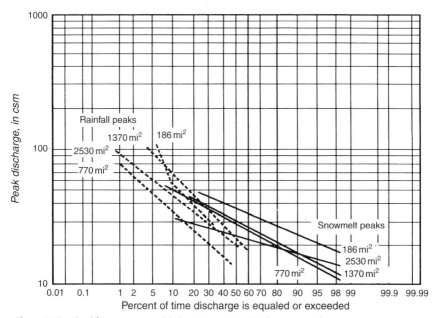

**Figure F-12a** Flood frequency curve. (Black, 1996. Reprinted with permission from Ann Arbor Press, Inc.)

**Flood fringe** A storage zone in which *flood water* is either ponded or moving slowly, thereby attenuating the *flood peak* as the *flood wave* proceeds down the watercourse. **Figure F-13** shows the plan and cross-sectional view of a *floodplain*, *floodway*, and flood fringe.

**Flood insurance rate map (FIRM)** A map developed by the United States Federal Emergency Management Agency (FEMA) as part of the minimum *floodplain* management standards for local governments.

**Flood mitigation, channel improvements for** Alterations to a particular *reach* of a water channel to speed the crest of a *flood* through that reach, thereby reducing the buildup of water and the resulting crest height. Aside from *flood-control* dams and *reservoirs*, the most common flood mitigation measures consist of cleaning, lining, straightening, or leveeing water channels. Most such channel improvements move the water more rapidly to the reaches beyond them, and there is always the possibility of creating bigger problems *downstream* than those solved. Constructing levees along a reach does not speed the passage of the *flood crest* but rather allows a higher crest to occur within the reach without overtopping the channel, spreading out in the adjacent *floodplain*, and causing damage. However, even with this type of improvement, the effects on downstream reaches must be considered. Cf. *Design flood; Flood control*.

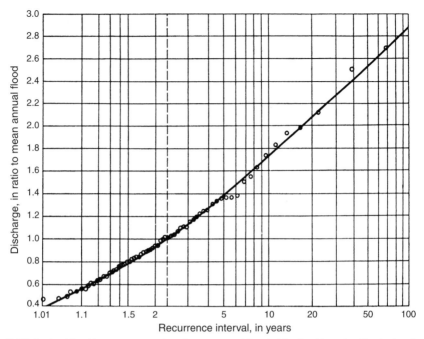

**Figure F-12b** Regional flood frequency curve where the recurrence interval of the flood is read on the abscissa. (Leopold, 1992. Reprinted with permission from Dover Publication, Inc.)

**Flood peak** Also called *flood discharge*. The highest value of the stage or *discharge* attained by a *flood*, i.e. *peak stage* or *peak discharge*. Cf. *Flood crest; Annual series; Flood plain.*

**Floodplain** The water surface elevation at specified locations along a watercourse profile during a *flood*. The floodplain is delineated by computing *flood runoff* rates and volumes, routing them through the *drainage* way, and computing the corresponding water surface profiles, as shown in **Figure F-13**. Once a *river* that has excavated a valley overflows its *banks*, and the decrease in velocity forces it to drop its *suspended load*, a floodplain will form in the valley. Cf. *Flood fringe; Floodway; Alluvial river.*

**Flood prediction** The study of basin characteristics, *precipitation* patterns, and *hydrographs* to predict the average frequency of a specific *flood* event by estimating the probable *discharge* that, on average, will be exceeded only once in a chosen period, e.g. a 50- or 100-year flood. One method of flood prediction is the use of *rainfall duration* and *rainfall frequency* data for a specific locality to estimate a flood of selected exceedance probability. Cf. *Flood forecasting.*

**Flood probability** The statistical determination that the occurrence of a *flood* of a specified magnitude will be equaled or exceeded within a given period. For example, a flood having a 10% probability would be called a 10-year flood. A 100-year flood is that magnitude of *flow* in an instant that, on average, can be expected to be observed once in 100 years at a specific location on a watercourse. Cf. *Flood frequency analysis; Flood damage analysis; Gringorten formula.*

**Flood profile** A graph of a watercourse in *flood stage* with the water surface elevation plotted as the ordinate, $x$, and distance measured *downstream* plotted on the abscissa, $y$. The flood profile can depict

- water elevation at a specified time
- *flood crests* during a particular flood

**Flood regulation** The reduction of downstream *flood peaks* through the use of *flood mitigation* reservoirs. Flood regulation, by retaining a portion of the *flood water* until it can be safely released, spreads the flood volume over a longer time and reduces the peak. *Reservoirs* for this purpose are usually sized to a projected *design flood* determined from analysis of historical flood records. Cf. *Flood damage analysis; Gringorten formula.*

**Flood routing** Determining the shape and timing of a *flood wave* at specified locations along a watercourse, typically a channel that has experienced moderate damage as a result of high water. Cf. *Dynamic equation.*

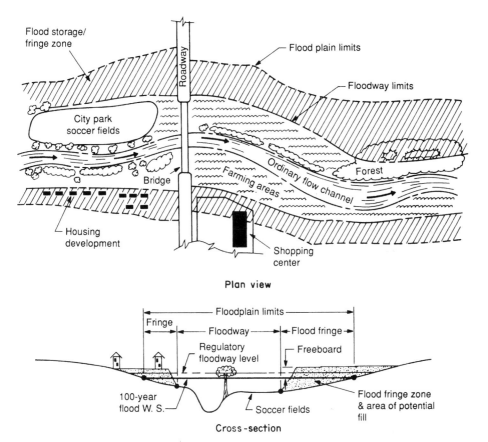

Plan view

Cross-section

Note: Minimum freeboard is typically set by local ordinance.
When it is not, suggest that freeboard be set using a
sensitivity analysis of flood profiles.

**Figure F-13** Map view and cross-sectional view of the floodplain, floodway, and flood fringe. (Maidment, 1993. Reprinted with permission from McGraw-Hill, Inc.)

**Flood runoff** A rise of water level that is due to a condition such as short-duration high-intensity *rainfall*, long-duration low-intensity rainfall, snowmelt, dam or levee failure, or a combination of these, and which exceeds the watercourse's *retention* capabilities. Cf. *Flood*; *Precipitation*.

**Flood series** A list of *flood* events in order of magnitude for a specified time period. A partial flood series comprises all floods above a selected base value, indicating the probability of such events being equaled or exceeded, e.g. two or more times per year. Cf. *Flood-frequency analysis*; *Gringorten formula*; *Recurrence interval*.

**Flood stage** The height or elevation of a water *gauge* at the lowest *bank* of a watercourse, e.g. *stream* or *river*, or the period during which a watercourse's *overflow* causes damage. Cf. *Flood frequency analysis*; *Gringorten formula*.

**Flood water** Excess water that has *overflowed* the natural or manmade confines of a watercourse. Cf. *Peak discharge*.

**Flood wave** A rise in stream stage that peaks in a crest before receding. Cf. *Flood crest*.

**Flood wave attenuation** See *attenuated flood waves*.

**Floodway** A zone exhibiting the greatest potential *flood* hazard within a *floodplain*, which is typically reserved for the passage of larger floods. **Figure F-13** illustrates a cross-section view of a floodway, floodplain and flood fringe.

**Flow** Fluid (i.e. gas or liquid), or fluid and its *load*, that is in motion. Flow is also loosely defined as the volume of liquid passing a given location in a unit of time. Cf. *Discharge; Peak flow; Reynolds number; Flow, steady; Turbulent flow; Uniform flow; Groundwater flow.*

**Flowage line** See *flow line.*

**Flow boundary** Also called a *flux* boundary. A limit or condition across which *groundwater flow* either enters or leaves a given area. Cf. *No-flow boundary; Boundary, flux; Darcy flux.*

**Flow channel** Also known as a streamtube. The channel or *flow path* that is formed between two adjacent *flow lines* in a *flow net* simulation. Cf. *Solution channel.*

**Flow coefficient** An experimentally determined proportionality constant, comparing the actual fluid velocity flow in a pipe or open channel to the theoretical velocity expected under certain assumptions. Cf. *Flow; Flow velocity.*

**Flow-duration curve/flood-duration curve** An expression of the relationship between magnitude of daily *flow*, in units of volume per time, and the number of time units, i.e. days, weeks, or months, during which that flow may be equaled or exceeded. A semi-logarithmic plot of how often a specified stream *discharge* is equaled or exceeded (**Figure F-14**) yields a straight line. When the data are plotted on natural scales (**Figure F-15**), the area under the curve shows the total volume of water that flowed past the *gauging station* during the time considered. The flow-duration curve is not a probability curve because discharge is related between successive time intervals. A flow-duration curve provides a graphical summary of watercourse variability. The analysis focuses on the minimum daily flows to evaluate minimum reliable water supplies, and the discharge characteristics are dependent on the season. Flow-duration curves are similar for *watersheds* that are geographically similar (i.e. in size, shape, and *facet*). Cf. *Flood-frequency analysis; Gringorten formula; Recurrence interval; Flood-frequency curve.*

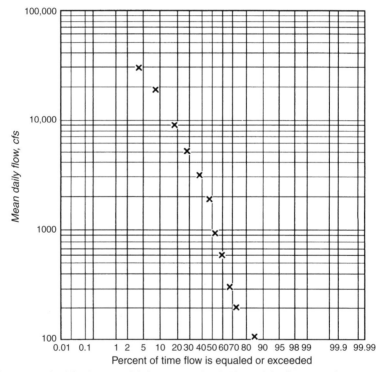

**Figure F-14** Flood duration curve. (Black, 1996. Reprinted with permission from Ann Arbor Press, Inc.)

**Flow equation** See *Darcy's law.*

**Flow frequency** Within a given period, the length of time or the number of times that a specified watercourse *flow* or *discharge* is expected to be exceeded. Cf. *Flood-frequency analysis; Flood-frequency curve; Gringorten formula; Recurrence interval.*

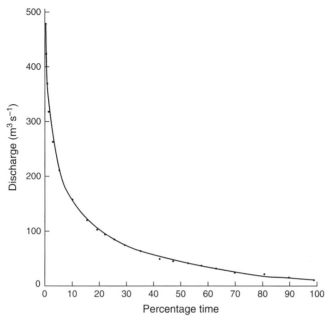

**Figure F-15** A flow duration curve for a particular point on a river in England. (Shaw, 1983. Reprinted with permission from Van Nostrand Reinhold Co. Ltd.)

**Flow hydrograph** A graph of the shape of a *flood wave* as a function of time. Cf. *Hydrograph; Unit hydrograph.*
**Flowing artesian well** See *artesian well; aquifer, artesian.*
**Flow layers, vertical** Within the *flow* of a watercourse, the zone in which each portion of the total *suspended load* is transported. The *sediment load*, expressed as a weight or volume passing through a given cross section of the watercourse per unit time, can be further broken down according to the vertical flow layer in which each separate category can be found. **Figure F-16** depicts the vertical flow layers for sediment load classifications.

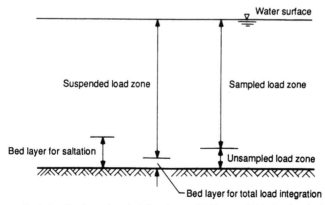

**Figure F-16** Sediment load classifications of vertical flow layers. (Maidment, 1993. Reprinted with permission from McGraw-Hill, Inc.)

Cf. *Bed load; Traction load.*

**Flow line** Also called flowage line. (1) A graphical depiction of the path that a particle of water travels under *laminar flow* conditions in a *porous medium*. (2) The position of the surface of a flowing fluid. (3) The hydraulic grade line in *open channel flow*. (4) The water-level contours of a body of water, e.g. a maximum or mean flow line of a *lake*. Flow lines are also used to represent the direction of *groundwater* movement toward *discharge*. If the *groundwater flow* is through an *isotropic* medium under steady-state conditions and the graph's horizontal and vertical scales are the same, the flow lines are perpendicular to equal *water table* elevations, *potential head* elevations, or *equipotential lines*: contour lines that connect points of equal *hydraulic head*. The commonly used contours of the *piezometric level* or of the water table represent the upper equipotential lines of the *aquifer*, assuming the air pressure above the aquifer is constant. Perpendicular to these equipotential lines is the upper flow line, which is the direction of upper aquifer *flow*, as shown in **Figures F-17a and F-17b**. Converging flow lines indicate an area of concentrated discharge.

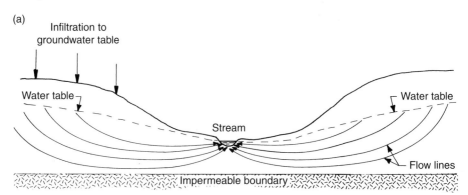

(a)

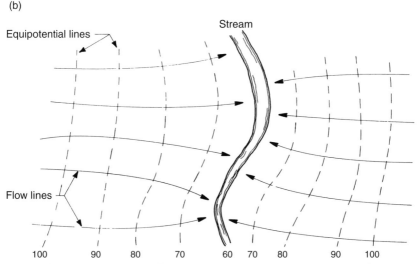

(b)

**Figure F-17a,b** a) A cross-sectional view of a stream valley, and b) the corresponding map view of a stream valley with the flow lines perpendicular to the dotted lines representing the points of equal groundwater elevation. (Driscoll, 1986. Reprinted by permission of Johnson Screens/a Weatherford Company.)

Cf. *Hydraulic grade line*.

**Flow line refraction, tangent flow** Varying zones of *hydraulic conductivity* result in refraction in *flowlines*, as the water seeks the shortest path through zones of lower hydraulic conductivity. Solute or *non-aqueous phase liquid (NAPL)* migrating with the *groundwater* can experience tangential flow at these interfaces. In petroleum geology, zones of tangential flow represent *hydrodynamic traps* creating oil reservoirs.

**Flow-meter logging** The measurement of horizontal and vertical *flow* of water. Conducted in both *streams* and down-hole in *boreholes* and *wells*. There are many types of flow meters, having impellers (*open channel flow*) to heat pulse (downhole). In measuring flow in a channel, metered flow readings are obtained at equally spaced distances across the channel and at depths to reduce interferences from friction from the riverbed and ice. Downhole measurements may be used to determine production rate from the various zones within a well or to locate *lost-circulation* zones or holes in the *well casing* outside the *screened intervals*. Horizontal flow is continuously recorded by a flow meter lowered into and retrieved from the well bore. The well is pumped from above the screened intervals, and a flow meter is lowered slowly past the screened intervals to take continuous recordings of vertical flow. The combined flow-meter readings represent the total flow up the well from the entire screened interval below the flow meter. The difference in flow between any two points in the well represents that zone's contribution to the well's total water production. Cf. *Float measurement of discharge*.

**Flow net** A graphical construction of *flowlines* and *equipotential lines* depicting the *flow paths* of water particles in two-dimensional steady *flow* through a *porous medium*. Flow nets are water-level maps or cross sections, with contour lines that connect points of equal *hydraulic head* or *piezometric head*. **Figure F-18** shows two-dimensional flow in which the

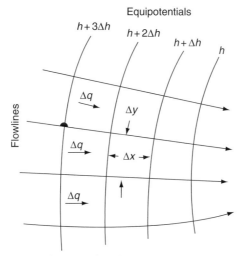

**Figure F-18** A flow net depicting two-dimensional flow. (Shaw, 1983. Reprinted with permission from Van Nostrand Reinhold Co. Ltd.)

change in *discharge* is constant between adjacent flow lines, i.e. no flow can cross them. The cells between these flow lines have the same change in *head* and therefore, must have the same length-to-width ratio. The cross section shows contours of equal hydraulic head, and flow lines for water flow direction are drawn at right angles to the contours, showing vertical components. A cross-sectional flow net shows that hydraulic head decreases with depth in a *recharge area* and increases with depth in a *discharge area*. Flow net construction assumes fairly *isotropic* conditions of *hydraulic conductivity*. For an isotropic *aquifer*, the flow lines and contours of piezometric head intersect at right angles. **Table F-2** shows the relationship of *boundaries* to the lines in a flow net.

**Flow path** The direction of water flow. Cf. *Flow line; Flow net; Flow system; Flow velocity; Groundwater flow*.

**Flow profile** The shape of the water surface, commonly known as the *backwater curve* of the *flow* in an open channel where the velocity slowly changes but the flow is assumed uniform for an incremental distance. Cf. *Backwater curve/dropdown curve; Flow velocity*.

**Flow rate** See *flow velocity*.

**Flow routing** Predicting the changing magnitude, speed, and shape of a *flood wave* as a function of time at a specified point or points along a watercourse, using equations derived for the purpose. Flow routing can characterize the *flow* resulting from a natural or artificially induced release of water thereby, enabling the necessary changes or modifications to protect property and loss of life from flooding. Flow routing has been classified as either hydrologic routing, in which the flow is computed as a function of time at one location, or hydraulic routing, in which the flow is computed as a function of time simultaneously at several cross-sectional areas along the flood route. Cf. *Flow hydrograph; Unit hydrograph; Floodway; Flood mitigation*.

**Flow, steady** The type of *flow* for which, at any point in the flow field, the magnitude and direction are constant in time. Cf. *Flow, unsteady; Equations of groundwater*.

**Table F-2** Line and boundary relationships in a flow net. (Sanders, 1998.)

| Type of boundary | Relationship of boundary to flow lines | Relationship of boundary to equipotential lines |
|---|---|---|
| **For cross-sectional or map-view flow nets** | | |
| Impermeable or no-flow boundary (e.g. cutoff wall, slurry trench, impermeable foundation of a structure, or other impermeable hydrostratigraphic unit) | Parallel | Perpendicular |
| Constant head (e.g. bed of a lake, bed of a constantly spilling reservoir of any type, a drain) | Perpendicular | Parallel |
| **For cross-sectional flow nets only** | | |
| Water table | Oblique (nonperpendicular, nonparallel) | Oblique (nonperpendicular, nonparallel) |

**Flow system** A *groundwater* network consisting of a *recharge area*, a *discharge area*, and a connecting *flow* pattern. Simple flow systems combine to form more complex patterns and relationships. A simple or local groundwater flow system consists of single recharge and discharge areas that result from well-defined topographic relief. *Recharge* occurs at a topographic high and *discharge* at a nearby low that may vary between a sharply defined location, such as a *spring* at the side of a valley, and a large area where *evapotranspiration* exceeds recharge by *precipitation*. When shown graphically, any two of the *flow lines* are always adjacent and can be intersected in only one direction. **Figure F-19** depicts such a one-flow system. Intermediate or regional groundwater flow systems have one or more local flow systems between

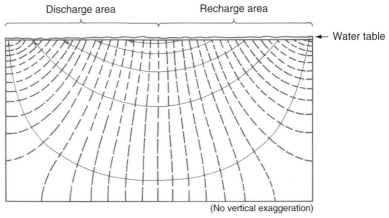

**Figure F-19** A regional one flow system. (Journal of Geophysical Research, 1962.)

their recharge and discharge areas and indicate areas where the surface topography has well-defined local relief. **Figure F-20** diagrams a series of local flow systems in a groundwater *drainage basin*, showing a regional flow system with the recharge area in the basin *divide* and the discharge area at the valley bottom. Cf. *Flow net*.

**Flow through cell** A device for measuring water quality parameters downhole or adjacent to the *well head* where the well discharge is directed to avoid contact with the atmosphere. A flow through cell is used to obtain measurements of temperature, specific conductance, *pH*, and oxidation-reduction potential (*redox*, or ORP). The benefit of the flow through cell is that interferences from sampling are reduced, providing representative measurements of the *groundwater quality* parameters. Cf. *Oxidation potential; Reduction potential*.

**Flow-through lake** A water body in which *groundwater* discharges into one side and *flows* out of another side. A flow-through *lake* is, therefore, neither a *discharge area* nor a *recharge area* for one *flow system*; rather, it is a boundary between two flow systems and is both a *discharge* area for the *upstream* system and the *recharge* area for the *downstream* system. Cf. *Boundary conditions*.

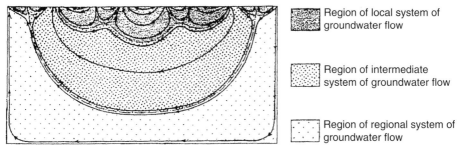

| | Region of local system of groundwater flow |
| | Region of intermediate system of groundwater flow |
| | Region of regional system of groundwater flow |

**Figure F-20** Local, intermediate, and regional systems of groundwater flow. (Freeze, 1979. Reprinted with permission from Prentice-Hall, Inc.)

**Flow, unsteady** Also called *transient flow* or non-steady flow. A *flow* for which the magnitude or direction changes with time at any point in the *flow* field. Cf. *Flow, steady; Unsteady-state flow; Equations of groundwater.*

**Flow velocity** Also called flow rate. A vector function indicating the rate and direction of *groundwater movement* per unit time through a *porous medium* perpendicular to the direction of movement. Flow velocity is expressed as the volume or mass passing through a cross section in a given time period, in dimensions of $L^3 \cdot T^{-1}$ or $M \cdot T^{-1}$. Refer to **Appendix A** for unit conversions. Using a form of the equation for *Darcy's law*,

$$Q = \frac{KA(h_1 - h_2)}{L}$$

where:

$Q = discharge\ [L^3 \cdot T^{-1}]$
$K = hydraulic\ conductivity\ [L \cdot T^{-1}]$
$A = cross\text{-}sectional\ area\ of\ flow\ [L^2]$
$(h_1 - h_2)/L = hydraulic\ gradient\ [L \cdot L^{-1}]$

combined with the standard continuity equation of hydraulics:

$$Q_{in} = V_1 A_1 = V_2 A_2 = Q_{out}$$

and knowing the hydraulic conductivity, $K$, hydraulic gradient, $(h_1 - h_2)/L$, and average *porosity*, $n$, the average *groundwater* flow velocity, $V_a$, can be calculated by the equation:

$$V_a = \frac{K(h_1 - h_2)/L}{n}$$

Determining the actual velocity of flow utilizes the porosity of the *aquifer* because flow only occurs through the interconnected *pore spaces* or *intersticies*, and not through the entire cross section. If the porosity is unknown, the average groundwater flow rate multiplied by the actual hydraulic gradient in the aquifer yields a velocity value that is greater than that calculated by the above equation. Cf. *Streamflow; Discharge; Specific discharge; Seepage velocity.*

**Flowage** The *flood water* of a watercourse, e.g. *river* or *stream*, or the act of flooding. Flooding of water, natural or induced, onto adjacent land.

**Flowage line** A contour line at the edge of a body of water, such as a *reservoir*, representing given water level. Cf. *Flow line.*

**Flowing artesian well** Also called a blow well, breathing well, or *blowing well*. A well that has sufficient *head* to bring the water up from the *aquifer* and above the ground surface without pumping. Cf. *Artesian well; Flowing well.*

**Flowing well** A well that yields or transports a fluid, e.g. water or oil, to the surface without requiring pumping. The *head* in a flowing well may be caused by pressure other than *artesian pressure*, such as gas pressure. The *potentiometric surface* of a flowing well is above the ground surface. The elevation of the potentiometric surface of the *aquifer* tapped by a flowing well must be assessed directly by allowing the water to rise to that level and then measuring its aboveground height or by attaching a pressure *gauge* to the top of the *well casing*. A gauge reading in units of pressure can be converted to a head measurement by the following equation:

$$\text{head above measuring point} = \frac{\text{gauge reading}}{\text{specific weight of water}}$$

Cf. *Artesian flow.*

**Fluid density ($\rho$)** A ratio of mass to volume, expressed in dimensions of $M \cdot L^{-3}$, which is dependent on temperature. For example, the fluid *density of water* at 10, 20, 30, and 40°C is 999.7, 998.2, 995.7, and 992.2 kg m$^{-3}$, respectively. Refer to **Appendix A** for conversions. When the fluid density, $\rho$, is a function of the *total dissolved solids* (TDS) within the fluid, the following equation applies:

$$\rho = \rho_0[1 + \beta_c(C - C_0)]$$

where:
$\rho_0 = $ *freshwater* density $[M \cdot L^{-3}]$
$\beta_c = 7.14 \times 10^{-4}$ L g$^{-1}$ within the *salinity* range from freshwater to seawater
$C = $ concentration of TDS $[M \cdot L^{-3}]$
$C_0 = $ TDS in freshwater $\rho_0$ $[M \cdot L^{-3}]$

Seawater with a TDS level of 35 g L$^{-1}$ has a fluid density of 1025 kg m$^{-3}$, whereas freshwater with 0 TDS has density values listed above. Cf. *Thalassic series; Density stratification.*

**Fluid potential** The mechanical energy of a unit mass of fluid at any point in space or time, which is proportional to the *total head* of the fluid. Fluid potential is calculated as the fluid *head* multiplied by the acceleration that is due to gravity. Cf. *Potential; Head; Groundwater energy.*

**Fluid pressure ($p$)** At point P in **Figure F-21**, the fluid pressure, $p$, is calculated as:

$$p = \rho g \psi + p_0$$

where:
$\rho = $ *fluid density* $[M \cdot L^{-3}]$
$g = $ gravitational constant $[L \cdot T^{-2}]$
$\psi = $ height of the liquid column above point P, or the *pressure head* $[L]$
$p_0 = $ atmospheric pressure or pressure at the standard state $[M \cdot L^{-1} \cdot T^2]$

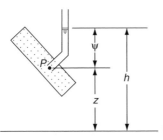

**Figure F-21** Seepage beneath a dam or levee through unconsolidated, homogenous isotropic sand. (Driscoll, 1986. Reprinted by permission of Johnson Screens/a Weatherford Company.)

Cf. *Fluid potential; Total head; Density of water.*

**Fluid storage** See *reservoir; coefficient of storage; pore space; interstices.*

**Fluid velocity** See *Flow velocity.*

**Fluidifier** An *admixture* used in *grouting* to increase the *grout* mixture's capacity to *flow* without having to add water.

**Fluidity** The opposite of *viscosity*. Fluidity expresses the ability of a substance to *flow*. Cf. *Flow velocity.*

**Flume** An artificial control mechanism with predetermined rating curves used to measure *flow* in open channels by restricting the flow, thereby causing an acceleration of the water and producing corresponding drops in the water levels, which can then be related to the *discharge*. Large flumes are required to measure flow in natural watercourses, but smaller flumes can be used to *gauge* flow from *wells*. The following conditions should be satisfied when using a flume to measure upstream flow:

- Bends in the stream or *piping* immediately *upstream* of the flume should be avoided.
- Water approaching the flume should be free of turbulence and waves.
- Because the flume restricts flow, the *banks* upstream should be high enough to contain increased depth.
- Because flow should be well distributed in the channel, irregular channel shapes should be avoided.
- The flume throat should be submerged as little as possible below the *downstream* backwater.

Cf. *Flume, cutthroat; Flume, Parshall.*

**Flume, cutthroat** A simple *flume* designed with a throat without parallel walls, which is used under both free-flow and submerged conditions to measure *drilling fluid* return. A cutthroat flume is accurate yet economical, and its flat floor design allows it to also be placed on a channel bed or inside a concrete-lined channel. **Figure F-22** shows a cutthroat flume.

F

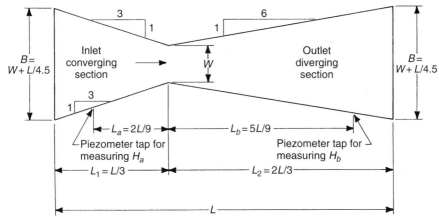

**Figure F-22** A typical cutthroat flume design. (Driscoll, 1986. Reprinted by permission of Johnson Screens/a Weatherford Company.)

Cf. *Flume, Parshall.*

**Flume, Parshall** A *flume* with a drop in the floor that causes *supercritical flow* through its throat, which is used in measuring well *discharge*. This type of flume was originally designed by R. L. Parshall of the US Soil Conservation Service (SCS). Under supercritical flow conditions, inertial forces are much greater than gravitational forces, resulting in rapid flow through the throat and a *hydraulic jump* that occurs well *downstream* of the Parshall flume. Under free-flow conditions, the hydraulic jump is not covered by the *backwater* downstream, simplifying flow rate calculations. When the water surface downstream is too high, however, it drowns the hydraulic jump in what is called submerged flow, making flow rate calculations more complicated and subject to more potential inaccuracy. **Figure F-23** shows a Parshall flume. Cf. *Froude number; Flume; Cutthroat.*

**Flush** A sudden increase in *stream* volume or *flow*.

**Fluvial** Of or pertaining to *rivers* and *streams*, existing, growing, or living in or near a river or stream. Cf. *Fluviatile; Drainage basin.*

**Fluvial lake** A standing body of water with detectable *flow*, or a water body that connects two other bodies of water with a distinguishable elevation difference, thereby creating flow from the higher to the lower. Cf. *Fluviatile lake.*

**Fluviatile** Resulting from or pertaining to *river* or *stream* action. Cf. *Drainage basin.*

**Fluviatile lake** A standing water body resulting from *river* or *stream* action. Cf. *Fluvial lake; Lateral lake.*

**Fluviation** The activity of a *stream* or *river*.

**Fluvioglacial** See *glaciofluvial.*

**Fluviometer/fluviograph** A measuring device capable of recording the rise and fall of water surface elevation in a *river* or *stream*. Cf. *Gauge; Float measurement of discharge; Flow meter; Chezy's formula.*

**Flux** A *transport* mechanism or movement of a substance through a bounding surface, usually of unit area. Refer to **Appendix A** for unit conversions. The transport of a constituent is measured as flux in units of quantity per area per time (in units of mass for mass flux, or volume for volume flux). Cf. *Darcy flux; Specific discharge; Advection; Diffusion; Boundary, flux.*

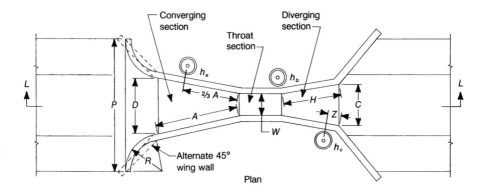

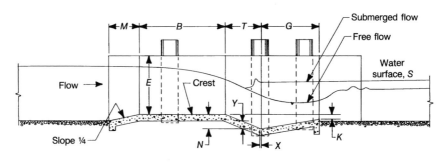

**Figure F-23** A Parshall flume. (Driscoll, 1986. Reprinted by permission of Johnson Screens/a Weatherford Company.)

**Flux boundary** See *boundary, specified-flow; boundary, flux; Darcy's flux.*

**Foam** See *drilling fluid/drilling mud, gel.*

**Forced convection** Also called an advective heat transfer regime. A type of convective heat transfer occurring only in moving *groundwater* in which fluid inflows and outflows are present. In the analysis of forced convection in the *transport* of heat by a natural *groundwater flow* system, density gradients and the resultant buoyancy force is subsidiary to the topographic driving force but is usually taken into account in numerical simulations. Fluid motion is due to the *hydraulic forces* acting on the boundaries of the system. Cf. *Free convection; Density current; Boundary conditions.*

**Force potential ($\phi$)** The total mechanical energy of the water at a specific point in an *aquifer*, consisting of the kinetic energy, elevation energy, and *pressure* at that point. Force potential is the driving impetus of *groundwater flow* and is equal to the *hydraulic head* times the acceleration that is due to gravity. The equation expressing these relationships under various assumptions of *steady-state flow and transient flow* is:

$$\phi = gz + \frac{P}{\rho} = gz + \frac{\rho g h_\text{p}}{\rho} = g(z + h_\text{p})$$

where:
$g$ = gravitational constant [L·T$^{-2}$]
$z$ = elevation of the center of gravity of the fluid above a reference elevation [L]
$P$ = pressure [M·L$^{-1}$·T$^{2}$]
$\rho$ = density [M·L$^{-3}$]
$h_\text{p}$ = height of the water column [L]

Because the elevation, $z$, plus the height of the water column, $h_p$, equals the hydraulic head, $h$, the force potential calculation can be simplified to:

$$\phi = gh$$

The total energy per unit mass of water is described by the *Bernoulli equation*. Cf. *Darcy's law; Hydraulic gradient; Equations of groundwater flow.*

**Forchheimer flow equation** The first non-linear polynomial expression for high flow rates, described by Forchheimer as:

$$i = aq + bq^2$$

where:

$i = $ *hydraulic gradient* $[L \cdot L^{-1}]$
$a = $ *laminar flow* coefficient $[T \cdot L^{-1}]$
$b = $ turbulence factor $[T^2 \cdot L^{-2}]$
$q = $ *discharge* $[L^3 \cdot T^{-1}]$

The parameters $a$ and $b$ are dependent on the properties of the medium and fluid, boundary configurations of the flow path, and on the *Reynolds number*, $R$. Based on experimentation, $a$ and $b$ are considered constants for Reynolds numbers between 1 and 50. For low flow rates, $aq >> bq^2$ and approaches linear flow, but at high flow, $bq^2 >> aq$ and flow becomes non-linear. Cf. *Froude number; Equations of groundwater flow; Laminar flow.*

**Ford** A shallow or narrow body of water that can be easily crossed by walking, riding, or using a mechanical vehicle. Cf. *Drift.*

**Form ratio (r)** The relationship of the mean depth of a watercourse or water body to its width as measured from *bank* to bank, indicating whether it is deep or shallow. If the watercourse or body is broad and shallow, the *hydraulic radius* can be substituted for the mean depth in the form ratio.

**Form roughness** See *bed roughness.*

**Formation packer** A device installed between the *borehole* wall and the *well casing* above a *screened interval* in a naturally developed well to prevent overlying materials, particularly fine-grained formation material, from sloughing down into the water-producing formation and thereby reducing the productivity of the well. Cf. *Well development; Well installation.*

**Formation sampling** The collection of materials from the *borehole*, typically during drilling to determine the nature of soils below ground surface.

**Formation stabilizer** Also called stabilizer sand. Sand, gravel, or other bolstering material placed between the *well casing* and the *borehole* wall to support the borehole and prevent caving of overlying fine-grained material into the screened portion of the well, either temporarily or for the long term. Formation stabilizers are also used to prevent premature caving during *well development* and to augment or maintain the *hydraulic conductivity* of the natural formation material. **Table F-3** details criteria for the selection of formation stabilizers.

**Table F-3** The selection of formation stabilizers for unconsolidated and semi-consolidated sediments. (Driscoll, 1986. Reprinted by permission of Johnson Screens/a Weatherford Company.)

| | | |
|---|---|---|
| **Type of aquifer** | Unconsolidated aquifers: alluvial and glaciofluvial sands and gravels, and beach deposits. | Semi-consolidated aquifers: dirty sandstones and siltstones, and sandy formations containing shells. |
| **Purpose** | Provide temporary support for the borehole walls next to the screen. | Permanently hold back the formation without providing any mechanical retention of small particles. |
| **Characteristics of stabilizer** | Grain-size distribution should be equal to or slightly larger than the original formation. | Grain-size distribution is usually greater than 12–13 times the 70%-retained size. |
| **Development** | Approximately 50 to 60% of the stabilizer is removed during natural development of the formation. | None of the stabilizer passes through the screen during development. |
| **Result** | The part of the formation stabilizer that remains next to the screen has a hydraulic conductivity similar to the natural formation so that flow is unimpeded. The stabilizer also plays a major role in preventing the migration of fine particles into the screen. | The formation cannot slump against the screen even if it becomes weakened over time. The porosity and hydraulic conductivity of the stabilizer are high, reducing drawdown in the immediate vicinity of the screen. |

**Formation water** Also called *native water*, an archaic term. The naturally occurring water residing within a formation. The term "formation water" is commonly used in drilling practices to differentiate this water from the fluid introduced during drilling. Cf. *Connate water*.

**Fossil ice** See *ground ice*.

**Fossil water** Water that is trapped in sediment pore spaces and buried with the rock during its deposition, or water trapped in undersea volcanic rock. Fossil water is generally not included in the *hydrologic cycle* or associated with it. Cf. *Connate water; Magmatic water; Juvenile water*.

**Fossilized brine** See *connate water*.

**Fouling** The accumulation of foreign material in a *filter* medium such as a *filtration* bed, *well screen, ion* exchanger, *filter pack*, or *well formation*, thereby retarding or impeding *flow* through the filter medium by clogging or coating surfaces.

**Fracture permeability** See *secondary permeability*.

**Fracture porosity** The volume of fracture-formed voids in a medium divided by the total volume of the medium. Fracture *porosity* may be naturally occurring, or the formation may be deliberately fractured with *explosives* or air pressure to increase formation *yield*. Cf. *Formation yield*.

**Fracture spring** See *spring, fracture*.

**Fracturing** See *hydraulic fracturing*.

**Frazil ice** One of the first types of ice to form in turbulent water as the temperature drops below freezing, consisting of small, suspended crystals. When the turbulence is not sufficient to overcome the buoyancy of the frazil ice, the crystals float to the surface to form sheet ice. Under some conditions, frazil ice may collect on the rocks of the *stream bed*, forming a "bottom ice" and thereby causing a small increase in *flow* stage. Cf. *Turbulent flow*.

**Free convection** A type of heat transfer or movement created by a density gradient such as a variation in water temperature, *salinity*, or suspended or dissolved constituents. In free convection, heat is transported through the system. Heat may enter the system through volcanic action and leave the system through thermal springs. The motion of the fluid is due to all forces acting on the system, including temperature gradients causing density variations, and is controlled by buoyancy effects. Cf. *Forced convection; Density current; Spring, thermal*.

**Free energy** Also known as Gibbs free energy where the thermodynamic function:

$$G = H - TS$$

where:
H = enthalpy
T = absolute temperature
S = entropy

It is the driving energy of a chemical reactions that occur between water and the solids and gases it contacts. Cf. *Osmotic pressure*.

**Free-flow carbonate aquifer** A carbonate ($CO_3$) formation in which the initial *groundwater* flow path is directed along *solution channels* controlled by bedding planes and fractures, and in which continued *flow* causes *dissolution* of the carbonate rock, thereby increasing the size and capacity of the *flow channels*. *Recharge* to a free-flow carbonate aquifer may occur through diffuse *percolation* of surface waters. *Sink holes* and solution channels to the surface may also capture surface *streams* and provide a more concentrated recharge.

**Free groundwater** See *aquifer, unconfined; water table aquifer*.

**Free moisture** See *free water*.

**Free surface** The uppermost fluid layer at which the *fluid pressure* is in equilibrium with the atmospheric pressure. The *water table* is the upper boundary of a *flow net*. A free surface must satisfy the *boundary conditions* of a water table ($h = z$) and of a *flow line*, intersecting *equipotential lines* at right angles. Cf. *Water level; Open channel flow*.

**Free water** Also called free moisture, or *gravitational water*. The portion of the *soil moisture* that is above the *field capacity* and is able to move through soil in response to gravity. Cf. *Infiltration water; Phreatic water*.

**Free water elevation/free water level/free water surface** Also called the *water table*, groundwater surface, and *groundwater elevation*. The elevation in relation to *sea level* at which the water pressure is zero with respect to atmospheric pressure, or the surface of any open water body or the water table at atmospheric pressure. Water pressure below the free water elevation or surface is greater than atmospheric pressure.

**Free water evaporation** *Evaporation* that occurs at the *free surface* at a temperature below the boiling point.

**Freeze-out lake** A *lake*, typically shallow, that is prone to becoming frozen for an extended period of time.

**Frequency analysis** The collection of data, such as extreme *rainfall* events and *flooding* to develop the best estimate of risk of the likelihood or probability of such events. The information is utilized in flood plain management and designs of flood control practices to reduce the loss of life and property damage costs. Cf. *Flood frequency analysis; Flow frequency; Flood duration curve; Gringorten formula; Recurrence interval*.

**Frequency of occurrence** See *recurrance interval*.

**Freshwater** *Surface water* or *groundwater* that is not composed of or influenced by *saltwater* or salt-bearing rocks, or water containing less than 1000 mg $L^{-1}$ of dissolved constituents. Freshwater may be considered *non-potable* or undesirable if it contains more than 500 mg $L^{-1}$ *total dissolved solids* (TDS) although many water supply systems, especially in the North American prairies, provide water above this criteria. Home water softening systems are usually employed where TDS is due to *hardness* to treat the water. According to the Venice classification system based on percent chlorine, freshwater contains 0.03% or less of chloride (Cl⁻), and the categories of *brackish water* contain successively more. Freshwater accounts for only 3.2% of the water on the earth. Cf. *Brine; Salinity; Chebotarev succession; Softening of water.*

**Freshwater estuary** A *bay* into which a *river* discharges a sufficient volume of *freshwater* to eventually exclude the *saltwater* that was present. Cf. *Estuary.*

**Freshwater lens** A lenticular zone of fresh *groundwater* overlying a saline water body, as is found below oceanic islands and along coastlines. Cf. *Ghyben–Herzberg lens; Saltwater intrusion; Salinity; Thalassic series.*

**Freundlich isotherm** A description of *sorption* at a specified temperature with partitioning of organic chemicals between the dissolved phase and the sorbed phase. An equilibrium representation of the ratio of a chemical adsorbed on soil particles to the amount of soil present is usually specified at a fixed temperature (*isotherm*). The generalized equation describing this relation is:

$$\frac{X}{M} = kc^{1/n}$$

**F**

where:
- $X$ = amount of chemical adsorbed [M]
- $M$ = amount of soil adsorbent [M]
- $c$ = concentration of the chemical in the solution in equilibrium with the soil [M·L⁻³]
- $k, n$ = empirically determined constants

A more useful expression is:

$$C^* = K_d C^b$$

where:
- $C^*$ = amount of the given *ion* adsorbed per unit weight of soil [M·M⁻¹]
- $K_d$ = slope of the isotherm, called the *distribution coefficient* [L³·M⁻¹]
- $C$ = *equilibrium concentration* of the ion in contact with the soil [M·L⁻³]

The amount of the ion adsorbed is proportional to the activity of the ion. The Freundlich isotherm is one of three *adsorption isotherms* whose relationships and associated equations are shown in **Figure F-24**. The adsorption isotherms are

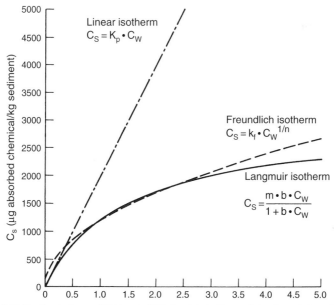

**Figure F-24** Sorption isotherms. (Maidment, 1993. Reprinted with permission from McGraw-Hill, Inc.)

133

extensively used in the study of mass *transport* of *solutes* in flowing *groundwater*. Both the Freundlich and the *Langmuir isotherm* equations are used when concentrations of organic chemicals are greater than $10^{-3}$ molar, as would occur in landfill *leachate* and waste processes.

Cf. *Activated carbon*.

**Friction head** Also called friction loss. The energy lost due to friction per unit weight of fluid. Refer to **Appendix A** for property conversions. Cf. *Dynamic head*.

**Friction loss** See *friction head*.

**Fringe water** See *capillary water; capillary fringe*.

**Froude number** A dimensionless number establishing the class or type of *flow*, according to energy criteria, in an open channel or watercourse. At a specified *discharge*, energy is a minimum at one depth, the critical depth $y_c$, as the energy of flow is a function of its depth and velocity. The Froude number, $Fr$, characterizing the flow is determined by the following equation:

$$Fr = \frac{v}{\sqrt{gv}}$$

where:

$v$ = velocity $[L \cdot T^{-1}]$
$g$ = gravitational constant $[L \cdot T^{-2}]$
$y$ = depth in open channel $[L]$

As shown in **Figure F-25**, when $Fr < 1$, flow is slow, gentle, or subcritical; when $Fr = 1$, depth $y$ = critical depth $y_c$ and flow is critical; and when $Fr > 1$, flow is fast or shooting and supercritical.

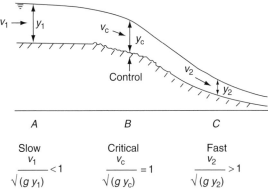

**Figure F-25** The character of flow as determined by the dimensionless Froude number. (Shaw, 1983. Reprinted with permission from Van Nostrand Reinhold Co. Ltd.)

Cf. *Reynold's number; Supercritical flow; Flume, Parshall; Open channel flow*.

**Fugacity** A measure of the tendency for a gas to expand or escape. Fugacity is the pressure at a given temperature where the properties of a non-ideal gas satisfy the equation for an ideal gas. Fugacity is defined as:

$$f_i = \gamma_i P_i$$

where:

$\gamma_i$ = the fugacity coefficient [unitless]
$P_i$ = the partial pressure for the component $i$ of the gas $[M \cdot L^{-1} \cdot T^2]$.

For an ideal gas, $\gamma_i = 1$.

**Fully penetrating well** A *water well* that penetrates, and whose *well screen* is within and open to the entire saturated thickness of an *aquifer* unit, allowing for the maximum well *yield*. A fully penetrating well that is screened across the entire thickness of a confined aquifer exhibits planar two-dimensional *groundwater flow* patterns. When viewed in a vertical cross section, *flow lines* toward the well are horizontal and *equipotential lines* are perpendicular. Cf. *Zone of saturation; Well yield; Sustained yield.*

**Funicular water** The *capillary water* in rock or within soil particles in the *unsaturated zone*, in which the *interstices* are completely filled with water bounded by a single closed meniscus. Cf. *Adhesive water; Pellicular water.*

**F**

**GAC** See *activated carbon.*

**Gage datum/gage height/gage pressure** See *gauge datum; gauge height; gauge pressure.*

**Gaining stream** An *effluent stream* receiving *groundwater discharge* from a *groundwater reservoir*, in which the stream *baseflow* increases or gains in the *downstream* direction as more water enters the *stream*. **Figure G-1** is a cross section of a *reach* in a gaining stream. The *hydraulic gradient* of an *aquifer* is toward a gaining stream, and the groundwater's rate of movement depends on the *hydraulic conductivity* of the stratum through which the water moves and on the depth of the water. A gaining stream during a *baseflow recession* may become a *losing stream*, or an *influent stream*, during a *flood*. This results from a reversal of the hydraulic gradient when the elevation of water in the channel is greater than the adjacent *water table* elevation. A gaining stream during a flood stage is shown in **Figure G-2**.

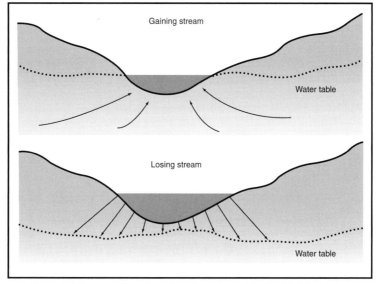

**Figure G-1** A cross-sectional view of a gaining stream. (Montana State University.)

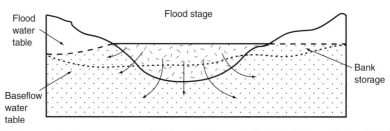

**Figure G-2** A former gaining stream in cross-sectional view at flood stage. (After Fetter 1994 and Montana State University.)

Cf. *Effluent stream; Influent stream.*

**Gallery** See *infiltration gallery.*

**Galvanic corrosion** A reaction between two metals placed in an electrolytic medium, in which the more active metal, as determined by their places in the *galvanic series of metals and alloys*, is anodic to the other and gives up metal. When two dissimilar alloys are placed in contact with an electrolytic medium such as water, as in the use of a low-carbon steel *well casing* with a stainless steel *well screen*, an electrical potential develops because one alloy dissolves, or corrodes, in the medium more readily than the other. Galvanic corrosion can cause premature failure of well casings. In *well installation* practices, metals with a significant voltage difference, or metals far apart in the galvanic series, are seldom joined.

If two dissimilar metals must be used together in a corrosive environment, they should be separated by non-conductive gaskets and insulated bolts.

**Galvanic series of metals and alloys** The relative potential for galvanic action between different metals can be estimated by their locations in the galvanic series as shown in **Table G-1**. The closer two metals are in the series, the less voltage they would produce if united in a galvanic cell. When two dissimilar metals are placed together in a conductive fluid, the metal closer to the bottom of the galvanic series (less noble) acts as the anode and will be corroded, while the metal nearer to the top of the series acts as the cathode and usually remains intact and non-corroded. In the case of a well having a *well screen* made of low-carbon steel and stainless steel, however, the electrochemical corrosion product from the anode is deposited on the cathode, blocking the screen openings and reducing *well yield*.

**Table G-1** The galvanic series of important metals and alloys. (Driscoll, 1986. Reprinted by permission of Johnson Screens/a Weatherford Company.)

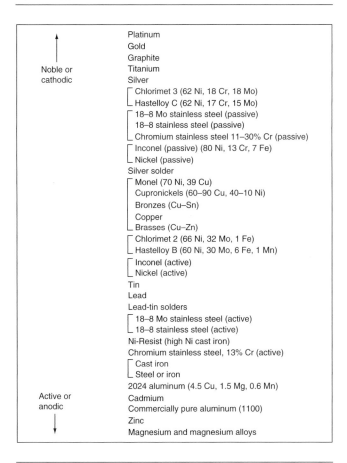

**Gamma attenuation** The weakening and back-scattering of gamma rays as they pass through a material such as soil, which is the basis for a radiological method of measuring *soil water* content. Naturally occurring radioisotopes in soils emit *gamma radiation*. Portable gamma attenuation equipment measures a 1-cm wide band, and the sensing device uses the difference between the natural terrestrial gamma radiation and that attenuated by soil water. Cf. *Soil water characteristic curve.*

**Gamma–gamma log** Also called a density log. A record of the *gamma radiation* received at a detector after gamma rays have been projected from a probe into the area surrounding a *borehole*. The gamma–gamma probe contains a source of gamma radiation, typically cesium-137, and one or more detectors. The gamma radiation emitted from the probe is attenuated as it passes through the borehole and surrounding rocks, and this *gamma attenuation* is proportional to the bulk density of the medium. If the grain density and *fluid density* are known, the *porosity* of the formation can then be calculated with the equation:

$$\text{porosity} = \frac{(\text{grain density}/\text{bulk density})}{(\text{grain density}/\text{fluid density})}$$

The bulk density measurement is obtained from the gamma–gamma log, the fluid density is $1\,\mathrm{g\,cm^{-3}}$ for most saturated fresh *groundwater* applications, and the grain or mineral density is obtained from a mineralogy text. The comparison of gamma–gamma logs before and after well completion can be used to locate cavities or unfilled annular space behind the *well casing* or to locate the top of the *cement* through the casing. Cf. *Geophysical exploration methods.*

**Gamma logging** Also called gamma-ray logging or natural-gamma logging. A radioisotope-based logging method used to identify lithology and stratigraphic correlations in *groundwater* investigations. The gamma logging technique either detects or creates unstable isotopes in the *borehole* area, and the total *gamma radiation* detected within that area is recorded. In water-bearing rocks, the naturally occurring radioactive isotopes detected by gamma logging are potassium-40 and the daughter products of the uranium- and thorium-decay series. Relative natural gamma intensities for different rock classifications are shown in **Figure G-3**. Knowledge of the local geology is required to identify numerous deviations from the typical response of the natural-gamma log when compared to the classification system presented. **Figure G-4** is a typical gamma log compared to a sequence of sedimentary rocks.

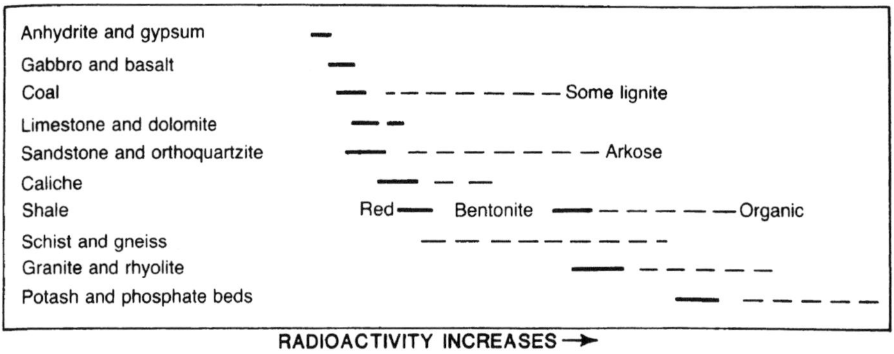

**Figure G-3** The relative radioactivity of common rock classifications. (Keys, 1989. Reprinted with permission from National Water Well Association.)

Cf. *Gamma-spectrometry logging; Gamma–gamma log; Geophysical exploration methods.*

**Gamma radiation** The electromagnetic radiation emitted from an atomic nucleus, about $10^{-10}$–$10^{-14}$ m in wavelength, accompanied by emission of alpha and beta particles. Gamma rays are similar to the X-rays emitted by radioactive substances but are shorter in wavelength.

**Gamma-ray log** The curve of the intensity of broad-spectrum, undifferentiated, natural *gamma radiation* emitted from the rocks in a *borehole*. A gamma-ray log can be used to distinguish and correlate shale from other types of rocks. Shale is typically rich in naturally radioactive elements. Cf. *Geophysical exploration methods; Gamma logging.*

**Gamma-ray logging** See *gamma logging.*

**Gamma-spectrometry logging** A method of identifying and quantifying the radioisotopes recorded by a gamma log. Gamma-spectrometry logging in a *borehole* provides information on lithology, as *gamma logging* does, and is also used to distinguish natural versus the artificial or induced radioisotopes migrating in *groundwater*, thereby making the interpretation of the gamma log a more accurate account of the borehole material. Cf. *Geophysical exploration methods.*

**Ganat** See *kanat/ganat/qanat.*

**Gas pycnometer** A laboratory device used for the measurement of *porosity.* Pycnometers operate using *Archimedes' principle* of fluid (gas) displacement and the technique of gas expansion to determine the true volume of a solid (and hence the volume of the voids). A gas, such as helium or nitrogen, is used to displace fluids and gases in *pore spaces.* Helium is considered the ideal gas to use for this purpose because of its small atomic dimension and hence it has the ability to penetrate into the pores approaching an Angstrom ($10^{-10}$ m). If this level of accuracy is not required, gases such as nitrogen may be substituted.

**Gas solution** See *Henry's law.*

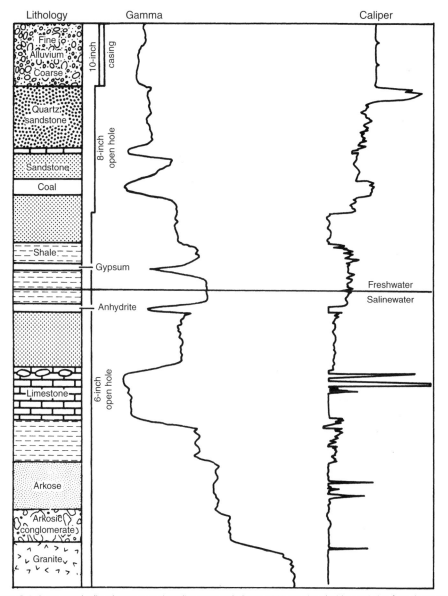

**Figure G-4** Gamma and caliper log response in sedimentary rock. (Keys, 1989. Reprinted with permission from the National Water Well Association.)

**Gas zone** Also called the gaseous area. When a *non-aqueous-phase liquid* (NAPL) enters the *unsaturated zone* and progresses downward, it leaves behind some residual liquid trapped in the *pore spaces* of the unsaturated zone by surface tension. A portion of the liquid volatilizes and forms a *vapor* extending beyond the initial NAPL. The NAPL extends downward by gravity movement, eventually reaching the saturated zone and spreads laterally along the *capillary fringe* if it is less dense than water (i.e. a *light non-aqueous-phase liquid*, LNAPL). The LNAPL residing on the top of the *water table* will continue to volatilize and extend the gas zone associated with the NAPL. **Figure G-5** illustrates NAPL movement and the resulting gaseous area or gas zone.

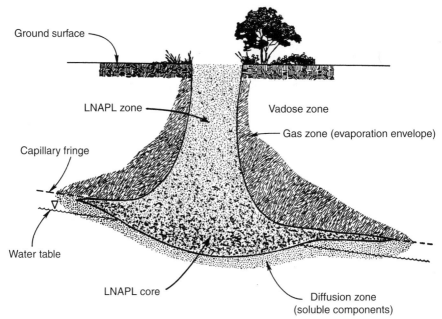

Ground surface

LNAPL zone

Vadose zone

Gas zone (evaporation envelope)

Capillary fringe

Water table

LNAPL core

Diffusion zone
(soluble components)

**Figure G-5** Schematic of LNAPL infiltration. (Maidment, 1993. Reprinted with permission from McGraw-Hill, Inc.)

**Gaseous area** See *Gas zone*.

**Gauge** A device used in hydraulic applications for measuring water surface elevations, *flow velocity, water pressure, precipitation*, etc. The measurement of *discharge* from a *stream, river*, or other channel requires locating the gauge on a straight section or *reach* of the channel. At this location, the *streamlines* of *flow* are nearly parallel, and water velocities are within the range that can be measured with a current meter. Cf. *Float measurement of discharge*.

**Gauge datum** Also gage datum. The elevation, usually *sea level* or an established arbitrary measuring point, to which the *gauge height* of a *stream* or other surface water body is referenced.

**Gauge height** Also gage height. The surface elevation at the gauge site of a surface water body, e.g. *lake* or *stream*, referred to *sea level* or an arbitrarily established *gauge datum* or elevation chosen for convenience. In stream-gauging operations, the height above the established datum can be called either the gauge height or *stage*, but the term "gauge height" is more appropriate. The term "stage" more accurately defines water levels in reference to the *bank* height of the river or stream.

**Gauge pressure** Also gage pressure. In *hydrology*, a *pressure* measured on a device that has been zeroed to atmospheric pressure. Equations in groundwater hydrology commonly set the atmospheric pressure, $p_0$, equal to zero, and work in gauge pressure.

**Gauging network** An organized system of *gauging stations* or measuring devices instituted for the collection of hydrologic data in order to evaluate *water resources*. The design of a gauging network depends on the information sought and its intended use, such as for planning, management, forecasting, or research. Cf. *Antecedent moisture content*.

**Gauging station** A selected area located on a surface water body, e.g. *stream, river, lake*, or *reservoir*, where observers or recording devices measure *gauge height* or similar *flow* parameters. Water surface elevations are recorded at gauging stations at designated time increments, typically every 15 min, and the information is stored on a data logger. The elevation datum is set by referencing at least three benchmarks located on stable ground. The most important aspect of choosing the location of the gauging station is that the hydraulic controls *downstream* are stable and reflect the changes in *discharge* and should be accessible at all times. The water levels at the gauging station are controlled by the hydraulic controls and determine the *stage-discharge relation*. A sequence of gauging stations constitutes a *gauging network*. Cf. *Basic data stations; Float measurement of discharge*.

**Gauze number** Slot openings in *well screens* are designated by numbers that correspond to the width of the openings in thousandths of an inch. A *slot size* designated as No. 10 is equal to an opening of 0.010 in. If the well screen is of small diameter and is covered with a wire mesh, the gauze number designates the number of openings in the mesh per inch. **Figure G-6** shows the comparison of the slot number and gauze number for common sizes of screen openings. *Sand pumping* problems may occur if too many slots are significantly different from the designated *slot size*.

**General circulation model (GCM)** See *hydrologic cycle*.

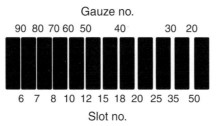

**Figure G-6** The gauze number for common sizes of screen openings. (Driscoll, 1986. Reprinted by permission of Johnson Screens/a Weatherford Company.)

**Gel** A jelly-like material that results from the coagulation of a *colloidal dispersion*, used in *drilling fluids*. A gel is a colloidal suspension in which a *shear stress* below a finite value fails to produce permanent deformation. A gel is more solid than a *sol*. The gel state occurs when the dispersed colloidal particles have a great affinity for the base fluid. The *gel strength* is of critical importance for the success of a drilling fluid. Cf. *Flocculation; Lost circulation; Insol; Guar gum.*

**Gel strength** A measure of a *drilling fluid's* ability to support suspended particles when the fluid is at rest. If the drilling fluid is made with clay particles, the alignment of the clay platelets with positive and negative charges joined together gives the fluid a plastic form and its gel strength properties. When sufficient agitation is applied to the drilling fluid by the fluid pump, however, the gel breaks down. Cf. *Flocculation.*

**General base level** See *ultimate base level.*

**General circulation model (GCM)** See *hydrologic cycle.*

**General head boundary condition** A *boundary condition* that can be applied in the computer model MODFLOW. General head boundary conditions are typically used to simulate *lakes* by assigning a head to a select set of nodes. If the *water table* rises above the specified head, water *flows* out of the *aquifer*. Conversely, if the water table is lower than the specified head, water *recharges* the aquifer. The flow rate in both cases is proportional to the head difference and the conductance. Cf. *Finite-difference technique; Finite-element technique.*

**Geochemical processes** The chemical changes and reactions that *groundwater* undergoes as it moves through the subsurface. These geochemical processes involve the chemical characteristics of *recharge* water, the solubilities of minerals encountered, the order in which chemicals are encountered and solubilized, and the actual reaction rates relative to groundwater movement. The ionic species $Na^+$, $Ca^{2+}$, $Mg^{2+}$, $HCO_3^-$, $SO_4^{2-}$, and $Cl^-$ typically constitute approximately 90% of the *total dissolved solids* (TDS). Cf. *Hydrochemical facies; Piper trilinear diagram; Chebotorev series; Electrochemical sequence.*

**Geochemical speciation** For solutions at chemical equilibrium, the distribution of elements among all the possible and feasible anions, cations, and complexes that would make up the *hydrochemical facies* can be calculated or geochemically speciated. The cations and anions in natural waters are generally at or near a state of local equilibrium (i.e. equilibrium continually shifts as different lithologies are encountered) once the dissolution of minerals from rocks has occurred. As *groundwater* moves through a geologic medium, it changes chemically as it comes in contact with different minerals and proceeds toward chemical equilibrium. Typically, the species Na, Ca, Mg, $HCO_3$, $SO_4$, and Cl constitute more than 90% of the *total dissolved solids* (TDS) in a groundwater sample. Cf. *Groundwater constituents.*

**Geodetic datum/sea level** See *mean sea level.*

**Geohydrochemical map** A specialized map that indicates both the geology of an area and the chemical characteristics of the *groundwater*. The geohydrochemical map is used to determine how the environment through which the groundwater migrates will affect the *water quality*. In conjunction with analysis of water samples to determine the chemical constituents, a geohydrochemical map can be used to trace flow paths and to classify the waters, e.g. bicarbonate water, sulfate water, or chloride water. Cf. *Geochemical speciation; Hydrochemical facies; Piper trilinear diagram.*

**Geohydrologic unit** The earth and water features, i.e. an *aquifer, confining bed*, or both, that as a unit from a distinct hydraulic system. Cf. *Hydrostratigraphic unit.*

**Geohydrology** In the engineering field, the study of the hydrologic or *flow* characteristics of subsurface waters. The term "geohydrology" is often used interchangeably with *hydrogeology* as the study of the interrelationships of the affects of geologic materials and processes with water.

**Geometric mean for hydraulic conductivity** A method to determine a representative value for a set of *hydraulic conductivity* data. Values for hydraulic conductivity for a given study area or aquifer can vary over several orders of magnitude. Using a simple arithmetic mean to obtain a representative value for hydraulic conductivity would scew the representative value to the high range of values. Determining a geometric mean provides a more representative value for the set of data. The geometric mean of a set of data $\{a_i\}_{i=1}^n$ is defined as:

$$G(a_1, \ldots, a_n) \equiv \left( \prod_{i=1}^n a_i \right)^{1/n}$$

Such that $G(a_1 a_2) = \sqrt{a_1 a_2}$, $G(a_1, a_2, a_3) = (a_1 a_2 a_3)^{1/3}$, and so on. Cf. *Harmonic mean of hydraulic conductivity.*

141

**Geologically bound water** Water that is out of circulation and is thought to have been in storage since the formation of the rock. The quantity of geologically bound water is unknown and inaccessible and not typically considered in *watershed management*. Cf. *Connate water; Formation water; Juvenile water; Native water*.

**Geophysical exploration methods** Also called logging. Investigation of subsurface materials, usually in search of resources such as hydrocarbons, mineral deposits, and water supplies, by geophysical techniques such as electric, gravity, magnetic, seismic, and thermal. *Groundwater* investigation generally occurs within the upper few hundred meters of the earth's surface, and the use of geophysical exploration methods has become increasingly popular. For example, boundaries between waters of differing chemical composition have been delineated using *surface resistivity* or electromagnetic conductivity methods. **Figure G-7** shows a comparison of the information obtained from a *borehole*

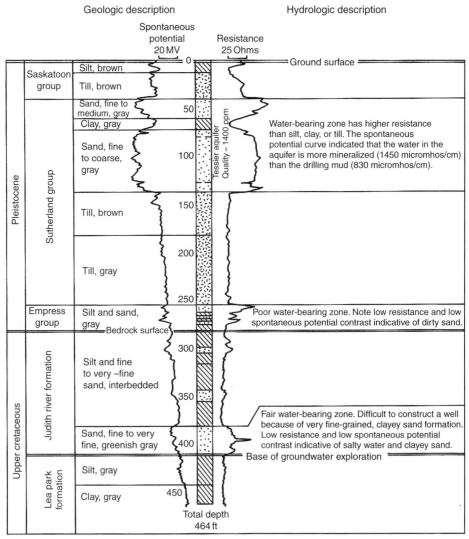

**Figure G-7** The hydrologic description based on information from the geologic log and electric logs. (Freeze and Cherry, 1979. Reprinted with permission from Prentice-Hall, Inc.)

well log and a geophysical investigation utilizing *spontaneous potential* and resistivity logs. For information on specific geophysical methods, see entries under each method. Cf. *Caliper log; Resistivity; Spontaneous potential logging; Electric logging; Electromagnetic resistivity; Gravity log; Ground-penetrating radar; Gamma–gamma log; Electromagnetic surveys.*

**Geopressured aquifer** An *aquifer* in which the *fluid pressure* exceeds the normal *hydrostatic pressure* of $9.79 \text{ kPa m}^{-1}$ (0.433 lb per square inch per foot) of depth.

**Geothermal energy** The heat generated from hot water within a *geothermal reservoir*. For this energy to be of practical use, the geothermal reservoir must have an adequate volume and *permeability* to ensure sustained delivery of fluid to wells, and temperatures must be in excess of 180°C. Hydrothermal convection systems, in which heat is transported by circulating fluids, lead to the existence of high-temperature fluids at shallow depth, which are ideal for pumping to the surface for use as geothermal energy.

**Geothermal gradient** The rate of increase in temperature in the earth with depth. The average geothermal gradient at the surface of the earth is approximately $25°C \text{ km}^{-1}$ of depth (or 2.5°C 100 m$^{-1}$). Beyond a certain point, the rate of temperature increase declines; otherwise, widespread mantle melting would occur. Geothermal gradient also changes with location, depending on regional heat flow, convective heat transport, and the *thermal conductivity* of the rocks in the area. The regional geothermal gradient directly affects the temperature of *groundwater*, and therefore, its chemical nature. The temperature of near-surface groundwater is typically constant at or near the mean annual air temperature. Generally, below depths of 30 m, geothermal heating may influence groundwater temperatures. As increasing water temperature generally increases the solubility of most minerals, groundwater at depth tends to be more mineralized. Cf. *Temperature gradient; Hydrothermal water; Temperature logging; Depth of circulation.*

**Geothermal reservoir** Any regionally localized setting where naturally occurring portions of the earth's internal heat flow are transported close enough to the earth's surface by circulating steam or hot water to be readily harnessed for use. An example is the Geyer's region in north western United States which is utilized to supply energy to the surrounding area. Cf. *Geothermal energy.*

**Geyser** A specialized hot spring resulting from the creation of steam by contact of *groundwater* with superheated rocks under conditions preventing free circulation, which generally produces an area of recurring eruptions of jets of hot water and steam above the ground surface. Cf. *Spring, thermal.*

**Geyser basin** A low-lying area containing *geysers* and usually numerous *springs*, steaming fissures, and vents, which are typically fed by the same *groundwater flow* in the area of superheated subsurface rocks.

**Geyser pipe** Also called a geyser shaft. A narrow outlet extending from the upper expression of a *geyser* to the water source or *geyser pool.*

**Geyser pool** A relatively shallow basin of heated water usually contained in a crater or mound of sinter at the top of a *geyser pipe.*

**Geyser shaft** See *geyser pipe.*

**Ghyben–Herzberg equation** See *Ghyben–Herzberg principle/Ghyben–Herzberg ratio.*

**Ghyben–Herzberg lens** The lens of *freshwater* that floats above saltwater in *hydrostatic equilibrium* as a result of the density difference between the two. The average specific density of seawater is $1.025 \text{ g cm}^{-3}$ compared to $1.0 \text{ g cm}^{-3}$ for freshwater. Cf. *Ghyben–Herzberg principle; Salinity.*

**Ghyben–Herzberg principle/Ghyben–Herzberg ratio** A formula presented in the late nineteenth century by W. Baydon-Ghyben and A. Herzberg to characterize the *freshwater/saltwater* interface in a coastal *aquifer.* The mathematical equation of the Ghyben–Herzberg principle is:

$$z_{(x,y)} = \frac{\rho_w}{\rho_s - \rho_w} h(x, y)$$

where:
$z_{(x, y)}$ = depth to the freshwater/saltwater interface below sea level at locations $x$, $y$ [L]
$\rho_w$ = density of freshwater [M·L$^{-3}$]
$\rho_s$ = density of saltwater [M·L$^{-3}$]
$h_{(x, y)}$ = elevation of the *water table* above sea level at locations $x$, $y$ [L]

The Ghyben–Herzberg principle, illustrated in **Figure G-8**, is valid only if the freshwater, interface, and saltwater are in equilibrium. The principle is also valuable in determining the location of the expected *freshwater lens* underlying an oceanic island. As shown in **Figure G-9**, at a depth of $z = 0$, the pressure at point A must equal the pressure at point B; therefore, the equation predicts that, if the density of seawater is 1025 kg m$^{-3}$ and freshwater density is 1000 kg m$^{-3}$, the distance of the freshwater lens below *sea level* is 40 times the freshwater's height above sea level at that area. The Ghyben–Herzberg principle can also be used in conjunction with the *Dupuit–Forchheimer equation* to describe the steady flow of *groundwater* in a coastal aquifer as:

$$\frac{\delta^2 h^2}{\delta x^2} + \frac{\delta^2 h^2}{\delta y^2} = \frac{-2w}{K(1 + G)}$$

where:
$w$ = *recharge* to the aquifer [L$^3$·T$^{-1}$]
$K$ = *hydraulic conductivity* [L·T$^{-1}$]
$G$ = equal to $\rho_w/(\rho_s - \rho_w)$ [M·L$^{-3}$]

143

(a)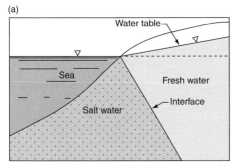

(b)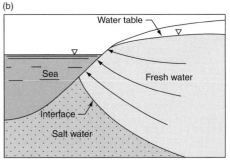

**Figure G-8** The position of the salt-water/fresh water interface (a) interface is stationary and there is no groundwater flow because the head created by the groundwater system equals the ocean head, and (b) groundwater head exceeds that of the ocean, therefore the groundwater flows into the ocean forcing the interface to move seaward. (Driscoll, 1986. Reprinted by permission of Johnson Screens/a Weatherford Company.)

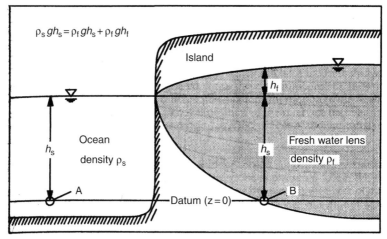

**Figure G-9** Demonstration of the fresh water lens using the Ghyben–Herzberg principle. (Maidment, 1993. Reprinted with permission from McGraw-Hill, Inc.)

**Giant's kettle** See *glacier well.*

**Glacial lake/glacier lake** A body of water either partially or entirely fed by glacial meltwater or formed within glacial ice by differential melting. Such a *lake* can often be formed by a morainal dam due to glacial retreat or can occupy a basin produced by glacial *erosion.* Cf. *Cirque lake; Kettle lake; Marginal lake; Ice-dammed lake.*

**Glacial mill** See *glacier well.*

**Glacier burst** See *glacier flood.*

**Glacier flood** Also called glacier burst. The sudden release of meltwater from a glacier or glacier-dammed lake, usually resulting in catastrophic flooding. A glacier flood can also be caused by the melting of a drainage channel or by subglacial volcanic activity. Cf. *Glacial lake.*

**Glacier well** Also called a moulin, glacial mill, pothole, or giant's kettle. A roughly cylindrical, nearly vertical hole or shaft in the ice of a glacier, scoured out by swirling meltwater as it pours down from the surface. Etymol: French, "mill," so called because of the loud roaring noise made by the falling water.

**Glaciofluvial** Also fluvioglacial. Flowing from a glacier, e.g. a meltwater stream, or formed by a glacier and/or a glacial stream, e.g. a glacial deposit.

**Glide** A gentle or slowly moving portion or *reach* of a shallow waterway. Cf. *Stream; River.*

**Governing equation for groundwater flow** In three dimensions, the governing equation for *groundwater flow* is:

$$\frac{\partial}{\partial x}\left(K_x \frac{\partial h}{\partial x}\right) + \frac{\partial}{\partial y}\left(K_y \frac{\partial h}{\partial y}\right) + \frac{\partial}{\partial z}\left(K_z \frac{\partial h}{\partial z}\right) = S_s \frac{\partial h}{\partial t} + W$$

where:
$K$ = hydraulic conductivity $[\text{L·T}^{-1}]$
$h$ = head $[\text{L}]$
$S_s$ = specific storage $[\text{L}^{-1}]$
$W$ = sterm for sources and sinks of water $[\text{T}^{-1}]$

Cf. *Equations of groundwater flow.*

**GPR** See *ground-penetrating radar.*

**Grab set** See *flash set.*

**Grade** The equilibrium or balance between a watercourse slope, *load*, volume, velocity, and characteristics. Equilibrium is attained between *erosion* and deposition through adjustments between the capacity of the watercourse to do work and the quantity of the work to be done. Watercourse grade should not be confused with watercourse *gradient*, which is a quantitative term. Cf. *Stream; River.*

**Grade level** The elevation at which an entire watercourse has attained a uniform *gradient*, i.e. when the watercourse's *longitudinal profile* is a straight line. Cf. *Stream; River.*

**Graded reach** A portion of a watercourse demonstrating a balance between *erosion* and deposition that may not be expressed throughout the entire watercourse profile. Cf. *Grade; Graded stream; Stream; River.*

**Graded stream** A watercourse that exhibits adjustability and a stage of stability indicating a balance between its transport capacity and the amount of material supplied to it. A graded *stream* has reached a condition of balance between *erosion* and deposition that is usually attained by mature *rivers*; equilibrium exists between the independent factors influencing the river, such as *discharge* and *sediment load*, and the dependent variables of *gradient, channel shape*, and channel *roughness*. The term "graded stream" relates to the stream's capacity and geomorphic character, as compared to the quantitative term "gradient," which can be determined for any watercourse or any *reach* along a watercourse. Cf. *Graded reach; Stream maturity; Steady-state stream.*

**Gradient** The steepness, topography, slope, or decrease in elevation of a *stream* bed as it flows over a horizontal distance. The *groundwater gradient, i*, is the change in *hydraulic head* along the *flow* path, which is mathematically stated as:

$$i = \frac{(h_1 - h_2)}{L}$$

where:
$h_1 - h_2$ = difference in hydraulic head $[\text{L}]$
$L$ = distance along the flow path between the points $h_1$ and $h_2$ $[\text{L}]$
Cf. *Hydraulic gradient; Stream gradient; Average linear velocity.*

**Gradient flow** Movement or *flow* occurring from areas of higher elevation or *head* to lower elevations. For streamflow, gradient flow is calculated as the *upgradient* stream elevation minus the *downgradient* stream elevation, divided by the distance between measurement points. Determination of the groundwater gradient, *i*, for a *porous medium* is based on measurements of the *total head* in wells or *piezometers* separated by a known distance and is calculated by the *gradient* equation.

**Grain** A unit of measurement of the *hardness* of water expressed in equivalents of $CaCO_3$. One grain per US gallon volume is equal to 17.1 parts per million (mg L$^{-1}$) of $CaCO_3$ by weight. See: **Appendix A – Conversions.**

**Grain angularity** The individual *grains* within a sample can be classified as either angular, subangular, subrounded, or rounded. The classification based on the shape of the grain will give an indication on the environment in which it came from. Well-rounded grains will indicate that the individual grain has either traveled great distance or within turbulent environment, whereas an angular fragment will have traveled less or deposited in tranquil situations, thereby not allowing much grain-to-grain contact and abrasion. Cf. *Bed roughness.*

**Grain roughness** See *grain angularity.*

**Grain size** A majority of prolific *aquifers* are comprised of subsurface sediments that can be classified by the size or the diameter of the individual grains. The two most common grain-size classification systems are the engineering classification system of the American Society of Testing Materials (ASTM), shown in **Table G-2**, and the geological or Unified Soil Classification System, shown in **Table G-3**. The engineering classification is based on *grain size distribution*, whereas the geological classification, typically used by sedimentologists, is based on the sediment size based on the limiting particle diameter.

**Table G-2**  Engineering grain size classification. (ASTM, 1994. ASTM Standards on Ground Water and Vadose Zone Monitoring.)

| Name | Size range (mm) | Example |
|---|---|---|
| Boulder | >305 | Basketball |
| Cobbles | 76–305 | Grapefruit |
| Coarse gravel | 19–76 | Lemon |
| Fine gravel | 4.75–19 | Pea |
| Coarse sand | 2–4.75 | Water softener salt |
| Medium sand | 0.42–2 | Table salt |
| Fine sand | 0.075–0.42 | Powdered sugar |
| Fines | <0.075 | Talcum powder |

**G**

**Table G-3**  Unified soil classification system. (Reprint from the United States Geological Society.)

| | | | | | | | | |
|---|---|---|---|---|---|---|---|---|
| Coarse grained. More than half of material is larger than no. 200 sieve size | GRAVELS Over 50 percent of coarse fraction larger than no. 4 sieve size (no. 4 = 1/4-in.) | CLEAN GRAVEL Little or no fines | Wide range in grain size and substantial amounts of all intermediate particle sizes. | | | | GW | Well graded gravel and well graded gravel with sand. |
| | | | Predominantly one size or a range of sizes with some intermediate sizes missing. | | | | GP | Poorly graded gravel and poorly graded gravel with sand. |
| | | GRAVELS WITH FINES Over 12 percent | Non-plastic lines (for identification procedures see ML below) | | | | GM | Silty gravel and silty gravel with sand. |
| | | | Plastic lines (for identification procedures see CL below) | | | | GC | Clayey gravel and clayey gravel with sand. |
| | SANDS Over 50 percent of coarse fraction smaller than no. 4 sieve size (no. 4 = 1/4-in.) | CLEAN SANDS Little or no fines | Wide range in grain sizes and substantial amounts of all intermediate particle sizes. | | | | SW | Well graded sand and well graded sand with gravel. |
| | | | Predominantly one size or a range of sizes with some intermediate sizes missing. | | | | SP | Poorly graded sand and poorly graded sand with gravel. |
| | | SANDS WITH FINES Over 12 percent | Nonplastic lines (for identification procedures see ML below) | | | | SM | Silty sand and silty sand with gravel. |
| | | | Plastic lines (for identification procedures see CL below). | | | | SC | Clayey sand and clayey sand with gravel. |
| Fine grained. More than half of material is smaller than no. 200 sieve size | | | Identification procedures of gradation smaller than no. 40 sieve size | | | | | |
| | | | Dry strength (crushing characteristics) | Dilatancy (reaction to shaking) | Toughness (consistency near plastic limit) | | | |
| | SILTS AND CLAYS Liquid limit less than 50 percent | | None to low | Rapid to slow | None | | ML | Silty, sandy, silty, clayey silt, or gravelly silt. |
| | | | Medium to high | None to very slow | Medium | | CL | Lean clay, sandy clay. of gravelly clay. |
| | | | Low to medium | Slow | Low | | OL | (Organic) silt, clay, sandy silt, sandy clay, gravelly clay, or gravelly silt. |
| | SILTS AND CLAYS Liquid limit 50 percent or more | | Low to medium | Slow to none | Slight to medium | | MH | (Elastic) silty, sandy silt, clayey silt, or gravelly silt. |
| | | | High to very high | None | High | | CH | (Fat) clay, sandy clay, silty clay, or gravelly clay. |
| | | | Medium to high | None to very slow | Low to medium | | OH | (Organic) clay, silt, sandy clay, silty clay, clayey silt, sandy silt, or gravelly silt and clays. |
| | HIGHLY ORGANIC SOILS | | Readily identified by color, oder, spongy feel and frequently by fibrous feature | | | | PT | Peat or other highly organic soil. |

Cf. *Sieve analysis.*

**Grain size distribution/Grain size distribution curve** The sediment size is measured by *sieve analysis,* which is then plotted on semi-logarithmic graph paper with the cumulative percent finer by weight plotted on the arithmetic scale and the *grain size* on the logarithmic scale. Plotting the variation in grain sizes for a granular sample as the total percentage passing each sieve produces a grain size distribution curve for that particular sediment, as shown in **Figure G-10**. Figure G-11 is a plot of the results of a sieve analysis for a fine- to medium-grained sand. The *uniformity coefficient* of a sediment is then based on the *effective grain size* as determined from the plot.

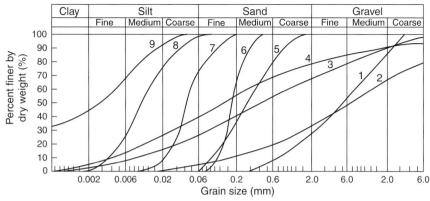

**Figure G-10** Grain size distribution curve plotted on semi-logarithmic paper. (Sen, 1995. Reprinted with permission from CRC Press, Inc.)

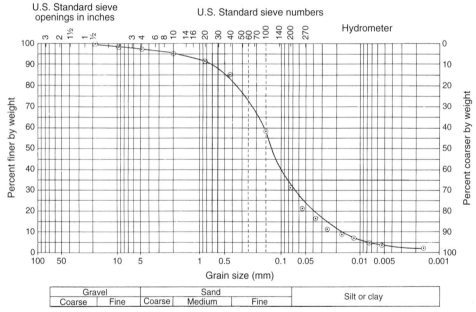

**Figure G-11** Example grading curve for design of monitoring well screens. (ASTM, 1994. ASTM Standards on Ground Water and Vadose Zone Investigations.)

Cf. *Fair-Hatch equation; Kozeny-carmen equation; Hazen method; Grain size gradation curve and conductivity.*

**Grain size graduation curve and conductivity** A relationship between the *grain size* of a *porous* granular sediment and its ability to conduct fluid. The plotted *grain size distribution* is used to determine the *effective grain size* of a sediment, which may be related to its *hydraulic conductivity* by the following equation:

$$K = Ad_{10}^2$$

where:
$K$ = hydraulic conductivity [$L \cdot T^{-1}$]
$A$ = coefficient [dimensionless, equal to 1.0 for a $K$ value in cm s$^{-1}$ and a $d_{10}$ value in mm]
$d_{10}$ = effective grain size [L]

**Grain size test** See *grain size distribution; grain size gradation and conductivity; sieve analysis*.

**Granular-activated carbon (GAC)** See *activated carbon*.

**Grapevine drainage pattern** See *drainage pattern, dendritic*.

**Graphing hydrochemical data** See *Stiff diagram; Piper trilinear diagram; Durov–Zaprorozec diagram; hydrochemical facies; hydrogeochemistry.*

**Gravel pack** In a water *well*, gravel or coarse sand that is placed opposite a water-producing zone and surrounding the *screened interval* to increase intake efficiency by creating a more permeable zone, or in oil production, gravel or coarse sand that is placed in a well opposite an oil-producing sand to prevent or retard the movement of loose and grains along with the oil into the *well bore*. Cf. *Filter pack; Production zone.*

**Gravimetric moisture content** The weight of the water within a *porous medium* divided by the weight of the solid, which can be used as a measure of the *soil moisture* in the aerated zone of the subsurface. The gravimetric moisture content, $\theta_m$, is obtained by weighing a soil sample of known volume, drying it until the final weight is constant, and applying the following equation:

$$\theta_m = \frac{m_w}{m_s} = \frac{m_t - m_s}{m_s}$$

where:
$m_w$ = *soil water* [M]
$m_t$ = soil sample weight [M]
$m_s$ = soil final weight after drying [M].

The percent of water-weight loss multiplied by the bulk density of the soil sample equals the percent volume loss, which can be translated to inches of water per depth. Refer to **Appendix A** for property conversions. Cf. *Volumetric moisture content; Zone of aeration; Antecedent precipitation index*.

**Gravitational potential energy ($E_g$)** The potential energy, or capacity to do work, resulting from the position of a fluid mass in relation to a specified datum. Gravitational potential energy, $E_g$, is defined by the equation:

$$E_g = mgz$$

where:
$m$ = mass of the fluid [M]
$g$ = gravitational constant [$L \cdot T^{-2}$]
$z$ = elevation of the center of gravity of the fluid above the reference elevation [L].

Cf. *Potential head.*

**Gravitational water** Also called gravity water, or *free water*. The water that moves freely through the soil under the influence of gravity. The movement of gravitational water can be limited by excess moisture in the underlying soil zones, by encountering the *water table*, or by impervious layers. Dry soils may also act as a barrier to gravitational movement of water because a continuous film of water may not be present, and at high tensions, the water in sand is held at contact points between large sand grains. Cf. *Field capacity; Soil water; Infiltration water; Phreatic water.*

**Gravity drainage** The gravity-driven movement of water within the *unsaturated zone*, or within an *aquifer*, the vertical flow resulting from the force of gravity, observed in *observation wells* or piezometers. Gravity drainage is also the predominant force responsible for the migration of *dense non-aqueous-phase liquid (DNAPL)* within the saturated zone. Gravity drainage is also responsible for the delayed-yield phenomenon in a dewatered zone. Cf. *Gravitational water; Free water; Field capacity; Hygroscopic moisture.*

**Gravity flow** Water flow solely under the influence of gravity as in *gravitational water* but it is also a cause of glacial movement in which the flow of the ice results from the downslope gravitational component in an ice mass resting on a sloping floor. Cf. *Groundwater gradient.*

**Gravity head** In *Bernoulli's equation*, the *total head h* along a streamline (parameterized by $x$) remains constant. As a result, *velocity head* can be converted into gravity head and/or *pressure head* (or vice-versa), such that the total head $h$ stays constant.

**Gravity potential of groundwater** The potential resulting from the position of *groundwater* or *soil moisture* above a specified datum. Cf. *Gravitational potential energy.*

**Gravity spring** See *spring, gravity.*

**Gravity survey/gravimetric survey** A record of minute differences in the earth's gravitational field caused by density differences of the surficial sediments and underlying rock. Changes in rock types associated with changes in *porosity* or grain density, *degree of saturation*, fault zones acting as *groundwater barriers*, and thickness of unconsolidated sediments have been investigated by direct measurement of density variations. **Figure G-12** shows the use of gravity measurements to locate buried bedrock valleys with the potential to yield high volumes of *groundwater*. Gravimetric surveys have been utilized at the ground surface for water well exploration; *borehole* gravity surveys are more commonly utilized for oil, gas, and mineral exploration.

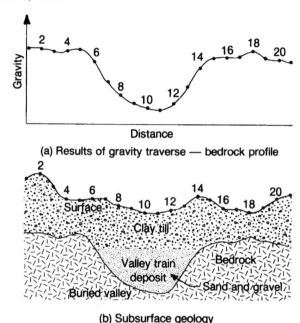

(a) Results of gravity traverse — bedrock profile

(b) Subsurface geology

**Figure G-12** The use of gravity measurements in locating valleys within the bedrock as potential water sources. (Driscoll, 1986. Reprinted by permission of Johnson Screens/a Weatherford Company.)

Cf. *Geophysical exploration methods.*
**Gravity water** See *free water; gravitational water.*
**Gray water** Waste water resulting from domestic use, e.g. household wash water.
**Green-Ampt infiltration model** A mathematical construct using *Darcy's law* to approximate the ponded *infiltration* into a deep *homogeneous* soil with a uniform initial water content. The Green-Ampt rate equation, ignoring the depth of ponding at the surface, is as follows:

$$f = K\left(1 + \frac{(\phi - \theta_i)S_f}{F}\right)$$

where:
$f$ = infiltration rate [L·T$^{-1}$]
$K$ = effective *hydraulic conductivity* [L·T$^{-1}$]
$\varphi$ = initial water content of the soil [L$^3$·L$^{-3}$]
$\theta_i$ = porosity [L$^3$·L$^{-3}$]
$S_f$ = effective suction at the wetting front [L]
$F$ = accumulated infiltration [L].

The Green-Ampt model also applies to *infiltration* under *rainfall* conditions; prior to ponding, the infiltration rate, $f$, equals the rainfall rate, $R$, which allows calculation of infiltration by the following equation:

$$f = R \quad \text{for} \quad t \leq t_p$$

149

$$f = K + \frac{KS_f(\phi - \theta_i)}{F} \quad \text{for} \quad t > t_p$$

where:

$t_p = F_p/R$

$F_p = [S_f(\varphi - \theta_i)]/(R/K - 1)$

**Figure G-13** shows the soil cross section in relation to the Green-Ampt model.

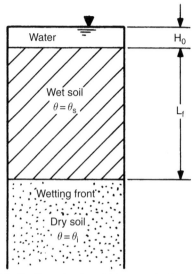

**Figure G-13** Soil horizons related to the Green-Ampt model to approximate ponded infiltration. (Maidment, 1993. Reprinted with permission from McGraw-Hill, Inc.)

**Gringarten–Witherspoon model** The first study evaluating *aquifer* parameters in a single fracture-well interface and derived curves for determining the hydraulic parameters from *aquifer test* responses. Their model, based on well-fracture intersections as shown in **Figure G-14**, is related to techniques for evaluating fracture effects on oil and gas wells.

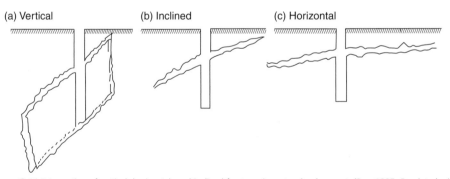

**Figure G-14** Intersection of vertical, horizontal, and inclined fractures in water development. (Sen, 1995. Reprinted with permission from CRC Press, Inc.)

**Gringorten formula** One of several equations in use to calculate the probability in a given year that an annual maximum *discharge* will equal or exceed a specified historical *peak flow*. For a *flood frequency analysis*, the annual maximum *flow* is calculated using a record of 24 annual maximum discharges from a continuous homogeneous-gauged river record. These peak flows are tabulated in decreasing order of magnitude and then ranked, *r*, by position on the chart with $r=1$ having the largest magnitude. The probability, $P(X)$, of an annual maximum equaling or exceeding an individual peak flow, $X$, in any given year, can be calculated for each value of $X$ (in units of $m^3 s^{-1}$) by the Gringorten formula:

$$P(X) = \frac{r - 0.44}{N + 0.12}$$

where:
$X$ = measurement of an occurrence of peak flow $[L^3 \cdot T^{-1}]$
$r$ = rank of $X$
$N$ = total number of data values.

Cf. *Flow frequency; Flood frequency curve; Recurrence interval; Design flood.*

**Gross precipitation** The amount of *rainfall* that is measured in an open area with little to no interference or is measured above a forest canopy.

**Ground ice** A body of frozen water, such as a lens or wedge, that has been enclosed in permanently frozen earth materials, often found at considerable depth, or ice of any origin that has formed on the ground or has been covered with soil. Cf. *Fossil ice.*

**G**

**Ground-penetrating radar (GPR)** An electromagnetic method used to map subsurface stratigraphy or search for *groundwater* contamination by transmitting repetitive electromagnetic pulses, with frequencies of 10–1000 MHz, into the ground. When the radiated energy encounters an interface between two materials of differing dielectric properties, the pulses are reflected back to the surface, and the voltage is recorded and displayed as a function of time, creating a continuous line profile as the GPR unit is pulled along the surface as shown in **Figure G-15**. GPR provides good vertical and lateral resolution but is limited in depth to approximately 15 m (50 ft), depending on the frequency used and the ground conductivity. GPR systems have been used successfully in hydrologic studies involving the location of abandoned waste-disposal sites in which the *leachate* has a dielectric potential. The technique has limited effectiveness in areas with near-surface clays.

**Figure G-15** Use of a ground-penetrating radar system. (Driscoll, 1986. Reprinted by permission of Johnson Screens/a Weatherford Company.)

Cf. *Geophysical exploration methods.*

**Groundwater** The excess *soil moisture* that saturates subsurface soil or rock and migrates downward under the influence of gravity. In the literal sense, all water below the ground surface is groundwater; in hydrogeologic terms, however, the top of this saturated zone is called the *water table*, and the water below the water table is called groundwater. **Figure G-16** shows a generalized classification of subsurface water. Under natural conditions, groundwater moves by gravity flow through rock and soil zones until it seeps into a *stream* bed, *lake*, or ocean, or discharges as a *spring*.

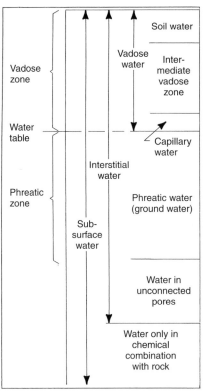

**Figure G-16** Subsurface water classifications. (Driscoll, 1986. Reprinted by permission of Johnson Screens/a Weatherford Company.)

Cf. *Phreatic water; Capillary water; Zone of saturation; Zones of water; Meteoric water; Juvenile water; Connate water; Gravitational water.*

**Groundwater artery** A confined subsurface zone of permeable material that is saturated, under *artesian pressure*, and surrounded by less-permeable material. *Groundwater* arteries typically occur in gravelly ancient *stream channels* buried by less-permeable silts. Cf. *Permeability.*

**Groundwater barrier** A natural or artificial subsurface obstacle inhibiting the lateral movement of *groundwater*, characterized by a substantial difference between the levels of the groundwater on opposite sides, thereby differentiating the obstacle from a *confining layer* or bed. Cf. *Aquiclude; Boundary conditions.*

**Groundwater basin** A subsurface area capable of retaining water that has reasonably well-defined boundaries and discernable areas of *recharge* and *discharge*. A groundwater basin can also be an *aquifer* or system of aquifers that may or may not be basin shaped. Cf. *Boundary conditions.*

**Groundwater budget** Also called a *hydrologic budget*, water budget, *water balance* groundwater inventory, or groundwater equation. A numerical accounting of the annual recharge to a *groundwater resource*. In a groundwater development program, such a budget is necessary to efficiently manage and utilize the resource. Many of the parameters that determine *recharge* to an *aquifer* are measured directly and some are computed from hydraulic characteristics and measured *potentiometric* data. The natural recharge to an undeveloped aquifer (one not being utilized) is determined by a water-budget analysis of the *recharge area* as follows:

Groundwater recharge = (precipitation + surface-water inflow + imported water + groundwater inflow) −

(evapotranspiration + reservoir evaporation + surface-water outflow + exported water

+ groundwater outflow) ± changes in surface-water storage

The above groundwater budget accounts for groundwater recharge from *precipitation, losing streams,* irrigation water, and unlined canals. In areas where the land surface of the recharge area is developed, additional components of recharge from such sources as agriculture, industry, and urban growth can be calculated by a supplemental water-budget analysis as follows:

Groundwater recharge = (industrial use + municipal use + domestic use + irrigation use) – (cooling-water evaporation + irrigation-water evapotranspiration + water exported in products + sewage discharge into surface waters)

Cf. *Groundwater inflow; Groundwater outflow.*

**Groundwater cascade** The subsurface of a waterfall where a near-vertical drop or *flow* of *groundwater* occurs over a *groundwater barrier.*

**Groundwater chemistry** See *hydrogeochemistry; geochemical processes; hydrochemical facies.*

**Groundwater, confined** See *aquifer, confined.*

**Groundwater constituents** The minerals dissolved by *groundwater* in the subsurface through which it travels, which determine the water's chemical nature. The slow *infiltration* and *gravity drainage* of surface water through the subsurface result in prolonged contact with minerals in the soil and bedrock. The water becomes saturated with the minerals it encounters, and in time a dynamic equilibrium is established. **Table G-4** lists the major (greater than $5 \, \mathrm{mg \, L^{-1}}$) cations and anions found in groundwater. Water is considered to be the universal solvent and has the ability to dissolve a greater range of substances than any other liquid.

**G**

**Table G-4**   Major ions in groundwater. (Adapted from Driscoll, 1986)

| Cations | Anions |
|---|---|
| Calcium ($Ca^{2+}$) | Bicarbonate ($HCO_3^-$) |
| Magnesium ($Mg^{2+}$) | Sulfate ($SO_4^{2-}$) |
| Sodium ($Na^+$) | Chloride ($Cl^-$) |
| Potassium ($K^+$) | Carbonate ($CO_3^{2-}$) |

Cf. *Total dissolved solids (TDS); Geochemical speciation.*

**Groundwater contamination** Typically all solutes introduced into *groundwater*, usually by man's activities, are referred to as contaminents, regardless of whether or not the concentration of the solute reach levels that cause degradation of the *water quality.*

**Groundwater data collection methods** Numerous direct and indirect techniques of obtaining information needed to properly understand a *groundwater resource*, whether for purposes of resource management or contaminant remediation. Groundwater data are collected through a given basin or within the area of contamination. **Table G-5** lists a variety of groundwater data collection methods and the associated advantages/disadvantages. The table is not all-inclusive but is a good representation of the more common methods.

**Table G-5**   Advantages and disadvantages of a variety of groundwater data collection methods. (Maidment, 1993. Reprinted with permission from McGraw-Hill, Inc.)

| Category | Commonly used methods | Advantages/disadvantages |
|---|---|---|
| Geophysics | Electromagnetics | Good for delineation of high-conductivity plumes |
| | Resistivity | Useful in locating fractures |
| | Seismic | Limited use in shallow soil studies |
| | Ground-penetrating radar | Useful in very shallow soil studies |
| Drilling | Augering | Poor stratigraphic data |
| | Augering with split-spoon sampling | Good soil samples |
| | Air/water rotary | Rock sample formation |
| | Mud rotary | Fills fractures—needs intensive development |
| | Coring | Complete details on bedrock |
| | Jetting/driving | No subsurface data |
| Groundwater sampling | Bailer | Allows escape of volatiles (operator-dependent) |
| | Centrifugal pump | Can produce turbid samples, increasing chance of misrepresentation contamination |
| | Peristaltic/bladder pump | Gives more representative samples |
| Soil sampling | Soil boring | Restricted to shallow depths |
| Aquifer tests | Pump test | Samples a large aquifer section |
| | Slug test | Does not require liquid disposal |

**Groundwater dating** See *dating of groundwater; carbon-14 dating; radiocarbon dating of groundwater.*

**Groundwater discharge** The release or flow of groundwater from the saturated zone or *aquifer* to the surface through *streams*, rivers, lakes, *springs*, and oceans. Cf. *Decrement; Discharge; Phreatic water; Baseflow; Runoff; Zone of saturation.*

**Groundwater divide** A location separating one *flow system* from another, such as the crest of a *water table* where *flow* occurs in opposite directions on either side of the crest as shown in **Figure G-17**. A subsurface groundwater divide often coincides with a high surface topographic feature. Water flows from topographically high areas or ridges, or areas of *recharge*, to lower elevations or valleys, or areas of *discharge*, which creates imaginary vertical no-flow boundaries beneath the ridges and valleys. These boundaries form groundwater divides where flow occurs in opposite directions.

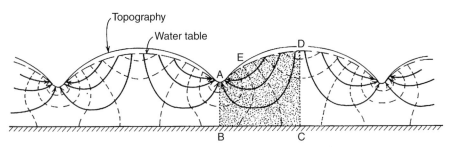

**Figure G-17** Flow lines diverge from recharge areas, such as mountain tips, and converge at discharge areas, such as valleys or rivers. Dotted lines AB and CD are the imaginary boundaries or groundwater divides. Region AE is the discharge area and ED is an area of recharge. (Freeze and Cherry, 1979. Reprinted with permission from Prentice-Hall, Inc.)

Cf. *Divide; Groundwater flow; Boundary, no-flow.*

**Groundwater elevation** *Groundwater* is generally always moving, vertically or horizontally, from higher *hydraulic head* in *recharge areas* to lower hydraulic head in *discharge areas*. Groundwater moves because there is a change in *head*, or a *hydraulic gradient* or slope in *potentiometric surface* of a groundwater system. Gravity is the driving force that moves water. Groundwater moves slowly (feet/yr or feet/day) that water can build up or mound in recharge areas before it affects or equilibrates in the system. The height of groundwater or the groundwater elevation is affected by the quantity of groundwater movement through a *porous medium*, which is defined by *Darcy's Law*. Cf. *Groundwater measurement.*

**Groundwater energy** The total *potential energy* in a *groundwater* mass, expressed as *head* in units of length. Groundwater energy has three components: *pressure*, velocity, and the *elevation head*. The sum of these energy potentials is mathematically expressed by the *Bernoulli equation*:

$$H = \frac{p}{\gamma} + \frac{V^2}{2g} + z$$

where:
$H$ = specific energy [L]
$p$ = *pressure head* [L]
$\gamma$ = specific weight of water
$V$ = velocity of *flow* [L·T$^{-1}$]
$g$ = gravitational constant [L·T$^{-2}$]
$z$ = elevation above a certain datum [L]

The elevation head, $z$, is the energy derived from the elevation of the water body; the *pressure head*, $p/\gamma$, is the energy contained in a water mass that can be attributed to the forces confining the water; and the *velocity head*, $V^2/2g$, is the energy component resulting from the movement of the water. Bernouilli's theorem is only valid for incompressible liquids. Cf. *Head, total; Head elevation.*

**Groundwater equation** See *groundwater budget; equations of groundwater flow.*

**Groundwater extraction** The removal of *groundwater* from an *aquifer* or a saturated zone for aquifer restoration, control of contaminant migration, water supply, construction *dewatering*, and seawater intrusion control. Cf. *Zone of saturation; Intrusion of salt water.*

**Groundwater facies** A mass of subsurface water that has a distinctive chemistry or character. The chemical composition of *groundwater* tends toward equilibrium at any point within the matric material at the prevailing conditions; therefore, it can be distinguished from adjacent water. Cf. *Hydrochemical facies; Hydrochemistry; Groundwater constituents.*

**Groundwater flow** The natural or artificially induced movement of subsurface water in a *zone of saturation.* In 1856 French engineer Henri Darcy recognized that the *flow* of water beneath the ground surface is analogous to pipeflow, and determined that the rate of flow through a column of saturated sand is proportional to the difference in *hydraulic head* at the ends of the column and inversely proportional to the length of the column; or, in the subsurface, that the rate of groundwater flow is proportional to the *hydraulic gradient.* His experiments led to the development of *Darcy's law,* which is expressed as the basic equation describing the flow of *groundwater.* **Figure G-18** is a schematic of a groundwater *divide* and associated *flow lines* of a groundwater flow system, in which the hydraulic head decreases with depth in areas of *recharge* and increases in areas of *discharge.* The determination of the *direction of groundwater flow* is analogous to a three-point problem in structural geology. **Figure G-19** illustrated a graphical method of determining the direction of flow and the hydraulic gradient.

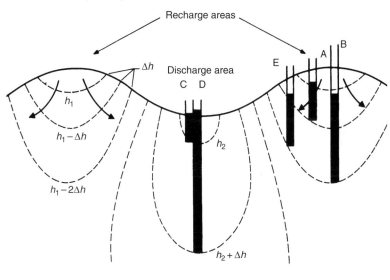

**Figure G-18** Groundwater flow system conceptual model. (Maidment, 1993. Reprinted with permission from McGraw-Hill, Inc.)

Cf. *Coupled flow; Chemical flow; Groundwater divide; Laminar flow; Transient flow; Governing equation of groundwater flow; Groundwater movement.*

**Groundwater flux** See *specific discharge.*

**Groundwater gradient** See *hydraulic gradient; Groundwater inflow.*

**Groundwater increment** See *recharge.*

**Groundwater inflow** A conceptual model used in most groundwater studies is the *hydrologic cycle,* but when a more quantitative approach is needed the conservation of mass or water budget analysis or *water balance* equation (Input − Output = change in storage) is used. **Table G-6** lists the most often used components for this quantitative approach. Cf. *Groundwater outflow; Groundwater budget; Groundwater recession curve.*

**Groundwater inventory** See *groundwater budget.*

**Groundwater lake** A surface water body resulting from the encounter of the upper surface of a *water table* or saturated zone with the surface of the ground. Cf. *Zone of saturation.*

**Groundwater law** See *water law.*

**Groundwater level** See *water table; groundwater elevation.*

**Groundwater measurement** A direct measurement for assessing *groundwater* is the depth to the *water table* in an unconfined aquifer or the location of the *piezometric surface* in a confined aquifer. The elevation of the top of the water within a well is obtained with a measuring device called a water-level recorder or electrical sounder. Cf. *Water level measurement; Aquifer, unconfined; Aquifer, confined.*

**Groundwater mining** The extraction or excessive pumping of *groundwater* at a rate that exceeds *recharge* of the basin, thereby creating a persistent decline in the groundwater level. Groundwater mining is often associated with areas in which the *aquifer* was recharged during previous wet climatic periods but no longer receives recharge or recharge at a much lower rate than previously occurred; the groundwater in such an area is mined as a local non-renewable resource. Cf. *Water table; Safe yield, Basin yield; Aquifer yield, Potential yield.*

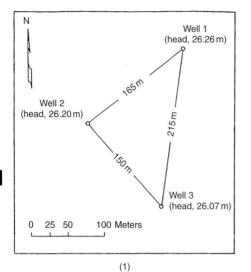

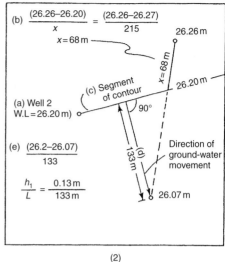

**Figure G-19** Both the direction of ground-water movement and the hydraulic gradient can be determined if the following data are available for three wells located in any triangular arrangement such as that shown on sketch 1:

1. The relative geographic position of the wells.
2. The distance between the wells.
3. The total head at each well.

Steps in the solution are outlined below and illustrated in sketch 2.

a. Identify the well that has the intermediate water level (that is, neither the highest head nor the lowest head).
b. Calculate the position between the well having the highest head and the well having the lowest head at which the head is the same as that in the intermediate well.
c. Draw a straight line between the intermediate well and the point identified in step b as being between the well having the highest head and that having the lowest head. This line represents a segment of the water-level contour along which the total head is the same as that in the intermediate well.
d. Draw a line perpendicular to the water-level contour and through either the well with the highest head or the well with the lowest head. This line parallels the direction of ground-water movement.
e. Divide the difference between the head of the well ($h_1$) and the distance between these wells ($L$) for the hydraulic gradient.

**Table G-6** Major Inflow and Outflow Components. (Weight, 2001. Reprinted with permission from McGraw-Hill)

| Inflow | Outflow |
| --- | --- |
| Precipitation | Evapotranspiration |
| Surface water | Surface water |
| Groundwater flux | Groundwater flux |
| Imported water | Exported water |
| Injection wells | Consumptive use |
| Infiltration from irrigation | Extraction wells |

**Groundwater monitoring** The growing demand for and dependence on *groundwater* increases the need to record the resource's availability and quality. Groundwater monitoring may be carried out to

- determine the *water quality* and chemistry of an area
- determine the water quality of a specific well designated for *water supply*
- determine the extent of groundwater contamination from a known source/sources
- provide regional information on depths to the *water table*

- provide information on *drawdown* in *piezomentric surface* due to regional or local pumping
- monitor a potential source of contamination to determine its potential effect on groundwater
- Groundwater samples can be collected from pumping or water supply wells, wells installed specifically for sampling such as *groundwater monitoring wells*, and *springs* or other points of *discharge*. Water samples are analyzed in a laboratory to determine the constituents and chemical make-up of the groundwater. Cf. *Water supply*.

**Groundwater monitoring well** A well installed for the specific purpose of sampling *groundwater* to determine its quality or of measuring its *flow* characteristics at a specific location. The specific design of a groundwater monitoring well depends on the nature of either the chemical or the physical parameters to be measured and on local *hydrogeologic* conditions. The installation of monitoring wells in areas under the jurisdiction of a regulatory agency may entail following specified installation and completion methods. During the installation of a groundwater monitoring well, it is vital to take precautions to prevent the introduction of materials that may affect or contaminate the *aquifer* that is to be sampled. **Figure G-20** is a well construction form and **Figure G-21** is an example of a completion diagram of a typical groundwater monitoring well.

G

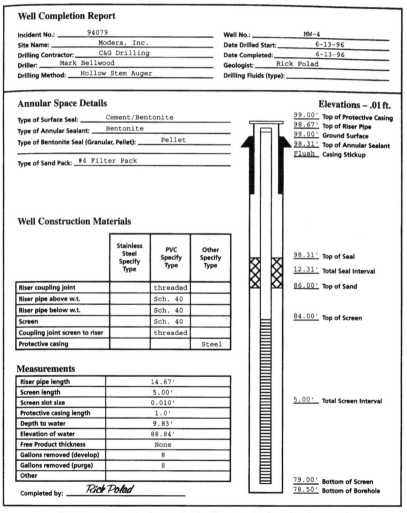

**Figure G-20** A sample of a completed well construction form. (Sanders, 1998.)

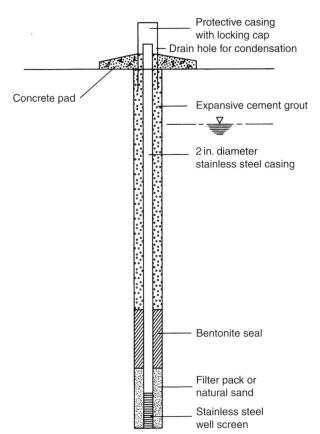

**Figure G-21** A sample of a well completion diagram. (ASTM, 1992. Special Technical Publication 1053, Ground Water and Vadose Zone Monitoring.)

Cf. *Well installation.*

**Groundwater mound** A dome-shaped *water table* or *potentiometric surface* typically formed in an area such as a lake perimeter, where the *unsaturated zone* is thinner and the water surface elevation builds up as a result of the downward *infiltration* of surface water, beneath landfills, beneath leaky canals and *lagoons*, etc. Groundwater mounds can form temporarily in the *downgradient* direction at the shoreline where lake water normally seeps out into the *groundwater*. A mound can inhibit *flow* and virtually back up the water system, causing a temporary halt of lake seepage.

**Groundwater movement** The *flow* of subsurface water in the saturated zone. Flow is generally calculated on a macroscopic scale rather than a microscopic scale, discounting the fine details of the subsurface strata. The equation for flow, *q*, expressed in units of velocity is as follows:

$$q = \frac{Q}{A}$$

where:
$v =$ *specific discharge* [L·T$^{-1}$]
$Q =$ flow rate [L$^3$·T$^{-1}$]
$A =$ cross-sectional area of porous material [L$^2$]

Cf. *Groundwater flow; Darcy's law; Governing equation of groundwater flow; Zone of saturation.*

**Groundwater network** See *flow system; groundwater divide; gauging network.*

**Groundwater outcrop** An area of *groundwater discharge* such as a *spring* or seep that can often be mapped through field observation. A groundwater outcrop might show clues of available water at the surface in the form of vegetation, saline soil, or a playa. Cf. *Groundwater outflow; Available moisture.*

**Groundwater outflow** The *discharge*, within a specified area such as a *drainage basin*, that is attributed to *groundwater* whether the *flow* is into a surface body of water, *spring, seepage face*, or other *groundwater outcrop*. The volume of water in groundwater outflow is used in the calculation of a *hydrologic budget* by the groundwater equation. Cf. *Groundwater budget; Interflow; Groundwater inflow*.

**Groundwater particle tracking** See *particle tracking*.

**Groundwater pollution** See *Groundwater contamination*.

**Groundwater, perched** See *aquifer, perched*.

**Groundwater pump and treat remediation** See *pump and treat systems; groundwater remediation*.

**Groundwater quality** The primary purpose of a water analysis and *groundwater monitoring* program is to monitor groundwater quality to determine the suitability of water for a proposed use such as domestic, agricultural, and industrial. Specific standards for groundwater quality are generally set by a governing body. In the United States, the *Environmental Protection Agency* (EPA) established *National Primary and Secondary Drinking Water Regulations* (NPSDW) under the provisions of the Public Health Service Act. Primary Drinking Water Regulations set *maximum contaminant levels* (MCL) for materials based on their potential to affect human health. Secondary Drinking Water Regulations suggest recommended maximum levels for contaminants that affect aesthetic and taste characteristics. A list of the currently established Primary and Secondary Drinking Water Regulations in the United States is presented in **Appendix B**. Cf. *Groundwater contamination*.

**Groundwater recession curve** The decreasing rate of *groundwater inflow*, shown on a portion of a *stream hydrograph*, that represents the time that surface *runoff* to the channel has ceased. The *stream baseflow* represents the withdrawal of groundwater from storage, or the groundwater recession, which is determined from the stream hydrograph and is shown to decay exponentially. Cf. *Overland flow; Interflow; Baseflow-recession hydrograph*.

**Groundwater recharge** See *baseflow recession hydrograph; basin yield; recharge; recharge rate; groundwater budget*.

**Groundwater region** An area designated according to the general geologic settings, *hydraulic characteristics*, and ranges of values of the dominant *aquifers*. Although *groundwater* exists in virtually all areas, it is not evenly distributed with respect to quantity or quality. In order for groundwater to be economically abundant, the local geologic conditions must permit the storage and transmission of large volumes of water, and climatic conditions must be favorable to *recharge* the aquifer. A *groundwater resource* is evaluated by *well yield; aquifer yield; basin yield, economic yield; and safe yield* .

**Groundwater remediation** Any hydrogeologic design for groundwater pollution control. The investigation and cleanup of *aquifers* became vital in the United States after 1984 because of environmental legislation related to the *Safe Drinking Water Act* (SDWA), the *Resource Conservation and Recovery Act* (RCRA), and the *Superfund Amendments and Reauthorization Act*. The first step in selecting the appropriate groundwater remediation process is to evaluate and characterize the contaminant source and the *hydrogeology* of the area. **Table G-7** lists some of the many groundwater remediation technologies in use today.

**Table G-7**   Groundwater remediation technologies.

| Conventional processes | Advanced processes |
|---|---|
| Coagulation, sedimentation, filtration | Activated alumina |
| Direct filtration | Adsorption |
| Diatomaceous earth filtration | GAC |
| Slow sand filtration | Powdered activated carbon |
| Lime softening | Resins |
| Ion exchange | Aeration |
| Oxidation-disinfection | Packed column |
| Chlorination | Diffused air |
| Chlorine dioxide | Spray |
| Chloramines | Slat tray |
| Ozone | Mechanical |
| Bromine | Cartridge filtration |
| Others | Electrodialysis |
| | Reverse osmosis |
| | Ultrafiltration |
| | Ultraviolet light (UV) |
| | UV with other oxidants |

Source: Dyksen, J.E., Hiltebrand, D.J., and Raczko, R.F., 1988, SDWA Amendments: Effects on the Water Industry. Journal Am. Water Works Assoc., v.80, no. 1. Copyright AWWA. Reprinted with permission

Cf. *Groundwater restoration; Groundwater contamination; Groundwater quality*.

**Groundwater replenishment** See *recharge*.

**Groundwater reservoir** The area of a viable *aquifer* that comprises all the subsurface materials constituting the saturated zone, including the *perched water*. Cf. *Groundwater resource; Zone of saturation*.

**Groundwater resources** A subsurface *water supply* that can be drawn upon for use. The exploitation of a *groundwater* resource is subject to supply and demand. When surface water supplies are abundant, groundwater is typically under-exploited; in populated arid regions, however, this valuable and crucial resource may be over-exploited. Groundwater resource management depends on the availability of the water supply and also on legal, political, and socioeconomic precedents and constraints. Cf. *Safe yield; Economic yield*.

**Groundwater restoration** Any of several measures implemented to restore a contaminated *groundwater basin* or *aquifer* to its precontaminated state. Groundwater may become contaminated through waste-disposal practices, spills, leaks, mine drainage, *saltwater intrusion*, poorly installed or *abandoned wells*, *infiltration* of contaminants from the surface, agricultural practices, highway deicing salts, and atmospheric contaminants transported by *precipitation*. The primary concern of groundwater restoration is to control or eliminate the source of contamination and then treat the contaminant plume itself.

**Groundwater ridge** A linear zone where the *water table* beneath an *influent stream* is elevated. Cf. *Divide; Groundwater divide*.

**Groundwater runoff** The excess water that enters the ground, becomes incorporated into a *groundwater basin* or *aquifer*, and is subsequently discharged into a *stream channel*. Cf. *Delayed runoff; Runoff; Influent stream; Discharge*.

**Groundwater sampling** A direct data collection method using *groundwater* specimens for the study and interpretation of subsurface conditions as part of a *hydrogeologic* investigation. A site sampling plan indicates the sampling frequency, locations, and possibly the statistical relevance of the samples that are collected. Samples that are indicative of site conditions are collected from *monitoring wells, springs*, and/or *piezometers* under strict quality assurance (QA) and quality control protocols (QA/QC). *Groundwater* is sampled after parameters such as *pH*, temperature, and specific conductance (as measured using a *flow through cell*) of an initial sample have equilibrated, indicating that the sample is from the formation of interest rather than a residual accumulation from the sample location. Cf. *Groundwater data collection methods*.

**Groundwater seepage** See *seepage/seepage force*.

**Groundwater storage** The water residing in the saturated zone, or the volume of water in an *aquifer*. Cf. *Specific storage; Storage coefficient; Specific yield*.

**Groundwater surface** See *water table; groundwater elevation*.

**Groundwater table** See *water table*.

**Groundwater transport** See *advection; dispersion; advection–dispersion equation; Darcy's law*.

**Groundwater trench** A longitudinal depression within a *potentiometric surface* or *water table* surface, resulting from the *discharge* of *groundwater* into a *stream* or drainage ditch.

**Groundwater, unconfined** See *aquifer, unconfined*.

**Groundwater velocity** See *groundwater flow; Darcy's law; velocity head*.

**Groundwater wave** An elevated area of the *water table* or *potentiometric surface* moving laterally in a wave motion, caused by the addition of a substantial quantity of water to the saturated zone within a short span of time. Cf. *Zone of saturation*.

**Groundwater yield** The maximum pumping rate that can be maintained in a *groundwater reservoir* while ensuring that water level declines are within acceptable limits. Cf. *Aquifer yield; Safe yield; Basin yield; Economic yield*.

**Grout** A mixture of water, neat *cement*, and various additives such as sand, *bentonite*, or hydrated lime (Ca(OH)$_2$) that is injected or poured into voids for the purpose of sealing. Grout is also available in a chip or pellet mixture, predominantly consisting of bentonite, that is hydrated after placement within the *well bore* as an effective upper seal for *groundwater monitoring wells*. Cf. *Well installation*.

**Grout curtain** A physical barrier constructed of sealant material for the purpose of controlling *groundwater flow*, typically to avoid leakage of subsurface waste or to avoid further spread of contamination. A grout curtain is installed by injecting different compounds such as *cement*, *bentonite*, silicate, and/or lignochrome *grouts* into *boreholes* drilled in the area to be isolated. Grout penetration varies depending on the subsurface material. Relatively closely spaced injection holes, e.g. 1.5 m, are commonly drilled in two or three staggered rows to ensure a continuous curtain, as shown in **Figure G-22**. The compounds fill the rock or soil pore spaces in the saturated zone and subsequently solidify to form the barrier. *Permeability* decreases and the soil-bearing potential increases after the grout hardens. A grout curtain is one of several methods (including a *slurry wall*, sheet piling, and *dewatering*) of isolating areas from groundwater flow. Cf. *Sheet piling; Slurry wall; Grouting*.

**Grouting** The mixture and placement by pressure injection of a fluid material, *grout*, into the void space of earth materials to reduce or eliminate their *permeability*, consolidate them, or increase their strength. Drilling contractors typically grout a hole or use grout to fill the *annulus* of a *well bore*. Low-viscosity grouting materials, e.g. a *bentonite*/gravel/clay *slurry*, are described in detail in American Standards and Testing Materials (ASTM) number D2144.

**Guar gum** A naturally occurring organic *colloid* obtained from finely ground seed of the guar plant, which is used as a *drilling fluid* additive. The physical properties of guar gum impart unusual hydration, gelling, and *viscosity*. The viscosity of an organic *drilling fluid* is rapidly reduced by adding acids, oxidizing agents, or enzymes to promote biodegradation. Cf. *Sol; Insol; Gel; Water of hydration*.

**Guelph permeameter** A device used to measure unsaturated *hydraulic conductivity* in the field. It is a *soil profile* measuring device. Use of the Guelph *permeameter* is more popular than other techniques, such as drained outflow measurement, the *piezometer* method, or the auger hole method, because it uses a smaller volume of water. Cf. *Flexible-wall permeameter*.

**G**

(a) No control measure

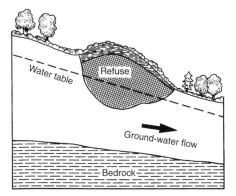

(b) Upgradient slurry wall to lower water table

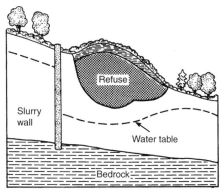

(c) Injection of grout to form a seal on sides and bottom

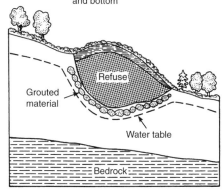

(d) Gradient control well to lower water table

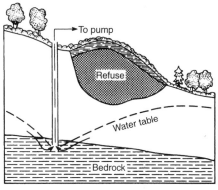

**Figure G-22** Methods of lowering the water table to prevent contact with waste material. (Fetter, 1994. Reprinted with permission from Prentice-Hall, Inc.)

**Gusher** See *geyser.*

**Gut** A very narrow passage or channel connecting two bodies of water, such as a small creek within a *marsh* or *tidal flat,* or a *tidal stream* connecting two larger waterways.

**Gypsum block** Also called a porous block. A relatively inexpensive pressure potential indicator, especially useful in the dry soil-water range or at potentials less than 100 kPa. Several different field and laboratory methods have been employed to measure water matrix potential; the principal measurement is the change in electrical resistance of the gypsum block with water content. Typically, *tensiometers* are used for these measurements in wet conditions in conjunction with gypsum blocks for dry conditions. Ceramic thermal conductivity sensors are gaining in popularity as an alternate pressure potential indicator. Cf. *Manometer; Thermocouple psychrometer.*

# Hh

**Half-flow interval** The length of time it takes for one-half of the annual *runoff* to exit a *watershed*. In hydrologic investigations, there are many ways of characterizing or analyzing runoff. Historically, half-flow interval was typically measured from the start of the record-keeping calendar or water year. It is now defined as the shortest interval, at any time during the year, in which one-half of the annual *runoff volume* leaves the watershed. Cf. *Quarter-flow interval; Hydrologic cycle.*

**Half-saturation constant ($k_{1/2}$)** The kinetic decay process of constituents in water where the concentration per unit time is the limiting reaction rate and where the concentration $C$ is equal to half, $k_{1/2}$, of the first-order rate coefficient $K$. The following equation for $k_{1/2}$ is typically applied to *nutrient* kinetics, especially in *eutrophication* research, and *monod kinetics* for biodegradation of organics:

$$\frac{dc}{dt} = \frac{k_s C}{k_{1/2} + C}$$

where:

$C$ = concentration of constituents [$M \cdot L^{-3}$]

$t$ = time [T]

$k_s$ = concentration/time of the limiting reaction rate when $C \gg k_{1/2}$ [$M \cdot L^{-3} \cdot T^{-1}$]

The negative sign indicates the kinetic decay or loss. Cf. *Governing equation for groundwater flow; Advection; Dispersion.*

**Halinity** The amount of chloride ($Cl^-$) in water, expressed as percent chlorinity. Halinity is used to classify water in the Venice system, as shown in **Table H-1**. Seawater is euhaline.

**Table H-1**   Table of halinity. (1994, Oxford University Press.)

| Zone | %Chlorinity |
|---|---|
| Euhaline | 1.65–2.2 |
| Polyhaline | 1.0–1.65 |
| Mesohaline | 0.3–1.0 |
| Alpha-mesohaline | 0.55–1.0 |
| Beta-mesohaline | 0.3–0.55 |
| Oligohaline | 0.03–0.3 |
| Freshwater | < 0.03 |

Cf. *Brine; Freshwater, salinity; Thalassic series.*

**Hantush inflection point method** A method of determining formation constants for a leaky artesian *aquifer* or semiconfined aquifer with no storage in the *confining layer* by plotting *drawdown* (arithmetic scale) versus time (logarithmic scale) on semilogarithmic graph paper. The Hantush inflection point method eliminates the use of type curves and results in an S-shaped plot, as shown in **Figure H-1**. One-half of the maximum drawdown is equal to the drawdown at the inflection point, $(h_0-h)_1$. After graphically determining the time $t_i$ at which $(h_0-h)_1$ occurs and determining the slope of the *drawdown curve* at the inflection point, $m_i$ (which is equal to the slope of the straight portion of the drawdown curve), the value of $f(r/B)$ can be calculated from the following equation:

$$f\left(\frac{r}{B}\right) = 2.3 \frac{(h_0 - h)_1}{m_1}$$

The value of $r/B$ can be determined from the function table (see **Appendix B**) by knowing the value of $f(r/B)$, and because $r$, the radial distance from the pumping well, is known, $B$ can be calculated. With the above information, *transmissivity*, $T$, and *storativity*, $S$, can then be calculated using the following equations:

$$T = \frac{QK_0\left(\frac{r}{B}\right)}{2\pi(h_0 - h)_{max}}$$

$$S = \frac{4t_i T}{2rB}$$

where:

$Q$ = *discharge* [$L^3 \cdot T^{-1}$]

$K_0$ = Bessel function [dimensionless]

162

Values of the functions $K_0(x)$ and $\exp(x)\ K_0(x)$ are given in **Appendix B**. If the thickness $b'$ of the confining layer is known, the conductivity $K$ may be determined by the following equation:

$$K' = \frac{Tb'}{B^2}$$

where:
$b' = $ thickness of the leaky layer [L]
$K' = $ vertical *hydraulic conductivity* of the leaky layer [L·T$^{-1}$]

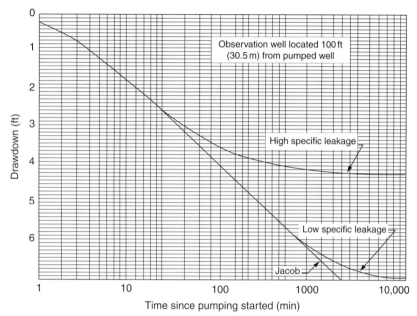

**Figure H-1** Drawdown graphs illustrating both low and high specific leakage. (Driscoll, 1986. Reprinted by permission of Johnson Screens/a Weatherford Company.)

Cf. *Hantush–Jacob formula; Aquifer, artesian.*
**Hantush-Jacob formula** Also called the Hantush–Jacob model or method. An equation, developed by M.S. Hantush and C.E. Jacob, describing the change with time in *hydraulic head* during pumping from a leaky *aquifer* with negligible storage in the *confining bed*. **Figure H-2** illustrates this concept. The initial pumping abstracts water from the main aquifer, but as pumping continues, other aquifers separated by semipervious layers above and below begin to contribute water to the pumped aquifer. Therefore, the pumped aquifer initially reacts similarly to a confined aquifer and is expected to have a time-drawdown variation similar to a *Theis-type curve*. When conditions are similar to those assumed by the *Theis method*, and *leakage* is vertical through the confining bed and proportional to *drawdown, head* in the supply bed is constant, and storage in the confining bed is negligible, the Hantush–Jacob formula for a leaky aquifer becomes:

$$h_0 - h = \frac{Q}{4\pi T}\,W(u, r/B)$$

where:
$h_0 - h = $ drawdown in the confined aquifer [L]
$Q = $ pumping rate [L$^3$·T$^{-1}$]
$T = $ *transmissivity* of the confined aquifer [L$^2$·T$^{-1}$]
$W(u, r/B) = $ *well function of u* for a leaky *artesian* well [dimensionless]
The values of the functions $W(u, r/B)$ for various values of $u$ are given in **Appendix B**, and the value $u$ is calculated by the following equation:

$$u = \frac{r^2 S}{4Tt}$$

163

where:
$r$ = radial distance from the pumping well to the *observation well* [L]
$S$ = storativity of the confined aquifer [dimensionless]
$t$ = time since pumping began [T].

The *leakage factor* is determined by the following equation:

$$B = \left(\frac{Tb'}{K'}\right)^{1/2}$$

where:
$B$ = leakage factor [L]
$b'$ = thickness of the leaky layer [L]
$K'$ = vertical *hydraulic conductivity* of the leaky layer [L·T$^{-1}$].

The *discharge* across the leaky layer, $q_L$, is determined by the following equation:

$$q_L = Q - q_s$$

where:
$Q$ = total discharge at time $t$ [L$^3$·T$^{-1}$]
$q_s$ = discharge from elastic storage in the confined aquifer at time $t$ [L$^3$·T$^{-1}$].

The discharge from elastic storage in the confined aquifer at the specified time, $t$, in days since pumping began, is calculated from the following equation:

$$q_L = Q^{(-Tt/SB^2)}$$

The Hantush–Jacob solution is reduced to the Theis solution when $K' = 0$ and $r/B = 0$, which occurs only when the *aquitard* is impermeable.

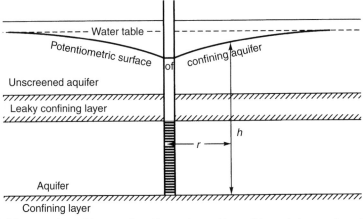

**Figure H-2** A fully penetration well in an aquifer with a semipermeable, confining or leaky upper layer. (Driscoll, 1986. Reprinted by permission of Johnson Screens/a Weatherford Company.)

Cf. *Jacob distance-drawdown straight line method; Jacob step-drawdown test; Jacob time-drawdown straight-line method.*
**Hantush-Jacob method** See *Hantush–Jacob formula.*
**Hantush partial-penetration method** Analysis of the effects of a well not screened through the entire thickness of the *aquifer*, by M.S. Hantush, regarding *drawdown* in an *observation well* during *pumping test* data collection. Calculating the value of horizontal and vertical *conductivities* of confining layers and storativity of the aquifer and confining layer is feasible if the well is not a *fully penetrating well* but only partially penetrating the aquifer, as stated in the simplifying assumptions, but caution must be taken because the time-drawdown curves may be different. A more distant observation well may exhibit greater drawdown than a closer well; drawdown curves for partial penetration may mimic a curve of downward *leakage* from storage through a thick, semipervious layer or a curve that is similar to the effect of a *recharge boundary*, sloping *water table* aquifer, or an aquifer of undulating thickness. The Hantush partial-penetration drawdown method compensates for the simplifying assumptions. Cf. *Screen interval; Storage coefficient; Jacob time-drawdown straight-line method.*

**Hard water** *Freshwater* that forms a scale upon *evaporation* and does not facilitate a soapy lather. Hard water contains more than 60% calcium carbonate ($CaCO_3$) or equivalent hardness-forming minerals. Cf. *Hardness; Softening of water*.

**Hardness** Also called total hardness. The amount of calcium and magnesium *ions* and, to a lesser degree, ions of other alkali metals, metals (e.g., iron), and hydrogen in *potable water*, expressed as parts per million (ppm), milligrams per liter (mpl) or, the *carbonate* and non-carbonate hardness as $CaCO_3$. (**See: Appendix A** for conversions.) For example, 40 ppm of Ca produces a water hardness of 100 ppm as $CaCO_3$. Water hardness, expressed in milligrams $CaCO_3$ per liter (L), can be classified as shown in **Table H-2**.

**Table H-2** Water hardness based on concentration of calcium and magnesium. For standardization purposes the water conditioning foundation classifies water hardness. (Driscoll, 1986. Reprinted by permission of Johnson Screens/a Weatherford Company)

| | |
|---|---|
| Very hard | $>180$ mg $L^{-1}$ |
| Hard | 120–180 mg $L^{-1}$ |
| Moderately hard | 60–120 mg $L^{-1}$ |
| Slightly hard | 9–60 mg $L^{-1}$ |
| Soft water | $<9$ mg $L^{-1}$ |

The above classification is based on the following equation:

$$\text{Hardness}\left(\frac{CaCO_3}{L}\right) = 2.497Ca + 4.118Mg$$

where:
$Ca = $ calcium content (mg $L^{-1}$)
$Mg = $ magnesium content (mg $L^{-1}$)

If the hardness of a water supply intended for human consumption is between 80 and 100 mg $L^{-1}$, it is common practice to add a softening agent. Cf. *Softening of water; Hard water*.

**Hargreaves equation** A temperature- or radiation-based equation used to estimate *evaporation* when the only data available is a temperature log. The Hargreaves equation is as follows:

$$E_{rc} = 0.0023 S_0\, \ddot{\delta}_T (T + 17.8)$$

$E = $ *potential evaporation* (mm day$^{-1}$)
$S_0 = $ water equivalent of extraterrestrial radiation for the location (mm day$^{-1}$]

$$S_0 = 15.392 d_r (\omega_s \sin \phi \sin \delta + \cos \phi \cos \delta \sin \omega_s)$$

$T = $ temperature (°C)
$\delta_T = $ difference between mean monthly maximum and minimum temperatures (°C)
$\omega_s = $ the sunset hour angle (rad)

$$\omega_s = ar \cos(-\tan \phi \tan \delta)$$

$\phi = $ latitude of the site ($+$ in northern hemisphere, $-$ in southern hemisphere)
$\delta = $ solar declination on day J (Julian calender) of the year (rad)

$$\delta = 0.4093 \sin\left(\frac{2\pi}{365}J - 1405\right)$$

$d_r = $ relative distance of the earth from the sun on day J

$$d_r = 1 + 0.033 \cos\left(\frac{2\pi}{365}J\right)$$

This method provides reasonable estimates because of the incorporation of the solar radiation term $S_0$. Cf. *Blaney–Criddle method; Thornthwaite equation; Temperature logging*.

**Harmonic mean of hydraulic conductivity** The harmonic mean yields the equivalent *hydraulic conductivity* for *groundwater flow* across a layered sequence of geologic units, each with differing hydraulic conductivity. Cf. *Geometric mean of hydralic conductivity*. The harmonic mean of $n$ observations $H(x_1, \ldots, x_n)$ is defined as:

$$\frac{1}{H} \equiv \frac{1}{n}\sum_{i=1}^{n}\frac{1}{x_i}$$

For example, when $n = 2$:

$$H(x_1 x_2) = \frac{2x_1 x_2}{x_1 + x_2}$$

and when $n = 3$:

$$H(x_1 x_2 x_3) = \frac{3x_1 x_2 x_3}{x_1 x_2 + x_1 x_3 + x_2 x_3}$$

For $n = 2$, the harmonic mean is related to the arithmetic mean $A$ and the geometric mean $G$ by the following equation:

$$H = \frac{G^2}{A}$$

Cf. *Geometric mean of hydraulic conductivity.*

**Hazen method** An empirical equation to estimate the *hydraulic conductivity* of sandy sediments based on the effective diameter of sediment grains as determined by a *grain-size distribution* curve. The Hazen equation is as follows:

$$K = C(D_{10})^2$$

where:
$K$ = hydraulic conductivity [$L \cdot T^{-1}$]
$D_{10}$ = *effective grain size* in which 10% of the particles are finer by weight [L]
$C$ = coefficient.

The Hazen method is valid if the effective grain size of the sand is between 0.1 and 3.0 mm; values for the coefficient $C$ are listed in **Table H-3**.

**Table H-3** The Hazen approximation coefficient. (Driscoll, 1986. Reprinted by permission of Johnson Screens/a Weatherford Company.)

| | |
|---|---|
| Very fine sand, poorly sorted | 40–80 |
| Fine sand with appreciable fines | 40–80 |
| Medium sand, well sorted | 80–120 |
| Coarse sand, poorly sorted | 80–120 |
| Coarse sand, well sorted, clean | 120–150 |

Cf. *Permeability; Fair-Hatch equation.*

**Head** The energy of a water mass or a water body, produced by elevation, pressure, or velocity. Head is interpreted as the height (above a datum) of the surface of a column of fluid that can be supported by the static pressure. Head can be viewed as shown in **Figure H-3** as the water-level elevation in a well, or in a flowing artesian well, as the height that water rises in a pipe terminating within the *aquifer*. Cf. *Elevation head; Hydraulic head; Pressure head; Stage, Total head.*

**Head capacity curve** The curve representing the relationship between a water pump's capacity (pumping rate) and the height to which this capacity is raised (*head*) above the surface of the water body (or *water table*). The depth of the pump below the surface of the water body or water table is not important unless it is far enough for frictional losses to become significant.

**Head-dependent boundary** Also called a head-dependent condition or a constant-head boundary. An aquifer boundary where the *boundary condition* is determined by head parameters. A head-dependent boundary usually takes the form of a fixed value of *hydrostatic head*, which might be determined by the level of a surface water body adjoining the *aquifer*.

**Head-dependent condition** See *head-dependent boundary.*

**Head, elevation** A latent form of energy in water due to an elevated position. Elevation head is expressed as the height of a given point in a column of liquid above a datum. As the water descends, the elevation head can be converted to velocity or *pressure energy*. Cf. *Head; Bernoulli equation; Velocity head; Head total; Groundwater energy.*

**Head loss** The part of the *head* energy that is lost due to the friction of flowing water. Cf. *Friction head.*

**Head, pressure** The energy present in a water mass resulting from forces exerted to contain the water. Energy of movement may result if containment is removed. Cf. *Bernoulli equation; Pressure head; Velocity head; Groundwater energy.*

**Head, total** The sum of *elevation head, pressure head*, and *velocity head* at a specific point in an *aquifer*. **Figure H-3** shows this relationship. *Total head* is often expressed as an energy potential in the irrotational form of the *Bernoulli equation*. Cf. *Groundwater energy; Darcy's law.*

**Head, velocity** The energy component of a water mass, which is due to the movement of the water mass. Cf. *Bernoulli equation; Groundwater energy.*

**Headcut** A physical process that extends the *drainage network* within a watercourse *drainage area* by the headward or up-valley *erosion* of a scarp.

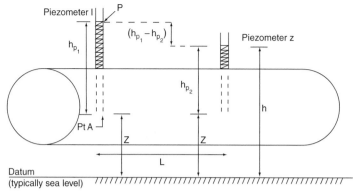

Pt A = point located at elevation Z, above a datum, at fluid pressure P.
P = fluid pressure at pt A.
z = (elevation head) elevation of the base of a piezometer above a datum.
$h_p$ = (pressure head) height of the water column in a piezometer.
h = (total head) sum of the elevation head (z) and pressure head ($h_p$).
$(h_{p_1} - h_{p_2})$ = the change in head (dh) over length L.

**Figure H-3** An apparatus consisting of a circular cylinder filled with sand to demonstrate Darcinian flow and showing the relationship between total head (h), elevation head (z) and pressure head ($h_p$). (After Fetter, 1994 and Freeze, 1979.)

**Heading** A horizontal tunnel penetrating fissures in an *aquifer*, constructed for water-supply purposes.

**Headpool** A body of water contained near the head of a *stream*.

**Headspace** The analysis or measurement of the air or *vapors* trapped within a *well casing* above the water, or in sample collection, the vapor between the meniscus and the sample cap. Many organic constituents in *groundwater* volatilize and collect in the air above the water. The amount of organic vapor in the headspace is proportional to the amount of these constituents dissolved in the water and their respective *vapor pressure*. Vapors trapped within a well casing are typically measured with an organic vapor analyzer (OVA) or similar device to assess the need for breathing protection. In groundwater sampling, if a sample is to be analyzed for volatile or semivolatile organic compounds, the sample container should be completely filled so that no air space remains. After collection, the sample vial should be inverted and tapped to see if a bubble appears, indicating that air is present, which would make the sample invalid.

**Headstream** A waterway, that is, the source, or one of the sources, supplying water to a larger *stream* or *river*, or the next higher-order stream. Cf. *Stream order*.

**Headwater** The source water or upper part of a stream, *river*, or contained watercourse in a *drainage basin*; or the water that is *upstream*. Cf. *Headstream*; *Stream order*.

**Heat capacity** The quantity of heat that raises the temperature of a system by 1 °C, expressed as calories (cal) of heat per degree (°C) Celsius. The heat capacity of water is 1 cal g$^{-1}$ °C$^{-1}$ in cgs units or 4186.8 J kg$^{-1}$ °C$^{-1}$ in SI units See: **Appendix A** for conversion units. Specific heat capacity refers to the heat capacity per unit mass of a given material. Heat capacity is treated as a constant for most thermodynamic calculations, but does vary with temperature and to a lesser extent with pressure.

**Heat flow, conductivity** See *depth of circulation; geothermal gradient; hydrothermal water; thermal conductivity; temperature logging*.

**Heat of condensation** See *latent heat of condensation*.

**Heat of fusion** See *latent heat of fusion*.

**Heat of hydration** When *Portland cement* is mixed with water, heat is liberated as a result of the exothermic reaction between the *cement* and water. The heat generated by the cement's hydration raises the temperature of concrete. A temperature rise of 55°C (100°F) have been observed with high cement content mixes. As a rule of thumb, the maximum temperature differential between the interior and the exterior concrete should not exceed 20°C (36°F) to avoid crack development. Cf. *Well installation*; *Grouting*.

**Heat of vaporization** See *latent heat of vaporization*.

**Heaving sands** A phenomenon observed during drilling (See: **Appendix C – Drilling Methods**) when sands below the *water table* tend to push up into the drill stem, either preventing further drilling or inhibiting sample collection. Saturated fine sands are typically more unstable, and sands under *artesian pressure* inhibit the drilling process even further. Heavy *drill muds* can counteract the pressures forcing the sands up the drill stem. A hollow-stem auger drill string with a *knockout plug* may also be useful with heaving sands, if sampling is not required. Cf. *Running sand*; *Quick conditions*.

**Hele-Shaw apparatus** See *parallel-plate model*.

**Hele–Shaw model** An immiscible or viscous fluid analogue model used to demonstrate *flow* and interaction of fluids with different densities, as occurs in *saltwater intrusion*. See *Parallel-plate model*; *Viscosity*.

**Helical flow** See *helicoidal flow*.

**Helicoidal flow** Also called helical flow. Fluid motion in a swirling or curled pattern near a bend in a watercourse, causing *erosion* of the concave outer *bank* and concurrent deposition along the convex inner bank.

**Henry's law** For gases that are slightly soluble in water, the relationship of the partial pressure of a gas at equilibrium dissolved in water to the *pressure* of that gas in the atmosphere. Henry's law is also stated as the partial pressure of a *solute* in a dilute solution being proportional to its molality; as such, it is applicable to gases of limited *solubility*, e.g., carbon dioxide ($CO_2$), oxygen ($O_2$), nitrogen ($N_2$), methane ($CH_4$), and hydrogen sulfide ($H_2S$). Significant mass can transfer between soil gases and *groundwater*. Under equilibrium conditions, the water-vapor partitioning is described by a linear relationship. Cf. *Henry's law constant*.

**Henry's law constant ($K_H$)** Also called the air–water partition coefficient. The ratio representing the equilibrium of the partial pressure of a compound in air to its concentration in water at a reference temperature. $K_H$ serves as the reference phase in water by the following equation:

$$c_a = K_H c_w$$

where:

$c_a$ = *solute* concentration in air (mol m$^{-3}$)
$c_w$ = *solute* concentration in water (mol m$^{-3}$)

Henry's law constant is dimensionless unless it is expressed as the ratio of the *vapor pressure*, $P_{vp}$, to the water *solubility*, S, for the constituent and therefore Henry's law constant, $K'_H$, has units of atm m$^3$ mol$^-$, as in the following equation:

$$K'_H = \frac{P_{vp}}{S}$$

**Table H-4** shows the vapor pressure, solubility, and Henry's law constant for many chemicals commonly identified as hazardous materials.

**Table H-4**  Common chemicals partitioning characteristics. (Maidment, 1993. Reprinted with permission from McGraw-Hill, Inc.)

| | Water solubility (mg L$^{-1}$) | Vapor pressure (atm) | Henry's law $K_H$ (atm · m$^3$ mol$^{-1}$) | Organic carbon $K_{oc}$ (L kg$^{-1}$) |
|---|---|---|---|---|
| Acetone | Infinite | 3.55E−01* | 2.06E−05 | 2.20E+00 |
| Aldrin | 1.80E−01 | 7.89E−09 | 1.60E−05 | 9.60E+04 |
| Atrazine | 3.30E+01 | 1.84E−09 | 2.59E−13 | 1.63E+02 |
| Benzene | 1.75E+03 | 1.25E−01 | 5.59E−03 | 8.30E+01 |
| Bis-(2-ethylhexyl)phthalate | 2.85E−01 | 2.63E−10 | 3.61E−07 | 5.90E+03 |
| Chlordane | 5.60E−01 | 1.32E−08 | 9.63E−06 | 1.40E+05 |
| Chlorobenzene | 4.66E+02 | 1.54E−02 | 3.72E−03 | 3.30E+02 |
| Chloroethane | 5.74E+03 | 1.32E+00 | 6.15E−04 | 1.70E+01 |
| DDT | 5.00E−03 | 7.24E−09 | 5.13E−04 | 2.43E+05 |
| Diazinon | 4.00E+01 | 1.84E−07 | 1.40E−06 | 8.50E+01 |
| Dibutyl phthalate | 1.30E+01 | 1.32E−08 | 2.82E−07 | 1.70E+05 |
| 1,1-Dichloroethane | 5.50E+03 | 2.39E−01 | 4.31E−03 | 3.00E+01 |
| 1,2-Dichloroethane | 8.52E+03 | 8.42E−02 | 9.78E−04 | 1.40E+01 |
| 1,1-Dichloroethene | 2.25E+03 | 7.89E−01 | 3.40E−02 | 6.50E+01 |
| 1,2-Dichloroethene *(trans)* | 6.30E+03 | 4.26E−01 | 6.56E−03 | 5.90E+01 |
| Dieldrin | 1.95E−01 | 2.34E−10 | 4.58E−07 | 1.70E+03 |
| Ethyl benzene | 1.52E+02 | 9.00E−03 | 6.43E−03 | 1.10E+03 |
| Methylene chloride | 2.00E+04 | 4.76E−01 | 2.03E−03 | 8.80E+00 |
| Methyl parathion | 6.00E+01 | 1.28E−08 | 5.59E−08 | 5.10E+03 |
| Naphthalene | 3.17E+01 | 3.03E−04 | 1.15E−03 | 1.30E+03 |
| Parathion | 2.40E+01 | 4.97E−08 | 6.04E−07 | 1.07E+04 |
| Phenol | 9.30E+04 | 4.49E−04 | 4.54E−07 | 1.42E+01 |
| Tetrachloroethene (PERC) | 1.50E+02 | 2.30E−02 | 2.59E−02 | 3.64E+02 |
| Toluene | 5.35E+02 | 3.70E−02 | 6.37E−03 | 3.00E+02 |
| Toxaphene | 5.00E−01 | 5.26E−04 | 4.36E−01 | 9.64E+02 |
| 1,1,1-Trichloroethane | 1.50E+03 | 1.62E−01 | 1.44E−02 | 1.52E+02 |
| Trichloroethene (TCE) | 1.10E+03 | 7.60E−02 | 9.10E−03 | 1.26E+02 |
| Trichloromethane (chloroform) | 8.20E+03 | 1.99E−01 | 2.87E−03 | 4.70E+01 |
| Vinyl chloride | 2.67E+03 | 3.50E+00 | 8.19E−02 | 5.70E+01 |
| o-Xylene | 1.75E+02 | 9.00E−03 | 5.10E−03 | 8.30E+02 |

Cf. *Octanol–water partition coefficient*.

**Hesitation set** *See False set.*

**Heterogeneity/heterogeneous** Also called non-uniformity. A medium or formation characteristic where the material properties vary from point to point, resulting in a lack of uniformity of the material properties and conditions. **Figure H-4** shows heterogeneity within a formation where *porosity, specific storage,* and *hydraulic conductivity* may be constant but the bed thickness changes spatially and therefore changes the *hydraulic characteristics* of the formation. Layered units within a sedimentary deposit may change in the vertical and/or the horizontal direction and can therefore be non-homogeneous, or heterogeneous, as well.

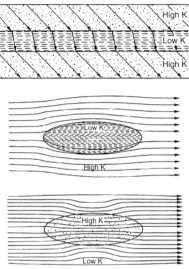

**Figure H-4** Flow crossing a boundary between materials of two different hydraulic conductivities is refracted at the boundary. (Sanders, 1998.)

Cf. *Isotropic; Anisotropic; Homogeneity.*

**High-level groundwater** *Perched water* separated from the *water table* by less-permeable materials such as ash beds, intrusive igneous rocks, or ice. High-level *groundwater* eventually joins the *basal groundwater* by flowing through, over, and around the lower permeable zone. Cf. *Groundwater elevation.*

**Hillside spring** See *spring, contact.*

**Hjulstrom effect** The noted difference between the *flow velocities* for deposition and *erosion* of fine-grained cohesive sediments. Silts and clays deposit only at low flow velocities, but, in contrast to larger-grained sediments, they require relatively very high velocities to be eroded once deposited, due to the cohesiveness of the finer grains. Cf. *Grain size analysis.*

**Hollow-stem auger** See **Appendix C – Drilling Methods**.

**Holomictic lake** An inland water body that, at least once per year, experiences a complete mixing or overturn of its water during circulation. Cf. *Dimictic; Monomictic; Thalassic series; Density stratification.*

**Holtan model** Holtan–Lopez modified the *Horton model* of *infiltration* based on the concept that the *soil moisture* storage, surface-connected *porosity,* and the root paths in soil are the dominant factors affecting the infiltration capacity of the soil. The Holtan equation for infiltration is as follows:

$$I = GIAS_a^{1.4}$$

where:
$I$ = infiltration rate [L·T$^{-1}$],
GI = growth index of the crop in percent maturity [between 0.1 and 1.0 during the season],
$A$ = infiltration capacity [L·T$^{-1}$],
$S_a$ = available *storage* in the surface layer A horizon [L]
The infiltration rate is a constant when the infiltration rate curve reaches an asymptote, thereby indicating a steady infiltration rate. **Table H-5** shows infiltration rates used in the above equation for each *hydrologic soil group.*

**Table H-5** Holton infiltration model parameter estimates. (Maidment, 1993. Reprinted with permission from McGraw-Hill, Inc.)
**Estimates of Vegetative Parameter A in Holtan Infiltration Model**

| Land use or cover | Basal area rating[a] | |
|---|---|---|
| | Poor condition | Good condition |
| Fallow[b] | 0.10 | 0.30 |
| Row crops | 0.10 | 0.20 |
| Small grains | 0.20 | 0.30 |
| Hay (legumes) | 0.20 | 0.40 |
| Hay (sod) | 0.40 | 0.60 |
| Pasture (bunch grass) | 0.20 | 0.40 |
| Temporary pasture (sod) | 0.20 | 0.60 |
| Permanent pasture (sod) | 0.80 | 1.00 |
| Woods and forests | 0.80 | 1.00 |

[a] Adjustments needed for "weeds" and "grazing."
[b] For fallow land only, poor condition means "after row crop" and good condition means "after sod."
Source: Ref. 30.

**Final Infiltration Rates by Hydrologic Soil Groups for Holtan Infiltration Model**

| Hydrologic soil group | $f_c \, (\mathrm{cm\,h^{-1}})$ |
|---|---|
| A | 0.76 |
| B | 0.38–0.76 |
| C | 0.13–0.38 |
| D | 0.0–0.13 |

Cf. *Runoff*.

**Homogeneity/homogeneous** A material or formation characteristic when the material properties are spatially identical or vary within a range an analyst would consider to be unimportant for the problem being evaluated. Wherever present, the *transmissivity* and storativity values would be the same throughout the formation. A consolidated unit would contain the same fracturing density, strike-and-dip joint sets, solution openings, degree of cementation, or *porosity* at all locations. Cf. *Anisotropic; Heterogeneity; Isotropic; Storage coefficient*.

**Homopycnal inflow** Water flow into a water body of the same density. Homopycnal inflow results in easier mixing. Cf. *Hyperpycnal inflow; Hypopycnal inflow; Density stratification*.

**Horizontal profiling** A *geophysical exploration method* used in *resistivity surveys* to determine lateral variations in formation resistivity. The use of horizontal profiling in hydrogeologic investigations is based on the premise that dry materials have a higher resistivity than the same materials when wet, and that gravel is more resistive, even when moist, than finer-grained materials such as clay, containing the same moisture content as the electrically charged surfaces of clays and silts, which are better conductors. Cf. *Electrical sounder; Schlumberger array; Wenner array*.

**Horizontal well** A well installed horizontally, or parallel to the ground surface, for the purpose of pumping or sampling fluid from producing strata. Horizontal wells are used in water production, vapor extraction, and remediation. Cf. *Kanat/ganat/qanat; Infiltration galleries*.

**Horseshoe lake** More commonly called an *oxbow lake*. A water body residing in a concave basin in the shape of a horseshoe.

**Horton analysis** R. E. Horton (1945) demonstrated that the *stream order* is related by geometric relationships to the number of streams, the channel length, and *drainage area*. **Figure H-5** is a semilogarithmic plot of stream order against the above-mentioned variables. Horton's system of stream order designation, Class I, is a *tributary* and increases to the highest classification number to the main stream or stem. The largest classifications were observed in the larger *watersheds* with the greatest number and extent of tributaries. Cf. *Bifurcation ratio*.

**Hortonian flow** See *Horton analysis; Horton model; Horton overland flow; surface runoff*.

**Horton model** An equation relating the *infiltration rate* or volume to time, as modified by soil parameters. The *infiltration capacity*, $I_p$, related to time is expressed as follows:

$$I_p = I_c + (I_0 - I_c)e^{-\beta t}$$

where:
$I_0 =$ maximum infiltration rate at the beginning of storm event [$\mathrm{L \cdot T^{-1}}$]
$I_c =$ constant rate of infiltration as soil becomes saturated [$\mathrm{L \cdot T^{-1}}$]
$\beta =$ parameter controlling rate of decrease in infiltration capacity [T]
$t =$ time since the beginning of event [T]

The values of $I_c$, $I_0$, and $\beta$, which are estimated from empirical infiltration data, are given in **Table H-6**. Horton's model applies only when the effective *rainfall intensity* is greater than $I_c$. In addition, the Horton model depends on specific soil and moisture conditions, which may limit its use. Cf. *Holtan model; Hydrologic soil group; Runoff*.

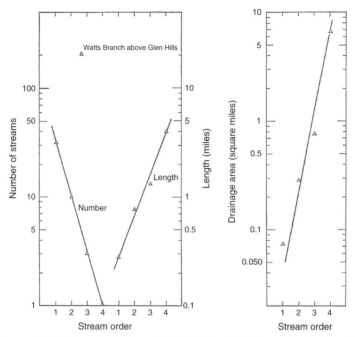

**Figure H-5** Horton analysis of drainage networks. (Leopold et al., 1992. Reprinted with permission from Dover Publications, Inc.)

**Table H-6** Horton infiltration model parameter estimates. (Maidment, 1993. Reprinted with permission from McGraw-Hill, Inc.)

| Soil and cover complex | $I_o$ (mm h$^{-1}$) | $I_c$ (mm h$^{-1}$) | $\beta$ (min$^{-1}$) |
| --- | --- | --- | --- |
| Standard agricultural (bare) | 280 | 6–220 | 1.6 |
| Standard agricultural (turfed) | 900 | 20–290 | 0.8 |
| Peat | 325 | 2–29 | 1.8 |
| Fine sandy clay (bare) | 210 | 2–25 | 2.0 |
| Fine sandy clay (turfed) | 670 | 10–30 | 1.4 |

**Horton overland flow** Also called Hortonian flow. A storm *runoff* process, named after Robert Horton, which occurs when the *precipitation* rate exceeds the *infiltration capacity* of the soils and the *depression storage* is substantially filled. **Figure H-6** shows the relationship between precipitation, infiltration, and *overland flow*. Cf. *Throughflow; Baseflow; Baseflow recession; Hydrograph; Interflow*.

**Horton runoff cycle** A *hydrograph* model of *baseflow recession curves* most used in *arid-zone hydrology*.

**Horton's stream order** See *Horton analysis*.

**Horton–Theissen method** A graphical method of estimating the mean *precipitation* over a given area. As shown in **Figure H-7**, a series of polygons is constructed with a *gauging station* at the centers, delimited by the perpendicular bisectors of lines between adjacent *gauges*, and the weight factors are determined by the percentage of the total area within each polygon. **Table H-7** is a sample calculation based on **Figure H-7** demonstrating that the mean precipitation is the total of the products. The Horton–Theissen method of estimation is valuable because it utilizes an established *gauge network*.

**Hot spring** See *spring, hot; spring, thermal*.

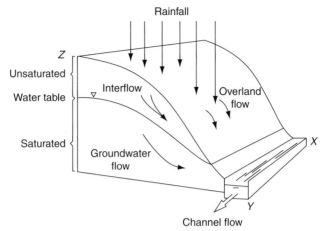

**Figure H-6** The relationship between precipitation as rainfall and hydrologic components. (Shaw, 1983. Reprinted with permission from Van Nostrand Reinhold Co. Ltd.)

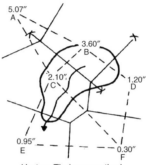

Horton–Theissen method

**Figure H-7** Calculation of areal precipitation by the Horton–Theissen method. (Black, 1996. Reprinted with permission from Ann Arbor Press, Inc.)

**Table H-7** Data for the calculation of mean precipitation in the Horton–Theissen method. (Black, 1996. Reprinted with permission from Ann Arbor Press, Inc.)

| Horton-Thiessen method | | | | |
|---|---|---|---|---|
| Gage No. ($\theta$) | Recorded precipitation (in) | Dod grid tally (No.) | Percent of total area (%) | Weighted precipitation (in) |
| A | 5.07 | - | - | - |
| B | 3.60 | 244 | 32.1 | 1.16 |
| C | 2.10 | 40S | 53.0 | 1.11 |
| D | 1.20 | 99 | 12.8 | .15 |
| E | .95 | 10 | 2.1 | .02 |
| F | .30 | – | – | – |
| Totals | | 760 | 100.0 | 2.44 |

**Humidity** The atmospheric water *vapor* content. Humidity may be a serious consideration in many hydrologic and hydro-geologic investigations. Field operations may require alterations because hot humid conditions could compromise worker safety, and equipment may need to be protected from excessive moisture. Design of certain hydrologic facilities depends on evaporative losses or cooling, both of which vary with humidity. Cf. *Evaporation; Absolute humidity; Relative humidity.*

**Hurst phenomenon** The suggestion that there is long-term persistence in the time series of hydrologic events in a system. The common assumption for hydrologic systems was that each event was independent of all others except the immediately preceding event. E. H. Hurst's long-term observations of the Nile River system suggest that a more persistent model showed relations between the whole series of hydrologic events. Cf. *Hydrologic cycle.*

**Hvorslev method/Hvorslev piezometer test** A means of calculating the *hydraulic conductivity* of a formation by using a partially penetrating *piezometer* (one that is not screened across the complete vertical extent of the *aquifer*) and a *slug test.* **Figure H-8** shows the geometry of piezometer installation for the Hvorslev method.

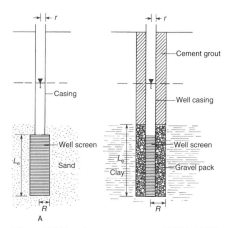

**Figure H-8** The piezometer geometry. (ASTM, 1992.)

Data from the Hvorslev slug test are plotted by calculating the *head* ratio of $h/h_0$ (where $h_0$ is the water level above *static water level* immediately after lowering the slug, and $h$ is the water level above static water level at some time $t$), versus time on a semilogarithmic scale (**Figure H-9**). The time–head ratio data plot on a straight line, and if the length of the screen is more than eight times the well screen radius, the hydraulic conductivity, $K$, is approximated by the following equation:

$$K = \frac{r^2 \, \ln(L/R)}{2LT_0}$$

where:
$K$ = hydraulic conductivity [L·T$^{-1}$]
$r$ = radius of the *well casing* [L]
$L$ = length of the *well screen* [L]
$R$ = the radius of the screen or *filter pack* [L]
$T_0$ = time taken for the water level to fall to 37% of the initial change [T], as shown in **Figure H-9**

Other formulas have been developed for other well geometries. Cf. *Cooper–Bredehoeft–Papadopulos method.*

**Hvorslev slug test** See *Hvorslev method/Hvorslev piezometer test; Slug test.*

**Hydration water** See *water of hydration.*

**Hydraulic action** The use of water pressure and the *hydraulic force* of moving water to mechanically loosen or remove material. Examples of hydraulic action are bank *erosion*, stream surges into rock fissures, and wave and current movement or surges.

**Hydraulic barrier** A blockage of *groundwater flow* resulting from an induced *hydraulic gradient* such as that produced by a line of *injection wells* or groundwater *extraction* wells. Cf. *ExtractionWells.*

**Hydraulic boundary** An interface within a *groundwater region* separating areas of different *hydraulic characteristics*, such as differences in *porosity, storativity, conductivity, transmissivity,* or any combination thereof. Cf. *Hydrogeochemical facies; Hydrogeologic boundary; Aquiclude; Hydraulic barrier; Storage capacity; Hydrogeochemistry.*

**Hydraulic capacitance** See *specific storage.*

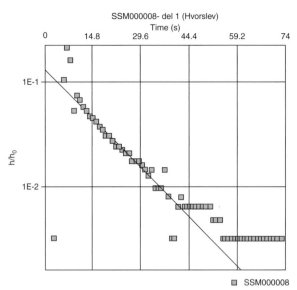

**Figure H-9** The Hvorslev plot of head ratio versus time. (Johansson, 2004.)

**Hydraulic characteristics** Parameters that aid in describing a specific section of a watercourse. Examples of hydraulic characteristics include width, depth, velocity, channel slope, *roughness*, and bed particle size. Cf. *Heterogeneity; Homogeneity; Hydraulic geometry.*

**Hydraulic conductivity (K)** The measure of the resistance to movement of water flowing through a *porous medium*, expressed in dimensions of $L \cdot T^{-1}$. Hydraulic conductivity within a saturated zone is generally substantially greater than that in the *unsaturated zone* because there is enough water in the *pore spaces* to satisfy adhesion and cohesion properties, allowing the excess water to flow freely. The following equation is used to calculate the hydraulic conductivity $K$:

$$K = \frac{Q}{A(\mathrm{d}h/\mathrm{d}L)}$$

where:
$Q = discharge\ [L^3 \cdot T^{-1}]$
$A = area\ [L^2]$
$(\mathrm{d}h/\mathrm{d}L) = gradient\ [L \cdot L^{-1}]$

Hydraulic conductivity is a function of the properties of the liquid as well as of the *permeability* of the medium through which it passes. The permeability, $k$, however, is solely a function of the medium. **Table H-8** lists typical ranges of hydraulic conductivities and permeabilities for sediments and rocks. The more viscous a fluid is, the slower is its velocity; therefore, hydraulic conductivity is directly proportional to the specific weight of the fluid $\gamma$ and inversely proportional to the *dynamic viscosity* of the fluid). Cf. *Darcy's law; Permeability coefficient; Porosity; Hydraulic diffusivity; Permeameter; Zone of saturation.*

**Hydraulic conductivity ellipsoid** A means of relating the principal *hydraulic conductivity* components $K_x$, $K_y$, and $K_z$ in the x, y, and z direction, to the hydraulic conductivity in any direction. This is described by the following equation for an ellipsoid (in Cartesian coordinates):

$$\frac{r^2}{K_s} = \frac{x^2}{K_x} + \frac{y^2}{K_y} + \frac{z^2}{K_z}$$

**Hydraulic containment** The modification of a *hydraulic gradient* by *groundwater* pumping, fluid injection, and/or installation of a cutoff wall or *grout curtain* for the purpose of controlling contaminants moving with the *groundwater flow*. Cf. *Groundwater mining.*

**Hydraulic corer/piston corer** A marine- or *lake*-sampling tube that penetrates by hydraulic pressure.

**Table H-8** Hydraulic conductivity and permeability ranges. (Sanders, 1998.)

| Sediment/Rock | Hydraulic conductivity (cm/sec) | Intrinsic permeability (cm$^2$) |
|---|---|---|
| Clay | $10^{-9}$ to $10^{-6}$ | $10^{-14}$ to $10^{-11}$ |
| Silt | $10^{-7}$ to $10^{-3}$ | $10^{-12}$ to $10^{-8}$ |
| Sand | | |
|   Fine or silty | $10^{-5}$ to $10^{-3}$ | $10^{-10}$ to $10^{-8}$ |
|   Course or well-sorted | $10^{-3}$ to $10^{-1}$ | $10^{-8}$ to $10^{-6}$ |
| Gravel | $10^{-1}$ to $10^{+2}$ | $10^{-6}$ to $10^{-3}$ |
| Till | | |
|   Dense/unfractured | $10^{-9}$ to $10^{-5}$ | $10^{-14}$ to $10^{-9}$ |
|   Fractured | $10^{-7}$ to $10^{-3}$ | $10^{-12}$ to $10^{-8}$ |
| Shale | | |
|   Intact | $10^{-11}$ to $10^{-7}$ | $10^{-16}$ to $10^{-12}$ |
|   Fractured/weathered | $10^{-7}$ to $10^{-4}$ | $10^{-12}$ to $10^{-9}$ |
| Sandstone | | |
|   Tightly cemented | $10^{-8}$ to $10^{-5}$ | $10^{-13}$ to $10^{-10}$ |
|   Loosely cemented | $10^{-6}$ to $10^{-3}$ | $10^{-11}$ to $10^{-8}$ |
| Limestone and dolomite | | |
|   Non-karst | $10^{-7}$ to $10^{-3}$ | $10^{-12}$ to $10^{-8}$ |
|   Reef or karst | $10^{-4}$ to $10^{+4}$ | $10^{-9}$ to $10^{-1}$ |
| Chalk | $10^{-6}$ to $10^{-3}$ | $10^{-11}$ to $10^{-8}$ |
| Anhydrite | $10^{-10}$ to $10^{-9}$ | $10^{-15}$ to $10^{-14}$ |
| Salt | $10^{-12}$ to $10^{-5}$ | $10^{-17}$ to $10^{-9}$ |
| Basalt | | |
|   Unfractured | $10^{-9}$ to $10^{-6}$ | $10^{-14}$ to $10^{-11}$ |
|   Fractured/vesicular | $10^{-4}$ to $10^{+3}$ | $10^{-9}$ to $10^{-2}$ |
| Unfractured igneous and metamorphic rocks | $10^{-12}$ to $10^{-8}$ | $10^{-17}$ to $10^{-13}$ |
| Fractured igneous and metamorphic rocks | $10^{-8}$ to $10^{-4}$ | $10^{-13}$ to $10^{-9}$ |

**Hydraulic current** The differences of water levels at opposing ends of a channel or watercourse producing a local continuous movement of water typically caused by a rising and falling tide.

**Hydraulic diffusivity (D)** An *aquifer* unit property or formation parameter calculated by dividing the *hydraulic conductivity* by the *specific storage*. Hydraulic diffusivity can also be determined by dividing the *transmissivity* by the storage coefficient as in the following equation:

$$D = \frac{T}{S} = \frac{K}{S_s}$$

where:
$T$ = transmissivity [L$^2$·T$^{-1}$]
$S$ = storage coefficient [dimensionless]
$K$ = hydraulic conductivity [L·T$^{-1}$]
$S_s$ = specific storage [L$^{-1}$]

Diffusivity has units of L$^2$·T$^{-1}$ and describes how *head* perturbances will propagate through a porous medium. Cf. *Laplace equation*.

**Hydraulic dispersion** See *Advection; Dispersion; Mechanical dispersion; Advection-dispersion equation*.

**Hydraulic element** A quantity within a specific stage and cross section of flowing water in a watercourse. Examples of hydraulic elements include depth of water, cross-sectional area, *hydraulic radius*, wetted perimeter, mean depth of water, velocity, *flow*, energy *head*, and friction factor.

**Hydraulic force** The capacity of moving water, without the addition of sediments, to erode. Cf. *Hydraulic action*.

**Hydraulic fracturing** Also called hydrofracturing. The fracturing of rock or sedimentary formations by pumping in pressurized fluid (usually water) and granular material (commonly sand). The purpose of hydraulic fracturing is to increase the *permeability* of a formation for production of water or petroleum, as well as for the cleanup of a contaminated *aquifer*, by creating artificial openings. The pressure of the injected fluid opens cracks and bedding planes, and the injected granular material provides a *permeable* path and maintains the enhanced openings after the pressure is released.

**Hydraulic friction** The resistance exerted by the surface of a watercourse to the flow of fluid as a result of the confining surface's roughness characteristics that reduce the fluid energy by drag. Stream characteristics such as turbulence and sinuosity are not considered functions of hydraulic friction.

**Hydraulic geometry** A method of describing a specific section of a watercourse showing the graphical relationships of plots of *hydraulic characteristics* by determining the shape of a natural watercourse as a simple power function of *discharge*. Watercourse adjustments include width, mean depth, mean velocity, *load*, slope, and water surface *gradient*. The relationship of these adjustments to *discharge* is expressed by the basic equations of hydraulic geometry, which are as follows:

$$B = aq^b$$
$$D = cq^f$$
$$v = kq^m$$

where:

$q$ = discharge [L$^3$·T$^{-1}$]
$B$ = channel top width [L]
$D$ = channel mean depth [L]
$v$ = mean velocity [L·T$^{-1}$]
$a, b, c, f, k, m$ = numerical constants.

Discharge is equal to $BDv$, therefore $a, c, k = 1$ and $b + f + m = 1$. When plotted on log–log graph paper, the above equations plot as a straight line and the exponents are the slope of the line while the coefficients are the intercept when $q = 1$. The values of the exponents as determined by various hydrologic investigations are presented in **Table H-9**. **Figure H-10**

**Table H-9** The hydraulic geometry interrelationship between high and low flow at stations A and B. (Linsley et al., 1982. Reprinted with permission from McGraw-Hill, Inc.)

|  | At station | | | Between stations | | |
|---|---|---|---|---|---|---|
|  | **b** | **f** | **m** | **b** | **f** | **m** |
| Average, midwestern states [8] | 0.26 | 0.40 | 0.34 | 0.5 | 0.4 | 0.1 |
| Ephemeral streams in semiarid United States [9] | 0.29 | 0.36 | 0.34 | 0.5 | 0.3 | 0.2 |
| Average, 158 United States stations [2] | 0.12 | 0.45 | 0.43 | | | |
| 10 stations on the Rhine River [2] | 0.13 | 0.41 | 0.43 | | | |
| Appalachian streams [10] | | | | 0.55 | 0.36 | 0.09 |
| Kaskaskia River, Ill. [11] | | | | 0.51 | 0.39 | 0.14 |

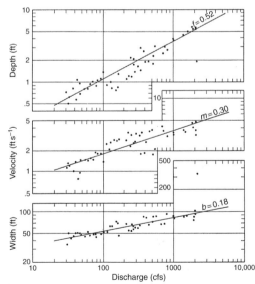

**Figure H-10** A river cross-sectional discharge with changes in width, mean depth, and mean velocity. (Leopold et al., 1992. Reprinted with permission from Dover Publication, Inc.)

shows "at-a-station curves" or plots of discharge at a given watercourse cross section versus adjustment variables or the channel characteristics. The average hydraulic geometry of channels is diagramed in **Figure H-11** along with a comparative graphic display of the channel characteristics.

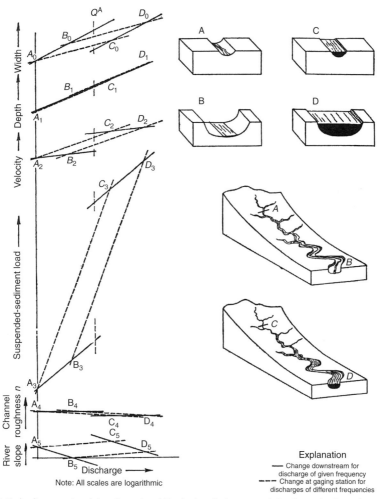

**Figure H-11** Hydraulic geometry of river channels: width, depth, velocity, suspended sediment load, roughness, and slope to discharge at station A and downstream B. (Leopold et al., 1992. Reprinted with permission from Dover Publication, Inc.)

**Hydraulic grade line** A line coinciding with the level of flowing water at any point along an open channel. In closed pipe systems, it is the level at which the water would rise in a *manometer* placed along the pipe. If the hydraulic grade line is above the crown of the pipe, then *pressure flow* conditions exist. If the hydraulic grade line is below the crown of the pipe, then open channel flow conditions exist. The hydraulic grade line is determined by subtracting the *velocity head* ($V^2/2g$) from the energy gradient (which represents the total kinetic plus *potential energy* in the system). Cf. *Flowline*; *Open channel flow*.

**Hydraulic gradient (I)** The change of total *head* per unit distance, at any point and in a specified direction; or the slope of a line representing the combined kinetic and potential energy along a water body. For steady, *uniform flow*, hydraulic gradient is equal to the slope of the water surface along the stream channel. The *gradient* is expressed by the following equation:

$$i = \frac{dh}{dL}$$

where:

$dh$ = change in head between two points that are relatively close together [L]

$dL$ = distance between the points [L].

The maximum *flow velocity* coincides with the area where the maximum fall in head per unit distance occurs. **Figure H-12** shows the determination of the hydraulic gradient from groundwater level data collected from *piezometers*. Negative values would indicate that the *flow* is in the direction of decreasing *hydraulic head*.

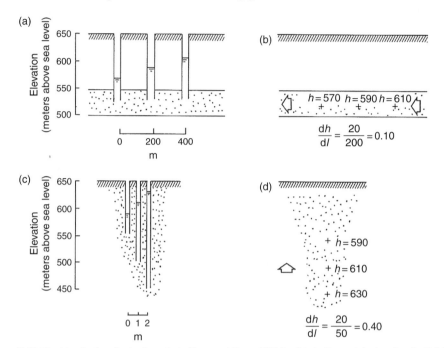

**Figure H-12** The determination of hydraulic gradients. (Freeze and Cherry, 1979. Reprinted with permission from Prentice-Hall, Inc.)

Cf. *Darcy's law; Pressure gradient.*

**Hydraulic head (h)** A specific measurement of water pressure above a geodetic datum. It is usually measured as a water surface elevation, expressed in units of length, at the entrance (or bottom) of a *piezometer*. In an aquifer, it can be calculated from the depth to water in a monitoring well or piezometer, and given information of the piezometer's elevation and screen depth. The hydraulic head can be used to determine a *hydraulic gradient* between two or more points. The fluid-driving force where the total mechanical energy per unit weight (equal to the sum of the kinetic, gravitational, and fluid *pressure energy*) in units of energy per unit weight (See: **Appendix A**) as shown in the following equation:

$$h = \frac{v^2}{2g} + z + \frac{P}{\rho g}$$

where:

$v$ = velocity [L·T$^{-1}$]

$g$ = gravitational constant [L·T$^{-2}$]

$z$ = elevation of the center of gravity of the fluid above the reference elevation, i.e., *elevation head* [L]

$P$ = *pressure* [M·L·T$^{-2}$]

$\rho$ = density of a fluid [M·L$^{-3}$].

The resultant hydraulic head is in units of length and represents the elevation of a body above a datum. In *groundwater flow*, the velocity components of energy are not significant, and therefore the term $v^2/2g$ is eliminated and the formula for hydraulic head in *groundwater* becomes:

$$h = z + \frac{P}{\rho g}$$

Because the weight of the overlying water per unit cross-sectional area is equal to the pressure, $P$, at any point, the second term can be replaced by the height of the water column as seen in the following equation:

$$h = z + \psi$$

where:

$\psi = pressure\ head$ [L].

**Figure H-3** displays all the components of head, i.e., the pressure head, $\psi$, and elevation head, $z$, equal to the *total head*. It is equal to the height above a given subsurface point of the *free surface* in a watercourse. Cf. *Hydraulic potential; Fluid potential; Artesian head; Fluid pressure.*

**Hydraulic jump** A change in fluid *flow* conditions accompanied by a stationary, abrupt, turbulent rise in the fluid level. In watercourse flow, particularly at the *upstream* or the *downstream* edge of man-made structures, abrupt changes in channel configuration may occur. Applying conservation of mass, momentum, and energy across one of these changes shows that a change in water level, or a hydraulic jump, is necessary to satisfy all three of the conditions simultaneously. Cf. *Flume, Parshall.*

**Hydraulic loading rate** The average *infiltration* into a *recharge basin* divided by time.

**Hydraulic mining** The substitution of water power for manpower to move vast quantities of soil, most often found in gold mining to separate the placer deposits from the gold. Unregulated hydraulic mining practices result in *loads* in excess of a *stream's capacity*, leading to extensive aggradation in the *downstream* region. Cf. *Hydraulic plucking.*

**Hydraulic percussion drilling** See **Appendix C – Drilling Methods.**

**Hydraulic permeability** See *permeability.*

**Hydraulic plucking** The removal of rock fragments resulting from stream *erosion* as water impacts cracks in rock surfaces. Cf. *Hydraulic mining.*

**Hydraulic potential** A measurable quantity at every point in a *flow system* where movement is always from regions of higher quantity to lower quantity regardless of location in space. The maximum hydraulic potential occurs at the crest of the *groundwater divide*, and minimum hydraulic potential occurs at the terminal base of *drainage* or the sea. Cf. *Hydraulic head; Fluid potential.*

**Hydraulic pressure** A pressure applied to any part of a confined fluid transmits to every other part with no loss. The pressure acts with equal force on all equal areas of the confining walls and perpendicular to the walls. Pressure (force/area) is usually measured in Pascal units, atmospheres, bars, or pounds per square inch. Refer to **Appendix A** for unit conversions. Cf. *Hydraulic head.*

**Hydraulic pressure test** See *packer test.*

**Hydraulic profile** A vertical section of the *potentiometric surface* in an aquifer.

**Hydraulic radius (R)** The ratio of a *stream's* cross section with its wetted perimeter or the water channel contact length in the cross-sectional area. This ratio is a measure of the frictional forces acting on a stream in proportion to the inertial forces or *channel efficiency* and is a controlling factor in *flow velocity*. The hydraulic radius, $R$, is defined as:

$$R = \frac{A}{2d + w}$$

where:

$A = area\ of\ cross\ section$ [$L^2$]
$d = channel\ depth$ [L]
$w = channel\ width$ [L]

The higher the ratio, the more efficiently the channel conveys water. The boundary of the channel experiences a frictional drag at its edges, while the body of the water in the channel is subjected to gravitational acceleration and possesses kinetic energy due to its motion. Cf. *Radius of influence; Form ratio; Manning's equation.*

**Hydraulic resistance (R_h)** The quantification of the retardation of vertical *flow* (as in an *aquitard*), which is the ratio of the distance traveled to the *hydraulic conductivity*. The hydraulic resistance is a measure of time. The length of the flow path is directly proportional to the effect of retarding or delaying flow: the longer the flow path is, the longer the flow is delayed. The hydraulic conductivity of the medium though which flow occurs is inversely proportional to the resistance: the greater the hydraulic conductivity, the less resistance to flow. The hydraulic resistance, $R_h$, is expressed as follows:

$$R_h = \frac{D}{K}$$

where:

$D = distance\ traveled$ [L]
$K = hydraulic\ conductivity\ of\ the\ medium$ [$L^2 \cdot T^{-1}$].

The hydraulic resistance is useful in estimations of *tracer tests* when the $R_h$ of the *aquifer* is the time required by the *tracer* to travel a given distance.

**Hydraulic routing** *Flow routing* is classified as either consolidated or distributed flow. Consolidated flow is hydrologic routing as the flow is computed as a function of time at one location along a water course. When the flow is distributed as in hydraulic routing the flow is calculated as a function of time simultaneously at many cross sections along the water course. Cf. *Dynamic equation; Attenuated flood wave; Flood routing; Flow routing.*

**Hydrochemical facies** The chemical character of water in hydrologic systems as a function of the lithology, flow path, and the solution kinetics within the system. The hydrochemical facies forms a distinct zone with dissolved concentrations that are within defined compositional categories. Although typically used in *groundwater* analysis, the use of facies in surface water is useful. One method of classification of hydrochemical facies has been based on the dominant anions and cations by means of *Stiff diagrams, Piper trilinear diagrams*, etc. **Figure H-13** combines the

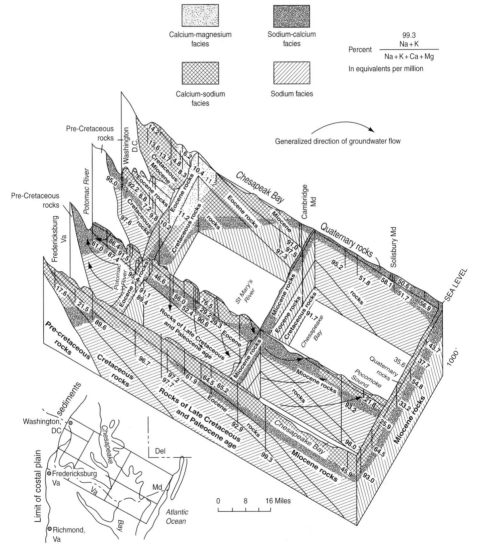

**Figure H-13** Fence diagram showing the hydrochemical cation facies. (Freeze and Cherry, 1979. Reprinted with permission from Prentice-Hall, Inc.)

geological fence diagram, illustrating rock types with the associated hydrochemical facies showing cation facies and *groundwater flow*. It is the chemical constituent and concentration in solution. Cf. *Electrochemical sequence; Hydrogeochemistry; Geochemical processes; Geochemical speciation; Groundwater facies.*

**Hydrochemistry** The study of the chemical constituents and processes in water. Dissolved constituents in water influence the media through which it flows, and analysis of the water composition provides clues to the *flow* path and geologic history. Cf. *Hydrogeochemistry; Geochemical processes; Geochemical speciation.*

**Hydrochloric acid** Often commercially known as *muriatic acid*. An inorganic acid, HCl, used extensively for removing mineral scale from *well screens* and *well casings* as well as for many other descaling applications outside the field of *hydrogeology*. When used as a well-scale remover, it is commonly diluted to about a 30% solution, and inhibitors may be added to reduce *corrosion* of the well components. To treat a well screen, HCl is usually pumped down a small-diameter pipe to the desired depth in the well. It is recommended to add enough acid to fill the screened length of the well, plus an additional volume of up to 50%. The appropriate quantities of acid needed for various sizes of wells are shown in **Table H-10**. Although HCl is very effective in removing mineral scale from wells, it is also very dangerous to handle and use. Contact with HCl can produce serious injuries to human tissue. Also, during its action as a descaling agent, chlorine gas is released, which can cause death or serious lung damage if inhaled.

**Table H-10**  Hydrochloric acid requirement for the treatment of incrusted screens. (Driscoll, 1986. Reprinted by permission of Johnson Screens/a Weatherford Company.)

| Screen diameter | | Amount of HCl acid (18° to 20° Baumé) per ft (0.3 m) of screen | |
|---|---|---|---|
| in. | mm | Gallons | Liters |
| 1½ | 38 | 0.11–0.14 | 0.42–0.53 |
| 2 | 51 | 0.20–0.24 | 0.76–0.91 |
| 2½ | 64 | 0.33–0.39 | 1.25–1.48 |
| 3 | 76 | 0.46–0.56 | 1.74–2.12 |
| 3½ | 89 | 0.63–0.75 | 2.38–2.84 |
| 4 | 102 | 0.81–0.98 | 3.07–3.71 |
| 4½ | 114 | 1.04–1.25 | 3.94–4.73 |
| 5 | 127 | 1.28–1.53 | 4.84–5.79 |
| 5½ | 140 | 1.54–1.85 | 5.83–7.00 |
| 6 | 152 | 1.84–2.21 | 6.96–8.36 |
| 7 | 178 | 2.50–3.00 | 9.5–11.4 |
| 8 | 203 | 3.26–3.92 | 12.3–14.8 |
| 10 | 254 | 5.10–6.12 | 19.3–23.2 |
| 12 | 305 | 7.35–8.82 | 27.8–33.4 |
| 14 | 356 | 10.0–12.0 | 37.9–45.4 |
| 16 | 406 | 13.1–15.7 | 49.4–59.4 |
| 18 | 457 | 16.5–19.8 | 62.6–75.1 |
| 20 | 508 | 20.4–24.5 | 77.2–92.7 |
| 22 | 559 | 24.7–29.6 | 93.5–112 |
| 24 | 610 | 29.4–35.3 | 111–133 |
| 26 | 660 | 34.5–41.4 | 131–157 |
| 28 | 711 | 40.0–48.0 | 151–182 |
| 30 | 762 | 45.9–55.1 | 174–208 |
| 32 | 813 | 52.2–62.7 | 198–237 |
| 34 | 864 | 59.0–70.7 | 223–268 |
| 36 | 914 | 66.1–79.3 | 250–300 |

**Hydrodynamic dispersion** See *dispersion*.
**hydrodynamic dispersion coefficient** See *dispersion coefficient*.
**Hydrodynamic traps** The movement of water can modify the geometry of an oil accumulation or even trap the oil and/or water in a location where it would otherwise escape, as in case D shown in **Figure H-14**.
**Hydrofracturing** See *hydraulic fracturing*.
**Hydrogenesis** The *condensation* of moisture within the air-filled *pore spaces* of surface soil or rock.
**Hydrogen ion concentration (pH)** The concentration of hydrogen *ions* in an aqueous solution, usually expressed as the negative logarithm (to the base 10) of the concentration. A neutral pH value of normal water is 7.0. A higher value of pH is basic, and a value less than 7.0 is acidic.

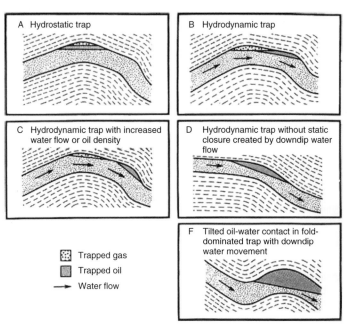

**Figure H-14** Schematic representations of traps for forming petroleum reservoirs. Illustrations of the qualitative effect of the amount and direction of water flow and oil density on hydrocarbon trap configuration. (A) Generalized hydrostatic trap. (B) Generalized hydrodynamic trap. (C) Hydrodynamic trap with increased water flow or oil density. (D) Hydrodynamic trap without static closure created by downdip water flow. (E) Same situation as in (D) but with updip water flow. (F) Tilted oil-water contact in fold-dominated trap with downdip water movement. (G) Tilted oil-water contact in fold-dominated trap with updip water movement. (Reprinted with permission of American Association of Petroleum Geologists).

**Hydrogeochemistry** The chemistry of *groundwater* and surface water and its relations to regional geology. The influence of the geochemistry of non-aqueous *aquifer* materials may be particularly important when investigating environmental groundwater chemistry and may even control the behavior of contaminants. Cf. *Hydrochemistry; Geochemical processes; Geochemical speciation; Hydrochemical facies.*

**Hydrogeologic boundary** The real-system hydraulic counterpart for instances where a *well* is not located in an *aquifer* of infinite areal extent but the *area of influence* encounters the edge of the aquifer or a source of *recharge*. Boundaries are either recharge or barrier boundaries. The *recharge boundary* is encountered in areas where the aquifer receives water, while barrier boundaries occur where the aquifer terminates by thinning, erosion, or encountering an impermeable or low-*permeability* formation. Real-system boundaries are simulated hydrologically by the use of *imaginary wells*. **Figure H-15a** and **b** shows cross sections of both recharge and boundary situations in a real system and the associated hydraulic modeling counterpart used to simulate the real world. The observed effect of boundaries on pumping is demonstrated by plotting *drawdown* as a function of time on semilogarithmic paper, as shown in **Figure H-16**. The rate of drawdown is retarded as the *cone of depression* encounters a recharge boundary and becomes zero if the pumping well is supplied entirely by recharge. The drawdown rate is accelerated and water levels decline more rapidly as the cone of depression encounters a barrier to *flow*. Cf. *Capture zone.*

**Hydrogeologic map** An interpretation of data within a specified area summarizing the topography, geology, hydrogeologic facies, geochemical, and water resource data. A hydrogeologic map is usually completed in the initial phase of a resource evaluation project or as a final product for an exploration program detailing trends across an area. Parameters may be displayed by contouring data values on a base map or by adding hydrochemical diagrams to the map. Cf. *Hydrochemical facies.*

**Hydrogeologic units** See *hydrostratigraphic units.*

**Hydrogeology** The study of the interrelationship of geologic materials and processes with water. The chemical, physical, and thermal interaction of water with the geologic medium and the transport of energy and chemicals with fluid *flow*. The lithology, stratigraphy, and structure of an area control the distribution and character of water bodies. Cf. *Geohydrology.*

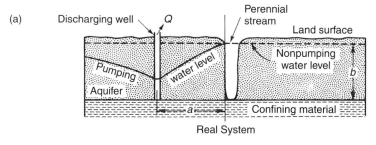

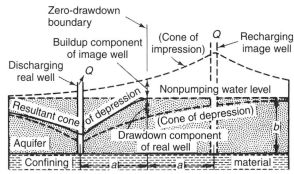

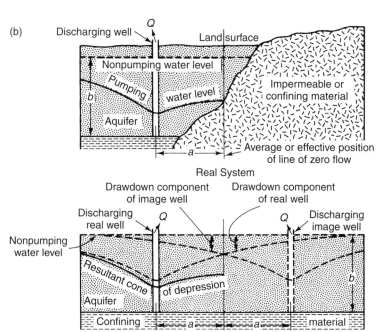

**Figure H-15** Real versus idealized aquifers bounded by a recharge source (a), and impermeable boundary (b). (USGS, 1962.)

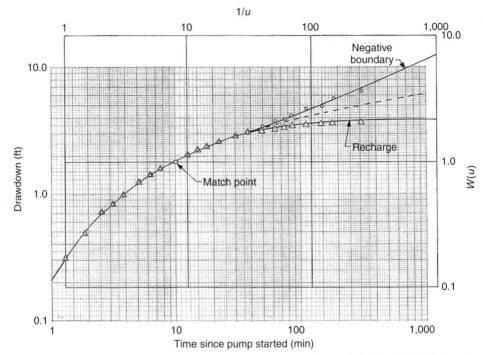

**Figure H-16** The effects of boundaries on drawdown-time thesis curve matching method. (Driscoll, 1986. Reprinted by permission of Johnson Screens/a Weatherford Company.)

**Hydrograph** A plot of variations of stage, *flow*, velocity, or other properties of water bodies as a function of time. Groundwater hydrographs plot changes in water levels or *hydraulic head* measured in a *monitoring well* versus time. A *stream* or river hydrograph may plot rate of flow or *discharge* at different times of the year. The river hydrograph of discharge versus time previously shown in **Figure B-8** displays the *hydrograph separation* and the relationship to *rainfall* in the catchment basin plotted above. Cf. *Baseflow-recession hydrograph; Recession curve.*

**Hydrograph separation** A plot of properties of water bodies as a function of time can be broken down into individual components, which are observed on the graph and indicate changes in data reflecting changes of the hydrologic system with time. Several techniques can be used to separate a *hydrograph* into two main components of *quick flow* or the surface *runoff* and *baseflow*. **Figure H-17** is a hydrograph of a *river* in a region experiencing a dry summer and can be separated into periods of *precipitation* and a period of *baseflow recession.*

**Hydrologic budget** A credit/debit system of accounting for water. *Groundwater budget* accounting includes inflow, outflow, and storage for a *hydrologic unit* such as an *aquifer*, a soil zone, a *lake*, or a *reservoir*. In surface water bodies, a hydrologic budget depicts the relationships between *evaporation, precipitation, runoff, seepage*, and the change in water storage. In subsurface water units, the budget considerations are for *recharge* from the surface, *discharge* to surface water bodies, and *leakage* to other aquifers. The *hydrologic equation* balances the budget. Cf. *Actual evapotranspiration (AE), Groundwater outflow.*

**Hydrologic cycle** The worldwide process, without a true beginning or end, that graphically or mathematically represents the movement of water from the oceans to the atmosphere, to land, and back to the oceans through the process of *evaporation, precipitation, infiltration, transpiration, overland flow, baseflow condensation*, and *runoff*. The hydrologic cycle is illustrated in **Figure H-18**. The annual quantities of the world *water balance* in the phases of the hydrologic cycle are presented in **Table H-11**. Cf. *Hydrologic budget; Hydrologic equation.*

**Hydrologic equation** A quantitative evaluation of the *hydrologic cycle* based on the law of mass conservation and used as an accounting system for a *hydrologic budget*. The hydrologic equation is simply expressed as follows:

$$\text{Inflow} - \text{Outflow} = \text{Change in } Storage$$

The equation allows for determining inequalities between system inflow and outflow and changes in storage. The parameter of inflow may include any means that adds water to the system: *precipitation, streams, groundwater*, and *runoff*.

Outflow may include *evapotranspiration*, outlet watercourses, and *groundwater seepage*. Both inflow and outflow must be measured over the same time period. A simplified water balance equation is as follows:

$$P = Q + E + \Delta S_S + \Delta S_G$$

where:

$P$ = precipitation $[L \cdot T^{-1}]$
$Q$ = runoff $[L \cdot T^{-1}]$
$E$ = evapotranspiration $[L \cdot T^{-1}]$
$\Delta S_S$ = change in storage in surface water [dimensionless]
$\Delta S_G$ = change in storage in groundwater [dimensionless].

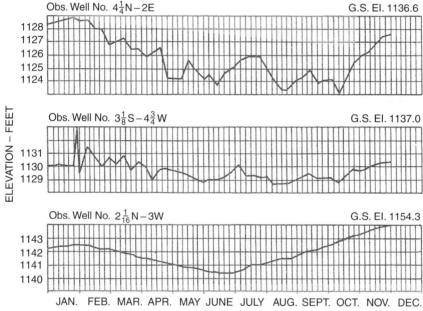

**Figure H-17** A plot of a river hydrograph. (U.S. Dept. of Interior, 1985.)

**Hydrologic head** See *Hydraulic head*.

**Hydrologic mapping** One of the fundamental tasks in hydrology involves mapping hydrological characteristics. This is done on the basis of hydrological series for different sites in a large *river* basin or different river basins within a region and/or climatological series and landscape information. Hydrological mapping follows a river from its *headwaters* to its mouth. The focus of modern hydrological mapping techniques respect *water balance* constraints as well as scaling properties of different hydrological characteristics. Since hydrology embraces the many subsystems relating to water, the maps may be constructed for each individually or may combine the features of more than one subsystem. Cf. *Hydrologic map*; *Hydrostratigraphic units*.

**Hydrologic modeling** See *Finite difference technique*; *Finite element technique*; *Finite volume technique*.

**Hydrologic network** An array of measuring or *gauging stations* producing data that, when combined, gives a realistic representation of the *hydrologic cycle* within the area covered by the array. The stations are commonly computer linked to a central data collection base and used for studying and forecasting hydrologic processes. Cf. *Gauging network*.

**Hydrologic properties** The governing parameters of a rock that determine its capacity to hold, transmit, or deliver water and the directions of maximum and minimum *permeability*. The hydrologic properties of a medium are *porosity*, *effective porosity*, *specific retention*, *specific yield*, permeability, *hydraulic conductivity*, *storage coefficient*, and *transmissivity*.

**Hydrologic region** An area with fixed boundaries in which the geology, climate, and topography are reasonably the same. Typically hydrologic regions are smaller areas within a larger *watershed* basin that facilitate the collection of the same hydrologic data each year for the purposes of historical analysis and prediction. Cf. *Watershed basin*.

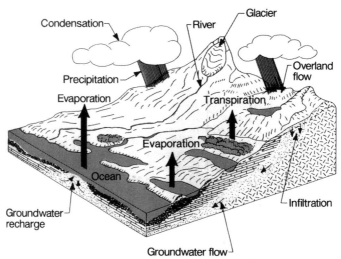

**Figure H-18** Schematic hydrologic cycle. (Driscoll, 1986. Reprinted by permission of Johnson Screens/a Weatherford Company.)

**Table H-11** Volume of water in the different phases of the hydrologic cycle. (Maidment, 1993. Reprinted with permission from McGraw-Hill, Inc.)

| Item | Area ($10^6$ km$^2$) | Volume (km$^3$) | Percent of total water | Percent of fresh water |
|---|---|---|---|---|
| Oceans | 361.3 | 1,338,000,000 | 96.5 | |
| Groundwater | | | | |
|   Fresh | 134.8 | 10,530,000 | 0.76 | 30.1 |
|   Saline | 134.8 | 12,870,000 | 0.93 | |
| Soil moisture | 82.0 | 16,500 | 0.0012 | 0.05 |
| Polar ice | 16.0 | 24,023,500 | 1.7 | 68.6 |
| Other ice and snow | 0.3 | 340,600 | 0.025 | 1.0 |
| Lakes | | | | |
|   Fresh | 1.2 | 91,000 | 0.007 | 0.26 |
|   Saline | 0.8 | 85,400 | 0.006 | |
| Marshes | 2.7 | 11,470 | 0.0008 | 0.03 |
| Rivers | 148.8 | 2,120 | 0.0002 | 0.006 |
| Biological water | 510.0 | 1,120 | 0.0001 | 0.003 |
| Atmospheric water | 510.0 | 12,900 | 0.001 | 0.04 |
| Total water | 510.0 | 1,385,984,610 | 100 | |
| Fresh water | 148.8 | 35,029,210 | 2.5 | 100 |

**Hydrologic routing** See *Hydraulic routing*.

**Hydrologic soil groups** Soils have been classified into four groups: A, B, C, and D. The groups are defined by their *infiltration* rates as follows:

- Group A soils have high infiltration rates and low *runoff* potential even when the soils are thoroughly wetted.
- Group B soils have moderate infiltration rates and typically consist of moderately deep, moderately well drained, fine to coarse texture soils.
- Group C soils have low infiltration rates when wetted with typically fine texture layer that impedes downward water movement.
- Group D soils have very low infiltration rates when wetted and high runoff potential consisting of clayey soils. Cf. *Holtan model; Horton model*.

**Hydrologic system** See *Hydrologic cycle*.

**Hydrologic unit** An area representing a distinct feature of a water-bearing, water-producing, or water-transmitting entity. Cf. *Hydrostratigraphic units*.

**Hydrology** The study or science of water including its circulation, distribution, and chemistry. Cf. *Hydrologic cycle*.

**Hydrolysis** A decomposing reaction in which the products resulting from the reaction of a substance in water react with the hydrogen or hydroxyl *ions* derived from the water and ionic disassociation. Hydrolysis refers to the natural combination of the constituents of rock with water in the chemical weathering process in which water acts as the *solvent*. Water can also transform *pollutants* by hydrolysis as in the breaking of the pollutants' chemical bonds to form new bonds, which is usually accelerated by acidic or basic conditions, and heat. Cf. *Hydrochloric acid.*

**Hydrometer** An instrument that measures the *specific gravity* of a liquid. A hydrometer can be used to estimate the *salinity* of seawater based on the density relating to the specific gravity.

**Hydrophilic** Having a strong affinity for water; literally, "water loving", referring to those *colloids* that swell readily in water yet do not tend to coagulate. A characteristic of polar compounds where the molecules are able to be wetted or solvated by water. Also a term used in reference to plants. Cf. *Hydrophobic; Solvent.*

**Hydrophobic** Having an aversion to water; literally, "water fearing" referring to those *colloids* that do not readily hydrate yet do coagulate easily. Molecules that resist wetting or solvation by water, i.e., non-polar compounds. Also a term used in reference to plants. Cf. *Hydrophilic; Solvent.*

**Hydroscopic capacity** See *Hygroscopic coefficient.*

**Hydroscopic water** See *hygroscopic moisture.*

**Hydrosphere** The water of the earth (the sphere), segregated from the lithosphere (rocks of the earth), biosphere (biota of the earth), or atmosphere (air of the earth). The hydrosphere incorporates all stages of water (liquid, solid, and *vapor*) that exist on or close to the earth's surface. Cf. *Hydrologic cycle.*

**Hydrostatic equilibrium** In a fluid when the *gravity head* and *pressure head* are in balance. Cf. *Elevation head; Total head.*

**Hydrostatic gradient** The change in *hydrostatic pressure* over a unit distance in any specific direction. In *groundwater*, the velocities are generally low enough that the direction of steepest negative *gradient* is often assumed to be the direction of groundwater movement. Because of sedimentary layering, the assumption can often be made that horizontal gradients are much greater than vertical gradients, and therefore the *flow* is mainly horizontal. The validity of this assumption should always be carefully assessed before depending on the results of its application.

**Hydrostatic head** The height of a vertical column of water. If applied to a unit cross section, the weight of the column is equal to the *hydrostatic pressure* at any given point. It is the *pressure head* or, as applied to water, the *static head*. Cf. *Artesian head.*

**Hydrostatic level** The level to which water rises in a well or *piezometer* under its *aquifer* pressure over the *screened interval* of the well. Hydrostatic level is the defining *potentiometric surface*. Cf. *Static water.*

**Hydrostatic pressure** The *pressure* exerted at the base of a water column. **Figure H-19** diagrammatically presents the following equation of hydrostatic pressure:

$$\rho = \gamma_w g h$$

where:

$\gamma_w$ = water density $[\text{M·L}^{-3}]$
$g$ = gravitational constant $[\text{L·T}^{-2}]$
$h$ = saturation depth $[\text{L}]$.

The direction of pressure is upward if the bottom of the representative volume is horizontal. Gravity forces are always greater than hydrostatic pressures, but the hydrostatic pressure is equal and with the same magnitude in every direction.

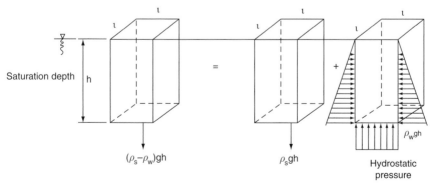

**Figure H-19** The hydrostatic pressure versus the bulk weight per area. (Sen, 1995. Reprinted with permission from CRC Press, Inc.)

Cf. *Artesian pressure; Elevation head; Pore pressure; Water pressure; Zone of saturation.*

**Hydrostratigraphic units** Also called hydrogeologic units. One or more geologic units over a wide lateral extent that is distinct from the surrounding body of rock or sediments and related by similar hydrologic parameters such as *hydraulic conductivity.* The hydrostratigraphic boundary may not correlate with the formation boundary; this may result in several geologic formations being combined into a single hydrostratigraphic unit such as an *aquifer,* or a single geologic formation being divided into several hydrostratigraphic units such as multiple aquifers with intervening *confining layers.* Cf. *Hydrochemical facies; Hydrogeologic boundary.*

**Hydrothermal system** A *groundwater flow* system where sufficient geothermal heat is added along the *flow path* between the *recharge* area and the *discharge* area to yield groundwater temperatures significantly above the mean annual ground surface temperature.

**Hydrothermal water** Subsurface water that is heated to a temperature high enough to be hydrogeologically significant, whether heated by radioactive decay, energy release, or by *geothermal gradient.* Cf. *Magmatic water; Metamorphic water.*

**Hydrothermal vent** Also called black smoker. A deep seafloor ejection of hot water, hydrogen sulfide, and other gases under *pressure* providing local nutrients, which would otherwise be unavailable because of depth below surface. Vents typically occur near spreading centers along oceanic ridges.

**Hyetal** Rain, *rainfall,* or a rainy region. For example, a hyetal interval would be the difference in rainfall between any two *isohyets.*

**Hyetograph** A depiction of *rainfall intensity* plotted versus time.

**Hyetometer** An instrument used to measure the amount of *rainfall;* a *rain gauge.*

**Hygroscopic capacity** See *hygroscopic coefficient.*

**Hygroscopic coefficient** Also called hygroscopic capacity. The ratio of the weight of water that can be absorbed by a mass of dry soil in contact with a saturated atmosphere until in equilibrium to the weight of the dry soil mass. The hygroscopic coefficient is expressed as a percentage. **Figure H-20** illustrates the relationship of the hygroscopic coefficient to soil moisture storage.

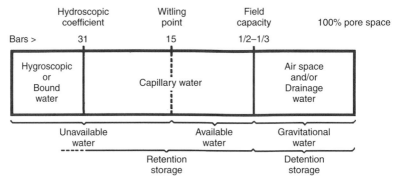

**Figure H-20** Soil moisture storage relations to hydroscopic coefficient. (Black, 1996. Reprinted with permission from Ann Arbor Press, Inc.)

Cf. *Capillary water; Hygroscopic water; Fixed moisture; Gravitational water; Retention storage; Detention storage; Zone of saturation.*

**Hygroscopic moisture/hygroscopic water** Also called hydroscopic water. The water held to soil particles by surface tension, generally greater than 31 bars or atmospheres, and which is at equilibrium with the adjacent atmosphere, thereby making it unavailable for plant use (**Figure H-20**). Water drains from soils by *gravity drainage* until the stress of gravity equals the surface tension. Gravity drainage ceases only when the hygroscopic water content is at the *specific yield.* A greater volume of hygroscopic moisture can be held by fine-grained material as a result of the greater surface area and the increased electrochemical forces displayed by most clays. Cf. *Hygroscopic coefficient.*

**Hygroscopicity** A soil's ability to absorb and retain water. Cf. *Hydrologic soil groups; Infiltration; Holton model.*

**Hyperhaline** See *Thalassic series.*

**Hyperpycnal inflow** Water that is denser than the water it is flowing into. Hyperpycnal inflow usually results in a *turbidity current.* Cf. *Hypopycnal inflow; Homopycnal inflow.*

**Hypersaline** *Salinity* significantly greater than that of seawater. Cf. *Thalassic series.*

**Hypolimnion** The bottom layer in a thermally stratified *lake,* often characterized by a stagnant, low-oxygen content, typically occurring in summer. Colder temperatures make deep-water layers more dense than water closer to the surface, thereby stabilizing the position of the hypolimnion. Organic material, or possibly oxidizable sediment, tends to deplete the oxygen in the bottom layer, resulting in stagnation. Physical disturbance of the lake by reduction in atmospheric temperature, or storm *flow* into the lake during autumn, may physically mix these layers, destroying the stability of *thermal stratification.* Cf. *Dystrophic lake; Epilimnion; Eutrophic.*

**Hypopycnal inflow** Water that is less dense than the body of water it flows into, such as *freshwater* flowing into *brine* or saltwater, as occurs in *delta* facies. The freshwater flows or "rides" above the denser saltwater. Cf. *Hyperpycnal inflow; Homopycnal inflow; Axial jet.*

**Hyporheic zones** A part of the landscape that contains water of both subsurface and *stream* channel origin. This zone becomes a dynamic bidirectional link that consists of multiple *flowpaths*. Stream and riverbeds contain material that is often coarse, that leak, exchanging water with the adjacent groundwater.

**Hypsographic curve** See *hypsometric curve.*

**Hypsometric curve** Also called a hypsographic curve. A cumulative frequency plot of ground surface elevations, either above *sea level* or above some other convenient elevation datum. Factors of interest in a *catchment area* or *drainage basin* often vary with elevation. By plotting the area frequency with which different elevations occur in a specific basin, different basins can be compared for similarities in elevation distribution. To facilitate interbasin comparison, the areas are usually plotted as a percentage of the total basin area. A typical hypsometric curve is shown in **Figure H-21**.

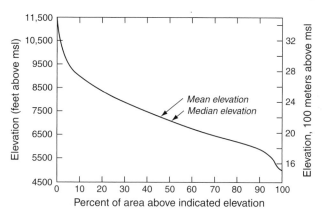

**Figure H-21** Typical hypsometric curve or area–elevation curve for a mature catchment basin. (Linsley et al., 1982. Reprinted with permission from McGraw-Hill, Inc.)

Cf. *Watershed basin*

**Hysteresis** A retardation of the effect when forces acting on a body are changed. The curves that define the relationship between moisture content and matric suction in an *unsaturated* soil are hysteretic, whereby the curve takes on a different shape depending upon whether the soil is wetting or drying.

**Image well theory** The *equilibrium well equations* assume *aquifers* of infinite areal extent. However, when a well is situated near a low-*permeability* boundary such as the edge of a paleochannel, the observed *drawdown* can be greater than predicted using the *Theis equation*. Further, when a well is located adjacent to a river, there may be significant *recharge* to the well resulting in less drawdown than predicted by the Theis equation. **Figure I-1** presents the image well

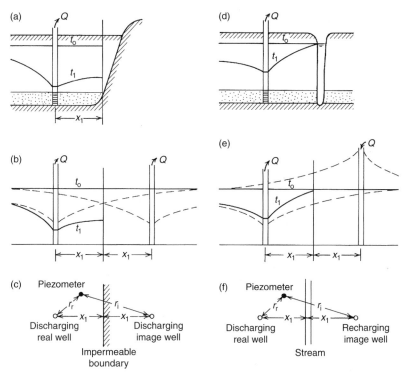

**Figure I-1** Image well theory. (a) Drawdown in the potentiometric surface of a confined aquifer bounded by an impermeable boundary; (b) equivalent system of an infinite extent; and (c) plan view. (Freeze and Cherry, Groundwater © 1979. Reprinted with permission by Prentice-Hall, Englewood Cliffs, NJ.)

theory to account from these deviations from the predicted conditions. Because the drawdown at any given point in an aquifer is the sum of the drawdowns created by the individual wells, the use of an imaginary well may be used to predict drawdowns. The drawdown in an aquifer bounded by an impermeable boundary is given by the following equation:

$$h_0 - h = \frac{Q}{4\pi T}[W(u_r) + W(u_i)]$$

where:

$$u_r = \frac{r_r^2 S}{4Tt}, \quad u_i = \frac{r_i^2 S}{4Tt}$$

$h_0$ = the static water level in the aquifer [L]
$h$ = the depressed head at the well [L]
$Q$ = the well discharge [L³·T⁻¹]
$T$ = *transmissivity* [L²·T⁻¹]
$r_r$ = radial distance from *pumping well* to measuring point or *piezometer* [L]
$r_i$ = radial distance from measuring point or piezometer to image well [L]
$W(u)$ = the well function of $u$.

For wells in an aquifer bounded by a constant head boundary

$$h_0 - h = \frac{Q}{4\pi T}[W(u_r) - W(u_i)]$$

Cf. *Well function of u; Boundary, head-dependent.*

**Imaginary wells** *Wells* used in the *image well theory* for the prediction of *drawdown*.

**Imhoff cone** A glass cone, as shown in **Figure I-2**, used for measuring the *total suspended solids* (TDS) in water. Because of the relatively small sample volume represented by an Imhoff cone, sediment concentration is determined by averaging the results of five samples taken at the following times during a *pump test*:

1. 15 min after start of test;
2. After 25% of the total pumping time has elapsed;
3. After 50% of the total pumping time has elapsed;
4. After 75% of the pumping time has elapsed; and
5. Near the end of the pumping test.

The amount of sediment collected at the bottom of the cone is weighed and represented as the weight of the volume of water sampled.

**Figure I-2** Imhoff cone for measuring total suspended solids in water. (Driscoll, 1986. Reprinted by permission of Johnson screens/a Weatherford company.)

**Immiscible flow** A condition where two fluids that are immiscible, such as oil and water, flow together. Since each fluid has its own wetting characteristics, density and *viscosity,* the transport relative to *groundwater flow* will be a result of these differences in viscosity and density. The migration of more viscous fluids will be retarded relative to groundwater flow. Immiscible fluids that are less dense than water will tend to float on the *water table*. More-dense fluids will have a tendency to sink in the *groundwater* under the influence of gravity. Cf. *Multiphase flow; Miscible flow.*

**Impeller** The part of a centrifugal pump that provides suction and lift to transport water. The volume of water that a centrifugal pump is capable of pumping is dependent on the design of the impellers and the speed at which they rotate. The motor determines the speed of rotation, while the area (opening) of the impeller controls the fluid movement. The equation relating impeller velocity and area is as follows:

$$Q = VA$$

where:
$Q$ = discharge [$L^3 \cdot T^{-1}$],
$V$ = velocity [$L \cdot T^{-1}$]
$A$ = area [$L^2$]

The capacity of the pump will vary directly with the impeller speed and diameter. The produced *hydraulic head* will vary with the square of the impeller speed and the square of the impeller diameter.

**Impermeable boundary** A boundary to groundwater *flow* that will not transmit water. An impermeable boundary may be found at the bottom of the lake where silts and clays form the bottom. This would unlikely transmit water. Another impermeable boundary could be fault thrust material impeding flow in any direction. Cf. *Barrier, aquifer effects.*

**Impervious boundary** *Aquifers* can rarely be approximated as of infinite areal extent, therefore, in a *pump test*, the *drawdown* exceeds that predicted by ideal conditions, indicating an impervious boundary.

**Impoundment** An artificial surface water body, used for temporary storage of liquids.

**Installation of well screens** See *well screen installation*.

**Inclinometer** Devices used in measuring the deviation from vertical, or plumb, for *wells*, as well as in measuring soil movement in earth engineering projects such as the construction of earth-fill dams. A number of inclinometer instruments can be employed to check for being plumb, including:

- Slope indicators
- gyrocompass
- plumb bob.

Slope indicator instruments require a grooved casing to align the downhole instrument. The instrument is lowered down the hole and readings are taken at depth intervals to measure the deviation from vertical. Gyrocompasses contain a compass and a gyro to measure the plumbness of the well. The controls can be mechanical- or computer-controlled. The gyrocompass is lowered downhole via a wireline to the select depths for measurement. They are computer-controlled and can take multiple measurements through the total depth of the borehole. Some use photographic film to record the measurements, while many relay direct readings to the computer on the surface. Shallow wells can use a special plumb bob arrangement, using a cylindrical 40-ft(12.2 m)-long dummy as the plumb bob or using a plumb bob alone. The dummy is 40 ft long since water well casing is typically in maximum 20 ft lengths, such that the dummy extends two casing lengths since bends in the well typically occur at casing joints. This arrangement is shown in **Figure I-3**.

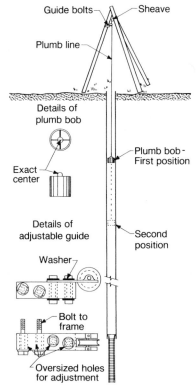

**Figure I-3** The straightness and plumbness of a boring/well can be determined by using a small tripod mounted over the top of a well bore and a plumb bob suspended in the casing. (Driscoll, 1986. Reprinted by permission of Johnson Screens/a Weatherford Company.)

Cf. *Alignment test*; *Plumbness.*

**Incrustation** The process whereby a crust or a coating is formed on or in another object typically relating to interference in the proper functioning of a *well screen*. Cf. *Entrance velocity.*

**Infiltration** The process by which water enters the subsurface. Cf. *Percolation.*

**Infiltration capacity** The maximum rate at which water can enter the soil at a particular point under a given set of conditions. The actual rate of infiltration ($f_i$) equals the infiltration capacity ($f_p$) only when the rainfall intensity minus the rate of retention equals or exceeds the infiltration capacity. Cf. *Green-Ampt model; Holton model; Depression storage; Horton model.*

**Infiltration galleries** One or more pipes laid horizontally, using conventional excavation techniques adjacent to or underneath a surface water body for *water supply* where the saturated thickness of the water-bearing zone is thin and near the surface. The typical depth of wells is less than 25 ft (7.6 m), since this is the practical depth limitation for conventional trenching machines. The criteria by which to locate an infiltration gallery adjacent to or underneath a surface water body are dependent upon a number of factors, which are as follows:

1. Galleries placed underneath surface water bodies have approximately twice the *yield* of those placed adjacent to the surface water. However, infiltration galleries are more expensive to construct. Over time, as sediment builds up in the lake or river, the *transmissivity* of the formation decreases, decreasing the yield.
2. Infiltration galleries placed adjacent to water bodies yield water of lower *turbidity* than those laid on or just below lake or riverbeds, as a result of the greater filtering of the water through sediments.
3. Maintenance of galleries adjacent to surface water bodies is easier and less expensive than those placed on or just below lake or riverbeds. Also, due to the sedimentation in lakes and rivers, there is a greater need for maintenance of galleries placed on beds (**Figure I-4**).
4. Lakes and rivers can be high-energy environments during storms, and when ice melt occurs, they can damage galleries installed on or below beds.

The orientation of the screen is always perpendicular to the direction of water flow to minimize *head loss*. As such, screens installed on or below riverbeds are perpendicular to the river *bank*, while infiltration galleries adjacent to a surface water body are placed parallel to the shoreline.

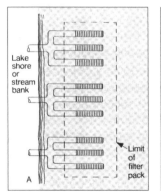

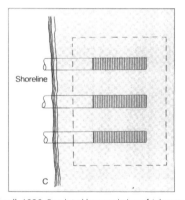

**Figure I-4** Screen arrangements for bed-mounted infiltration galleries. (Driscoll, 1986. Reprinted by permission of Johnson Screens/a Weatherford Company.)

Cf. *Zone of saturation.*

**Infiltration rate** *Rainfall* reaching the ground surface infiltrates the surface soils at a rate that decreases with time until a constant rate is achieved. If at any time the rate of rainfall exceeds the *infiltration capacity* of the soil, excess water will pond on the soil surface. Cf. *Horton model; Hydrologic soil groups; Depression storage.*

**Infiltration test** A determination of the *percolation* rate of water into the subsurface. An infiltration test is usually performed to determine whether a soil is suitable for a *septic system*. Cf. *Borehole infiltration test; Infiltration capacity; Infiltration rate.*

**Infiltrometer** A device used to measure the *infiltration* rate of water through soils. It consists of two open-ended metal cylinders that are driven concentrically into the ground and then partially filled with water. As water drains into the soils, water is added to the cylinders to keep the liquid level constant. By measuring the amounts of water added to each cylinder, the operator is able to calculate the *infiltration rate* of the soil. From the infiltration rate, *hydraulic conductivity* can be calculated. Cf. *Fixed ring infiltrometer; Ring infiltrometer.*

**Inflatable packers** Typically used for casing repairs, isolating zones of rock formations for injection testing (also known as *packer testing* a formation), and hydrofracturing. Inflatable packers have larger expansion ratios than do casing hangers used in water well completions. The packer, usually made from neoprene, is inflated with a gas (air, nitrogen), water, or a solidifying liquid for permanent installations. Cf. *Packer.*

**Influent stream** A *stream* or a *river* that recharges groundwater. Cf. *Effluent stream; Bank storage; Gaining stream.*

**Injection wells** Although conventional wells are used for the withdrawal of water, injection wells are used for various purposes, although at some point in their lifetime may be used for water withdrawal. Injection wells are used for:

- *recharge wells* for *water supply,*
- *groundwater* control,
- *saltwater intrusion* control
- *solution mining*
- waste disposal, and
- geothermal energy.

In addition to taking special care in the construction of injection wells, special attention must be paid to water chemistry interactions, *air entrainment*, thermal interference, and *sand pumping*. Because *groundwater* within a short distance of the *water table* is typically *anaerobic*, the addition of oxygenated water can result in biofouling, chemical reactions producing undesirable mineral precipitates, potentially clogging *well screens* and formations. Sand pumping is a significant problem, for sand at concentrations as low as $1\,mg\,L^{-1}$ can clog injection wells within a short period of time.

Clogging of screens is the most serious problem in the operation of injection wells. Particular attention needs to be paid to the screen open area and screen length for optimal conditions. The average well screen *entrance velocity* should be designed for $0.05\,ft\,s^{-1}$ ($0.015\,m\,s^{-1}$). This results in the screen length being twice as long for an injection well as for a withdrawal well for the same volume of water. See: **Appendix A** for conversion factor.

When recharge wells are used for Water Supply to slow or prevent *overdrafts* of groundwater, *artificial recharge* of groundwater increases the rate at which water infiltrates from the land surface to the *aquifer*. When water is introduced into an injection well, a cone of *recharge* is formed that is similar, but in reverse to a *cone of depression* surrounding a pumping well. Injection pressures must be controlled so that the formation is not fractured in weakly consolidated stratified sediments. In unconsolidated formations, the injection pressure must never result in the cone of recharge extending to the land surface or subsurface structures, for *quick soil conditions* will result. The ground will become unstable, lose its bearing capacity, resulting in possible foundation problems. Further, there is the potential that shallow wells may rise from the ground under such conditions. For poorly consolidated coastal plain sediments, fracturing can occur at 0.5 psi ft$^{-1}$ (11.3 kPa m$^{-1}$) depth. For crystalline rock, up to 1.2 psi ft$^{-1}$ (27.1 kPa m$^{-1}$) depth is possible before fracturing occurs. In unconsolidated sediments, the injection pressures should not exceed $0.2h$, where $h$ is the depth from the ground surface to the top of the screen or *filter pack*. The equations governing pumping to an injection well are the same as for water withdrawal wells. For a confined aquifer:

In SI units (American units can be found in **Appendix A**)

$$Q = \frac{Kb(h_w - H_0)}{0.366 \log(r_0/r_w)}$$

where:

$Q$ = rate of injection (m$^3$ day$^{-1}$)
$K$ = hydraulic conductivity (m day$^{-1}$)
$b$ = aquifer thickness (m)
$h_w$ = head above the bottom of aquifer while recharging (m)
$H_0$ = head above the bottom of aquifer when no pumping is taking place (m)
$r_0$ = radius of influence (m)
$r_w$ = radius of injection well (m)

In English units

$$Q = \frac{Kb(h_w - H_0)}{0.366 \log(r_0/r_w)}$$

where:

$Q$ = rate of injection (gpm)
$K$ = hydraulic conductivity (ft day$^{-1}$)
$b$ = aquifer thickness (ft)
$h_w$ = head above the bottom of aquifer while recharging (ft)
$H_0$ = head above the bottom of aquifer when no pumping is taking place (ft)
$r_0$ = radius of influence (ft)
$r_w$ = radius of well (ft)

In an unconfined aquifer, the equation becomes:
In SI units

$$Q = \frac{K\left(h_w^2 - H_0^2\right)}{0.733 \log(r_0/r_w)}$$

In English units

$$Q = \frac{K\left(h_w^2 - H_0^2\right)}{141.4 \log(r_0/r_w)}$$

The equations for pumping and recharge wells suggest that the rate of recharge would equal the pumping capacity of a well. In practice, the recharge rates are less than the pumping *yield*. This is because the formation around the well can quickly become plugged with sand, entrained air, and buildup of biomass and chemical precipitate.

Artificial recharge may provide the following benefits:

1. By maintaining water levels in an aquifer, the operating costs are reduced by reducing the lift requirements in a well.
2. As above, the need for deepening wells as the aquifer becomes exploited is reduced or eliminated.
3. Maintains a dependable supply of water.
4. Allows for municipal expansion through increased and dependable water supplies.
5. Prevents *saltwater intrusion* into coastal areas.
6. Prevents intrusion of connate *brines* from deeper formations.
7. Prevents land subsidence through maintaining groundwater levels.
8. Used in natural-gas storage fields for maintaining pressures for gas withdrawal.

The California Department of Water Resource in the United States has listed the following requirements for a successful recharge project:

1. The basin must be suitable from the standpoint of storage capacity and *transmissivity* of aquifers.
2. Adequate recharge water must be available.
3. Infiltration rates must be maintained at adequate levels.
4. The recharge water must be compatible with the existing groundwater and be injected at temperatures consistent with natural conditions.

Infiltration wells are used to prevent saltwater intrusion: In coastal areas, exploitation of *groundwater resources* has resulted in the migration of the salt/freshwater boundary inland. This has resulted in increased chloride ($Cl^-$) concentrations in the drinking water of coastal communities. **Figure l-5** illustrates three schemes to control saltwater intrusion. Two of these schemes use artificial recharge as a means of controlling saltwater intrusion.

Solution Mining: Solution mining is accomplished by injecting water into the ore body. Depending on the application, the water being injected may be chemically treated to aid in solutioning the ore. Solution mining of mineral salts, the water is treated only insofar as the water that is used is recirculated; such treatment involves the removal of the mineral salts as part of a closed-circuit solutioning process. Both injection and

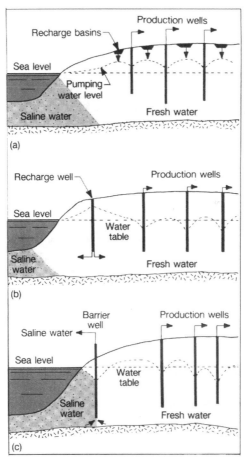

**Figure I-5** (a) Use of artificial recharge in the area of production wells in an unconfined coastal aquifer; the recharged water maintains the water table above sea level to prevent saltwater intrusion. (b) Use of injection wells to form a pressure ridge to prevent saltwater intrusion in an unconfined coastal aquifer. (c) Use of pumping wells at the coastline to form a trench in the water table, which acts as a barrier against further saltwater encroachment. (Driscoll, 1986. Reprinted by permission of Johnson Screens/a Weatherford Company.)

*recovery wells* are used in this process. Solution mining eliminates the need for large excavations and tailings disposal, requiring less capital and labor costs. Once the ore body is defined, a network of injection and *extraction wells* are drilled and constructed. **Figure I-6** presents three injection production well layout approaches. The overall scheme is to provide for a central production well surrounded by injection wells, laid out in a five or seven-spot arrangement.

Injection wells have been used for Waste Disposal Hazardous and toxic wastes are disposed into aquifers containing saltwater brines. In the United States, it is now illegal to inject hazardous or toxic wastes above or into underground sources of drinking water. *Deep-well injection* for disposal of wastes is expensive, and only those waters that cannot be treated at the surface may be injected. Injection of wastes must be placed at least 1320 ft (433 m) below a *potable* aquifer, and the intervening strata should consist in part of tight clay or other material that is impervious to upward migration of fluids or wastes.

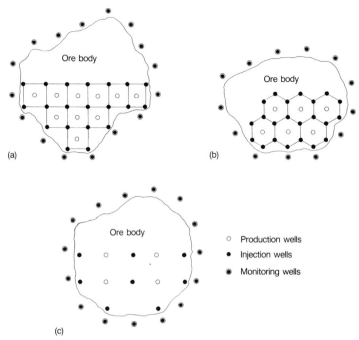

**Figure I-6** Well arrangements used in solution mining: (a) five-spot pattern, (b) seven-spot pattern, and (c) line-drive pattern. Distances between wells vary, but 50 ft (15 m) is typical. (Driscoll, 1986. Reprinted by permission of Johnson Screens/a Weatherford Company.)

**Insol** Derived from the word insoluble. A term associated with *drilling fluids* prepared with *polymers*. One of the desired properties of a drilling fluid is to prevent fluid loss in a borehole. As such, the drilling fluid coats the borehole wall. The plugging of the soil pores is caused by the plastering of the insol (insoluble) particles onto the borehole wall. Each insol is surrounded by a viscous film of fluid, and as insols build up on the borehole wall, fluid is prevented from leaving the borehole (**Figure I-7**).

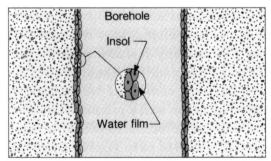

**Figure I-7** Plugging caused by the coating of insoluble (insol) particles onto a borehole wall. Each insol is surrounded by a viscous polymeric film. In general, polymeric films are much thinner than filter cakes composed of clay particles. (Driscoll, 1986. Reprinted by permission of Johnson Screens/a Weatherford Company.)

**Interface probe** An interface probe is lowered down a *borehole* or *well casing*. The first sounding usually a high pitched beep followed by a second sounding when it is lowered further is an indication the probe first encounter a hydrocarbon and the second 'beep' was the depth to water. A well-sounding device is used to detect and/or measure the thickness of *non-aqueous-phase liquid (NAPL)* in a well. Interface probes come in various forms, using differences in electrical conductivity or hydrophobicity to differentiate the non-aqueous-phase liquid from water. Cf. *Water level measurement; Groundwater measurement; Electrical sounder; Electrical conductance.*

**Interfacial tension** The adhesion forces (tension) between the liquid phase of one substance and another substance. The interaction occurs at the surfaces between the two, i.e., at their interfaces. This effect results in capillary action in pore spaces and results in the formation of a meniscus representing the interfacial tension between the two liquids. Cf. *Surface tension.*

**Interference** A condition that occurs when two *wells* are pumping within the same *aquifer* and are close to each other, such that the *radius of influence* from each well overlaps. As such the resultant *drawdown* in this overlapping area becomes the sum of the *drawdown* from the individual wells at this point. Cf. *Image well theory.*

**Interflow** Infiltrated water may move horizontally in the unsaturated zone if layers of soil with a lower vertical *hydraulic conductivity* exist in the subsurface. In some *drainage basins*, this interflow may be substantial and significantly contributes to the basin's total *streamflow*. Cf. *Groundwater inflow; Groundwater outflow.*

**Intermediate belt** The zone of vadose water that lies between the soil belt and the *capillary fringe*. Water in the *pore spaces* is held by molecular attraction, and at times of groundwater *recharge* water may also be migrating downward to the *water table*. The intermediate zone may be absent or may be several hundred feet thick, depending on the local geology, topography, and climate. Cf. *Belt of soil water; Zones of water; Vadose zone.*

**Intermediate flow system** Groundwater flow systems have been characterized as having local, intermediate, and regional flow systems. As such, the definition of where these systems exist is site- and scale-dependent. Local flow systems are considered to encompass the groundwater flow regime that receives recharge from highlands and discharges in the adjacent lowlands. The regional flow system would encompass basin-wide or large groundwater flow regimes. The intermediate flow system would encompass flow regimes between the local and regional (**Figure I-8**).

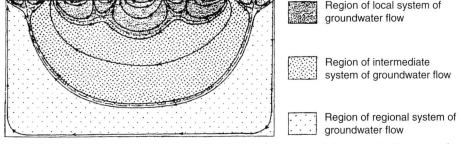

Figure I-8 Local, intermediate, and regional systems of groundwater flow. (Freeze, 1979. Reprinted with permission from Prentice-Hall, Inc.)

**Intermediate water** Water found between 200 and 2000m deep in the oceans.

**Intermediate zone** See *intermediate belt.*

**Intermittent lake** Lakes that disappear seasonally are called intermittent lakes and are typical of karstic terrain. Also known as seasonal lakes. Cf. *Playa Lake; Dry Lake.*

**Intermittent pumping** Pumping on a non-continuous basis. Cf. *Pump; Pumping test.*

**Interstices** See *Pore spaces.*

**Intrinsic bioremediation** The natural biodegradation process that occurs in the subsurface whereby microorganisms metabolize or facilitate the breakdown of organic chemicals. Compounds such as fuel hydrocarbons, ketones, and alcohols are directly metabolized by natural microorganisms in the subsurface. Moreover, chlorinated aliphatic and aromatic hydrocarbons are degraded through a cometabolic process, whereby the microorganisms do not utilize the chlorinated compounds as substrate, but through the oxidation of other organic compounds create the appropriate redox conditions for the dehalogenation of the chlorinated compounds. Cf. *Electrochemical sequence.*

**Intrinsic permeability** The portion of *hydraulic conductivity* ($K_i$) which is representative of the properties of the porous medium along function of size of the openings through which the fluid moves

$$K_i = Cd^2$$

where
  $C$ = dimensionless constant
  $d$ = mean pore diameter (L)
  $K_i$ = hydraulic conductivity ($L^2$)

Relationship between hydraulic conductivity and intrinsic permeability is as follows:

$$K = K_i \left( \frac{\gamma}{\mu} \right)$$

where
  $\gamma$ = specific weight of fluid
  $\mu$ = dynamic viscosity of fluid (g/cm sec)

Or

$$K = K_i \left( \frac{\rho g}{\mu} \right)$$

where
  $\rho$ = density of fluid (g/cm$^3$)
  viscosity units = centipoise (0.01 g/(cm sec))
  g = acceleration constant due to gravity (9.80665 m/s$^2$)
  Cf. *Permeability; Permeability, absolute; Effective Permeability; Dynamic viscosity coefficient.*

**Intrusion of salt water** Along coastal areas where groundwater is pumped from aquifers, gradients are set up that induce the flow of salt water from the sea inland. **Figure I-9** illustrates equilibrium, non-pumping conditions for the

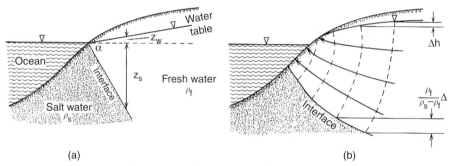

**Figure I-9** Saltwater intrusion interface in an unconfined coastal aquifer (a) under hydrostatic conditions and (b) under steady-state seaward flow. (Freeze and Cherry, 1979. Reprinted with permission from Prentice-Hall, Inc.)

salt/fresh water interface at a coastal setting. Under equilibrium hydrostatic conditions, the mass of a unit column of water extending from the water table to the interface is balanced by a unit column of water to the same depth on that interface. From **Figure I-9**

$$\rho_s g z_s = \rho_f g (z_s + z_w)$$

or

$$z_s = \frac{\rho_f}{\rho_s - \rho_f} z_w$$

For $\rho_f = 1.0$ and $\rho_s = 1.025$

$$z_s = 40z_w$$

where:
$\rho_s$ = specific gravity of salt water $[M \cdot L^{-3}]$
$\rho_f$ = specific gravity of fresh water $[M \cdot L^{-3}]$
$g$ = gravitational constant $[L \cdot T^{-2}]$
$z_s$ = elevation head of salt water $[L]$
$z_w$ = elevation head of fresh water $[L]$

This relationship is often called the *Ghyben–Herzberg* equation. In most real situations, the *Ghyben–Herzberg* equation underestimates the depth to the fresh/saltwater interface. Further, in reality, the interface is gradual, for there is some mixing of the waters.

**Inverse problem** In the calibration of numerical groundwater flow models, the inverse problem involves identifying a set of model parameters (e.g., hydraulic conductivity and recharge rates) that produce hydraulic head values that reasonably match the observed values of hydraulic head measured in wells or piezometers. Two approaches are adopted to solve the inverse problem: (1) trial-and-error calibration and automated inversion schemes that use, for example, a weighted least squares formulation and (2) and optimization algorithm to estimate a set of model parameters.

**Ion** An element or a compound that contains an electrical charge. That is, it is not electrically neutral, having gained or lost an electron.

**Ion exchange** A process for *softening water* whereby calcium, magnesium, iron, manganese, and strontium are replaced by sodium. Water is passed through a synthetic *cation exchange* medium (resin) containing the sodium ions. When expended, the resin is regenerated by passing *brine* through the medium. The exchanged ions are replaced by sodium. Multiple regenerations are possible with the resin usually lasting for years. Cf. *Base exchange*.

**Iron bacteria** Iron-bearing waters, which are under reducing conditions, favor the growth of iron bacteria. The most common genus are Crenothrix, Gallionella, and Leptothrix, of which Crenothrix is the most common, with a filamentous nature forming large gelatinous masses. These masses can break loose from pipes or *well screens* and can clog the system. The bacteria have the ability to convert soluble ferrous iron ($Fe^{2+}$) to insoluble ferric iron ($Fe^{3+}$), creating a buildup of precipitate. Iron bacteria can be controlled with a *bactericide*.

**Iron incrustation** Caused by the sudden change from ferrous ($Fe^{2+}$) to ferric ($Fe^{3+}$) iron. Ferrous iron is the more soluble form of iron under normal pH conditions (up to 50 mg $L^{-1}$ in oxygen-deficient conditions). Ferric iron is almost completely insoluble under alkaline or acidic conditions. When water with a pH of 7–8.5 is aerated, almost all of the iron becomes insoluble. This creates incrustation problems with treatment systems such as air strippers or spray aeration systems. Ferrous iron can also combine with carbonate ($CO_3^{2-}$) ions that often contribute to the plugging of well screens.

**Iron in water** Iron is one of the most common elements in the earth. Concentrations of up to 5 mg $L^{-1}$ iron are normally found in *groundwater*, with the iron content increasing with anaerobic conditions. US EPA *drinking water standards* (secondary, non-enforceable) for iron are 0.3 mg $L^{-1}$, because iron can stain plumbing fixtures, stain clothes during washing, incrust *well screens*, clog pipes, and generally have an objectionable taste.

**Isoelectric point** Also known as the point of zero charge ($z_{pc}$). The ability of soils to adsorb ionic and polar molecules is a result of the electrical charges on the mineral grains. The net surface charge is pH-dependent. At the isoelectric point, the charges are neutral and the soil mineral's ability to adsorb ionic or polar molecules is neutralized. The greater the difference in pH from the isolectric point, the greater the ability of the mineral to adsorb ionic or polar molecules. **Table I-1** presents the isoelectric point from various soil minerals.

**Table I-1**  Selected isoelectric points

| Substance | $pH_{z_{pc}}$ | Substance | $pH_{z_{pc}}$ |
|---|---|---|---|
| Quartz | 1–3 | Kaolinite | ≤2–4.6 |
| $SiO_2$ gel | 3.5 | Montmorillonite | ≤2–3 |
| Feldspars | 6–7 | Gibbsite | ~9 |
| Hematite | 4.2–6.9 | β-$MnO_2$ | 4.6–7.3 |
| Goethite | 6–9 | δ-$MnO_2$ | 1.5–2.8 |

Cf. *Cation exchange capacity.*
**Isohales** Contour lines of constant *salinity*.

**Isohyetal method** The average depth of *precipitation* over a specified area, an individual storm, a specified season, or on an annual basis is a requirement in many hydrologic investigations. The average depth can be determined using three different methods: (a) arithmetically averaging the gauge amounts in the area of investigation, (b) the *Thiessen polygon method*, and (c) the isohyetal method, where gauging stations and precipitation amount is plotted on a map, as shown in **Figure I-10**. Contours of equal precipitation, or isohyets, are drawn. The average precipitation for an area is

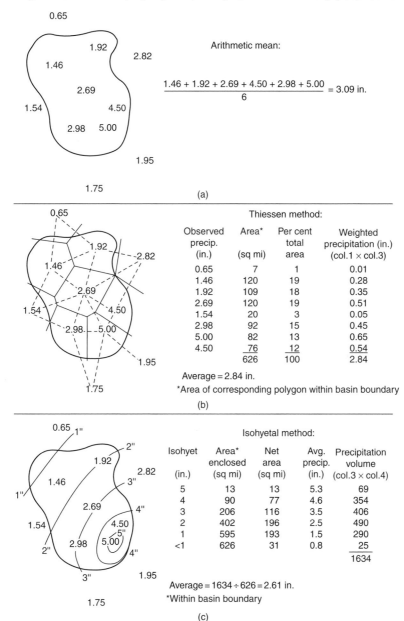

Arithmetic mean:

$$\frac{1.46 + 1.92 + 2.69 + 4.50 + 2.98 + 5.00}{6} = 3.09 \text{ in.}$$

(a)

Thiessen method:

| Observed precip. (in.) | Area* (sq mi) | Per cent total area | Weighted precipitation (in.) (col.1 × col.3) |
|---|---|---|---|
| 0.65 | 7 | 1 | 0.01 |
| 1.46 | 120 | 19 | 0.28 |
| 1.92 | 109 | 18 | 0.35 |
| 2.69 | 120 | 19 | 0.51 |
| 1.54 | 20 | 3 | 0.05 |
| 2.98 | 92 | 15 | 0.45 |
| 5.00 | 82 | 13 | 0.65 |
| 4.50 | 76 | 12 | 0.54 |
| | 626 | 100 | 2.84 |

Average = 2.84 in.

*Area of corresponding polygon within basin boundary

(b)

Isohyetal method:

| Isohyet (in.) | Area* enclosed (sq mi) | Net area (sq mi) | Avg. precip. (in.) | Precipitation volume (col.3 × col.4) |
|---|---|---|---|---|
| 5 | 13 | 13 | 5.3 | 69 |
| 4 | 90 | 77 | 4.6 | 354 |
| 3 | 206 | 116 | 3.5 | 406 |
| 2 | 402 | 196 | 2.5 | 490 |
| 1 | 595 | 193 | 1.5 | 290 |
| <1 | 626 | 31 | 0.8 | 25 |
| | | | | 1634 |

Average = 1634 ÷ 626 = 2.61 in.

*Within basin boundary

(c)

**Figure I-10** Averaging the amount of precipitation using three different methods (a) Arithmetic mean (b) Thiessen method and (c) Isohyetal method. (Linsley, 1982. Reprinted with permission from McGraw-Hill; Inc.)

calculated by weighing the average precipitation between isohyets by the area between isohyets, totaling the results, and dividing by the total area. Cf. *Horton-Theissen method*.

**Isopach map** Contour maps showing the thickness of geologic units.

**Isostatic equilibrium** Equilibrium resulting from equal *pressure* on all sides. In geologic terms, this is the equilibrium of the earth's crust resulting from the gravitational forces acting on rock masses of unequal *specific gravity*.

**Isotherm** Lines connecting points of equal temperature. Cf. *Freundlich isotherm; Adsorption isotherm*.

**Isotropic** Having the same physical properties in every direction. In *hydrogeology*, the term "isotropic" usually means having equal *hydraulic conductivity* in the $x$, $y$, and $z$ directions as shown in **Figure I-11**.

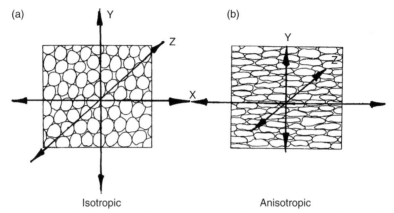

**Figure I-11** An aquifers grain size, shape, and orientation determines the isotropic or anisotropic nature. The hydraulic conductivity in an isotropic environment (a) will be equal in the X, Y and Z directions, while in an anisotropic environment (b) the hydraulic conductivity will be greater in the X direction. (After Fetter, 1994.)

Cf. *Anisotropic; Heterogeneity*.

**Isotropic medium** A medium whose properties that are directionally dependent are equal in all directions (e.g., *hydraulic conductivity*).

**Jacob distance–drawdown straight-line method** The Jacob distance–drawdown straight-line method may be used if the *drawdown* in two or more *observation wells* is recorded. The drawdown must be measured simultaneously and, according to the *Theis non-equilibrium equation*, varies with the distance from the *pumping well*. Using semilogarithmic paper, drawdown is plotted on the arithmetic scale as a function of the distance from the pumping well on the logarithmic scale. The line drawn through the data points for the closest wells is extended to the zero–drawdown intercept line, $r_0$. This is the distance at which the pumping well does not affect the water level (past the extent of the *radius of influence* of the pumping well). **Figure J-1** shows this variation of the Jacob straight-line method of solution of *pumping test* data for a fully confined aquifer. The *transmissivity*, $T$, and *storage coefficient*, $S$, are then calculated by the following equation:

$$T = \frac{2.3Q}{2\pi\Delta(h_0 - h)}$$

and

$$S = \frac{2.25Tt}{r_0^2}$$

When the units of measure are feet, minutes, and gallons, the equations are written as:

$$T = \frac{70Q}{\Delta(h_0 - h)}$$

and

$$S = \frac{Tt}{640r_0^2}$$

where:

$Q$ = constant rate of pumping (gpm)
$(h_0 - h)$ = drawdown per log cycle of distance (ft)
$t$ = time since pumping began (min)
$r_0$ = projection of the straight line, back through the zero–drawdown axis (ft).

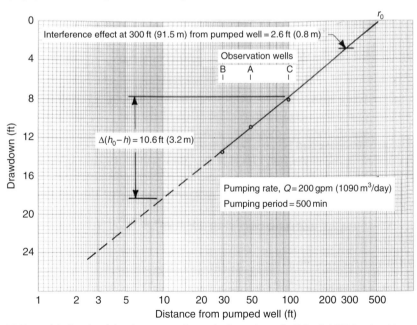

**Figure J-1** The straight line plot of drawdown versus distance in observation wells. (Driscoll, 1986. Reprinted by permission of Johnson Screens/a Weatherford Company.)

Cf. *Aquifer, confined.*

**Jacob equation/Jacob method of analysis** The *Theis non-equilibrium well equation* for non-steady radial flow in a confined areal extensive *aquifer*, for situations when u (value of the *well function*) is <0.05 was simplified by Cooper and Jacob in 1946. The Jacob equation replaces the exponential integral function, $W(u)$, with a logarithmic term simplifying the analysis, although the assumptions for non-equilibrium remain the same. Variations of the Jacob method include the time–drawdown method, the distance–drawdown straight line method, and variations of the two. The Jacob equation determines *transmissivity* and storage coefficient using the *Jacob straight line method* and is valid when the time of pumping, t, is sufficiently large and the distance from the center of the pumped well to a point where the drawdown is measured, r, is sufficiently small. Therefore, straight-line plots of drawdown versus time can be drawn after sufficient time has elapsed. A table of values of u & w/u, see **Appendix B**.

**Jacob–Hantush method** See *Hantush–Jacob formula*.

**Jacob step-drawdown test** To evaluate well characteristics, a specific type of *pumping test*, called a step test, is employed. Different methods have been developed and applied to *aquifer tests*, resulting in numerical calculations of the aquifer parameters. Step tests are employed to determine the effects of well emplacement on *groundwater* movement. The Jacob step-drawdown method aids in determining the *well loss coefficient* from data concerning two successive intervals. **Figure J-2** is a graph of sample time–drawdown data on semilogarithmic paper for

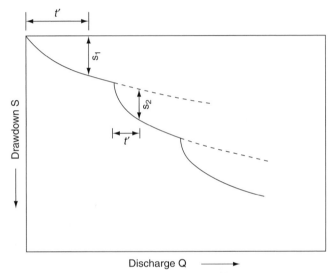

**Figure J-2** Determining well loss coefficients on a graph of time–drawdown data on semilog paper for genuine drawdown increments in each step. (Sen, 1995. Reprinted with permission from CRC Press, Inc.)

drawdown increments in each step. Jacob theorized that the *drawdown* in a well is composed of two parts, as stated in the following equation:

$$s_w = AQ + CQ^2$$

where:
$s_w$ = total energy loss or drawdown [L]
$A$ = undisturbed *aquifer loss* coefficient
$C$ = well loss coefficient [dimensionless]
$Q$ = *discharge* [L$^3$·T$^{-1}$]

The total energy loss, or drawdown, in the well at any time is composed of the aquifer loss and well loss. The Jacob expression for step-drawdown analysis is as follows:

$$C_i = \frac{(\Delta s_{wi}/\Delta Q_i) - (\Delta s_{w(i-1)}/\Delta Q(i-1))}{\Delta Q_i + \Delta Q_{(i-1)}}, \quad i = 1, 2, \ldots, j$$

The remaining terms are explained in **Figure J-2** and **Table J-1**.

**Table J-1** The drawdown increments in each step is entered and then the remaining table entrants are calculated, which gives the individual C estimations for each step and then an average value representing the well loss factor. (Sen, 1995. Reprinted with permission from CRC Press, Inc.)

| Step no. (1) | $Q_i(\text{m}^3/\text{s})$ (2) | $\Delta Q_i(\text{m}^3/\text{s})$ (3) | $\Delta S_{wi}(\text{m})$ (4) | $\Delta S_{wi}/\Delta Q_i$ (s/m$^2$) (5) | $C_i$ (from Equation 9) (s$^2$/m$^5$) (6) |
|---|---|---|---|---|---|
| 0 | 0 | 0 | 0 | 0 | |
| 1 | $Q_1$ | $Q_1$ | $S_{w1}$ | $S_{w1}/Q_1$ | $C_1$ |
| 2 | $Q_2$ | $Q_2' = Q_2 - Q_1$ | $S_{w2}$ | $S_{w2}/Q_2'$ | $C_2$ |
| 3 | $Q_3$ | $Q_3' = Q_3 - Q_2$ | $S_{w3}$ | $S_{w3}/Q_3'$ | $C_3$ |
| 4 | $Q_4$ | $Q_4' = Q_4 - Q_3$ | $S_{w4}$ | $S_{w4}/Q_4'$ | $C_4$ |
| 5 | $Q_5$ | $Q_5' = Q_5 - Q_4$ | $S_{w5}$ | $S_{w5}/Q_5'$ | $C_5$ |

**Jacob time–drawdown straight-line method** A graphical method of obtaining *transmissivity* and storativity estimates from *pumping test* data for a fully confined *aquifer* of large areal extent. C.E. Jacob noted that after the pumping well has been running for an extended time, straight-line plots of *drawdown* versus time can be obtained. In the Jacob time–drawdown straight-line method, field-data points of drawdown in an *observation well* are plotted as a function of time on semilogarithmic paper; then, a straight line is drawn through the field-data points and extended backward to the zero–drawdown axis (**Figure J-3**). The transmissivity, $T$, and storativity, $S$, are determined by using the graphical

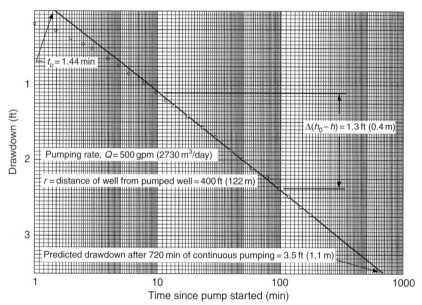

**Figure J-3** Data of drawdown versus time plotted on semilogarithmic graph paper. (Driscoll, 1986. Reprinted by permission of Johnson Screens/a Weatherford Company.)

method of solution. The semilogarithmic plot of $s(r,t)$ versus $\ln(t/r^2)$ allows $T$ and $S$ to be estimated from the slope and intercept of the resulting straight line and the following equations:

$$T = \frac{2.3Q}{4\pi\Delta(h_0 - h)}$$

and

$$S = \frac{2.25Tt_0}{r^2}$$

or if there are multiple observation wells

$$S = 2.25T\left(\frac{t}{r^2}\right)$$

where:

$T$ = transmissivity [$L^2 \cdot T^{-1}$]
$Q$ = *discharge* [$L^3 \cdot T^{-1}$]
$(h_0 - h)$ = drawdown per log cycle of time [L]
$t$ = time since pumping began [T]
$r$ = distance to the observation well [L]

If the drawdown and distance are in feet and time is in minutes, the above equations become:

$$T = \frac{35Q}{\Delta(h_0 - h)}$$

and

$$S = \frac{Tt_0}{640r^2}$$

where:

$Q$ = constant rate of pumping (gpm)
$(h_0 - h)$ = drawdown per log cycle of time (ft)
$r$ = distance to the observation well (ft)
$t_0$ = projection of the straight line, back through the zero–drawdown axis (min)
$t$ = time since pumping began (min)

The resulting transmissivity is in square feet per day (ft$^2$ day$^{-1}$), and the distance to the pumped well, $r$, is in feet. For example, in **Figure J-3**, $T$ is 1400 ft$^2$ day$^{-1}$, and $S$ is 0.000017. Referred to **Appendix A** for units. In the absence of an observation well, both the Theis and Jacob methods can be utilized by plotting the time–recovery data from the pumped well. After pumping has ceased, water-level measurements are recorded during the recovery period. Cf. *Theis solution*.

**Jet flow** Also called shooting flow. A *streamflow* characterized by water plunging or moving in forceful surges, typically occurring in stream *reaches* of high velocity along inclined stretches, at a waterfall, or where a *turbulent stream* enters a standing water body.

**Jet grouting** Drilling technique utilizing horizontal and vertical high-speed water jets to excavate soils and produce hard impervious columns by pumping *grout* through the horizontal nozzles that jets and mixes with foundation material as the drill bit is withdrawn.

**Jetted well** A *well* installed by exerting a high-*pressure flow* of water into the subsurface. Use of the jetted well technique is limited to very shallow wells, generally less than 30 ft below the ground surface, and in unconsolidated sandy sediments. Cf. *Driven well; Well installation*.

**Jetting in well development** A technique used to loosen and remove fine-grain formation material adjacent to the *well screen* with high-velocity water jets directed horizontally from the inside of the well screen out into the formation. A jetting tool typically has four nozzles attached to a pipe carrying high-pressure water. The apparatus is lowered to the bottom of the well screen and then slowly rotated and raised approximately 0.3 m (1 ft) every 5–15 min, loosening fine material from the adjacent formation. The fine-grain material settles to the bottom of the *well casing* and then removed by bailing or *airlift pumping*. The jet nozzle fitting may have a check-valve provision to allow the pump to be used for *intermittent pumping* of the well. Cf. *Well development; Bail-down procedure*.

**Jenkins–Prentice fractured-medium aquifer test model** An equation derived by Jenkins and Prentice in 1982 to calculate *transmissivity* and storativity in a fractured medium such as a *homogeneous* aquifer bisected by a single vertical fracture of extensive length. The simplifying assumptions for the Jenkins–Prentice model, are the same for any individual fracture model development. As depicted in **Figure J-4**, the *flow* in the *aquifer* is linear toward a single vertical fracture. *Drawdown* in *observation wells* in the rock matrix is directly related to the distance of these wells from the vertical fracture. The equation for drawdown, $s$, in a *pumping well* is expressed as follows:

$$s = \frac{\sqrt{\pi}Q}{2L\sqrt{TS}}\sqrt{t}$$

where:

$s$ = drawdown in the pumping well [L]
$t$ = time [T]
$Q$ = constant discharge from the well [$L^3 \cdot T^{-1}$]
$L$ = fracture length, assumed to be very long [L]
$T$ = transmissivity [$L^2 \cdot T^{-1}$]
$S$ = storativity [dimensionless]

The above equation is valid when utilizing consistent units, and when drawdown is plotted versus time on log–log paper, it yields a straight line with slope of 1/2. The simplified recovery-drawdown equation following the pumping period for linear flow becomes independent of distance:

$$s(t, t') = \frac{Q}{2LT} \frac{4T}{S} \sqrt{(t - t')}$$

where:
$t$ = time since pumping began [T]
$t'$ = recovery period time since pumping stopped [T]

The recovery data plotted as $s(t, t')$ versus the square root of $(t - t')$ yield a straight line in a linear *flow regime*. See: **Appendix A** for conversion units.

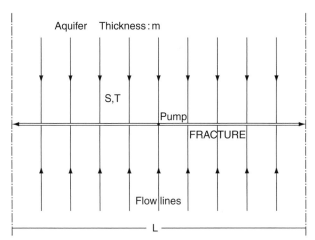

**Figure J-4** Transmissivity and storativity for flow in a fractured-medium aquifer that is linear toward an extensive single vertical fracture is determined by using the Jenkins–Prentice model. (Sen, 1995. Reprinted with permission from CRC Press, Inc.)

**Joint inversion** A mathematical technique used to determine rock property data from geophysical measurements. Geophysical measurements have been traditionally used to define changes in lithology. It is beyond the scope of this document to describe in detail joint inversion algorithms used to calculate rock properties. In general, joint inversion algorithms use two or more independent geophysical measurements to determine rock properties. For example, both P and S wave velocities can be used to determine fluid content within clastic rock formations. P waves are sensitive to changes in pore fluid and may travel at reduced velocities with changes in pore fluid. S waves are not sensitive to pore fluids. These differences in P and S waves can be useful as a direct hydrocarbon indicator in clastic rocks.

**Joint probability analysis** In general, an analysis to predict the probability of occurrence of events in which two or more partially dependent variables simultaneously take high or extreme values. Several different environmental variables are potentially important in design and assessment of *flood* and coastal defenses, including waves, *tides*, surges, river flows, *precipitation*, swell, and wind. Joint probability analysis of *flood discharge* uses transition probability matrices or a large number of simulations with random values drawn from assumed distributions. It is a general approach used to estimate a flood of selected exceedance probability from rainfall intensity–duration–frequency data for a designated locality.

**Jokulhlaup** A sudden, short, but often violent increase in the meltwater *stream* discharge from a glacier or an ice cap, generally resulting from subterranean volcanic activity. A lake may develop above the heat source and then overflow or breach the embankment to produce a torrent of meltwater. Flow velocity can reach 7–8 m s$^{-1}$, and the *discharge* may attain 100,000 m$^3$ s$^{-1}$, which is comparable to rates of flow of the Amazon River in Brazil.

**Junior water rights** In a United States court of law, "water right" is the legal right to use water and is universally recognized as real property with constitutional protection against deprivation without due process of law. A water right may be sold, leased, abandoned, or severed from the lands to which it is attached. A water right is also defined in relation to other users, and the holder of the right only acquires the right to use a specific quantity of water under specified conditions. A junior water right is that of a more recent holder than those of the senior rights holders and is not met until the senior rights have been satisfied.

**Juvenile water** Also called magmatic water. Original water formed as a result of magmatic processes. Juvenile water has not yet entered the *hydrologic cycle* and is therefore found relatively close to the point of origin. Water, gas, or ore-forming fluids are called juvenile if derived from a magma, as opposed to fluids of surface, *connate*, or meteoric origin. Magma can contain large quantities of magmatic water; for example, a magma body with a density of 2.5 g mL$^{-1}$ containing 5% water (M/M), a thickness of 1 km, and an areal extent of 10 km$^2$ contains approximately $1.25 \times 10^9$ m$^3$ of water, a portion of which may never have been in the atmosphere. Cf. *Resurgent; Resurgent vapor; Meteoric water.*

**J**

# Kk

**Kalman filter technique** A general mathematical method for integrating noisy data. In hydrogeologic evaluations, this is a method that provides for parameter estimation, addressing field data that deviates from the basic assumptions of aquifer test models. Conventional models for confined *aquifer tests*, e.g., the *Theis* and the *Cooper–Bredehoeft–Papadopulos method*, do not account for deviations from the basic assumptions of aquifer test models. Other methods to solve parameter estimation problem include general minimization procedures, weighted least-squares method, and the Bayesian decision-theoretic approach.

**Kanat/ganat/qanat** A method devised in 800 B.C. in Persia to deliver water from a mountainous water-bearing zone to typically arid or semiarid areas. **Figure K-1** shows a cross-sectional view of a kanat water delivery system. Kanat system benefits are the following:

- gravitational flow, eliminating the need for a power source
- minimal evaporative losses
- a dependable, sustainable water supply naturally protected from surface pollution

If the *water table* is somewhere above the level of the area to be supplied, the water can be tapped because the loss of *hydraulic head* incurred in the tunnel *flow* is less than that in the *aquifer*; the water flows into the kanat and down the constructed potential gradient.

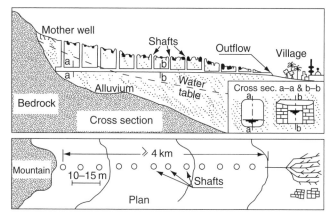

**Figure K-1** A cross-sectional and map view of a kanat (qanat) water delivery system. The well shafts are typically 300–400 m apart, up to 100 m deep, and diameters of at least 75 cm. (Sen, 1995. Reprinted with permission from CRC Press, Inc.)

Cf. *Springs; Driven well; Jetted well.*

**Kaolinite** A common high-alumina clay mineral ($Al_2Si_2O_5[OH]_4$) of the kaolin group, used in freshwater *drilling fluids* because of its *hydration potential*. Kaolinite does not expand as much with varying water content as do other more commercially available clays. In addition, kaolinite does not exchange aluminum for iron or magnesium within its structure.

**Karst lake** Also called karst pond, sink lake, and solution lake. A body of water contained within a closed depression or a sinkhole resulting from karstification.

**Karst spring** A *spring* that emerges from a karst area. Karst is a subsurface and topographic environment formed by *dissolution* features in limestone, dolomite, or gypsum, and characterized by caves, sinkholes, and typically extensive underground *drainage*. In areas of high *flow*, a karst spring may become a karst river. Cf. *Springs*.

**Kemmerer sampler** A specialized device used to collect water at a given depth in both open bodies of water and *groundwater* wells. The Kemmerer sampler is lowered in an open position to the desired depth and then a weight is lowered sliding on a cable, triggering a spring-loaded device to close the sampler and thereby collect the water sample. Cf. *Groundwater sampling*.

**Kennedy's critical velocity** The rate ($L \cdot T^{-1}$) of fluid flow in an open channel at which silt is neither picked up nor deposited. Cf. *Critical flow; Laminar flow; Turbulent flow; Reynolds number; Belanger's critical velocity; Unwin's critical velocity.*

**Kielland table** A chart enabling estimates of the *activity coefficients* for less-common *ions* (at ionic strengths below 0.1) using the *Debye–Huckel equation*. *Solute* concentrations are expressed as activities in the law of mass action. In *groundwater*, there are six major ions with significant concentrations. The activity coefficients of these major ions are determined graphically and are typically less than 1. **Table K-1** gives the values of parameters to be substituted in the Debye–Huckel equation for the less-common ions.

209

**Table K-1** Kielland table for ion-activity coefficients for the Debye–Huckel equation: $\log\gamma = -Az^2\,\sqrt{I}/1+\mathring{a}B\sqrt{I}$. (Freeze and Cherry, 1979. Reprinted with permission from Prentice-Hall, Inc.)

### Values of the Ion-Size Parameter å for Common Ions Encountered in Natural Water:

| $a \times 10^8$ | Ion |
|---|---|
| 2.5 | $NH_4^+$ |
| 3.0 | $K^+$, $Cl^-$, $NO_3^-$ |
| 3.5 | $OH^-$, $HS^-$, $MnO_4^-$, $F^-$ |
| 4.0 | $SO_4^{2-}$, $PO_4^{3-}$, $HPO_4^{2-}$ |
| 4.0–4.5 | $Na^+$, $HCO_3^-$, $H_2PO_4^-$, $HSO_3^-$ |
| 4.5 | $CO_3^{2-}$, $SO_3^{2-}$ |
| 5 | $Sr^{2+}$, $Ba^{2+}$, $S^{2-}$ |
| 6 | $Ca^{2+}$, $Fe^{2+}$, $Mn^{2+}$ |
| 8 | $Mg^{2+}$ |
| 9 | $H^+$, $Al^{3+}$, $Fe^{3+}$ |

| | Parameters A and B at 1 Bar | |
|---|---|---|
| Temperature (°C) | A | $B\,(\times 10^{-8})$ |
| 0 | 0.4883 | 0.3241 |
| 5 | 0.4921 | 0.3249 |
| 10 | 0.4960 | 0.3258 |
| 15 | 0.5000 | 0.3262 |
| 20 | 0.5042 | 0.3273 |
| 25 | 0.5085 | 0.3281 |
| 30 | 0.5130 | 0.3290 |
| 35 | 0.5175 | 0.3297 |
| 40 | 0.5221 | 0.3305 |
| 50 | 0.5319 | 0.3321 |
| 60 | 0.5425 | 0.3338 |

**Kinematic viscosity (ν)** The ratio of a fluid's *absolute (dynamic) viscosity* or viscosity coefficient (the measure of its resistance to *shear stress*) to its density at room temperature. The viscosity of a fluid is its resistance to motion during fluid *flow*, whether in open channels, in *aquifers*, or driven by density differences. Kinematic viscosity, $\nu$, is calculated by the following equation:

$$v = \frac{\mu}{\rho}$$

where:
$\mu$ = absolute (dynamic) viscosity [$M \cdot L^{-1} \cdot T$]
$\rho$ = density [$M \cdot L^{-3}$]

Cf. *Eddy viscosity*; *Coefficient of kinematic viscosity.*

**Kinematic wave** Describes flood wave motion without considering the influence of mass and hydrodynamic force. The 1D kinematic wave model is defined by the following coupled equations:

$$\text{Continuity}: dQ/dx + dA/dt = q$$
$$\text{Momentum}: S_o = S_f$$

where:
$Q$ = discharge [$L^3 \cdot T^{-1}$]
$A$ = cross-sectional area [$L^2$]
$q$ = lateral inflow [$L^3 \cdot T^{-1}$]
$S_o$ = channel bottom slope [$L \cdot L^{-1}$]
$S_f$ = frictional slope [$L \cdot L^{-1}$]

In theory, the kinematic wave should advance downstream with its rising limb getting steeper. However, the size of the wave does not become longer or attenuated. The kinematic wave model can be solved by using the finite difference scheme or the method of characteristics. For a problem with simpler boundary and initial conditions, kinematic wave routing can be performed analytically.

**Knockout plug** A stopper, usually made of wood, that covers the open leading end of a hollow-stem auger, which can be removed (or "knocked out") for sample collection or well installation. If a hollow-stem auger is not to be utilized during the drilling operation, a more durable plug, typically steel, is used in place of the knockout plug. Cf. **Appendix C – Drilling Methods**.

**Kozeny–Carmen equation** A formula used to predict the *hydraulic conductivity*, *K*, based on empirical correlations with *grain size*. The Kozeny–Carmen equation is as follows:

$$K = \frac{\rho g}{\mu} \frac{n^3}{(1-n)^2} \frac{d_m^2}{180}$$

where:
$K$ = hydraulic conductivity
$\rho$ = density [M·L$^{-3}$]
$g$ = gravitational constant [L·T$^{-2}$]
$\mu$ = *viscosity* [M·L$^{-1}$·T]
$n$ = *porosity* [L$^3$·L$^{-3}$]
$d_m$ = mean grain size [L]

Cf. *Hazen method; Fair-Hatch equation.*

**Kozeny equation** A method of reducing the effect of pumping from a confined aquifer when the well penetrating the *aquifer* is only partially screened. When the intake section of a well fully penetrates a confined aquifer, the *groundwater flow* path to the well is radial. Partially penetrating well-intake areas, however, force water to move along paths longer than the radial flow lines of a *fully penetrating well*, as shown in **Figure K-2**. *Head loss* is increased by the longer flow paths and reduced cross-sectional area; therefore, for a given *yield*, the *drawdown* in the pumping well is greater. **Figure K-3** shows the relationship of partial penetration and attainable *specific capacity* in a *homogeneous* confined aquifer, based on the Kozeny equation:

$$\frac{Q/s_p}{Q/s} = L\left(1 + 7\sqrt{\frac{r}{2bL}}\cos\frac{\pi L}{2}\right)$$

where:
$Q/s_p$ = specific capacity of a partially penetrating well [L$^3$·T$^{-1}$·L$^{-1}$ or L$^2$·T$^{-1}$]
$Q/s$ = maximum possible specific capacity of a well fully penetrating the aquifer [L$^2$·T$^{-1}$]
$r$ = radius of the *well screen* [L]
$b$ = thickness of the aquifer [L]
$L$ = length of the well screen [L]

The equation assumes the following:

• The fraction of penetration, $L/b$, is small (less than 1/2).
• The well radius is small compared to the screen length.
• The aquifer thickness is large.

The Kozeny equation is utilized in unconfined aquifers when the *drawdown* is less than 20–30% of the original saturation thickness.

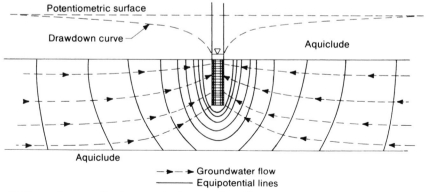

**Figure K-2** Flow patterns resulting from pumping from a well partially penetrating a confined aquifer deviate from the radial patterns associated with fully penetrating wells. (Driscoll, 1986. Reprinted by permission of Johnson Screens/a Weatherford Company.)

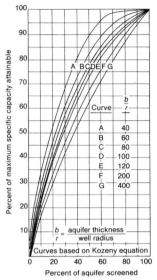

**Figure K-3** The graphical relationship developed from the Kozeny equation of partial penetrations and expected attainable specific capacity in homogeneous confined aquifers. (Driscoll, 1986. Reprinted by permission of Johnson Screens/a Weatherford Company.)

**K-packer** A neoprene rubber *packer* placed at the top of a telescoping *well screen* assembly to prevent sand from entering the space between the *well casing* and the top of the telescoped well screen. Packers are commonly composed of rubber, as is the K-packer, or of lead. The flexible K-packer is attached to a steel coupling, sealing the casing tightly to the screen. A lead packer requires the use of a *swedge block* to expand it, as shown in **Figure K-4**.

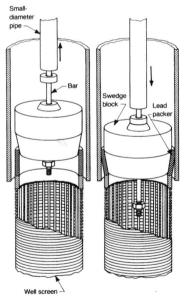

**Figure K-4** Two common types of packers used are the neoprene rubber or K-packer and a lead packer. (Driscoll, 1986. Reprinted by permission of Johnson Screens/a Weatherford Company.)

**Lacustrine** Also lacustral or lacustrian. Produced by, formed in, referring to, deposited in, growing in, or inhabiting a *lake*.

**Lade** The beginning or mouth of a *river* or other watercourse.

**Lag/lag time** A measure from the beginning to the peak, e.g., the catchment response time, in a rainfall and river *hydrograph*; or, the time from the start of *precipitation* until a watercourse starts to rise. The relationship is shown in **Figure L-1**. The measurement of lag time begins from the center of gravity of the effective rainfall to the center of gravity of the direct surface *runoff*. In *geophysical logging*, lag, $L'$, is related to the speed at which *nuclear logs* are to be run. Lag is measured in feet as the distance that a detector moves during one time constant.

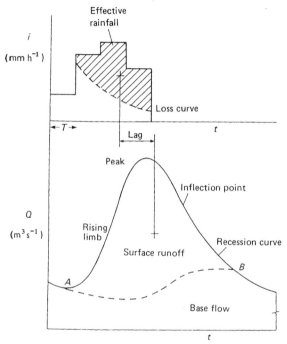

**Figure L-1** The graphical relationship between rainfall and a river hydrograph with the associated lag time between the start of the precipitation and the river rise. (Shaw, 1983. Reprinted with permission from Van Nostrand Reinhold Co. Ltd.)

Cf. *Basin lag.*

**Lagoon** A shallow body of water (natural or manmade) that is enclosed, or nearly enclosed, between the mainland and a saltwater body (natural), i.e., a coastal water body. A natural lagoon can be formed either from saltwater influx or within a basin of a hot spring. Cf. *Spring, hot; Salinity.*

**Lagrangian current measurement** The measurement of water movement by tracing particle paths over a sustained time period. In Lagrangian current measurement, a device is released and allowed to drift passively, and its course is followed and plotted. The device may be a neutrally buoyant float for subsurface current tracking or a buoy or a drift pole for surface-water movement.

**Lake** A *freshwater* or saline inland body of water enclosed within a depression of the earth's surface, either natural or man-made. Lakes can be classified on the basis of size, chemical, or biological variables (such as *oligotrophic* or *eutrophic*), or being an open system (connected with a flowing *stream*) or a closed/isolated system. Naturally occurring lakes are typically found in topographically low areas where the *water table* intersects the ground surface, or when the near-surface materials exhibit low *hydraulic conductivity* causing a perched condition forming an aboveground body of water. Regional *groundwater* strongly influences the hydrologic regime of a lake. If the lake has an obvious inlet and outlet, it may be stream-fed and/or drained, but a lake may occur without an aboveground inlet or outlet and be supplied by groundwater or by *precipitation* alone. Large permanent lakes in relatively low topographic areas are typically fed

213

by the regional groundwater system, whereas small permanent lakes in upland areas are usually *discharge* points for the local *flow system*. Cf. *Epilimnion; Hypolimnion; Density currents; Perched water.*

**Lake gauge** A device for measuring and sometimes recording water surface elevations within a *lake.*

**Lamellar flow** Liquid *flow* in layers that are capable of gliding over or past adjacent layers. Cf. *Laminar flow.*

**Laminar flow** Fluid movement in which the particle paths are smooth, horizontally straight, and parallel to the channel walls because the fluid *viscosity* is damping out *turbulent flow*. In laminar flow, the forces of gravity and friction dominate, and flow in open water channels occurs at low velocities that increase away from the watercourse bed. Laminar and turbulent flow, or a combination of the two, (see **Figure L-2**) has also been observed and quantified in surface flow that is spatially varied and unsteady as a result of variation in *rainfall, infiltration* rate, and topography. The depth of *overland flow* is typically very small but the volume can be large. If the overland flow is laminar, it is assumed that the slope is very small, and the depth increases rapidly to accommodate increased turbulent flow. *Groundwater flow* on a macroscopic scale is normally assumed to be laminar, moving slowly and steadily through interconnected rock/soil interstices, allowing the water particles to move along in regular, steady flow paths, or streamlines, without crossing other paths. Laminar flow conditions are assumed in most groundwater modeling and aquifer test interpretations. Laminar flow conditions are assumed to exist in an *aquifer* during pumping, forming the basis for well hydraulics where *drawdown* is directly proportional to the pumping rate; if laminar flow conditions are not present, then this relationship is not valid. On the microscopic scale, movement may be quite different. Further, turbulent groundwater flow may be expected in karst settings, in fractured bedrock, at the base of a dam, etc. In fluid mechanics, the dimensionless *Reynolds number* is used to determine whether flow is laminar at low velocity or turbulent at high velocity.

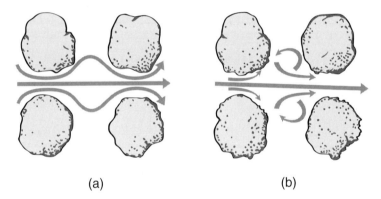

(a)                                         (b)

**Figure L-2** A schematic representation of a) laminar and b) turbulent flow in a granular deposit. (Driscoll, 1986. Reprinted by permission of Johnson Screens/a Weatherford company.)

Cf. *Critical velocity; Laminar velocity; Turbulent velocity; Open channel flow.*

**Laminar head loss** The first-order term plus the second-order turbulent component to express the *drawdown* in a *well* from a well efficiency *step-drawdown test* for the purpose of determining optimum *pumping rates*. The laminar term, $BQ$, has been attributed to the *aquifer loss* and the turbulent term, $CQ^2$, to the well loss or *head loss* from inefficiency. In actuality, both terms include loss from both the aquifer and the well; therefore, the ratio of the laminar head loss to the total head loss, expressed as a percentage, is computed from the step-drawdown test data by utilizing the following equation:

$$L_p = \frac{BQ}{BQ + CQ^2} \times 100$$

where:
$L_p$ = percentage of the total head loss that is attributable to laminar flow [%]
$BQ$ = first-order laminar component
$CQ^2$ = second-order turbulent term
$Q$ = *discharge* [L$^3$·T$^{-1}$]

If the drawdown divided by the discharge is plotted versus the discharge, as in **Figure L-3**, the intercept and the slope on the resulting straight line are the values of $B$ and $C$, respectively. The value of $L_p$ is not a measure of the well efficiency.

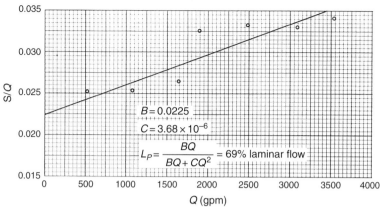

**Figure L-3** The values for *B* and *C*, intercept and slope, in the step-drawdown equation are determined from the above graph of *s/Q* versus *Q*. (Driscoll, 1986. Reprinted by permission of Johnson Screens/a Weatherford Company.)

**Laminar velocity** In a watercourse, that rate of water movement below which only *laminar flow* occurs and above which either laminar or *turbulent flow* occurs. In fluid mechanics, the dimensionless *Reynolds number* is used to determine whether flow is laminar at low velocity or turbulent at high velocity. Cf. *Turbulent velocity*.

**Land pan** A device used to measure *free water evaporation* by using shallow, unpainted, galvanized metal pans (122 cm in diameter and 25.4 cm deep) placed on supports to allow air to circulate freely. The depth of water in the pan is kept between 18 and 20 cm, and daily records are collected on the depth of water, volume of water added to replace evaporated water (giving pan evaporation), and volume of water added as *precipitation*. The data is used to calculate the daily evaporation using the *hydrologic budget*.

**Landslide lake** An interior water body formed by a mass movement of earth material that dams or blocks a formerly free-flowing watercourse.

**Land subsidence** Settlement of the land surface primarily as a result of excessive *groundwater* withdrawals in regions containing compressible sediments such as clays. As *fluid pressures* decline due to water withdrawals from *aquifers*, fluid pressure reductions in any adjacent lower-*permeability* units cause the effective stress to increase and the lower-permeability, more compressible units to then consolidate. The cumulative effect of this process is subsidence of the ground surface.

**Langelier index** An analytical method, based on calcium carbonate ($CaCO_3$) equilibrium in water, used to predict the *saturation pH*, $pH_s$, to determine the incrusting or corrosive tendency of the water. If *dissolved oxygen* (DO) is present and the water has the capacity to dissolve $CaCO_3$, then the water will likely be corrosive to steel. If the water is supersaturated with $CaCO_3$, incrustants will likely form. The Langelier index equation is as follows:

$$\text{Langelier index} = pH - pH_s$$

where:
$pH$ = actual pH of water
$pH_s$ = saturation pH

If the measured pH is less than the calculated $pH_s$, the water has a negative Langelier index and will dissolve $CaCO_3$ and is likely to corrode steel. If the pH is greater than the calculated $pH_s$, the index is positive and supersaturated with $CaCO_3$-forming *incrustations*. The saturation pH is determined using nomographs. Cf. *Ryznar stability index*.

**Langmuir isotherm** For a chemical system at equilibrium, a mathematical expression that assumes a non-linear function in relating the mass of *solute* sorbed to the dissolved solute concentration. Three types of *isotherms* are applicable to the *sorption* of organic chemicals: the Langmuir, *Freundlich*, and *linear isotherms*, in relationship to each other as shown in **Figure L-4**. When organic concentrations are greater than $10^{-3}$M, as for landfill *leachate* or waste from industrial processes, the Langmuir or Freundlich isotherms are useful. The plot of the sorbed concentration of the *ion* in question divided by the amount adsorbed is plotted versus the concentration of the ion in solution in contact with soil. If there is a maximum capacity of the soil to sorb the ion, a Langmuir adsorption isotherm can be used to describe the process. The Langmuir isotherm is given by the following equation:

$$\frac{C}{C^*} = \frac{1}{\beta_1 \beta_2} + \frac{C}{\beta_2}$$

where:
$C$ = equilibrium concentration of the ion in contact with the substrate [$M \cdot L^{-1}$]
$C^*$ = amount of the ion adsorbed per unit weight of soil [$M \cdot M^{-1}$]

215

$\beta_1 =$ adsorption constant related to the binding energy $[M \cdot L^{-1}]$
$\beta_2 =$ adsorption maximum for the soil $[M \cdot L^{-1}]$

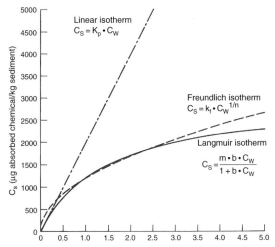

**Figure L-4** Sorption isotherms. (Maidment, 1993. Reprinted with permission from McGraw-Hill, Inc.)

The slope of the line divided by the intercept is $\beta_1$, the binding energy constant, and the maximum ion adsorption for the substrate is the reciprocal of the line. The Langmuir isotherm is applicable to both anions and cations, and the data may plot as two straight lines: one for the lower ion concentration and the other for the higher concentration with a lower slope. **Figure L-5** shows a two-line plot for phosphorous adsorbed on a calcareous substrate.

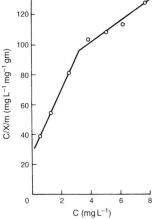

**Figure L-5** A plot for the sorption of phosphate adsorbed on calcareous outwash displaying a two-surface isotherm. (Fetter, 1977.)

Cf. *Activated carbon.*

**Laplace equation** A partial differential equation of *groundwater* movement describing *steady-state flow* through a *homogeneous, isotropic* medium. The Laplace equation, stated below, determines the value of the *hydraulic head, h,* at any point in a three-dimensional flow field:

$$\frac{\partial^2 h}{\partial x^2} + \frac{\partial^2 h}{\partial y^2} + \frac{\partial^2 h}{\partial z^2} = 0$$

216

In the case of steady state, saturated two-dimensional flow for a unit thickness, the expression $(\partial^2 h/\partial z^2)$ is dropped. The Laplace equation can often be solved graphically by constructing a flow net displaying the streamlines and equipotentials. Cf. *Equations of groundwater flow*.

**Latent heat of condensation** The heat released by the *condensation* of water vapor into liquid or solid form. The amount of thermal or heat energy within a substance is measured in calories, and the latent heat of condensation for water is 590 cal g$^{-1}$. Meteorologically, the transformation from solid directly to vapor, and vapor directly to solid, is called *sublimation*. Cf. *Latent heat of fusion; Latent heat of transition; Latent heat of vaporization*.

**Latent heat of fusion** The amount of heat required to change 1 g of ice to liquid water at the same temperature. The addition of 80 cal of heat is required to melt 1 g of ice at 0°C. Cf. *Latent heat of transition*.

**Latent heat of transition** The heat required to initiate a phase change to a higher energy state, i.e., from solid to liquid and from liquid to gas, as in the latent heat of melting or, in reverse, the latent heat of crystallization. Latent heat measurements are expressed in joules per mole (J mol$^{-1}$) or in calories. See: **Appendix A** for conversions. Cf. *Latent heat of condensation; Latent heat of vaporization; Latent heat of fusion*.

**Latent heat of vaporization** The amount of heat absorbed by a substance without a change in temperature that occurs while changing from liquid to liquid or solid state. The input of 590 cal of heat energy is required to *evaporate* or vaporize 1 g of water at 15°C. The water vapor retains this energy until it condenses and releases the energy as *latent heat of condensation*.

**Lateral consequent stream** A *tributary* of a *subsequent* or *subconsequent stream* that flows down the flank of an anticline or a syncline.

**Lateral dispersivity** See *transverse dispersion*.

**Lateral inflow** Any volume of water added to a water body by *groundwater* inflow, *overland flow, interflow*, or small *springs* and seeps. *Flow* in a watercourse is a distributive process because the rate, velocity, and depth vary at different cross sections along the channel. The volumetric flow rate of a *stream* or *river*, Q, in dimensions of L$^3$·T$^{-1}$, (Refer to **Appendix A** for conversions) increases *downstream* from lateral inflows. Lateral inflows are not only significant additional sources of water, but may also introduce additional *solute* masses dissolved from the surrounding area.

**Lateral lake** A *fluviatile lake* formed by silting up of the channel of the main stream in the valley of a *tributary* impounding the water of the tributary.

**Lateral storage** See *bank storage*.

**Lateral stream** A watercourse that flows along an edge or is situated on, directed toward, or comes from the side.

**Lattice drainage pattern** See *drainage pattern*.

**Lava-dam lake** Impounded water resulting from the obstruction of a watercourse by a volcanic flow.

**Law of basin areas** Statement of the direct geometric relationship between *stream order* and the mean *basin area* of stream order within a specified *drainage basin*. As postulated by Schumm in, 1956 the relation is a linear correlation of the logarithm of *mean basin area* regressed on stream order; there is a linear relation, the positive regression coefficient being the logarithm of the basin area ratio.

**Layered heterogeneity** Commonly found in layered sedimentary rocks, unconsolidated *lacustrine*, and marine deposits in which the *hydraulic conductivities* of the individual beds or layers of the formation are *homogeneous* but the conductivity of the formation as a unit is *heterogeneous*, as pictured in **Figure L-6**.

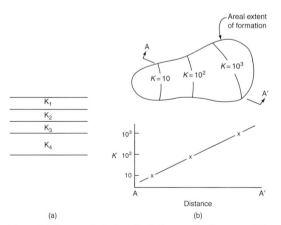

**Figure L-6** Comparison of layered heterogeneity to trending heterogeneity. (Freeze and Cherry, 1979. Reprinted with permission from Prentice-Hall, Inc.)

**Leachate** The solution formed by the dissolution of contaminants, or *leaching* by water as it *percolates* through wastes. Leachate forms when water *infiltrates* areas such as feedlots, landfills, or pesticide applications, and mixes with contaminated water or with dissolved compounds in solid waste. Leachate formation may result in deeper percolation of hazardous substances into subsurface soil, *groundwater*, and eventually *surface water*. Cf. *Freundlich isotherm; Langmuir isotherm*.

**Leachate collection system** A series of drains installed in an area receiving *leachate* so that it may be collected and treated or transported to avoid affecting subsurface and eventually surface-water systems.

**Leaching** The removal or extraction of soluble constituents from surface and subsurface materials. The soluble constituents may be nutritive or degradative (e.g., mineral salts/organic matter), and naturally or artificially occurring. Leaching can dissolve or remove these constituents by *percolation* through either natural (*rainfall*) or artificial (*irrigation*) means. Cf. *Acid mine water*.

**Leaching requirement** The addition of irrigation water in excess of the volume required for crop hydration; it is used as a method of soil salinization control. High levels of salt in the crop root zone can become toxic to plants. Dissolved salt in irrigation water tends to accumulate in soils as *evapotranspiration* consumes soil water. The excess irrigation water flushes or leaches the salt through the crop root zone.

**Lead** An artificial watercourse originating or leading from a mine, a mill, or a *reservoir*.

**Lead packer** A type of sand-tight seal placed between the top of a telescoped screen assembly and the outer *well casing* and attached directly to either the top of the *well screen* or the top of a *riser pipe*, as shown in **Figure L-7** in comparison to a *K-packer*.

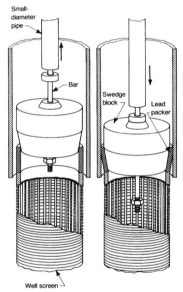

**Figure L-7** Two common types of packers used are the neoprene rubber or K-packer and a lead packer. (Driscoll, 1986. Reprinted by permission of Johnson Screens/a Weatherford Company.)

Cf. *Swedge block*.

**Leakage/leakage factor** Water *flow* across layers with different *hydraulic conductivities*. Leakage is the resistance of a semipervious layer to upward or downward leakage in a leaky aquifer. The leakage factor, $L$, can be calculated as follows:

$$L^2 = \frac{K_U m_U m_L}{K_L}$$

where:
$m_U, m_L$ = thickness of the recipient and source layers [L]
$K_U, K_L$ = hydraulic conductivities of the recipient and source layers [L·T$^{-1}$]

218

The above equation is demonstrated in **Figure L-8**. High values of $L$ indicate a greater resistance of the source strata to flow or negligible leakage. **Table L-1** is a leakage classification based on the above equation. In the *Hantush–Jacob formula* for leaky aquifers, the leakage factor, $B$, is calculated as follows:

$$B = \left(\frac{Tb'}{K'}\right)^{1/2}$$

where:
$T = transmissivity\ [\text{L}^2\cdot\text{T}^{-1}]$
$b' = thickness\ of\ the\ leaky\ layer\ [\text{L}]$
$K' = vertical\ hydraulic\ conductivity\ of\ the\ leaky\ layer\ [\text{L}\cdot\text{T}^{-1}]$

The solution assumes that leakage through the confining bed is vertical and proportional to the *drawdown*, the head in the layer leaking is constant, and storage in the confining bed is negligible.

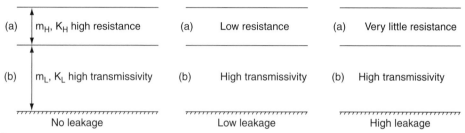

**Figure L-8** The relationship between leakage-hydraulic resistance and transmissivity. (Sen, 1995. Reprinted with permission from CRC Press, Inc.)

Table L-1   Leakage factors

| Leakage factor | Classification |
| --- | --- |
| $L<1000$ | Potential leakage |
| $1000<L<5000$ | Moderate leakage |
| $5000<L<10{,}000$ | Low leakage |
| $L<10{,}000$ | Negligible leakage |

Cf. *Aquifer, leaky confined; Layered heterogeneity.*
**Leakage coefficient ($L_c$)** The rate of *leakage* passing through a unit area of the lower surface of a semipervious layer under the condition that the *head* drop is equal to unity. The leakage coefficient, $L_c$, is the reciprocal of *hydraulic resistance*, $R_h$, and is calculated by the following equation:

$$L_c = \frac{1}{R_h} = \frac{K_u}{m_u} = \frac{T}{L^2}$$

where:
$L_c = leakage\ coefficient\ [\text{T}^{-1}]$
$R_h = hydraulic\ resistance\ [\text{T}]$
$K_u = hydraulic\ conductivity\ of\ the\ semipervious\ layer\ [\text{L}\cdot\text{T}^{-1}]$
$m_u = thickness\ of\ the\ semipervious\ layer\ [\text{L}]$
$T = transmissivity\ [\text{L}^2\cdot\text{T}^{-1}]$
$L^2 = leakage\ factor\ [\text{L}^2]$
**Leakance** See *Leakage/leakage factor.*
**Leaky aquifer** See *Aquifer, leaky confined.*
**Leaky confined aquifer** See *Aquifer, leaky confined.*
**Leaky confining layer** See *Aquitard.*
**Left bank** The side of a watercourse that is to the left of an observer facing *downstream*. Cf. *Bank.*

**Leibnitz's rule** The *drawdown* increment with respect to time to obtain the temporal changes in the drawdown during an *aquifer test* can be differentiated, leading to the following equation:

$$\frac{ds(r,t)}{dt} = \frac{Q}{4\pi T}\frac{e^{-r^2 S/4rT}}{t}$$

where:
$s(r,t) =$ drawdown [L] as a function of $r$ and $t$
$\quad t =$ time [T]
$\quad Q =$ discharge [L$^3$·T$^{-1}$]
$\quad T =$ transmissivity [L$^2$·T$^{-1}$]
$\quad S =$ storativity [dimensionless]

The equation for Leibnitz's rule is similar to the *distance–drawdown* equation but incorporates the expansion of the *cone of depression* with time, as shown in **Figure L-9**. The slope of the time–drawdown plot is represented by the left-hand side of the equation and, theoretically, is inversely and non-linearly related to the aquifer parameters. The terms of the equation, with the exception of $T$ and $S$, are determined from aquifer test results. Local changes in these transmissivity and storativity values are estimated by the slope-matching method.

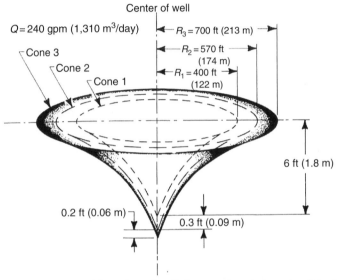

Center of well

$Q = 240$ gpm (1,310 m$^3$/day)     $R_3 = 700$ ft (213 m)

Cone 3     $R_2 = 570$ ft (174 m)

Cone 2     $R_1 = 400$ ft (122 m)

Cone 1

6 ft (1.8 m)

0.2 ft (0.06 m)

0.3 ft (0.09 m)

**Figure L-9** Relationship between cone of depression, radius of influence, and constant pumping. (Driscoll, 1986. Reprinted by permission of Johnson Screens/a Weatherford Company.)

Cf. *Jacob step-drawdown test; Jacob time-drawdown straight-line method.*
**Level** An open *reach* of a watercourse; or, the elevation, *water table*, of a water body's surface. Cf. *Stream.*
**Level of saturation** See *water table.*
**Level recorder** A method of recording water elevations in a *well*. This may be performed manually using a calibrated tape with a water sensor at the end, which gives either a auditory or visual indication when water is encountered as the probe is lowered down a well. The depth to water is then read off the calibrated tape. Automated methods involve the use of floats connected to strip chart recorders or *pressure transducers* coupled with data loggers. Cf. *Electrical sounder.*
**Lifting head** The vertical distance between the water level that is being pumped and the delivery point. Cf. *Dynamic head; Total head; Performance curve.*
**Ligand** An atom, an *ion*, or a functional group that is bonded to one or more central (usually metal) ion(s) forming a complex. The central atom or group of atoms will have a positive charge, and the ligands will bond by compensating that charge with their own negative charge or characteristics. As the central atom has a specific charge, it has a maximum potential number of ligands (as each must donate at least one electron into the charge compensation equation) that it can bond to. Common ligands are F, Cl$^-$, Br$^-$, I$^-$, CO, and RC:CR compounds with double carbon bonds, benzene, cryptates, crown ethers, and OH$^-$. In biochemistry, a ligand will refer to a small molecule that binds to a larger macromolecule, whether or not the ligand actually binds at a metal site or not. This is probably a carryover from the large number of binding studies on oxygen transport proteins, such as hemoglobin, where the ligand indeed did bind at a metal site, an expansion of the term to a more general case of binding.

**Light non-aqueous-phase liquid (LNAPL)** An immiscible fluid that has a lower density than that of water will float on top of the heavier fluid. In *groundwater*, an example of a light non-aqueous-phase liquid (LNAPL) is a contaminant such as gasoline, which is less dense than water and therefore tends to float on top of the saturated zone once it has infiltrated the *unsaturated zone*. *Dense non-aqueous-phase liquid* (DNAPL), by contrast, tends to sink to the base of the *aquifer* or to the top of an impermeable zone. **Figure L-10** shows the result of density differences in fluids in a groundwater system. Once an LNAPL reaches the *capillary fringe*, it tends to spread laterally, allowing for greater distribution: as a gas zone in the upper layer, where the more volatile compounds will migrate into *pore spaces*, and as a diffusion zone in the lower layer, where the more soluble portion will be disseminated. The lens of LNAPL, where present, may also depress the groundwater surface.

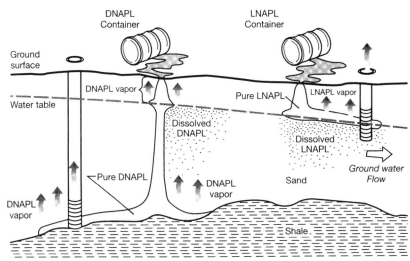

**Figure L-10** The migration character of LNAPL and DNAPL as a result of density differences. (Sanders, 1998.)

Cf. *Non-Aqueous-Phase Liquid (NAPL)*.

**Liman** A shallow, muddy enclosure at the mouth of a watercourse, protected by a seaward barrier, as is an *estuary* or a broad *freshwater* bay off the sea.

**Lime softening** The application of hydrated lime $(Ca(OH)_2)$ to water to precipitate calcium carbonate $(CaCO_3)$ and/or magnesium hydroxide $(Mg(OH)_2)$ in order to reduce the water's *hardness*. Once coagulated, these inorganic materials are removed by sedimentation and *filtration*. Cf. *Softening of water*.

**Limewater** Naturally occurring water in which a greater portion of the *total dissolved solids* (TDS) consists of calcium bicarbonate $(CaCO_2)$ or calcium sulfate $(CaSO_4)$.

**Limiting conditions** General solutions for *groundwater movement* obtained by incorporating assumptions based on the model used and conditions encountered. Specific solutions of *groundwater* movement require additional knowledge of the limiting conditions associated with the *aquifer* being studied. Limiting conditions include the initial state of the *piezometric surface* of the groundwater and the *boundary conditions* affecting cross-flow.

**Limnal/limnic** Related to *freshwater* bodies, as in a *lake* or lakes.

**Limnology** The study of the physical, chemical, and biological characteristics of inland bodies of water, including all enclosed water bodies such as *lakes*, ponds, pools, as well as *rivers*, *streams*, bogs, and *wetlands*.

**Line of seepage** See *seepage line*.

**Line source** A contaminant release that spreads or disperses in a line and diffuses only in the $x-y$ plane perpendicular to the line of release and uniformly in the $z$ direction. A line source can be either instantaneous or continuous, as shown in **Figure L-11**.

**Line swabbing** See *Swabbing*.

**Lineament** A linear feature greater than 300 m (1000 ft) in length, appearing on an aerial or satellite photo, which is used in fracture trace analysis. The fracture traces represent zones of increased *porosity* and *hydraulic conductivity* in *carbonate*, igneous, and metamorphic rocks.

**Linear flow** See *Laminar flow*.

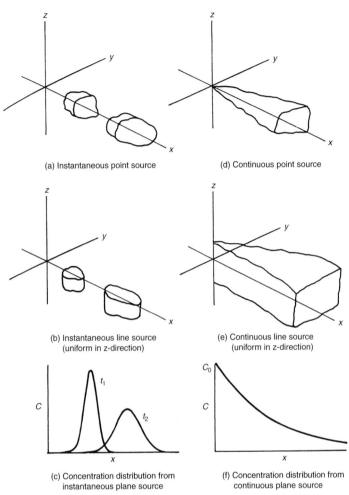

**Figure L-11** Comparison between instantaneous and continuous source of contaminant. (Maidment, 1993. Reprinted with permission from McGraw-Hill, Inc.)

(a) Instantaneous point source

(d) Continuous point source

(b) Instantaneous line source (uniform in z-direction)

(e) Continuous line source (uniform in z-direction)

(c) Concentration distribution from instantaneous plane source

(f) Concentration distribution from continuous plane source

**Linear isotherm** For a chemical system at equilibrium, a mathematical expression that relates the mass of solute sorbed to the dissolved solute concentration by assuming a linear relationship between the sorbate concentration and the solute concentration. Three types of *isotherms* are applicable to the *sorption* of organic chemicals: the linear, *Langmuir*, and *Freundlich isotherms*, shown in relationship to each other in **Figure L-4**. The linear isotherm is an accurate representation of partitioning for organic chemicals at low concentration in natural waters and the sorbed concentration and expressed by the following equation:

$$C_s = K_p C_w$$

where:

$C_s$ = sorbed concentration, or mass of contaminant per unit mass of sediment $[M \cdot M^{-1}]$
$K_p$ = partitioning coefficient, or volume of water per mass of sediment $[L^3 \cdot M^{-1}]$
$C_w$ = dissolved concentration, or mass of contaminant per volume of water $[M \cdot L^{-3}]$

The above equation is also called the linear expression of the Freundlich isotherm when the slope of the line, $b$, is equal to 1.0. The retardation coefficient $K_d$, is related to $K_p$, and is defined as

$$K_d = \frac{\text{mass of solute on the solid phase per unit mass of solid phase}}{\text{concentration of solute in solution}}$$

**Links** The curved sections along a watercourse, including the ground along that section.

**Linn** Water running over a precipice or a steep edge and pooling below (e.g., waterfall, cascade, and cataract).

**Liquid-water content** The portion of the total volume of wet snow that is liquid. The liquid-water content of snow is not the same as the equivalent water represented by the snow.

**Liquidity index** An indication of the consistency of a soil at its natural *moisture content*. The liquidity index is the moisture content minus the *plastic limit* moisture content, divided by the *plasticity index* at the liquid limit.

**Lithia water** A mineral water containing lithium bicarbonate ($LiHCO_3$), lithium chloride ($LiCl$), or other lithium salt.

**Lithologic log** See *Driller's log.*

**Littoral** Immediately surrounding and on the outer edge of a water body, such as a *lake*, pond, *river*, or ocean. The littoral zone, or the shoreline, is the area to which most biota restrict their activities.

**Littoral lake** The area surrounding an enclosed water body, or the shoreward portion of a *lake*, that is capable of supporting rooted vegetation.

**Littoral rights** See *Riparian rights/Riparian doctrine.*

**Live stream** See *Perennial stream.*

**Livingston atmometer** A device used to measure *evaporation*, consisting of porous spheres on the ends of glass capillary tubes. The porous spheres of a Livingston atmometer are exposed to ambient and radiative energy levels and the lower ends of the tubes are immersed in distilled water. The difference in the changes in water level is an indication of the evaporation between readings. Cf. *Atmometer.*

**LNAPL** See *Light, non-aqueous-phase liquid.*

**Load** The total amount of material that is either carried or moved by natural agents such as wind, water, or glaciers. The load or sediment concentration, $C$, may be calculated either as the total mass, $M$, in a volume of water, $V$

$$M = \frac{C}{V}$$

or as the mass flow rate, $L$, in dimensions of $M \cdot T^{-1}$, in water with a flow rate of $Q$, in dimensions of $L^3 \cdot T^{-1}$

$$L = CQ$$

Refer to **Appendix A** for property conversions or unit changes.

The concentration of sediment, calculated by multiplying the number of days having a given range of *discharge* by the mean discharge of that range, increases with discharge and then decreases at the higher discharges. **Figure L-12** shows this relationship. Although the flows at the high end of the graph are effective in *erosion* and transportation, they are infrequent. As a result, their contribution to the total load is meager compared to the lower discharges.

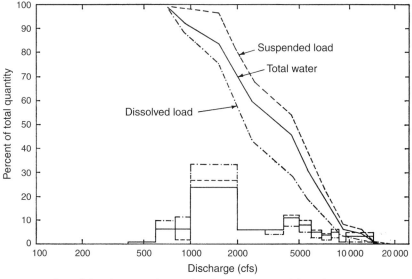

**Figure L-12** Comparison of the relative contribution of discharge rates to total flow of water, suspended load, and dissolved load. (Leopold et al., 1992. Reprinted with permission from Dover Publication, Inc.)

Cf. *Sediment load; Bed load; Dissolved load.*

**Loaded stream** A contained watercourse that is at its maximum sediment-carrying or transporting capability. Cf. *Stream; Load.*

**Loading function** A term frequently used in *groundwater*, non-point pollution calculation procedures for estimating the average annual *load*, and the storm event load, of a *pollutant* from an individual land-use category. (See **Figure L-11**). The most widely used loading functions in the United States are the Environmental Protection Agency (EPA) Screening Procedures or the EPA Water Quality Assessment Methodology, which describe calculation procedures to estimate non-point loads for urban and non-urban areas.

**Local base level** See *Temporary base level*.

**Local flow system** See *Flow system*.

**Log–log-type curve matching** The determination of *transmissivity*, $T$, and storativity, $S$, from a plot of $h_0-h$ and $t$ and $W(u)$ versus $1/u$-type curves, as shown in **Figure L-13**. Log–log type curve matching is performed as follows:

1. By plotting the dimensionless theoretical response, or *type curve*, of $W(u)$ versus $1/u$ and the time–drawdown values, $h_0-h$, versus $t$, on log–log graph paper of the same size and scale.
2. By superimposing the field curve on the type curve until most of the observed data points fall on the type curve.
3. By reading off the paired values of $W(u)$, $1/u$, $h_0-h$, and $t$ at an arbitrary match point and calculating $u$ from $1/u$.
4. By calculating $T$ and $S$ from the following relationships:

$$T = \frac{AQW(u)}{h_0 - h}$$

$$S = \frac{uT_t}{Br^2}$$

where:

$Q$ = pumping rate $[L^3 \cdot T^{-1}]$
$W(u)$ = *well function of u* [dimensionless]
$h_0-h$ = recorded drawdown [L]
$u = r^2 S/4Tt$
$t$ = time [T]
$A$, $B$ = coefficients dependent on the units used

For SI units of meters and seconds, $A = 0.08$ and $B = 0.25$. For units of feet, days, gpm, or gal day$^{-1}$ per ft, $A = 114.6$ and $B = 1.87$; or, if imperial gallons are used, $B = 1.56$. Refer to **Appendix A** for unit conversions.

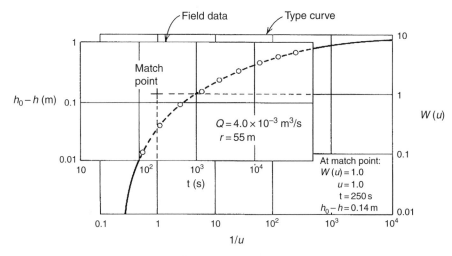

$$T = \frac{QW(u)}{4\pi(h_0-h)} = \frac{(4.0\times10^{-3})(1.0)}{(4.0)(3.14)(0.14)} = 0.0023 \text{ m}^2/\text{s} \ (15\,700 \text{ U.S. gal day}^{-1}\text{ft}^{-1})$$

$$S = \frac{4uTt}{r^2} = \frac{(4.0)(1.0)(0.0023)(250)}{(55.0)^2} = 7.5\times10^{-4}$$

**Figure L-13** The $W(u)$ versus $1/u$-type curve for the determination of $T$ and $S$ using the log–log curve-matching procedure. (Freeze and Cherry, 1979. Reprinted with permission from Prentice-Hall, Inc.)

**Logging** See *geophysical exploration methods.*

**Longitudinal consequent stream** A *consequent stream* whose flow direction is determined by the plunge of a fold, as in a synclinal trough.

**Longitudinal dispersion** The mixing, or *dispersion*, of a contaminant in the direction of fluid *flow. Advective transport* characterizes the movement of a contaminant at the average velocity of the fluid. Superimposed on the advective process, there is a dispersive process caused by velocity variations in the fluid on a scale smaller than that captured in the advective term. Longitudinal dispersion causes a fraction of the contaminant to spread ahead of the average advective position and a second fraction to move with a lower than average velocity, and to lag behind the average advective position in a fluid. Cf. *Dispersion coefficient.*

**Longitudinal dispersion coefficient** The proportionality constant, $D_L$, that accounts for the *dispersion* of a *solute* in the *downgradient* or longitudinal direction. The longitudinal dispersion coefficient has dimensions of $L^2 \cdot T^{-1}$. The following equation shows the relationship between the coefficients of *mechanical dispersion* and the linear groundwater velocity:

$$D'_L = \alpha_L v$$

where:

$D'_L$ = mechanical component of the longitudinal dispersion coefficient $[L^2 \cdot T^{-1}]$
$\alpha_L$ = longitudinal dispersivity of the medium $[L]$
$v$ = linear groundwater velocity $[L \cdot T^{-1}]$.
Cf. *Peclet number.*

**Longitudinal flow** One-dimensional *flow* that is the simplest form of flow geometry in *groundwater* investigations. Longitudinal flow is a simplification used as an assumption in an alluvium fill of a valley that has only a single-layer, longitudinal velocity component, which is treated as a one-dimensional flow problem.

**Longitudinal profile** The cross-sectional view of a watercourse as seen from the head or source area down to the mouth or area of *discharge.* The size of a watercourse increases in the *downstream* direction as a result of discharge from its tributaries. As the channel width and depth increases, the bed particle size tends to decrease and the *gradient* generally flattens, thereby making the longitudinal profile concave. Cf. *Tributary.*

**Longitudinal stream** A watercourse that has adjusted to the regional geology, a *subsequent stream*, that *flows* along the strike of the underlying formation. Cf. *Stream.*

**Losing stream** A watercourse that has a permeable section through which water can be transmitted to *groundwater.* Water drains from the losing *stream* into the ground when the bottom of the channel is higher than the local *water table.* **Figure L-14** shows a cross section of a losing or *influent stream.* The rate of water loss is a function of the depth of the water and the *hydraulic conductivity* of the channel bottom.

**L**

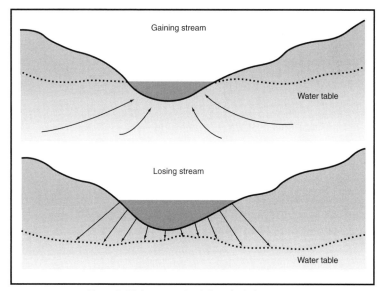

**Figure L-14** A cross-sectional view of a losing stream.

Cf. *Gaining stream; Intermittent stream.*

**Lost circulation** When *drilling fluid*, which is pumped into the *borehole* during borehole advancement for the purpose of removing borehole cuttings and cooling and lubricating the drill bit, exits or is lost to the formation instead of returning to the surface. This occurs when the *drill string* encounters or penetrates a cavity, crevice, fault, fracture, or other subsurface opening into which the drilling fluid can escape. Noticing areas of lost circulation are critical when monitoring for contaminants or production fluid, as they are potentially areas of preferential *flow*. If significant amounts of fluid are lost downhole, drilling must be stopped and the zone of lost circulation must be "plugged" by adding *grout*, *cement*, or other polymeric *gel* to the borehole. Once the area is sealed, the drill string can be advanced through the plugging agent. Cf. *Flow-meter logging*.

**Lost stream** A watercourse that has dried up in an arid region or disappeared below ground in a karst area. Cf. *Sinking stream; Karst spring*.

**Louver screen** Also called a bridge-slot screen. *Well screens* in which the openings are arranged in rows that are either at right angles or parallel to the screen axis. Large blank spaces must be left between openings for stability and, therefore, the percentage of *open area* is limited. Because the slotted openings tend to become blocked by formation of material during *well development* procedures, the louver or bridge-*slot screen* is used almost exclusively in filter-packed wells. Cf. *Gauze number*.

**Lower zone of moisture** The arbitrary separation of moisture storage into the upper zone of moisture and lower zone in the determination of the functional relationship of *actual* and *potential evapotranspiration*. The depletion of moisture from the lower zone occurs when the upper zone of moisture has been depleted. At this point, the rate of *evapotranspiration* is assumed to be proportional to the moisture available in the lower zone. This assumption, coupled with observations of *precipitation* and total *runoff*, allows for the derivation of the daily values of evapotranspiration by accounting procedures. Cf. *Belt of soil moisture*.

**Low flow** The least-sustainable discharge during *base runoff* of a watercourse. Low flows are relatively long events; only regulated *rivers* drop to low levels and recover quickly by releases, and have specific characteristics of *discharge*, duration, and frequency. *Low-flow frequency curves*, constructed from random data sequences, allow for the determination of the probability of occurrence of a specific low-flow event. Cf. *Drainage basin; Flash hydrologic behavior*.

**Low-flow frequency curve** The magnitude and occurrence of *low flows* during a specified time period, displayed graphically. Cf. *Frequency analysis*.

**L**

**Low-flow sampling technique** A technique for sampling *monitoring wells* used to minimize the disturbance to water samples caused by the act of sampling. Historically, monitoring wells have been sampled using a *bailer*. A bailer is lowered into a *well*, allowed to fill, retrieved, and the sample bottles are filled from the bailer. In performing the sampling, the bailer may be allowed to drop into the well, creating splashes, resulting in losses of volatile compounds from the sampled groundwater. Further, rapid filling of the bailer can result in losses of volatile compounds from the sample *groundwater*. Low-flow sampling techniques involve the use of a peristaltic pump to obtain samples to minimize losses of volatile compounds in transferring the water from the well to the bailer to the sample bottles. Another purpose of low-flow sampling is to minimize disturbance of colloidal material in the formation surrounding the well screen, providing a better indication of "true" mobile concentration of metals in groundwater.

**Lysimeter** A device used to obtain a direct estimate of the *actual evapotranspiration* (AE) and water lost by deep *percolation*. A watertight container is driven into an area of undisturbed soil and associated representative vegetative cover. A sealing base with a drain pipe is secured to the bottom, and a weighing device is established underneath. The change in the weight of the soil and vegetation is recorded and indicates the amount of water retained by the system. *Evaporative* measurement is calculated from the following equation:

$$E_t = S_i + P + I - S_f - D$$

where:
$S_i$ = volume of initial soil moisture [L$^3$]
$S_f$ = volume of final soil moisture [L$^3$]
$P$ = *precipitation* into the lysimeter [L$^3$]
$I$ = irrigation water added to the lysimeter [L$^3$]
$D$ = excess moisture drained from the soil [L$^3$]

The accuracy of the measurement depends on the accurate reproduction of the soil type, profile, moisture content, vegetation, and on the sensitivity of the weighing device. Large block samples are required to detect small changes in soil moisture storage. Measuring devices with sealed bottoms are also called tanks or evapotranspirometers, and devices with pervious bottoms or mechanisms to maintain negative pressure are lysimeters or suction lysimeters. Suction lysimeters with porous porcelain tips, or cups with negative pressure applied to draw soil moisture into the cup, are used to sample soil moisture in situ and can be used in low-permeability sediments below the *water table*. Another potentially serious limitation when using a suction lysimeter to analyze *soil moisture* for chemistry or solute isotopes is that the negative pressure applied can influence the water chemistry by vaporizing volatile compounds, or causing carbonate compounds to precipitate. Cf. *Belt of soil moisture*.

# Mm

**Macropore flow** In soils, root holes, animal burrows, and desiccation or cooling, cracks represent macropores that can lead to rapid flow rates, especially under very wet conditions. In complex *unsaturated zones*, macropore structure and texturally contrasting layers may have competing influences on *groundwater flow*.

**Magmatic water** See *Juvenile water*.

**Magnetometers** A geophysical instrument used to detect the presence of buried ferrous materials. In groundwater investigations, these instruments are used to detect the location of metal objects. As such they are used to detect buried pipe, and in environmental work, they can be used to detect shallow buried drums, underground storage tanks, and metal debris in landfills. Magnetometers will not give the depth of the buried object, nor will it provide information on site stratigraphy.

**Maintenance frequency for wells** *Wells*, like any other manmade device, require maintenance to continue to operate efficiently. Listed in **Table M-1** are the maintenance frequencies in various types of *aquifers* for wells constructed to locally acceptable design and construction standards.

**Table M-1** Maintenance frequency requirements for municipal wells set in differing aquifer types

| Aquifer Type | Most Prevalent Well Problems | Major Maintenance Frequency Requirement (Municipal) |
|---|---|---|
| Alluvial | Silt, clay, sand intrusion; iron precipitation; incrustation of screens; bio-fouling; limited recharge, casing failure | 2–5 years |
| Sandstone | Fissure plugging; casing failure; sand production; corrosion | 6–10 years |
| Limestone | Fissure plugging by clay, silt and carbonate scale | 6–12 years |
| Basaltic lavas | Fissure and vesicle plugging by clay and silt; some scale deposition | 6–12 years |
| Interbedded sandstone and shale | Low initial yields; plugging of aquifer by clay and silt; fissure plugging limited recharge; casing failure | 4–7 years |
| Metamorphic | Low initial yield; fissure plugging by silt and clay; mineralization of fissures | 12–15 years |
| Consolidate Sedimentary | Fissure plugging by iron and other minerals; low to medium initial yields | 6–8 years |
| Semi-consolidated and consolidated sedimentary | Clay, silt, sand intrusion; incrustation of screens in sand and gravel wells; fissure plugging of limestone aquifers in the interbedded sand, gravel, marl, clay silt formations; bio-fouling; iron precipitation | 5–8 years |

Estimates of major maintenance frequencies are based on the following assumptions:

1. Wells are being pumped continuously at the highest sustained yield they are capable of producing.
2. Major maintenance is required when the sustained yield decreases to 75 percent of the initial yield.
3. Major maintenance is considered to represent a cost expenditure of approximately 10 percent of the total current replacement cost. Minor maintenance is excluded.
4. Wells are designed in accordance with current practices, not necessarily in accordance with best available technology.

**Manganese incrustation** The precipitation of manganese resulting from the release of dissolved carbon dioxide ($CO_2$) in water entering a well.

**Manganese in water** Manganese is an element commonly found in groundwater under *anaerobic* conditions. Manganese in deep groundwater may reach concentrations as high as $2–3$ mg $L^{-1}$. Manganese precipitates as soluble manganese bicarbonate, which is converted to insoluble manganese hydroxide ($Mn(OH)_2$), when it reacts with atmospheric oxygen. Manganese bicarbonate ($Mn_2HCO_3$) precipitates out of solution as a black sooty deposit when carbon dioxide ($CO_2$) is liberated from the water in the vicinity of the *well screen*. Such precipitation can virtually cement a poorly designed well point in the ground, making removal and replacement difficult.

To avoid staining and objectionable taste, water quality criteria limit manganese in drinking water to less than $0.5$ mg $L^{-1}$. Manganese can be kept in solution by adding a small amount of sodium hexametasulfate as a *chelating agent* to the water, delaying precipitation. Additional stabilizing chemicals must be added before the water is exposed to air.

**Manning's equation** The Manning equation (a.k.a. the Manning formula) is used to describe steady uniform flow of water in prismatic open water channels. The Manning equation is derived from the *Chezy formula* or equation. The Chezy equation is as follows:

$$V = \sqrt{RS}$$

where:
$V$ = average cross-sectional velocity $[L \cdot T^{-1}]$
$C$ = Chezy coefficient, defined below [unitless],
$R$ = the hydraulic radius of the conduit [L]
$S$ = the hydraulic head losses per unit weight per unit length, or the slope of the water surface that is parallel to the channel bottom [L]

Substituting

$$C = \frac{C_m}{n} R^{1/6}$$

such that

$$V = \frac{C_m}{n} R^{2/3} S^{1/2}$$

The value of $C_m$ is 1.0 using SI units and 1.49 when using English units. The roughness coefficient is represented by $n$. Channel efficiency hydraulic radius. Cf. *Chezy formula; Darcy-Weisbach resistance coefficient.*
**Manning's roughness coefficient** Values of the roughness coefficient "$n$" are presented in **Table M-2**.

**Table M-2**  Manning roughness factors for various natural and manmade features (from Streeter and Wylie (1975) and Albridge and Garrett (1973))

| Boundary Material | Manning $n$ |
|---|---|
| Planed wood | 0.012 |
| Unplanned wood | 0.013 |
| Finished concrete | 0.012 |
| Unfinished concrete | 0.014 |
| Cast iron | 0.015 |
| Brick | 0.016 |
| Riveted steel | 0.018 |
| Corrugated metal | 0.022 |
| Rubble | 0.025 |

| | Median Size of Bed Material (mm) | Straight Uniform Channel | Smooth Channel |
|---|---|---|---|
| Sand | 0.2 | 0.012 | |
| | 0.3 | 0.17 | |
| | 0.4 | 0.20 | |
| | 0.5 | 0.22 | |
| | 0.6 | 0.23 | |
| | 0.8 | 0.25 | |
| | 1.0 | 0.26 | |
| Rock cut | | | 0.025 |
| Firm soil | | 0.025–0.032 | 0.020 |
| Coarse sand | 1.2 | 0.026–0.035 | |
| Fine gravel | | | 0.024 |
| Gravel | 2–64 | 0.028–0.035 | |
| Coarse gravel | | | 0.026 |
| Cobble | 64–256 | 0.030–0.050 | |
| Boulder | >256 | 0.040–0.070 | |

**Manometer** A liquid hydrostatic measuring instrument usually limited in application to measuring pressures near at or near atmospheric pressure. A simple instrument consisting of a clear tube with one end inserted into the fluid and the other end of the tube open to the atmosphere placed at various locations in a pipe to determine the pressure drop with flow along the length of the pipe. Cf. *Tensiometer; Gypsum block; Thermocouple psychrometer.*
**Marsh** A type of *wetland* dominated by herbaceous plants such as rushes, reeds and sedges, subject to frequent or continuous inundation. Woody plants may be low-growing shrubs. A marsh is different from a swamp, which in North America is the term for a wetland dominated by trees. Coastal marshes may be associated with *estuaries* and are also along waterways between coastal barrier islands and the inner coast. The estuarine marsh, or tidal marsh, is often based on soils consisting of sandy bottoms or bay muds. The water may be fresh, *brackish*, or saline. Decomposition of plant materials below water often produces marsh gas, which may begin to burn by self-ignition making Cf. *Fen; Wetland; Bog; Tide.*

**Marsh funnel viscosity** More commonly known as funnel viscosity. This is the number of seconds required for 1 quart (946 mL) of a given fluid to flow through a Marsh funnel.

**Maui-type well** See *basal tunnel*.

**Maximum contaminant level (MCL)** The maximum concentration of a defined chemical in drinking water, according to the World Health Organization (WHO), US Environmental Protection Agency (EPA), or other regulatory body. Listed in **Appendix B** are the primary and secondary MCLs as defined by the US EPA for drinking water. In the United States, primary criteria are required by law for the water to be considered potable. Secondary criteria are desirable, but not enforceable by law. Cf. *Water quality.*

**MCL** See *maximum contaminant level*.

**Mean river level** The average height of the surface of a *river* at any point for all stages of the *tide* over a 19-year period.

**Mean sea level** The average *sea level* for all stages of the *tide* over a 19 year period, usually determined from hourly height readings from a fixed reference level. Cf. *Geodetic datum.*

**Meander/Meandering streams** *Stream* channels are described by the shape the water course assumes from the *upstream* towards the *downstream* direction and fall into three flow patterns: meandering *braided* or straight. The meandering stream flows in large symmetrical bends or loops with the median length of the meander approximately 1.5 times the valley length. The meander wavelength is typically from seven to eleven times the channel width and the amplitude or width of the meander belt varies considerable being controlled more by the *bank* material characteristics, usually from ten to twenty times channel width. Cf. *Oxbow lake*; *Channel shape.*

**Measuring drawdown in wells** Performing *pumping tests* and monitoring the influence of pumping involve measuring the *drawdown* of the water table or piezometric head in a well. This can be accomplished manually, or using several indirect and automated methods. Drawdown data can be obtained from both the pumping well and appropriately placed *observation wells*. In pumping wells, however, turbulence caused by the pumping can affect the measurements. This can be taken care of by installing measurement tubes or *piezometers* around the outside of the well casing from which to take the measurement, as illustrated in **Figure M-1**. Typically, drawdown measurements for the purposes of

**M**

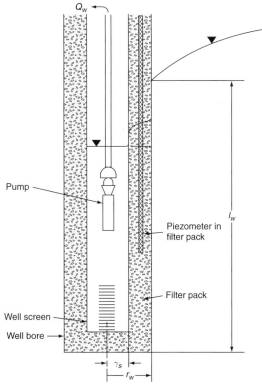

**Figure M-1** Use of a piezometer installed outside a well casing to measure head loss and well efficiency during pumping. (Powers, J. P., 1981. Reprinted with permission by John Wiley and Sons, Inc.)

performing a pump test are obtained by using *pressure transducers* and water level data-gathering instruments as pictured in **Figure M-2**. Before the advent of such devices, and if these are not available, drawdown measurements can be obtained manually. **Figure M-2** also shows an example of the *electrical sounder* for measuring the depth to water. Prior to beginning the pumping test, all of the observation wells and the pumping well must have an initial depth to water measurement from which to determine the drawdown. At the start of the pumping test, all measurements to depth to water from the individual observation wells and pumping well must be synchronized to occur at the same scheduled time. This is usually done by synchronizing the watches of all the persons performing the measurements. Typically, once the watches of all persons performing these measurements are synchronized, a truck horn is sounded (or if visible to all, the truck headlights are flashed to signal the start of the testing the measurement).

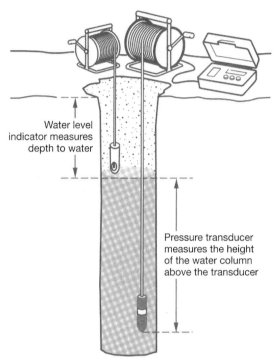

Water level indicator measures depth to water

Pressure transducer measures the height of the water column above the transducer

**Figure M-2** Pressure transducers measure the height of the water column above the transducer. (Sanders, 1998.)

**Mechanical methods removing incrustants** Techniques for scrubbing or prying *incrustation* from wells and screens. Typically, acid treatments are performed to remove incrustants from well screens. However, several mechanical methods are useful in preparing for acid treatments or as the primary method of treatment. Wire brushes or other mechanical scrapers can remove precipitate on the inside of a well. The loosened material is removed through bailing, *airlift pumping*, or other means. Mechanically removing the precipitated material reduces the quantity of acid necessary to improve *well yield*. Controlled blasting can also be used to fracture the incrusting matrix. This can be effective where the incrusting material is located in the formation outside of the well screen. Combined with acid treatments can be particularly effective to improve well yield and is provided by specialized service companies. Cf. *Well maintenance*.

**Mechanical dispersion** A component of hydrodynamic *dispersion*. Two processes, mechanical dispersion and molecular dispersion, result in the spread of contaminants over a greater area than *advection* or linear groundwater velocity alone would produce. **Figure M-3** illustrates other factors of dispersion. The following equation shows the relationship between the coefficients of mechanical dispersion and the linear groundwater velocity:

$$D'_L = \alpha_L v$$

where:

$D'_L$ = mechanical component of the *longitudinal dispersion coefficient* $[L^3 \cdot T^{-1}]$
$\alpha_L$ = longitudinal dispersivity of the medium $[L]$
$v$ = linear groundwater velocity $[L \cdot T^{-1}]$

M

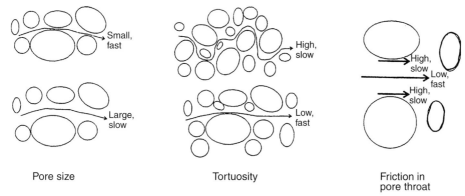

Pore size          Tortuosity          Friction in
                                        pore throat

**Figure M-3** Hydrologic factors that are capable of spreading a contaminant in addition to adjective forces. (Air Force Center for Environmental Excellence, 1995.)

Cf. *Advection–dispersion equation; Breakthrough; Peclet number.*

**Mechanical mixing** One of the fate and transport mechanisms that the pollutant or contaminant plume undergoes during transport. The pollutants are spread by *advection* and mechanical mixing as the plume migrates through the geologic material and tends to take on an ellipsoidal shape, as illustrated in **Figure M-4**. **Figure M-5** is a schematic of mixing on a microscopic scale.

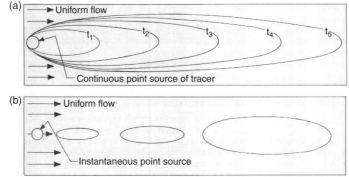

**Figure M-4** The plume-spreading characteristics for a continuous point source and an instantaneous point source. (Driscoll, 1986. Reprinted by permission of Johnson Screens/a Weatherford Company.)

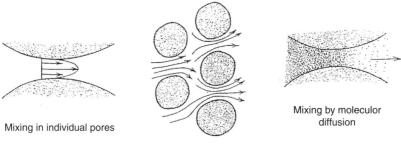

Mixing in individual pores

Mixing of pore channels

Mixing by moleculor diffusion

**Figure M-5** Schematic of dispersion and mixing on a microscopic scale. (Freeze and Cherry, 1979. Reprinted with permission from Prentice-Hall, Inc.)

Cf. *Mechanical dispersion; Advection–dispersion equation; Breakthrough.*

**Mean sea level** See *sea level*. Cf. *Geodetic datum*

**Mesohaline** See *Thalassic series*.

**Mesotrophic lake** The mesotrophic lake is intermediate in most characteristics between the *oligotrophic* and *eutrophic lakes*. The water is moderately clear with phosphorus and chlorophyll concentrations between those characteristic of oligotrophic and eutrophic lakes. Mesotrophic lakes usually have some scattered weed beds, and within these beds the weeds are usually sparse. Production of plankton is intermediate, with some organic sediment accumulating and as a result some loss of oxygen in the lower waters. Depending on the depth of the lake, the oxygen may not be entirely depleted except near the bottom. Cf. *Dystrophic lake*.

**Metahaline** See *Thalassic series*.

**Metalimnion** See *thermocline*.

**Meteoric water** *Groundwater* which originates in the atmosphere and reaches the *zone of saturation* by *infiltration* and *percolation*.

**Meteoric waterline** The relationship between deuterium ($^2$H) and $^{18}$O concentrations in water from global precipitation surveys is as follows:

$$\delta^2 H\text{‰} = 8\delta^{18}O\text{‰} + 10$$

where:

‰ = parts per thousand

When water evaporates from soil or surface water bodies under natural conditions, the residual water becomes enriched (where this enrichment is represented by $\delta$) with $^{18}$O and $^2$H. This relative degree of enrichment is different from the enrichment that occurs during condensation. This relationship and deviations thereof can be used in hydrogeologic studies, including the quantity of groundwater *baseflow* to lakes and rivers and the effects of *evaporation* on infiltration. This has also been used to identify groundwater bodies with different recharge areas in mountainous terrains. The ratio of $\delta^{18}O/\delta^2H$ for precipitation that has partially evaporated is greater than the ratio for normal precipitation as determined from the above equation.

**Methods to join casing** Since there are various materials used for casings, there are various methods to join the casing sections together. Steel casings are joined by threading or welding, although threads are generally not available in casings exceeding 12 in. (305 mm) in diameter. At this size, it becomes extremely difficult to align the threads and turn the casing sections. Plastic casing has either threads, cam-locking lugs, or screws, or can be joined by solvent welding, although solvent welding is not permitted in environmental applications for the construction of monitor wells using plastic casing. The solvents from the weld can leach in the water and result in false-positive readings of groundwater contamination. Cam locking is not common in well construction due to the cost of machining the cams, but is typically found in such casing applications as slope Indicators, where the grooves within the casing need to be aligned to a high degree of precision. Most fiberglass casing is threaded together. In some instances, fiberglass is joined by slip joints, or a flexible key that locks the male and female ends together. Cf. *Polyvinyl chloride casing*; *Monitoring well*.

**Mining** See *groundwater mining*.

**Miscible flow** Refers to *flow* situations where two fluids are soluble in each other. The most typical example of this in natural *flow systems* occurs where waters of different chemistry mix, such as seawater and *freshwater*, or freshwater and contaminated water, or ethanol and water. Cf. *Immiscible flow*; *Multiphase flow*.

**Misfit stream** A *stream* whose meanders are either to large or to small to have eroded the valley in which it flows. Cf. *Braided stream*; *Straight stream*; *Paleochannels*; *Paleohydrology*.

**Mix water for drilling fluids** The water used to make up the *drilling fluid* needs to meet several criteria so as to effectively react with drilling muds, or to minimize impacts on *water quality* within a well. Mix water should never be taken from *wetlands*, swamps, or small lakes near the well site. These sources of water may have iron- or sulfate-reducing bacteria present, which can grow within a well and result in clogging of the *well screen* and formation. It is a good practice to chlorinate all water prior to use as a mix for the drilling fluid or makeup water. Mix water should be chlorinated to a concentration of 50–100 mg L$^{-1}$, depending on the manufacturer's recommendation for the additive being used. The chlorine concentration can fluctuate because of the instability of chlorine in solution, such as in *sodium hypochlorite* solutions. This can be easily determined in the field measured using chlorine paper and should be maintained at 10 mg L$^{-1}$ during drilling. The *pH* should be maintained above 6.5 to maintain the performance of clay additives. *Soda ash* [sodium carbonate ($Na_2CO_3$)] can be used to reduce hardness, by precipitating calcium carbonate before the drilling fluid is mixed. Under normal circumstances, 0.5–3 lb (0.2–1.4 kg) of soda ash per 100 gal (0.4 m$^3$) is sufficient to soften the mix water. Cf. *Soda ash*.

**Mixoeuhaline** See *Thalassic series*.

**Modified non-equilibrium equation, Jacob** This variation of the Theis equation for non-equilibrium flow was developed by Cooper and Jacob in 1946. In its simplest form, the Theis equation can be written as follows:
In SI units

$$s = \frac{1}{4\pi}\frac{Q}{T}W(u)$$

where:

s = drawdown in m at any point in the vicinity of a well discharging at a constant rate (m)
$Q$ = pumping rate (m$^3$ day$^{-1}$)
$T$ = coefficient of transmissivity of the aquifer (m$^2$ day$^{-1}$)
$W(u)$ = well function of $u$ and represents an exponential integral

$$u = \frac{r^2 S}{4Tt}$$

where:

$r$ = distance from the center of a pumped well to a point where the drawdown is measured (m)
$S$ = coefficient of storage [dimensionless]
$T$ = coefficient of transmissivity (m$^2$ day$^{-1}$)
$t$ = time since pumping started (days)

In English Units

$$s = \frac{15.3QW(u)}{T}$$

where:

s = drawdown at any point in the vicinity of a well discharging at a constant rate (ft)
$Q$ = pumping rate (gpm)
$T$ = coefficient of transmissivity of the aquifer (ft$^2$ day$^{-1}$)
$W(u)$ = well function of $u$ and represents an exponential integral

In the $W(u)$ function, $u$ is equal to

$$u = \frac{360r^2 S}{Tt}$$

where:

$r$ = distance ft from the center of a pumped well to a point where the drawdown is measured (ft)
$S$ = coefficient of storage [dimensionless]
$T$ = coefficient of transmissivity (ft$^2$ day$^{-1}$)
$t$ = time since pumping started (days)

Cooper and Jacob reasoned that if $u$ is sufficiently small, the Theis non-equilibrium equation can be reduced to the following without significant error.

In SI units

$$s = \frac{264Q}{T} \log \frac{0.3Tt}{r^2 S}$$

In English units

$$s = \frac{0.183Q}{T} \log_{10}\left(\frac{2.25Tt}{Sr^2}\right)$$

where $u$ is less than approximately 0.05; the modified non-equilibrium equation yields essentially the same result as the Theis equation, previously shown. For ease of analysis of pump tests, the pumping rate is held constant, such that $Q$, $T$, and $S$ are all constants. The modified non-equilibrium equation, which is valid when $u$ is less than 0.05, shows that the drawdown, $s$, varies with $\log_{10}(t/r^2)$. From this, two important relationships can be derived:

1. If $r$ is constant, $s$ becomes a function only of time, such that it varies with $\log C_1 t$, where $C_1$ represents all of the constant terms stated above.
2. At a given time $t$, $s$ varies as a function of $r$, such that $s$ varies with $\log C_2 / r^2$, where $C_2$ represents all of the constants defined above, as well as the specific value of $t$.

The data from a pump test can be easily analyzed using these simplified relationships to obtain values of transmissivity and storage. To analyze data using the Cooper and Jacob method, the drawdown versus time data are plotted on a semilogarithmic plot. As shown in **Figure M-6**, most of the data points fall on a straight line. During the first 10 min of a pump test, the value of $u$ is much larger than 0.05, and so these values do not fall on a straight line and hence the modified non-equilibrium equation does not apply.

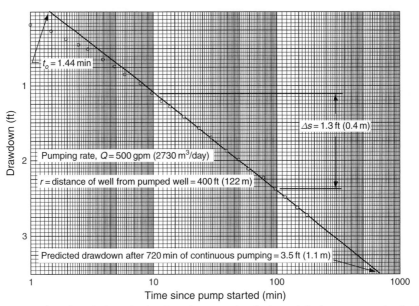

**Figure M-6** Use of semilogarithmic to plot drawdown data versus time for the analysis of a pump test using the Cooper and Jacob method. (Driscoll, 1986. Reprinted by permission of Johnson Screens/a Weatherford Company.)

$T$ and $S$ can be determined as follows:
In SI units

$$T = \frac{2.3}{4\pi}\frac{Q}{\Delta s} = \frac{0.183Q}{\Delta s}$$

where:
$\Delta s$ = slope of the time–drawdown graph over one log cycle (m)

In English units

$$T = \frac{264Q}{\Delta s}$$

where:
$\Delta s$ = the slope of the time–drawdown graph over one log cycle (ft)

To determine $S$ in SI units

$$S = \frac{2.25Tt_0}{r^2}$$

To determine $S$ in English units

$$S = \frac{0.3Tt_0}{r^2}$$

where:
$t_0$ = intercept of the straight line at zero drawdown (days).

**Moisture content** The quantity of water in a mass/volume of soil, sewage, sludge, or samples thereof; expressed in percentage by weight of water in the mass/volume. The volumetric moisture content is determined by:

$$\theta = \frac{V_W}{V_T}$$

where:
$\theta$ = the volumetric moisture content [unitless],
$V_T$ = the total unit volume [$L^3$]
$V_W$ = the volume of water [$L^3$]

As with *porosity*, $\theta$ is usually reported as a decimal fraction or percent. In unsaturated conditions, $\theta$ is less than the porosity, $n$; while under saturated conditions, $\theta = n$.
Cf. *Antecedent precipitation index; Gravimetric moisture content; Volumetric moisture content; Plastic limit; Liquidity index.*

**Moisture potential ($\psi$)** The negative pressure due to the soil-water attractive forces which increases with decreasing amounts of *soil moisture*. The *infiltration* of water into the *unsaturated zone* is dependent on both the gravity potential (z) and the moisture potential ($\psi$) also called the soil-water potential ($\psi_{sw}$). Either the moisture potential or the gravity potential will predominate depending on the soil-moisture content. If the soil is very dry the moisture potential will be much greater than the gravity potential, but if the moisture content is close to the *specific retention* the gravity potential will be greater. Cf. *Soil moisture storage capacity.*

**Molality** This is the number of moles of *solute* dissolved in 1 kg mass of solution. One mole of a compound is the equivalent of one molecular weight. In SI units, the symbol for this quantity is $m_b$, where b denotes the solute. Refer to **Appendix A** for unit conversions.

**Molarity** This is the number of moles of *solute* in 1 L of solution. In SI units, this is designated as mol $L^{-1}$. Moles per liter (mol $L^{-1}$) is a permitted symbol for molarity and is commonly used in groundwater studies

$$\text{molarity} = \frac{\text{(milligrams per liter)}}{1000 \times \text{formula weight}}$$

or

$$\text{molarity} = \frac{\text{(grams per cubic meter)}}{1000 \times \text{formula weight}}$$

**Mole fraction** Denoted as $X_B$, is the ratio of the number of moles of a given solute species to the total number of moles of all components in solution.

$$X_B = \frac{n_B}{n_A + n_B + n_C + n_D + \cdots}$$

where:
$n_B$ = the moles of solute
$n_A$ = the moles of solvent
$n_C, n_D, \ldots$ = the moles of other solvents.

**Molecular diffusion** Dispersal or spreading of a chemical resulting from the kinetic activity of the ionic or molecular constituents. Ionic species moves in the direction of lower concentrations and is proportional to the concentration *gradient* as described by *Fick's law.* Cf. *Diffusion; Diffusive model of dispersion; Peclet number.*

**Molecular dispersion** See *mechanical dispersion.*

**Monitoring well** A well installed for the specific purpose of sampling *groundwater* to determine the quality of the groundwater or its *flow* characteristics at a specific location. **Figure M-7** is a completion diagram of a typical monitoring well. The specific design of the well depends on whether it is to be used for measuring physical or chemical parameters, the nature of those parameters, or the hydrogeologic condition. In an area overseen by a regulatory agency, the installation of monitoring wells may entail installation and completion methods specified in that jurisdiction. During the installation of a monitoring well, it is vital to take precautions to prevent the introduction of materials that may impact or contaminate the *aquifer* to be sampled.

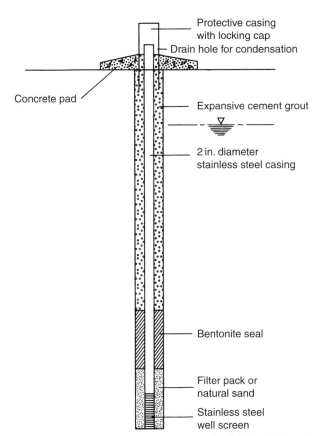

**Figure M-7** A sample of a well completion diagram. (ASTM, 1992. Special Technical Publication 1053, Ground Water and Vadose Zone Monitoring.)

Cf. *Well maintenance.*

**Monitored natural attenuation** The use of *natural attenuation* processes as part of overall site remediation. Natural attenuation processes (biodegradation, *dispersion*, sorption, and *volatilization*) affect the fate and transport of *pollutants* in all hydrologic systems. In the United States, US Environmental Protection Agency (EPA), Office of Solid Waste and Emergency Response defines monitored natural attenuation as the "reliance on natural attenuation processes (within the context of a controlled and monitored cleanup approach) to achieve site-specific remedial objectives within a time frame that is reasonable compared to other methods."

**Monod kinetics** Used to model the kinetics of biodegradation of organic compounds and associated biomass growth. The depletion rate of substrate is described by the following equation:

$$\frac{dC}{dt} = -q_{max} \frac{C}{K_s + C} X$$

The time dependence of biomass growth is described by the following equation:

$$\frac{dX}{dt} = q_{max} Y \frac{C}{K_s + C} X - bX$$

where:

$C$ = substrate concentration (mg L)

$t$ = time (h)

$q_{max}$ = the maximum substrate utilization rate per unit biomass ($mg_{substrate}\ mg_{biomass}^{-1}$ [protein] $h^{-1}$)

$K_s$ = half saturation coefficient (mg $L^{-1}$)

$X$ = biomass concentration ($mg_{protein}\ L^{-1}$)
$Y$ = yield coefficient ($mg_{protein}\ mg_{substrate}^{-1}$)
$b$ = endogenous decay rate ($h^{-1}$)
  Cf. *Hall saturation constant; Nutrient; Governing equation for groundwater flow.*

**Monomictic** *Holomictic lakes* that mix from top to bottom during one mixing period each year. Monomictic lakes may be subdivided into cold and warm types. Cold monomictic lakes are covered by ice throughout much of the year. During summer months, the surface waters remain at or below 4°C and lack significant thermal stratification, mixing thoroughly from top to bottom. These lakes are typical of cold climate regions (e.g., much of the arctic). Ice prevents these lakes from mixing in winter. Warm monomictic lakes are lakes that never freeze and are thermally stratified throughout much of the year. The density difference between the warm surface waters (the *epilimnion*) and the colder bottom waters (the *hypolimnion*) prevents these lakes from mixing in summer. During winter, the surface waters cool to a temperature equal to the bottom waters. Lacking significant thermal stratification during winter months, these lakes mix thoroughly during this period from top to bottom. These lakes are widely distributed from temperate to tropical climatic regions. Cf. *Dimictic; Thermocline.*

**Montmorillonite** Also called sodium montmorillonite used in *bentonite*. The clay most commonly commercially available and used for *drilling mud*. Sodium montmorillonite has the greatest *viscosity*-building characteristic of any clay because the sheets of atoms making up the structure of the clay particles are thinner and separate more readily than other clays. As a result, sodium montmorillonite can swell up to ten times its original volume when exposed to water. Cf. *Attipulgite clay; Flocculation.*

**Moulin** See *Glacial well.*

**Mud** See *drilling fluid/drilling mud.*

**Mud cake** See *filter cake.*

**Mud pit** The fluid used during drilling is usually mixed adjacent to the drilling rig, typically in excavated and lined pits or via portable containers that meet the needs of reverse circulation drilling in which the mud will lift the cuttings from the borehole as the drill bit is advanced. **Figure M-8** depicts the various pit placement options to maximize efficiency.

M

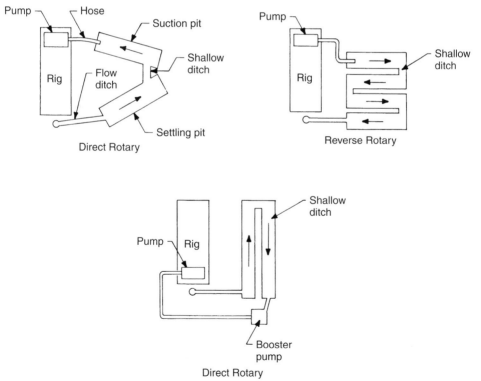

**Figure M-8** Placement schemes for mud pits relative to a rotary drill rig. (Driscoll, screens 1986. Reprinted by permission of Johnson a Weatherford Company.)

The design of the mud pit should consider the ability to store adequate volumes of drilling fluid and to act as an effective settling basin for cuttings. The pit can be constructed in two sections: the settling section and the suction section. A deeper pit is preferable to a longer pit to reduce the velocity of the drilling fluid and hence allow the cuttings to settle out. **Figure M-9** lists mud pit capacities and dimensions.

---

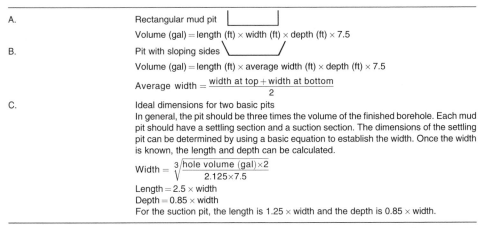

A.  Rectangular mud pit
    Volume (gal) = length (ft) × width (ft) × depth (ft) × 7.5

B.  Pit with sloping sides
    Volume (gal) = length (ft) × average width (ft) × depth (ft) × 7.5

    $$\text{Average width} = \frac{\text{width at top} + \text{width at bottom}}{2}$$

C.  Ideal dimensions for two basic pits
    In general, the pit should be three times the volume of the finished borehole. Each mud pit should have a settling section and a suction section. The dimensions of the settling pit can be determined by using a basic equation to establish the width. Once the width is known, the length and depth can be calculated.

    $$\text{Width} = \sqrt[3]{\frac{\text{hole volume (gal)} \times 2}{2.125 \times 7.5}}$$

    Length = 2.5 × width
    Depth = 0.85 × width
    For the suction pit, the length is 1.25 × width and the depth is 0.85 × width.

---

**Figure M-9** Typical mud pit capacities and dimensions. (Driscoll, 1986. Reprinted by permission of Johnson Screens/a Weatherford Company.)

**M**

Cf. *Filter cake; Drilling fluid.*

**Multiphase flow** Different fluids of differing viscosities densities, *solubilities*, and partitioning characteristics with the *aquifer* matrix will migrate at different rates with *groundwater flow*. The movement of oil, gas with groundwater is just one example of immiscible multiphase flow. To model such systems, it is necessary to consider the different Darcy equations (*Darcy's Law*) for each of the fluids flowing simultaneously through the *porous media*. From the vector analysis presented in **Figure M-10**, it can be seen that as a result, not only are the magnitude of the flow vectors different for the various fluids, but so are the directions as a result of the differing *fluid densities*.

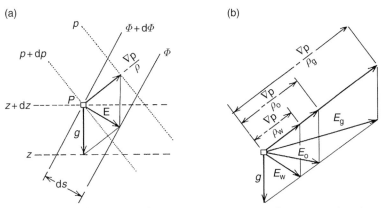

**Figure M-10** (a) Components of the impelling force $E$ acting on a unit mass of water at a point $P$ in a steady-state groundwater flow system and (b) impelling forces on water, oil, and gas in a three-phase steady-state flow system. (Freeze and Cherry, Groundwater© 1979. Reprinted with permission by Prentice-Hall, Englewood Cliffs, NJ.)

Cf. *Immiscible flow; Miscible flow.*

**Muriatic acid** A commercial grade of hydrochloric acid used in *well maintenance*.

# Nn

**NAPL** See *Non-aqueous-phase liquid*.

**National Contingency Plan** The policy for federal response actions in the United States under the *Comprehensive Environmental Response, Compensation Liability Act* (CERCLA), wherein has established a hazard-ranking system and procedures and standards for responding to uncontrolled releases of hazardous substances and *pollutants*. The plan is a regulation subject to regular revision. Cf. *Superfund Amendments and Reauthorization Act (SARA)*.

**National Pollutant Discharge Elimination System (NPDES)** A national program in the United States established under the *Clean Water Act* (CWA) and administered by the individual states that establishes water quality criteria for discharges to waters of an individual state. This system provides for issuing, revoking and reissuing, terminating, and enforcing permits. Further, procedures are also established for imposing and enforcing treatment standards.

**National Priorities List (NPL)** Under *Comprehensive Environmental Response Compensation and Liability Act* (CERCLA) in the United States, this is a list and ranking of uncontrolled hazardous waste sites designated as requiring cleanup. The purpose of the list is to prioritize cleanup efforts and inform the public of the most serious waste sites. The US *Environmental Protection Agency* (EPA) periodically revises the list as new sites are added or deleted.

**Native water** An archaic term referring to water that resides in a particular basin or *formation water*. Water is in constant movement in hydrologic systems by *discharge* from *groundwater* to surface water, by *infiltration* from the surface to groundwater *flow* systems, and by *evaporation* and *precipitation*. In deep groundwater systems, the residence time of the water may be very long (e.g., $10^5$–$10^6$ years) or present when the particular strata is formed, known as *connate water*, which is the preferred term.

**Natural attenuation** The processes that occur in the subsurface that retard or attenuate the movement of contaminant plumes in groundwater. These in situ processes include biodegradation, *bacterial degradation, dispersion*, dilution, *sorption, volatilization*, and chemical or biological stabilization, transformation, or destruction of contaminants.

**Natural development** A well constructed without artificial *filter* material placed around the screen. The native *aquifer* material is allowed to collapse into the *screened interval* after the drill casing is removed, such as in the use of telescoping screens or the *pull-back method of screen installation*. Cf. *Well development*.

**Natural gradient tracer test** See *tracer test*.

**Natural load** The *sediment load* carried by a *river* or a *stream* under normal *flow* conditions. Cf. *Flow layers, vertical*.

**Natural–gamma logging** See *gamma logging*.

**Navier–Stokes equations** Equations describing the motion of a fluid resulting from the forces acting on a small element of the fluid in open channel or pipe *flow*. Since water is considered an incompressible fluid, the equations are as follows for each Cartesian coordinate:

$$-\frac{1}{\rho}\frac{\partial}{\partial x}(p + \gamma h) + v\nabla^2 u = \frac{du}{dt}$$

$$-\frac{1}{\rho}\frac{\partial}{\partial y}(p + \gamma h) + v\nabla^2 v = \frac{dv}{dt}$$

$$-\frac{1}{\rho}\frac{\partial}{\partial z}(p + \gamma h) + v\nabla^2 w = \frac{dw}{dt}$$

where:

$v\,A\phi\upsilon\xi\nu$ = kinematic viscosity, assumed to be constant [$L^2 \cdot T^{-1}$]

$\quad t$ = time [T]

$\quad \rho$ = density [$M \cdot L^{-3}$]

$\quad p$ = *pressure* [$M \cdot L^{-2}$]

$\quad \gamma h$ = *elevation head* term [L]

$\mathbf{u}, \mathbf{v}, \mathbf{w}$ = force vectors in the $x$, $y$, $z$ directions, respectively [M].

The operator $\nabla^2$ is defined as follows:

$$\nabla^2 = \frac{\partial^2}{\partial x^2} + \frac{\partial^2}{\partial y^2} + \frac{\partial^2}{\partial z^2}$$

Cf. *Darcy's equation*.

**Nephelometric Turbdity Unit (NTU)** A light-scattering method used for measuring turbidity in low-turbidity waters. The light transmitted through the sample is measured at 90° and 270° to the incident beam. The units of measurement are the Nephelometric Turbidity Units. Cf. *Turbid; Roil*.

**Nernst equation** Mathmatically defines oxidation potential, Eh:

$$Eh(volts) = Eh^0 + \frac{2.3RT}{nF}\log\left(\frac{[oxidant]}{[reductant]}\right)$$

where:

Eh = oxidation potential expressed in volts (millivolts are the commonly used units)

$Eh^0$ = standard or reference condition at which all substances are at unit activity [ ]

$$Eh^0 = \frac{RT}{nF}\ln K$$

where:

$R$ = universal gas constant ($0.0821$ l bar $K^{-1}$ mol$^{-1}$)

$T$ = temperature (°K)

$n$ = number of electrons transferred in the half reaction

$F$ = Faraday constant ($9.65 \times 10^4$ C mol$^{-1}$)

$K$ = equilibrium constant for a given reaction [ ]

where:

$$K = \frac{[D]^d[E]^e}{[e]^n[B]^b[C]^c}$$

For the general half reaction

$$bB + cC + ne = dD + eE$$

Cf. *Eh-pH diagram; Oxidation potential.*

**Net infiltration** The amount of water available to *recharge* groundwater from *precipitation* after *evaporation* and *transpiration* is factored out.

$$\text{Infiltration}_{net} = \text{Precipitation} - \text{Evaporation}$$

Cf. *Inflow; Outflow.*

**Net positive suction head (NPSH)** The suction head at the inlet portion of a pump needed to eliminate boiling of the water under the reduced pressure conditions found near the *impellers*. If the condition of net positive suction head (NPSH) is not met, *cavitation (vaporization)* results in severe pitting of the impellers and pump housing, shortening the effective life of the pump. The equation for calculating available NPSH is as follows:

$$NPSH = H_a + H_s - H_f - H_{vp}$$

where:

$H_a$ = absolute head on the liquid surface of the water [L]

$H_s$ = elevation of the liquid above or below the impeller eye while pumping [L]. If the level is above the eye, $H_s$ is positive, if below, $H_s$ is negative.

$H_f$ = friction head losses in the suction piping [L]

$H_{vp}$ = absolute vapor pressure of the liquid at the pumping temperature [L]

The atmospheric pressure at various altitudes and the vapor pressure of water are listed in **Appendix B**.

**Net rainfall** The quantity of *precipitation* measured in unit length (i.e., millimeters or inches), which fall during a given precipitation event. See: **Appendix A** for other units of conversion.

**Neuman condition** Also known as a Neuman boundary. In numerical and analytical *groundwater modeling*, solving the differential conditions requires specifying *boundary conditions*. A Neuman boundary establishes a condition where there is a specified flux along a boundary. A *no-flow boundary* is a special case of this type. Cf. *Boundary, flux; Boundary, specified-head.*

**Neutron–activation log** A geophysical technique, sometimes referred to as an activation log, that can identify elements present in the borehole fluid and in adjacent rocks under a wide variety of borehole conditions. Neutron activation produces radioisotopes from stable isotopes; the parent or stable isotope may be identified by the energy of the gamma radiation emitted and its half-life. **Table N-1** lists some common stable isotopes that are readily activated by thermal-neutron capture. If a daughter nuclide can be identified, this concentration can be determined from the measurements of the gamma activity. Quantitative neutron activation analysis is unlikely to be as accurate as laboratory analysis, but can still be potentially useful in hydrology.

**Table N-1**  Activation data for some common isotopes for neutron activation logging. (Keys, 1989. Reprinted with permission from the National Water Well Association.)

| Parent isotope | Daughter isotope | Counts per second per gram after 2-min irradiation | Half-life | Energy of major gamma peaks (MeV) |
|---|---|---|---|---|
| Aluminum-27 | Aluminum-28 | $2.7 \times 10^4$ | 2.3 min | 1.78 |
| Chlorine-37 | Chlorine-38 | $8.1 \times 10^2$ | 37.5 min | 2.16, 1.63 |
| Potassium-41 | Potassium-42 | $1.9 \times 10^2$ | 12.4 h | 1.53 |
| Magnesium-26 | Magnesium-27 | $3.1 \times 10^2$ | 9.5 min | 0.85, 1.02 |
| Manganese-55 | Manganese-56 | $1.2 \times 10^4$ | 2.58 h | 0.84, 1.81, 2.13 |
| Sodium-23 | Sodium-24 | $2.1 \times 10^2$ | 15.0 h | 1.37, 2.75 |
| Silicon-30 | Silicon-31 | 5.9 | 2.6 h | 1.26 |

Based on a normal nuclide abundance, a flux of $10^8$ neutrons per square centimeter per second, and a 10% counting efficiency.

**Neutron-gamma log**  A downhole geophysical logging technique used primarily as an indicator of total *porosity* under saturated conditions. It can also be used in determining the amount of moisture and in doing so indicate the depth of the *water table*. It is also used to log *moisture content* and change in moisture content above the water table. Neutrons emitted from a radioactive source in the sonde collide with various particles in the formation, lose energy, and are captured before impinging on a detector in the sonde. Most of the energy lost by neutrons is through collisions with hydrogen ions, a principal component of water. If the energy loss is large, the amount of hydrogen in the formation is large and there the saturated porosity is large. A common source of neutrons is americium-241/beryllium. Neutrons are emitted when alpha particles from americium-241 impinge on the beryllium. The neutron source can be used in cased or open holes, filled with fluid, or dry. Care must be exercised in the interpretation if the hole has been freshly drilled with *drilling fluid*. Cf. *Geophysical exploration methods; Zone of saturation.*

**Neutron soil moisture meter**  A probe for determining *moisture content* that operates on the same principle as neutron logging. Most commonly used in earthworks such as road building coupled with a nuclear densometer for measuring compaction and with the neutron probe to determine moisture content. Cf. *Geophysical exploration methods.*

**Newtonian fluid**  A fluid that deforms proportionately to an applied stress, such as water, is an example of a Newtonian fluid. Higher-viscosity fluids such as *drilling fluids* with clay additives perform as plastic materials. That is, they do not begin to deform until a significant amount of stress is applied. A *plastic fluid* may become Newtonian when the yield point is reached, i.e., when the relationship between stress and strain is more or less constant. **Figure N-1** illustrates the relationship between stress and strain for plastic and Newtonian fluids.

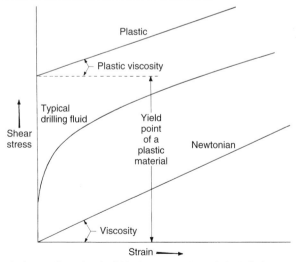

**Figure N-1**  A stress–strain diagram illustrating the field where Newtonian and plastic fluids respond to stress. Newtonian fluids, such as water, will deform proportionately to the stress applied. A plastic substance begins to deform only after a certain amount of stress is applied. The force required to induce a flow is a measure of the viscosity of the fluid. (Streeter, V. L. et al., 1975. Reprinted with permission from McGraw-Hill).

Cf. *Non-Newtonian flow.*

**Next-Generation Weather Radar (NEXRAD)** A useful tool for compiling *precipitation* estimates used in the United States. The databases available are through the National Weather Service of the National Oceanographic and Atmospheric Administration.

**Nitrification** The inorganic process whereby oxygen is consumed in *groundwater* through the oxidation of ammonia ($NH_4^+$), producing nitrate ($NO_3^-$). This is an important reaction associated with such activities as septic tanks, manure storage, and feedlots. Nitrification is illustrated in the following reaction:

$$NH_4^+ + O_2 = NO_3^- + 2H^+ + H_2O$$

Cf. *Septic systems.*

**Non-aqueous-phase liquid (NAPL)** A liquid, typically organic, having measurable aqueous *solubility,* such that when released to the subsurface and after migration to the *water table* has the ability to be present in an undissolved form. As such, liquids of this nature have the potential of representing continuous sources of *groundwater* pollution, slowly dissolving over time. Non-aqueous-phase liquid, especially when denser than water (*dense non-aqueous-phase liquid*, DNAPL), has the ability to migrate through the water table in a direction contrary to the prevailing *hydraulic gradient*, leaving residual liquid in the soil pores, representing long-term source of *groundwater pollution*. Cf. *Light Non-Aqueous-Phase Liquid (LNAPL).*

**Non-artesian groundwater** See *Aquifer, unconfined.*

**Non-consumptive use** Water that is used for other purposes besides that of direct consumption. Examples are industrial use, irrigation, pumping to control *saltwater intrusion*, water produced in groundwater *pump and treat remediation systems* that is not used for human consumption.

**Non-Darcian flow** *Darcy's law* is a linear law relating *specific discharge* and *hydraulic gradient*. There is an upper and lower limit to this linear relationship. In fine-grained materials of low *permeability*, laboratory evidence suggests that there may be a threshold hydraulic gradient below which *flow* does not take place. There has been little practical importance to investigate this threshold gradient. It has been recognized and accepted for years that at very high rates of *flow*, the linear relationship between flow and hydraulic gradient breaks down. The upper limit is identified with the aid of the *Reynolds number*, *Re*, a dimensionless number used in fluid mechanics to distinguish between *laminar flow* at low velocities and *turbulent flow* at high velocities. The Reynolds number for flow through *porous media* is defined as:

$$R_e = \frac{\rho v d}{\mu}$$

where:

$\rho$ = fluid density [$M{\cdot}L^{-1}{\cdot}T$]
$\mu$ = fluid viscosity [$L^2{\cdot}T^{-1}$]
$v$ = specific discharge [$L{\cdot}T^{-1}$]
$d$ = representative length dimension for the porous medium, such as mean grain diameter or mean pore diameter [L]

From this, Darcy's law is valid based on the average grain diameter not exceeding a Reynolds number of between 1 and 10. **Figure N-2** illustrates the range of validity of Darcy's law.

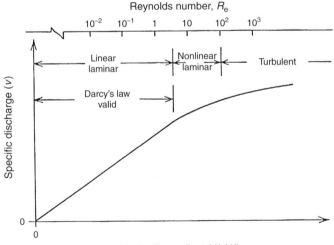

**Figure N-2** Range of validity of Darcy's law. (Freeze and Cherry, 1979. Reprinted with permission by Prentice-Hall.)

**Non-equilibrium-type curve** See *Theis nonequilibrium-type curve.*

**Non-equilibrium well equation** Also known as the Theis non-equilibrium well equation, which is used to predict *drawdown* at any time after pumping begins from a *well*. The basic assumptions are the following:

1. The water bearing formation is *homogeneous* and *isotropic*, with the *hydraulic conductivity* equal in all directions.
2. The formation is of uniform thickness and of infinite areal extent.
3. The formation receives no *recharge* from any source.
4. The well is a *fully penetrating well* receiving water from the full thickness of the formation.
5. Water released from *storage* is discharged instantaneously from the well when the *head* is lowered.
6. The *pumping well* is 100% efficient.
7. All water removed from the well comes from *aquifer* storage.
8. *Laminar flow* exists throughout the well and the aquifer.
9. The water table or *potentiometric surface* has no slope (static flow conditions).

These assumptions are identical to those for the *equilibrium well equations*, except the water levels within the *cone of depression* need not have reached equilibrium.
The simplest form of the Theis equation is as follows:
In SI

$$s = \frac{1}{4\pi}\frac{Q}{T}W(u)$$

In FPS

$$s = \frac{15.3QW(u)}{T}$$

where:
   $s$ = drawdown in a well discharging at a constant rate (m or ft)
   $Q$ = pumping rate (m$^3$ day$^{-1}$ or gpm)
   $T$ = coefficient of transmissivity of the aquifer (m$^2$ day$^{-1}$ or ft$^2$ day$^{-1}$)
   $W(u)$ = "*the well function of u*" (an exponential integral).
   Cf. *Well function of u* ($W(u)$).

**Non-flowing artesian well** A *well* tapping an *aquifer* whose *artesian pressure* has been reduced through *overpumping* or seasonal fluctuations in *piezometric head* that was at one time flowing. Cf. *Aquifer, artesian.*

**Non-flowing well** A *well* that relies on the use of pumps to produce water. The *piezomteric head* in the well is less than the elevation of the top of the well. Cf. *Elevation head; Total head.*

**Non-Newtonian fluid** A fluid having an initial yield stress, which must be exceeded to cause a continuous deformation. In non-Newtonian fluids, there is a non-linear relationship between the magnitude of applied stress and the rate of deformation. **Figure N-3** presents a rheological diagram illustrating the relationship between applied stress and rate of deformation for *Newtonian*, non-Newtonian, ideal plastic, and *thixotropic* substances. Cf. *Drilling fluids; Gel strength.*

**Non-point source** A source of *pollution* that cannot be traced back to a single location. In contaminant hydrogeology, sources of pollution are classified as point and non-point sources. Examples of non-point sources include fertilizers and pesticides that have been spread on farm fields, or chloride (Cl$^-$) pollution from the application of deicing salts on roadways in winter. Cf. *Plume.*

**Non-potable** *Water quality* determined through chemical analysis to be below the criteria (chemical, radiological, biologic, taste, odor, and turbidity) considered acceptable for drinking. See: **Appendix B** for a list of the *maximum contaminant levels (MCLs)*. Cf. *Potable water.*

**Non-pumping well** A non-specific term that is rarely used. Such terms as an *artesian well*, an *observation well*, a *pressure relief well*, or a *monitoring well*, which are more specific and specify the use of the well, are the preferred terms.

**Non-steady flow** See *unsteady-state flow/unsteady flow.*

**Nonsteady radial groundwater flow** See *Unsteady radial groundwater flow.*

**Nonsteady-state groundwater flow** See *unsteady-state groundwater flow.*

**Non-uniform flow** *Flow* for which the velocity, *discharge*, magnitude, and direction varies along the length of a watercourse, or in hydrogeology, flow path. There may be convergence or divergence and spatially varying gradients. Cf. *Uniform flow.*

**Non-uniformity** See *heterogeneity/heterogeneous.*

**Normal depth** In *open-channel flow*, this is the depth of steady *uniform flow*. In open-channel flow, velocity varies with depth as a result of friction or drag on the channel walls, ice cover, and to some extent with the atmosphere. There is no analogy in flow through *porous media*.

**N**

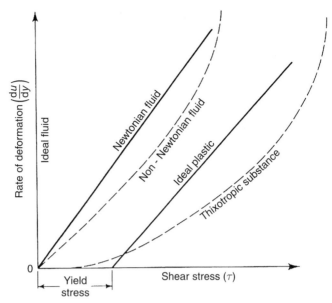

**Figure N-3** Rheological diagram showing the relationship between shear stress and rate of deformation for various substances. (Streeter, V. L. and Wylie, E. B., Fluid Mechanics © 1975. Reprinted with permission by McGraw-Hill Book Company, New York, NY.)

**Normal moisture capacity** See *Field capacity; Moisture content.*

**NPL** See *National Priorities List.*

**NPSH** See *Net positive suction head.*

**NTU** See *Nephelometric turbidity unit.*

**Numerical model** A simulation of *groundwater flow* utilizing electronic computers that may incorporate contaminant transport utilizing finite difference or finite element methods to discretize segments of an *aquifer* system. Also known as a computer model. Simulations can be performed in two or three dimensions. Two-dimensional models can be oriented vertically, as a slice through an aquifer, or horizontally to simulate conditions over an area of interest.

**Nutrient** A mineral or substrate that supports organisms. Common nutrients that support microbial life are nitrogen, oxygen, phosphates, and carbon.

**Oasis** Within an arid region, an isolated area that supports vegetation, or where *springs* or *seepage* exist throughout the year. An oasis is usually located in a depressed area in which the *water table* is close to the ground surface and within the reach of plant roots. Water transported from a deep-solution cavity network in an arid karst region may also form an oasis. Cf. *Kanat*.

**Obsequent stream** A *stream*, or a valley in which a stream flows, oriented in a direction opposite to that of the original *consequent stream* or the dip in the underlying strata. An obsequent stream is a *tributary* to a subsequent stream that developed independent of, and subsequent to, the original topography.

**Observation well** A *non-pumping well* used to collect chemical or physical water data and to evaluate *aquifer* parameters or fluctuations, especially during *pumping tests*. An observation well is employed, for example, to monitor changes in water level in an unconfined aquifer, or changes in the *potentiometric level* and pressure in a confined aquifer. The observation well is used for piezometric measurements instead of the pumping well because of the turbulence created by pumping, and more insight is gained using data from multiple distances from the pumping well. An observation well is usually distinguished from a *piezometer* by having larger-diameter casing, but in practice the two have been used interchangeably. Chemical and water-level parameters measured in the observation well during a pumping test are best recorded with an electronic logger, especially if parameters are anticipated to change rapidly. It is important not to situate the observation well too close to the test pumping well, and the screens of the observation well and the pumping well should be similar in their placement. Observation wells for unconfined aquifers are usually sited not further than 30–90 m from the pumped well and for well-stratified confined aquifers, typically within 90–200 m. The distances should be determined using preliminary calculations with non-equilibrium equations for the anticipated hydraulic conductivity based on geologic knowledge. The further the observation wells are sited from the test pumping well, the longer the pumping test must run in order to observe changes in *head* caused by pumping; otherwise, *boundary conditions* within the aquifer would go unnoticed. The number of observation wells used in an aquifer study is dependent on the parameters being observed and the budget. Cf. *Aquifer, confined*; *Aquifer, unconfined*.

**Occlusion** See *Absorption*.

**Octanol–water partition coefficient ($K_{ow}$)** The ratio of the concentration of a *solute* in octanol ($C_8H_{18}O$) to its concentration in water. Values of the octanol–water partition coefficient, $K_{ow}$ correlate well with bioconcentration factors in aquatic organisms and *adsorption* to soil or sediment. The $K_{ow}$ is used to calculate the adsorption or *distribution coefficient*, $K_d$. The $K_{ow}$ and the *solubility* of an organic compound may be calculated or are usually available in tables or publications. **Figure O-1** graphically displays the relationship between the octanol–carbon partition coefficient, $K_{oc}$, and $K_{ow}$.

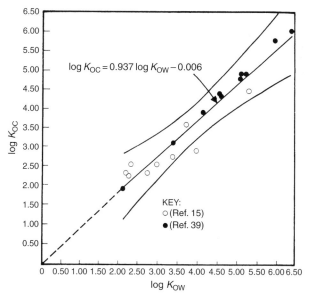

$$\log K_{OC} = 0.937 \log K_{OW} - 0.006$$

KEY:
○ (Ref. 15)
● (Ref. 39)

**Figure O-1** Graphical relationship between $K_{oc}$ and $K_{ow}$ for a coarse silt. (Maidment, 1993. Reprinted with permission from McGraw-Hill, Inc.)

Cf. *Organic carbon content*.

**Octapent gauge** See *Rain gauge*.

**O'Donnell model** In *catchment* area modeling, whenever the processes are described mathematically, such as *evaporation*, and are established as a series of interlinked processes and storages for which *water budgets* are determined. The O'Donnell model is built around the following four interdependent storage types that vary in volume with time:

  - *surface storage, R*: the volume of *rainfall* retained on the surface or intercepted by vegetation
  - *channel storage, S*: the volume of water in surface streams, *rivers*, and *lakes*
  - *soil moisture storage, M*: the volume of water contained in the unsaturated soil layers
  - *groundwater storage, G*: the volume of water in the deeper saturated zone below the *water table*

**Figure O-2** shows O'Donnell's conceptual catchment model for *discharge* at a *gauging station*. The model inputs and the interactions between the various storages result in the calculated total discharge at a gauging station, $Q$, as described by the following equation:

$$Q = Q_S + Q_B$$

where:
$Q_S$ = surface storage [L$^3$]
$Q_B$ = groundwater storage [L$^3$]

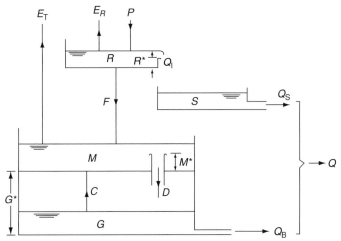

**Figure O-2** The O'Donnell conceptual catchment model for discharge at a gauging station. (Shaw, 1983. Reprinted with permission from Van Nostrand Reinhold Co. Ltd.)

**Off-channel storage** See *dead storage*.

**Offset stream** A *stream* course displaced either laterally or vertically by faulting.

**Oligohaline** The *brackish water* chloride (Cl$^-$) zone in which the percent chlorinity is 0.03–0.3. Cf. *Halinity; Thalassic series*.

**Oligomictic** Considered to be thermally stable. In an *oligomictic lake*, mixing is rare and usually occurs only during abnormal cold spells. Oligomictic lakes are small in area but exceptionally deep, e.g., Lake Tahoe in California, or are tropical lakes with very high surface temperature (20–30°C) that maintain stable *thermal stratification*. Cf. *Dimictic*.

**Oligotrophic lake** A *lake* deficient in plant nutrients and with low biological productivity. As a result of the relatively small amounts of organic matter, abundant dissolved oxygen (DO) exists in the *hypolimnion* or bottom layers of an oligotrophic lake, typically to a greater extent than in the *epilimnion*. Oligotrophic lakes are typically very deep; plankton blooms are rare and littoral plants are usually scarce. Oligotrophic lakes are considered geologically younger than *eutrophic* lakes, or are less modified by weathering and erosion products. Cf. *Dystrophic lake; Eutrophic; Epilimnion; Mesotrophic lake*.

**One-dimensional flow** The simplest type of *flow* geometry utilized in *groundwater* analysis. One-dimensional flow typically involves using analytical equations to simulate groundwater movement along a *flow path* or transport experiments in laboratory columns. Cf. *Groundwater energy*.

**One-hundred-year flood** The instantaneous magnitude of *flow* that can be expected at a frequency of once in a 100-year average at any specified point on a *stream*. The probability of the 100-year flood occurring is 1 in 100, or a 0.01 probability. There are several designated floods of any magnitude, but the 100-year flood is most often used for engineering designs, insurance regulations, and *floodplain* management.

**Open area** The amount of area in a *well screen* that allows *flow* into the *well*, usually expressed as a percentage of the casing area in the screened interval. The open area typically ranges between 1 and 40% of the total casing area. The percentage of open area achievable depends on screen construction factors such as material, slot size and configuration, and casing diameter. Representative open areas for various screen sizes, slot types, and materials are listed in **Appendix B**.

**Open-channel flow** The movement of water that occupies hollows in the ground surface such as valleys and gullies. Water *flows* in a well-defined channel as long as there is a *free water* surface and under-gravity flow. Open-channel flow, shown in **Figure O-3**, is characterized by the velocity pattern seen in a cross-sectional area of the channel as *uniform*, non-uniform, *laminar flow*, *turbulent flow*, critical, or *steady-state flow*.

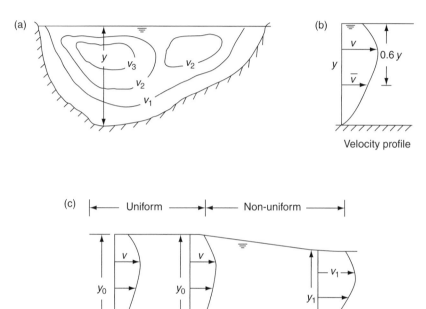

**Figure O-3** (a) Cross-sectional view of open-channel flow and velocity changes. Cross-sectional and vertical section velocity contours. (Shaw, 1983. Reprinted with permission from Van Nostrand Reinhold Co. Ltd.). (b) Vertical uniform and non-uniform flow. (Shaw, 1983. Reprinted with permission from Van Nostrand Reinhold Co. Ltd.)

Cf. *Dead storage; Gravitational water.*

**Open-end well** A *well* bore that is cased to its terminal depth and open to the water-bearing unit only at bottom; or a production or *monitoring well* that does not have a slotted screen interval for the intake of water. Cf. *Piezometer.*

**Open hole** An uncased well bore or *borehole*, or the bottom portion of a cased well where the borehole extends beyond the depth at which the casing terminates. Wells of this nature are typically constructed in hard rock, and not in sediments.

**Open lake/open water** An inland body of water with a surface free of ice or overabundance of vegetation; or a *drainage lake* or effluent lake in which the surface elevation is controlled by the lake *effluent*.

**Open system** A cell or a basin in dynamic equilibrium in which inflow of materials is unrestricted but the form or character of the system remains unchanged. Both energy and mass freely cross the boundaries of an open system. A surface-water example of an open system is a section of a river channel that, over a period of time, receives an inflow of sediment and water that is discharged *downstream* while the channel itself remains essentially unchanged. A subsurface example is organic degradation in the vadose zone, where oxygen consumed in the process can be readily replaced. Cf. *Closed system.*

**Open water evaporation** See *evaporation.*

**Operation stations** The *gauging stations* in a streamflow network that are required for streamflow forecasting, project operation, water allocation, and other day-to-day project requirements. Typically, every major *stream* and all main *tributaries* within the investigation area are gauged near their mouth, or area of *discharge.* The operation stations, or network of stations, are sited and operated for as long as their purpose exists. Cf. *Basic data stations.*

**Optimal yield** An optimal yield must be determined to establish a sound water management policy in a watershed basin, and is a plan that best meets a set of economic and/or social objectives associated with the water use. Through decision analysis, optimal yield is obtained by minimizing the expected value of the loss function. The optimal yield varies over time either monthly or seasonally because the probability density function for *rainfall* is time dependent. *Groundwater* withdrawal exceeding the safe yield can create adverse effects such as land subsidence and *seawater intrusion.* Minimizing these negative impacts depends on regulating the rate of groundwater withdrawal. Cf. *Basin yield; Overpumping.*

**Order of streams** See *stream order.*

**Organic carbon content ($f_{OC}$)** In the subsurface environment, a non-polar organic compound has an affinity to the solid organic matter or organic carbon of the soil matrix, when the compound is *hydrophobic.* The organic carbon content, $K_{OC}$, is the most important property of soil or aquifer material that affects the *sorption* of undissociated organic chemicals in the soil/water matrix. If organic carbon is readily available in the subsurface, the sorption of non-polar organic compounds increases. Organic carbon is also found in the dissolved phase in the subsurface. The typical ranges of concentration of dissolved organic carbon are 1–6 mg $L^{-1}$ for streams and *rivers,* 0.3–32 mg $L^{-1}$ for *groundwater,* and 0.10–27 mg $L^{-1}$ for seawater. The *distribution coefficient,* $K_d$, is related to the organic carbon content fraction, $f_{OC}$, by the following equation:

$$K_d = K_{OC} * f_{OC}$$

where:
$K_{OC}$ = organic carbon partitioning coefficient [$L \cdot M^{-1}$]
$f_{OC}$ = fraction of total organic carbon content (grams of organic carbon per gram of soil).

**Table O-1** lists several values of $f_{OC}$ and the partitioning characteristics for several chemicals. Refer to **Appendix A** for conversion units. Organic carbon content is measured by evaporating water from the sample aliquot and combusting the residue left behind. The organic carbon content is then determined from the quantity of carbon dioxide produced from the combustion.

**Table O-1** Typical values of the fraction of organic carbon in aquifer materials. (Maidment, 1993. Reprinted with permission from McGraw-Hill, Inc.)

| Aquifer material | Value (%) | Reference |
|---|---|---|
| Borden aquifer Canada (sand) | 0.028 | 59 |
| Silt clay | 16.2 | 56 |
| Sandy loam | 10.8 | 56 |
| Silt clay | 1.7 | 56 |
| Silt loam | 1.0 | 56 |
| Silt loam | 1.9 | 20 |
| Silt clay loam | 2.6 | 70 |
| Silt clay loam | 1.8 | 70 |
| Traverse City, Mich. (sand and gravel) | 0.008 | 82 |
| Granger, Ind.(sand) | 0.1 | 82 |
| Fine sand | 0.087 | 100 |
| Silt loam | 1.6 | 83 |

**Original formation water** See *Connate water.*

**Original stream** See *Consequent stream.*

**Oxidation-reduction potential (ORP)** Also known as *redox potential.* See *Redox potential.*

**Orthotropy** The preferred *flow* in a *porous medium* through channel sets that are mutually perpendicular. Orthotropy is a special case of *anisotropy.* Cf. *Isotropy.*

**Osmosis** The chemical movement of fluid in response to a concentration *gradient.* Cf. *Electro-osmosis; Osmotic pressure; Reverse osmosis.*

**Osmotic potential** See *osmotic pressure.*

**Osmotic pressure** The pressure required to move solutions of differing concentrations that are separated by a semipermeable membrane. The *solvent,* for example, water, moves to equalize osmotic pressure within the system. Chemical forces drive a less-concentrated solution through a semipermeable membrane toward the more-concentrated solution until the liquid head on the concentrated side is able to stop the transfer of liquid through the membrane. Movement is a result of the *free energies* and ceases once the free energy is the same on both sides of the membrane. **Figure O-4a** illustrates two solutions of different concentrations in equilibrium, with the osmotic pressure indicated by the liquid head needed to maintain that equilibrium. **Figure O-4b** demonstrates how the application of *pressure* greater than osmotic pressure can be used to drive liquid from an area of high concentration across the membrane to an area of low concentration. This process of reversing the *flow* with pressure, and effectively producing a liquid with a lower chemical concentration, is called *reverse osmosis*.

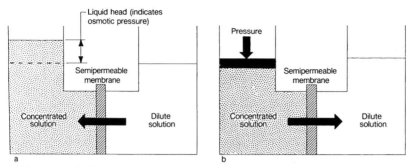

**Figure O-4** (a) Solutions of differing concentrations flow until osmotic pressure on the dilute solution and the liquid head on the concentrated solution side balances. (b) The method of reversed osmosis by applying pressure to the high-concentration solution. (Driscoll, 1986. Reprinted by permission of Johnson Screens/a Weatherford Company.)

Cf. *Water potential.*

**Ott recorder** A five-channel paper tape recorder for a series of river stage measurements taken automatically at fixed time intervals.

**Outcrop spring** See *Spring, contact.*

**Outfall** The naturally occurring outlet or mouth of a watercourse, or the volume of water leaving one system or basin. Outfall is typically the lowest portion of a waterway or body of water that enters into a larger body of water, e.g., a *lake* or a *river* entering the sea, or streamflow out of a *watershed basin.*

**Outpost well** See *Discovery well.*

**Outside recharge** The addition of water to a *groundwater* aquifer that occurs outside the particular unit or subunit under consideration. The term "outside recharge" is often applied when the *recharge* to a particular *hydrostratigraphic unit* is not vertically down from above the unit, but from a different unit, horizontally adjacent, or from a surface water body at one horizontal extreme of the unit.

**Overbank deposits** Overbank deposits occur when *floodwaters* are in excess of strem *channel capacity.*

**Overflow bank** See *Overflow; Overbank deposits.*

**Overdevelopment** *Well development* beyond the point of improving *aquifer* characteristics in the vicinity of the *well,* typically caused by *overpumping.* Overdevelopment may cause physical damage to the well installation by removing too much fine materials from the aquifer or the less-permeable layers above or below. The removal of fine materials from a layer or formation that does not have larger grains in grain-to-grain contact causes settlement of the formation. The most common result of overdevelopment is damage to the well itself; uneven adjustment of the materials around a well can result in broken casings and necessitate redrilling and reinstallation of the well. A more severe, but less common, result of overdevelopment is surface ground settlement causing damage to nearby structures or utilities. Cf. *Grain size distribution.*

**Overdraft** The excessive depletion of streamflow or overaggressive *groundwater* pumping and withdrawal above the *safe yield.* Overdraft can result not only in the depletion of groundwater and surface-water reserves but also in the intrusion of undesirable quality water, induced *infiltration,* and land *subsidence.* Cf. *Groundwater mining.*

**Overflow/overbank flow** Water received in excess of the channel capacity that flows over the *banks* of the channel or levee, inundating all or parts of the surrounding *floodplain.* Cf. *Design Flood.*

**Overflow channel** Also called a spillway. A glacial drainage channel cut by meltwater escaping from a preglacial lake, typically U-shaped and not integrated within the local *drainage pattern,* or a man-made or artificial channel that conveys water overflowing from a water basin.

**Overflow spring** See *Spring, overflow.*

**Overflow stream** A temporary waterway resulting from a *river* that has overflowed its *banks* and exceeded its *bank storage,* or from the *transport* of *effluent* from a *lake* to another lake or water body in the *downgradient* direction.

**Overflow well** See *artesian well.*

**Overland flow** Also called saturation overland flow, sheet flow, or sheetwash. *Precipitation* that does not immediately penetrate the land surface drains across the surface before discharging into a *stream, river,* or other surface water body. **Figure O-5** demonstrates the interrelationship of the hydrologic components resulting from *rainfall.* Surface water can result from overland flow or from *groundwater* that has seeped into the stream or river bed. Overland flow is spatially varied and unsteady as a result of variations in supply (rainfall) and rate of depletion (*infiltration*) coupled with variations

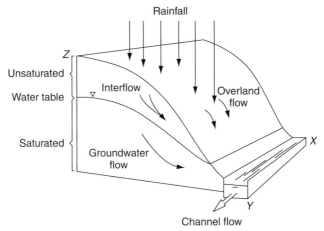

Channel flow

**Figure O-5** Hydrologic components resulting from precipitation. (Shaw, 1983. Reprinted with permission from Van Nostrand Reinhold Co. Ltd.)

introduced by topography; clear-cutting of forests and destruction of vegetation, for example, increase the occurrence and volume of surface runoff and overland flow. Overland flow becomes an important participant in the formation of *flood peaks* if it occurs in significant quantity. Although the term "overland flow" is used synonymously with "surface runoff," the strictly technical meaning varies depending on the inclusion of the amount of precipitation falling on the stream system itself, whereas overland flow does not include this precipitation. Overland flow becomes part of the volume of total runoff after entering another water body. Cf. *Baseflow; Interflow; Horton overland flow*.

**Overloaded stream** A *stream* having a suspended sediment load that is in excess of its capacity, thereby forcing it to deposit a portion of the *load*. Cf. *Sediment load*.

**Overpressured system** A confined *aquifer* system in which the subsurface *fluid pressure* exceeds the *hydrostatic pressure* for that depth. Since such conditions exist at groundwater discharge zones also, an understanding of the flow conditions is also necessary for proper interpretation. Cf. *Underpressure system; Aquifer, confined*.

**Overpumping** A method for removing *fines* from an unconsolidated water-bearing formation around a pumping or an *observation well*, by which the pumping rate is set higher than the *optimal yield* or the rate at which the well will be pumped when in service. *Bridging* (**Figure O-6**) is a common drawback of overpumping in *well development*. When

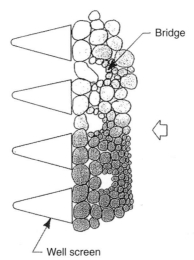

**Figure O-6** Bridged sand grains as a result of well development by overpumping. (Driscoll, 1986. Reprinted by permission of Johnson Screens/a Weatherford Company.)

pumping at high rates during the development process, sand grains *bridge* and prevent fines from entering the *well screen*; the bridge then collapses after pumping ceases or when the well is placed in service, allowing the entry of the fine constituents. Overpumping may also cause fines to compact around the well screen, causing blockage. Other well development methods, including *backwashing, surging,* and *air development,* are viable alternatives and are often used in conjunction with overpumping the formation. Cf. *Fineness modulus.*

**Oxbow lake** The standing water body created when a meander loop in a *stream* or a *river* is abandoned by the stream during channel alignment. *Young streams* in the first stages of development are relatively straight channeled, but as the stream ages, the *gradient* or topographic slope decreases, and a meandering configuration develops. In a *flood stage,* the increased streamflow can overtop the meander *banks,* changing the stream's channel to a more direct course and isolating a meander to create an oxbow lake. An aerial photograph illustrating this process is shown in **Figure O-7.**

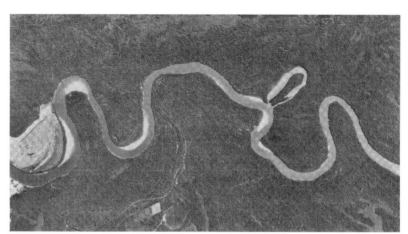

**Figure O-7** An areal photograph of a meandering stream with an associated oxbow lake. (Linsley et al., 1982. Reprinted with permission from McGraw-Hill, Inc.)

**O**

Cf. *Horseshoe lake; Fluviatile lake.*

**Oxidation** A chemical reaction that increases the oxygen content of a compound. Cf. *Oxidation potential.*

**Oxidation potential (Eh)** The oxidation potential of an aqueous solution, in a *oxidation-reduction reaction* (also known as redox reaction), is called the Eh. A redox reaction has an electrical potential caused by the transfer of electrons. The transfer of electrons from one compound to another, gaining oxygen or losing hydrogen or electrons, is the oxidation portion of a redox reaction. The *Nernst equation* is used to calculate Eh:

$$\text{Eh} = E^0 = \frac{RT}{nF} \ln K_{sp}$$

where:
$E^0$ = standard potential (V)
$R$ = gas constant [0.00199 kcal (mol °K)$^{-1}$]
$T$ = temperature (°K)
$F$ = Faraday constant (23.1 kcal V$^{-1}$)
$n$ = number of electrons
$K_{sp}$ = equilibrium equation

Standard potential, $E^0$, has been measured for many reactions at 25°C and 1 atm. The sign of the potential is positive if the reaction is oxidizing and negative if reducing. Oxidation potential is measured with an ion electrode meter. If the Eh and pH of the aqueous solution are known, the stability of minerals in contact with the water can be determined. The stability relationship is represented by an *Eh–pH diagram.* Water is stable only within a part of the Eh–pH diagram (**Figure O-8**).

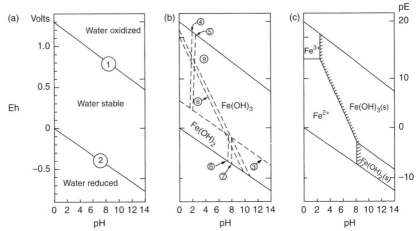

**Figure O-8** *Eh–pH* diagram for 25°C and 1 bar. (a) Stability field for water; (b) construction lines for the Fe–H₂O system; (c) completed diagram showing stability fields for major dissolved species and solid phases. (Freeze and Cherry, 1979. Reprinted with permission by Prentice-Hall, Inc.)

Cf. *Chebotarev's succession.*

**Oxidation-reduction reactions** Also known as Redox reactions. An oxidizing chemical change, where an elements positive valence is increased (electron loss) simultaneously with the reduction (electron gam) of an associated element.

**Oxygen demand** The measure of the amount of oxidizable material present in a stream, *lake*, or other body of water. Oxygen demand is also used as a measure of the organic *pollution* within a water body. Cf. *Biological Oxygen Demand (BOD); Chemical Oxygen Demand (COD).*

**Oxygen, dissolved** See *Dissolved oxygen.*

**Oxygen sag** The fall and recovery in the curve of the dissolved oxygen (DO) percentage saturation levels in bodies of water *downstream* from a *pollution* source.

**Packer** A device that provides seals with a *borehole*, or within a well, usually constructed of neoprene rubber and is either inflatable or self-sealable. They are used in performing downhole tests, for providing seals in telescoping screens, and in multipoint monitoring systems. Inflatable packers are used in downhole tests to seal off select zones in rock for the purpose of conducting a *packer test* (determinations of *hydraulic conductivity*), or for the purpose of sampling *groundwater* within discrete zones. A packer assembly (**Figure P-1**) is lowered down a drill hole to the

**Figure P-1** Inflatable packers are used as permanent or temporary sealing devices to isolate portions of the well bore. (Driscoll, 1986. Reprinted by permission of Johnson Screens/a Weatherford Company.)

**P**

select zone to be tested. Packers can be inflated through a number of means: by applying pressure on the packer assembly using the drill rig or through pressurized gas (e.g., air and nitrogen). Once the zone is sealed off, water can be injected under pressure to determine hydraulic conductivity, or pumped out for sampling groundwater from discrete zones. Multipoint *groundwater monitoring* systems consist of a number of sample ports set on the same casing with one borehole (**Figure P-2**). The sample ports typically have a port for the collection of water samples and a pressure transducer to obtain piezometric measurements. These sample ports are isolated within the borehole using inflatable packers. Because of the permanent nature of these installations, these packers are inflated using a material that expands when wet. The permeability of the membrane is such that it allows sufficient installation time before the packer begins to swell. Permanent packers are used in well construction to seal the well screen to the well casing. **Figure P-3** shows the placement of a packer in the construction of a well in unconsolidated sands using a telescoping screen, also known as the *pull-back method of screen installation*. This method makes use of the natural sand for the well filter material. Two types of packers are used in this application, neoprene rubber (also called *K packers*) and lead. The use of lead packers is falling out of favor due to concerns of lead finding its way into the drinking water. Neoprene rubber packers do not require inflation; however common practice calls for the use of multiple packers to compensate for variations in pipe diameter and finish of the interior of the pipe. Further, for wells deeper than 300 ft (91.5 m), constructed with welded pipe, multiple packers are recommended because of the potential that weld slag, weld beads, and the roughness of the pipe interior can damage the lips of the packer as it is being put in place. Typically the lips of the packer are lubricated with petroleum jelly to minimize damage to the *K* packer as the screen is lowered through the casing. Cf. *Packer test*.

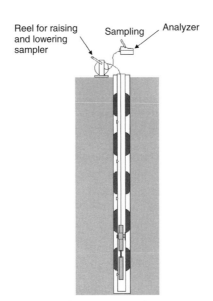

**Figure P-2** Illustration of multipoint groundwater monitoring system.

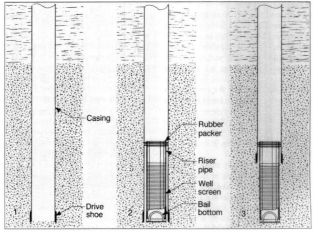

**Figure P-3** Basic operations in setting a well screen by the pull-back method. (1) Driving, bailing, or lowering (rotary method) the casing to full depth of the well, (2) lowering the screen inside the casing, and (3) pulling the casing back to expose the screen to the aquifer. (Driscoll, 1986. Reprinted by permission of Johnson Screens/a Weatherford Company.)

**Packer tests** Also known as hydraulic pressure tests are primarily conducted in rock. The test is conducted by forcing water into the formation or zone to be tested. The zone of rock is isolated through the use of one or more *packers*. There are generally two types of tests: single or double packers. A single packer setup tests the entire boring from the packer to the bottom of the hole. This technique is more commonly performed while drilling the hole, with each successive zone tested before the next core run. The double-packer arrangement isolates a specific zone of rock between the packers. The hydraulic pressure-testing system consists of a Moyno-type pump, a *flow meter*, a pressure *gauge*, and the packer assembly. Prior to testing, the flow meter and the pressure gauge should be calibrated. Further, the entire testing assembly should be checked on the ground surface to assure that all perforations in the pipe are unobstructed. It is also a good idea to determine the pumping capacity of the entire assembly prior to testing. For highly permeable zones, the quantity of *flow* into the zone to be tested should be limited by the *permeability* of the zone, rather than by the limits of the equipment. Packer permeability tests may be carried out above or below the *water table*, but are more successful below the water table. In a packer test, a length of borehole ($L$) is isolated between inflatable packers, and fluid is injected into the test interval. Injection is continued until the flow rate stabilizes. The *hydraulic head* within the injection zone and the flow rate $Q$ are measured. Excess hydraulic head above the pretest measurements is calculated ($\Delta h_w$). *Hydraulic conductivity* is determined using the following equation:

$$K = \frac{Q}{2\pi L \Delta h_w} \ln \frac{L}{r_w}$$

For $L \gg r_w$

where:
$Q$ = flow rate for the test [$L^3 \cdot T^{-1}$]
$L$ = length of the section of borehole over which the test was performed [L]
$K$ = hydraulic conductivity [$L \cdot T^{-1}$]
$h_w$ = excess hydraulic head above the pretest measurement [L]
$r_w$ = radius of borehole [L]

Sources of Error in Testing
Some of the common sources of error in testing are the following:
Ensure that the test zone is under saturated conditions

- At flow rates greater that $0.06\,L\,s^{-1}$ (1 gpm), the friction between the water and the pipe can cause a significant reduction in *head* (the sum of water pressure between the level of the gauge and the water level of the stratum). These head or friction losses need to be accounted for when analyzing the data.
- The equations for analyzing the data assume *laminar flow* within the rock formation. When the flow reaches turbulent conditions, the relationships are no longer valid. To maintain laminar flow, *gauge pressures* are kept to within 100–170 kPa (15–25 psi) during the test.
- When the pressure acting on the test zone exceeds the strength of the rock combined with the overburden pressures, fractures may be enlarged by the introduction of water under pressure, yielding erroneous readings of flow, and hence permeability. A rule of thumb for the maximum allowable test pressure is $22.6\,kPa\,m^{-1}$ ($1\,psi\,ft^{-1}$) above the *water table* and $12.9\,kPa\,m^{-1}$ ($0.57\,psi\,ft^{-1}$) below the water table. This assumes an overburden density of $2300\,kg\,m^{-3}$ ($144\,lb\,ft^{-3}$).
- The *borehole* should be cleaned out prior to testing. Cuttings from drilling cat plug fractures in the zone to be tested result in improper seals between packers and borehole walls.
- To provide adequate sealing on the borehole walls, the minimum length of the expanded gland of the packer should be five times the diameter of the boring. For NX core, this is 38 cm (15 in.).
  Cf. *Flow velocity*.

**Paleochannels** Also known as buried *stream* deposits. In a hydrogeologic sense, this provides for long, narrow *aquifers*, typically bounded by silts and clays as a result of the deposition of *overbank deposits*.

**Paleofloods** Major *floods* that have occurred beyond the historical record found from old newspaper reports or archive municipality reports. Evidence of paleofloods are seen in geological, geomorphological or botanical information.

**Paleohydrology** The study of ancient surface-water-flow systems with particular emphasis on features developed by *paleofloods, paleaochannels, misfit streams*, etc.

**Pan coefficients, evaporation** A measure of the potential *evaporation* that can be expected for a given location. This is used in *water balance* analysis to determine the amount of water available for *recharge* from *precipitation* for a given basin or aquifer system. Natural evaporation is measured as the rate of loss of liquid water from the surface or as the rate of gain of water vapor by the atmosphere. The use of the evaporation pan measures the loss of liquid from either an assumed or created closed system. The evaporation from a pan can differ significantly from the surrounding area. Using empirical pan coefficients accommodate these sometimes large differences.

**Parallel-plate model** Also known as the Hele–Shaw apparatus. Before *numerical modeling*, modeling of groundwater *flow* systems was performed using various techniques; physical models such as electrical analogue models and in this case, a model that involved two closely spaced plates containing a viscous material, usually oil. The first apparatus of this kind was developed by Hele–Shaw in 1897. Until the advent of numerical modeling and the digital computer, the Hele–Shaw analog had been used extensively to simulate two-dimensional *laminar flow* of water through porous media. The Hele–Shaw apparatus is also used to calculate *hydraulic conductivity* and flow in a single fracture as a function of aperture. The average velocities of viscous flow between two parallel plates may be expressed as follows:

$$u_{ave} = -K_m \frac{\partial h_m}{\partial x}$$

where:
$u_{ave}$ = average velocity [$L \cdot T^{-1}$]
$K_m$ is defined by the following equation:

$$K_m = \frac{b^2}{12\mu} \gamma$$

P

and:

$$h_m = z = \frac{p}{\gamma} + \text{constant}$$

These equations are directly analogous to *Darcy's law* and the equation of *fluid potential*. **Figure P-4** presents a diagram of a Hele–Shaw apparatus. Because the viscosity of most oils and glycerin is temperature-dependent, these models require being set up in temperature-controlled rooms. The spacing of the plates was in the order of a millimeter or less for light oils or water to limit the *Reynolds number* to 1000 to maintain laminar flow between the two plates.

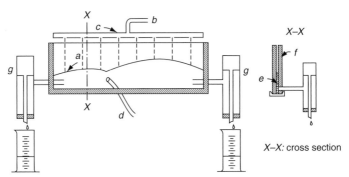

**Figure P-4** Hele–Shaw apparatus: (a) groundwater table; (b) supply of rain; (c) sprinkler; (d) discharge tube; (e) viscous liquid; (f) transparent plates; and (g) vertically adjustable overflow. (Davis and DeWiest, 1966. Reprinted with permission by John Wiley and Sons, Inc., NY.)

**Parshall flume** See *Flume, Parshall.*

**Partial-duration flood series** A list of all *flood peaks* for a given watercourse that exceed a chosen base storage or *discharge*, without regard for annual frequency. This list is compiled to develop a *partial-duration series (PDS)*, which is used to calculate a *recurrence interval.*

**Partial-duration series (PDS)** The ranking of *flood* events for a given watershed or *stream*, from which to determine the *recurrence interval.* The analysis using a PDS, sometimes referred to as the *peaks-over-threshold (POT)* includes all peak's of *flood events* above a truncation or threshold level. The benefit of using PDS instead of just choosing the largest event of each year (annual maximum series) are that relatively long and accurate data are available. The PDS should yield more accurate estimates of extreme quantities. Cf. *Annual series.*

**Partial penetration** A condition that exists where the intake portion of a *well* does not extend to the deepest portion of the *aquifer* (**Figure P-5**); or in a confined aquifer, where the *well screen* does not extend throughout the aquifer thickness. In either situation, water does not flow radially to the well, and vertical components of *flow* are induced. As a result, the *pressure* varies with depth, and it can be seen that with the variation in *drawdown* with depth, the *yield* of the well is

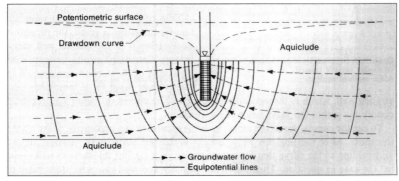

**Figure P-5** Flow patterns resulting from pumping from a well partially penetrating a confined aquifer deviate from the radial patterns associated with fully penetrating wells. (Driscoll, 1986. Reprinted by permission of Johnson Screens/a Weatherford Company.)

decreased. **Figure P-6** presents a relationship between partial penetration and *specific capacity* for wells in a *homogeneous* confined aquifer. The Kozeny equation, developed in 1933, was used to develop this diagram:

$$\frac{Q/s_p}{Q/s} = L\left[1 + \sqrt[7]{\frac{r}{2bL}}\cos\left(\frac{\pi L}{2}\right)\right]$$

where:

$Q/S_p$ = specific capacity of a partially penetrating well [m³ day⁻¹ per m (gpm ft⁻¹)]

$Q/s$ = maximum possible specific capacity of a fully penetrating well [m³ day⁻¹ per m (gpm ft⁻¹)]. The ratio of these two parameters presents the relative production of a partially penetrating well to a fully penetrating well.

$r$ = well radius [m (ft)]

$b$ = aquifer thickness [m (ft)]

$L$ = well screen length as a fraction of the aquifer thickness

To use the curves presented in **Figure P-6**, the screen length, $L$, is expressed as a percentage of the aquifer thickness. The ratio of $b/r$ is determined. From there, the percent of the aquifer screened is located on the x-axis. This point is traced up to the appropriate curve, and the corresponding percentage of maximum specific capacity is determined.

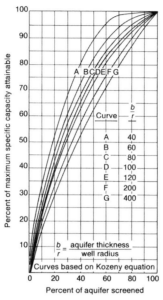

**Figure P-6** The graphical relationship developed from the Kozeny equation of partial penetrations and expected attainable specific capacity in homogenous confined aquifers. (Driscoll, 1986. Reprinted by permission of Johnson Screens/a Weatherford Company.)

**Particle tracking** A technique used in numerical modeling to depict the movement of solute or to trace *groundwater movement* to understand the geometry of the *flow system*. Cf. *Tracer*.

**Partition coefficient** See *Distribution coefficient*.

**Passive diffusion sampler** A device that is placed in a *monitoring well* to provide a time-weighted average concentration of contaminants in the groundwater being monitored at that location. In its simplest form, it represents a bladder containing dionized water, the membrane of which allows the diffusion of the contaminants from the groundwater into the dionized water. When equilibrium conditions are achieved, the water chemistry inside the sampler is equal to the groundwater chemistry. The water within the membrane is then analyzed using conventional chemical methods appropriate for the analyte. The advantage of the passive diffusion sampler is that there is no *purging* of the well to obtain representative samples reducing labor costs associated with this. The device is left in the monitoring well for a preset period of time and then retrieved. As a result, there is no *purge water* to dispose after sampling. Cf. *Peeper (dialysis membrane samplers)*.

**Passive film** A *corrosion* analysis term, having application to stainless steel *well screens*. Passive films provide corrosion protection and act as hydrated oxide film having a *gel*-like structure. Films on stainless steel are about 1.5–3 nm thick and contain chromium and iron-rich areas ($Cr^{+3}$, $Fe^{+2}$, and $Fe^{+3}$) that are incorporated into layers having both hydroxide ($OH^-$) and $O^{-2}$ *ions*. It has also been observed through X-ray photon electron spectroscopy that hydrogen-bonded water may also be present in the film. Corrosion resistance of passivated stainless steel is almost completely controlled by the presence of the passive film and is sensitive to the presence of certain ions, such as chloride ions (Cl⁻). Corrosion protection is reduced when the passive film is worn through mechanical action such as sand passing through the screen under pumping conditions or damaged during installation. Cf. *Well maintenance.*

**Pasteurization treatment for iron bacteria** This method involves circulating hot water within a *well* to control the growth of bacteria. Water is introduced down the well at temperatures of 80°C (176°F) until the return water reaches the same temperature. At temperatures of 45°C (113°F), the bacteria are dispersed, and at a temperature of 54°C (129°F), the bacteria are killed. This treatment is effective for small-diameter wells, but is ineffective for treating the bacteria within the formation. In large-diameter wells, the treatment becomes uneconomic. Cf. *Well maintenance.*

**Pathline** See *flow path.*

**Pathogenic** Capable of causing disease, when pertaining to water purification. Cf. *Shock chlorination.*

**PDS** See *partial-duration series.*

**PE** See *potential evapotranspiration (PE); actual evapotranspiration (AE).*

**Peak discharge (q)** The estimation of peak discharge, also known as peak flow, is the greatest overall economic importance in the application of flood estimation. The point on a *stream* or *river hydrograph* representing the highest *flow* from a single (or cumulative) *precipitation* event (**Figure P-7**). The traditional approach for estimating floods on small drainage basins use the following formula:

$$q = CiA$$

where:

$q$ = peak discharge (ft³/s)
$C$ = *runoff coefficient* (ratio of runoff to rainfall) dimensionless
$i$ = *rainfall intensity* (in/h)
$A$ = drainage basin area (A)

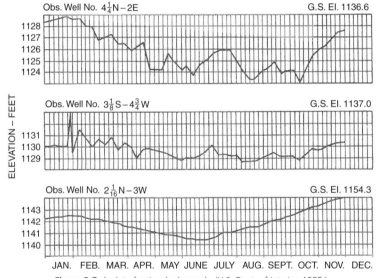

**Figure P-7** A plot of a river hydrograph. (U.S. Dept. of Interior, 1985.)

Cf. *Baseflow recession hydrograph; Recession curve; Storm hydrograph; Rational equation; Flood forcasting.*

**Peak flow** See *peak discharge.*

**Peak stage** The highest point in the *flood stage*, in a given water course for which *discharge-flow* curves have been developed. Cf. *Flood forcasting.*

**Peak runoff** See *runoff; rational equation.*

**Peaks over threshold (POT)** A database of *flows* of a water course over a predefined threshold to give on average five peaks a year above the selected threshold. The year of data may be a calendar year or a water year, depending upon *precipitation* patterns. Cf. *Partial Duration Series (PDS).*

**Peclet number** A parameter used to define *dispersion* (mechanical mixing) of the solute front as it migrates through *porous medium*. At very low groundwater velocities, *diffusion* is the dominant mixing mechanism, whereas at higher velocities, dispersion is the dominant mixing mechanism. Laboratory experiments have established a relationship between the influence of diffusion and *mechanical dispersion*. As illustrated in **Figure P-8**, the dimensionless parameter, $vd/D^*$, is known as the Peclet number. The average particle diameter is defined by $d$. "$v$" is the average linear water velocity and $D^*$ is the diffusion coefficient. The exact shape and relation between the Peclet number and the ratio $D_l/D^*$ (the ratio of the *dispersion coefficient* to the *diffusion coefficient*) depends on the nature of the medium and the fluid used in the experiment. The Peclet number is also used as a design criterion for numerical simulations of the *advection dispersion equation*.

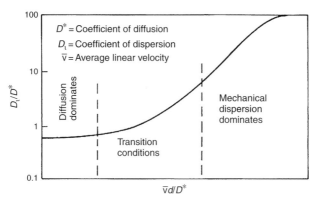

**Figure P-8** Relation between the Peclet number and the ratio of longitudinal dispersion coefficient and the coefficient of molecular diffusion in a sand of uniform-sized grains. (Freeze and Cherry, 1979. Reprinted with permission from Prentice-Hall, Englewood Cliffs.)

**Peeper (dialysis membrane sampler)** A variation on a *passive diffusion sampler* consisting of screening material stretched over the mouth of a jar. The dialysis membrane sampler allows water to move across the membrane. The screening material made from regenerated cellulose are compatible with most analytes, including metals and volatile organic compounds (VOCs). Disadvantages with these samplers are that they biodegrade and cannot be left in place very long, they are more difficult to use than passive diffusion samplers because the material is manufactured only in flat sheets, and they must be filled on-site. Cf. *Adhesive water; Attached groundwater.*

**Pellicular water** Films of *groundwater* adhering to particles or *pore spaces* above the *water table*. Cf. *Adhesive water; Attached water; Capillary fringe; Vadose zone.*

**Penetration test** See *split-spoon samplers.*

**Perched water** A zone of local saturation, which occurs above the *water table*. Found where groundwater accumulates in top of lower-*permeable* stratum above the water table. This can be found in clay-rich glacial drift, where sand lenses and stringers within a clay matrix result in the accumulation of water. The upper surface of this water is called a perched water table. This situation is illustrated in **Figure P-9**. Cf. *Aquifer, perched.*

**Perched water table** See *Perched water.*

**Percolation/percolates** The process by which water seeps through soil without following a definite channel. Cf. *Absorption; Infiltration; Leaching; Seepage/seepage force.*

**Perennial stream** Also known as a permanent stream. A *stream* or *reach* of the stream that maintains a continuous flow throughout the year and whose upper surface is generally lower than the regional *water table* to which the stream belongs.

**Perennially frozen ground** See *permafrost.*

**Performance-monitoring network** A *groundwater*-monitoring network established to evaluate the effectiveness of a groundwater engineering effort, such as a *saltwater intrusion* barrier, a contaminant plume recovery system, and a natural attenuation monitoring system. Cf. *Plumes.*

**Performance curves** Also called a power curve or a pump curve. A graphical depiction of the range of efficient operation. The curves show the *pumping rate* versus the lift that the pump is capable of. **Figure P-10** depicts a performance curve for various horsepower pumps. The flat horizontal portion of the curve represents the range of inefficient operation. To determine the most appropriate horsepower rating of the required pump, estimations of *well yield* and required lift are made. Cf. *Lifting head.*

**Peristaltic pumps** A low–flow, rotary-positive displacement pump typically used for chemical feed systems, obtaining samples from shallow-*monitoring wells*, or for the filtering of samples for analysis. It is basically a positive displacement-type pump where flexible tubing is inserted into a housing and a rotating cam squeezes the flexible tubing to provide the positive displacement. Peristaltic pumps have the ability to lift water from as deep as 8.5 m (28 ft).

**P**

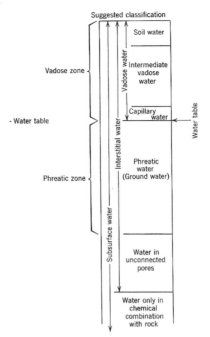

**Figure P-9** Classification of subsurface water. (Davis, 1966. Reprinted with permission from John Wiley & Sons, Inc.)

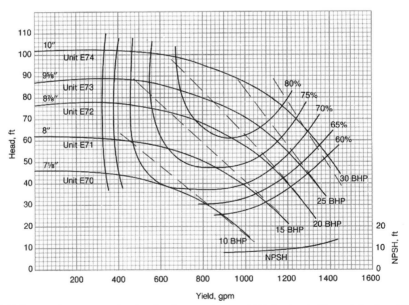

**Figure P-10** Pump performance at various impeller diameters. (Powers, 1981. Reprinted with permission from John Wiley and Sons, Inc., NY.)

**Permafrost** Perennially frozen ground. Soil temperatures significantly below 0°C (32°F) are required to initiate the change in state of the *pore water* into ice. This is dependent on the salt content of the water, the *grain size distribution* of the soil, the soil mineralogy, and the soil structure. The hydrogeologic importance of permafrost resides in the changes in *permeability* that result when the pore water freezes. What would normally be considered permeable sediments can become low permeable to impervious *aquitards* at temperature slightly below 0°C. This greatly modifies *recharge/discharge* characteristics in the Arctic regions.

**Permanent stream** See *perennial stream*.

**Permeability/permeable** The property or capacity of a porous media (soil, rock, or sediment) for transmitting a fluid. It is also defined as the relative ease of fluid flow under differential pressure through a media. Cf. *Hydraulic conductivity*.

**Permeability, absolute (*k*)** The capacity of a porous rock, sediment, or soil to transmit a fluid such as water or a gas at 100% saturation. Permeability is a measure of the relative ease of fluid *flow* under unequal pressure. In consulting work, permeability and *hydraulic conductivity*, *K*, tend to be used interchangeably when referring to water transmission through natural materials. Hydraulic conductivity, however, is a function of properties of both the *porous medium* and the fluid passing through it, whereas permeability, or what has also been called absolute and intrinsic permeability, is representative of the medium alone, is independent of the transmitted fluid and has dimensions $L^2$. The equation for relating permeability, *k* and hydraulic conductivity, is as follows:

$$K = \left(\frac{k\rho g}{\mu}\right)$$

where:
$K$ = hydraulic conductivity $[L \cdot T^{-1}]$
$k$ = permeability $[L^2]$
$\rho$ = fluid density $[M \cdot L^{-3}]$
$g$ = gravitational constant $[L \cdot T^{-2}]$
$\mu$ = dynamic viscosity $[M \cdot L^{-1} \cdot T]$

Permeability is a function of the size and shape of the pore openings through which the fluid moves. The larger the square of the mean pore diameter, *d*, the lesser the resistance to flow. The cross-sectional area of a pore is also a function of the shape of the opening. The overall effect of the shape of the pore spaces can be described by a dimensionless constant called the shape factor, *C*. Using this constant, permeability is expressed by the following equation:

$$k = Cd^2$$

The dimensions for *k* are area, $L^2$, and the coefficient *C* has the following typical values:

| | |
|---|---|
| Very fine sand, well graded | 40–80 |
| Fine sand with appreciable fines | 40–80 |
| Medium sand, poorly graded | 80–120 |
| Coarse sand, well graded | 80–120 |
| Clean coarse sand, poorly graded | 120–150 |

In unconsolidated sediments, the permeability is a function of the grain size, *d*, which increases or decreases the surface area in contact with fluid. For sedimentary deposits

- as the grain size increases, so does the permeability, because of larger pore openings.
- A well-graded deposit has a decreased permeability because the voids tend to be filled with the finer material.
- A well-graded coarse sediment is less permeable than a well-graded fine sediment.

In consolidated sediments, the permeability is due to the primary openings created during the rock's formation and the secondary openings created after the rock was formed. *Primary* and *secondary permeability* depend on the size of the openings (fractures, solution channels, and bedding planes), the degree of connectivity, and the amount of open space. The rock's *porosity* affects its permeability. The above table presents the typical ranges of permeability for unconsolidated and consolidated material. Permeability is a function of the medium only, and therefore has a dimension of area, $L^2$; in the petroleum industry, the permeability is usually expressed in *darcies*. Cf. *Effective permeability*; *Hydraulic conductivity*; *Intrinsic permeability*; *Relative permeability*; *Zone of saturation*.

**Permeability coefficient** See *Coefficient of permeability*.

**Permeameter** Saturated hydraulic conductivity (K) is usually measured in the laboratory on undisturbed soil samples using a constant-head permeameter method and/or a falling-head permeameter method (**Figure P-11**).

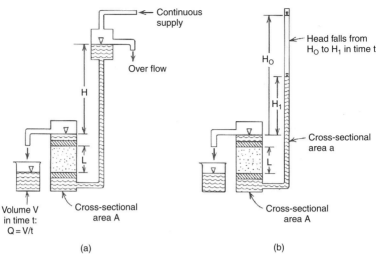

**Figure P-11** Permeameters: (a) constant-head and (b) falling-head. (Freeze and Cherry, 1979. Reprinted with permission from Prentice-Hall, Inc., NY.)

- Constant-head permeameter method. In a constant-head permeameter, a soil sample is enclosed within a cylindrical tube, with porous plates at either end. In performing these tests, it is important that no air become entrapped within the test assembly or the sample. This is accomplished using de-aired water, with the sample saturated from below. If disturbed samples are being tested, they need to be compacted in the test cell prior to testing, and carefully saturated from the bottom of the sample upward prior to testing. The sample length $L$, cross-sectional area $A$, and the constant-head differential $H$ across the sample are known, such that a simple application of *Darcy's law* can be used to determine hydraulic conductivity:

$$K = \frac{QL}{AH}$$

where:
$K$ = hydraulic conductivity of the sample [L·T$^{-1}$]
$Q$ = volumetric *discharge* through the sample [L$^3$·T$^{-1}$]
$L$ = length of the sample [L]
$A$ = cross-sectional area of the sample [L$^2$]
$H$ = constant-head differential across the sample [L]

- Falling-head permeameter the head is allowed to fall from an initial head $H_0$ to $H_1$ over a period of time $t$. The hydraulic conductivity is calculated from the following equation:

$$K = \frac{aL}{At} \ln\left(\frac{H_0}{H_1}\right)$$

where:
$a$ = cross-sectional area of the tube [L$^2$]

Permeameters can be quite accurate provided that a sample has been returned to its field condition. Every sample becomes at least partially disturbed during collection and transport to the laboratory. The original packing density must be reestablished if the measurements are to be representative of percolation rates found in nature.

The laboratory permeameters fall into 2 categories: rigid-wall permeameters and flexible-wall permeameters. In a rigid-wall device the soil is contained in a rigid cylinder and soils having low saturated conductivity side wall flow may occur and affect results. The flexible wall permeameter the soil volume is confined with a flexible latex membrane which prevents side wall flow. The most commenly used technique to measure saturated conductivity in the field is the *guelph permeameter* which uses a smaller volume of water compared with other techniques.

**pH** The negative logarithm of the hydrogen ion activity. In a liquid state, water undergoes the following equilibrium dissociation:

$$H_2O \Leftrightarrow H^+ + OH^-$$

Therefore, from the law of mass action, this can be restated as:

$$K_w = \frac{[\text{H}^+][\text{OH}^-]}{[\text{H}_2\text{O}]}$$

where the square brackets denote activity.

The activity of water is defined as unity at 25°C at 1 bar. Water has equal $\text{H}^+$ and $\text{OH}^-$ activities at 25°C and pH 7 ($[\text{H}^+] = [\text{OH}^-] = 1.0 \times 10^{-7}$). At lower temperatures, the equality of $\text{H}^+$ and $\text{OH}^-$ activities occurs at higher pH values. For example, at 0°C, the equality of activities occurs at a pH of 7.53, while at 50°C, the equality of activities occurs at a pH of 6.63.

**Phreatic surface** Also called the *potentiometric surface*. An imaginary surface, represented by the *piezometric pressures* within an *aquifer*. In practice, it is represented by contoured piezometric pressure readings obtained from wells and *piezometer*. In unconfined aquifers under horizontal *flow* conditions, this is typically synonymous with the *water table*. Cf. *Aquifer, unconfined*.

**Phreatic water** Water below the ground surface, also called *groundwater*, which is at or above atmospheric pressure. Water in the subsurface can be divided into two zones: *phreatic* and *vadose zones* (see **Figure P-9**). Phreatic water can freely enter wells and is available for human use. Cf. *Zone of saturation*.

**Phreatophyte** A water-loving organism such as a plant that thrives in environments where there is excess available water. These plants' roots may extend below the *water table* and hence may represent a major consumptive use of water in arid regions. Examples of phreatophytes include willows and bull rushes.

**Physical plugging** The trapping of fine sediment from the *aquifer* in the *well screen* slot openings. Fine particle movement to the well screens results from the following:

- a poorly designed *filter pack*
- improper screen placement
- improper screen size opening
- insufficient or improper *well development*
- *corrosion* of the screen or *well casing*
- *overpumping*
- excessive pump cycling
- chemical precipitation in the screen slots.

As physical plugging occurs, the *entrance velocity* of the water under pumping conditions increases, eroding the screen openings. This in turn allows sediment to enter the well, which can result in pump failure. Cf. *Sand pumping; Well maintenance*.

**Phytoremediation** A soil and *groundwater* remediation technique that utilizes vegetative root systems to treat contamination. In groundwater applications, phytoremediation may provide a hydraulic barrier for plume and groundwater flow. Further, uptake from root systems may transfer pollutants to the leaves, which in turn may be released to the atmosphere through *evapotranspiration*. In soil remediation applications, phytoremediation may be used as a means of removing contaminants from the soil through the harvesting of plants. The plants remove the pollutants from the soil via uptake through the root system, and the pollutants become incorporated into the plant material. The harvested plants have lower concentrations of pollutants than the soil, reducing volume and concentration in the media to be disposed. Research is ongoing in this field, and it is also believed that redox conditions may be modified in the root zone, which may also have beneficial remediation effects. Cf. *Plume*.

**Piezometer** A tube or a pipe in which the elevation of water can be determined. It must be open to water *flow* at the bottom and to the atmosphere on top. The intake is typically a section of slotted pipe or commercially available *well screen*. The device is called a piezometer in the field and is analogous to a *manometer* in the laboratory. Since the piezometric pressure acts on the length of the screen, the point of measurement in a piezometer is at the midpoint of its screen, not at the water level.

**Piezometer nest** A set of *piezometers* arranged so that groundwater *gradient* (and hence *flow*) can be determined in three dimensions. Piezometer screens at difference depths provide information on vertical flow, while horizontal separation of the piezometers provides information on *horizontal gradients*, as illustrated in **Figure P-12**, where the horizontal gradient is determined from piezometer installations. Cf. *Hydraulic gradient; Direction of groundwater flow*.

**Piezometric surface** See *Potentiometric surface*.

**Piezometric head** Energy contained by fluid because of its pressure, usually expressed in units of length or elevation relative to a datum.

**Piezometric level** Undeveloped and non-penetrated confined *groundwater* is usually under pressure because of the weight of the overburden and the *hydrostatic head*. If a *well* penetrates the *confining layer*, water will rise to this level, the piezometric level, the artesian equivalent of the *water table*. If the piezometric level is above ground level, the well discharges as a *flowing well, artesian well*, or a spring. Cf. *Aquifer, confined*.

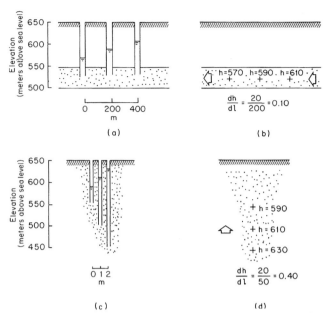

**Figure P-12** The determination of hydraulic gradients. (Freeze, 1979. Reprinted with permission from Prentice-Hall, Inc.)

**Pipe-base well screens** A type of *well screen* that consists of two sets of openings: an outer opening consisting of wire-wound screen (called the wrap-on-pipe screen) covering perforations or holes drilled into the pipe (called the structural core). A pipe-base screen is illustrated in **Figure P-13**. The hydraulic performance of the well screen depends primarily on the percentage open area of the pipe base, which is usually low. These types of screens are often specified in oil field work where it may become necessary to retrieve the screen from great depths. In some parts of the world, this type of well screen is used for water wells, but their efficiency is low and they are also susceptible to corrosion. The structural core is usually constructed from carbon steel, while the wire-wound wrap-on-screen is usually constructed of stainless steel. Stainless steel in contact with carbon steel pipe always results in some electrolytic or *galvanic corrosion* of the carbon steel pipe. This can be prevented by constructing the pipe-base screen using the same metal.

**Figure P-13** A pipe-base screen is constructed by placing a continuous wire-wound screen around a perforated pipe base. This results in an exceptionally strong screen, typically specified for oil wells or deep-water wells. Both steel and plastic materials are used in this type of well construction. (Driscoll, 1986. Reprinted by permission of Johnson Screens/a Weatherford Company.)

Cf. *Well design.*

**Piper trilinear diagram** A graphical method for representing the distribution of major *ions* in *groundwater*. In a Piper trilinear diagram, the percentage of cations and anions is plotted on separate triangular graphs as shown in **Figure P-14**. The intersection of the lines extended from the two sample points on the triangular graphs to the central diamond-shaped graph gives a point that represents the major ion composition on a percentage basis. This is used for visually describing differences in major ion chemistry. **Figure P-15** illustrates the *hydrochemical facies* of groundwater. **Figure P-16** presents the general chemical evolution of major ion distributions in a sedimentary basin.

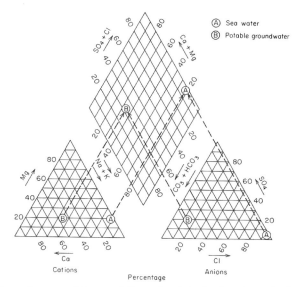

**Figure P-14** Piper trilinear diagram used for presentation and analysis of major anion and cation data in groundwater. Concentrations are presented in milliequivalents per liter. (Freeze and Cherry, 1979. Reprinted with permission by Prentice-Hall, Englewood Cliffs, NJ.)

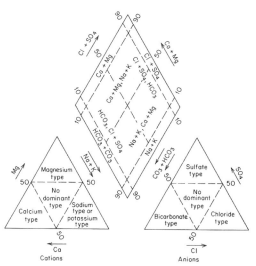

**Figure P-15** Classification diagram for anion and cation facies in terms of major ion percentages. (Freeze and Cheery, 1979. Reprinted with permission from Prentice-Hall, Inc.)

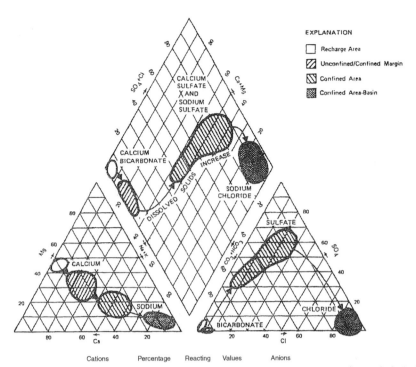

**Figure P-16** General chemical evolution of major ion distribution in groundwater with increasing flow path. (United States Geological Survey, 1982.)

Cf. *Fingerprint diagram; Stiff diagram; Durov-zaprorozec diagram.*

**Pipe-size screens** A *well screen* that is sized to have the same inside diameter as the corresponding standard pipe. See **Table P-1** for pipe sizes.

**Table P-1** Pipe size Johnson well screens. (Driscoll, 1986. Reprinted by permission of Johnson Screens/a Weatherford Company.)

| Screen size | | Inside diameter | | Outside diameter | | OD of female threaded end | |
|---|---|---|---|---|---|---|---|
| in. | mm | in. | mm | in. | mm | in. | mm |
| 2 | 51 | 2 | 51 | 2⅝ | 67 | 2¾ | 70 |
| 3 | 76 | 3 | 76 | 3⅝ | 92 | 3¾ | 95 |
| 4 | 102 | 4 | 102 | 4⅝ | 117 | 4¾ | 121 |
| 5 | 127 | 5 | 127 | 5⅝ | 143 | 5¾ | 146 |
| 6 | 152 | 6 | 152 | 6⅝ | 168 | 7 | 178 |
| 8 | 203 | 8 | 203 | 8⅝ | 219 | 9¼ | 235 |
| 10 | 254 | 10 | 254 | 10¾ | 273 | 11⅜ | 289 |
| 12 | 305 | 12 | 305 | 12¾ | 324 | 13⅜ | 340 |
| 14 | 356 | 13⅛ | 333 | 14 | 356 | | |
| 16 | 406 | 15 | 381 | 16 | 406 | | |
| 20 | 508 | 18¾ | 476 | 20 | 508 | | |
| 24 | 610 | 22¾ | 578 | 24 | 610 | | |
| 30 | 762 | 28¾ | 730 | 30 | 762 | | |

Cf. *Well design.*

**Piping** Also called tunnel erosion. A phenomenon that occurs when *seepage* forces or the *hydraulic gradient* exceed the critical *gradient* below a formation or an earth-fill dam. Piping can occur in man-made structures and in nature as well. When the seepage forces exceed the critical gradient, *quick conditions* can result in soils. The critical gradient is defined as:

$$i_{crit} = \frac{h_{crit}}{L} = \frac{\gamma - \gamma_w}{\gamma_w} = \frac{\gamma_b}{\gamma_w}$$

where:

$\gamma$ and $\gamma_w$ = the soil bulk weight and the water unit bulk weight, respectively [M]

$L$ = length of the vertical *flow path* [L],

$H$ = *groundwater* head causing flow [L]

Piping occurs in the extreme quick condition when the soil is washed away at the toe of a dam or a levee to create a cavity, which then attracts further flow and *erosion*. This can lead to the collapse of a dam or a levee if left untreated such as occurred at the Teton Dam in Idaho in 1976 and along sections of the Mississippi River in 1994. Cf. *Quick soil conditions; Pressure-relief wells.*

**Piracy** See *Beheaded stream; Diversion; Stream capture.*

**Pitless adaptors** An assembly for the underground discharge of water from a well. Prior to the development of the pitless adaptor, wells were installed in pits for domestic or public supply to protect the discharge lines from freezing, creating potentially unsanitary conditions. This device, illustrated in **Figure P-17**, attaches directly to the well casing and extends the *well casing* above the ground surface.

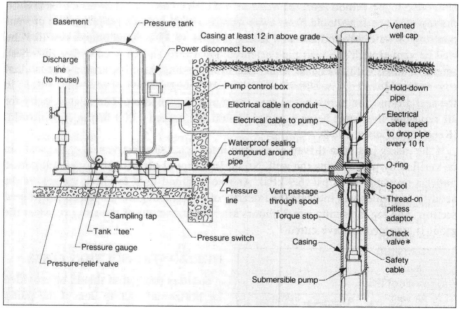

**Figure P-17** Pitless adaptor and piping to a pressure tank installed at a basement or crawl space of a house. Well is equipped with a submersible pump. (Michigan Department of Public Health.)

**Plane jet** A jet of water exiting from a nozzle forming a plane. A method of well development or maintenance using the pressure of water exiting from the nozzle. Cf. *Axial jet.*

**Plastic limit** Cf. *Moisture content; Liquidity index.*

**Plastic fluid** A flowing behavior of a material, such as seen in putty, that occurs after the applied stress reachs a critical value called the yield value. A solid body undergoes a permanent change in shape or size when subjected to the yield value. Cf. *Newtonian fluid; Non-Newtonian fluid.*

**Plasticity index** The percent difference between *moisture content* of soil at the liquid and plastic limits. The plastic limit occurs at the point of transition between the plastic and semi soild state. Cf. *Plastic fluid; Newtonian fluid.*

**Plumbness and alignment** It is desirable that a water *well* be both plumb, or perpendicular, and straight. In practice, however, wells tend not to be perfectly straight and can go out of plumb. Conditions that cause a *borehole* to go out of plumb are the following: (1) cobbles or variations in the hardness of subsurface strata, (2) inconsistent pressure on the drill bit, and (3) trueness of the *well casing* and drill pipe. It is difficult to install casing and pump turbines in a borehole that makes twists and turns. Too much misalignment out of plumb may affect the operation and life of some pumps. Cf. *Alignment test; Inclinometer.*

**Plumes** Also called a zone of dissolved *contaminants*. Typically a plume will originate from a source area and migrate in a *downgradient* direction traveling with *groundwater flow*. The geologic material and chemical composition of the plume will have a significant impact on the path and time of travel. The type of plume that may develop will be affected by the density of the contaminant (**Figure P-18a**), and also the spreading characteristic. **Figure P-19** shows the differences in an instantaneous point source or a continuous source. **Figure P-18c** illustrates the gravity effects.

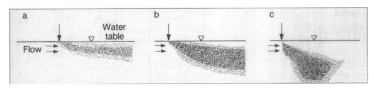

**Figure P-18** The effect of density of a contaminant on the spread of the plume. (Driscoll, 1986. Reprinted by permission of Johnson Screens/a Weatherford Company.)

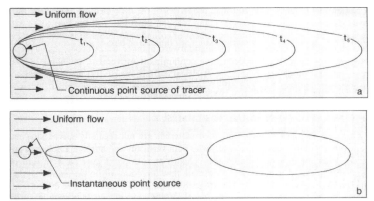

**Figure P-19** The plume-spreading characteristics for a continuous point source and an instantaneous point source. (Driscoll, 1986. Reprinted by permission of Johnson Screens/a Weatherford Company.)

Cf. *Contamination; Dense non-aqueous-phase liquid; Dispersion; Groundwater restoration; Natural attenuation; Mechanical mixing.*

**Plume stabilization wells** Also know as plume containment *wells*. These are wells usually installed at the front of a *contaminant* plume for the purpose of stopping its migration. Cf. *Capture zone; Injection wells.*

**Playa lake** A typically shallow, *intermittent lake* occurring in an arid or semiarid region covering a dry, vegetation-free, flat area near the lowest part of an undrained desert basin, a playa. It is an intermittent lake that leaves or forms the "playa" upon evaporation. Cf. *Dry lake.*

**Point-of-use water treatment systems** A treatment system installed to treat water from domestic water *wells*. The various treatment methods available and the problems treated are summarized in **Table P-2.**

**Point source** See *Plumes.*

**Pollutant** Also called a *contaminant*. An anthropogenic or a naturally occurring compound introduced by man-made or natural conditions into a particular resource, such as *groundwater*. The level of contamination or concentration of the pollutant then restrict the potential use of the resource. Cf. *Plumes; Light non-aqueous-phase liquid (LNAPL); Dense-non-aqueous-phase liquid (DNAPL).*

**Pollution** A condition where the *pollutant* concentration exceeds levels that restrict the potential use of *groundwater*. Cf. *Groundwater pollution.*

**Polygorskite** See *Attapulgite clay.*

**Polyhaline** See *Thalassic series.*

**Polymer** A long-chain chemical compound that consists of many small molecular units (monomers) combined together. Strictly speaking, a polymer is a substance that results from the union of two or more molecules of the same chemical, linked end to end into another compound having the same elements in the same proportion but a larger molecule with hence a larger molecular weight. Polymers typically have a high molecular weight and consist of several thousand monomers. It is when these long chains become tangled that they form a strong film. It is this strong film characteristic that makes them useful as a *drilling fluid*, either on their own or as a bentonite additive. When hydrated, the chemical mixture thickens to provide a mixture that can be used to maintain fluid return within the *borehole*. Further, the control of fluids into the borehole can reduce bottomhole pressure. Additional features of polymers are the following:

- Reduce torque and friction on the drill bit and rods.
- Rapid settling of drill cuttings at the surface, allowing for more rapid recirculation, hence smaller mud tanks are required over natural drilling muds.
- Some polymers are compatible with *brackish* and *saline* formation waters.
- No *filter cake* buildup on the rock cores to mask the samples.

**Table P-2**  Processes for effective removal of drinking water contaminants by point-of-use systems

| General groundwater problem | Contaminants or dissolved solids | Processes | | | | | | |
|---|---|---|---|---|---|---|---|---|
| | | Cation exchange | Chlorination | Sediment | Taste & Odor | Reverse osmosis | Distillation | Ultra-violet |
| Particulates | Sand | | | × | | | × | |
| | Silt | | | × | | | × | |
| | Rust Particles | | | × | | | × | |
| Inorganics | Arsenic | × | | | | × | × | |
| | Barium | × | | | | × | × | |
| | Cadmium | × | | | | × | × | |
| | Calcium | | | | | × | × | |
| | Chromium | × | | | | × | × | |
| | Copper | × | | | | × | × | |
| | Iron | × | | | | × | × | |
| | Lead | | | | | × | × | |
| | Magnesium | × | | | | × | × | |
| | Manganese | × | | | | × | × | |
| | Mercury | | | | | × | × | |
| | Radium 226 & 228 | | | | | × | × | |
| | Selenium | | | | | × | × | |
| | Silver | | | | | × | × | |
| | Sodium | | | | | × | × | |
| | Strontium 90 | × | | | | × | × | |
| | Zinc | | | | | × | × | |
| | Chloride | | | | × | × | × | |
| | Chlorine | | | | | | × | |
| | Nitrates | | | | | | × | |
| | Sulfates | | | | | × | × | |
| | Sulfides | | | | | × | × | |
| Organics | Benzene | | | | × | × | × | |
| | Trihalomethanes | | | | × | × | × | |
| | Polychlorinated Biphenyls | | | | × | × | × | |
| | Petroleum Solvents | | | | × | × | × | × |
| | Pesticides | | | | × | × | × | × |
| | Herbicides | | | | × | × | × | × |
| | Tannin (Humic Substances) | | | | | × | × | |
| | Odors | | | | × | × | × | |
| | Swampy Taste | | | | × | × | × | |
| Biological | Algae | | × | | | × | × | |
| | Bacteria | | × | | | × | × | |
| | Viruses | | | | | | × | |

**Polyphosphates** Adding a small amount of polyphosphate before or during *well development* provides considerable help in the removal of clays naturally occurring in the *aquifer* and clays introduced as part of the *drilling fluid*. Polysulfate disperses

P

the clay particles in the formation, making them easier to remove by groundwater pumping. Two types of polyphosphates are used in well development: crystalline and glassy. Crystalline polyphosphates are sodium acid polyphosphate (SAPP), tetrasodium polyphosphate (TSPP), and sodium tripolyphosphate (STP). A glassy polyphosphate is sodium hexametaphosphate (SHMP) that is readily available and as a result is widely used in developing wells. Commonly, about 6.8 kg (15 lb) of a polyphosphate should be used for each 0.4 m$^3$ (100 gal) of water in the screen. However, polyphosphates also promote bacterial growth, and as a result, 0.9 kg (2 lb) of *sodium hypochlorite* (bleach) at a 3–15% solution should be added to every 0.4 m$^3$ (100 gal) to control this growth. Polyphosphates do not easily mix with cold water and therefore should be well mixed before introduction into the well. The mix water may be heated to promote the dissolution of the polyphosphates. If SHMP is used, care must be taken for the polyphosphate can become glassy in the well and under certain conditions may plug the formation and the screen. For example, an overly rich solution of SHMP may form glassy precipitates, which are gelatinous masses that are very difficult to remove when the solution comes in contact with cold *groundwaters* found in northern latitudes.

**Polyvinylchloride (PVC) casing** A polymer of vinyl chloride that is insoluble in most organic solvents and is used in molded rigid products such as *well casings* and *well screens*. Cf. *Acrylonitrile butadiene styrene (ABS) casing*.

**Pond** See *Recharge basin; Depression storage*.

**Pore channels** *Pore spaces* that are highly interconnected, providing conduits of *flow*. Pore channels may consists of joins within the soil fabric or macropore size *pore spaces* that are connected. Cf. *Porosity*.

**Pore pressure** Also called pore water pressure. Soils are predominantly two-phase systems, composed of relatively flexible particles (soil grains), and the pores of which contain a relatively incompressible fluid. When the soil is saturated with *free water*, rather than adsorbed water, the *pore water* has a positive pressure equal to the *hydraulic head* of the *free water*. The pore pressure $u$ at a point O in a soil mass at a depth $z$ below the ground surface and a depth $h$ below the *water table* will then be given by the following equation:

$$u = \gamma_w h$$

where:

$u$ = pressure at a point O in a soil mass at depth $z$ below the ground surface [M·L$^{-2}$]
$\gamma_w$ = unit weight of water [M·L$^{-3}$]
$h$ = depth of point O below the water table [L]

This is illustrated in **Figure P-20**.

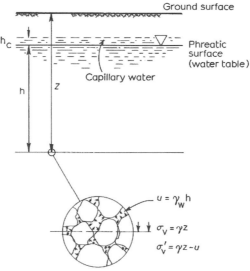

**Figure P-20** Representation of pore water pressure at a point below the water table. (Attewell and Farmer, 1976. Reprinted with permission by Chapman & Hall New York, NY.)

Cf. *Effective stress*

**Pore spaces** Soil consists of solid, liquid and gaseous phases. The solid phase, representing approximately 50% by volume, contains mainly minerals of varying sizes as well as organic compounds. The remaining pore space contains either liquids or gas. There are three pore types: Micropores, Mesopores and Maropore.

Micropores (Size $<0.2$ μm) are filled with water that are adsorbed onto surfaces of clay molecules. These pores are too small for plant use without great difficulty. Water held in micropores is important to the activity of microbes creating moist anaerobic conditions that can cause either the oxidation or reduction of molecules in the crystalline structure of the soil minerals.

- Mesopores (Size 200 μm–0.3 μm) are ideally always full or contain liquid for successful plant growth. The properties of mesopores are studied by soil scientists for agriculture and irrigation purposes.
- Macropores (Size $>2$ mm) are too large to have any significant capillarity and are full of air at field capacity. Macropores can be caused by cracking, division of aggregates, the action of plant roots, and zoological exploration. Cf. *Porosity.*

**Pore volume** The volume occupied by the pores, either saturated or unsaturated, in porous media such as soils. Cf. *Zone of saturation; Unsaturated zone.*

**Pore water/pore fluid** Water or fluid that occupies the pore space of a *porous medium.* Cf. *Adsorbed water; Interstices; Pure pressure.*

**Pore water pressure** See *pore pressure.*

**Pore water velocity** See *seepage velocity.*

**Porosity (n)** The quantity of water that an aquifer can hold in storage is determined by its porosity. The porosity of aquifer materials is a function of the size, shape and arrangement of mineral grains. The porosity of earth materials is the percentage of the rock or soil that is void of material, known as the *pore spaces,* and is mathematically calculated by:

$$n = \frac{100V_v}{V}$$

n = porosity (%)
$V_v$ = volume of void space ($L^3$)
V = volume of earth material ($L^3$)

The porosity is also a function of the formation of the earth material.

Primary porosity is the void space originating when the rocks were formed which include vesicles, intergranular pores, unconformities and the isotropic or anisotropic nature of soil grains.

Secondary porosity is the void space created after the rock is formed including faults, fractures, solution cavities and disruption of soil by plants and animals. Cf. *Homogeneity; Heterogeneity.*

**Porosity, effective** See *Effective porosity.*

**Porous block** See *Gypsum block.*

**Porous medium/porous** A matrix of material (typically soil) that has a measurable volume of pores or open areas that are available for fluids and/or gases. Cf. *Zone of saturation; Void spaces; Pore spaces.*

**Potash Lake** An *alkali lake* whose waters contain a high content of dissolved potassium salts. Cf. *Soda Lake.*

**Potable water** Water fit for human consumption. Cf. *Maximum Contaminant Levels (MCLs); Safe Drinking Water Act (SDWA).*

**Potassium permanganate** A strong oxidizing agent with the chemical formula $KMnO_4$, which is an excellent bactericide. Potassium permanganate is available as a dry purplish-colored solid that is relatively inexpensive and relatively safe to use. It provides good control of iron bacteria. A solution of 1000–2000 mg $L^{-1}$ permanganate dissolved in enough water to fill the *well screen* has been found to achieve excellent results. A 1000 mg $L^{-1}$ solution is equivalent to 0.38 kg (0.83 lb) in 0.4 $m^3$ (100 gal) of water. The overall effectiveness of the oxidant treatment is enhanced through mechanical action through *surging* and *jetting* to loosen the biomass plugging material. Cf. *Polyphosphates; Sodium hypochlorite.*

**Potential energy** See *Gravitational potential energy; Groundwater Energy; Head; Elevation head.*

**Potential evaporation** Also known as evaporative power, evaporation capacity, evaporative capacity and evaporativity. A measure of the degree to which the weather of an area is favorable to *evaporation,* under the existing atmospheric conditions from a chemically pure water surface at the temperature of the lowest layer of the atmosphere.

**Potential evapotranspiration (PE)** The amount of water that would be removed from the land surface by both *evaporation* and *transpiration* if there is available water to meet this demand. Potential evaporation varies by season, and is expressed in units of water depth. Cf. *Evapotranspiration; Actual evapotranspiration.*

**Potential head** See *Hydraulic head.*

**Potential yield** The maximum theoretical amount of water that can be produced from a well.

**Potentiometric level** See *Potentiometric surface.*

**Potentiometric surface** Also known as piezometric surface, isopotential level, and pressure surface. When confined aquifers are tapped by wells, water from the aquifer will rise within the well casing representing an imaginary surface which can be represented by contours of *hydraulic head.* This map of contoured hydraulic heads provides an indication of horizontal *groundwater* flow. However, *groundwater* flow is three-dimensional, and interpretations of flow based on a potentiometric surface map alone may be misleading. Cf. *Aquifer, confined; Confining bed.*

**Pothole** See *Glacier well.*

**Power curve** See *performance curve.*

**Precipitation** The falling of water in liquid or solid form that occurs when the atmosphere becomes saturated with water vapor. The water condenses and falls out of solution (i.e., precipitates). Two processes, either cooling of the air or an input of greater moisture, can lead to the air becoming saturated. Precipitation that reaches the surface of the earth can occur as *rain*, freezing rain, drizzle, *snow*, hoar frost, ice pellets, and hail. Virga is precipitation that begins falling to the earth but *evaporates* before reaching the surface. Precipitation is a major component of the water cycle and is responsible for depositing most of the fresh water on the planet. Approximately 505,000 km³ (121,000 cu mi) of water falls as precipitation each year, 398,000 km³ (95,000 cu mi) of it over the oceans. Given the Earth's surface area, that means the globally averaged annual precipitation is about 1 m (39 in.), and the average annual precipitation over oceans is about 1.1 m (43 in). Cf. *Horton-Theissen method; Water balance.*

**Precipitation gauge** Also known as rain gauge, udometer, pluviometer, fluviograph. An instrument used by meteorologists and hydrologists to gather and measure the amount of liquid precipitation (as opposed to solid precipitation that is measured by a snow gauge) over a set period of time. Most gauges generally measure the precipitation in units of length, such as millimeters, inches, or centimeters. Cf. *Rain gauge; Stream gauge; Gauge.*

**Preferential flow path** Represents the path of least resistance for a fluid moving through a *porous medium*. The preferential flow path has implications in constructing a *flow net* for heterogeneous systems, as the *flow lines* are longest through more permeable formations. Cf. *Macropore flow; Paleochannel.*

**Premature stiffening** See *False set.*

**Pressure** Force per unit area. See **Appendix A** for property conversions.

**Pressure energy** If water is allowed to move under gravity, the *pressure* or energy is converted to velocity or pressure energy. This is expressed by the following equation:

$$H = \frac{P}{\gamma} + \frac{V^2}{2g} + z$$

where:
$H$ = *hydraulic head* [L]
$P$ = pressure [M·L⁻²]
$\gamma$ = specific weight of water [M·L⁻³]
$V$ = velocity of *flow* [L·T⁻¹]
$g$ = gravitational constant [L·T⁻²]
$z$ = elevation above a certain datum [L]
In this equation, $z$ is the *elevation head*.

**Pressure head** Also known as static pressure head or simply static head, but not to be confused or substituted for static head pressure. The internal energy of a fluid due to the pressure exerted on its container. Pressure head is shown in **Figure P-21**

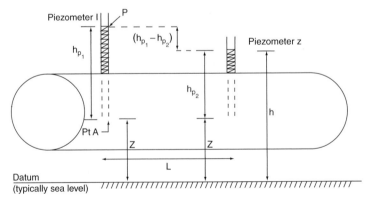

Pt A = point located at elevation Z, above a datum, at fluid pressure P.
P = fluid pressure at pt A.
z = (elevation head) elevation of the base of a piezometer above a datum.
$h_p$ = (pressure head) height of the water column in a piezometer.
h = (total head) sum of the elevation head (z) and pressure head ($h_p$).
$(h_{p_1} - h_{p_2})$ = the change in head (dh) over length L.

**Figure P-21** An apparatus consisting of a circular cylinder filled with sand to demonstrate Darcinian flow and showing the relationship between total head (h), elevation head (z) and pressure head ($h_p$). (After Fetter, 1994 and Freeze, 1979.)

and is mathmatically expressed as:

$$\psi = \frac{p}{\gamma} = \frac{p}{\rho g}$$

where:
$\psi$ = pressure head (L)
$p$ = fluid pressure ($M_{force}/L^2$)
$\gamma$ = the specific weight ($M_{force}/L^3$)
$\rho$ = the density of the fluid ($M/L^3$)
$g$ = acceleration due to gravity ($L/T^2$).
Cf. *Hydraulic gradient; Darcy's law; Gravity head; Total head.*

**Pressure-relief screens** Used to relieve pressure in the *filter pack* under pumping conditions where the *drawdown* outside of the well may be significantly above the pumping level inside the *well casing*. This condition is most likely to occur in tight or highly stratified formations and can create strong upward *flow* in the filter pack. This strong upward flow can lift filter material to the pump intake, or lift heavy lead slip *packers*. As shown in **Figure P-22**, a short pressure-relief screen is installed in the *riser pipe*, just above the bottom of the casing. This allows pressures to be relieved through the screen, rather than the filter pack.

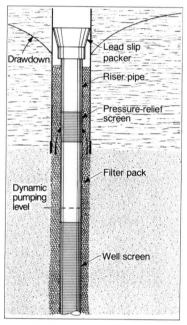

**Figure P-22** A pressure-relief screen installed inside the bottom of a casing to relieve large differential pressures. (Driscoll, 1986. Reprinted by permission of Johnson Screens/a Weatherford Company.)

**Pressure-relief wells for dams and levees** Dams or levees are typically constructed of impervious or semipermeable material overlying the natural permeable alluvial material. The *pressure* from the water dammed behind the structure is transmitted landward from the standing water. Under high-water conditions, this creates a landward zone where the hydraulic pressure is higher than the ground surface, creating a zone of *heaving sands* or *quick conditions* (**Figure P-23**). The principal requirements of a pressure-relief well system are the following:

1. The wells must penetrate to the layers of sediment underlying the dam or levee that have the highest *transmissivity.*
2. The wells must intercept a major part of the subsurface *seepage*. This may involve closer spacings in more transmissive bodies.
3. Well must be efficient to minimize head losses.
4. The well screens should be designed to minimize fine-particle movement into the screened zone.
5. Wells should be resistant to chemical attack and biological fouling.

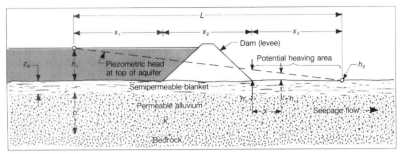

**Figure P-23** Cross section through a dam or a levee showing the transmission of hydraulic head landward, resulting in heaving or quick conditions. (Driscoll, 1986. Reprinted with permission by Johnson Division, St. Paul, MN.)

Pressure-relief well systems can be designed through the use of streamflow lines and *equipotential* analysis to construct a *flow net* as illustrated in **Figure P-24.**

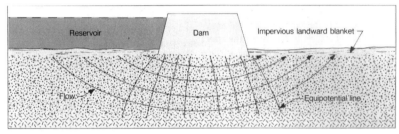

**Figure P-24** Seepage beneath a dam or a levee through unconsolidated, homogeneous isotropic sand. (Driscoll, 1986. Reprinted by permission of Johnson Screens/a Weatherford Company.)

The unit flow volume $q$ through each section along the levee or dam can be determined from the following equation:

$$q = \frac{N_f}{N_e} K h$$

where:
$N_f$ = number of *flow channels* in the flow net [unitless]
$N_e$ = number of potential drops along each channel [unitless]
$K$ = *hydraulic conductivity* [L·T$^{-1}$]
$h$ = total loss of head [L]

The total flow through a section of width $w$ is given by the following equation:

$$Q = TIw$$

where:
$T$ = transmissivity [L$^2$·T$^{-1}$]
$I$ = hydraulic gradient [L]

The head loss is measured between the seepage entrance and the seepage exit. To evaluate the potential for soil heaving, or quick conditions, it is necessary to determine the effect of the reservoir pressure on the surface materials. The upward pressure on a soil column resulting from reservoir pressure is given by the following equation:

$$F_w = (H_1 - H_2)\gamma_w$$

where:
$F_w$ = reservoir pressure [M·L$^{-2}$]
$\gamma_w$ = specific gravity of water [M·L$^{-3}$]
$H_1$ = head at the bottom of the soil column [L]
$H_2$ = head at the top of the soil column [L]
Cf. Piping.

**Pressure transducer** An electronic device that measures pressure and transmits this information to a data collection/storage device. Pressure transducers are used in gathering water level information during pump tests and for long-term monitoring of *groundwater elevation* data. Cf. *Electrical sounder; Measuring drawdown in wells; Level Recorder.*

**Primary drinking water regulations** See *Maximum contaminant level* (MCL).

**Primary permeability** See *permeability, absolute (k).*

**Primary porosity** See *porosity.*

**Priming pumps** The purpose of priming pumps is to expel the air that may be present in the pumping chamber. This is to prevent leakage past pistons, valves, and other working parts.

**Prior appropriation right** Also called the Appropriation doctrine. In the United States, a surface-water law by which those who intend to use a particular water source place themselves on a priority list for a particular source as in "first in time, first in right." An appropriator must divert the water from its natural channel and make beneficial use of the water in order to establish and maintain their right to use. Prior appropriation right law is typically thought of as a system for areas where there may not be enough water for all *Riparian Rights* owners. Cf. *Rule of capture; Rule of reasonable use.*

**Probability** See *Recurrance interval.*

**Procedure for acid treatment** The introduction of a strong acid solution into a *well* to remove chemical *incrustation* by dissolving the precipitate, allowing it to be pumped from the well. The most commonly used acids in well rehabilitation are hydrochloric acid (or muriatic acid) (HCl), sulfamic acid ($H_3NO_3S$), and hydroxyacetic acid ($C_2H_4O_3$). When using any liquid acid, personal protective equipments in the form of rubber gloves, chemical-resistant coverall (Saranex®), and protective face shield should be used. Further, a breathing apparatus should be worn by personnel working near the well. To minimize reaction to the acid, all tanks and *tremie pipe* should be constructed of plastic or black iron. As a safety precaution to take care of spills, a large quantity of water, or a water tank with a mixture of sodium bicarbonate, should be available to dilute or neutralize the acid. The breathing space above the well should be well ventilated, for the fumes generated by the addition of acid can be lethal. The acid solution should be introduced into the well screen area via a tremie pipe. This allows the acid to react with the zone to be treated without becoming spent in reacting with the chemical precipitate in the upper part of the well. It is recommended that the well be treated in 5-ft intervals from the bottom-up. Acid should be added in quantities to fill each 5 ft section of *screen interval.* Pelletized forms of sulfamic acid are commercially available, which are dropped into the casing and go into solution throughout the entire column of water in the well. Follow manufacturer's specifications when using pelletized or granular forms of sulfamic acid. After the acid is placed in the well, a full casing volume of water is added, forcing the added acid into the filter material and into the formation. This is to dissolve chemical precipitant therein. The water is followed by mechanical agitation, surging, and/or jetting to remove dissolved and suspended incrustants. It can never be assumed that the acid diffused evenly into the formation, and zones within the formation around the well may be partially or completely clogged. The acid solution will take the path of least resistance, and it can be difficult, if not impossible, to diffuse the chemical solution to all areas to remove unwanted deposits. If iron and or manganese incrustants are present, and the pH of the treating fluid is 3 or less, the use of chelating agents is recommended. At pH levels that are this low, these metals are for insoluble precipitates that settle out and reduce the effectiveness of the treatment. Common chelating agents are citric, phosphoric, and tartaric acid. Typical ratios of chelating agent to acid are as follows:

- 1.8 kg (4 lb) of chelating agent to 3.8 L (1 gal) of 20N Baumé muriatic acid.
- 0.5 kg (1 lb) chelating agent to 6.8 kg (15 lb) granular sulfamic acid.

The well is allowed to sit after mechanical agitation until the pH returns to between 6.5 and 7. The time required for this is of course well- and formation-specific and may require up to 15 h. It is a good practice to develop the well after treatment to remove particulates from fouling the screen, pump intakes, and pump bowls. Cf. *Well maintenance.*

**Production well** Term is typically used in the oil and gas industry but may also be substituted for *water well.* In petroleum engineering a well is drilled to obtain gas or liquid hydrocarbons from a reservoir bed within a stratigraphic series in an oil province.

**Production zone** The soil or rock stratum of a water-, oil-, or gas-bearing zone that will produce these fluids or gas in economic quantities when penetrated by a well.

**Propping agents** A material, typically sand or small particles of high-strength plastic, used in hydrofracturing to increase *well yield* by keeping the fractures open. Fractures formed during hydrofracturing tend to heal, negating the intended increase in yield. Propping agents are introduced into the *well bore* using a high-pressure pump. Or if a high-pressure pump cannot be used, the sand is placed into the well bore before pressurization. The act of pressurization then causes the propping agents to be distributed into the formation. Cf. *Hydrofracturing.*

**Pull-back method of screen installation** Also known as a telescope screen or well installation. The method involves installing the casing to the total depth of the *well,* lowering (telescoping) the *well screen* to the bottom of the casing and pulling back the casing, leaving the screen in place. This method uses natural formation material as the filter pack. The basic operation is illustrated in **Figure P-3**. After the screen is pulled back to the desired depth, a *swedge* block is used to expand the lead packer attached to the top of the telescoped screen (**Figure P-25**). This method of installing wells that was popular before direct rotary drilling methods became more common.

**Pump** A mechanical device used to transfer liquids, gases, or solids in suspension. Also used for compressing or attenuating gases. Cf. *Check valve; Flow meter logging; Impeller; Lifting head; Pumping test; Rawhiding; Sand pumping; Step-drawdown tests; Suction lift limits; Well design; Well maintenance.*

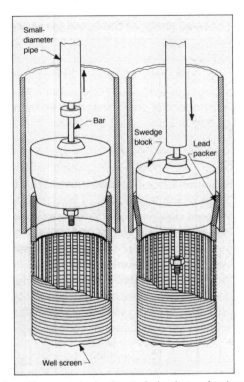

**Figure P-25** A swedge block is used to expand a lead packer attached to the top of a telescoped screen assembly. (Driscoll, 1986. Reprinted with permission by Johnson Division, St. Paul MN.)

**P**

**Pump curve** See *performance curve.*

**Pump and treat system** In environmental *groundwater restoration* projects, by which contaminated groundwater is pumped from an aquifer, treated, and either reinjected into the *aquifer* or discharged to a surface water body, or publically owned treatment works. In the United States, this has historically been the presumptive remedy in groundwater treatment; however, as more and more is learned about the fate and transport of anthropogenic substances in the subsurface, this technique is losing favor to in situ methods for environmental restoration projects. The basic reason for this is that many anthropogenic substances partition into the aquifer matrix, and pumping the groundwater treats only a small portion of the released compound. This method still may be economical in instances such as (1) the source of the release cannot be accessed; (2) it is too deep for other forms of treatment; (3) to control the migration of groundwater plumes; or (4) where the contaminant is highly mobile, such as in the case of methyl-*tertiary*-butyl ether. Cf. *Point-of-use treatment; Plumes.*

**Pumping cone** See *Cone of depression.*

**Pumping rate** The quantity of fluid that a pump or *well* is discharging in units of volume per unit time. Typical units of measurement are gallons per minute, cubic meters per second, liters per minute, and cubic feet per minute. Refer to **Appendix A** for conversions.

**Pumping test** Also called an *aquifer* test, a borehole *permeability* test, *step* test, or a *hydraulic conductivity* test. Pumping tests are used for two purposes: (1) to determine the performance characteristics of a *well* and (2) to determine the hydraulic characteristics of an aquifer. To ensure that the best-quality data is obtained, the tests need to be carried out in a methodical manner, carefully recording time, *discharge*, and depth measurements. A pumping test is facilitated through the use of pressure transducers, data loggers, and computers. However, even with improved technology, there are some basic steps that need to be performed to ensure quality data. Several days before the test, the well should be pumped for several hours to determine the following:

- The maximum anticipated *drawdown.*
- The quantity of water produced at various pump speeds and drawdown.
- The best method to measure the *yield* (e.g., *flow meter, weir, flume,* and critical orifice).

- Where to discharge the pumped water to avoid *recharge*, or in the case of environmental work, where to store the contaminated water.
- Whether the *observation wells* are located such that sufficient drawdown is observed to produce useable data.

Before beginning a pumping test, ensure that the water levels have returned to normal (pretest) *static water levels*. The accuracy of the drawdown data is dependent on the following:

- Maintain constant well discharge throughout the test.
- Measure the drawdown in the pumping well and several observation wells.
- Program the monitoring equipment to take readings at the appropriate time intervals.
- Determine how changes in barometric pressure and *tidal influences* affect drawdown data.
- Compare recovery data with drawdown data.
- Confined aquifers should have 24 hours of drawdown and 24 hours of recovery (this may be project-/site-specific).
- Unconfined aquifers should have 72 hours of drawdown and 72 hours of recovery (this may be project-/site-specific).
- *Step-drawdown* tests should be conducted over a 24-hour period for either type of aquifer.
- Pumping rates should be measured and recorded periodically, and adjusted if needed. Variations in pumping rates can result in erratic data. Pumping rates can change as a result of variations in line voltage, variations in air temperature, humidity, or gasoline mixture in the generator.

Cf. *Aquifer, confined; Aquifer, unconfined.*

**Pumping water level** Also called the dynamic water level. The water level in a well when pumping is occurring. For *artesian wells*, it is the above-ground water level occurring from a flowing well.

**Purging** See *well development*.

P

# Qq

**Qanat** See *kanat*.

**Quality of groundwater** The suitability of *groundwater* for a particular use, as indicated by the total concentration of dissolved minerals or solids. The *total dissolved solids* (TDS) in groundwater can be calculated by adding the concentration of the individual *ions* in the sample. In the field, TDS can be approximated by measuring the *specific conductance*. **Table Q-1** is a simple classification of groundwater quality based on TDS.

**Table Q-1**   Groundwater classification based on total dissolved solids (TDS). (Driscoll, 1986. Reprinted by permission of Johnson Screens/a Weatherford Company. Originally from Freeze and Cherry, 1979.)

| Quality classification | Total dissolved solids (mg l$^{-1}$) |
|---|---|
| Freshwater | 0–1000 |
| Brackish water | 1000–10,000 |
| Saline water | 10,000–100,000 |
| Brine water | >100,000 |

Groundwater characteristics are affected by *percolation* and *groundwater flow* through the subsurface. **Table Q-2** lists the major (greater than 5 mg L$^{-1}$), minor (between 0.01 and 10.0 mg L$^{-1}$), and trace (less than 0.1 mg L$^{-1}$) dissolved constituents in a typical groundwater sample.

**Table Q-2**   Major, minor, and trace dissolved constituents commonly found in groundwater samples. (Davis and Dewiest, 1966. Reprinted with permission from John Wiley & Sons, Inc.)

| Major constituents (greater than 5 mg L$^{-1}$) | | Copper | Rubidium |
|---|---|---|---|
| Bicarbonate | Silicon | Gallium | Ruthenium |
| Calcium | Sodium | Germanium | Scandium |
| Chloride | Sulfate | Gold | Selenium |
| Magnesium | | Indium | Silver |
| Minor constituents (0.01–10.0 mg L$^{-1}$) | | Iodide | Thallium |
| Boron | Nitrate | Lanthanum | Thorium |
| Carbonate | Potassium | Lead | Tin |
| Fluoride | Strontium | Lithium | Titanium |
| Iron | | Manganese | Tungsten |
| Trace constituents (less than 0.1 mg L$^{-1}$) | | Molybdenum | Uranium |
| Aluminum | Bromide | Nickel | Vanadium |
| Antimony | Cadmium | Niobium | Ytterbium |
| Arsenic | Cerium | Phosphate | Yttrium |
| Barium | Cesium | Platinum | Zinc |
| Beryllium | Chromium | Radium | Zirconium |
| Bismuth | Cobalt | | |

Cf. *Chebotarev's succession*; *Water quality*; *MCLs*.

**Quality of precipitation** The air quality and wind patterns of an area through which rainfalls determine the dissolved substances that are picked up by the water particles and thereby affect the purity of the *precipitation*. Atmospheric water accumulates carbon dioxide ($CO_2$), nitrates ($NO_3^-$), trace organic compounds, and inorganic forms of phosphorus and sulfur, especially in industrialized areas. The chemicals picked up by precipitation can become a pollution *non-point source* that can eventually be washed into streams and *lakes*, carried down into the *groundwater* system, or deposited directly into surface water bodies.

**Quality of runoff** Water traversing the land surface during or immediately after a storm event dissolves constituents in the soil and transports *suspended materials* encountered along its *flow path*, and thereby chemically alters the purity of stormwater runoff. Materials transported by *runoff* are often major contributors to water pollution as *non-point sources*. Common transported materials are *sediment loads*, microbial populations, organic compounds, pesticides from agricultural areas, refined petroleum compounds, dissolved metals, and mineral salts. These constituents of runoff may be in solution, suspended, or absorbed to the transported sediment load.

**Quarter flow interval** The length of time it takes for one-quarter of the annual *runoff* to exit a *watershed*. This can take place over any time period in a year (calendar or water year) and is the shortest interval at any time of the year in which one-quarter of the annual rainfall leaves the watershed. Cf. *Half flow interval*; *Hydrologic cycle*.

**Quasi-equilibrium** A state of approximate equilibrium, for a cross-sectional area in a *reach* or section of a *stream*, providing a smooth *longitudinal profile* even though the stream may still be down cutting its base, or in *groundwater*, a *steady-state flow* system in which insignificant changes occur in *head* or in the *drawdown* occurring with time in the *cone of depression*.

**Quasi-steady-state drawdown** The *drawdown* that occurs when the well discharge is kept constant. The productivity of wells with quasi-steady-state drawdowns can be compared, provided that the same time duration is considered.

**Quasi-steady-state flow** The *flow* that occurs when variations of groundwater levels with time are negligible. Because of the inherent variation in subsurface conditions, true *steady-state flow* in nature is virtually impossible. However, groundwater situations considered to be quasi-steady-state flow can be solved if they are considered as steady-state flow, which is mathematically simpler.

**Quick conditions** A significant reduction in the bearing capacity of a soil that is caused by water flowing upward with sufficient velocity to decrease the soil's intergranular pressure or particle-to-particle attraction. Quick conditions are often mistakenly called quicksand. In an *aquifer*, fluid pressures force groundwater to flow from the aquifer through the underlying sediments. If the subsurface pressures are greater than the saturated weight of the surface materials, the excess pressure causes localized heaving of the soil; these areas are called sand boils and are characterized by quick conditions. **Figure Q-1** is a cross section through a structure demonstrating the transmission of the static *hydraulic head* down-river or landward of the structure. The water/soil friction resulting from water flowing through the subsurface causes an energy transfer or *head loss* from the water to the soil. This seepage force, which is the energy transfer, produces quick conditions. Quick conditions also occur naturally in streambeds fed by groundwater *baseflow*, and in aquifers associated with groundwater discharge when the upward *hydraulic gradient* is great enough to reduce the bearing capacity of the soil.

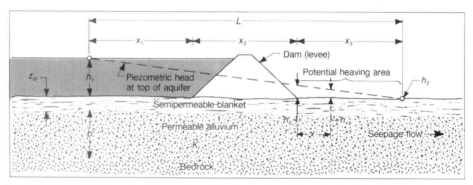

**Figure Q-1** Cross section through a dam or a levee showing the transmission of hydraulic head landward, resulting in heaving or quick conditions. (Driscoll, 1986. Reprinted by permission of Johnson Screens/a Weatherford Company)

**Q**

**Quick flow** The rapid runoff of "new water" during and after *rainfall*, which consists of *interflow* and saturated *overland flow*. Quick flow is a segment of a storm rainfall that moves rapidly to a stream channel by surface runoff or interflow, producing a *flood wave* in the channel. Water that has not infiltrated into the catchment soils is the quick flow or the surface runoff component of the peak on a storm *hydrograph*, which immediately follows rainfall. The hydrograph can be separated into two main components: quick flow or surface runoff and *baseflow*. In some literature, quick flow is substituted for storm flow, which is a combination of surface and subsurface *runoff*, channel and *bank storage*, and may also include a rapid response from the *groundwater reservoir*. Cf. *Interflow; Runoff; Hydrograph*.

**Quick set** See *Flash set*.

**Quick soil condition** The condition created when the effective pressure between solids is reduced to zero as a result of an upward seepage force and the soil exhibits properties of a liquid.

**Quicksand** An area of fine-grained, well-rounded sand that lacks a tendency for the individual grains to adhere. A quicksand area is usually saturated with water flowing upward through the pore spaces, forming a soft, shifting, semiliquid, and highly mobile mass that yields to *pressure*. Objects on the surface of a quicksand tend to sink because the sand surface has lost its bearing capacity and therefore cannot support the load. Cf. *Heaving sand; Quick sand*.

**Race** A narrow or constricted section in a watercourse through which a strong, rapid current flows. A race may occur either naturally, when *tides* meet at a headland between two *bays*, or artificially, by converting water to or away for industrial use. Cf. *Water race*.

**Radial drainage pattern** See *drainage pattern*.

**Radial flow** *Groundwater flow* in an *aquifer* in the direction of a pumping vertical well. Radial flow has inherent to it the assumptions of transient and *steady-state flow*, as well as those relating to pumping from confined and unconfined aquifer. The equation for the radial flow of groundwater is derived from *Darcy's law* and is written as follows:

$$Q = (2\pi rb)K\left(\frac{dh}{dr}\right)$$

where:

$Q$ = pumping rate $[L^3 \cdot T^{-1}]$
$r$ = radial distance from circular section to well $[L]$
$b$ = aquifer thickness $[L]$
$K$ = hydraulic conductivity $[L \cdot T^{-1}]$
$dh/dr$ = hydraulic gradient [dimensionless]

Water flows from every direction toward a *pumping well*. As water moves closer to the pump source, the cylindrical area of the water becomes smaller, causing its velocity to increase (**Figure R-1**). The effective stress induced by pumping is greater in the vertical direction than in the horizontal direction. The horizontal hydraulic gradient induced by pumping results in the decrease of *hydraulic head* in the aquifer around the well.

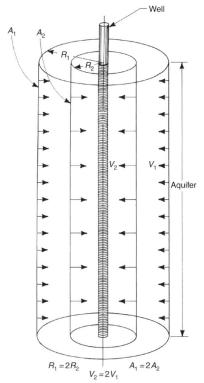

**Figure R-1** Imaginary cylindrical surfaces get smaller as water converges on a well. (Driscoll, 1986. Reprinted by permission of Johnson Screens/a Weatherford Company.)

Cf. *Cone of depression*; *Thiem equations*; *Aquifer, confined*; *Aquifer, unconfined*.

R

**Radial well** An *infiltration gallery* modification in which the *well screen* portion of a vertical caisson placed near a water body is extended horizontally. Cf. *Radial flow*; *Cone of depression*.

**Radiation log** See *radioactivity log*.

**Radioactive contaminant** See *water quality*.

**Radioactive dating** See *Carbon-14 dating*; *Dating of groundwater*; *Radiocarbon dating of groundwater*; *Tritium in groundwater*; *Tracer*.

**Radioactive spring** See *spring; radioactive*.

**Radioactive tracer** A radioactive isotope used to track or monitor the movement and pathway of water. Non-radioactive chemical analogues are often used for the same objectives as radioactive tracers if their presence can be monitored. Examples of non-radioactive tracers are stable isotopes of common elements, such as strontium.

**Radioactivity log** Also called a radiation log. A well log generated from the response of rocks in a well bore, either cased empty or fluid-filled, to bombardment with gamma rays (neutrons) emitted from a logging sonde. Cf. *Neutron activation log*; *Gamma–gamma log*.

**Radiocarbon dating of groundwater** A method for determining the age of organic material that is applicable to samples up to approximately 70,000 years old. As the Earth's upper atmosphere is bombarded by cosmic radiation, atmospheric nitrogen is broken down into an unstable isotope of carbon–carbon 14 ($^{14}$C). The unstable isotope is brought to earth by atmospheric activity, such as storms, and becomes fixed in the biosphere. The half-life of $^{14}$C is 5730 years. The $^{14}$C radioactivity is expressed as a percentage of the original using the following equation:

$$A_C = QA_0 2^{(-t/T_C)}$$

where:
$A_C$ = measured $^{14}$C radioactivity $[T^{-1}]$
$A_o$ = activity at the time the sample was isolated $[T^{-1}]$
$Q$ = adjustment factor accounting for diluted carbon [dimensionless; range 0.5–0.9]
$t$ = age of the sample [T]
$T_C$ = half-life of $^{14}$C [5730 years]

The values of $Q$ and $A_o$ are estimated, and they depend on the carbonate equilibria of *groundwater* in an *open system* exposed to an infinite amount of $CO_2$. Cf. *Carbon-14 dating*; *Dating of groundwater*.

**Radiogenic isotope** See *Radioactive tracer*.

**Radiohydrology** The study of the interactions (i.e., processing, extraction, and disposal) of radioactive elements or radioactive materials associated with waste processes, within the hydrogeologic regime.

**Radius of curvature** In the geometry of watercourse *meanders*, the empirical relations between channel width and meander length and between channel width and the radius of curvature, as shown in **Figure R-2**.

R

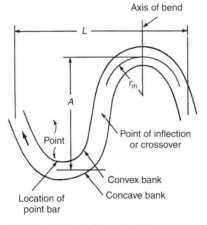

Axis of bend

$L$ = Meander length (wave length)
$A$ = Amplitude
$r_m$ = Mean radius of curvature

**Figure R-2** A meander section and sketch of terms. (Leopold et al., 1992. Reprinted with permission from Dover Publication, Inc.)

The graphical relation between meander length or wavelength, amplitude, radius of curvature, and channel width plots linearly, as shown in **Figure R-3a** and **b**.

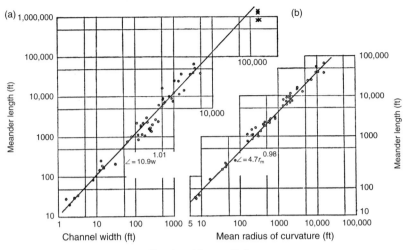

Meanders of rivers and in flumes
Meanders of Gulf stream
Meanders of glacier ice

**Figure R-3a and 3b** Graphical relation of meander length (a) to width and (b) to radius of curvature in channels. (Leopold et al., 1992. Reprinted with permission from Dover Publication, Inc.)

Cf. *Channel shape.*

**Radius of influence ($r$ or $R_0$)** The horizontal distance from the center of a *well bore* to the furthest point of noted pumping-induced change in the *aquifer*. In other words, the radius of influence is the distance from the pumping center to the point where the *drawdown* is zero. The radius of influence is greater for *cones of depression* in confined aquifers than for unconfined aquifers; representative field averages vary from 300 to 30 km for a confined aquifer and from 50 to 500 m for an unconfined aquifer. Pumping water from a well over an extended time draws water from the aquifer storage at greater and greater distances from the well bore until equilibrium is established. As the cone expands, the radius of influence increases. **Figure R-4** shows the relation between cone of depression, radius of influence, and time during constant rate pumping.

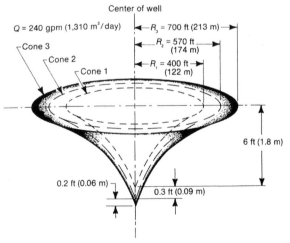

**Figure R-4** Relationship between cone of depression, radius of influence, and constant pumping. (Driscoll, 1986. Reprinted by permission of Johnson Screens/a Weatherford Company.)

Cf. *Area of influence of a well; Hydraulic radius; Jacob distance–drawdown straight-line method; Aquifer, confined; Aquifer, unconfined.*

R

**Rain** See *rainfall.*
**Rain gauge** See *Precipitation gauge.*

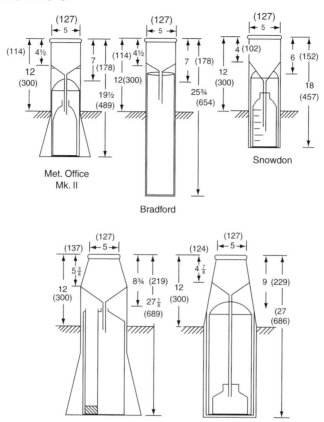

Met. Office
Mk. II

Bradford

Snowdon

Octapent          Seaththwaite

**Figure R-5** Standard types of rain gauges. (Shaw, 1983. Reprinted with permission from Van Nostrand Reinhold Co. Ltd.)  **R**

**Raindrop** The form that water attains as it travels in the atmosphere. When air moves upward, a water drop forms and remains in the atmosphere until enough drops coalesce to overcome the air friction and buoyant forces or until a downdraft propels them toward the earth's surface. Raindrop size varies from 0.4 to 7 mm and averages 1 mm. Refer to **Appendix A** for unit conversions. The drop breaks up if it exceeds the terminal velocity (the limiting rate of fall in still air). The *terminal velocity of raindrops* is from 170 to 900 cm s$^{-1}$, which are speeds that can be achieved in as little as 30 ft of free fall, e.g., dropping from tree leaves. Downdrafts can propel raindrops with sufficient force to cause significant damage to unprotected surface soils. Cf. *Rainfall; Precipitation.*

**Rainfall** A form of *precipitation,* that is, the quantity or measure of water discharged from the atmosphere within a given time period, typically expressed as the resulting depth of water within the area receiving the water. Rainfall often constitutes the greatest portion of the precipitation in an area. The worldwide range of maximum annual rainfall varies from 0 in some deserts to over 25,400 mm or 1000 in. in monsoonal conditions. The amount of rainfall received over a *catchment basin* is calculated by taking *rain gauge* samples from representative point locations to determine the total volume (L$^3$), or the equivalent areal depth (L), over the catchment area. The chemical composition of rainfall is highly variable, but the major dissolved constituents are sulfate (SO$_4^{2-}$), bicarbonate (HCO$_3^-$), chloride (Cl$^-$), nitrate (NO$_3^-$), calcium (Ca$^{2+}$), sodium (Na$^+$), magnesium (Mg$^{2+}$), ammonium (NH$_4^+$), potassium (K$^+$), and hydrogen (H$^+$) ions. Atmospheric gases and inorganic impurities also occur in rainfall. The natural pH of rainfall where the air is not significantly contaminated ranges from 5 to 6 and is controlled by the concentration of dissolved carbon dioxide (CO$_2$). Cf. *Precipitation rate; Terminal velocity of raindrops.*

**Rainfall area** The geographic area receiving *rainfall*. Cf. *Catchment area*.

**Rainfall density** The areal distribution of *rainfall*. Rainfall density aids in the determination of the number of *rain gauges* and *gauging stations* that are required to adequately monitor an area. **Table R-1** gives the minimum number of rain gauges for monthly average rainfall estimates. The design of a *gauging network* requires careful consideration of such variables as areas with low rainfall totals but high *rainfall intensity*, which would require adequate sampling because of the potential of serious flooding.

**Table R-1** Anticipated number of rainfall gauges to calculate the estimated average rainfall. (Shaw, 1983. Reprinted with permission from Van Nostrand Reinhold Co. Ltd.)

| Square miles | Square kilometres (approx.) | Number of rain gauges |
|---|---|---|
| 10 | 26 | 2 |
| 100 | 260 | 6 |
| 500 | 1300 | 12 |
| 1000 | 2600 | 15 |
| 2000 | 5200 | 20 |
| 3000 | 7600 | 24 |

Cf. *Rainfall frequency*.

**Rainfall distribution coefficient** A quantity derived by dividing the maximum *rainfall* received at a specified location within a storm area by the mean rainfall amount within the *drainage area* or basin.

**Rainfall duration** A fundamental observation in the characteristics of *rainfall*, recorded in units such as minutes, hours, and days. The greater the rainfall duration, the greater the rainfall amount and the lower the *rainfall intensity* for a given total volume of rain. As shown in **Figure R-6**, the critical duration of rainfall for the estimation of a *design flood* is the plot of *peak discharge* versus rainfall duration.

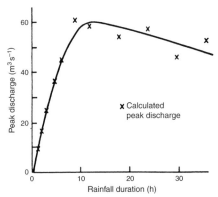

**Figure R-6** Plot of peak discharge versus duration for design rainfall. (Maidment, 1993. Reprinted with permission from McGraw-Hill, Inc.)

Cf. *Flood prediction*.

**Rainfall erodibility factor (R)** An evaluation of the potential for each storm to effect soil *erosion*, which begins when *raindrops* impact the ground surface and dislodge soil particles. The rainfall erodibility factor, $R$, for each storm is summed over time increments (hours) according to the following equation:

$$R = 0.01 \sum EI$$

where:
$E$ = kinetic energy [E•L$^{-2}$; typically reported as MJ ha$^{-1}$ or mega-Joules per hectare]
$I$ = *rainfall intensity* [L•T$^{-1}$; typically reported in mm h$^{-1}$].

The kinetic energy, $E$, is calculated by the following equation:

$$E = (916 + 331 \log I)$$

The average annual rainfall erosion index for the United States can be determined from **Figure R-7**.

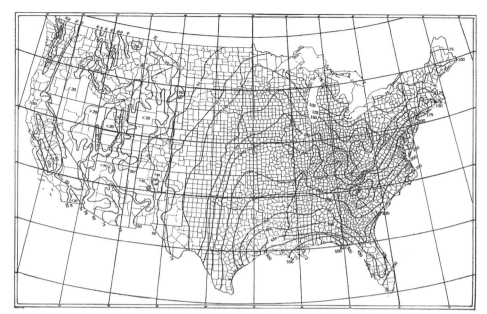

**Figure R-7** Determination of the average annual rainfall erosion index for the United States (Maidment, 1993. Reprinted with permission from McGraw-Hill, Inc.)

**Rainfall excess** Also called effective rainfall. The volume of *rainfall* that is available for *direct runoff* when the *rainfall intensity* exceeds the rate at which water can enter the soil. Cf. *Horton overland flow.*

**Rainfall frequency** The measurement of the incidence of *rainfall*. In the analysis of rainfall and *precipitation*, which includes snowfall, daily *rain gauges* allow for the calculation of rainfall totals without information as to the time of occurrence. Different climatic regimes experience significant differences in seasonal rainfall quantities. The determination of annual rainfall frequency requires many years of data collection to establish a representative pattern. Cf. *Precipitation gauge.*

**Rainfall intensity** The volume of *rainfall* measured in a specified time or the amount of rainfall divided by the *rainfall duration*, expressed in dimensions of $L\,T^{-1}$. Refer to **Appendix A** for unit conversions. Recorded rainfall is shown in a *hyetograph*, which represents the time pattern of rainfall intensity. Cf. *Rainfall frequency; Rain gauge; Precipitation gauge; Horton model.*

**Rainfall penetration** The depth below the soil surface of *infiltration* from a *rainfall* event.

**Rainfall recorder** A *rain gauge* that collects data on the quantity of *rainfall* and the time period(s) of its occurrence. This data is used to determine *rainfall frequency.*

**Rainwash** A hillslope process consisting of the loosening of surface material by *rainfall* and the subsequent transfer of that material across the surface prior to concentration into a watercourse, i.e., sheet erosion, the movement of material by gravity. Cf. *Overland flow.*

**Rainwater** Water falling as *rain* that has not yet acquired any soluble matter from soil. Cf. *Precipitation.*

**Rainwater train** A series of uniformly spaced waves in *runoff* that occurs as part of *overland flow*. A rainwater train is typically associated with heavy *rainfall.*

**Ranney collector** See *Collector well.*

**Ranney Well** See *Collector well.*

**Raoult's law** For oil–water partitioning, contaminant transport in *unsaturated flow* states that the aqueous-phase concentration is equal to the aqueous-phase *solubility* of the constituent in equilibrium with the pure constituent phase, multiplied by the mole fraction of the constituent in the oil phase. This law is represented by the following equation:

$$K_{\mathrm{o}} = \frac{\omega_k \sum_{j=1}^{N} (c_{\mathrm{o}j}/\omega_j)}{S_k \gamma_k}$$

where:

$K_{\mathrm{o}}$ = aqueous-phase concentration [$\mathrm{M}{\bullet}\mathrm{L}^{-3}$]

$k$ = one species out of the $N$ species that make up the oil phase

285

$\omega_j$ = molecular weight of $j$th constituent $[M \bullet mol^{-1}]$
$c_{oj}$ = concentration of the $j$th constituent in the oil phase $[M \bullet L^{-3}]$
$S_k$ = solubility of $k$ in water $[M \bullet L^{-3}]$
$\gamma_k$ = *activity coefficient* of the $k$th species

For ideal solutions, the activity coefficient is 1. As the composition of the oil phase changes, $K_o$ also changes because of *dissolution, volatilization,* and eventual degradation of constituents. Raoult's law also describes the process of volatilization, or phase partitioning between a liquid and a gas, for *solvents* in terms of equilibrium theory. Cf. *Henry's law; Sorption isotherm.*

**Rapid flow** Where the mean velocity is equal to or greater than *Belanger's critical velocity,* a *Froude number* is greater than 1, and a velocity is greater than the celerity of a gravity wave. Cf. *Critical velocity; Tranquil flow.*

**Rapids** A swift, *turbulent flow* or current of water, or the part of a watercourse where the velocity is greater than that of the surrounding water, typically where the surface is obstructed. The break in slope at rapids is less than that defining a *waterfall* and may be a series of small consecutive slope breaks or steps, possibly resulting from a sudden steepening of the *gradient.*

**Rate of infiltration** See *Infiltration rate.*

**Rating curve** Also called a *stage–discharge* curve. A graph of the measured *discharge* of a watercourse and its *stage* at various stage heights used to determine the empirical relationship between the watercourse's stages and discharges (**Figure R-8**). Once a rating curve has been established for a watercourse, the discharge for any measured stage can be estimated from the graph, which is useful because empirical discharge measurement is costly and slow.

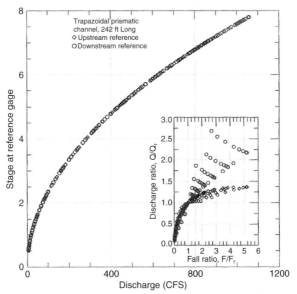

**Figure R-8** Stage–discharge rating curve. (Schmidt, 2001.)

**Rating table** A chart that displays data obtained from a *rating curve.*

**Rational equation/rational formula** An equation that predicts the amount of *runoff,* or peak flow, expected from a given storm. The rational equation multiplies *peak discharge* (rainfall per hour) by the *drainage area* and by a *runoff coefficient,* as follows:

$$Q_p = CIA$$

where:

$Q_p$ = *peak runoff rate* $[L^3 \bullet T^{-1}]$
$C$ = runoff coefficient [dimensionless]
$I$ = average *rainfall intensity* $[L \bullet T^{-1}]$
$A$ = drainage area $[L^2]$

The runoff coefficient, $C$, is dependent on *drainage basin* characteristics. **Table R-2** gives values of $C$ for different land uses with different *infiltration* rates. The rational equation applies to events for which the period of *precipitation* exceeds the time of concentration, which is the time required for the runoff from an impervious *watershed* to equal the precipitation rate.

286

**Table R-2**  The value of C, runoff coefficient, used in the rational equation that predicts the amount of runoff expected from a given storm. (ASCE Manuals and Reports on Engineering Practice no. 37 and WPCF Manual of Practice No. 9, 1969.)

| Description of Area | C |
|---|---|
| Business | |
|    Downtown | 0.70–0.95 |
|    Neighborhood | 0.50–0.70 |
| Residential | |
|    Single-family | 0.30–0.50 |
|    Multiunits, detached | 0.40–0.60 |
|    Multiunits, attached | 0.60–0.75 |
| Residential suburban | 0.25–0.40 |
| Apartment | 0.50–0.70 |
| Industrial | |
|    Light | 0.50–0.80 |
|    Heavy | 0.60–0.90 |
| Parks, cemeteries | 0.10–0.25 |
| Playgrounds | 0.20–0.35 |
| Railroad yard | 0.20–0.35 |
| Unimproved | 0.10–0.30 |
| | |
| *Character of surface* | |
| Pavement | |
|    Asphalt and concrete | 0.70–0.95 |
|    Brick | 0.70–0.85 |
| Roofs | 0.75–0.95 |
| Lawns, sandy soil | |
|    Flat, up to 2% grade | 0.05–0.10 |
|    Average, 2–7% grade | 0.10–0.15 |
|    Steep, over 7% | 0.15–0.20 |
| Lawns, heavy soil | |
|    Flat, up to 2% grade | 0.13–0.17 |
|    Average, 2–7% grade | 0.18–0.22 |
|    Steep, over 7% | 0.25–0.35 |

Cf. *Manning's equation.*

**Rawhiding**  Also called *backwash.* A *well development* procedure of *surging,* alternately lifting and dropping a column of water a significant distance above the pumping water level. The reversal of *flow* through the *well screen* caused by rawhiding agitates the sediment, removes the finer fraction in the sediment, rearranges the remaining formation particles, and breaks down the *bridging* between large particles. Cf. *Air development; Overpumping.*

**Raw water**  Water entering the first stage in a treatment facility or water of a water supply collected from a natural or impounded body of water.

**Rayleigh number**  The ratio of buoyancy forces to viscous resistance. The onset of convention occurs when the Rayleigh number reaches a critical value. The Rayleigh number is defined as:

$$Ra \equiv \frac{g\alpha\Delta T d^3}{\nu\kappa}$$

where:

$g$ = gravitational constant [L•T$^{-2}$]

$\alpha$ = (volume) thermal expansion coefficient [°K$^{-1}$]

$\Delta T$ = change in temperature [°K]

$d$ = the width of the zone being heated [L]

$\nu$ = kinematic viscosity [L$^2$•T$^{-1}$]

$\kappa$ = thermal diffusivity [L$^2$•T$^{-1}$]

L = much greater than the thickness of the zone being heated

Cf. *Buoyant density; Viscosity.*

**Reach**  Uniform *discharge,* depth, area, and slope in a specified length of a *stream,* such as that between two *gauging stations.* A specified section of a restricted watercourse. Cf. *Yazoo stream; Dynamic equation.*

**RCRA**  See *Resource Conservation and Recovery Act (RCRA).*

**Reactive aggregate** A common cause of concrete cracking resulting in significant damage to concrete structures world-wide. Alkali-aggregate reaction is a chemical reaction between certain types of aggregates and hydroxyl ions ($OH^-$) associated with alkalis in the cement. Usually, the alkalis come from the Portland cement but they may also come from other ingredients in the concrete or from the environment. Concrete deterioration caused by alkali-aggregate reaction is generally slow, but progressive. There are two types of alkali-aggregate reaction: alkali-silica reaction and alkali-carbonate reaction.

The alkali-silica reaction is the most common form of alkali aggregate reaction and results from the presence of siliceous aggregates in the concrete, such as some granites, gneisses, volcanic rocks, greywackes, argillites, phyllites, hornfels, tuffs, and siliceous limestones. The product of the alkali-silica reaction is a gel that absorbs water and increases in volume. Pressure generated by the swelling gel ruptures the aggregate particles and causes cracks to extend into the surrounding concrete.

With the alkali-carbonate reaction, certain dolomitic limestone aggregates react with the hydroxyl ions in the cement (or other sources such as de-icing salts) and cause swelling. The swelling of the limestone particles causes the concrete to expand and crack. Cf. *Alkali aggregate; Reaction aggregate; Cement grout.*

**Real-time forecasting** Short-term predictions, the knowledge of *rainfall* and *runoff* from 1 to 24 h in the future. Real-time forecasting is a meteorological concern, for which the information is applied by a hydrologist to model and then predict the subsequent behavior of a basin.

**Reasonable use rule** See *Rule of reasonable use.*

**Recession** The lowering of water levels in surface water bodies or *groundwater* within a specified area. A decline in the system's natural output, e.g., a declining *water table* after a rise resulting from *recharge.* A *groundwater recession curve,* as seen in a stream *hydrograph,* results from the withdrawal of groundwater from storage and denotes the *baseflow* component of the watercourse. Cf. *Recession curve.*

**Recession curve** Also called the falling limb of a *hydrograph.* The decrease in *runoff* rate after rainfall or *precipitation,* as plotted on a hydrograph. **Figure R-9** shows a recession curve and a method of recession analysis. The recession curve can be defined by the following function:

$$q_1 = q_0 K_r$$

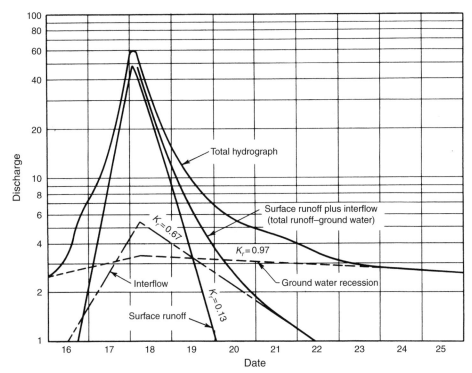

**Figure R-9** Hydrograph showing method of recession analysis. (Linsley et al., 1982. Reprinted with permission from McGraw-Hill, Inc.)

where:
$q_0 = flow$ at any time $[L^3 \cdot T^{-1}]$
$q_1 = $ flow one time unit later $[L^3 \cdot T^{-1}]$
$K_r = $ recession constant less than unity

When plotted on semilogarithmic paper, as in **Figure R-9**, a stream recession slope gradually decreases; that is, the $K_r$ value increases because with time, water will be added from stream channels, surface soil, and *groundwater storage*. The lowest portion of the recession is the characteristic $K_r$ for *groundwater*, assuming that both *interflow* and surface runoff have ceased. Cf. *Direct runoff; Base runoff; Base flow recession; Depletion curve.*

**Recharge** The addition of water to the saturated zone, either naturally (by *rainfall, precipitation,* or *runoff*) or artificially (by *spreading zones* or *injection*). The recharge amount in natural recharge of confined and unconfined aquifers is controlled by the same factors:

- the amount of precipitation available after runoff;
- the vertical *hydraulic conductivity;*
- the *transmissivity* of the aquifer;
- the vertical infiltration through vadose zone.

Recharge to a confined aquifer occurs in areas where the *confining bed* is absent or compromised. Recharge can also occur if the *hydraulic gradient* across a leaky confining bed is in a direction that promotes flow into the aquifer. Areas of recharge and discharge are often defined by the use of a *flownet diagram*. Cf. *Artificial recharge; Aquifer, confined; Aquifer, unconfined; Hydrologic budget; Zone of saturation.*

**Recharge area** The area receiving water from *infiltration* and *percolation*, beneath which the vertical components of *hydraulic head* exist and *groundwater* moves deeper within the *aquifer*. Recharge areas typically have several physiographic features distinguishing them from *discharge areas*; for example, topographically high locations (as compared to discharge areas in low areas) and a thicker *unsaturated zone* between the land surface and the *water table*. As shown in **Figure R-10**, *flow lines* typically diverge from recharge areas and converge in discharge areas.

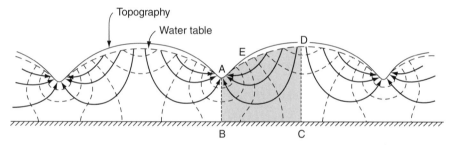

**Figure R-10** Flow lines diverge from recharge areas, such as mountain tips, and converge at discharge areas, such as valleys or rivers. Dotted lines AB and CD are the imaginary boundaries or groundwater divides. Region AE is a discharge area and ED is an area of recharge. (Freeze and Cherry, 1979. Reprinted with permission from Prentice-Hall, Inc.)

**R**

Cf. *Catchment area.*
**Recharge basin** Also called a recharge pit or pond. A depression constructed to replenish *groundwater* supply. Surface water is captured in artificial or modified naturally occurring basins during spring *runoff* and slowly *recharges* the *groundwater basin* in the subsequent days and months. The recharge basin method is shown in **Figure R-11**. Cf. *Artificial recharge; Spreading zone.*
**Recharge boundary** The limit of the extent of an *aquifer* system that contributes water to the system when the system is pumped, when a water body recharges a pumped aquifer. The recharge boundary is a type of *hydrogeologic boundary* used to simulate *groundwater flow* to a *well* not located in an aquifer of infinite areal extent. **Figure R-12** displays a "real" system recharge boundary versus its vertical counterpart. The region where the aquifer is replenished, such as a watercourse within the *aquifer basin*, is at a recharge boundary.
**Recharge pit** See *Recharge basin.*
**Recharge rate (W)** The net *groundwater* recharge divided by the time interval. Recharge rate, W, is given by the following equation:

$$W = \frac{G}{\Delta t}$$

where:

$G$ = groundwater recharge volume or deep *percolation* volume [L³]

$\Delta t$ = time interval [T]

The application of the *water balance* equation over an extended time interval in which the water storage change is negligible, i.e., 1 year or more, allows for the calculation of the groundwater recharge rate.

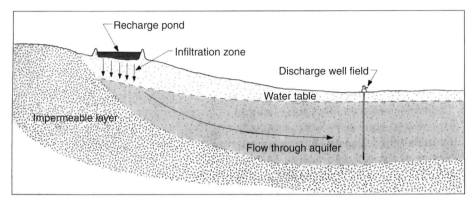

**Figure R-11** Recharge of an unconfined aquifer from a recharge basin. (Driscoll, 1986. Reprinted by permission of Johnson Screens/a Weatherford Company.)

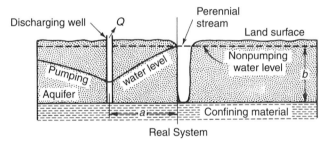

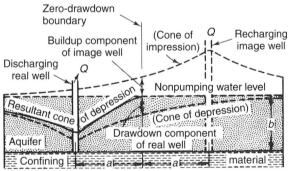

**Figure R-12** A real versus idealized aquifer bounded by a recharge source. (USGS, 1962.)

**Recharge well** The injection of water into one or more *aquifers*. Recharge wells are used to either slow or stop *overdraft* of the groundwater system. **Figure R-13** illustrates the difference between *artificial recharge* via basins and wells. Both methods increase the *infiltration* of water from the land surface to the groundwater system.

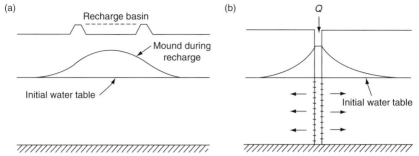

**Figure R-13** Examples of expected effects on the initial water table with recharge through basins (a) and wells (b). (Domeninco, 1990. Reprinted with permission from John Wiley & Sons, Inc.)

Cf. *Injection well; Barrier well; Saltwater intrusion.*
**Reclaimed water** Treated wastewater that is recycled for irrigation, *recharge*, or other reuse.
**Reconsequent** See *Resequent stream.*
**Recovery** A rise in water level in a *well* or *borehole* that occurs once *discharge* from within the *area of influence* ceases and the well and *aquifer* water levels rise toward prepumping levels. **Figure R-14** shows *drawdowns* during the pumping period and *residual drawdown* remaining during the recovery period.

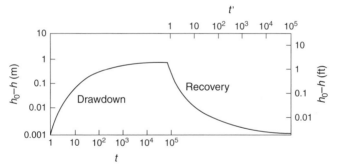

**Figure R-14** Diagram of expected recovery in hydraulic head after pumping has ceased. (Freeze and Cherry, 1979. Reprinted with permission from Prentice-Hall, Inc.)

Cf. *Aquifer test; Recovery test.*
**Recovery curve** See *Recovery; Recovery test; Residual drawdown graph.*
**Recovery test** Also called a build-up test. The measurement of the rate of *fluid pressure*, or the return to *static water level*, in a well or *borehole* after a specified time of either pumping or injection. A recovery test is an important part of an *aquifer test*, and the results are used to calculate *transmissivity* and storage from the slope of the recovery curve. **Figure R-15** is a graph of a typical *drawdown* and *recovery* plot for a well pumped at a constant rate. A semilogarithmic plot of *residual drawdown* versus pumping duration and time since pumping ceased is used in the *Jacob equation*. Recovery test data are most reliable if obtained from an *observation well* located close enough to the pumped well to demonstrate significant changes in drawdown during pumping.
**Recovery well** A *well* installed for the specific purpose of intercepting and extracting contaminants or contaminated water from *groundwater*. When it is part of a *pump-and-treat system* of wells, a recovery well may also be called an *extraction well*. A recovery well differs from a production well in the placement of the *screened interval*. The *well screen* in a production well intercepts the most productive zones within the *aquifer*, whereas a recovery well is designed to intercept the zone that contains the contaminant(s). The hydrologic design of the recovery well must be considered along with the physical and chemical properties of the constituents to be recovered. Cf. *Monitoring well.*
**Rectangular drainage pattern** See *drainage pattern.*

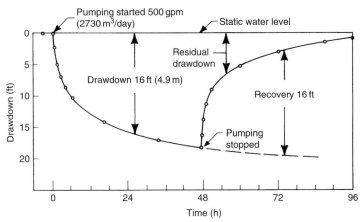

**Figure R-15** Graph of a typical drawdown and recovery plot at constant rate of pumping. (Driscoll, 1986. Reprinted by permission of Johnson Screens/a Weatherford Company.)

**Recurrence interval ($T_r$)** Also called the return period or frequency of occurrence. The average time between hydrologic events equal to or greater than a certain magnitude. The recurrence interval is the average interval in years within which a given event will be equaled or exceeded, and is the reciprocal of frequency. Recurrence interval is a way of expressing the exceedance probability of a specified event. To describe a *design flood*, the recurrence interval, $T_r$, can be substituted for probability, $p$. As seen in the following equation, $T_r$ is the reciprocal of $p$:

$$p = \frac{1}{T_r}$$

or

$$T_r = \frac{1}{p} = \frac{(n+1)}{m}$$

where:
$n$ = number of years
$m$ = rank of event

The estimation of *peak discharge, Q,* in *flood frequency analysis* is likely to be equaled or exceeded on average recurrence interval of $T$ years. The intervals may vary considerably around the average value of $T$; a record may indicate a peak discharge of a 25-year event, $Q_{25}$, occurring over much longer or shorter intervals, or even in successive years. When plotting a series of *peak flows* as a *cumulative frequency curve,* a probability or recurrence interval is associated with each peak. Cf. *Annual flood series; Flood frequency analysis; Gringorten formula; Partial-duration series; Bankfull discharge.*

**Redox potential** Also known as oxidation-reduction potential (ORP). The voltage difference at an inert electrode immersed in a reversible oxidation-reduction system; it is the measurement of the state of oxidation of the system. See *Eh.* Cf. *Nerst equation; Oxidation potential; Reduction potential.*

**Reduction potential** The potential drop involved in the reduction of a positively charged ion to a neutral form or to a less highly charged ion; or of a neutral atom to a negatively charged ion. See *Eh.*

**Redox reactions** See *Oxidation-reduction reactions.*

**Reduction factor** The coefficient used in the *Forchheimer* equation to determine the *discharge* from a well that does not penetrate the entire thickness of an unconfined *aquifer.* This reduction factor equation is as follows:

$$Q_P = \pi K \frac{h^2(R) - [h(r) - t]^2}{\ln\left(\frac{R}{r}\right)} \alpha$$

For equation parameters, refer to the *Dupuit–Forchheimer equation.* The reduction factor, $\alpha$, is expressed as:

$$\alpha = \left(\frac{L}{H}\right)^{1/2} \left(2 - \frac{L}{H}\right)^{1/4}$$

where:
$H$ = saturation thickness before discharge [L]
$L$ = distance from bottom of *well screen* to *piezometric level* prior to pumping [L]

For a *fully penetrating well*, $L = H$, and the reduction factor equation is equivalent to the Dupuit–Forchheimer equation for unconfined aquifer discharge. Cf. *Thiem equation*.

**Regime canal** A non-scouring and non-silting irrigation channel shown to be analogous to a *graded stream* or to a watercourse in *quasi-equilibrium*.

**Regimen** The distribution of *runoff* over time. A regimen is a method of characterizing runoff to aid in the management or prediction of *streamflow*.

**Regional flow system** A method of dividing a groundwater study area according to *flownet* analysis, as shown in **Figure R-16**.

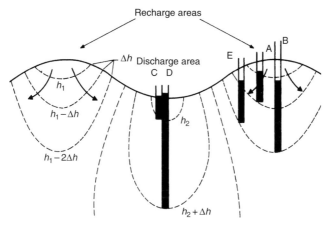

**Figure R-16** Groundwater flow system conceptional model. (Maidment, 1993. Reprinted with permission from McGraw-Hill, Inc.)

Cf. *Intermediate flow system*.

**Regional frequency analysis** See *Flood frequency analysis*; *Flow frequency*; *Flood duration curve*; *Gringorten formula*.

**Regrading stream** A contained watercourse that is aggrading (building up) certain *reaches* or specified parts of its profile while degrading (downcutting or eroding) other reaches or parts. Cf. *Aggrading stream*; *Degrading stream*; *Stream load*.

**Regulation** The manipulation of *flow* in a watercourse by artificial means.

**Rehabilitation of wells** See *Well maintenance*.

**Rejuvenated stream** Also called a revived *stream*. A watercourse that is revived, typically by uplift, after reaching maturity, thereby renewing its former erosive capabilities. Cf. *Stream maturity*.

**Rejuvenated water** Formerly isolated water that is returned to the terrestrial water supply as a result of compaction and metamorphism.

**Relative humidity (RH or $H_r$)** The actual volume of water vapor in a given volume of air divided by the amount that would be present if the air were saturated at the same temperature. Relative humidity, RH or $H_r$, is expressed as a percentage and is calculated by the following equation:

$$RH = 100\frac{e}{e_s}$$

where:

$e =$ *vapor pressure* $[M{\bullet}L^{-2}{\bullet}T]$
$e_s =$ saturation vapor pressure $[M{\bullet}L^{-2}{\bullet}T]$

Relative humidity can also be approximated from the air temperature and *dew point* using the following equation:

$$H_r \approx 100\left(\frac{112 - 0.1T + T_d}{112 + 0.9T}\right)^8$$

where:

$T =$ temperature (°C)
$T_d =$ dew point (°C)

Cf. *Absolute humidity*; *Humidity*.

**Relative permeability ($k_{rw}$)** The ratio of the *permeability* at a given water content to its saturated value. Fluids in a multifluid system have a tendency to interfere with one another during flow. Relative permeability describes the relation between *effective* permeability and the permeability at 100% saturation, i.e., the ratio of a fluid at partial saturation and permeability at 100%, which ranges from 0 at a low saturation to 1.0 at 100% saturation.

293

**Relative water content** See *Liquidity index*.

**Relict lake** A remnant water body persisting after the extinction of a larger body of water or a body of water that results after uplift and separation from the sea.

**Relief–length ratio** A mathematical comparison of the difference in elevation within the area in which a given watercourse is contained and the distance between two points of interest along that watercourse. The reciprocal of the relief–length ratio indicates the concavity of a watercourse's *longitudinal profile*. **Figure R-17** shows a typical non-dimensional diagram of the relief, length, and the resulting concavity of a watercourse. The usual concavity of the longitudinal profile results from the *downstream* increase in *discharge*. The relief–length ratio also influences the movement of sediments within a *drainage basin*. **Figure R-18** illustrates the ratio's relationship to the mean annual sediment yield in a typical basin in the western United States.

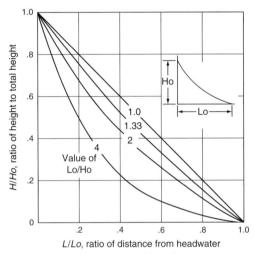

**Figure R-17** Profile of the concavity of a watercourse. (Leopold et al., 1992. Reprinted with permission from Dover Publications, Inc.)

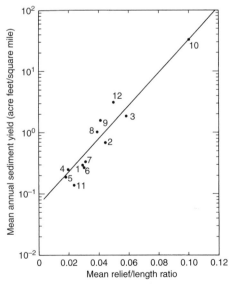

**Figure R-18** The relief–length ratio's influence on the mean annual sediment yield. (Schumm, 1977. Reprinted with permission from John Wiley & Sons, Inc.)

**Relief well** Also called an absorbing well or drainage well. A well installed to relieve excess *hydrostatic pressure*. Relief wells may be employed in *dewatering* soils or, when situated on the landward side of dams and levees, in preventing *blow-outs* during high water. Cf. *Pressure-relief well system for dams and levees.*

**Renewed consequent stream** See *resequent stream.*

**Replenishment** The addition of water to a surface water body, *groundwater reservoir*, or basin. *Precipitation* supplies the continuous replenishment that is required to maintain the *hydraulic head* of a water body in a confined system that extends downdip from a *water table* in an outcrop area. In an unconfined *aquifer*, the water table is a subdued expression of the overlying topography, providing the hydraulic head, creating *flow*. Cf. *Recharge; Depletion; Depletion curve.*

**Representative elementary volume** The smallest volume of geologic media that is considered representative of the entire domain being studied. The size of the representative elementary volume (REV) is related to how locally correlated the property is on the microscopic scale. The REV is considered to be in the vicinity of 100–1000 grain diameters, large enough for averages to be statistically meaningful but small enough to avoid *heterogeneity.*

**Resequent stream** Also called a reconsequent stream or renewed reconsequent stream. A *tributary* of a *subsequent stream* that flows downdip of the underlying formation, and in the same direction as the original *consequent stream* but at a lower level. Cf. *Obsequent stream.*

**Reservoir** In general, reservoir is any body, natural or man-made, where fluids accumulate or are stored. This may be a lake, pond, *detention* or *retention basin*, or *recharge basin*. It is a term used in oil exploration and production applications, and surface water hydrology, but rarely in groundwater hydrology, since in groundwater hydrology flow is mainly through interconnected pore spaces. In groundwater hydrology, *karst* features or large fractures may provide for groundwater reservoirs.

**Residence time** The amount of time that water resides within its source area, *aquifer*, watercourse, or water body before moving to another phase of the *hydrologic cycle*. Residence times vary depending on location and climatic conditions. The residence time for a watercourse such as a mountain *stream* or a *river* could be a few days, whereas the residence time for isolated water bodies may be decades. The residence time for storage would be equal to the total amount of stored water divided by the amount of *flow* into or out of *storage*, because long-term storage is neither increasing nor decreasing. Cf. *Water balance; Native water; Connate water.*

**Residual drawdown ($\Delta s'$)** The difference between the original *static water level* or prepumping water level, $h_0$, and the depth to water at any given instant during the *recovery* period after pumping is discontinued, $h'$. The residual drawdown non-equilibrium equation is reduced to the following form:

$$h_0 - h' = \Delta s' = \frac{2.3Q}{4\pi T}\log\frac{t}{t'}$$

where:

$\Delta s'$ = residual drawdown

$Q$ = *discharge* $[L^3 \bullet T^{-1}]$

$T$ = *transmissivity* $[L^2 \bullet T^{-1}]$

$t$ = time since pumping started $[T]$

$t'$ = time since pumping stopped $[T]$

In an *aquifer test*, this represents the difference between the original static water level in an observation well and the level at a given instant during the recovery period.

Cf. *Recovery test; Residual-drawdown graph.*

**Residual drawdown graph** A plot of the *recovery* period during an *aquifer test*, which is an inverted image of a *drawdown curve*, with the shape of the curve determined by the physical characteristics of the *aquifer*. **Figure R-14** shows the recovery plot representing the *residual drawdown* in the well during the recovery period. The differences between the original *static water level* and the depth to water at given instants during recovery are the points on the recovery curve.

**Residual saturation** Fluid held by capillary forces in the narrowest section of the soil pore space, typically where water is the wetting fluid and air is the non-wetting fluid (**Figure R-19**). Because a fluid at residual saturation is not connected across the network of pores, the fluid is not capable of *flow*. Under conditions of residual saturation, NAPL contaminant migration becomes discontinuous and can be immobilized by capillary forces.

**Resistance** See *Hydraulic resistance.*

**Resistance factor** See *Darcy – Weisbach resistance coefficient.*

**Resistivity logging** See *Electric logging.*

**Resistivity survey** The electrical resistance measured in ohms, which varies with the composition of the strata and their water content. The apparent resistivity is measured by passing an electric current into the ground via current electrodes and recording the drop in potential across a pair of electrodes, as shown in **Figure R-20**. *Electrical resistivity* decreases with increasing water content in porous rocks. A resistivity survey is a relatively cheap and reliable method of determining the depth to the *water table*, or the depth of bedrock material.

**Resource Conservation and Recovery Act (RCRA)** A management system enacted by the United States, in 1976, to regulate the monitoring, investigation, and corrective actions at all hazardous-materials treatment, storage, and disposal facilities. Enacted as groundwater protection, the Resource Conservation and Recovery Act (RCRA) has provisions for the management of solid hazardous waste from the time it is produced to its ultimate disposal; these RCRA provisions are often called the "cradle-to-grave" concept. Cf. *Comprehensive Environmental Response; Compensation and Liability Act.*

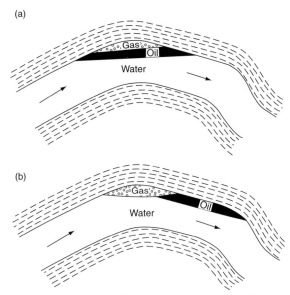

**Figure R-19** Example of how a liquid not capable of flow can become trapped, a residual saturation. (Domenico, 1990. Reprinted with permission from Wiley & Sons Inc.)

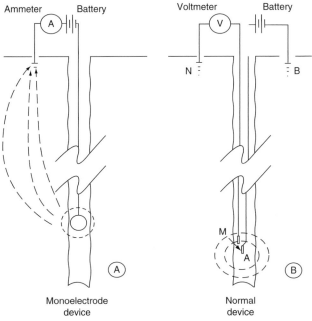

**Figure R-20** Example of an early model, monoelectrode device to the current electrical resistivity recording device where an electric current is fed from the surface to the downhole electrode. (Welenco, 1996. Reprinted with permission from Welenco Inc.)

**Resurgence** Cf. *Debouchment; Spring, rising.*

**Resurgent** Derived from sources on the earth's surface, in the atmosphere, or from the country rock next to a magma body (as in magmatic or juvenile water or gases). Cf. *Juvenile water.*

**Resurgent vapor** *Vapor* or gases from *groundwater* that have volatilized from contact with hot rock or magmatic gas resulting from dissolved or assimilated country rock. Cf. *Juvenile water.*

**Retained water** The water that is isolated in interstices or held by molecular attraction, thereby remaining in the soil or rock after *gravity drainage.*

**Retardation** The reaction on an absorbed solute front of a water body when the front of the reactive *solute* spreads out but travels behind the front of a non-reactive solute. **Figure R-21** is a concentration profile depicting the retarded species. Understanding solute retardation is critical in the investigation of contaminants in *groundwater.*

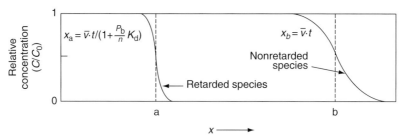

**Figure R-21** Concentration profile depicting the retarded species. (Freeze and Cherry, 1979. Reprinted with permission from Prentice-Hall, Inc.)

Cf. *Retardation equation; Retardation factor.*

**Retardation equation** A formula used to calculate the *retardation* of a solute front relative to the bulk mass of water. When the partitioning of a contaminant is well described by the *distribution coefficient,* the retardation equation is:

$$\frac{v}{v_c} = 1 + \frac{\rho_b}{\eta} K_d$$

where:
$v$ = average linear velocity of *groundwater*
$v_c$ = velocity of the relative concentration, $C/C_0$ $[L \bullet T^{-1}]$
$\rho_b$ = bulk mass density $[M \bullet L^{-3}]$
$n$ = *porosity* [dimensionless]
$K_d$ = distribution coefficient describing the partitioning of the adsorbed species $[L^2 \bullet M^{-1}]$

This equation has been used to determine the retardation of the center of mass of a contaminant moving from a point source while undergoing *adsorption.* With minimal information, the retardation equation can begin to explain the distribution of constituents and potentially predict future dispersal in the subsurface. Cf. *Retardation factor; Plume.*

**Retardation factor ($R_f$)** A quantification of the inhibition or delay of solute movement that is due to particular soil types or chemical processes. Numerically, the retardation factor, $R_f$, is described by the following equation:

$$R_f = 1 + \frac{\rho_s K_d (1 - n)}{n}$$

where:
$\xi \rho_s$ = density of the solid $[M \bullet L^{-1}]$
$K_d$ = *distribution coefficient* $[L^2 \bullet M^{-1}]$
$n$ = *porosity* [dimensionless]

The reciprocal of $R_f$ is the relative velocity, $v_c/v$. The results are used to interpret the *breakthrough curves* for chemicals and to quantify the total *sorption* process. Some chemical compounds have greater mobility than that predicted by hydrophobic sorption theory. Although a compound may be strongly sorbed on solid organic carbon, transport of the compound can be facilitated by cosolution with dissolved organic compounds moving at groundwater velocities. This process is explained by a modified *retardation equation,* describing $R_f$ as:

$$R_f = 1 + \frac{K_{oc} f_{oc} \rho_s [(1 - n)/n]}{1 + K_{DOC} DOC \times 10^{-6}}$$

where:
$K_{OC}$ = *octanol–carbon partition coefficient*
$f_{OC}$ = fraction of the total organic carbon content (grams of organic carbon per gram of soil)

$\rho_s$ = density of the solid [M•L$^{-1}$]

$n$ = porosity [dimensionless]

DOC = dissolved organic carbon content [M•L$^{-3}$]

$K_{DOC}$ = distribution coefficient describing the partitioning of the adsorbed species [L$^2$•M$^{-1}$]

Cf. *Adsorption; Adsorption isotherms; Distribution coefficient.*

**Retention** The part of *precipitation* in a *drainage basin* that does not become part of the surface *runoff* within a specified time frame. Retention is the difference between the total precipitation and the total runoff, and represents *evaporation, transpiration,* and *infiltration* for a specified period. Cf. *Specific retention.*

**Retention basin** Used to manage stormwater runoff to prevent flooding and downstream erosion and improve water quality in an adjacent river, stream, lake or bay. It is essentially an artificial lake with vegetation around the perimeter and includes a permanent pool of water in its design (also known as a wet pond or wet detention basin). It is distinguished from a *detention basin*, sometimes called a dry pond, which temporarily stores water after a storm, but eventually empties out at a controlled rate to a downstream water body. It also differs from an *infiltration basin* or *recharge basin*, which is designed to direct stormwater to groundwater through permeable soils. Retention basins are frequently used for water quality improvement, groundwater recharge, flood protection, aesthetic improvement or any combination of these. Sometimes they act as a replacement for the natural absorption of a forest or other natural process that was lost when an area is developed. As such, these structures are designed to blend into neighborhoods and viewed as an amenity. Cf. *Detention basin; Recharge basin.*

**Retention capacity** The maximum *retention* attainable under given conditions.

**Retention storage** The capillary *pore water* in soil that is held by gravity and is available to plants. Water is held in retention storage at high tensions, whereas water is held in *detention storage* at nearly atmospheric tensions. Cf. *Field capacity; Capillary water.*

**Retention time** The duration for which water or wastewater is held at a given facility or *catchment area.*

**Retention, surface** See *Adsorption; Detention storage; Depression storage; Detention basin; Retention storage.*

**Return flow** Also called return water. A component of *throughflow* that drains across the land surface before reaching a stream. Prior to entering the watercourse, this water appears as *overland flow*. Cf. *Interflow.*

**Return period** See *Recurrence interval.*

**Return water** See *Return flow.*

**Reverse-circulation rotary drilling** See: **Appendix C – Drilling Methods**.

**Reverse osmosis** A water treatment process utilizing *pressure* to force water through a semipermeable membrane for purification. *Osmosis* is a spontaneous process in which the *flow* of water across a membrane is normally toward areas of higher *solute* concentration. **Figure R-22a** and **b** illustrates the differences between solutions of differing concentrations flows until *osmotic pressure* on the dilute-solution side and the liquid head on the concentrated-solution side become balanced, i.e., osmosis and reverse osmosis. The membrane design maximizes the surface area relative to the volume to minimize early fouling by colloidal material.

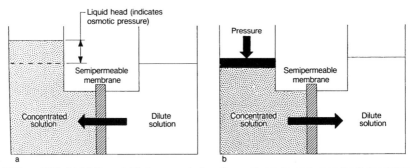

**Figure R-22a and 22b** (a) Solutions of differing concentrations flow until osmotic pressure on the dilute solution and the liquid head on the concentrated solution side balances. (b) The method of reversed osmosis by applying pressure to the high-concentration solution. (Driscoll, 1986. Reprinted by permission of Johnson Screens/a Weatherford Company.)

Cf. *Desalination; Point-of-use water treatment systems.*

**Reverse rotary drilling** See: **Appendix C – Drilling Methods**.

**Reverse-type curve** See *Theis non-equilibrium-type curve.*

**Reversed consequent stream** A *consequent stream* flowing contrary to the underlying geologic formation, i.e., part of a captured consequent stream. Cf. *Obsequent stream; Resequent stream.*

**Reversed stream** A watercourse for which the *flow* pattern has been reversed as the result of glacial action, landslides, regional or local uplift, or *stream capture.*

**Revived stream** See *Rejuvenated stream*.

**Reynolds' critical velocity** See *Critical velocity*; *Reynolds number*; *Belager's critical velocity*; *Kennedy's critical velocity*; *Unwin's critical velocity*.

**Reynolds number (R)** Also called Reynolds critical velocity. A dimensionless ratio relating four factors that determine whether fluid *flow* will be linear (*laminar flow*) or non-linear (*turbulent flow*). The Reynolds number is applicable to flow in a closed system where the *free surface* is not considered. The equation for the Reynolds number, *R*, expresses the ratio of inertial to viscous forces during flow as:

$$R = \frac{\rho v d}{\mu}$$

where:

$R$ = Reynolds number [dimensionless]
$\rho$ = *fluid density* $[M \bullet L^{-3}]$
$v$ = *discharge velocity* $[L \bullet T^{-1}]$
$d$ = diameter through which fluid moves [L]
$\mu$ = *fluid viscosity* $[M \bullet T^{-1} \bullet L]$

The value of *d* is a representative length dimension for the *porous medium*, often the average grain diameter, mean pore dimension, or some function of the square root of its *permeability*. In *groundwater flow*, the Reynolds number has been used to validate the Darcy equation; *Darcy's law* has been shown by experimentation to be valid only when the resistive forces of *viscosity* predominate, which occurs when *R* is less than 10, which corresponds to velocities commonly observed in most *groundwater* systems. For laminar *open-channel flow*, *R* is less than about 500. Cf. *Darcy–Weisback formula*; *Moody diagram*.

**Rhombohedral packing** A configuration in which the equidimensional spheres or sand grains in one layer align in the hollows formed by four adjacent spheres or grains of the lower layer of spheres or grains. This results in a rhomboid, as shown in **Figure R-23**. In the determination of *porosity*, the porosity value associated with rhombohedral packing is 25.95% as compared to 47.65% with cubic packing.

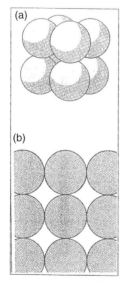
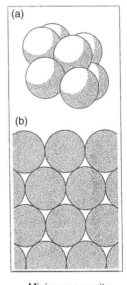

Maximum porosity          Minimum porosity

**Figure R-23** (a) Three dimensional view and (b) cross-sectional view of spheres illustrating a maximum porosity of 47.6% with cubic packing and minimum porosity of 25.9% with rhombohedral packing. (Welenco, 1996.)

Cf. *Isotropic*; *Anisotropic*.

**Richard's equation** A parabolic, non-linear equation describing water *flow* in the *unsaturated zone*. The Richard's equation is the *pressure head*-based equation for transient flow through unsaturated *porous media*:

$$\frac{\partial}{\partial x}\left[K(\psi)\frac{\partial \psi}{\partial x}\right] + \frac{\partial}{\partial y}\left[K(\psi)\frac{\partial \psi}{\partial y}\right] + \frac{\partial}{\partial z}\left[K(\psi)\left(\frac{\partial \psi}{\partial z}+1\right)\right] = C(\psi)\frac{\partial \psi}{\partial t}$$

299

where:

$x, y, z$ = Cartesian coordinates
$K$ = hydraulic conductivity [L•T]
$\psi$ = pressure head [L]
$C$ = specific moisture capacity [unitless]
$t$ = time [T]

**Rift lake** A water body formed in a rift valley, typically elongated in shape. Cf. *Sag pond*.

**Right bank** The stream boundary on the right-hand side of an observer who is facing *downstream*. Cf. *Bank*.

**Rill** A very small stream of water. The term "rills" typically refers to the first channels formed by *runoff*, which carry water only during storms. Newly formed rills initially develop in a parallel pattern in the predetermined slope without materially altering the natural slope. These "shoestring" rills can become part of a drainage net by micropiracy (the capture of small drainage channels by a larger channel). The ridges between initially separated adjacent rills break down and the rills join as part of the *drainage pattern's* evolution. It has been demonstrated mathematically that the *drainage density* of a rill, $D$, in dimensions of $L•L^{-1}$, increases with slope, $S$, expressed as a percentage, according to the following equation:

$$D = 0.909 + 22.42S$$

**Rill flow** The movement of water from surface *runoff* in very small, irregular channels. Cf. *Rill*.

**Ring infiltrometer** A metal ring of approximately 12-in. (0.3 m) diameter or greater that is inserted in soil and flooded with more than 2 cm of water. The rate of *infiltration* is determined from the time taken by the water in the ring infiltrometer to seep into the soil and is calculated in centimeters (inches) per hour. Also information on sorptivity and steady-state infiltration rate can be obtained. Cf. *Fixedring infiltrometer*.

**Ripa** Legal term for a *bank* of a stream or a *river*.

**Riparial** See *riparian*.

**Riparian** Also riparial. Legal term for existing on the sides or *banks* of a watercourse or body of water. Cf. *Riverain*.

**Riparian rights/riparian doctrine** A United States doctrine ensuring the right of land owners to the reasonable use of water in a water body or watercourse abutting their land, with respect to other riparians. An owner of property adjacent to a surface water body has the first right to use and withdraw the water, and water is to be used for natural purposes; each user of the water is to return the *flow* to the original channel unimpaired in quality. When applied to *groundwater*, the riparian doctrine allows the owner of land above a *groundwater resource* to affect the quantity available for other uses; the owner is entitled to reasonable use of the water, which is interpreted on a case-by-case basis. Cf. *Prior appropriation right; Rule of capture; Rule of reasonable use*.

**Riparian water loss** The *evapotranspiration* of water along a watercourse via vegetation on the *banks*.

**Riser pipe** A section of pipe that is attached to the top of a *well screen* to provide extra safety during well installation, to ensure maximum utilization of the screen intake area, and to prevent the loss of the screen below the *well casing* if the casing is unintentionally pulled back too quickly beyond the screen intake area. **Figure R-24** shows the basic operations in setting a well screen in a *pull-back method of screen installation* with a riser pipe.

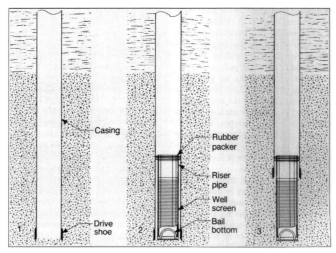

**Figure R-24** Basic operations in setting a well screen by the pull-back method. (1) Driving, bailing, or lowering (rotary method) the casing to full depth of the well, (2) lowering the screen inside the casing, and (3) pulling the casing back to expose the screen to the aquifer. (Driscoll, 1986. Reprinted by permission of Johnson Screens/a Weatherford Company.)

**R**

**Rising-head test** An in situ method of estimating the *hydraulic conductivity* of an *aquifer* or *confining bed* by removing a slug of water from a *well casing* to cause an instantaneous drop in the water level. The rate of the *recovery* to the initial or *static water level* is correlated with hydraulic conductivity, i.e., a faster return to the initial level indicates higher hydraulic conductivity. The rising-head test is one of two types of *slug test*, the other being a *falling-head test*. Cf. *Aquifer test*; *Hvorslev method*.

**Rising limb of a hydrograph** The portion of a graphical plot of *discharge* that shows an increase from the beginning of a hydrologic event, e.g., a storm, to its crest or peak. The character of the event causing this increasing discharge determines the shape of the rising limb of the *hydrograph*.

**Rithron** The portion of a watercourse that displays high velocities and *turbulent flow* and that is shallow and relatively cold.

**River** A permanent or seasonal natural *freshwater* surface watercourse or *stream*, typically characterized by a large *discharge*, which is confined within a channel and *flows* toward another watercourse, water body, or the sea. **Appendix-B** gives *water quality* parameters used to describe rivers and streams. Cf. *Alluvial River*; *Channel shape*; *River Channel pattern*.

**River basin** The total areal extent of the topography that is drained by a given *river* and its *tributaries*. Cf. *Drainage basin*.

**River breathing** The observable water-level fluctuation in a *river*.

**River capture** See *beheaded stream; diversion; stream capture*.

**River channel pattern** The size and shape of the surface area containing a watercourse, which is a function of the watercourse *flow*. Increasing *discharge* alters the size of a *stream* or a *river*, but the shape of the channel tends to remain the same as the *banks* give way and the central portion enlarges. **Figure R-25** shows several river channel patterns with their respective relationships to sediment supply and texture, channel *gradient*, and stability.

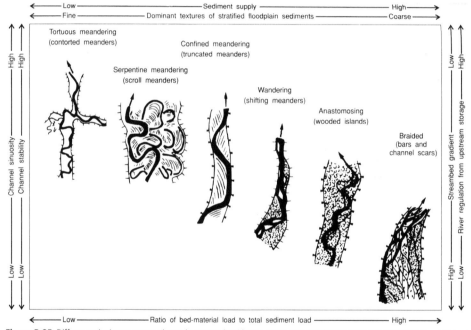

**Figure R-25** Differences in river or stream channel patterns. (Maidment, 1993. Reprinted with permission from McGraw-Hill, Inc.)

Cf. *Stream roughness; Stream channel shape*.

**River gauge** An instrument that measures and records the height of a lake or the streamflow of a river over time. The US Geological Survey (USGS) operates more than 7,000 gages across the United States. The instrument is housed in a shelter, referred to as a gauging station. The gauging station can be located over the water, on shore, or attached to a bridge over a large river. A gauging station may also include equipment that measures water-quality conditions such as temperature, pH, dissolved oxygen, and dissolved chemicals, or weather conditions such as air temperature, precipitation, and wind speed. Cf. *Stream gauge; Precipitation gauge; Gauging station*.

**River gauging** The measurement of a watercourse, whether a small *stream* or a larger *river*, by continuously monitoring one variable such as water level, which is related to *discharge* calculated from other variables such as velocity and depth.

River gauging should be conducted in a short stretch of channel where the discharge variation will not continuously modify the cross-sectional area, and in a site with a stable bed profile. River gauging is analogous to rain gauging. Cf. *Rain gauge; Stream gauge; Precipitation gauge.*

**River piracy** See *beheaded stream; diversion; stream capture.*

**River profile** See *longitudinal profile.*

**River regime** *Flow* records for a natural watercourse that cover 20–30 years and provide a representative distinctive seasonal pattern. A river regime can be estimated from the climate of the region because the main climatic features of temperature and *rainfall* directly affect the catchment *runoff*. Simple river regimes are represented by plots of monthly mean *discharges* from January through December, represented as proportions of the mean of the 12 monthly values.

**River shed** The *drainage basin* of a river or a watercourse. Cf. *River basin; River channel pattern.*

**River stage** The elevation of the surface of a watercourse at a station above an arbitrary zero datum, which is typically either mean *sea level* or (more commonly) a point slightly below the point of zero *flow*. **Figure R-26** shows a typical method of measuring river stage with a staff gauge.

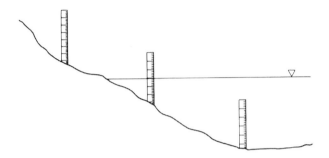

**Figure R-26** Illustration on a typical method of measuring river stage with a sectional staff gauge. (Linsley et al., 1982. Reprinted with permission from McGraw-Hill, Inc.)

Cf. *River gauging.*

**River system** Also called a water system. The watercourse main branch and all associated *tributaries.*

**Riverain** Existing on or near a *bank* of a watercourse, and much wider in areal extent than *riparian.*

**Robbery** See *Diversion; Beheaded stream; Stream capture.*

**Rock basin lake** A glacial water body formed in a rock depression. Cf. *Paternoster lake.*

**Roil** A small section exhibiting swift *turbulent flow* within a watercourse. Cf. *Turbidity; Nephalometric Turbidity Unit (NTU).*

**Rorabough method** A procedure used for calculating steady-state *drawdown* in pumping and *observation wells* near a *recharge boundary.* The influence of a recharge boundary depends on its *abstraction*, duration, and distance to the pumping well. *Hydrologic boundaries* or discontinuities influence *groundwater flow* and drawdown by the addition of water. The Rorabough equation for the drawdown in the pumping well, $s_{ro}$, is:

$$s_{rp} = \frac{2.30Q}{2\pi T} \log \sqrt{4a^2 + r_w^2 - 4ar_r \cos\left(\frac{\phi}{r_w}\right)}$$

and the equation for the drawdown in the *observation well*, $s_{ro}$, is:

$$s_{ro} = \frac{2.30Q}{2\pi T} \log \frac{2a - r_w}{r_w}$$

where:
$Q = discharge\ [\mathrm{L^3 \cdot T^{-1}}]$
$T = transmissivity\ [\mathrm{L^2 \cdot T^{-1}}]$
$a =$ distance from pumping well to the recharge boundary [L]
$\phi =$ angle between the line joining the pumping and image wells and another line joining the pumping and observation wells [L]
$r_w =$ radius of the main well [L]

**Figure R-27** shows the Rorabough method setup, demonstrating the variables in the equation.

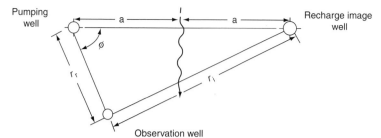

**Figure R-27** Rorabough method used for calculating steady-state flow in pumping and observation wells near a recharge boundary. (Sen, 1995. Reprinted with permission from CRC Press, Inc.)

Cf. *Steady-state flow.*

**Rotary drilling** See: **Appendix C – Drilling Methods**.

**Roughness** The property of a wetted surface that affects the energy available for *evaporation* or a channel function affecting *flow velocity.* Cf. *Roughness coefficient; Dispersion; Bed roughness; River channel pattern.*

**Roughness coefficient** A factor of the *average linear velocity* of *flow* in an open channel indicating the effect that the surface of the channel has on the energy loss in the flowing water. Cf. *Manning's formula; Manning's roughness coefficient.*

**Routing** See *Flow routing; Flood routing; Attenuated flood wave.*

**Rubber set** See *False set.*

**Rule of capture** Also called Absolute ownership rule, or English rule of capture. A common law doctrine in the United States giving the landowner rights to all the *groundwater* he can find beneath his own land, in the same manner that he owns the surface vegetation, no matter how much interference it may cause with a similar right of his neighbor. Cf. *Rule of reasonable use.*

**Rule of reasonable use** Also called the American rule of reasonable use, Reasonable use doctrine, American Doctrine or Reasonable use rule. A US law stating that landowners have the right to withdraw percolated *groundwater* to the extent that they must exercise their rights reasonably in relation to the similar rights of others. Also according to this rule, the landowner's use of groundwater is limited to the amount necessary for some useful or beneficial purpose in connection with the overlying land. The determination of "reasonableness" is made in a court of law on a case-by-case basis. Cf. *Correlative rights; Prior appropriation; Rule of capture.*

**Run** A specified section of a watercourse or a small, swiftly flowing *stream* of water.

**Runnel** Also called a brook, rivulet, or streamlet. A very small watercourse.

**Running sand** An unstable layer of sand or sandy soil that occurs below the *water table* or in a confined *aquifer.* During drilling operations, running sands collapse into the *borehole* as soon as drilling tools are removed, filling the hole with sediment and thereby making it difficult to set a *well screen.* If running sand is under *artesian pressure,* it may even be forced up through the drill bit and into the drill stem while drilling is in progress. Heavy *drilling muds* are used to counteract the pressures that are forcing the sand upward; alternatively, a hollow-stem auger is used in conjunction with a *knockout plug* to hold the hole open and prevent running sand from moving up the auger. Cf. *Heaving sand; Quicksand.*

**Running water** Water flowing in a confined channel, as in a watercourse. Cf. *Channel shape.*

**Runoff** The portion of *precipitation* that eventually appears in surface watercourses or water bodies. Runoff is typically expressed as the average depth of *flow* over the *watershed* in which precipitation occurred. After reaching the ground, a water particle enters a watercourse (such as a *stream*) through *overland flow, interflow,* or *groundwater flow.* Overland flow is also called "surface runoff," although surface runoff technically includes the portion of precipitation that falls directly into a surface depression that transmits water, whereas overland flow is the portion of runoff that travels over the ground surface and eventually discharges into a water body. Surface runoff is an important factor in streamflow only during occurrences of heavy or high-intensity rains. The runoff data used in storm analyses are dependent on *rainfall intensity;* for basins of $250 \, \text{km}^2$ $(100 \, \text{mi}^2)$, refer to **Appendix A** for unit conversions, or more, it is common to use an average intensity as determined by amount and duration. The production of runoff is not uniform over an entire basin in most areas, and most runoff distribution maps use a plot of *mean annual runoff.* Cf. *Direct runoff; Base runoff; Groundwater runoff; Runoff cycle; Infiltration; Delayed runoff; Drainage coefficient; Alluvial-dam lake; Antecedent runoff conditions.*

**Runoff coefficient** The proportion of *precipitation* that becomes *runoff,* expressed as a constant between 0 and 1. The runoff coefficient is determined by the climatic conditions and physiographic characteristics of the *drainage basin.* **Table R-3** is an abbreviated version of **Table R-2** of runoff coefficients used in the *rational equation* that is used to predict the amount of runoff expected from a given storm.

**Table R-3** An abbreviated table of values for C, runoff coefficient, used in the national equation that predicts the amount of runoff expected from a given storm. (Black, 1996. Reprinted with permission from Ann Arbor Press, Inc.)

| Surface | Value of c |
|---|---|
| Level terrane not affected by snow | 0.200 |
| Rolling farmland, long narrow valleys | 0.333 |
| Uneven terrane, wide valleys | 0.500 |
| Rough, hilly country, moderate slopes | 0.667 |
| Steep, rocky ground, abrupt slopes | 1.000 |

**Runoff curve number (CN)** A dimensionless expression of potential *retention* standardized by the Soil Conservation Service (SCS). The value of the runoff curve number, CN, depends on the soil and the hydrologic condition of the land surface. The graphical solution of the SCS runoff equation using the cumulative *direct runoff*, Q, and the cumulative *rainfall*, P, is shown in **Figure R-28**.

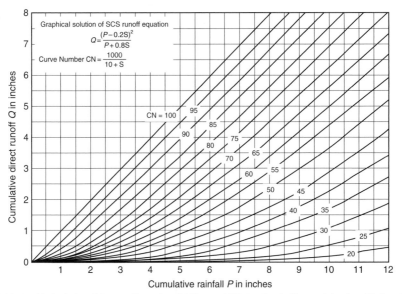

**Figure R-28** Graphical solution of the runoff equation. (Maidment, 1993. Reprinted with permission from McGraw-Hill, Inc.)

Cf. *Antecedent moisture content.*

**Runoff cycle** The portion of the *hydrologic cycle* between *precipitation* and subsequent *evapotranspiration* and/or *discharge* through watercourse channels. **Figure R-29** is a diagram of hydrologic factors during a uniform storm in an idealized, dry, spatially *homogeneous* basin. The shaded area represents the portion of precipitation that eventually becomes a part of the watercourse. At the onset, the rate of interception is high but the available storage capacity and *depression storage* is quickly depleted. The rate of surface runoff, which starts at zero, increases slowly and then approaches a relatively constant percentage of the precipitation rate. Cf. *River channel pattern.*

**Runoff intensity** Also called runoff rate. The depth of *rainfall*, or the volume of water on the land surface resulting from *precipitation*, per hour. Runoff intensity is dependent on storm characteristics, rainfall amount, *rainfall intensity*, and *rainfall duration*, and on the initial moisture conditions of the soils within the *catchment basin*. Cf. *Runoff volume.*

**Runoff mechanism** Stormwater runoff mechanisms, for example, include *infiltration*, saturation *overland flow*, and subsurface storm flow or *interflow*.

**Runoff rate** See *runoff intensity.*

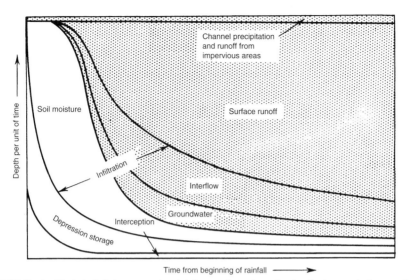

**Figure R-29** Graph of hydrologic factors encountered during a uniform storm showing the increase in the rate of surface runoff. (Linsley et al., 1982. Reprinted with permission from McGraw-Hill, Inc.)

**Runoff volume** The quantity of water from a specified *precipitation* event that reaches surface watercourses or water bodies during a specified time. *Runoff* volume is typically expressed in hectare-meter *(acre-feet)* or as depth in millimeters (inches) over the *drainage area*. Refer to **Appendix A** for unit conversions. **Figure R-30** graphically presents the changes is accumulated rainfall, losses and runoff typically encountered during a uniform storm.

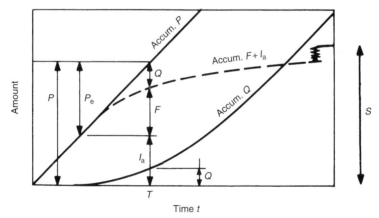

**Figure R-30** Graphical representation of the volume of runoff during a uniform storm. (Maidment, 1993. Reprinted with permission from McGraw-Hill, Inc.)

Cf. *Surface runoff.*
**Ryznar stability index** An analytical method to determine the incrusting or corrosive tendencies of water, developed by X.X. Ryznar by adapting the *Langelier index* based on *saturation pH*, $pH_s$. The Ryznar stability index is used to predict the reaction of a metal, such as that in a well pipe, to saturated subsurface conditions. The equation to determine the Ryznar stability index value, $I$, for a water sample is:

$$I = S - C - pH$$

where:

$I$ = Ryznar stability index [dimensionless]
$S$ = function of *total dissolved solids* (TDS)
$C$ = function of calcium ion ($Ca^{2+}$) concentration and methyl orange alkalinity
pH = acidity or alkalinity of the water sample.

The water sample is analyzed for pH, TDS, methyl orange alkalinity, and calcium *hardness*. The values for $S$ and $C$ are based on the water sample analysis as plotted in the graphs in **Figure R-31a and b**, and then substituted into the above equation. A water is considered corrosive if $I$ is greater than 7 and incrusting if $I$ is less than 7.

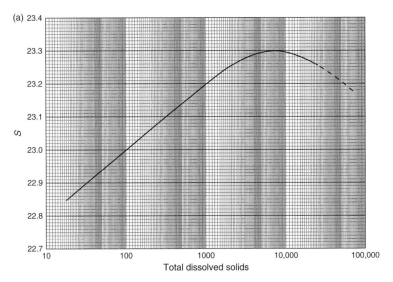

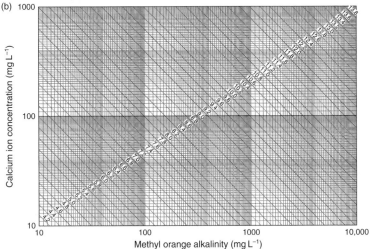

**Figure R-31a and 31b** (a) The $S$ value as a function of TDS in the Ryznar stability indexes. (b) The $C$ value as a function of calcium ion concentration and methyl orange alkalinity. (Driscoll, 1986. Reprinted by permission of Johnson Screens/a Weatherford Company.)

**Safe drinking water** See *Maximum contaminant level*. Cf. *Potable water*; *Safe Drinking Water Act (SDWA)*.

**Safe Drinking Water Act (SDWA)** A 1974 and amendments of 1977 in the United States established separate regulations for surface, underground, and bottled drinking water supplies; required the *Environmental Protection Agency (EPA)* to set *Maximum Contaminent Levels (MCLs)* for certain chemical and bacteriological *pollutants*; provided for each state to assume primary enforcement authority; and established emergency powers and citizen's enforcement powers. Cf. *Maximum Contaminant Level (MCL)*; *Water quality*.

**Safe yield** The amount of water that can be withdrawn from a *groundwater basin* on an annual basis without producing an undesired effect. The undesired effect must balance economic and/or social objectives with which the water is to be used. Any withdrawal in excess of the safe yield is an *overdraft*. Overdraft may result in the depletion of the groundwater reserves, lowering the *water table* or *potentiometric head*, which would increase pumping costs, or result in the intrusion of water with undesirable quality. Cf. *Economic yield*; *Intrusion of saltwater*; *Aquifer yield*.

**Sag pond** A body of water, typically small and often intermittent, formed where active or recent fault movement, at or near the groundwater table, has impounded *drainage* or changed the local *groundwater flow* that brings it to the surface, i.e., enclosed depression or sag that captures and stores water.

**Saline** Waters from inland seas that have a composition that differs from that of the ocean.

**Saline, contamination** The presence of chloride ions in an *aquifer* from an anthropogenic source or influence, e.g., *saltwater intrusion*.

**Salinity** The concentration of *total dissolved solids* (TDS) in water or more specifically, the total concentration of dissolved minerals or inorganic salts in water. In ocean studies, it is a measure of the amount of salt (in grams) dissolved in 1000 g of seawater, presented in parts per thousand (ppt ‰).

| Freshwater | Brackish | Saline | Brine |
|---|---|---|---|
| <0.05% | 0.05–3% | 3–5% | 75% |
| <0.5 ppt | 0.5–30 ppt | 30–50 ppt | 750 ppt |

The technical term for saltiness in the ocean is *halinity*, from the fact that halides – chloride specifically, are the most abundant anions in the mix of dissolved elements.

**Salinity current** The difference in density between salt water and fresh water may result in differences in water flow between the two at say, the mouth of a river. Cf. *Density current*; *Turbidity current*.

**Salt accumulation** The increase in the salt or chloride ($Cl^-$) concentration as groundwater migrates through a sedimentary basin. Cf. *Salinity*.

**Salt lake** See *Alkali lake*.

**Saltation** The rolling forward of sediment particles by a current in conditions of sediment transport where the sediment clast is too heavy to remain in suspension. The particle size that can be saltated depends upon the velocity of the current and the density of the sediment clast. Cf. *Traction*; *Sediment load*.

**Saltwater, contamination** See *Intrusion of saltwater*.

**Saltwater intrusion** Also called seawater intrusion. The encroachment of saltwater into groundwater, which exists as a natural condition along any coastline as lighter freshwater is depleted through pumping or drought. Control of saltwater intrusion was initially controlled by drilling wells inland, which only resulted in further saltwater intrusion. Current measures to control this include ponding of surface water runoff or treated water from sewage treatment plants on the surface and letting it infiltrate to resupply the fresh water. Other measures include the use of deep recharge wells where a potentiometric surface is created to prevent seawater intrusion and to allow pumping below sea level landward of the barrier. Different control schemes are illustrated in **Figure S-1**. Cf. *Intrusion of saltwater*.

**Sample splitter** Also called a sand separator. A device used in air drilling to collect samples of the formation being drilled after the cuttings pass through a cyclone. Typically, a three-tier sampler is used to separate out the various sediment sizes. The samples are then collected in a tubular bag for analysis. Typically, 5 ft (1.52 m) of sample bag will be filled for every 20 ft (6.1 m) drilled. Cf. *Grain size distribution*.

**Sampling plan** Also called a site sampling plan or SSP. A document defining the sampling frequency, locations, analysis, and possibly the statistical relevance of samples to be collected from an area being investigated for soil and/or water quality.

**S**

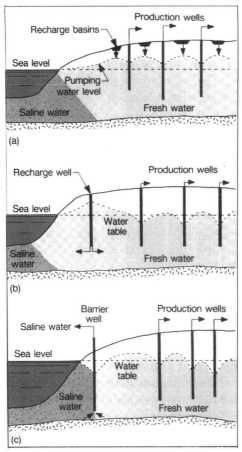

**Figure S-1** (a) Use of artificial recharge in the area of production wells in an unconfined coastal aquifer; the recharged water maintains the water table above sea level to prevent saltwater intrusion. (b) Use of injection wells to form a pressure ridge to prevent saltwater intrusion in an unconfined coastal aquifer. (c) Use of pumping wells at the coastline to form a trench in the water table, which acts as a barrier against further saltwater encroachment. (Driscoll, 1986. Reprinted by permission of Johnson Screens/a Weatherford Company.)

**Sand content** Also called *total suspended solids (TDS)*. The amount of sediment that is present in water. Sand content can have detrimental effects on pumps, fittings, and discharge nozzles on irrigation systems. Limits on the quantity of sediment in water must take into account its use. The US EPA and the National Water Well Association have recommended the following limits:

1. $1\,mg\,L^{-1}$ – water in direct contact with or used in the processing of food and beverages.
2. $5\,mg\,L^{-1}$ – water used in homes, institutions, municipalities, and industries.
3. $10\,mg\,L^{-1}$ – water for sprinkler irrigation systems, industrial evaporative cooling systems, and any other use where a moderate amount of sediment is not especially harmful.
4. $15\,mg\,L^{-1}$ – water for flood-type irrigation.

The concentration of sediment can be estimated by using a large container such as a 55-gal drum (200 L), a centrifugal sampler, or an *Imhoff cone*. The use of an Imhoff cone may not be representative of higher-yielding wells. Whereas, estimation of sediment content through collection of samples in a drum tends to be inaccurate. Sand content is usually measured at the time of a *pumping test*, and it is recommended that five samples be averaged over the period of the pump test at the following intervals:

1. Fifteen minutes after the start of the test
2. After 25% of the total projected pumping time period

3. After 50% of the total project pumping time period
4. After 75% of the total projected pumping time period
5. Near the end of the test.

The recommended volume of water to obtain representative samples in gallons per minute is as follows: multiplying the flow rate times 0.05. For yields over $0.06 \, m^3 \, s^{-1}$ (1000 gpm), the test volume should be 190 L (50 gal). While for yields less than $1.3 \times 10^{-3} \, m^3 \, s^{-1}$ (20 gpm), the test volume should be 19 L (5 gal). Cf. *Grain size distribution.*

**Sand drain** A vertical man-made channel backfilled with sand. Originally conceived for the purpose of improving soil consolidation by relieving *pore pressure*, sand drains have been used to improve the interconnectivity of *braided stream deposits* for the purpose of *pump and to treat systems* for *groundwater remediation* and for the purpose of relieving pressure buildup in soils at the toe of an earth-filled dam.

**Sand, hydraulic properties of** Sand formations are typically the most valuable storage sites for *groundwater*. Of these, beach sand deposits had the potential to store the greatest quantity of groundwater. *Void space* in new beach sand can be as high as 25–40%; this volume is reduced over time through consolidation. Other processes that reduce the void radio are chemical precipitation and heating. Some individual sandstone formations are quite extensive; for example, the St. Peter Sandstone, located in the central United States, covers more that $751,000 \, km^2$ ($290,000 \, mi^2$) and averages 24–49 m (80–160 ft) thick.

**Sand joint for removal of well screens** A technique commonly employed in instances where physical removal of a well is required. **Figure S-2** shows how a sand joint is made within the well screen to pull it out. A burlap bag is tied around a pulling pipe, lowered down the *borehole*, and filled with angular sand. The pipe is lowered to about 70% of the screen length. Sand is poured to fill two-thirds of the screen length. If necessary, to prevent *bridging*, the sand may be washed into the burlap bag. The pulling pipe is slowly pulled out of the well to create the sand lock. Using constant pressure and avoiding jerking of the pipe, the well is then pulled from the borehole. At the surface, the sand joint can be washed from the screened interval.

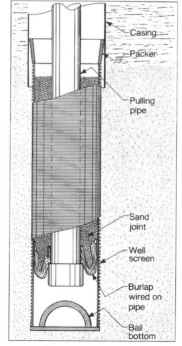

**Figure S-2** The use of a sand joint to remove telescoping screens. Angular sand placed between pulling pipe and screen becomes locked when lift is applied to the pulling pipe. (Driscoll, 1986. Reprinted by permission of Johnson Screens/a Weatherford Company.)

**Sand pumping** The introduction of a certain quantity of sand into a *well* during pumping, which can be a result of poor *well design* or inadequate *development*; some wells pump a certain quantity of sand. Other wells may begin to pump sand after several months or years of service. As a well ages, localized *corrosion* or mineral precipitation can result in increased *entrance velocities*. Sand grains, mobilized by the higher entrance velocities, can erode portions of the *well screen*, pump intakes, bowls, and vanes. Continued *erosion* of the screen can result in larger particles entering the well and increased damage to the well. Cf. *Effective grain size.*

**Sand separators** See *Sample splitter.*

**Sanitary protection measures** The *well completion* procedures and techniques employed to ensure safe drinking water from *wells*. Sanitary protection measures include *grouting*, sealing casing, horizontal suction lines, *pitless adaptors*, and determining the proper location for a well. Cf. *Maximum Contaminant Levels (MCLs); Safe Drinking Water Act (SDWA).*

**Saprolite** Soil formed by rock that has weathered in situ. Saprolitic soils tend to retain the structure of the parent rock, such that relict faulting, and joint patterns from the parent rock are also observed in the soils. This feature can be significant in hydrogeologic evaluations for preferential *flow* zones may be present.

**SAR** See *Sodium adsorption rate.*

**SARA** See *Superfund Amendments and Reauthorization Act (SARA).*

**Satellite imagery** A technology enabling the visualization of topographic features from space and is used (among other purposes) to determine likely sources of *groundwater*. Through satellite imagery, the identification of lineaments (linear traces longer than one mile) can be used to discern fault and fracture traces. In igneous and metamorphic rocks, this can help determine where zones of higher *permeability* are present that would permit *flow* to wells. "LandSat" photos can be used to differentiate vegetation types, which in turn can be used to infer soil types.

**Saturated pore space** See *zone of saturation.*

**Saturated thickness** See *Zone of saturation.*

**Saturated zone** See *Zone of saturation.*

**Saturation** See *Zone of saturation.*

**Saturation index** See *Common-ion effect*

**Saturation overland flow** See *Overland flow.*

**Saturation pH** The *pH* conditions at which calcium carbonate ($CaCO_3$) begins to precipitate and form *incrustation* on a *well screen* and in the filter material just outside of the screen. Generally, water with a pH 7.5 or greater tends to form precipitates in areas where there is a pressure drop, which occurs under pumping conditions in the screen slots and in the filter or formation material just outside of the screen. The tendency to precipitate $CaCO_3$ increases with temperature, which can be a problem where wells are used for thermal storage in aquifers, groundwater heat pumps, and where heating the subsurface is being used for *groundwater remediation* purposes, such as in the use of steam or hot-water injection, or where electrical or thermal conduction methods are used to heat up the *groundwater*. Two well-known methods, the *Langelier index* and the *Ryznar stability index*, have been developed to predict the incrustation or corrosive tendencies of a particular water. Langelier was the first to develop a method for predicting the saturation pH, termed $pH_s$, which is based on $CaCO_3$ equilibrium values, taking into account dissociation factors for carbonic acid ($H_2CO_3$), bicarbonate ($HCO_3^-$) and carbonate ($CO_3^{2-}$), and the theoretical solubility of calcium carbonate. The Langelier index is defined as:

$$\text{Langelier Index} = pH - pH_s$$

A negative Langelier index indicates that the water will dissolve $CaCO_3$ and will likely be corrosive to steel in the presence of *dissolved oxygen*. A positive Langelier index indicates supersaturation with calcium carbonate and incrustation will likely form on well screens and in the formation just outside of the screen. To more accurately reflect these tendencies, Ryznar further developed the Langelier index into the Ryznar stability index, I, as:

$$I = S - C - pH$$

where:

$I$ = Ryznar stability index

$S$ is a factor determined from **Figure S-3a** and the known TDS of the sample. $C$ is a factor determined from **Figure S-3b** and the known methyl orange alkalinity and calcium ion concentration (0.4 × calcium hardness). For a given water sample having:

$$pH = 7.0$$
$$\text{Total dissolved solids} = 400 \text{ mg L}^{-1}$$
$$\text{Methyl orange alkalinity} = 200 \text{ mg L}^{-1}$$
$$\text{Calcium hardness} = 125 \text{ mg L}^{-1}$$

note that the calcium ion concentration for the Ryznar stability index equation is $0.4 \times 125 = 50 \text{ mg L}^{-1}$.

1. The value of $S$ from **Figure S-3a** is 23.12.
2. The value of $C$ from **Figure S-3b** is 8.0.
3. The Ryznar stability index, $I$, is calculated as $23.12 - 8.0 - 7.0 = 8.12$.

A Ryznar index greater than 7 indicates corrosive water, whereas a Ryznar index lower than 7 indicates incrusting water. Thus, for the example presented, the water is corrosive.

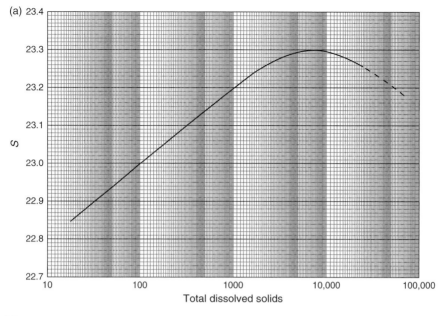

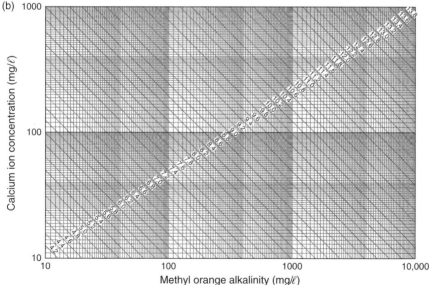

**Figure S-3** (a) The S value as a function of total dissolved solids (TDS) in the Ryznar stability indexes. (b) The C value as a function of calcium ion concentration and methyl orange alkalinity. (Driscoll, 1986. Reprinted by permission of Johnson Screens/a Weatherford Company.)

**Scale effects in estimating hydraulic conductivity** In estimating *hydraulic conductivity*, there are differences in whether this is done on a core sample or on a formation evaluation. Intuitively, there will be averaging on formation-scale estimations. However, differences in *anisotropy* can be quite pronounced. Core samples of clay and shale seldom show horizontal ($K_x$) to vertical ($K_z$) anisotropy ($K_x$:$K_z$) of greater than 10:1, and typically this is less than 3:1.

In estimating hydraulic conductivity on a formation-scale basis, where cyclic sedimentary deposition has taken place, it is not uncommon for regional anisotropy values on the order of 100:1 or greater.

**Scaling** The precipitation of calcium carbonate ($CaCO_3$) as a result of an increase in temperature. Scaling is typically observed in heating or boiling *hard water*. Cf. *Corrosion; Incrustation; Well maintenance.*

**Schlumberger array** An electrode configuration used in a *surface geophysical method* for measuring earth resistivity where the current electrodes are spaced much farther apart than the potential electrodes (**Figure S-4**). For depth

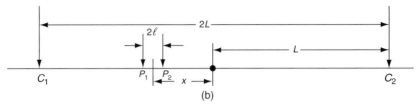

**Figure S-4** Electrode spread for a Schlumberger array. (Telford, et al., 1968. Reprinted with permission of Cambridge University Press)

probing, the potential electrodes remain fixed, while the current electrode spacing is expanded symmetrically about the center of the spread. As a result, for depth probing for only two electrodes need move about the point at which the depth probing is taking place. Lateral exploration may be performed in either of the following two ways. (1) The current electrodes are fixed large distances apart (typically 360 m or 1000 ft apart), and the potential electrode pair is moved between them with a fixed spacing. The apparent resistivity is plotted as the midpoint between the potential electrode pairs. (2) The other layout is similar to the *Wenner array* in that the electrode spacing remains fixed, and the survey is performed by moving the whole array along the survey line in suitable steps. This is not typically done, for it is less convenient, requiring the movement of all four electrodes to the next station to perform the survey. The potential electrodes may be moved offline from the current electrodes to perform a two-dimensional mapping. The apparent resistivity is determined as follows:

$$\rho_a = \frac{\pi L^2}{2L}\left(\frac{\Delta V}{I}\right)$$

where:

$\rho_a$ = apparent resistivity (ohms)
$L$ = half the distance between the two stationary current electrodes (m)
$2L$ = the distance between the two charged electrodes (m)
$\Delta V$ = applied voltage (V)
$I$ = current (A)
Cf. *Surface resistivity*.

**Screen** See *Well screen*.

**Screen installation** See *well screen installation*.

**Screen slot size** See *Slot openings*.

**Screened interval** See *well screen length*.

**Sea level** The elevation datum to which the elevation or altitude to which most objects are referenced. Used interchangeably with the term *mean sea level* to account for the natural variations in sea level. In 1929, sea level was replaced by the National Geodetic Vertical Datum (NGVD) for vertical control surveying in the United States. To establish the NGVD, mean sea level was measured at 26 tide gauges: 21 in the United States and 5 in Canada. The NGVD of 1929 was subsequently replaced by the North American Vertical Datum of 1988 (NAVD 88) based on the General Adjustment of the North American Datum of 1988. Cf. *Bouguer anomaly value*.

**Seal** See *Sealing abandoned wells; Sealing the wellhead*.

**Sealing abandoned wells** Closing off *wells* that are no longer in use. The principal purpose of sealing an *abandoned well* is to return the *aquifer*, to the extent practical, to a state before the well was installed. This is done to eliminate physical hazards, seal off a conduit for *groundwater* contamination, maintain confined head conditions (where applicable), conserve aquifer yield, and prevent the potential of poor-quality water from migrating from one aquifer to another. Laws and regulations on the sealing of abandoned wells in the United States vary from state to state. In many states, monitoring wells (especially those installed through concrete surfaces) may be abandoned by cutting off the *riser pipe* below the concrete and *grouting* the well riser and screen into the ground. Abandonment of water supply wells typically requires the installation of adequate seals. If the casing has not been grouted, it may be removed using the *sand joint technique*, hydraulic jacks, vibration hammer, or in extreme cases explosives may be used to loosen and remove casing.

If casing or liners cannot be removed, it may be necessary to perforate the casing or liners to ensure adequate sealing in these zones. In unconfined conditions, the objective is to prevent *percolation* of surface water through the casing or outside the casing between the casing wall and the well bore wall. In confined conditions, the intent of the sealing is to maintain the *confining pressure* of the aquifer. **Figure S-5** (bridge seal) shows the setup of well seals for confined conditions.

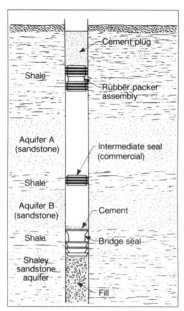

**Figure S-5** A bridge seal commonly used to abandon a producing well. (Driscoll, 1986. Reprinted by permission of Johnson Screens/a Weatherford Company.)

Intermediate seals are placed between confined water-transmitting zones having different static heads to prevent migration from one zone to another. *Bridge seals* are cement or weighted wood plugs placed beneath major aquifers. If desired, or where required by law, the lower part of the well bore may be filled with disinfected fill, such as chlorinated pea gravel. Top seals are placed above the aquifers and form the base for neat cement–bentonite *grout* to seal to the surface. Cf. *Abandoned well.*

**Sealing the wellhead** Completing the procedures necessary to maintain the integrity of a *well* after construction. One requirement for properly sealing a wellhead is to have the casing extend at least 12 in. (0.3 m) above the floor or ground elevation. If located in a *floodplain*, the casing should extend 2 ft (0.3–0.6 m) above the highest flood elevation. *Monitoring wells* located in thoroughfares, of course, cannot have casing extending above the ground. For this instance, special "traffic-rated" well boxes are installed. In all instances, a welded, threaded, flange cap or a compression seal (sanitary seal) must be fixed to the top of the well. **Figure S-6** shows a proper wellhead seal. The ground or floor area should slope away from the wellhead in all directions to prevent the ponding of water around the well. In pumped wells, a vent line should be installed with the vent opening pointed down. Further the opening of the vent should be covered with a fine-mesh screen or filled with fiberglass wool to prevent sucking foreign material into the well under pumping conditions.

**Seawater** See *Salinity.*

**Seawater intrusion** See *Intrusion of saltwater.*

**Secondary drinking water regulations** Guidelines that specify the aesthetic qualities of drinking water such as taste, odor, color, and appearance. Lack of adherence to these regulations deters acceptance of drinking water from public water supplies. Secondary drinking water regulations cover chloride, color, copper, corrosivity, foaming agents, iron, manganese, odor, pH, sulfates, TDS, and zinc. Cf. *Maximum Contaminant Level (MCL).*

**Secondary maximum contaminant levels** See *Maximum Contaminant Level (MCL).*

**Secondary permeability** Also called fracture permeability. In describing the characteristics of rock, the term for *permeability* that is associated with fractures in the rock matrix, rather than with interconnected *pore spaces*. Cf. *Permeability, absolute (k).*

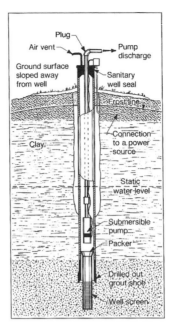

**Figure S-6** Proper sealing and venting of well. The groundwater surface around the top of the casing has been graded to slope away in all directions. (After US Environmental Protection Agency, 1973.)

**Second type boundary** Also known as a constant flux boundary. A *boundary condition* established in *numerical modeling* where there is a specified flux or *flow* across a model boundary.

**Secondary porosity** See *porosity*.

**Sediment concentration, allowable** See *Sand content*.

**Sediment load** The *total suspended solids (TDS)* present in a *stream* or a *river*. The sediment load is expressed as a weight or volume passing through a given cross section per unit time. Cf. *Loaded stream; Bed load; Flow layers, vertical*.

**Sediment-size analysis** See *grain size distribution; Uniformity coefficient; Grain size*.

**Seepage/seepage force** The slow movement of water through small openings and spaces in the surface of unsaturated soil into or out of a body of surface or subsurface water. Water *flows* past a soil grain in response to the *hydraulic potential* or energy per unit mass of flowing fluid by the difference in *hydraulic head* ($\Delta h$) between the front of the soil grain and back of the grain. The force exerted on the grain is the seepage force which is exerted in the direction of flow expressed as:

$$F = \rho g \, \Delta h \, A$$

where:

$A$ = cross sectional area of the grain
$\rho$ = mass density of water

If the cross sectional area encompasses many grains the expression of the seepage force during vertical flow is directly proportional to the *hydraulic gradient*. In areas of downward percolating groundwater, the seepage forces act in the same direction of gravity, but in areas of upward flowing water they oppose gravity. If the upward directed seepage force at a *discharge area* in a flow system exceeds the downward directed gravity force soil grains will be carried away potentially causing severe damage. Cf. *Piping*.

**Seepage face** **Figure S-7** illustrates this concept in a cross-sectional view, in which BC is a constant-head boundary, and DC is an impermeable boundary. If there is no *infiltration*, then AB will act as an impermeable boundary. The *water table* as defined by EF intersects the outflow boundary at exit point E, forming the seepage face at ED.

**Seepage line** *Streams* may gain or lose water through the streambed along a *seepage* line. *Baseflow* to perennial streams changes with corresponding changes in the elevation of the *water table*, where increased baseflow from groundwater would represent a *gaining stream*, and *recharge* to groundwater from the streambed would represent a *losing stream*. Cf. *Seepage velocity*.

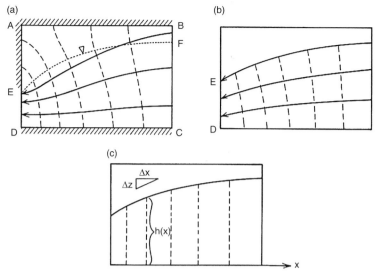

**Figure S-7** Development of a seepage face on a free outflow boundary. (a) Saturated–unsaturated flow net; (b) free-surface flow net; and (c) Dupuit–Forchheimer flow net. (Freeze and Cherry, 1979. Reprinted with permission by Prentice-Hall, Inc.)

**Seepage meter** Used to measure discharges of groundwater at a sediment–water interface. This has traditionally been performed using the drum method, whereby the lower portion of a metal drum with a port for collecting water is inserted upside down into the sediments, and the seepage is measured as the amount of time required for a plastic "bag" fitted to the sampling port to fill with groundwater (**Figure S-7a**). This method has been improved upon using an ultra-sonic flow-measuring device (**Figure S-7b**) located on a buoy that also facilitates sampling the groundwater.

**Seepage surface** See *Seepage face; seepage line; Dupuit–Forchheimer equation.*

**Seepage velocity ($V_s$)** The average speed at which groundwater moves through the interconnected pore spaces of a porous medium. Seepage velocity is defined by the following equation:

$$V_s = \frac{KAi}{n}$$

where:
$V_s$ = seepage velocity [L•T$^{-1}$]
$K$ = hydraulic conductivity [L•T$^{-1}$]
$i$ = hydraulic gradient [L•L$^{-1}$]
$A$ = cross-sectional flow area [L$^2$]
$n$ = porosity [%]

**Seiches** The increase in water level in a water body resulting from a storm occurring on another side. This is a common occurrence on the great lakes in North America. For instance, as thunderstorms travel across Lake Michigan, winds at the leading edge of the storm can "push" water ahead of it. When the storm passes to the east shore, the high water that is pushed ahead of the storm then "sloshes" back to the west side of the lake, creating a seiche. Cf. *Attenuated flood wave.*

**Self-potential** See *Spontaneous potential.*

**Semiconfined aquifers** See *Aquifer, leaky confined.*

**Septic system** A system that contains a filter that uses sedimentation processes to separate solid domestic waste materials from liquids. Wastes collected from the household drain to the *septic tank*, typically constructed of concrete or fiberglass, buried below the ground and containing a series of holes to allow water to drain. Septic tanks become very biologically active due to the presence of oxidizable organic material and the fresh supply of oxygenated water from household uses. Solids settle out, setting up a biologically active material or sludges in the subsurface. The liquids pass through the biologically active sludges, and from there *recharge* the *groundwater.* There are specific state, federal, or local ordinances governing the setback distance between a septic system and a water production well.

**Septic tank** The tank component of a *septic system*, which contains the solids and sludges that result from treating waste water. These must be cleaned out on a periodic basis.

**Settling pit** See *Mud pit.*

315

**Settling velocity** Also referred to as the falling velocity. The rate [L•T$^{-1}$] (Refer to **Appendix A** for unit conversions) at which *suspended solids* within a liquid subside or fall and are deposited.

**Shale catcher** See *Shale trap.*

**Shale trap** Also known as a shale catcher. A rubber formation packer mounted on the casing of a well above the *well screen* to prevent clay or shale from falling into the annular space between the *borehole* wall and the screen during well installation (**Figure S-8**).

**Figure S-8** A rubber formation packer (shale trap) mounted on the casing above the screen prevents clay from falling into the annulus around the screen. (Driscoll, 1986. Reprinted by permission of Johnson Screens/a Weatherford Company.)

**Shear stress** The stress that acts tangential, a given vector acting at Right angles to a given radius of a given circle, to a plane through any given point within a given body or an object.

**Sheet erosion** The detachment of land surface material with water acting as an agent of erosion by raindrop impact and thawing of frozen grounds and its subsequent removal of *overland flow.* The cohesiveness of soil particles is removed by the impact of raindrops. Cf. *Precipitation.*

**Sheet flow** See *Overland flow.*

**Sheetwash** See *Overland flow.*

**Shock chlorination** A term reserved for the use of chlorine solutions greater than 1000 mg L$^{-1}$ for disinfection. Cf. *Pathogenic; Corrosion; Sanitary protection measures.*

**Shooting** See *Explosives, use of.*

**Shooting flow** See *Jet flow.*

**Shroud** A shield usually constructed of rubber, or synthetic neoprene rubber, used at the *wellhead* during drilling to keep return fluid from mud, water, or air-rotary drilling operations from spraying up the *well bore.*

**Shut-in head** Also known as shut-in pressure. The *pressure* that builds up inside an *artesian well* when it is capped or in situations such as a *packer test,* the excess pressure that is confined within the zone of interest.

**Shut-in pressure** See *Shut-in head.*

**Shut-off head** In a situation where a pump runs when a valve is closed, or a *pressure* setting has been set higher than the range of the pump, the pump continues to pump until it reaches a point where it can no longer pump any water; a point of zero production or the *shut-in head.* The pump will recirculate water, creating friction and heat, which may eventually cause the water to boil. This in turn results in a lack of coolant for the pump, which may cause damage to the bearings.

**Sichardt equation** An empirical equation relating *radius of influence* from a pumping *well* and *drawdown* within that well. This equation is based on numerous pump tests and provides a good approximation of the radius of influence at steady-state conditions. In SI units it is:

$$R = 3000s\sqrt{K}$$

where:
$R$ = radius of influence (m)
$s$ = drawdown (m)
$K$ = *hydraulic conductivity* (m s$^{-1}$)

When used in conjunction with the *Thiem equation,* the value of $R$ can be used to determine the approximate *well yield.*

**Sieve analysis** Determination of the particle size distribution of soil, sediment, or crushed rock by measuring the percentage of particles that are retained on standard-sized sieves. **Figure S-9** presents the various sieve groups that are suitable for the various classes of unconsolidated sediments. Cf. *Grain size distribution; Effective grain size.*

**Sieve test** See *Sieve analysis; Grain size distribution; Grain size gradation and conductivity.*

**Site sampling plan (SSP)** See *sampling plan.*

S

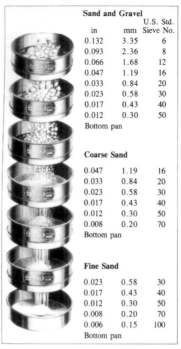

| Sand and Gravel | | |
|---|---|---|
| in | mm | U.S. Std. Sieve No. |
| 0.132 | 3.35 | 6 |
| 0.093 | 2.36 | 8 |
| 0.066 | 1.68 | 12 |
| 0.047 | 1.19 | 16 |
| 0.033 | 0.84 | 20 |
| 0.023 | 0.58 | 30 |
| 0.017 | 0.43 | 40 |
| 0.012 | 0.30 | 50 |
| Bottom pan | | |

| Coarse Sand | | |
|---|---|---|
| 0.047 | 1.19 | 16 |
| 0.033 | 0.84 | 20 |
| 0.023 | 0.58 | 30 |
| 0.017 | 0.43 | 40 |
| 0.012 | 0.30 | 50 |
| 0.008 | 0.20 | 70 |
| Bottom pan | | |

| Fine Sand | | |
|---|---|---|
| 0.023 | 0.58 | 30 |
| 0.017 | 0.43 | 40 |
| 0.012 | 0.30 | 50 |
| 0.008 | 0.20 | 70 |
| 0.006 | 0.15 | 100 |
| Bottom pan | | |

**Figure S-9** Various sieve groups or percentages of particles retained on standard-sized sieves. (Driscoll, 1986. Reprinted by permission of Johnson Screens/a Weatherford Company.)

**Sinkhole** A natural depression in the surface topography caused by the removal of soil and/or bedrock by water. Mechanisms of formation may include the gradual removal of slightly soluble bedrock (such as limestone) by percolating water, the collapse of a cave roof, or a lowering of the water table. They may occur gradually or suddenly and are found worldwide. Sinkholes also form from human activity, such as the collapse of abandoned mines in urban areas due to water main breaks or sewer collapses when old pipes give way. They can also occur from the overpumping of groundwater and subsurface fluids. Sinkholes have been used for centuries as disposal sites for various forms of waste. As a result, groundwater pollution may be associated with old skinholes.

**Slot openings** The slot openings in a *well screen* must strike a balance between maximizing the *open area* to permit water flow into the *well* and providing the strength to withstand the hydraulic and formation forces acting on the screen. Further, the selection of the screen slot size is also determined by the grain size of the aquifer. *Well screen open areas* range from 1% for perforated pipe to more than 40% for wire-wound screen. Slot configuration controls how much development energy can reach the formation. Cf. *Gauze number; Louver screen.*

**Slotted pipe** Notched or perforated pipe that can be used for *well screens*. Slots are cut with a saw (typically used in PVC pipe), oxyacetylene (cutting) torch, or punched with a chisel-and-die casing perforator that is used once the casing is set in the ground. Limitations of perforated pipe are the following:

1. Openings cannot be closely spaced;
2. Low percentage of open area;
3. Significant variation in slot size of *slot openings* varies significantly;
4. Difficult to produce openings small enough to control fine or medium particle size; and
5. Slotted steel pipe tends to corrode more quickly than continuous slot screens.

Cf. *Gauze number; Louver screen.*

**Slough** A *lake* or a pond where most of the water present is *groundwater*-derived. As such, there is little flushing, and these bodies of water tend to be *eutrophic*.

**Slow drainage** A condition occurring during the early stages of a *pumping test*, which can influence the validity of the results of *drawdown* monitoring. Slow drainage results where there are significant differences between the horizontal and vertical *hydraulic conductivity*, which can occur in glacial drift where there may be coarse sand and gravel layer present within a silty or clayey zones above the *aquifer* being tested. This condition typically lasts for a matter of hours or days before the slope of the *drawdown curve* matches the true aquifer characteristics (**Figure S-10**).

**S**

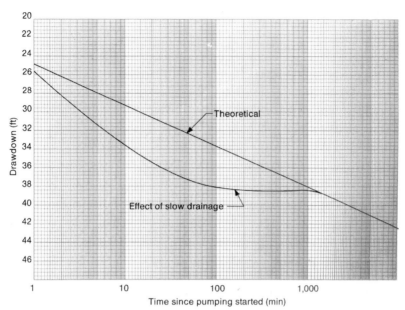

**Figure S-10** Time–drawdown curve showing the effect of slow drainage on the early part of the curve. (Driscoll, 1986. Reprinted by permission of Johnson Screens/a Weatherford Company.)

Cf. *Delayed drainage*.

**Slug test** Also known as a *bail-down test*, bailer test or borehole permeability test. A single-well test to determine the *hydraulic conductivity* of a formation, by which a slug of known volume is inserted into a well and the water level response to insertion of the slug (rising head) and removal of the slug (falling head) is monitored (**Figure S-11**).

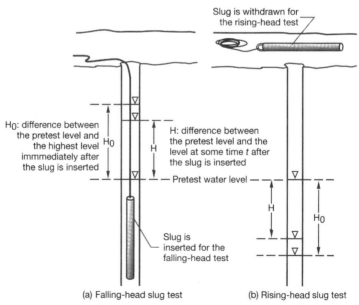

**Figure S-11** Schematic illustrating the performance of a slug test in a well. (Sanders, 1998.)

The most well-known techniques for analyzing the data from these tests are that of Hvorslev, Bouwer and Rice, and Cooper, Bredehoft, and Papadopolous. The basic principle is that as the slug is inserted or removed, *flow* is induced into or out of the formation that is screened. The response rate in water level in the well is reflective of the hydraulic conductivity. There are commercially available computer programs for the analysis of these data. Cf. *Rising-head test; Falling-head test; Hvorslev method.*

**Slurry** A suspension of finely divided particles of cement and/or clay in water. Cf. *Grout; Mud pit; Drilling mod.*

**Slurry wall** A barrier installed for the purpose of *groundwater* control for constructing underground structures such as subway stations or for the containment of groundwater contamination. Slurry walls consist of *bentonite*, water, and backfill materials placed in deep trenches. A mixture of soil material and hydrated bentonite can be placed as deep as 30.5 m (100 ft). The bottom of the wall is tied to an impermeable formation to hydrologically isolate the area to be sealed off. Under natural conditions, a slurry wall will last 20–40 years. However, in cases where isolation of groundwater contamination is required, the contaminant may react with the slurry material, shortening the life of the wall. Cf. *Dewatering; Grout curtain.*

**SMOC** See *Standard Mean Ocean Chloride.*

**SMOW** See *Standard Mean Ocean Water.*

**Snow chemistry** Compared to *rainfall*, snowmelt tends to be higher in *total dissolved solids (TDS)* as a result of particulates falling out of the atmosphere while the snow is accumulating on the ground that dissolve when the snow melts. Rainwater and melted snow in non-urban areas tend to have pH values between 5 and 6 from contact with carbon dioxide ($CO_2$) in the earth's atmosphere, resulting in the formation of carbonic acid ($H_2CO_3$).

$$H_2O + CO_2 = H_2CO_3$$

In industrial areas, the pH of *precipitation* can be below 5.7, frequently as low as 3 or 4 as a result of air borne particulate sulfur and sulfate compounds emitted from factories, mine-processing plants, and coal- or petroleum-fired electric-generating stations. As a result, snowmelt and rainwater are dilute, slightly to moderately acidic, oxidizing agents that can alter the chemistry of soils and geologic material within which they percolate.

**Soda ash** Sodium carbonate ($Na_2CO_3$). Soda ash can be used to soften *hard water* and to soften it for use in *drilling mud* by precipitating the calcium as calcium carbonate ($CaCO_3$). The following relationship can be used to estimate the quantity of soda ash required for softening:

$$A\,(mg/l) = 0.0095 \times hardness(CaCO_3)$$

where:

$A$ = soda ash per 100 gal of mix water (lb/100 gal)

Typically, 0.5–3 lb (0.2–1.4 kg) of soda ash per 100 gal (0.4 m$^3$) is sufficient to soften the mix water. After mixing, the pH of the water should be checked to ensure that it is 7.5 or higher.

**Soda lake** An ephemeral lake in arid locations that evaporates quickly leaving behind alkaline evaporates (including sodium carbonate and sodium bicarbonate). A soda lake, located in San Bernardino County, California, is part of what remains of the ancient Ice Age Lake Mojave, located at the terminus of the Mojave River and has no outlet to the sea. In North America, soda lakes are found in Nevada and California. Cf. *Alkali lake; Potash lake.*

**Sodium adsorption ratio** A term that was developed to measure the effect of sodium on irrigated soils. When sodium-rich water is applied to soil, some of the sodium is taken up by the clay, which results in the clay releasing calcium and magnesium to the water. The reaction, known as base exchange, alters the physical characteristics of the soil and can result in alkali soils and retarded growth of the cultivated plants. The clay that undergoes the base exchange becomes sticky and slick when wet and has low *permeability.* When dry, large clods are formed, which are difficult to cultivate. The sodium adsorption ratio (SAR) is determined from the following equation:

$$SAR = \frac{Na}{\sqrt{\frac{Ca+Mg}{2}}}$$

The amounts of sodium, calcium, and magnesium, expressed in *milliequivalents per liter* (mequiv L$^{-1}$), are determined from a water analysis. An SAR of 18 or greater indicates that base exchange may occur in soils, whereas SAR values less than 10 indicates there is a little chance of base exchange occurring.

**Sodium hypochlorite** Also known as household bleach. A chemical with the formula NaOCl, which is commonly used in *well* treatment for disinfection. Sodium hypochlorite is commonly available in solutions of 5–15%. Pure NaOCl is explosive and as such cannot be handled safely. In solution, sodium hypochlorite is unstable and tends to break down with time. Over a 6-month storage period, a 10% solution can lose 20–50% of its useful chlorite. Cf. *Polyphosphates.*

**S**

**Softening of water, processes for** Using either lime softening or ion exchange to reduce the *hardness* of water by 95–100%. The application of hydrated lime ($Ca(OH)_2$) results in the precipitation of calcium carbonate ($CaCO_3$) and/or magnesium hydroxide ($Mg(OH)_2$). Most alkaline waters contain bicarbonate ($HCO_3^-$); thus, the precipitation of $CaCO_3$ and $Mg(OH)_2$ necessitates the reaction of $CO_2$ and $HCO_3$ to form $CO_3^{2-}$:

$$Ca(HCO_3)_2 + Ca(OH_2) \rightarrow 2CaCO_3 \downarrow + 2H + 2O$$

$$CO_2 + Ca(OH)_2 \rightarrow CaCO_3 \downarrow + H_2O$$

$$2Ca(OH)_2 \rightarrow Mg(OH)_2 \downarrow + 2CaCO_3 \downarrow$$

The materials precipitated in lime softening are removed through sedimentation (1–4-hours settling time) followed by *filtration*. In water softening by ion exchange, sodium ions replace the undesired ions as water is passed through a cation exchange resin wherein sodium is exchanged for calcium, magnesium, iron, manganese, and strontium. Water is passed through a cation exchange resin, wherein the sodium ions are exchanged for the undesired ions. When all of the sodium is exhausted, the resin is regenerated using a *brine*, which displaces the accumulated calcium, magnesium, iron, or other ions. Ion exchange resins typically last for years and are not adversely affected by the ion exchange process. Cf. *Ion exchange*.

**Soft water** Water containing less than 75 mg/l of calcium carbonate ($CaCO_3$). The hardness classification of water is shown in **Table S-1**.

**Table S-1** Hardness classification of water based on the amount of $CaCO_3$ present. (Walton, 1970. Reprinted with permission from McGraw-Hill, Inc.)

| Hardness as $CaCO_3$ (mg/l) | Water Class |
|---|---|
| 0–75 | Soft |
| 75–150 | Moderately hard |
| 150–300 | Hard |

Cf. *Softening of water, processes for; Hardness.*

**S**

**Soil leaching** The removal of soluble constituents from the soil. Leaching often occurs with soil constitutents such as nitrate fertilizers with the result that nitrates migrate to potable ground water.

**Soil moisture** The quantity (volume) of water occupying a given volume of soil, expressed as a percentage. Three soil moisture regions reside within the *vadose zone*. Up to 30 feet (10 m) below the soil surface in the area penetrated by roots is the soil water region, which is highly variable. Where the *water table* is deep, an intermediate region exists where moisture concentration is constant at the field capacity of the soil or rock. Above the water table and below the soil water region, moisture is raised by *capillarity* into the *capillary fringe* region. This region may have a vertical extent of a few inches to several feet depending on the pore size of the soil. Cf. *Antecedent precipitation index; Gravimetric moisture content; Volumetric moisture content.*

**Soil moisture storage capacity** Also known as soil water storage capacity. For irrigation of crops, the soil water storage (SWS or SMS) capacity is defined as the total amount of water that is stored within the plant's root zone. This is determined by soil texture and the crop rooting depth. A deeper rooting depth results in a larger volume of water stored in the soil and therefore a larger reservoir of water for the crop to draw on between irrigations. This allows the farmer to determine how much water to apply at one time and how long to wait between each irrigation. For example, the amount of water applied at one time on a sandy soil, which has a low soil water storage capacity, would be less than for a loam soil, which has a higher soil water storage capacity, assuming the crop's rooting depth is the same for both soils. Applying more water to the soil than can be stored results in a loss of water to deep percolation and leaching of nutrients beyond the root zone. In some cases, leaching of salts is desirable and extra irrigation would be desired.

Determination of soil water storage (SWS) and the maximum soil water deficit:

Step 1. Determine the effective crop rooting depth, *RD (m)*, **Table S-2**.

**Table S-2**   Effective rooting depth of mature crops for irrigation (From the Ministry of Agriculture, Food and Fisheries, British Columbia, Canada)

| Shallow (0.45 m, 1.5 ft) | Medium shallow (0.60 m, 2 ft) | Medium deep (0.9 m, 3 ft) | Deep (1.20 m, 4 ft) |
|---|---|---|---|
| Cabbages | Beans | Brussel sprouts | Asparagus |
| Cauliflower | Beets | Corn (sweet) | Blackberries |
| Cucumbers | Blueberries | Eggplant | Grapes |
| Lettuce | Broccoli | Kiwifruit | Loganberries |
| Onions | Carrots | Peppers | Raspberries |
| Radishes | Celery | Squash | Sugar beets |
| Turnips | Potatoes | Saskatoons | Tree fruits |
| | Peas | Tree fruits | (spacing 4 m × 6 m) |
| | Strawberries | (spacing 2 m × 4 m) | |
| | Tomatoes | | |
| | Tree fruits | | |
| | (spacing 1 m × 3 m) | | |

Step 2. Determine the available water storage capacity of the soil, $AWSC$ (mm m$^{-1}$), **Table S-3**.

**Table S-3**   A guide to available water storage capacities of soils (From the Ministry of Agriculture, Food and Fisheries, British Columbia, Canada)

| Textural Class | Available water storage capacity (AWSC) | | |
|---|---|---|---|
| | in. water/in. soil | in. water/ft soil | mm water/m soil |
| Clay | 0.21 | 2.5 | 200 |
| Clay loam | 0.21 | 2.5 | 200 |
| Silt loam | 0.21 | 2.5 | 208 |
| Clay loam | 0.20 | 2.4 | 200 |
| Loam | 0.18 | 2.1 | 175 |
| Fine sandy loam | 0.14 | 1.7 | 142 |
| Sandy loam | 0.12 | 1.5 | 125 |
| Loamy sand | 0.10 | 1.2 | 100 |
| Sand | 0.08 | 1.0 | 83 |

Step 3. Calculate the total soil water storage, *SWS* (mm).

$$SWS \text{ (mm)} = RD \text{ (m)} \times AWSC \text{ (mm m}^{-1}) \tag{1}$$

Step 4. Determine the availability coefficient of the water to the crop, AC (%), **Table S-4**.

**Table S-4**   Availability coefficient (AC%) of water to crop

| Availability Coefficients | |
|---|---|
| Crop | Maximum Percent (%) |
| Peas | 35 |
| Potatoes | 35 |
| Tree Fruits | 40 |
| Grapes | 40 |
| Tomatoes | 40 |
| Other crops | 50 |

**S**

Step 5. Calculate the maximum soil water deficit, MSWD (mm).

$$MSWD = SWS \text{ (mm)} \times AC \text{ (\%)} \tag{2}$$

Cf. *Actual Evapotranspiration (AE); Potential Evapotranspiration (PE).*

**Soil sniffing** A term used in environmental site investigations to denote the practice of screening the soils using an organic vapor analyzer for the presence of volatile organic compounds. Soils are retrieved from the sampler, placed in a plastic bag, sealed, and allowed to warm. The probe of the organic vapor analyzer is then inserted into the bag, and the readings of the vapor concentrations in the headspace of the bag are taken.

**Soil water** Water residing in the *unsaturated zone*. In the upper soils of the unsaturated zone, or *belt of soil water*, water is lost from the soils through *transpiration* or by direct *evaporation*. Cf. *Available moisture; Field capacity*.

**Soil water characteristic curve** The soil water characteristic curve (SWCC) is fundamental to hydrological characterization and required for most analyses of water movement in unsaturated soils. Further, the SWCC is also used in characterizing the shear strength and compressibility of unsaturated soils. The unsaturated hydraulic conductivity of soil is often estimated using properties from the SWCC and the saturated hydraulic conductivity. This method applies only to soils containing two pore fluids: a gas and a liquid. The liquid is usually water and the gas is usually air. Other liquids may also be used, but caution must be exercised if the liquid being used causes excessive shrinkage or swelling of the soil matrix. Cf. *Gypsum block; Manometer; Unsaturated zone*.

**Sol** A soluble particle. In hydrogeology, a commonly used sol is the polysaccharide *guar gum*, which when dispersed in a polymeric *drilling fluid* provides fluid loss control. This fluid loss control is due to both the sols and the *insols* (the insoluble cell wall residue of the gum). Each insol particle is covered with a relatively thick, viscous film of fluid produced by the dissolved sol. These particles form a layer on the *borehole* wall that controls fluid loss from the borehole and is thinner than the functionally equivalent layer formed by clay additives. Cf. *Lost circulation; Gel; Flocculation*.

**Solar evaporation** Evaporation caused by solar radiation.

**Solid-shaft electric motors** A motor used to drive vertical turbine pumps. A solid-shaft electric motor is used for shallow-well pump installations, for the shaft needs to be turned to thread directly into the line shaft, which can be unwieldy, given the weight of a solid shaft.

**Solid-stem augers** A tool used to advance holes in stable formations. A solid-stem auger is not typically used for well drilling, but more commonly for foundation investigation activities.

**Solifluxion** The slow downhill movement of soil or scree material as a result of the alternate freezing and thawing of *pore water*.

**Solubility** For a given compound, the maximum mass of the compound that can be dissolved in a *solvent* such as water. Solubility typically refers to aqueous solubility and is usually expressed in milligrams per liter or as a percent. See conversions in **Appendix A-1**.

**Solute** An organic or inorganic constituent dissolved within a *solvent*. The amount of solute present in solution is usually expressed as mass per unit volume, e.g., milligrams per liter (mg L$^{-1}$). It is also becoming common to express concentration on a molar basis, such as millimoles per liter (mM L$^{-1}$). See **Appendix A-1** conversions.

**Solution channel** See *Karst*.

**Solvent** The liquid in which a *solute* is dissolved. In *hydrology* and *hydrogeology*, the solvent is usually water.

**Solvent welding** The joining of plastic casing (PVC, ABS, and CPVC) using a solvent-based cement. Solvent welding is not recommended for the construction of *monitoring wells* for environmental purposes because the *solvents* can leach into the groundwater, contaminating the water samples obtained from the wells. Solvent welding may be used in the construction of *piezometers*, suction lysimeters, pore pressure-monitoring devices, or supply wells where there is minimal concern over solvent contamination of groundwater (e.g., irrigation wells and *dewatering wells*). Solvent welding typically done with "bell-end"-type pipe and the pipe manufacturers' recommendations should be followed. Typically, this involves selecting the proper cement for the pipe, cleaning the pipe prior to welding, applying the cement, and joining the bell and spigot sections together.

**Sonic logging** See *acoustic velocity logging*.

**Sonic tool** See *Acoustic log*.

**Sorption** A class of process by which one material partitions to another. *Absorption* refers to the process of the penetration of one material into another. *Adsorption* refers to the collection of one material on another's surface.

**Sorption isotherm** See *Freundlich isotherm; Langmuir isotherm*.

**Sounder** See *electrical sounder*.

**SP** See *spontaneous potential logging*.

**Spatial correlation** See *correlation length*.

**Specific capacity** The quantity of water produced in a well per unit *drawdown*. Specific capacity is commonly expressed as gallons per minute per foot (gpm ft$^{-1}$), or cubic meter per second per meter (m$^3$ s$^{-1}$ per m). See **Appendix A-1** conversions.

**Specific discharge** Also called filter velocity, *Darcy flux* or Darcian velocity. In a permeameter, the specific discharge, $v$, is defined as:

$$v = \frac{Q}{A}$$

where:

$Q$ = steady-state flow of water through the permeameter $[L^3 \cdot T^{-1}]$

$A$ = cross-sectional area of the permeameter $[L^2]$

Given the dimensions of $Q$ and $A$, $v$ has the dimensions of velocity $[L \cdot T^{-1}]$. Cf. *Discharge velocity*.

**Specific electrical conductance** The electrical conductance of a cubic centimeter of any substance compared with the electrical conductance of the same volume of water. The unit of conductance is mhos, the inverse of ohms, used to express resistance. For water, units of conductance are expressed as mhos $\times 10^{-6}$ (micromhos) or micro-Siemens ($\mu S$; SI). Chemically pure water has a very low electrical conductance, and as such is a good insulator. The dissolution of a minute amount of mineral matter will increase the conductance of water. However, the specific conductance for the same concentration of different minerals differs. A concentration of 100 mg $L^{-1}$ of sodium chloride will result in a higher conductance than 100 mg $L^{-1}$ of calcium chloride since 100 mg of sodium (atomic weight 22.99) results in a greater number of ions dissolved in *groundwater* compared with calcium (atomic weight 40.08). There is a strong correlation between the concentration of *TDS* and specific conductance. For most waters, the specific conductance times a factor of 0.55–0.75 provides a reasonable estimate of the concentration of dissolved solids in milligrams per liter. The factor for saline waters is usually higher than 0.75, while for acidic waters, the value is less than 0.55.

**Specific gravity** The mass of a specified volume of any substance relative to the same volume of water at standard temperature and pressure. Cf. *Baume gravity*.

**Specific moisture capacity** The unsaturated storage property of a soil. **Figure S-12** presents the characteristic curves relating the hysteretic curves for the wetting and drying of soils. The specific moisture capacity is defined as:

$$C = \frac{\partial \theta}{\partial \Psi}$$

where:

$C$ = specific moisture capacity $[L^{-1}]$

$\partial \theta$ = change in moisture content [unitless]

$\partial \Psi$ = change in pressure head $[L]$.

Because of the hysteresis and the non-linear relationship, $C$ is not a constant. In the saturated zone, the moisture content $\theta$ is equal to the *porosity*, a constant such that $C = 0$.

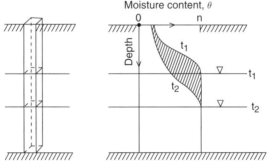

**Figure S-12** Concept of specific yield viewed in terms of the unsaturated moisture profiles above the water table. (Freeze and Cherry, 1979. Reprinted with permission from Prentice-Hall, Inc.)

**Specific retention** The ratio of the volume of water that is retained on the solid matrix (rock or soil), relative to that volume of solid. This value is typically stated as a percentage. Cf. *Adhesive water; Attached groundwater*.

**Specific speed ($N_s$)** Specific speed is the speed in revolutions per minute (rpm) at which a given impeller would operate if reduced proportionately in size to deliver 1 gpm (5.5 m$^3$ day$^{-1}$) against a total dynamic head of 1 ft (0.3 m) (Colt Industries, 1974, as presented in Driscoll, 1986). In impeller design, specific speed is used to compare one type of impeller or impeller system to another. $N_s$ is determined from the following equation:

$$N_s = \frac{\text{rpm} \times \text{gpm}}{H^{0.75}}$$

where:

$H^{0.75}$ = head per stage in ft (m)

The *head* and capacity for the pump are selected by determining the highest efficiency point of the largest-diameter impeller in the pump. As a result, different conditions of head and capacity are accommodated by adjusting the impeller speeds. Impellers for high heads usually have low specific speeds, whereas impellers for low heads have high specific speeds. A pump with low specific speed will avoid *cavitation* problems by producing a greater suction lift than one with

323

higher specific speed. Slower speeds are used when the suction lift is more than 15 ft (4.6 m), as this requires a larger pump. If the suction lift is low, or if there is a positive head on the suction end of the pump, the specific speed may increase resulting in a reduction in the required pump size.

**Specific storage ($S_s$)** The volume of water that a unit volume of *aquifer* releases from *storage* under a unit decline in *hydraulic head*. The water that is produced is a result of two mechanisms: (1) compaction of the aquifer caused by the increase in *effective porosity* ($\sigma_e$) and (2) the expansion of water resulting from the reduction in pressure ($p$). Compaction of the aquifer is controlled by aquifer compressibility ($\alpha$) and the expansion of the water by fluid compressibility ($\beta$). Specific storage is defined as:

$$S_s = \rho g(\alpha + n\beta)$$

where:
$g$ = the acceleration of gravity
$n$ = porosity
$S_s$ has dimensions of $L^{-1}$, which follows as $S_s$ is defined as a volume per volume per unit decline in head, measured in units of length.
Cf. *Hydraulic diffusivity; Storativity.*

**Specific yield ($S_y$)** The volume of water that is released from storage in an *aquifer* per unit drop in the *water table* under pumping conditions. The specific yield ($S_y$) of an unconfined aquifer is higher than the storativity of a comparable confined aquifer: the typical range of $S_y$ is 0.01–0.30, whereas typical values for storativity range from 0.005 to 0.00005. The difference is in unconfined aquifers; an actual *dewatering* of the soil pores occurs, whereas releases from storage in confined aquifers represents the effects of water expansion and aquifer compaction as the *fluid pressure* changes. Thus, unconfined aquifers are more efficient for exploitation by wells for the yield realized is greater per unit change in head over less-extensive areas. **Figure S-12** depicts a unit drop in a water table with the crosshatched area representing the specific yield. Cf. *Air-space ratio; Aquifer, confined; Aquifer, unconfined.*

**Specified flow boundary** See *boundary, specified head; flow boundary.*

**Specified head boundary** See *boundary, specified head; flow boundary.*

**Spillway** See *Overflow channel.*

**Split-spoon samplers** A device used to collect soil samples for geotechnical and environmental investigations. As illustrated in **Figure S-13**, a spoon is typically 18–24 in. long, with an outside diameter of 1.5 in. Samples are obtained in a

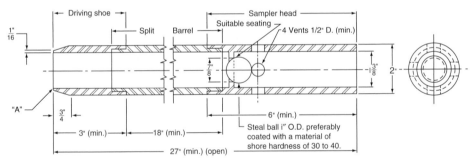

**Figure S-13** Standard split-spoon (barrel) sampler assembly used for sampling of soils. (Reprinted with permission from American Society for Testing of Materials.)

procedure known as a standard penetration test, where the spoon is inserted into a *borehole* and is driven using a standard 63.6-kg (140 lb) hammer, dropped a free fall distance of 0.91 m (30 in.), or the mechanical hammer delivering equivalent energy. The number of blows to penetrate 18–24 in. (46–61 cm) is counted and recorded for each 6-in. (15 cm) interval. The standard penetration test value, $n$, is the sum of the number of blows to penetrate the second 12 in. (30 cm); the number of blows for the first 6 in. is not included, for this may be indicative of slough in the borehole and not representative of natural conditions. If a 24-in. sampler is used, the last 6 in. of penetration may be influenced by the spoon becoming overfilled. The $n$ values are used to determine changes in stratigraphy. The split spoon is split open and samples are retrieved for physical logging and laboratory testing (e.g., grain size, chemical analysis for environmental *contaminants*). The spilt spoon is cleaned out and reused for sampling deeper intervals. Cf. *Grain size analysis.*

**Spontaneous potential (SP) logging** A downhole technique for determining changes in lithology by measuring naturally occurring electrical potentials that exist at contacts between different types of geologic materials. These spontaneous potentials, also called self–potentials, result from chemical and physical changes that occur between a clay layer and a sand layer. For spontaneous potential logging, a sonde (or tool) is lowered into an uncased *borehole* filled with *drilling fluid*. The sonde contains an electrode connected to one terminal of a millivolt meter and recorder. The other terminal of the millivolt meter is connected to a ground terminal at the surface, typically the *mud pit*. No external source of current is applied to this circuit. Current flow results from the electrochemical action between the drilling fluid and the formation, or the formation water. The measured potential drop is that between the fluid column and the downhole electrode and the surface electrode. **Figure S-14** presents an idealized log showing both resistivity and SP.

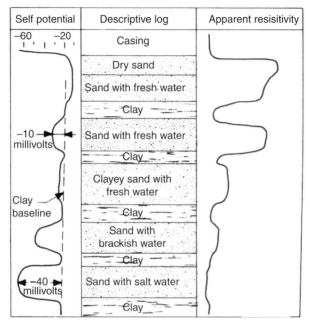

**Figure S-14** Idealized spontaneous potential and resistivity curve showing electric log response corresponding to changes in stratigraphy. Illustrated is a sequence of alternating sand and clay strata. Note the increase in spontaneous potential with increasing salt content in the lower sand unit, with a corresponding decrease in resistivity. (Driscoll, 1986. Reprinted by permission of Johnson Screens/a Weatherford Company)

Cf. *Geophysical exploration methods.*

**Spreading Zone** Under *artificial recharge* (**Figure S-15**), infiltrating water spreads from the footprint of the basin under which the *recharge* which is taking place, known as the spreading zone. The spreading zone is encountered in the *infiltration* zone (i.e., the *vadose zone*) above the *water table*.

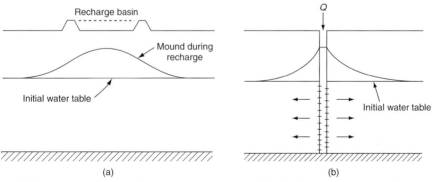

**Figure S-15** Illustration of the spreading zone above the water table that takes place during artificial recharge. (Doneminco, 1990. Reprinted with permission from John Wiley & Sons, Inc.)

**Springs** A *groundwater discharge*, either ephemeral or continuous, at the ground surface. The observation of flowing water usually differentiates a spring from a seep, which may simply be a moist area. Vegetation or chemical precipitates can provide clues as to the presence of springs and seeps. Vegetation includes salt-tolerant phreatophytes such as willow, cottonwood, mesquite, salt grass, and greasewood. Highly *saline* groundwater springs can result in the formation of saline soils, playas, salinas, and salt precipitates. **Figure S-16** presents several different types of springs, and many others are described below.

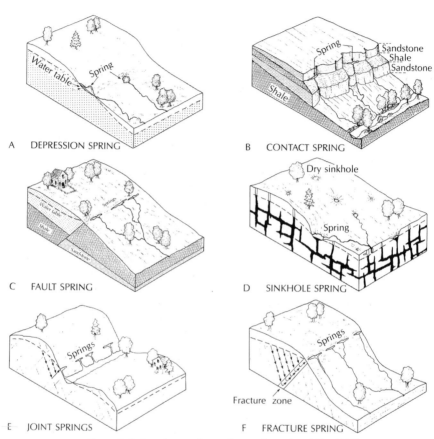

A   DEPRESSION SPRING   B   CONTACT SPRING

C   FAULT SPRING   D   SINKHOLE SPRING

E   JOINT SPRINGS   F   FRACTURE SPRING

**Figure S-16** Several types of naturally occurring springs. (GSA, 2005.)

**Spring, artesian** An area where water issues from the ground at a potentiometric head that is greater than the elevation head at that point, artesian springs are flowing springs where the water rises above ground level. Cf. *Artesian*.

**Spring, barrier** The emergence of *groundwater* at the ground surface as a result of its flow being diverted over or underneath an *impermeable* barrier in an *aquifer* system.

**Spring, boundary** Also referred to as a border spring or alluvial-slope spring. A special gravity spring in which water is emitted from the lower slope of an alluvial cone.

**Spring, contact** Also known as a hillside spring, outcrop spring, or hardpan spring. A gravity flow of water from a *groundwater* source to the land surface from permeable strata overlying less permeable or impermeable stata that prevent or delay *percolation*.

**Spring, depression** Also known as a dimple spring. Gravity flow of water onto the land surface sloping to the *water table*.

**Spring, dike** Naturally flowing *groundwater* issuing at a contact between a dike composed of impermeable material, such as basalt or quartz, and a *permeable* rock into which the dike was intruded.

**Spring, fault** Also known as a fault-dam spring. Free-flowing *groundwater* onto the land surface issuing from a previously faulted area that brought a permeable bed into contact with a less-permeable layer.

**Spring, fault-dam** See *spring, fault*.

**Spring, filtration** Naturally flowing *groundwater* that flows or percolates from numerous small openings in permeable material whose discharge may be quite variable.

**Spring, fissure** Natural flow of *groundwater* issuing from a crack or joint in bedrock and may be flowing at several different locations along the joint.

**Spring, fracture** Natural flow of *groundwater* issuing from a joint or other fractures in bedrock and may be flowing at several different locations along the fracture.

**Spring, gravity** A surface expression or outcropping of the *groundwater*. Natural flow of groundwater from the point where the *water table* and the land surface intersect.

**Spring, hot** Natural flow of groundwater whose temperature is greater than normal human-body temperature (generally greater than 45–50°C; 110–120°F). The water is heated below the ground surface in volcanically active areas.

**Spring, karst** Natural flow of groundwater to the surface from underground *Karst* features.

**Spring, overflow** In areas where a permeable deposit dips below an impermeable stratum, *groundwater* overflows onto the land surface at the edge of the impermeable contact. A special type of contact spring.

**Spring, periodic** *Groundwater* that ebbs and flows above ground due to natural siphoning, typically emanating from solution channels.

**Spring, radioactive** *Groundwater* naturally flowing above the ground surface that has a detectable emission of radiation resulting from a radioactive decay process.

**Spring, rising** Also known as a resurgent spring, where the water level or flow rate is increasing with time.

**Spring, talus** *Groundwater* emanating above the ground within a talus slope, typically at the base. The water may not be classified as part of the *groundwater basin* because it may have originated from recent rainfall seeping into the slope and collected at a less-permeable strata.

**Spring, terrace** *Groundwater* naturally rising above the ground surface that is built up through deposition of material from flowing water forming a series of terraces or basins.

**Spring, thermal** Also known as hot spring or warm spring. *Groundwater* that naturally flows above the ground and whose temperature is higher than the mean local atmospheric temperature. Typically located in volcanic areas.

**Spring, tubular** *Groundwater* naturally flowing from circular or semicircular fissures, solution channels, or lava tubes.

**Stable isotopes** A non-radioactive isotope or a radioactive isotope having an extremely long half-life. Examples of stable isotopes include $^{18}O$ and $^{2}H$.

**Stablizer sand** See *formation stabilizer*.

**Stagnation point (in a flow system)** In a *groundwater flow* system described by a series of *equipotential lines* and *flow lines*, a point in the flow system where the *hydraulic gradient* is flat, and hence there is no movement of water. Since the *water table* fluctuates, a stagnation point may be a temporary condition.

**Stagnant water** Non-flowing water, generally characterized as being *eutrophic*, potentially under sulfate- or methane-reducing conditions if sufficient organic carbon is present. Cf. *Ficks First Law*; *Brackish water*; *Thalassic series*.

**Standard mean ocean chloride (SMOC)** A standard established for the concentration of chloride ($Cl^-$) in the ocean, used as a reference point in geochemical and isotopic studies.

**Standard mean ocean water (SMOW)** A standard used to compare isotopes of oxygen and hydrogen in water. The isotope ratios are expressed in delta units $\delta$ as per mille (parts per thousand or ‰) differences relative to this standard:

$$\delta\ (\text{‰}) = \left[ R - \frac{R_{\text{sample}}}{R_{\text{standard}}} \right] \times 1000$$

where $R_{\text{sample}}$ and $R_{\text{standard}}$ are the isotopic ratios, $^{2}H/^{1}H$ or $^{18}O/^{16}O$ of the sample and the standard, respectively. The accuracy of the measurement is typically better than $\pm 0.2$ ‰ and $\pm 2$ ‰ for $\delta^{18}O$ and $\delta^{2}H$, respectively.

**Stage** Cf. *Rating curve*; *Stage-discharge relation*.

**Stage-discharge curve** See *Rating curve*.

**Stage-discharge relation** Rating curves relating river stage ($y$; i.e., elevation) with discharge ($Q$). Stage-discharge rating curves for flow in rivers and channels are established by concurrent measurements of stage, $y$, and discharge, $Q$, through velocity measurements. These measurements need to cover a range of flows for a given gauging station and are fitted graphically or statistically to yield rating curves. Stage-discharge ratings are generally treated as following a power curve of the form given by the following equation, $Q = c\,(a + y)^{\alpha}$; in which $\alpha$ is an index exponent, and $a$ and $c$ are constants. Cf. *Rating curve*.

**Static head** The maximum height (pressure) a pump can provide.

**Standard penetration test** See *Split spoon samplers*.

**Static water level** The level of water in a well that is not being affected by the action of withdrawal or injection of groundwater. Cf. *Static head*; *Residual drawdown*.

**Steady-state flow** A condition at which, at any point in a flow field, the magnitude and the direction of *flow velocity* are constant with time. Cf. *Equations of groundwater*; *Darcy's law*; *Unsteady-state flow*; *Flow-steady*.

**S**

**Steady-state stream** In surface water hydrological modeling, evaluating consumption of water, irrigation and human consumption inflow of treated wastewater are evaluated to determine residual impacts in the last year of a user-specified pumping duration. Steady state is attained when the overall model residual approaches zero. A 25-year model run is generally sufficient to provide a good estimate of residual distributions at steady state. Cf. *Graded stream.*

**Steam stripping** A water treatment technique where steam is used to remove volatile organic compounds in a *countercurrent-packed tower.* In common applications of *air stripping*, air is forced countercurrent to the water flow, but for less-volatile, moderately strippable compounds, such as acetone, steam stripping may be employed as means of heating the water, increasing the *Henry's law* constant of the compound, enabling treatment of the water to remove the compounds. Cf. *Pollutant.*

**Step-drawdown test** A procedure for examining the performance of a well under conditions of *turbulent flow.* Conducted to determine well efficiency, optimum pumping rates, and to separate the laminar and turbulent components of flow. In a step-drawdown test, a well is pumped at several successively higher pumping rates and the *drawdown* for each rate (or step) is recorded. Typically, the entire test takes place over the course of one day, using five to eight pumping steps, with the well pumped at each successive pumping rate for 1 or 2 h. Calculations are simplified if the well is pumped for the same amount of time for each successive rate. It is desirable to have the water level in the pumping well return to the static water level before the next successive pumping; however, time does not usually permit this. The data from a step drawdown test can be used to determine the proportion of laminar and turbulent flow from the following equation:

$$s = \frac{264Q}{T} \log\left(\frac{0.3Tt}{r^2 S}\right)$$

where:
$s$ = drawdown [L]
$T$ = *transmissivity* [L$^2 \bullet$T$^{-1}$]
$t$ = time since pumping began [T]
$r$ = *radius of influence* [L]
$S$ = *storativity* [dimensionless]

This can be shortened to:

$$s = BQ$$

where $B$ is derived from the following equation:

$$B = \frac{264}{T} \log\left(\frac{0.3Tt}{r^2 S}\right)$$

From this, the value of $B$ is time-dependent; however changes in $B$ after a certain period of time become small, and this term may then be considered constant. When turbulent flow to a well is induced (as would result from rapid drawdown in the vicinity of the well where the gradient would be highest), drawdown in the well may be more accurately depicted as having a component of laminar flow and a turbulent component such that:

$$s = BQ + CQ^2$$

In this equation, the laminar term ($BQ$) is the *aquifer loss* (i.e., losses in *hydraulic head* attributable to flow through the aquifer) and the turbulent term ($CQ^2$) is the *head loss* attributable to inefficiency. In the real world, this is not always the case, in that the laminar term always contains a major portion of the well losses, while the turbulent term occasionally includes some aquifer losses. A graphical method has been developed to determine $B$ and $C$. Dividing the preceding equation by $Q$ yields:

$$\frac{s}{Q} = CQ + B$$

Plotting $s/Q$ against $Q$ results in a straight line, with slope $C$ and intercept $B$. This is illustrated in **Figure S-17**. The ratio of laminar head loss to total head loss is determined by the following equation:

$$L_p = \frac{BQ}{BQ + CQ^2} \times 100$$

S

where:

$L_p$ = percentage of the total head loss attributable to laminar flow

*Specific capacity* can also be defined as:

$$\frac{Q}{s} = \frac{T}{264 \log \frac{0.3Tt}{r^2 S}}$$

For example, using typical values of $t = 1$ day, $r = 0.5$ ft, $T = 30,000$ gpd ft$^{-1}$ and $S = 1 \times 10^{-3}$ for a confined aquifer and $S = 7.5 \times 10^{-2}$ for an unconfined aquifer, the specific capacity of a confined aquifer is given by:

$$\frac{Q}{s} = \frac{T}{2000}$$

For an unconfined aquifer, the specific capacity is:

$$\frac{Q}{s} = \frac{T}{1500}$$

To determine well efficiency, the 1-day specific capacity is compared to the theoretical specific capacity.

$$\left(\frac{Q}{s}\right)_{\text{theoretical}} = \frac{T}{2000}$$

The actual specific capacity is defined as $Q/s$. Therefore, well efficiency is:

$$E = \frac{(Q/s)_{\text{actual}}}{(Q/s)_{\text{theoretical}}}$$

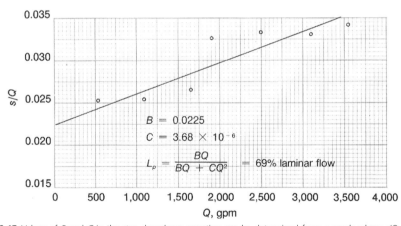

**Figure S-17** Values of $B$ and $C$ in the step-drawdown equation can be determined from a graph where $s/Q$ is plotted against $Q$. (Driscoll, 1986. Reprinted by permission of Johnson Screens/a Weatherford Company.)

Cf. Laminar head loss; Pumping test.

**Step test** See *Jacob step-drawdown test*.

**Stiff diagram** A technique for depicting the major ions of inorganic water chemistry such as calcium ($Ca^{2+}$), sodium ($Na^+$), potassium ($K^+$), magnesium ($Mg^{2+}$), chloride ($Cl^-$), bicarbonate ($HCO_3^-$), carbonate ($CO_3^{2-}$), and sulfate ($SO_4^{2-}$). A stiff diagram such as **Figure S-18** provides a pictorial version of the data for comparison with other water types.

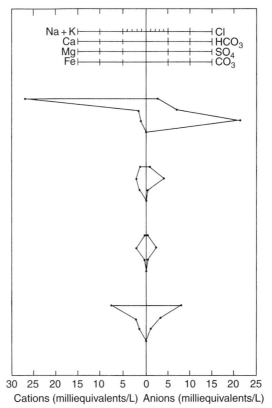

**Figure S-18** Presentation of major ion data using a stiff diagram. (Freeze and Cherry, 1979. Reprinted with permission by Prentice-Hall, Inc.)

Cf. *Piper trilinear diagram; Durov–Zaprorozec diagram; Hydrochemical facies.*

**Storage** A concept in stream hydrology relating groundwater discharge and recharge in the vicinity of a stream. Also known as (river) bank storage. During a flood period, river stage increases induce infiltration of stream water into the adjacent river banks or an aquifer; subsequent declines in river stage cause a reverse motion of the infiltrated water, that is, the base flow of the river results from bank storage. The (bank) storage zone is the part of aquifer where groundwater is replaced by stream water during the flood stage of the river. Cf. *Antecedent moisture conditions.*

**Storage capacity** See *Soil moisture storage capacity.*

**Storage coefficient (S)** See *storativity (S).*

**Storativity (S)** A dimensionless quantity also called storage coefficient and coefficient of storage. It is the volume ($L^3$) of water that a *permeable* unit releases from or takes into *storage* per unit surface area ($L^2$) of the unit per unit change in *hydraulic head.* For a confined aquifer of thickness $b$, the storativity is equal to the product of the *specific storage* ($S_s$) and the aquifer thickness, mathematically represented by:

$$S = S_s b$$

The hydraulic head for a confined aquifer is often displayed as a *potentiometric surface.* **Figure S-19** illustrates the concept of storativity. In an unconfined aquifer the storativity is equivalent to the *specific yield* ($S_y$), or more accurately the specific yield is considered to be the "unconfined storativity". Both representations of storativity are shown in **Figure S-19.** Cf. *Aquifer, confined; Aquifer, unconfined.*

**Stormwater** Precipitation from a relatively short-lived but intense rain event. Design of flood control facilities take into account these relatively short duration but intense events. Cf. *Retention basin; Detention basin.*

**Stormwater runoff** Water resulting from storms in rivers, gullies, streets, and other locations where water accumulates.

**Straight-line method** See *Jacob distance–drawdown; Jacob time–drawdown* and *step-drawdown test.*

**Straight stream** A stream flow path or watercourse with a sinuosity of less than 1.25 (median length times the valley length). Cf. *Meander; Braided stream.*

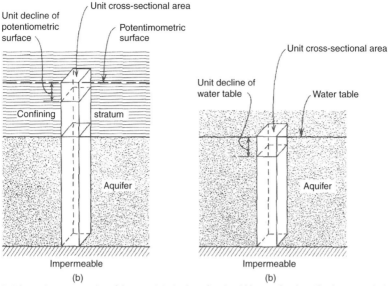

**Figure S-19** Schematic representation of the storativity in a) confined and b) unconfined aquifer. (Freeze and Cherry, 1979. Reprinted with permission from Prentice-Hall, Inc.)

**Stream** A conduit in the water cycle, with a current confined within a bed and stream *banks* which acts as a mechanism for groundwater *recharge* or *discharge* and also serves as a corridor for fish and wildlife migration. Stream is an umbrella term used in the scientific community for all flowing natural waters, regardless of size. Depending on its location, size, local vernacular or certain characteristics, a stream may be referred to as brook, beck, burn, creek, crick, kill, lick, rill, river, syke, bayou, rivulet, or run. The biological habitat in the immediate vicinity of a stream is called a riparian zone. Cf. *Downstream; Braided stream; Straight stream; Meander; Stream order.*

**Streambed** Materials that are part of or lying within a watercourse. The streambed materials, as part of the channel through which water is transported, are inherently part of the geologic or stratigraphic record of the local and possibly upgradient area and are an integral part of the dynamics of river mechanics. The streambed consists of two major divisions of elements: discrete particles that affect stream flow as a function of the size and shape of the particles themselves and the aggregates forming the structures in the streambed. **Figure S-20** displays the physical differences as well as their mode of travel between the stream's bed load, suspended load, and dissolved load.

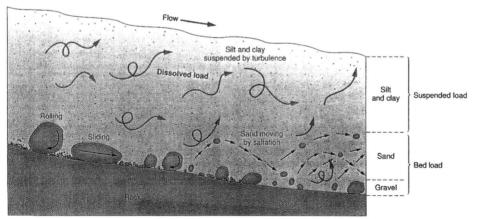

**Figure S-20** The stream's bed load, suspended load, and dissolved load. (McGeary, 2001. Reprinted with permission from McGraw-Hill, Inc.)

**Stream baseflow** *Groundwater* seepage into a *stream* channel. During most of the year, stream flow is composed of both groundwater *discharge* and land surface *runoff*. When groundwater provides the entire *flow* of a stream, stream baseflow conditions are said to exist. Cf. *Groundwater recession curve*.

**Stream capacity** The total amount of sediment a stream is able to transport, which usually corresponds to the stream *gradient* along a stream profile.

**Stream capture** Also called stream piracy. Diversion effected by a stream eroding headward so as to tap and lead off or capture the waters of another stream. The natural diversion of stream flow into the channel of another stream flowing at a lower level and having greater erosional activity. Cf. *Abstraction; Consequent stream*.

**Stream flow** The water that constitutes stream flow reaches the *stream channel* by several paths from a point from where it first reaches the earth, or enters the *hydrologic cycle*, as *precipitation*. For hydrologic investigations involving the streamflow rate its critical to determine and distinguish between the components of total flow. Examples of components of streamflow are *overland flow*, *infiltration*, inflow, outflow, interflow and base flow.

**Stream gauge** See *River gauge*.

**Stream gradient** The ratio of decline in elevation in a *stream* per unit distance, usually expressed as feet per mile or meters per kilometer. High *gradient* streams tend to have steep, narrow V-shaped valleys and are referred to as young streams. Low-gradient streams have wider and less rugged valleys, with a tendency for the stream to meander. Cf. *Gradient flow; Meandering*.

**Stream hydrograph/streamflow hydrograph** Streamflow hydrographs depict two very different types of contribution from a watershed. The hydrograph is constructed with the months of the year on the x-axis and stream *discharge* $(L^3/t)$ on the y-axis. The stream discharge for the hydrograph is the total of the fast response to short term changes in the subsurface flow systems in hillslopes adjacent to *stream channels* (subsurface storm flow, groundwater flow) and slow responses to long term changes in the watersheds flow system (baseflow). Cf. *Hydrograph separation*.

**Streamlines** Fluid flow is described in general by a vector field in three (for steady flows) or four (for nonsteady flows including time) dimensions. Streamlines are a family of curves that are instantaneously tangent to the velocity vector of the flow. By definition, streamlines defined at a single instant in a flow do not intersect. This is so because a fluid particle cannot have two different velocities at the same point.

**Stream load** See *Load; Loaded stream; Sediment load; Bed load; Dissolved load*.

**Stream maturity** Cf. *Rejuvenated stream*.

**Stream meanders** See *meander/meandering streams*.

**Stream order** A method of numbering the *streams* within a single *drainage basin*. The first-order stream would be the smallest unbranched stream or mappable *tributary*. The second-order stream would be the stream into which the first order empties. A sketch of two differing stream basins is presented in **Figure S-21**, demonstrating how the number of streams and stream order may affect drainage densities.

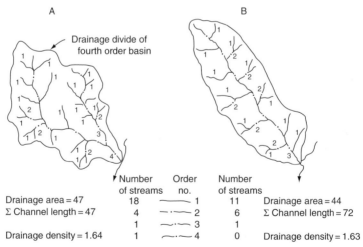

**Figure S-21** How the number of streams and stream order in different basins may affect drainage densities. (Leopold et al., 1992. Reprinted with permission from Dover Publications, Inc.)

Cf. *Basin area; Basin order; Bifurcation ratio; Law of basin area; Horton analysis*.

**Stream stage** See *River stage.*

**Stream tube** See *Flow channel.*

**Stream velocity** In hydrologic studies, gauge height is calibrated to stream velocity and/or stream discharge. A section of a stream or river is selected that is relatively straight and the laminar water flow is present. A transect perpendicular to the stream is established and an imaginary box is placed around each spot on the stream transect. Flow is determined for each box and summed for all boxes. Flow associated with each box is calculated by multiplying the width of the box at each spot (typically 1 m) by stream depth (measured at each velocity measuring point) by the velocity determined using a flow meter. The flow of each box is in cubic meters per second ($m^3 s^{-1}$) and the flow of each box is added together to give total flow.

$$\text{Total flow} = (W_1 \times SD_1 \times SV_1) + (W_2 \times SD_2 \times SV_2) + (W_n \times SD_n \times SV_n)$$

Average stream velocity is calculated by dividing total flow by the cross-sectional area of your transect. The cross-sectional area is determined by calculating a cross-sectional area for the box at each spot of your transect and then summing the cross-sectional areas.

$$\text{Average stream flow} = \frac{\text{total flow}}{(W_1 \times SD_1) + (W_2 \times SD_2) + (W_n \times SD_n)}$$

**Stress** The application of a force.

**Structural Lake** See *Tectonic Lake.*

**Subconsequent stream** Also known as a secondary *consequent stream*. A *tributary* of a *subsequent stream*, flowing parallel to or down the same slope as the original consequent stream. It usually develops after the formation of a subsequent stream but in a direction consistent with that of the original consequent stream. Cf. *Lateral consequent stream.*

**Sublimation** The conversion of a solid to a *vapor* (or gas) without first changing phase to a liquid. Cf. *Latent heat of condensation.*

**Submergence** The state of being covered with or immersed in water.

**Subsequent stream** Also know as longitudinal stream or strike stream. A *stream* that *flows* in the direction of the strike of the underlying strata and is subsequent to the formation of the *consequent stream* of which it is a *tributary.* Cf. *Obsequent stream.*

**Subsidence** Differential settlements resulting from excessive groundwater withdrawals. Pumping induces horizontal *hydraulic gradients* toward a well that result in a decrease in *hydraulic head* near the well. Effective stresses are increased at these points, resulting in aquifer compaction. The feature that is common to all instances of subsidence is that there is a layer of unconsolidated or poorly consolidated sediments within an interbedded aquifer–aquitard system. The actual strata that are compacted are the aquitard, for the compressibility of clay is 1–2 orders of magnitude greater than the compressibility of sand. Since the *hydraulic conductivity* of clay is orders of magnitude less than that of sand, the drainage of water from the clay is much slower, and as a result the compaction process is much slower.

**Subsoil** Generally, the soil beneath the organic soils or topsoil found at the surface.

**Successful wildcat** See *discovery well.*

**Suction lift, limits of** Suction lift limits of conventional vacuum pumps are typically between 20 and 25 ft (6.1–7.6 m), but can be increased through the use of slurping or dual-phase extraction, whereby a liquid ring pump draws a water–air mixture. The *specific gravity* of the water–air mixture is less than that of water alone, and as such the suction lift can be increased. Cf. *Pumping test; Shut-off head; Specific speed; Well design; Check valve; Lifting head.*

**Sulfamic acid** Also known as aminosulfonic, amidosulfonic, and amidosulfuric acid. A dry, white granular material with the formula $H_3NO_3S$ that produces a strong acid when mixed with water. Sulfamic acid is primarily used for treating calcium and magnesium incrustants. It is generally less effective on iron and *manganese incrustants*, but can dissolve iron with the addition of rock salt. Sulfamic acid has handling and safety benefits over *hydrochloric acid* (HCl). In its dry form, it is relatively safe to handle, does not give off fumes, and will not irritate dry skin. Available in pelletized, granular, and powdered forms, dry $H_3NO_3S$ is relatively safe to handle. In situations where the *well screens* are deep within the water column, the pellets are heavier and sink to the zone to be treated. Sulfamic acid may be marketed under the trade name of Nu-Well®. **Table S-5** presents the amount of Nu-well to treat a moderately plugged 1 ft (0.3 m) section of screen. Cf. *Well maintenance.*

**Sulfate reduction** The biologically mediated reduction of sulfate ($SO_4^{2-}$) to sulfide in *groundwater*. As water *flows* through soil–rock systems, oxygen is consumed through the oxidation of organic matter. Oxidation can still occur after the dissolved oxygen in water is consumed, but the oxidizing agents are nitrate ($NO_3^-$), manganese dioxide ($MnO_2$), ferric iron ($Fe^{3+}$), and $SO_4^{2-}$. Cf. *Electrochemical sequence.*

**Supercritical flow** Also known as *rapid flow* or supersonic flow. Flow of a fluid over a body at speeds greater than the speed of sound in the fluid, and in which the shock waves start at the surface of the body. Cf. *Froude number.*

S

**Table S-5**  Quantity of Nu-Well required to treat a moderately plugged 0.3-m (1 ft) section of screen

| Screen diameter | | Screen capacity | | Nu-Well Required | |
|---|---|---|---|---|---|
| mm | in | $L\,m^{-1}$ | $gal\,ft^{-1}$ | $kg\,m^{-1}$ | $lbs\,ft^{-1}$ |
| 38 | $1\,^{1}/_{2}$ | 1.2 | 0.1 | 0.3 | 0.2 |
| 51 | 2 | 2.5 | 0.2 | 0.6 | 0.4 |
| 76 | 3 | 5.0 | 0.4 | 1.3 | 0.9 |
| 102 | 4 | 8.7 | 0.7 | 2.4 | 1.6 |
| 127 | 5 | 12.4 | 1.0 | 3.9 | 2.6 |
| 152 | 6 | 18.6 | 1.5 | 5.5 | 3.7 |
| 203 | 8 | 32.3 | 2.6 | 9.7 | 6.5 |
| 254 | 10 | 50.9 | 4.1 | 15.2 | 10.2 |
| 305 | 12 | 73.2 | 5.9 | 21.9 | 14.7 |
| 356 | 14 | 99.3 | 8.0 | 29.8 | 20.0 |
| 406 | 16 | 129 | 10.4 | 38.9 | 26.1 |
| 457 | 18 | 164 | 13.2 | 49.2 | 33.0 |
| 508 | 20 | 202 | 16.3 | 60.8 | 40.8 |
| 559 | 22 | 246 | 19.8 | 73.6 | 49.4 |
| 610 | 24 | 292 | 23.5 | 87.5 | 58.7 |
| 711 | 28 | 397 | 32.0 | 119 | 80.0 |
| 762 | 30 | 455 | 36.7 | 137 | 91.8 |
| 813 | 32 | 519 | 41.8 | 156 | 104 |
| 864 | 34 | 586 | 47.2 | 176 | 118 |
| 914 | 36 | 657 | 52.9 | 197 | 132 |

**Superfund Amendments and Reauthorization Act (SARA)** Also called Superfund. The 1986 reauthorization of the *Comprehensive Environmental Response Compensation Liability Act* (CERCLA), which had been established by the United States Congress in 1981 to identify, prioritize, and cleanup abandoned sites or areas where released hazardous substances have migrated offsite in an uncontrolled manner. CERCLA gave rise to the definition of Superfund sites as entire regions containing multiple sources of pollution, each contributing to a much larger groundwater plume, such as the San Fernando or San Gabriel Valley Superfund sites in the greater Los Angeles area in California. This added structure to the program including specific deadlines for the more important Superfund activities, linked remedial action closely to the administrative record, and encouraged more state involvement in the Superfund process.

**Superposition** See *Image well theory*

**Supersaturation** A situation where there is an excess of ionic constituents. Supersaturation can be determined using the saturation index. Consider the reaction of constituents B and C to produce D and E:

$$bB + cC \Leftrightarrow dD + eE$$

where $b, c, d,$ and $e$ are the mole fractions of constituents B, C, D, and E, respectively. The reaction quotient of this reaction can be expressed as:

$$bB + cC \Leftrightarrow dD + eE$$

At equilibrium, the equilibrium equation is expressed as:

$$K_{eq} = \frac{[D]^d[E]^e}{[B]^b[C]^c}$$

The saturation index is expressed as:

$$S_i = \frac{Q}{K_{eq}}$$

Where $S_i$ is greater than 1, supersaturation conditions exist and the tendency is for the precipitate to be formed from the solution. At $S_i < 1$, dissolution will occur.

**Supersonic flow** See *Supercritical flow.*

**Supplied water** The self- or third-party provision of water of various qualities and quantities to different users.

**Surface geophysical methods** Geophysical exploration techniques employed at the ground surface for mapping or vertical profiling to define variations in lithology, ore deposits, or buried features.

**Surface resistivity** A surface *geophysical exploration method* involving mapping of the *electrical resistivity* of subsurface strata. Surface resistivity can also be used for *electrical profiling* or electrical drilling.

**Surface retention** See *Detention storage; Depression storage; Retention storage; Surface storage.*

**Surface runoff** See *Runoff; Overland flow.*

**Surface storage** Also known as initial detention or surface retention. The part of *precipitation* retained temporarily at the ground surface as interception or *depression storage* so that it does not appear as *infiltration* or subsurface *runoff* either during the *rainfall* period or shortly thereafter. Cf. *O' Donnell model.*

**Surfactant** A substance when dissolved has the ability to reduce the surface tension of the liquid in which it is dissolved. Used to promote the dissolution of *non-aqueous-phase liquids* into *groundwater*, to produce foam in air-based drilling fluids, and to disaggregate clays during *well development.*

**Surge block** Also called a surge plunger. In *well development*, a block that is inserted into the well at the end of a drill rod. On the downstroke, water is forced outward into the formation, while water, silt, and sand are then pulled into the well on the upstroke. **Figure S-22** is an illustration of a surge block.

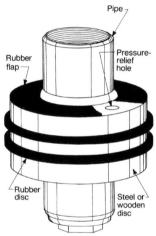

**Figure S-22** Illustration of a surge block and its use in well development. The typical surge block consists of two rubber disks sandwiched between three steel or wooden disks. On the downstroke, water is forced outward into the formation. Water, silt, and fine sand are pulled through the well screen on the upstroke. (Driscoll, 1986. Reprinted by permission of Johnson Screens/a Weatherford Company.)

**S**

**Surge plungers** See *Surge block.*

**Surging** A *well development* technique using a *surge block.*

**Suspended load** The sediment, or *sediment load*, carried in suspension by a *river* or *stream*. Cf. *Traction; Bed load; Bed material.*

**Suspended solid** Insoluble material that either floats on the surface of or is held within the water or other liquid column. A suspended solid could be colloidal, dispersed, coagulated, flocculated, and physically held in suspension by liquid flow or agitation. Cf. *Flocculation.*

**Sustained yield** *Groundwater* basins that contain vast reserves of water that may be pumped with a planned withdrawal of water at a rate that can be sustained over a long period.

**Swabbing** Also called line swabbing. A surge-type *well development* technique, which typically involves a rubber-flanged mud scow or *bailer* that is lowered into the well and raised at about 3 ft s$^{-1}$ (0.9 m s$^{-1}$) with no attempt to reverse the *flow*. As the scow or the bailer is raised, high pressure is created near the top of it, forcing water into the formation. Low pressure is created at the bottom of the scow as it is drawn up, drawing water and sediment into the well. Swabbing is used to clean fine material from deep wells drilled in consolidated rock aquifers. **Figure S-23** illustrates a swabbing tool.

**Swale** A low-lying area that acts as drainage way in times of *precipitation*. A swale is typically above the *water table* and its flow is discontinues after the precipitation event is over. The relatively short length differentiates it from an *ephemeral river.*

**Swamp** See *wetland.*

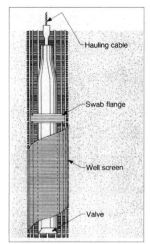

**Figure S-23** Illustration of the use of a swabbing tool. As the swab is pulled upward at about $1 \, \text{m s}^{-1}$ ($3.1 \, \text{ft s}^{-1}$), high-pressure conditions at the top of the swab force water into the formation. Low-pressure conditions at the base cause flow of sand, silt, and water back into the borehole. Primarily used in unconsolidated conditions. (Driscoll, 1986. Reprinted by permission of Johnson Screens/a Weatherford Company.)

**Swedge block and bar** Tools used to set telescoping casing fitted with a *lead packer* into a telescoped screen assembly. The bar allows a driving stroke of approximately 0.3 m (1 ft) to expand the lead packer. **Figure S-24** illustrates the use of a swedge block and bar.

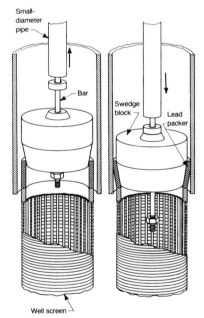

**Figure S-24** A swedge block is used to expand a lead packer attached to the top of a telescoped screen assembly. (Driscoll, 1986. Reprinted by permission of Johnson Screens/a Weatherford Company.)

S

**Swedging** The use of a *swedge block and bar* to set a *lead packer* in a telescoped screen installation.

**Swing joint** In a *well-point* system, a device that allows *well points* to be easily connected to the vacuum header for the extraction of *groundwater*. **Figure S-25** illustrates a standard well-point swing and a no-tool swing.

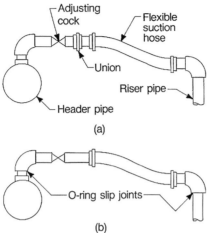

(a)

(b)

**Figure S-25** Two basic types of well-point connections: (a) standard well-point swing and (b) no-tool swing. The standard swing configuration is preferred because the well point and the header can always be easily disconnected, even if the header and the well point are not properly spaced. (Powers, P. A., CONSTRUCTION DEWATERING© 1981. Reprinted with permission by Wiley-Interscience.)

**Tag line** A cable set across the width of a watercourse perpendicular to *flow*, as shown in **Figure T-1**, used to identify locations where current velocity meter readings are to be taken.

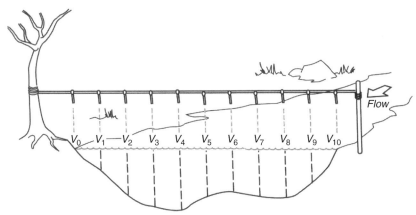

**Figure T-1** The tag line stretched across the stream at marked points where velocity is to be measured. Velocity is measured at the edge of each section and not in the center. (Sanders, 1998.)

Cf. *Stream gauge.*

**Tail** The downgradient or *downstream* section of a small water body or watercourse. The term "tail" especially refers to a relatively calm *reach* or section that follows a more turbulent section.

**Tangential flow** See *Flowline refraction.* Cf. *Streamline.*

**Talus spring** See *Spring; talus.*

**Talweg** See *thalweg in groundwater; Thalweg in streams.*

**Tangent law of refraction** In a steady-state and transient *groundwater flow* system, *flow lines* bend at the interface of materials of differing *permeabilities.* The refraction pattern follows a law of tangents similar to Snell's law of light refraction involving sines. Cf. *Steady-state flow; Unsteady-state flow.*

**Tapoon** A dam installed at the subsurface of a dry wash to either increase *recharge* to nearby *wells* or impound water for later use.

**Tarn** A landlocked water body such as those occurring in tracts of marsh, bog, or swamp, or in a glacial gorged basin, which are small and typically deep with sheer sides.

**Tattletale screen** See *telltale screen.*

**Taylor series approximation** See *Finite difference technique.*

**TDS** See *Total Dissolved Solids (TDS).*

**Tectonic lake** Also called a structural lake. A basin produced predominantly by crustal movement, e.g., upwarping causing the impoundment of a *drainage basin* or water accumulating in a graben.

**Telescope installation** See *Pull-back method of screen installation.*

**Telescope-size screen** A screen designed to be placed in a well by "telescoping" it through the *well casing.* The diameters of telescope-size screens allow them to be lowered freely through the standardized pipe used as *well casing.* The outer-diameter size of the screen allows for just enough clearance for it to pass through the selected casing. The screen size selected is dependant on the size of the well casing or inside diameter of the drill pipe. **Table T-1** provides telescope-size screen dimensions used in relation to the casing size.

**Telltale screen** Also called a tattletale screen. A short section of screen placed in the *annular space* above the production screen of a well, used in *well installation* to determine whether the desired fill level has been reached. As the *filter pack* and water are pumped into the *annulus,* the return flow passes through the production and telltale screens. When the annular space around the production screen is filled up, the return water is then observed only in the telltale screen, indicating that the *filter pack* has been installed to the desired level.

**Telluric water** Water formed at high temperature and pressure by combining hydrogen with the oxygen of the atmosphere.

**Tempe cell** The name given to a short soil column used for determining soil–water retention curves. Tempe cells are similar to flow cells, but flow cells are more flexible in their use. Flow/Tempe cells come in various lengths and diameters, with highly permeable nylon membranes, which are readily replaceable and require less time for soil to come to equilibrium during water retention experiments. Flow cells can be used for determining soil water retention curves, performing solute transport experiments, as well as for determining the saturated and unsaturated hydraulic conductivities of soils.

**T**

**Table T-1** The screen dimensions and inside/outside screen diameters for several telescope-size well screens. (Driscoll, 1986. Reprinted by permission of Johnson Screens/a Weatherford Company.)

| Nominal Casing Size | | Screen Outside Diameter | | Screen Inside Diameter | | Pipe-Size Threaded Fittings | | |
|---|---|---|---|---|---|---|---|---|
| in | mm | in | mm | in | mm | in | | mm |
| 3 | 76 | $2^3/_4$ | 70 | 2 | 51 | 2 | M or F | 51 |
| 4 | 102 | $3^3/_4$ | 95 | 3 | 76 | 3 | M or F | 76 |
| 5 | 127 | $4^3/_4$ | 121 | 4 | 102 | 4 | M or F | 102 |
| 6 | 152 | $5^5/_8$ | 143 | $4^7/_8$ | 124 | 5 | M or F | 127 |
| 8 | 203 | $7^1/_2$ | 191 | $6^5/_8$ | 168 | 6 | M or F | 152[a] |
| 10 | 254 | $9^1/_2$ | 241 | $8^5/_8$ | 219 | 8 | M or F | 203[a] |
| 12 | 305 | $11^1/_4$ | 286 | $10^3/_8$ | 264 | 10 | M or F | 254[a] |
| 14 | 356 | $12^1/_2$ | 318 | $11^3/_8$ | 289 | 12 | M or F | 305[a] |
| 16 | 406 | $14^1/_4$ | 362 | $13^1/_8$ | 333 | | | |
| 18 | 457 | $16^1/_4$ | 413 | 15 | 381 | | | |
| 20 | 508 | $18^1/_4$ | 464 | 17 | 432 | | | |
| 24 | 610 | $22^5/_8$ | 575 | $20^3/_4$ | 527 | | | |
| 30 | 762 | $27^7/_8$ | 708 | 26 | 660 | | | |
| 36 | 914 | $31^7/_8$ | 810 | 30 | 762 | | | |

[a] Special threads are used when connecting multiple screen sections to maintain ID dimensions.

**Temperature logging** A downhole logging technique utilizing a sonde equipped with a thermocouple or other device to record temperature at depth. Temperature logging is especially useful in investigating fracture flow in bedrock, as *recharge* from the surface (especially during winter in northern climates) is at a lower temperature than the *groundwater*. The technique has also been used to identify fractures through which groundwater contributes to recharging. Temperature, along with *conductivity* and usually *pH*, is continuously monitored during *pumping tests* to help determine whether water is breaching in from an *aquifer* other than the one being tested. Because temperature is sensitive to aquifer depth, temperature logging can be used to distinguish different water systems. A constant temperature is an indication that the pumping test is restricted to a single aquifer. Groundwater is often temperature-equilibrated with the subsurface material in which it resides (subject to convective influences); therefore, temperatures measured in wells or *springs* provide general information on the *depth of circulation* when used with the following equation:

$$\text{depth} = \frac{T_{\text{measured}} - T_{\text{surface}}}{\Delta T/100}$$

where:
$T_{\text{measured}}$ = temperature of the spring or well water [°]
$T_{\text{surface}}$ = local average annual surface temperature [°]
$\Delta T$ = local *geothermal gradient* established by geophysical studies

The values obtained from the above calculation are the minimum depths of circulation expected if the water samples are collected from a low-discharging sampling port, as the water may cool during ascent. Contours of equal temperature measured from groundwater wells have been used to determine recharge from a *river system* because of the tendency of the seasonal variations in aquifer temperature to lag behind the seasonal changes measured in river temperature. Temperature measurements are also critical to the determination of temperature-dependent rate coefficients; reaction rates tend to increase with increasing temperatures. The temperature-dependent physical properties of water are listed in **Appendix B**. Cf. *Thermal-neutron log; Thermal stratification; Spring, thermal.*

**Temporary base level** Any elevation, other than *sea level*, below which *erosion* can no longer reduce the land area. The temporary base level is typically controlled by a resistant layer. Cf. *Local base level; Ultimate base level.*

**Temporary hardness** More appropriately called carbonate hardness versus permanent hardness as non-carbonate hardness. See *carbonate, carbonate hardness, hard water, lime softening, softening of water.*

**Temporary lake** See *intermittent lake; playa lake; dry lake.*

**Temporary stream** See *intermittent stream.*

**Tensiometer** A device that measures soil–water tension or soil suction, used in the field to determine the negative *fluid pressure*, $\psi$, in the *unsaturated zone*. A typical tensiometer is a tube filled with water that is sealed at the top and has a *porous*, permeable ceramic cup or tip at the bottom. Water *flows* out through the bottom into the soil being tested and creates a vacuum at the top; the more the water flowing into the soil, the greater the vacuum. The tensiometer is connected to a pressure-measuring transducer, *manometer*, or vacuum gauge to determine soil suction or the relative measure of *soil moisture*. The measurements are collected over time until equilibrium is attained. The equilibrium pressure remains

T

steady until the moisture conditions in the soil change. Cf. *Gravimetric pressure plate or membrane; Thermocouple psychrometer.*

**Tension** A condition allowing for the existence of interstitial water at a pressure less than the atmospheric pressure. Tension exists when *fluid pressures* above the *water table* are negative with respect to the local atmospheric pressure. Air may also exist in the voids above the water table, but there the air pressure is equal to the atmospheric pressure. Cf. *Moisture potential; Matric potential.*

**Tension barrier** An effective boundary in strata, delineated by an abrupt change in soil pore size, that prevents downward movement of water. As shown in **Figure T-2**, if a clay layer overlying a pure sand is saturated, the water from the clay cannot percolate down because there is no *gradient* for movement. If the pure sand layer is above the clay, relatively little water moves from the sand to the clay; although the water is under strong tensional forces, the distinct difference in tension between the two layers creates a tension barrier, and without the gradient, the water does not move.

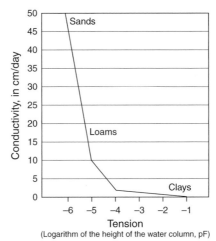

**Figure T-2** An abrupt change in soil pore size prevents downward movement of water. The graph shows the relationship among capillary conductivity, tension, and soil particle sizes. (Black, 1996. Reprinted with permission from Ann Arbor Press, Inc.)

Cf. *Retention storage; Detention storage.*

**Tension-saturated zone** The *capillary fringe* in which water is held by forces of adhesion and cohesion. The pores within the tension-saturated zone are saturated, but the *pressure head* is less than the atmospheric pressure.

**Terminal velocity of raindrops** The limiting rate of water falling as *precipitation* under the action of gravity. The terminal velocity of raindrops is dependent on *raindrop* size, water temperature (affecting *viscosity*), height of fall, and vertical air movement. The calculation of the velocity, $v(D)$, can be simplified as:

$$v(D) = 3.86D^{0.67}$$

where:
$D =$ raindrop diameter (between 0.8 and 4.0 mm) [L]

The terminal velocity range is approximately 170–900 cm s$^{-1}$. Refer to **Appendix A** for unit conversions. Both the fall velocity and the size distribution of raindrops are critical in the determination of precipitation rate, which influences what happens once the raindrops encounter the land surface. Precipitation occurs during updrafts and therefore tends to remain aloft until drops coalesce and become large enough to overcome air friction and buoyant forces, or until there is a downdraft. If there is too much friction, the drops break up.

**Terrace spring** See *spring; terrace.*

**Terrain conductivity** A *geophysical exploration method* using electromagnetic waves to map variations in electrical resistance or *conductivity* in the subsurface. Terrain conductivity is typically employed in mineral exploration and *groundwater* investigations. In contaminant investigations, terrain conductivity has been used to delineate plumes of higher conductivity in groundwater. Cf. *Surface conductivity; Schlumberger arrays; Wenner array.*

**Tertiary sewage treatment** The removal of residual biological oxygen demand (BOD), phosphate, odor, and taste in the final stage of purification in wastewater treatment. Cf. *Wastewater treatment.*

**T**

**Test reach** A section of a watercourse between two *gauging stations* that have sufficient length to determine the *gradient* within that section.

**Test well** A well dug or drilled in search of a viable water source or with the intent of measuring water elevations, collecting water samples for chemical analysis, conducting *aquifer tests*, or conducting geophysical surveys of subsurface formations.

**Texture of drainage network** See *drainage pattern*.

**Thalassic series** A method of classifying *brackish waters* and freshwater by their percent *salinity* in the following zones.

| | |
|---|---|
| >300% | |
| | Hyperhaline |
| 60–80% | |
| | Metahaline |
| 40% | |
| | Mixoeuhaline |
| 30% | |
| | Polyhaline |
| 18% | |
| | Mesohaline |
| 5% | |
| | Oligohaline |
| 0.5% | |

Cf. *Chebotarex's succession; Halinity*.

**Thalweg in groundwater** Also spelled talweg. A path of *groundwater flow* below a surface watercourse or valley, usually flowing in the same direction as its surface counterpart.

**Thalweg in streams** Also spelled talweg. A *longitudinal profile* or line connecting the lowest or deepest points along a watercourse, line of maximum depth, median line, or channel line.

**Theis curve** See *Theis non-equilibrium-type curve; Theis non-equilibrium equation/Theis equation/Theis solution*.

**Theis method** A graphical technique used to estimate values of *transmissivity* and storativity by comparing plots of field data (time versus *drawdown*) collected during an *aquifer test* to the *Theis non-equilibrium-type curve* of the well function. At least one non-pumping *observation well* is necessary to calculate storativity by the Theis method. Cf. *Theis non-equilibrium equation; Theis non-equilibrium-type curve; Thiem equations; Jacob straight-line method*.

**Theis non-equilibrium equation/Theis equation/Theis solution** An analytical calculation of the *drawdown, s*, caused by a pumping well. In addition to the basic assumptions about hydraulic conditions listed in **Appendix B**, Theis made the following assumptions for *flow* in a confined aquifer:

- The *aquifer* is fully confined and has no source of *recharge* or *leakage* into it.
- The aquifer water is compressible and is released instantaneously as the *head* is lowered.
- The well fully penetrates the aquifer and is pumped at a constant rate.
- The water in storage in the well bore is ignored.

**Figure T-3** is a conceptual model showing the *hydraulic characteristics* or parameters of an aquifer as determined by the Theis solution. The well *discharge, Q*, is constant, the *hydraulic head* is constant at a time of $t = 0$, and the drawdown, *s*, expressed as hydraulic head is $h - h_0$. The Theis equation for the drawdown becomes:

$$s(r, t) = \frac{Q}{4\pi T} \int_u^\infty \frac{e^{-z}}{z} \, dz = \frac{Q}{4\pi T} W(u)$$

where:

$r$ = radial distance to the *observation well* [L]
$S$ = aquifer storativity [dimensionless]
$T$ = aquifer *transmissivity* [$L^2 \cdot T^{-1}$]
$t$ = time since pumping began [T]
$W(u)$ = *well function of u*

and for which

$$u = \frac{r^2 S}{4Tt}$$

The *well function of u* is an integral equation with a numerical evaluation that follows a series expansion. The value for the well function is determined from Table W-5. The Theis equation has been simplified by C. E. Jacob, who observed that for values of *u* less than 0.05, the Theis equation can be approximated by the *Jacob straight-line method* as shown in **Figure T-4**. Cf. *Hantush–Jacob equation; Jacob method; Thiem equations*.

341

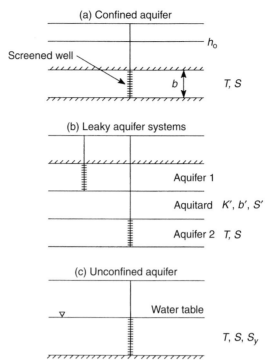

**Figure T-3** Hydraulic characteristics or parameters of an aquifer as determined by the Theis solution. (Maidment, 1993. Reprinted with permission from McGraw-Hill, Inc.)

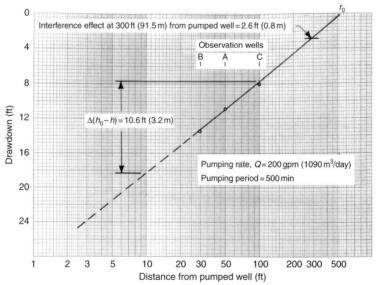

**Figure T-4** The straight line plot of drawdown versus distance in observation wells. (Driscoll, 1986. Reprinted by permission of Johnson Screens/a Weatherford Company.)

**Theis non-equilibrium-type curve** The plot of the *well function of u* on the *y*-axis of three-log-cycle graph paper versus $1/u$ on the *x*-axis of five-log-cycle, as shown in **Figure T-5**. The *argument of the well function, u,* describes the theoretical relationships between *observation well* distance, time since the beginning of pumping, and the storativity and *transmissivity* of a confined aquifer. Field data collected for the observation well during the *aquifer test* are plotted, also on 3 × 5 log–log graph paper, with *drawdown* on the *y*-axis versus time on the *x*-axis. The field data plot is compared to the type curve to obtain a match point. The match point where $W(u) = 1$ and $1/u = 1$ is chosen to simplify the calculations. The values of drawdown and radius or time/radius$^2$ can be substituted into the following Theis equation:

$$T = \frac{Q}{4\pi(h_0 - h)} W(u)$$

where:

$T$ = transmissivity $[L^2 \cdot T^{-1}]$
$Q$ = test pumping rate $[L^3 \cdot T^{-1}]$
$W(u)$ = value of $W(u)$ at the match point
$h_0 - h$ = drawdown value at the match point [L]

Once the value of $T$ is known, the storativity, $S$, can be calculated using the following equation:

$$S = 4Tu\left(\frac{t}{r^2}\right)$$

where:

$S$ = storativity [dimensionless]
$u$ = constant [dimensionless]
$t$ = time since pumping began [T]
$r$ = radial distance to the observation well [L]

Commercially available software now allows this analysis to be done on a computer screen.

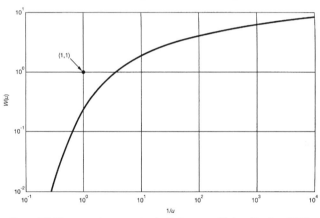

**Figure T-5** The reverse-type curve for the Theis non-equilibrium. (Sanders, 1998.)

Cf. *Theis non-equilibrium equation; Theis method; Thiem equation.*

**Thesis Non-equilibrium well equation** See *Non-equilibrium well equation.*

**Theis-type curve** For a confined aquifer, a logarithmic plot of the well function, $W(u)$, versus the argument $u$, in which $u$ describes the theoretical relationships among observation well distance, time since the beginning of pumping, and the aquifer *storativity* and *transmissivity*. Based on the results of an *aquifer test,* the Theis-type curve is used to determine aquifer properties, drawdown, or pumping rate. Cf. *Theis non-equilibrium equation; Theis method; Thiem equation; Area relations of catchments.*

**Thermal conductivity** A measure of the ability of the basic property of rocks to conduct heat, or the time rate of heat flow by conduction through a unit thickness, across a unit area. The strata's ability to conduct heat is dependent on composition and geometry of matrix, *porosity, pore medium* and varies with temperature. Values of thermal conductivity for geologic materials range over approximately one order of magnitude. Cf. *Geothermal gradient; Depth of circulation; Hydrothermal water; Temperature logging.*

**Thermal-neutron log** A neutron–neutron type of radioactive log for which the downhole detector records slowed neutrons with thermal energy levels of about 0.025 electron volts (eV).

**Thermal pressure/aquathermal pressure** *Pressure* created by thermal expansion of fluids in a medium that is less expansive. Cf. *Temperature logging.*

**Thermal spring** See *spring; thermal.*

**Thermal stratification** The layering produced by the changes in temperature at different depths in a water body, resulting in horizontal layers of different densities. Cf. *Density stratification; Temperature logging; Thalassic series.*

**Thermal water** Water, above or below the ground surface, that has a temperature significantly above the mean annual air temperature.

**Thermocline** Also known as metalimnion. A temperature gradient in which the temperature decrease with depth is greater than that of the overlying and underlying water and is usually associated with water density. Cf. *Thermal stratification; Thalassic series.*

**Thermocouple psychrometer** Used to measure water potential (*head*) in soils and leafy plants. Cf. *Manometer; Tensiometer.*

**Thermoplastic casing** Plastic casing materials typically formed under an extrusion process involving heating. Thermoplastic casing is commonly composed of ABS (as in *acrylonitrile butadiene styrene casing*), polyvinyl chloride (PVC), or chlorinated polyvinyl chloride (CPVC). Other more exotic materials such as Kynar™ and teflon may also be used in rare situations. Plastic *well casing* has been employed since the late 1940s in waters in which steel corrodes rapidly. Thermoplastic casing is resistant to *corrosion* and acid treatment, and its light weight, relatively low cost, and ease of installation make it useful where high strengths are not required. **Table T-2** presents the standard specifications for thermoplastic casing as set by the American Standard Testing Materials F 480-76.

**Table T-2** The wall thickness and tolerances for thermoplastic water well casing pipe as set by ASTM F480-76. (Driscoll, 1986. Reprinted by permission of Johnson Screens/a Weatherford Company.)

| Nominal Pipe Size | SDR 26 | | SDR 21 | | SDR 17 | | SDR 13.5 | |
|---|---|---|---|---|---|---|---|---|
| | Minimum | Tolerance | Minimum | Tolerance | Minimum | Tolerance | Minimum | Tolerance |
| 2 | – | – | 0.113 | +0.020 | 0.140 | +0.020 | 0.176 | +0.021 |
| A 2 ½ | – | – | 0.137 | +0.020 | 0.169 | +0.020 | 0.213 | +0.026 |
| 3 | – | – | 0.167 | +0.020 | 0.206 | +0.025 | 0.259 | +0.031 |
| 3 ½ | – | – | 0.190 | +0.023 | 0.235 | +0.028 | 0.296 | +0.036 |
| 4 | 0.173 | +0.021 | 0.214 | +0.026 | 0.265 | +0.032 | 0.333 | +0.040 |
| 5 | 0.214 | +0.027 | 0.265 | +0.032 | 0.327 | +0.039 | 0.412 | +0.049 |
| 6 | 0.255 | +0.031 | 0.316 | +0.038 | 0.390 | +0.047 | 0.491 | +0.058 |
| 8 | B 0.332 | +0.040 | 0.410 | +0.049 | C 0.508 | +0.061 | – | – |
| 10 | 0.413 | +0.050 | 0.511 | +0.061 | 0.6312 | +0.076 | – | – |
| 12 | 0.490 | +0.059 | 0.606 | +0.073 | 0.750 | +0.090 | – | – |

[A]The minimum is the lowest wall thickness of the well casing pipe at any cross section. All tolerances are on the plus side of the minimum requirements.
[B]Dimensions below the line meet or exceed Schedule 40 in SDR 13.5, 17, 21 and 26
[C]Dimensions below the line meet or exceed Schedule 80 in SDR 13.5, and 17

**Thiem equations** The formulas developed to determine the hydraulic parameters of steady *radial flow* in either a confined or an unconfined *aquifer*. The Thiem equations relate *drawdown* in several *observation wells* to the *discharge* of a pumping well, using information collected during an *aquifer test* under the following conditions and assumptions:

- The pumping well is screened only within the aquifer being tested.
- All observation wells are screened only in the aquifer being tested.
- The entire aquifer thickness is screened by both observation and pumping wells.
- The pumping well is pumped at a constant rate.
- There is no change in drawdown with time, i.e., equilibrium has been attained.



For confined flow, the following additional assumption is required to validate the Thiem equation:

- The aquifer being tested is confined at the top and bottom.

**Figure T-6** illustrates the above conditions and assumptions. The Thiem equation for *transmissivity*, *T*, of a confined aquifer is:

$$T = \frac{Q}{2\pi(h_2 - h_1)} \ln\left(\frac{r_2}{r_1}\right)$$

where:

$T$ = transmissivity $[L \cdot T^{-1}]$
$Q$ = *discharge* $[L^3 \cdot T^{-1}]$
$h_2$ = *head* above a datum at radial distance $r_2$ from the pumping well (saturated thickness) [L]
$h_1$ = head above a datum at radial distance $r_1$ from the pumping well (saturated thickness) [L]

For unconfined flow, the following is added to the general assumptions:

- The unconfined aquifer being tested is underlain by a horizontal *aquiclude*.

The conditions and assumptions for an unconfined aquifer are also shown in **Figure T-6**. The Thiem equation for *hydraulic conductivity*, *K*, of an unconfined aquifer is:

$$K = \frac{Q}{\pi\left(h_2^2 - h_1^2\right)} \ln\left(\frac{r_2}{r_1}\right)$$

where:
$K$ = hydraulic conductivity [L·T$^{-1}$]

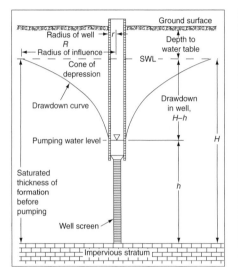
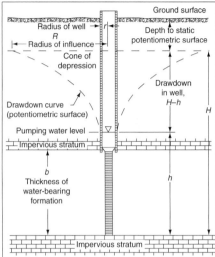

**Figure T-6** A well in an unconfined (a) and confined (b) aquifer displaying the terms used in the equilibrium equation. (Driscoll, 1986. Reprinted by permission of Johnson Screens/a Weatherford Company.)

Cf. *Theis non-equilibrium equation*; *Theis method*; *Theis non-equilibrium-type curve*; *Dewatering*.

**Thiessen polygon method** A graphical method of computing the mean areal *precipitation* based on the assumption that for any point in the *watershed*, actual *rainfall* is equal to the observed rainfall at the closest *gauging station*. As shown in **Figure T-7**, the Thiessen polygons are set up with the boundaries of the polygons formed by the perpendicular bisectors of the lines adjoining adjacent gauges. The relative areas formed in the Thiessen polygon network are used to adjust non-uniform gauge distribution by utilizing a weighing factor for each precipitation gauge.

**Thiessen–Scheel–Diesselhorst equation** The calculation of water density, $\rho_0$, as a function of the water's temperature, $T$, in degree Celsius, from the following equation:

$$\rho_0 = 1000\left[1 - \frac{T + 2.889414}{508929.2(T + 68.12963)}(T - 3.9863)^2\right]$$

**Figure T-8** shows the curve for temperatures in degrees Celsius (°C). Equation assumes that no solutes are dissolved in the water affecting its density. Cf. *Temperature logging*.

**Third-type boundary** This refers to the boundary condition for numerical contaminant transport modeling, establishing the continuity of mass flux at the domain boundary.

**Thixotropy** The *gel strength* that develops as a *drilling fluid* sets. When left standing in a *borehole* or a mud pit, a drilling fluid with clay additives gains strength as increasing numbers of clay platelets align themselves. If the resulting gel strength is too high, excessive pump pressure may be required to resume circulation, so that the drilling fluid may be forced into fractured or weak formations. Pump agitation within the mud pit can be used to restore the drilling fluid to its original *viscosity*. Thixotropy is also a characteristic of many paints and varnishes. Cf. *Newtonian fluid*; *Non Newtonian fluid*; *Plastic fluid*.

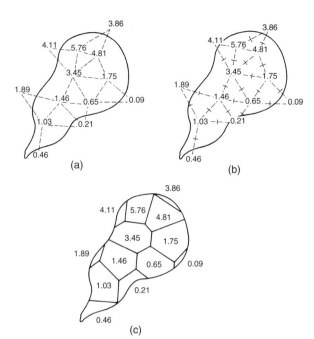

**Figure T-7** Thiessen polygons (a) with rain gauge stations connected with lines, (b) using perpendicular bisector of each line, and (c) formed around each station using extended bisectors. (Fetter, 1994. Reprinted with permission from Prentice-Hall, Inc.)

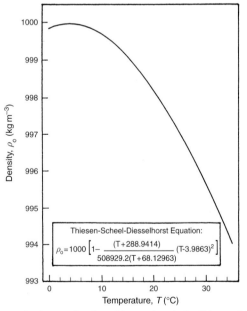

**Figure T-8** Graph of pure-water density as a function of temperature in the Thiessen–Scheel–Diesselhorst equation. (Maidment, 1993. Reprinted with permission from McGraw-Hill, Inc.)

**Thornthwaite equation** A meteorologically based formula for estimating *potential evapotranspiration (PE)* based on mean monthly air temperature, and month, while ignoring the influence of vegetation.

$$ET_{adj} = 1.6\left[10\frac{t_c}{I}\right]^a$$

where:
$ET_{adj}$ = adjusted monthly potential evapotranspiration
$t_c$ = mean monthly temperature in °C

$$I = \sum i_m = \sum \left[\frac{t_c}{5}\right]^{1.5}$$

$I$ = the heat index for a given year
$a = 6.7 \times 10^{-7}$

Although the Thornthwaite equation is incorrect, it provides a reasonably accurate estimate of annual potential evapotranspiration. Both the *Hargreaves equation* and *Blaney–Criddle method* can replace the Thornthwaite equation when more accurate estimates are needed. Cf. *Evapotranspiration; Actual evapotranspiration.*

**Thread** The line connecting points of maximum current velocity along the surface of a watercourse. Cf. *Channel line.*

**Three-point problem** See *Groundwater flow; Direction of groundwater flow; Hydraulic gradient.*

**Throughfall** The water from *precipitation* that reaches the ground through plant cover or that drips onto the ground from branches, leaves, or snowmelt. The throughfall amount depends on the canopy density. As shown in **Figure T-9**, the portion of throughfall dripping off the canopy is called leaf drip.

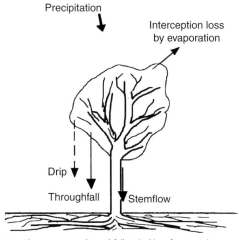

**Figure T-9** Diagram of the interception process or throughfall typical in a forest environment. (Black, 1996. Reprinted with permission from Ann Arbor Press, Inc.)

Cf. *Throughflow.*

**Throughflow** The portion of *runoff* that infiltrates and percolates rapidly into the soil, typically through large openings or macropores such as cracks, roots, and animal boreholes; or water that encounters a less-permeable zone and travels laterally in a temporary saturated zone. Throughflow differs from other flows by its larger magnitude and rapid response and therefore reaches a watercourse quickly. Cf. *Throughfall; Infiltration; Percolation.*

**Tidal estuaries** A semienclosed coastal body of water with one or more rivers or streams flowing into it, with a free connection to the open sea. Estuaries are typically the tidal mouths of rivers and are often characterized by sedimentation carried in from terrestrial runoff and, frequently, from offshore. They are characterized as having brackish water. Estuaries are often given names like bay, sound, fjord, and so on. Estuaries are often associated with high levels of biological diversity. Cf. *Drouned coast.*

**Tidal influence** Periodic inundation of seawater twice daily due to the rising and falling of the tides. This action can influence groundwater flow, contaminant, and nutrient transport in coastal areas.

**Tides** The cyclic rising and falling of Earth's ocean surface caused by the gravitational forces of the moon and the sun acting on the oceans. Tides cause changes in the depth of the marine and estuarine water bodies and produce oscillating currents known as tidal streams, making prediction of tides important for coastal navigation. The changing tide produced at a given location is the result of the changing positions of the moon and sun relative to the Earth coupled with the effects of Earth's rotation and the bathymetry of oceans, seas, and estuaries.

**Time average equation** See *Acoustic velocity logging.*

**Time–drawdown graph** See *Jacob time–drawdown straight-line method.*

**Time domain reflectometry (TDR)** A method of soil moisture measurement based on the change in dielectric constant of the soil with changes in soil moisture.

**Time step** The time interval at which measurements are made in a *pumping test.* Water levels in *observation wells* at the beginning of the pumping test change rapidly, and therefore measurements are made more frequently. As water is removed from storage, the drawdown area expands, the aquifer response tends toward steady state, and the changes in water levels are more gradual; as a result, the time step between measurements can then be increased.

**Topographic adjustment** The ground surface change when the *gradient* of a *tributary* is consistent with the gradient of the main stream.

**Topographic divide** See *divide; groundwater divide.*

**Torrent** A violent and typically sudden onset of rushing water as seen in a flooded area, mountain waterway, or a rapid water-level rise in a watercourse due to increased *rainfall* or snowmelt.

**Tortuosity** For a watercourse, the ratio of its length as measured along the middle to its axial length, or for *groundwater,* the length of the actual *flow path* of interest divided by the straight-line distance between the beginning and the end of the path.

**Tortuous flow** See *turbulent flow.*

**Total dissolved solids (TDS)** The solid material in a solution, whether ionized or not. Total dissolved solids (TDS) in natural waters range from less than 10 mg $L^{-1}$ (for rain or snow) to not more than 300,000 mg $L^{-1}$ for some *brines.* Total dissolved solids is measured by boiling the water away in a given sample and weighing the solids remaining. Total dissolved solids may also be approximated by measuring the *specific electrical conductance.* Water for domestic and industrial uses should contain less than 1000 mg $L^{-1}$ TDS; water for agricultural use should contain less than 3000 mg $L^{-1}$ TDS. Water use should not be decided by the criterion of TDS, but rather by the concentrations of individual *ions* present. Cf. *Brine; Brackish water; Chebtarer's sucession; Dissolved solids.*

**Total dynamic head** The sum of all *head losses* that must be overcome to pump water from its source to a desired location. Total dynamic head includes the *lifting head* and the *friction head,* which are losses experienced in a piping network.

**Total hardness** See *hardness.*

**Total head** The sum of the *elevation head,* the *pressure head,* and the *velocity head* see **Figure T-10.** In *groundwater,* the velocity head in calculations of total head is typically negligible and therefore ignored. Total head is calculated by adding the elevation of the base of a *piezometer* and the height of the water column in the piezometer (assuming a datum of *sea level*).

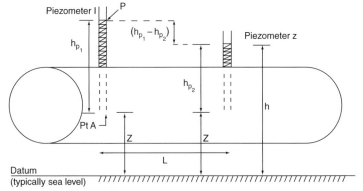

Pt A = point located at elevation Z, above a datum, at fluid pressure P.
P = fluid pressure at pt A.
z = (elevation head) elevation of the base of a piezometer above a datum.
$h_p$ = (pressure head) height of the water column in a piezometer.
h = (total head) sum of the elevation head (z) and pressure head ($h_p$).
$(h_{p_1} - h_{p_2})$ = the change in head (dh) over length L.

**Figure T-10** An apparatus consisting of a circular cylinder filled with sand to demonstrate Darcinian flow and showing the relationship between total head (h), elevation head (z) and pressure head ($h_p$). (After Fetter, 1994 and Freeze, 1979.)

Cf. *Fluid potential; Hydraulic head; Static head; Piezometric pressure; Potentiometric surface; Hydraulic gradient.*

T

**Total head loss**  Cf. *Laminar head loss*.

**Total load**  The graph in **Figure T-11** shows the relationship of *discharge* rates to the total *flow* of water and the *suspended* and *dissolved load* contributing to the total load. The graph shows that although very high flows [about 10,000 ft$^3$ s$^{-1}$ (cfs) or more] effectively erode and *transport* material, they occur infrequently and transport only about 9% of the total measured load.

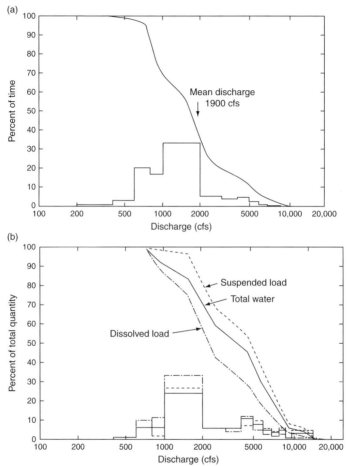

**Figure T-11**  Histograms of water versus load: (a) duration curve of percentage of time at various discharge rates are equaled or exceeded and (b) relative contribution of various discharge rates to total flow of water, suspended load, and dissolved load. (Leopold et al., 1992. Reprinted with permission from Dover Publication, Inc.)

Cf. *Stream bed*.

**Total porosity**  See *porosity*.

**Total precipitation**  See *Precipitation*.

**Total runoff**  See *runoff*.

**Trace element**  An element dissolved in minute quantities in water, typically in concentrations less than 1 mg in 1 L of water (1 mg L$^{-1}$).

**Tracer**  A material or a substance that is introduced into a chemical, biological, or physical medium and used to track or follow a travel or flow path and/or determine the travel time or mean velocity. Common tracers include rhodamine WT uranine, and fluorescent dyes Cf. *Dating of groundwater; Radioactive tracer; Radiocarbon dating of groundwater*.

**Tracer test**  A test in which a *tracer* is introduced into a *flow system* and then tracked to define a flow path. Cf. *Hydraulic resistance; Dating of groundwater*.

**Tract**  A portion of a watercourse, e.g., a mountain tract.

**Traction** A process of material *transport* by which particles in air or water are moved on, near, or immediately above and parallel to a bottom surface by rolling, sliding, dragging, pushing, or *saltation*. Traction is a mode of sediment transport. Cf. *Suspension*.

**Traction current** Movement in standing water that transports particles along and in contact with the bottom surface of a watercourse. Cf. *Turbidity current*.

**Traction load** See *bed load*.

**Tranquil flow/streaming flow** May also be called subcritical flow. Tranquil or streaming flow is called subcritical flow when the mean velocity of *turbulent flow* is less than the critical flow. Cf. *Rapid flow; Reynolds number; Critical velocity*.

**Transient condition** A situation in which the magnitude or direction of *flow* changes with time.

**Transient flow** See *unsteady-state flow/unsteady flow*.

**Transient state flow** This represents a condition where at any point in a flow field, the magnitude or direction of the flow velocity changes with time. Cf. *Flow, unsteady*.

**Transit-time logging** See *Acoustic velocity logging*.

**Transmissibility coefficient** See *transmissivity*.

**Transmission constant** The ability of a permeable medium to conduct a fluid under *pressure* or the *discharge* of groundwater through a cross-sectional area under a *hydraulic gradient* of 1.

**Transmissivity ($T$)** Also known as the coefficient of transmissivity. The rate of movement of water at the prevailing *kinematic viscosity* through an *aquifer* of unit width under a unit of *hydraulic gradient*. Transmissivity, $T$, is a function of the liquid and the aquifer, as is *hydraulic conductivity*, $K$, but also incorporates the saturated thickness along with the properties of the contained liquid. Cf. *Hydraulic diffusivity; Zone of saturation*.

**Transpiration** The process by which water is absorbed by plants and *vapor* is given off through their leaves; also, the volume of water moved by this process within a specified time. Plant roots extract water from the soil while leaves lose or transpire water vapor into the atmosphere; therefore, the depth at which transpiration affects the *soil moisture* is determined by the depth of the plant's root system. Roots typically penetrate approximately 2 m into the soil, although some *phreatophytes* have much deeper roots. In the United States on a country-wide basis, approximately 21 in. of the 30 in. of *precipitation* received annually are returned to the atmosphere through *evaporation* and transpiration (*evapotranspiration*).

**Transport** The movement of *solute*, suspended matter, or heat in either a liquid or a *porous medium* by *advection, diffusion, conduction*, or *convection*. Cf. *Erosion*.

**Transporting erosive velocity** The rate of movement, in a confined area or channel, of water that is capable of carrying silt in the *load* and scours the channel bed.

**Transverse dispersion** The spread of a *solute* that occurs normal to the direction of movement of *groundwater flow*. Cf. *Dispersion; Longitudinal dispersion*.

**Treatment train** Methods performed in series to treat a given matrix. For example, a treatment train for *groundwater* may involve *filtration* to remove *suspended solids*, followed by *air stripping* to remove volatile organic compounds, then followed by *carbon adsorption* (usually called carbon polishing) to remove less-strippable organic compounds.

**Trellis drainage pattern** See *drainage pattern*.

**Tremie pipe** A small-diameter pipe used for the placement of the *filter pack* and *grout* during well construction. The tremie pipe is lowered into the annular space between the *well casing* and the *borehole* wall to the desired depth of placement, the filter or grout material is introduced, and the tremie pipe is raised as the annular space fills. This method minimizes particle separation and *bridging*, resulting in fairly uniform placement of the materials. The diameter of the tremie pipe varies depending upon the material to be placed and the annular space available. Material can be pumped directly through the tremie pipe or poured through a funnel arrangement.

**Tributary** Also called a *branch*, affluent feeder, confluent feeder, feeder, or tributary *stream*. A watercourse that flows into, feeds, or joins a larger watercourse or water body. Tributary *discharge* and *sediment load* contributions to the main channel can shift or force a watercourse to a different position in a valley. Cf. *Distributary*.

**Tributary stream** See *tributary*.

**Trilinear diagram** See *Piper trilinear diagram*.

**Trip blank** In water quality sample, a sample of deioinized water is typically prepared as a quality assurance/quality control protocol. This called a trip blank since the deionized water should be pure. This is prepared to determine if the sample set became contaminated during shipment, as part of the shipment process. Typically trip blanks are only prepared for analysis of volatile organic compounds.

**Tritium in groundwater** A radioactive hydrogen isotope, $^3H$, containing two neutrons and one proton in the nucleus (regular hydrogen has one of each), which can be used to date waters 50 years old or younger. Before 1952, the $^3H$ content of *rainfall* ranged from 1 to 10 tritium units (TU). Atmospheric testing of thermonuclear devices in 1952 greatly increased the $H^3$ content of rain and surface water; subsequent TU values varied over a wide range with time and geographic position, with a high of 2937 TU recorded at Ottawa, Canada, in March 1963. As a result, $H^3$ measurements may be used to date relatively recent waters. The bomb–tritium impact was especially high between 1961 and 1975, reaching a maximum during 1963. Tritium values have been decreasing with a half-life of 12.4 years; $^3H$ in most of the pre-1954 water will be below trace concentrations in 40–60 years. It has been proposed that when the $^3H$ in *groundwater* is no longer measurable, the $^3H$ isotopic decay product tritiogenic helium ($^3He$) could be used as the substitute for the initial $^3H$ concentration.

**T**

**Trophogenic zone** The upper portion of a continental water body, e.g., a *lake*, where photosynthesis occurs, converting inorganic matter to organic matter. Cf. *Euphotic zone;Tropholytic zone.*

**Tropholytic zone** The deeper portions of continental water bodies, e.g. *lakes*, where it is too deep for light to penetrate and photosynthesis to occur and where organic matter therefore accumulates. Also known as the benthic zone. Cf. *Profundal;Trophogenic zone.*

**Tropical lake** A continental water body in which the surface water temperature is consistently greater than 4°C.

**Tunk** The main watercourse or channel of a system of *tributaries.*

**Tube well** See *driven well.*

**Tubular spring** See *spring; tubular.*

**Tunnel erosion** See *piping.*

**Turbid** Appearing muddy, cloudy, or opaque with suspended material, as occurs in a sediment-laden water body or watercourse. Cf. *Roil;Turbidity; Nephelometric Turbidity Unit (NTV).*

**Turbidity** Fluid that is not clear because of suspended matter. Turbidity is the degree to which the suspended particles interfere with light penetrating the fluid. Cf. *Diatomaceous earth filter.*

**Turbidity current** Also called a suspension current. Material that is set in motion within the surrounding air, water, or other fluid by density differences, e.g., a sediment that has been stirred up and causes some water to be denser than the surrounding or overlying sediment-free or clear water. Cf. *Turbid; Turbidity flow; Density current; Salinity current.*

**Turbidity flow** A dense, gravity-induced movement of a fluid and suspended material.

**Turbulence** See *turbulent flow.*

**Turbulent diffusion** Molecular diffusion as a transport mechanism in surface water is rarely an important process because of the velocity fluctuations caused by *turbulent flow.* Turbulent transport mechanisms could be analyzed as an *advective* process if it could be described exactly as a function of time and space. Ordinarily, turbulent transport or diffusion can be described only statistically. Diffusivities are usually estimated from tracer experiments or other empirical data.

**Turbulent flow** Also called turbulence or tortuous flow. The movement of a fluid displaying *heterogeneous* mixing or non-parallel *flow lines,* a condition in which inertial forces predominate over viscous forces and *head loss* is not linearly related to velocity. The character of fluid flow is expressed by the dimensionless *Reynolds number,* $N_R$, determined from the following equation:

$$N_R = \frac{vd}{(\mu/\rho)}$$

where:
$v =$ velocity $[L \cdot T^{-1}]$
$d =$ depth $[L]$
$\mu =$ viscosity $[L^2 \cdot T^{-1}]$
$\rho =$ density $[M \cdot L^{-3}]$

Turbulent flow occurs when the Reynolds number is greater than 10. The increase in the *sediment load* in a watercourse dampens the water turbulence and alters the resistance to *flow.* Cf. *Critical flow; Critical velocity; Laminar flow; Tranquil flow; Eddy; Open channel flow.*

**Turbulent head loss** *Flow* occurring around the *well screen* as a result of *head losses.* Turbulent head loss reduces well efficiency.

**Turbulent stream** *Turbulent flow* conditions are defined as occurring when the *Reynolds number* is greater than 10. Turbulent flow in a stream results in erosional processes that change the characteristics of the stream itself. If we consider an erodible bottom sheared by a turbulent stream, where the strength of the flow is sufficiently high, sediment grains get detached from the bed. In turn, the moving grains interact with the flow and decelerate it, so that the flow can only erode a limited quantity of sediment. As a result, a stream under turbulent flow conditions does not exhibit uniform flow as erosion, sowing of flow, and re-deposition occur. Cf. *Turbulent flow.*

**Turbulent velocity** The rate of movement of fluid in a watercourse above which *turbulent flow* dominates and below what would be either *laminar flow* or turbulent flow. Cf. *Laminar velocity.*

**Turnover** The time or season, typically fall and/or spring, of uniform vertical temperature in a water body when convective circulation or overturn occurs.

**Two-and-eight-tenths depth method** A means of channel velocity measurement by which a current meter or a *flow meter* is placed in a cross-sectional area where the depth or *stage* and width of the channel is stable. Meter readings are then collected in that cross-sectional area at two-tenths and eight-tenths of the depth to calculate the true average in the vertical.

**T**

# Uu

**Ultimate base level** Also called general base level. The lowest possible water level or the base level for a *stream* in *sea level*. If this level is projected inland as an imaginary surface below the stream, the imaginary line is the ultimate base level for that stream.

**Ultraviolet-light water purifiers** Ultraviolet is that part of the electromagnetic spectrum ranging in wavelength from 40 to 4000 Å units, mainly in the 3000–4000 range just beyond the high-energy (violet) end of the visible-light band of the solar spectrum. Bacteria, viruses, and spores, in domestic water supplies can be eradicated by exposing the water to short-wave ultraviolet radiation waves on the order of 2000 to 2950 Å. The wavelength is produced by low-pressure mercury vapor lamps at a typical dosage of 635,000 µWs cm$^{-2}$ (MWS/cm$^2$) with the retention time of at least 15 s. The number of lamps and retention time determines the actual amount of radiation received, which is determined by the degree of purification required.

**Unavailable water** The water existing in *soil pores* in an amount below the wilting point. Water that is present below the ground surface but cannot be utilized by plants and trees because of the water's strong attraction to the soil particles. Cf. *Unavailable moisture.*

**Unconcentrated flow** See *Overland flow.*

**Unconcentrated wash** See *Sheet erosion.*

**Unconfined aquifer** See *Aquifer, unconfined.*

**Unconfined groundwater** See *Aquifer, unconfined.* Cf. *Phreatic water; Unconfined water.*

**Unconfined steady-state flow equation** A method used to estimate the height of the *water table* and the volumetric flow, at an intermediate location between two *monitoring wells* where the water table elevation is known in an unconfined aquifer. The simplified equation simulates *steady-state flow* in an unconfined aquifer using the *Dupuit-Forchheimer assumptions.* In an unconfined aquifer, the elevation is equal to the *hydraulic head, h,* and the boundary becomes part of the solution when predicting the hydraulic head distribution in the aquifer. As shown in the conceptual model in **Figure U-1**,

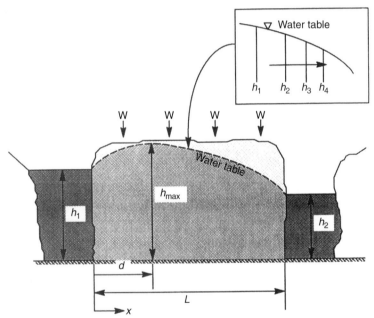

**Figure U-1** Conceptual model for unconfined flow to a stream using the Dupuit approximation. (Maidment, 1993. Reprinted with permission from McGraw-Hill, Inc.)

the following equation solves for the height of the water table, the *discharge* per unit width at any distance from the origin ($Q_x$), and the location of the *groundwater divide* (d). The equations based on Dupuit-Forchheimer assumptions are:

$$h^2 = h_1^2 - \frac{(h_1^2 - h_2^2)x}{L} + \frac{w}{K}(L - x)x$$

$$Q_x = \frac{K(h_1^2 - h_1^2)}{2L} - w\left(\frac{L}{2} - x\right)$$

$$d = \frac{L}{2} - \frac{K}{w}\left(\frac{(h_1^2 - h_2^2)}{2L}\right)$$

where:
$h$ = height of the water table [L]
$L$ = distance between $h_1$ and $h_2$ [L]
$w$ = recharge rate at the water table [L·T$^{-1}$],
$Q_x$ = discharge per unit width [L$^3$·T$^{-1}$]
$K$ = *hydraulic conductivity* [L·T$^{-1}$]
$d$ = the location of the groundwater divide [L]

**Underfit stream** A watercourse typically resulting from drainage changes, i.e., *stream capture*, glaciers, or variations in climate, that appears to be too small to have eroded the valley in which it currently flows. Cf. *Misfit stream*.

**Underflow** *Groundwater* movement through subsurface soil or beneath a structure. Typically used to indicate groundwater flow below a *stream* or a riverbed, moving in the same *downstream* direction. Also indicating groundwater flow under a dry stream channel in an arid region or the erosional effect of water undercutting a structure. Cf. *Underflow conduit*.

**Underflow conduit** A deposit capable of transporting fluids below a surface *stream* channel that is limited on the bottom and sides by relatively impermeable strata, simulating flow through a pipe. The below-ground channel transports *groundwater* in the same general direction as the above-ground stream. Cf. *Underflow*.

**Underground Injection Control (UIC)** A United States "*Safe Drinking Water*" program that regulates the use of *wells* that pump fluids below ground, either for *recharge* or disposal purposes.

**Underground stream** Water flowing in a distinct channel below the ground surface typically thought to be flowing as a distinct current. This type of *flow* is characteristic of fracture, joint, or *karst* flow. The term has been inappropriately used to indicate an unconfined or *water table* aquifer.

**Underground water** See *Groundwater*.

**Underpressure system** The *fluid pressure* within the *hydrogeologic system* is less than the *hydrostatic pressure*. Cf. *Overpressure system*.

**Uniform channel** A watercourse displaying a constant cross-sectional view, i.e., same geomorphology.

**Uniform flow** The velocity and *discharge*, magnitude and direction, of *flow* do not vary along the length of a watercourse; or in a groundwater *flow path*, the velocity and the direction of flow are the same at all points in the field of flow. The flow in which there is no convergence or divergence. Typically used as a simplifying assumption in hydraulics. Cf. *Groundwater flow*.

**Uniformity coefficient ($C_u$)** A numerical expression of the sorting or grading of a sediment, which is the ratio of the *grain size* that is 60% finer by weight, $D_{60}$, to the grain size that is 10% finer by weight, $D_{10}$. The *porosity* of sediments is affected by the distribution of grain size and the shape of the grains. Poorly sorted sediments will have lower *porosities*. The uniformity equation is written as follows:

$$C_u = \frac{D_{60}}{D_{10}}$$

**U**

Generally, a sediment with a $C_u$ less than 4 is well-sorted or poorly graded and if $C_u$ is greater than 6, it is poorly sorted or well-graded. The uniformity coefficient, defines the sieve size on which 40%, by weight, of grains retained divided by the 90% retained size. The lower the value, the more uniform the grading of the sediment between the 40 and 90% retained. Cf. *Effective grain size; Grain size distribution; Grain size distribution curve.*

**Uniformness of aquifer** Depth variations within the *aquifer basin* will determine the geometrical dimensions of the groundwater *reservoir* being studied. Knowledge of the areal and cross-sectional geometry of the aquifer is required to evaluate movement and volume of groundwater. **Figure U-2** shows a cross-sectional view of a *recharge* phenomenon in a *groundwater basin* of uniform depth over the *area of influence*. Groundwater equations use the assumption of uniformness of the aquifer, or uniformity, allowing the use the *transmissivity* concept throughout the uniform range, as shown in **Figure U-2**.

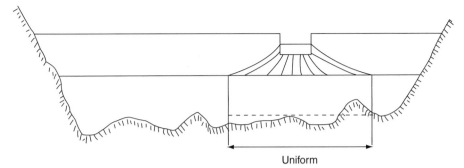

Uniform

**Figure U-2** Cross-sectional geometry of an aquifer demonstrating the use of uniformity of an aquifer. (Sen, 1995. Reprinted with permission from CRC Press, Inc.)

**Unit hydrograph** The graph of surface *or direct runoff* resulting from effective rainfall falling in a unit of time, which produces uniformly in space and time over the total *catchment area*. The unit hydrograph method of flood estimation was introduced in 1932. The unit hydrograph, as shown in **Figure U-3**, is a rainfall–runoff relationship, where the volume surface runoff is given by the area under the hydrograph and is equivalent to the depth of effective rainfall over the catchment area. Once a unit hydrograph of specified duration has been derived for a catchment area and/or specified storm type, then for any sequence of effective rainfalls in the same duration, an estimate of the surface runoff can be

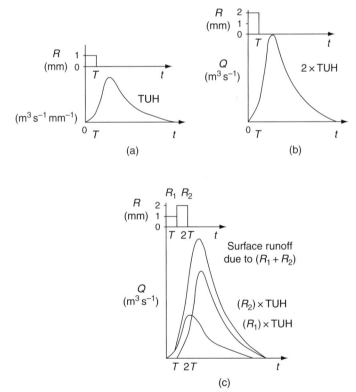

**Figure U-3** The rainfall–runoff relationships displayed in unit hydrographs. (Shaw, 1983. Reprinted with permission from VanNostrand Reinhold Co. Ltd.)

determined. Variable characteristics of storms, such as rainfall duration, time intensity pattern, *rainfall distribution*, and amount, cause variations in the shape of a basin hydrograph; therefore, there in no one typical hydrograph for any one basin even though the basins characteristics (shape, size, slope, etc.) are relatively invariable. The best unit hydrograph is compiled from the average of several unit hydrographs of several storms of approximately equal duration. **Figure U-4** demonstrates the derivation of a unit hydrograph developed from a storm of uniform intensity, desired duration, and relatively large runoff. The *baseflow* is separated from the direct runoff and the volume of the direct runoff is determined. The ordinates of the direct-runoff hydrograph are divided by observed runoff depth and the adjusted ordinates form a unit hydrograph.

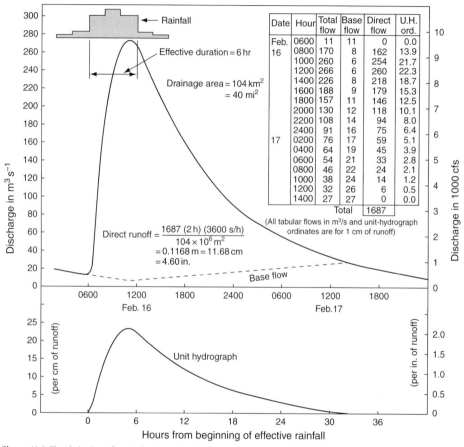

| Date | Hour | Total flow | Base flow | Direct flow | U.H. ord. |
|------|------|-----------|-----------|-------------|-----------|
| Feb. | 0600 | 11 | 11 | 0 | 0.0 |
| 16 | 0800 | 170 | 8 | 162 | 13.9 |
|  | 1000 | 260 | 6 | 254 | 21.7 |
|  | 1200 | 266 | 6 | 260 | 22.3 |
|  | 1400 | 226 | 8 | 218 | 18.7 |
|  | 1600 | 188 | 9 | 179 | 15.3 |
|  | 1800 | 157 | 11 | 146 | 12.5 |
|  | 2000 | 130 | 12 | 118 | 10.1 |
|  | 2200 | 108 | 14 | 94 | 8.0 |
|  | 2400 | 91 | 16 | 75 | 6.4 |
| 17 | 0200 | 76 | 17 | 59 | 5.1 |
|  | 0400 | 64 | 19 | 45 | 3.9 |
|  | 0600 | 54 | 21 | 33 | 2.8 |
|  | 0800 | 46 | 22 | 24 | 2.1 |
|  | 1000 | 38 | 24 | 14 | 1.2 |
|  | 1200 | 32 | 26 | 6 | 0.5 |
|  | 1400 | 27 | 27 | 0 | 0.0 |
|  | Total |  |  | 1687 |  |

(All tabular flows in m³/s and unit-hydrograph ordinates are for 1 cm of runoff)

$$\text{Direct runoff} = \frac{1687\,(2\,\text{h})\,(3600\,\text{s/h})}{104 \times 10^6\,\text{m}^2}$$
$$= 0.1168\,\text{m} = 11.68\,\text{cm}$$
$$= 4.60\,\text{in}.$$

Effective duration = 6 hr

Drainage area = 104 km² = 40 mi²

**Figure U-4** The derivation of a unit hydrograph. (Linsley et al., 1982. Reprinted with permission from McGraw-Hill Inc.)  **U**

Cf. *Rainfall runoff; Flood forcasting; Flood prediction; Flood-frequency curve; Gringorten formula; Flow hydrograph.*

**Unsaturated flow** Water infiltrating into the unsaturated zone is defined by the relation of the *gravity potential, Z,* and *moisture potential, ψ*. The moisture potential is the negative pressure resulting from the soil–water attraction, which increases when soil moisture decreases. When the moisture content of the soil approaches the specific retention, the gravity potential predominates unsaturated flow. If the soil is dry, the moisture potential may be several orders of magnitude greater than the gravity potential. Although the unsaturated *hydraulic conductivity, K(θ),* is not a constant, *Darcy's law* for flow is valid. The $K(\theta)$ is a function of the *volumetric water content, θ.* The soil moisture flows downward by gravity flow through interconnected pores that are filled with water. The greater the number of pores that are filled, the greater the rate at which water moves downward, or as $\theta$ increases $K(\theta)$ increases. The total potential in unsaturated flow, $\phi$, is the sum of the moisture potential and the elevation head and is determined using the following equation:

$$\phi = \varphi(\theta) + Z$$

**Figure U-5** demonstrates the relationship between $K$, the hydraulic conductivity, and $\theta\,K$, the volumetric water content for a typical clay.

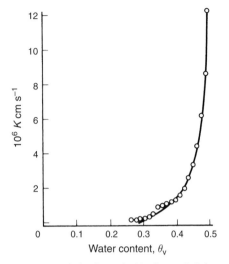

**Figure U-5** Graph of water content versus hydraulic conductivity for a typical clay material. (Constantz, 1982.)

Cf. *Unsaturated hydraulic conductivity.*

**Unsaturated hydraulic conductivity** See *Unsaturated flow; Hydraulic conductivity.*

**Unsaturated zone** The area below the ground where the *void spaces* are filled with a mixture of water under pressure less than atmospheric, which includes water held by *capillarity* and air (gases) under atmospheric pressure. Subsurface water is classified into groups according to the amount of water filling the *void spaces* below the ground surface. The unsaturated zone lies above the saturated zone and is below the ground surface. This zone is also referred to as the *zone of aeration* or the *vadose zone*. The unsaturated zone can be further divided into the *belt of soil water*, the *intermediate belt*, and the *capillary fringe*. Cf. *Phreatic zone; Zones of water; Zone of saturation; Void spaces.*

**Unsteady radial groundwater flow** This condition occurs during the initial phases of pumping groundwater to a well. Under unsteady or transient flow conditions, groundwater flow to a well is faster near the well than at increasing distance. Further, flow to the well is assumed to be homogeneous and isotropic, and therefore drawdown around the well will be radially symmetric. Under these conditions, it is convenient to express the problem in terms of polar coordinates (theta and radius) rather than x and y. Since velocities increase closer to the well bore is approached, pumping a fully confined aquifer at a steady discharge rate $Q$ creates a drawdown that never reaches steady state as water is produced by compressing the aquifer rather than recharge. In order to keep squeezing the aquifer, it is necessary to keep dropping the head, and the cone of depression continually expands. See *theis nonequilibrium equation/theis equation/theis solution* and *theis nonequilibrium-type curve* for an analytical solution to this condition. Cf. *W(u).*

**Unsteady-state groundwater flow or unsteady flow** Unlike *steady-state* groundwater flow, the direction and/or the magnitude of *specific discharge* changes with time. The following properties characterized unsteady-state flow:

1. In a *flownet*, either the *streamline* or the *equipotential line* or both change their position with the passage of time.
2. The discharges at different sections have different values.
3. Points 1 and 2 imply that the specific discharge magnitude and/or direction change with time.
4. The groundwater level fluctuates with time.
5. There is not always continuous external recharge but even if there is, it may not be enough to balance the amount of water abstracted from the aquifer.
6. Changes occur in the storage within the aquifer.

The water movement alone is not enough to represent the groundwater flow. In addition to this, the storage capacity should be known. Natural conditions of unsteady-state flow include mounding of the water table below recharge areas, recharge to a river. In steady state, the hydraulic head will vary with distance only, while hydraulic head varies with time also in unsteady-state flow. Cf. *Nonsteady flow; Steady flow; Transient flow; Equations of groundwater.*

**Upconcavity** The consistent decrease in gradient in the *downstream* direction as profiled in the channels of *streams*.

**Upconing** The most common cause of upconing is evident in areas where a *freshwater* aquifer overlies a *saltwater* zone. High pumping rates will have a tendency to draw the underlying salt water up into the freshwater aquifer. **Figure U-6** depicts upconing of saline water induced by a pumping well.

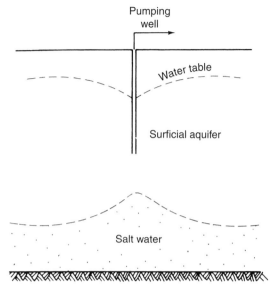

**Figure U-6** Overpumpage can cause salt water intrusion resulting from upconing. (Bruington, 1969.)

Cf. *Coning.*

**Uphole velocity** See *Annular velocity.*

**Uplift** Beneath a structure or formation, the *hydrostatic force* of water will tend to displace or disturb structures. When used in relation to drilling materials, such as during *grouting*, uplift refers to the vertical displacement of the formation material because of the grout injection.

**Unwin's critical velocity** Cf. *Critical velocity; Kennedy's velocity.*

**Upstream** A streams *gradient* usually decreases in the *downstream* direction or at the lower elevation. The upstream direction is usually at higher elevations in the headwater or beginning of the stream.

**Upwelling** The process by which water rises from a deeper to a shallower depth, usually as a result of divergence of offshore currents.

**Urban drainage** Hydrologic design for urban areas is challenging because the well-studied principles of *watershed* hydrology often cannot be applied directly to urban hydrology. *Floodplain* management practices differs from urban or stormwater drainage, which includes the design of storm sewers, detention and retention facilities, and water quality enhancement facilities.

**Urban runoff** The accumulative *direct runoff* originating from city streets and gutters, typically collected via stormwater collection basins. Urban runoff usually contains varied trash and organic and bacterial waste.

**U**

**Vadose zone** One of the two zones in which water occurs in the subsurface. The other zone is the *saturated zone*. The vadose zone consists of three separate types of water: soil water, intermediate vadose water, and *capillary water*. In the vadose zone, water is held in soils by molecular attraction and capillarity, which act against the forces of gravity. Molecular attraction on the one hand results in water providing a thin film on the surface of soil particle. Capillarity on the other hand holds water in the pore spaces between the soil grains. When the water-holding capacity of the capillary forces is exceeded, the water begins to percolate downward under the force of gravity. Soil water has particular importance in agriculture, for it provides moisture for plant growth. Water is lost from this zone through the processes of evaporation, transpiration, and percolation to deeper zones. The intermediate zone exists between the soil water zone and the zone of capillary water. Most of the water in this zone is moving downward, with some being retained in soil pores or as a film on soil grains. In humid regions, this zone can be quite thin or non-existent. In dry regions, little water passes through this zone. The capillary fringe is the area where the water from the *water table* is drawn upwards through capillary forces. The thickness of this zone is a function of the grain size of the soils in this zone. Water has the potential to migrate upward of 3 m (10 ft) in fine-grained, well-sorted sediments. The various zones are illustrated in **Figure V-1**.

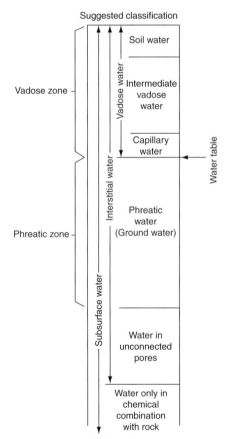

**Figure V-1** Classification of subsurface water. (Davis and DeWiest, 1966. Reprinted with permission from John Wiley & Sons, Inc.)

Cf. *Phreatic zone; Average pore water velocity in the vadose zone.*

**Valence** The capacity of an *ion* to react, unite, or interact with something else. For example, the charge of $-4$ for the complex ion $SiO_4^{-4}$ is its valence and four electrons are available to unite with cations.

**Vanes** Used in centrifugal or more specifically turbine pumps. Diffuser vanes in turbine pumps provide gradually increasing passages for the water to pass. Since water is considered an incompressible fluid and there is conservation of energy, the velocity of the water is reduced, while the pressure increases. It is this increase in pressure that makes turbine pumps useful in deep-lift operations.

**van Genuchten parameters** Parameters for an equation commonly used as a functional expression that relates matric suction to moisture content for unsaturated soils. The van Genuchten equation was developed to predict soil water content in the vadose zone as a function of depth, $h$:

$$\theta(h) = \theta_r + \frac{\theta_s - \theta_r}{[1 + (\alpha h)^n]^m}$$

where:

$\theta_r$ = residual soil water content $[L^3 \cdot L^{-3}]$
$\theta_s$ = soil water content at saturation $[L^3 \cdot L^{-3}]$
$\alpha$ = van Genuchten parameter $[L^{-1}]$
$n$ = van Genuchten parameter [unitless]
$m = 1 - 1/n$

The van Genuchten parameters have been estimated from field data for different soil types. Cf. *Brooks and Corey method; Buckingham–Darcy equation.*

**Vapor** The gaseous state of water.

**Vaporization** Cf. *Cavitation.*

**Vapor pressure** Also called the partial pressure of a vapor. **Table V-1** lists vapor pressures of water at various temperatures. The molecules are more than 10 times further apart in water vapor than in the liquid phase, causing the intermolecular forces to be much less. With an increase in temperature, the molecular separation increases. This separation increases greatly during evaporation and the energy that is absorbed is called the *latent heat of vaporization* of water, $\lambda$. The latent heat of vaporization, where the temperature $T_s$ is in °C, is:

$$\lambda = 2.501 - 0.002361 T_s$$

Approximately 2.5 million joules are required to evaporate 1 kg of water. **Figure V-2** depicts the exchange of water molecules between liquid and vapor.

**Table V-1**  Vapor pressure of water

| Temperature | | Absolute vapor pressure | | | |
|---|---|---|---|---|---|
| °F | °C | psia | kPa | ft of water | m of water |
| 32 | 0 | 0.09 | 0.62 | 0.20 | 0.06 |
| 40 | 4.4 | 0.12 | 0.83 | 0.28 | 0.09 |
| 50 | 10.0 | 0.18 | 1.24 | 0.41 | 0.13 |
| 60 | 15.6 | 0.26 | 1.79 | 0.59 | 0.18 |
| 70 | 21.1 | 0.36 | 2.48 | 0.89 | 0.27 |
| 80 | 26.7 | 0.51 | 3.52 | 1.2 | 0.37 |
| 90 | 32.2 | 0.7 | 4.83 | 1.6 | 0.49 |
| 100 | 37.8 | 0.95 | 6.55 | 2.2 | 0.67 |
| 110 | 43.3 | 1.28 | 8.83 | 3.0 | 0.91 |
| 120 | 48.9 | 1.69 | 11.7 | 3.9 | 1.19 |
| 130 | 54.4 | 2.22 | 15.3 | 5.0 | 1.52 |
| 140 | 60.0 | 2.89 | 19.9 | 6.8 | 2.07 |
| 150 | 65.6 | 3.72 | 25.6 | 8.8 | 2.68 |
| 160 | 71.1 | 4.74 | 32.7 | 11.2 | 3.41 |
| 170 | 76.7 | 5.99 | 41.3 | 14.2 | 4.33 |
| 180 | 82.2 | 7.51 | 51.8 | 17.8 | 5.43 |
| 190 | 87.8 | 9.34 | 64.4 | 22.3 | 6.80 |
| 200 | 93.3 | 11.5 | 79.3 | 27.6 | 8.41 |
| 210 | 98.9 | 14.1 | 97.2 | 33.9 | 10.3 |

**V**

**Variable density log** See *acoustic televiewer (ATV) logging.*
**Variable-head permeameter** See *permeameter.*

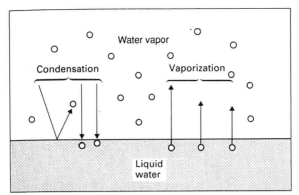

**Figure V-2** The molecular exchange between water and vapor. (Maidment, 1993. Reprinted with permission from McGraw-Hill, Inc.)

**Velocity head** This is the result of the motion of a fluid (kinetic energy). Velocity head can be expressed as:

$$h = \frac{v^2}{2g} \tag{1}$$

where:

$v$ = velocity (ft s$^{-1}$, m s$^{-1}$)
$g$ = acceleration of gravity (32.174 ft s$^{-2}$, 9.81 m s$^{-2}$)
Cf. *Groundwater energy; Total head; Elevation head; Pressure head; Darcy's law.*

**Velocity of groundwater** The velocity of groundwater is determined by combining Darcy's law with the standard continuity equation of hydraulics. Basically, the standard continuity equation is based on the conservation of mass in dynamic systems, such that whatever flows into a system must flow out. Darcy's Law can be written as follows:

$$Q = \frac{KA(h_1 - h_2)}{L}$$

Substituting $Q = VA$

$$VA = \frac{KA(h_1 - h_2)}{L}$$

and by eliminating terms, $V$ becomes:

$$V = \frac{K(h_1 - h_2)}{L}$$

Since groundwater flow occurs through the pores, the groundwater velocity will actually be greater than that indicated above. The actual groundwater flow (also known as the seepage velocity) is a function of the hydraulic conductivity, hydraulic gradient, and the porosity:

$$V_s = \frac{(K(h_1 - h_2)/L)}{n}$$

where:

$Q$ = volumetric flux through the porous media [L$^3$·T$^{-1}$]
$K$ = the hydraulic conductivity [L·T$^{-1}$]
$h_1$ = the head at a downgradient measuring point [L]
$h_2$ = the head at an upgradient measuring point [L]
$L$ = the horizontal distance between the points where the head is measured [L]
$n$ = the porosity.

**V**

**Velocity uphole** See *annular velocity*.

**Venice system** See *halinity; Chebotarov's succession; Brine; Thalassic series*.

**Vertical leakage** Occurs in a confined aquifer when the *piezometric pressure* is reduced, typically under pumping conditions. During a *pump test*, vertical leakage modifies the time–drawdown curve, typically at the latter portion of the test as illustrated in **Figure V-3**.

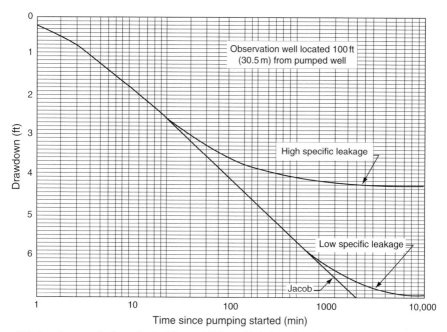

**Figure V-3** Drawdown graphs illustrating both low and high specific leakage. (Driscoll, 1986. Reprinted by permission of Johnson Screens/a Weatherford Company.)

**Vertically averaged concentration** Concentration of parameters of concern in groundwater averaged over the length of the sampled water column. The transport of contaminants in groundwater can occur through distinct zones such as macropores or paleochannels. If the well screen from which a sample is collected for characterizing the groundwater contamination is greater than the thickness of the zone responsible for the contamination transport, the measured concentration may reflect contributions from lesser-contaminated zone and hence be averaged relative to the main zone of transport.

**Vesicle** A void space created in volcanic rock by the escape of hot gases as the rock cools.

**Vibrating wire piezometer** A type of piezometer that converts pore pressure to an electronic signal that can be read with the manufacturer's readout device. A vibrating wire piezometer (VWP) converts water pressure to a signal via a diaphragm, a tensioned steel wire, and an electromagnetic coil. A change in pore pressure causes a change in the tension of the wire. The vibration of the wire in the proximity of the coil generates a frequency signal that is transmitted to a readout device. The advantages of a VWP are the following:

- high resolution
- simplified installation since sand filter and bentonite seals are not required
- rapid response to changes in pore pressure and
- typically equipped with a temperature sensor.

**Vibrocore** A soil-sampling device that vibrates a soil coring tool into the ground for the purposes of collecting a sample.

**Viscometer** An instrument that measures the amount of *shear stress* between an inner stationary cylinder and a rotating outer cylinder (**Figure V-4**). The annular space is filled with the fluid of interest. The shear stress produced from the rotation of the outer cylinder is directly proportional to the viscosity.

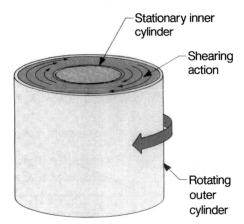

**Figure V-4** A viscometer measures the amount of shear stress (drag) that develops between a stationary and a rotating cylinder when the volume between the two is filled with a fluid. As the outer cylinder rotates, the amount of shear stress produced is directly proportional to the viscosity of the fluid. (Driscoll, 1986. Reprinted by permission of Johnson Screens/a Weatherford Company.)

**Viscosity** The property of a fluid that allows it to resist relative motion and shear deformation during *flow*. The more viscous the fluid, the greater the *shear stress* at a given velocity. There are two measures of viscosity: dynamic and kinematic. Newton's law of viscosity relates shear stress and the velocity gradient:

$$\tau = \mu \frac{dv}{dy}$$

where:
$\tau$ = the shear stress [$M \cdot L^{-1} \cdot T^2$]
$dv/dy$ = the velocity gradient [unitless]
$\mu$ = the dynamic viscosity [$M \cdot L^{-1} \cdot T^2$]

The *kinematic viscosity* $v$ [$L^2 \cdot T^{-1}$] is given by:

$$\nu = \frac{\mu}{\rho}$$

where:
$\rho$ = the fluid density [$M \cdot L^{-3}$]

The density and viscosity of fluids are functions of the temperature and *pressure*. Because the range of temperature and pressure in *groundwater* is not extreme, it is common practice to use the following measurements as constants at 15°C: $\rho$ = 1.0 g cm$^{-3}$, $\mu$ = 1.124 cP. Cf. *Fluid pressure; Aggregation; Newtonian fluid.*

**Void ratio** A soil mechanics term that is closely related to porosity (*n*). Void Ratio (*e*) is defined as the volume of voids in a mixture divided by the volume of solids:

$$e = \frac{V_v}{V_s}$$

where:
$e$ = the void ratio [unitless]
$V_v$ = volume of voids [$L^3$]
$V_s$ = volume of solids. [$L^3$]

*The void ratio, e, is related to n by the following equation:*

$$e = \frac{n}{1 - n}$$

or

$$n = \frac{e}{1 + e}$$

**Void spaces** Pore spaces typically occupied by air or other gases. Cf. *Unsaturated zone; Zone of saturation.*

**Volatilization** Also called vaporization. The conversion of a liquid to a vapor.

**Volume of water in casing or hole** **Table X in Appendix B** presents a compilation of the volume of water in casing or in a cylindrical hole.

**Volumetric moisture content** The volume of water present in the pore space divided by the total volume of the sample:

$$\theta = \frac{V_w}{V_t}$$

where:

$\theta$ = the volumetric moisture content [unitless]

$V_W$ = the volume of water in the sample [$L^3$]

$V_T$ = the total volume of the sample [$L^3$]

For saturated flow, the volumetric moisture content is equal to the *porosity* of the sample; for unsaturated flow in the vadose zone, $\theta$ is less than the porosity. Cf. *Gravimetric moisture content; Soil moisture; Anticedent precipitation index.*

**V**

**Wash-down method of installation** Also called jetting. In the wash-down method, the casing is first installed to the desired depth and grouted in place. After the *grout* is set, the *cement* plug at the bottom of the casing is drilled out and a pilot hole is drilled to the desired depth at the bottom of the screen. A self-closing bottom fitting or a backpressure valve is mounted at the bottom of the screen (**Figure W-1**). The top of the screen is connected to a wash pipe, which is usually the drill pipe. Water or a light-weight *drilling fluid* is pumped through the wash line, and the jetting action loosens and removes the sediment.

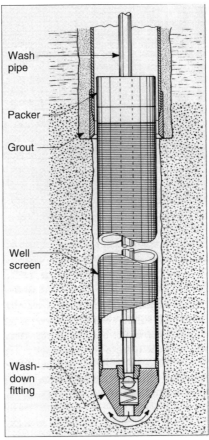

**Figure W-1** A wash-down bottom with spring-loaded valve permits washing a screen into place. Space around the lead packer allows return flow outside the well screen. (Driscoll, 1986. Reprinted by permission of Johnson Screens/a Weatherford Company.)

**Wash load** See *bed-material load*; *Sediment load*; *Flow layers, vertical*.

**Water balance** Also called a water budget or *hydrologic budget*. The quantification of the *recharge–discharge* interrelationships within a *watershed basin*. In a watershed the *divides* in surface water and *groundwater* coincide, where fluxes into or out of the watershed over an annual period are equal to the *precipitation* as shown in the following equation:

$$P = Q + E + \Delta S_S + \Delta S_D$$

where:

$P$ = average annual precipitation [L]

$Q$ = average annual runoff [L]

$E$ = average annual evapotranspiration [L]

$\Delta S_S$ = the change in storage of the surface water reservoir during the annual period [L]

$\Delta S_G$ = the change in storage of the groundwater reservoir (both saturated and unsaturated) during the annual period [L]

Averaging over many years of observation, the change in groundwater and surface water is usually assumed to be zero, such that $\Delta S_S = \Delta S_S = 0$ and the water balance can be simplified to:

$$P = Q + E$$

The values for $Q$ and $E$ are usually given as a unit of length (e.g., inches or centimeters) such that their units are consistent with those for $P$, measured on an annual basis. Cf. *Basin accounting; Groundwater budget.*

**Water-bearing formation** Any lithospheric layer that includes water. Cf. *Aquifer; Hydro stratigraphic units.*

**Water budget** See *water balance; Hydrologic budget; Groundwater budget.*

**Water of capillarity** See *capillary water.*

**Water conservation** The protection, development, and efficient management of water resources for beneficial use.

**Water content** Also known as free-water content and liquid water content. The liquid water present in a sample of *snow* expressed as a percent by weight. The water content in percent of water equivalent is 100 minus the quality of snow.

**Watercourse** A natural channel through which water may run or does run/flow. Cf. *Stream, river.*

**Water crop** See *water yield.*

**Water fall** A perpendicular or steep descent of a *stream*, where it crosses an outcrop of resistant rock over hanging softer rock that has been eroded, or *flows* over the edge of a plateau or cliffed coast. Cf. *Rapids; Turbulent flow.*

**Water equivalent** The depth of water that would result from the melting of a snow pack or of a snow sample; therefore, the water equivalent of a new snowfall is the same as the amount of *precipitation* represented by that snowfall.

**Water horizon** See *zones of water.*

**Water law** Legislation in North America is established at the state or provincial level governing the development and management of *groundwater resources.* Variations exist in traditions, rights, and statutes of water law. The two main doctrines of water law are appropriation and *riparian rights*, summarized in **Table W-1**.

**Table W-1** Summary and comparison of doctrines of appropriation and riparian rights

| Appropriation | Riparian |
|---|---|
| 1. Beneficial use, independent of land ownership, is the basis of water right. | 1. Land ownership is the basis of the water right. Water may be used for any reasonable purpose. |
| 2. Priority of use is the basis of allocation between rival claimants. Rights of appropriation are not equal. | 2. Co-sharing equality is the basis of allocation between rival claimants. |
| 3. Rights are to a definite quantity of water. | 3. Rights not fixed to a definite quantity of water. |
| 4. Water may be used on non-riparian land. | 4. Use of water may be restricted to riparian land. |
| 5. Right may be lost by non-use or abandonment. | 5. Right does not depend on use and is not subject to abandonment. |
| 6. There is no natural flow requirement. | 6. There is a qualified right to natural flow in some jurisdictions. |

Groundwater laws are constantly changing as a result of occurrences of groundwater *pollution.* The trends in groundwater law are that the following:

1. The riparian doctrine, with its concept of absolute property rights in water, is becoming obsolete.
2. The principle of the appropriation doctrine, which suggests that ownership of water rests in the public collectively, is becoming a more widely accepted view.
3. The right to appropriate water is being decided on whether the greater public good is being served.
4. All surface and subsurface water resources are being handled by a single governmental agency with expertise in all aspects of the *hydrologic cycle.*

**W**

European water laws are in a state of flux as well resulting from the formation of the European Union. Several sovereign states may occupy a given *watershed basin.* Within the sovereign state, there may be different agencies that govern water management and quality. A water framework directive (WFD) has been established for all waters in the European Union on the scale of river districts (equivalent to watersheds), where an authority implements integrated management programs. Aquifers are linked to the most appropriate river district. The purpose of the WFD is to provide 'good status' for all ground and surface waters in quantity and quality by 2015. This is done through integrated water management of quality, quantity, surface and groundwater exploitation and preservation. Presented below in **Table W-2** is the milestone schedule for implementing the WFD.

**Table W-2**   Water Framework Directive (WFD) schedule

| Deadline | Implementation Steps | Articles of the WFD |
|---|---|---|
| 2000 | Adoption of the WFD | Directive 200/60/EC |
| 2003 | Transposition of the directive in the national legislation | Art. 24 |
| | Designation of the competent authorities in river basin districts | Art. 3 |
| 2004 | Register of the characteristics of the river basin district | Art. 5 |
| | Register of protected areas | Art. 6 |
| 2006 | Implementation of the monitoring program | Art. 8 |
| 2009 | Publication of the management plan | Art. 13 |
| | Publication of the program of measures | Art. 11 |
| 2010 | Implementation of full-cost recovery of water uses | Art. 9 |
| 2012 | Entry into force of the program of measures | Art. 12 |
| | Combined approach of the emission controls for point and diffuse sources | Art. 10 |
| 2013 | Repeal of a set of former EU water directives | Art. 22 |
| 2015 | **Good Status for all waters** | Art. 4 |
| 2019 | Review of the directive | Art. 19 |
| 2024 | Prohibition of emissions of priority hazardous substances | Art. 16 |

Cf. *Beneficial use*; *Correlative rights*; *Prior appropriation*; *Rule of capture*; *Rule of reasonable use*.

**Water of hydration** Water that is present in a definite amount and attached to a compound to form a hydrate. Water can be removed, that is, by heating, without altering the composition of the compound. Cf. *Heat of hydration*.

**Water level measurement** The act of sounding a well or a *piezometer* to determine the depth of water within it. Water level measurement can be accomplished using a device that lights up or emits a sound when contact with the water is made. If an *electric sounder* is unavailable, a weighted-measuring tape-lowered downhole may be chalked to record the water level or jiggled so as to listen for making contact with the water surface. Cf. *Groundwater measurement; Interface probe*.

**Water level recovery** The second stage of a *pumping test* in which the rate of the water's return to its prior level is monitored. This data is used to corroborate the *drawdown* data to analyze *transmissivity* and *storativity*.

**Water loss** See *Evapotranspiration*.

**Water-air partitioning coefficient** See *Henry's law constant*.

**Water pressure** The force per unit area at some point within a body of water. Water pressure occurs in one of two situations: (1) open condition, where the water is open to the atmosphere, and (2) a closed condition, such as water in a pipe or in a confined *aquifer*. Pressure in open conditions usually can be approximated as the pressure in "static" or nonmoving conditions (even in the ocean where there are *waves* and currents), because the motions create only negligible changes in the pressure. Such conditions conform with principles of fluid statics. The pressure at any given point of a nonmoving (static) fluid is called the *hydrostatic pressure*. Closed bodies of fluid are either "static," when the fluid is not moving, or "dynamic," when the fluid can move. Cf. *Aquifer, confined*.

**Water purification** The process in which undesirable impurities are removed or neutralized as in chlorination, filtration, *ion exchange*, and distillation.

**Water quality** A set of chemical parameters found in water. The primary use of water quality analysis is to determine the suitability of water for a proposed use. Three main classes of use are domestic (household), agricultural, and industrial. In the United States, water quality criteria are designated by the *Safe Drinking Water Act (SDWA)*. Within this, there are both primary and secondary *drinking water standards*. The primary standards are enforceable by law, whereas the secondary standards act as guidelines and are not enforceable. These criteria are listed in **Appendix B** *Maximum Contaminant Levels (MCLs)*. Cf. *Bacteriological water quality standards*.

**Water retention curve** The relationship between the water content ($\theta$) and the soil water potential ($\psi$) used to predict the soil water storage, water supply to the plants (field capacity) and soil aggregate stability. The water holding capacity of any soil is due to the porosity and the nature of the bonding in the soil. The hysteretic effect of water filling and draining of pores, different wetting and drying curves may be distinguished. This curve is characteristic for different types of soil and is also called the soil moisture characteristic. The general features of a water retention curve can be seen in **Figure W-2**, in which the volume water content, $\theta$, is plotted against the matric potential, $\Psi_m$. At potentials close to zero, a soil is close to saturation and water is held in the soil primarily by capillary forces. As $\theta$ decreases, binding of the water becomes stronger, and at small potentials (increasing negativity, approaching wilting point) water is strongly bound in the smallest of pores, at contact points between grains and as films bound by adsorptive forces around particles. Sandy soils will involve mainly capillary binding and will therefore release most of the water at higher potentials, while clayey soils, with adhesive and osmotic binding, will release water at lower (more negative) potentials. At any given potential, peaty soils will usually display much higher moisture contents than clayey soils, which would be expected to hold more water than sandy soils.

**W**

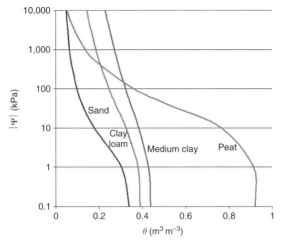

**Figure W-2** Water retention curves for sand, clay loam, medium clay and peat.

Cf. *Brooks and Corey model.*

**Water sampling** Any method of collection of water for analysis, whether from surface-water or *groundwater* sources.

**Watershed** See *watershed basin.*

**Watershed basin** A *drainage basin* or *catchment area* in which the surface waters, from natural or manmade water collection systems, drain. Watersheds drain into other watersheds in a hierarchical form, larger ones breaking into smaller ones or subwatersheds with the topography determining where the water flows. Cf. *Hydrologic region; Artesian basin.*

**Watershed management** The planned manipulation of one or more of the factors of the environment of a natural or disturbed *drainage* so as to effect a desired change in or maintain a desired condition of a water resource. Comprehension of complex interrelationships between vegetation, soil, climate, geologic forces, and the impact of civilizations are imperative to a watershed management plan.

**Water supply** A division of engineering concerned with the development of sources of supply, transmission, distribution, and treatment of water. Cf. *Groundwater reservoir.*

**Water table** Also called groundwater level, groundwater surface, groundwater table, or level of saturation. In an unconfined *aquifer*, the level at which the *piezometric* pressure is equal to atmospheric pressure. Below this level, all *pore spaces* are filled with water. The location of the water table is determined in the field by the level at which water stands in a shallow well open along its length and penetrates surficial deposits just deeply enough to encounter standing water. Cf. *Aquifer, unconfined.*

**Water temperature** The average temperature of groundwater in the mid-latitudes is approximately 10°C (52°F). In the far northern and southern areas of the world, groundwater temperature can be close to freezing and can be permanently frozen (*permafrost*). In the tropical latitudes, water temperature can be as high as 25–30°C (75–81°F). Generally, within a given geographic area, shallow groundwater temperatures are consistent within a few degrees, and below a depth of approximately 10 m approximate the mean annual temperature of the region.

**Water treatment** See *Point-of-use water treatment systems; Air sparging; Air stripper; Water scrubber.*

**Water well** An excavation or structure created in the ground by digging, driving, boring, or drilling to access water from an *aquifer*. The well may consist of structural material (casing) to keep the *borehole* open, and an open area for the water to enter the well. The well water is drawn into the well via an electric submersible pump or a mechanical pump (e.g., from a water-pumping windmill). Some water wells are simply areas excavated in the ground surface and remain open to atmospheric and lithologic units. This water can also be drawn up using containers, such as buckets that are raised mechanically or by hand. Wells can vary greatly in depth, water yield of completion, water quality, and methods of completion. Cf. *Production well.*

**Water-vapor partitioning coefficient** See *Henry's law constant.*

**Water race** See *Race*

**Water yield** Also called water crop. The *precipitation* minus the *evapotranspiration* or the *runoff* from a delineated *drainage basin*. The water yield includes the *groundwater outflow* into the basin's watercourses plus that portion that bypasses the *gauging stations* and leave the basin through underground routes.

**Water yielding** A general term applied to formations that transmit water, or *aquifers.*

**Waterborne disease** Illness resulting from the spread of pathogenic agents through the consumption of, or exposure to, water containing bacteria. The leading causes of waterborne disease are as follows:

- contamination of water by overflow or seepage of sewage;
- contamination of springs;
- contamination of water by chemicals;
- contamination of water from surface runoff; and
- contamination of water by flooding.

All water that percolates into the subsurface acquires and transports surface material to the groundwater system. Rain and melting snow pick up carbon dioxide ($CO_2$), pesticides, herbicides, fertilizers, minerals, bacteria, and inorganic compounds such as oxides of sulfur and nitrogen. *Percolation* through landfills and septic fields picks up bacteria, viruses, and toxic substances. Further, in industrial areas, *percolating water* may leach industrial chemicals into the water table. As a result, any well has the potential to become biologically or chemically contaminated.

**Wave** A disturbance that moves through or over the surface of a liquid, e.g. the sea, from one point to other points without any permanent displacement.

**Weirs** A vertical baffle that restricts the total *flow* of water in an open or closed channel and represents a simple method to measure the flow of water. The crest of a weir is the bottom of a notch or the level to which water must rise for flow to occur. Weirs with rectangular or triangular openings are used for measuring flow in open channels. The relationship between *head* and *discharge* of a weir varies according to the shape of the weir. Francis' equation for a rectangular weir without end contractions is:

$$Q = 1495Lh\sqrt{h}$$

where:
$Q$ = discharge (gpm)
$L$ = length of the crest (ft)
$h$ = head on weir (ft)

It is important to measure $h$ as far *upstream* as possible to eliminate the effect of the increase in velocity as the water spills over the weir. **Figure W-3** illustrates a rectangular weir without end contractions.

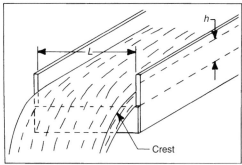

**Figure W-3** A rectangular weir without end contractions. (Driscoll, 1986. Reprinted by permission of Johnson Screens/a Weatherford Company.)

**Well** See *water well; production well.*
**Well bore** See *Borehole; Bored well.*
**Well casing** A casing, typically a section of pipe, that is used to prevent a *borehole* from collapsing, to suspend a *well screen*, or to provide access for downhole pumping equipment. Cf. *Acrylonitrile Bufadine Styrene (ABS); Polyvinyl Chloride (PVC); Casing.*
**Well cuttings** See *drillers log.*
**Well design** The process of specifying materials and dimensions for a well. A good design should follow the following principal objectives:

- Obtain the highest yield with the minimum *drawdown*, consistent with *aquifer* capability.
- Proper protection from *contamination*.
- Obtain good-quality water.
- Produce water that is sand-free.
- Provide for a long life of 25 years or more.
- Provide for reasonable short-term and long-term maintenance costs.

**W**

The design engineer must have in mind not only the objectives listed above, but also common sense and the needs of the well owner. A suburban homeowner should not pay for a 300-gpm well when his need could be satisfied with a 15-gpm well. Every well consists of two main elements: the casing portion and the intake portion. The *well casing* houses the pumping equipment and provides a conduit to bring the water to the surface for use. The intake portion in unconsolidated and semiconsolidated formations is generally screened to prevent sediment from entering the well and to provide a structural barrier between the well and the formation. In consolidated formations, the intake portion may be open borehole; however, in rock formations such as sandstone, screens are often employed in anticipation of the formation's deterioration with time. The following information are necessary for well design:

- stratigraphy of the aquifer and overlying sediments;
- *transmissivity* and storage coefficient values for the aquifer;
- current and long-term *water balance* conditions for the aquifer;
- *grain size distribution* for unconsolidated sediments;
- identification of rock or mineral types in consolidated formations; and
- water quality information.

If other wells are present in the area, this is a good starting point to evaluate appropriate designs. Well records are maintained by federal, state, and provincial agencies. Standard well design procedures involve selecting the casing dimensions and materials, the well depth and the *well screen* dimensions, slot size and materials. These aspects of well design include the following:

1. The casing diameter must accommodate the pump with enough clearance for installation and efficient operation. **Table W-3** presents recommended casing diameters for anticipated *well yields*.

**Table W-3**   Recommended well diameters for various pumping rates

| Anticipated well yield | | Nominal size of pump bowls | | Optimum size of well casing | | Smallest size of well casing | |
|---|---|---|---|---|---|---|---|
| gpm | m³ day⁻¹ | in. | mm | in. | mm | in. | mm |
| <100 | <545 | 4 | 102 | 6 ID | 152 ID | 5 ID | 127 ID |
| 75 to 175 | 409 to 954 | 5 | 127 | 8 ID | 203 ID | 6 ID | 152 ID |
| 150 to 350 | 818 to 1,910 | 6 | 152 | 10 ID | 254 ID | 8 ID | 203 ID |
| 300 to 700 | 1,640 to 3,820 | 8 | 203 | 12 ID | 305 ID | 10 ID | 254 ID |
| 500 to 1,000 | 2,730 to 5,450 | 10 | 24 | 14 OD | 356 OD | 12 ID | 305 ID |
| 800 to 1,800 | 4,360 to 9,810 | 12 | 305 | 16 OD | 406 OD | 14 OD | 356 OD |
| 1,200 to 3000 | 6,450 to 16,400 | 14 | 356 | 20 OD | 508 OD | 16 OD | 406 OD |
| 2,000 to 3,800 | 10,900 to 20,700 | 16 | 406 | 24 OD | 610 OD | 20 OD | 508 OD |
| 3,000 to 6,000 | 16,400 to 32,700 | 20 | 508 | 30 OD | 762 OD | 24 OD | 610 OD |

2. The selection of casing materials is based on *water quality*, depth, cost, diameter, and the appropriate regulations. Typical casing materials are carbon steel, thermoplastic (e.g., polyvinyl chloride, PVC), fiberglass, and galvanized steel. **Table W-4** presents a comparison of well casing materials.

Well design is also dependent upon the available drill rig. **Appendix C** presents some drilling terms and methods.

**Well development** Procedures to maximize *well yield* by repairing the damage done to the formation through drilling and to alter the physical characteristics of the *aquifer* near the *borehole* to improve the *flow* of water to the well. Drilling alters the physical characteristics of the aquifer in the vicinity of the *borehole*, which can restrict the flow of water, for example, through the invasion of *drilling fluids*, to the formation or compression of unconsolidated sediments around the borehole. Well development:

- removes fine materials from the pore spaces around the borehole;
- increases the natural *porosity* and *permeability* of the formation in the vicinity of the borehole;
- removes the filter cake or drilling-fluid film that has invaded the formation; and
- creates a zone of graded sediment around a screen in a naturally developed well to stabilize the formation.

Well Development Methods:

*Overpumping:* Pumping at a rate higher than the rate at which the well will be pumped when in service.
*Backwashing:* Reversal of flow through the screen to remove the finer fractions from the well, usually through *surging*. This can be done through pumping of water, mechanical surging, or airlift methods.
Cf. *Allowable; Sediment concentration; Aquifer development; Air development.*

**W**

**Table W-4** Comparison of well casing materials

| Material | Specific gravity | Tensile strength (psi) | Tensile modulus ($10^5$ psi) | Impact strength (ft–lb/in.) | Upper temperature limits (°F) | Thermal expansion ($10-6$ in./in. °F) | Heat transfer (Btu-in./h-ft² °F) | Water absorption (wt%/24 h) |
|---|---|---|---|---|---|---|---|---|
| ABS | 1.04 | 4.500 | 3.0 | 6.0 | 180 | 5.5 | 1.35 | 0.30 |
| PVC | 1.40 | 8.000 | 4.1 | 1.0 | 150 | 3.0 | 1.10 | 0.05 |
| Styrene Rubber | 1.06 | 3.800 | 3.2 | 0.8 | 140 | 6.8 | 0.80 | 0.15 |
| Fiberglass Epoxy | 1.89 | 16,750 | 23.0 | 20.0 | 300 | 8.5 | 2.30 | 0.20 |
| Asbestos Cement | 1.85 | 3,000 | 30.0 | 1.0 | 250 | 1.7 | 0.56 | 2.0 |
| Low-Carbon Steel | 7.85 | 35,000 (yield)[a] 60,000 (ultimate) | 300.0 | [b] | 800–1,000 | 6.6 | 333.0 | Nil |
| Type 304 Stainless Steel | 8.0 | 30,000 (yield) 80,000 (ultimate) | 290.0 | [b] | 800–1,000 | 10.1 | 96.0 | Nil |

[a]Yield strength is the tensile stress required to produce a total elongation of 0.5% of the gauge length as determined by an extensometer. Expressed in psi.
[b]Because testing methods for steel and other materials are not the same and the three results are not comparable, the impact strength values for steel are not shown. In any event, the actual strength of steel is so high relative to the demands of water well work that in can be ignored in design considerations.

**Well drilling methods** See: *Appendix C – Drilling Methods*.

**Well efficiency** See *Step-drawdown test*.

**W(u) (well function of u)** In 1915, Theis developed a breakthrough in hydrologic methodologies by utilizing an analogy to heat flow for the equation describing transient *groundwater flow* to a well. The equation for *non-steady-state groundwater flow* to a *well* in radial coordinates is:

$$\frac{\partial^2 h}{\partial r^2} + \frac{1}{r}\frac{\partial h}{\partial r} = \frac{SS}{T}\frac{\partial h}{\partial t}$$

The *Theis solution*, written in terms of *drawdown*, is:

$$h_0 - h(r,t) = \frac{Q}{4\pi T}\int_u^\infty \frac{e^{-u}\,du}{u}$$

where:

$$u = \frac{r^2 S}{4Tt}$$

The various parameters are illustrated in **Figure W-4**. The exponential integral shown above is well-known in mathematics and is easily solved. The term within the integral is known as the well function $u$, $W(u)$. On substituting:

$$h_0 - h = \frac{Q}{4\pi T}W(u)$$

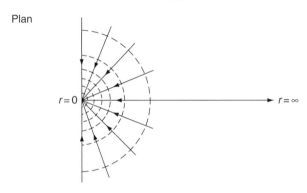

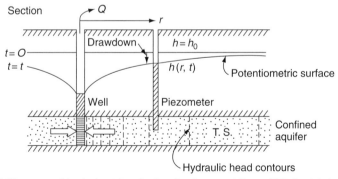

**Figure W-4** Radial flow to a well in a horizontal confined aquifer. (Freeze and Cherry, 1979. Reprinted with permission by Prentice-Hall, Inc.)

Values of $W(u)$ versus $u$ are presented in **Appendix B**, while **Figure W-5a** shows the graphical relationship between $W(u)$ and $1/u$, which is known as the *Theis non-equilibrium-type curve* and **Figure W-5b** presents field data that is overlaid on the type curve to determine a "match point" for the drawdown and time, for solving the above equations. It is typically convenient to select a point equal to $W(u)$ of say "1" and $1/u$ of 10 for the calculation.

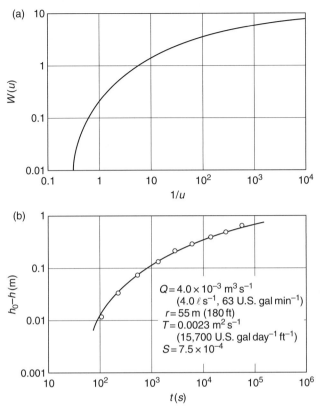

**Figure W-5** (a) Theoretical curve of $W(u)$ versus $1/u$. (b) Calculated curve of $h_0 - h$ versus $t$. (Freeze and Cherry, 1979. Reprinted with permission by Prentice-Hall, Inc.)

Cf. *Image well theory; Equilibrium well equations.*

**Well head** The component at the top of a well used to suspend casing strings and provide sealing functionality for wells. The primary component of a water well head system is the casing head. A wellhead serves numerous functions such as (1) casing suspension, (2) casing pressure isolation when multiple casing strings are used, (3) well access, (4) pump attachment, and (5) tubing suspension. The tubing is a removable pipe installed in the well for transmitting water. Cf. *Well design.*

**Well hydraulics** The *well function of u*, $W(u)$ is the derivation of the equation of *radial flow* to a well with the *Theis solution* to the integral relating the change in *head* to the *flow* of water from the well. This provides the basis for defining well hydraulics of an idealized system in which the strata overlying and underlying a confined *aquifer* are completely impermeable. In reality, there is leakage between water-transmitting zones. **Figure W-6** presents a schematic of a two-aquifer "leaky" system. The solution to this was developed in 1955 by Hantush and Jacob, and can be written in a form analogous to the Theis equation (see $W(u)$) as:

$$h_0 - h = \frac{Q}{4\pi T} W\left(u, \frac{r}{B}\right)$$

for which:

$$\frac{r}{B} = r\sqrt{\frac{K'}{K_1 b_1 b'}}$$

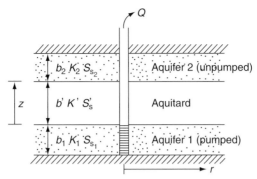

**Figure W-6** Schematic diagram of a two-aquifer "leaky" system. (Freeze and Cherry, 1979. Reprinted with permission by Prentice-Hall, Inc.)

**Figure W-7** presents a plot of $W(u, r/B)$ (the leaky-well function) versus $1/u$. If the *aquitard* is impermeable, then $K' = 0$ and the *Hantush–Jacob formula* reduces to the Theis solution. In unconfined aquifers, there is a vertical component of flow induced in the flow system as water is *drawdown* under pumping conditions. Under pumping conditions in unconfined aquifers, there is a delayed response in *piezometers* to pumping, and three distinct periods or segments of response have been noted. During the first segment, an unconfined aquifer reacts in the same manner as a confined aquifer. The effects of gravity drainage are shown in the second segment, where there is a decrease in the time–drawdown relative to the Theis-type curve as a result of gravity drainage resulting from the falling water table in the vicinity of the well. In the later stages of pumping, the time–drawdown response again appears to conform to confined conditions. The mathematical solution to the relationship of flow of a well and drawdown can be expressed in a form analogous to the Theis solution:

$$h_0 - h = \frac{Q}{4\pi T} W(u_A, u_B, \eta)$$

where:

$$u_A = \frac{r^2 S}{4Tt}$$

where:
$S =$ elastic storativity [dimensionless]
$T =$ transmissivity $[L^2 \cdot T^{-1}]$
$t =$ time [T]

In the later stages of pumping:

$$h_0 - h = \frac{Q}{4\pi T} W(u_B, \eta)$$

where:

$$u_B = \frac{R^2 S_y}{4Tt}$$

where:
$S_y =$ specific yield responsible for the delayed release of water to the well.

Under *anisotropic* conditions, having horizontal *hydraulic conductivity* $K_r$ and vertical hydraulic conductivity $K_z$, $\eta$ is defined as:

$$\eta = \frac{r^2 K_z}{b^2 K_r}$$

If the aquifer is *isotropic*, then $K_z = K_r$ and $\eta = r^2/b^2$. Other modifications could also be considered to the Theis equation, e.g., partial penetration, anisotropy in pumped aquifers, and multiaquifer systems.

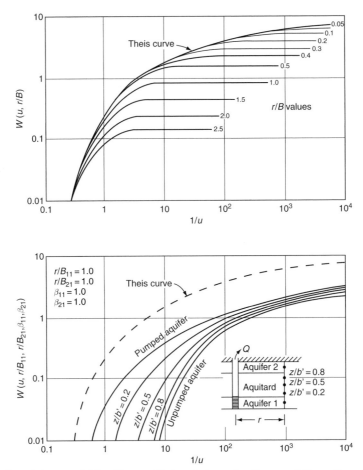

**Figure W-7** Theoretical curves of $W(u, r/B)$ versus $1/u$ for a leaky aquifer. (Freeze, and Cherry, Groundwater© 1979. Reprinted with permission by Prentice-Hall, Englewood Cliffs, NJ.)

Cf. *Aquifer, confined; Aquifer, unconfined; Aquifer, leaky.*

**Well inefficiency** See *step-drawdown test.*

**Well installation** See *Groundwater monitoring well; Packer; Washdown method of installation; Pull back method of screen installation.*

**Well log** The recorded observations, either manual or electronic, of subsurface material encountered during the emplacement of a well or a *borehole.* Cf. *Well logging.*

**Well logging** The measurement, recording, and/or analysis of the geologic strata that are encountered during or after drilling a *borehole.* Well logging is performed either directly through cores, sampling, or observation of drill cuttings or indirectly through *geophysical exploration methods.* Cf. *Borehole geophysics; Gram size analysis; Split-spoon samples.*

**Well loss** *Head loss* in a *well* due to well inefficiency. Cf. *Step drawdown test.*

**Well maintenance** Wells, like any other man-made device, require maintenance. Water, being the universal solvent, has the ability to dissolve and precipitate a wide variety of compounds in the subsurface. Further the presence of bacteria in the subsurface can further result in changes in geochemical conditions, which can result in the deposition of biomass and precipitates. Well maintenance generally entails chemical treatments or physical treatments, although for many problems, a combination of multiple treatments is usually required. Types of well problems include the following:

- reduction in *well yield*, which can be caused by chemical *incrustation* or biofouling of the *well screen* and the formation materials around the intake portion of the well;
- plugging of the formation around the well screen by fine particles;
- *sand pumping*, usually attributable to poor *well design* or inadequate development;
- structural collapse of the *well casing* or screen often due to acidic waters; and
- pump conditions resulting from poor design or corrosion.

Causes of Chemical Incrustation: *Carbonate* incrustation forms where there is a pressure release, as occurs at the interface where the water enters a well. Calcium carbonate ($CaCO_3$) can be carried in solution in proportion to the amount of dissolved carbon dioxide ($CO_2$), which varies with *pressure*; the higher the pressure, the higher the concentration of $CO_2$. Under pumping conditions, the *water table* is drawn down to produce the necessary *gradient* for *flow* to the well. The greatest pressure change under these conditions occurs in the immediate vicinity of the well. As pressure decreases, $CO_2$ is released and the equilibrium with $CaCO_3$ is disturbed, resulting in the precipitation of carbonate as illustrated below:

$$Ca(HCO_3)_2 \xrightarrow{-\Delta P} CaCO_3 \downarrow + CO_2 \uparrow + H_2O$$

where $-\Delta P$ is the negative change in pressure. The solubility of $Ca(HCO_3)_2$ on the left side of the equation is about 1300 mg $L^{-1}$, whereas the solubility of $CaCO_3$ on the right side of the equation is about 13 mg $L^{-1}$. In waters that are rich in iron and manganese under pumping conditions, velocity-induced pressure changes can result in the formation of insoluble iron and manganese hydroxides. These hydroxides have a gel-like consistency that have the ability to eventually harden. The reactions induced through pumping are:

$$Fe(HCO_3)_2 \xrightarrow{-\Delta p} Fe(OH)_2 \downarrow + 2CO_2$$

With aeration during pumping, the reaction proceeds further, as indicated below:

$$4Fe(OH)_2 + 2H_2O + O_2 \rightarrow 4Fe(OH)_3 \downarrow$$

The solubility of ferric hydroxide is less than 0.01 mg $L^{-1}$. Ferric hydroxide is a reddish-brown deposit similar to rust; hydrated ferrous hydroxide forms a viscous black sludge. A similar reaction occurs for manganese bicarbonate ($Mn(HCO_3)_2$):

$$2Mn(HCO_3)_2 + O_2 + 2H_2O \rightarrow 2Mn(OH)_4 + 4CO_2 \uparrow$$

and eventually produces manganese oxide ($MnO_2$), which is also black or dark brown.

Treatment of Incrustation: From the *electrochemical sequence*, we know that the presence of iron and manganese in the *groundwater* can result from bacterially mediated redox reactions. These redox reactions can be localized as a result of materials used during drilling, such as biodegradable drilling fluids. In this instance, the problems can be taken care of through the use of bactericides such as sodium hypochlorite, chloride dioxide, or other bactericides. Other methods include

- designing the well to have the maximum-possible inlet area to reduce the *entrance velocity*;
- developing the well thoroughly to remove as much of the material introduced during drilling as possible;
- reducing the pumping rate; possibly having a trade-off with longer pumping periods;
- increasing the number of wells to decrease the rate of pumping from an individual well; and
- frequent maintenance, including application of a bactericide.

Cf. *Procedure for acid treatment*; *Passive film*; *Pasteurization treatment for iron bacteria*; *Physical plugging*; *Propping agents*; *Calcium hypochlorite*; *Dissolved oxygen*.

**Well point** Also called a drive point. A *well screen* with a narrowed tip at the end designed for being driven into the ground, or more commonly, a component of *well point system* for *dewatering* where the water is removed by applying a vacuum. Well points can either be driven or installed via a predrilled *borehole*. Typically they have short screens and are restricted to depths of approximately 6 m (20 ft), the practical limitation on the suction lift. However, the typical design basis for a suction lift system is to limit the systems to 4.6 m (15 ft). Cf. *Well point system*.

**W**

**Well point system** A series of closely spaced wells, connected to a header for *suction lift*. Operation of a well point system entails the use of a central pump that applies a vacuum to lift the water. Well point systems are typically used for *dewatering* applications in the following conditions:

- areas where the desired dewatering level is within the suction lift of the system, i.e., 4.6 m (15 ft) of the surface;
- formations with low *hydraulic conductivities*, e.g., fine sands and silts; and
- shallow formations that overlie impervious formations.

Cf. *Dewatering well.*

**Well rehabilitation** See *well maintenance.*

**Well screen** Part of a complete *well design* that acts as a filtering device used to keep sediment from entering a well. The screened portion of a completed well allows water, oil, or other liquid to enter the well from the saturated zone, prevents unwanted sedimentation within the well, and serves structurally to support unconsolidated *aquifer* material. Proper well screen selection, e.g. *gauze number, slot openings, open area, louver screen*, affects the hydraulic efficiency of a well and the long-term cost and use to the well owner. Cf. *Pipe base well screen; Pipe size screens.*

**Well screen installation** The exact procedures to be followed during installation depend on the nature of the *aquifer* materials, the method of drilling, the *borehole* dimensions, the hydraulic conditions, and the casing and *well screen* material. Cf. *Well screen length; Groundwater monitoring well; Pull-back method of screen installation.*

**Well screen length** The optimum well screen length is based on a number of factors and aquifer conditions. *Aquifers* are typically *heterogeneous*, and in such conditions, the well screen should be placed to intercept those zones of higher *hydraulic conductivity* that will produce more water. The recommended screen length for typical hydrogeologic settings are described below.

Homogeneous unconfined aquifers: Typically the bottom one-third to one-half of an *aquifer* of less than 46 m (150 ft) thick is optimum. In deep, thick aquifers, as much as 80% of the aquifer thickness may be screened to obtain higher *specific capacity.* There are, however, competing considerations in the selection of screen length. The maximum practical *drawdown* in an unconfined aquifer is two-thirds the thickness of the water-bearing zone. The longer the screen length, the greater the specific capacity, while the shorter the screen length, the greater the available drawdown. The greater the available drawdown, the greater the achievable pumping rate.

Non-homogeneous unconfined aquifers: Screen length recommendations for these conditions are similar to those for *homogeneous* unconfined aquifers, except that the screen is situated to intercept the zone(s) of higher hydraulic conductivity. A general rule of thumb is to screen the bottom one-third of the producing formation. Homogeneous confined aquifers: Here 80–90% of the aquifer thickness should be screened, assuming that water levels under pumping conditions do not fall below the top of the aquifer. Maximum available drawdown in confined aquifers should be the distance from the potentiometric surface to the top of the aquifer.

Non-homogeneous confined aquifers: In this case, 80–90% of the most permeable zone(s) should be screened.

**Well screen open area** The total spaces in the screened interval of a well through which water can flow into the well. In order to maximize water entry, the well screen open area should be as large as possible, while maintaining the strength of the screen. At present, the most efficient well screen design is the wire-wound screen.

**Water scrubber** A system where gases are in contact with water, as in spraying or bubbling through, to washout traces of water-soluable components of the gas stream. Cf. *Air sparging; Air stripping.*

**Well seal/sealing of the well head** After construction a welded, threaded, or flanged cap or compression seal must be fixed to the top of the well to prevent foreign material from entering the well. Regardless of the type of seal, the watertight casing should extend at least 12 inches (305 mm) above the ground elevation. The ground around the top of the casing, whether natural or manmade, should slope away from the *well head* to prevent pending around the *well casing.* If the well is located in areas known to occasionally *flood*, the top of the casing should be set one to two feet (0.3 to 0.6 m) above the highest recorded flood level.

**Well sounder** See *electrical sounder; interface probe.*

**Well yield** The maximum safe volume of water discharged per given unit of time, usually measured in gallons per minute or cubic meters per day. For other units for conversion see **Appendix A**. Cf. *Basin yield; Aquifer yield; Well development; Sustained yield; Yield.*

**Wenner array** Used in surface resistivity mapping. This is one of the most popular arrays used for resistivity surveys and the most simple one. In the Wenner array, the electrodes are uniformly spaced in a line as illustrated in **Figure W-8**.

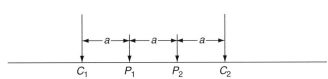

**Figure W-8** Electrode spread in a Wenner array. (Telford et al., 1968. Reprinted with permission of Cambridge University Press.)

The apparent resistivity measured using this array is:

$$\rho_a = \frac{2\pi a \Delta V}{I}$$

where:

$\rho_a$ = apparent resistivity (ohms)
$\Delta V$ = change in potential (V)
$I$ = current (A)
$a$ = distance between electrodes [L]

Cf. *Schlumberger array; Electric profiling.*

**Wetland** Areas of *marsh, fen,* peatland, or water, whether natural or artificial, permanent or temporary, with waste that is static or flowing, fresh, brackish, or salt, including areas of marine water, the depth of which at low *tide* does not exceed 6 m (Ramsar Conventin). Cf. *Fen; Bog.*

**Wettability** Describes the relative preference of a material to be covered by a certain liquid or phase. Wettability can be affected by minerals present in the pores. Quartz, sands, and sandstones have a greater tendency to become water wet than carbonates.

**Wildcat well** See *discovery well.*

**Withdrawal well** See *dewatering well.*

**Wrap-on-pipe screen** See *pipe-base well screens.*

**Wylie equation** See *Acoustic velocity logging.*

**W**

**Yazoo stream** A long *reach* or portion of a *tributary* flowing parallel to the main *stream* before joining at a *deferred junction*. A yazoo stream may result when a stream is forced to flow along the base of a natural levee formed by the mainstream or *river*. Cf. *Deferred tributary; False stream*.

**Yield** The rate of *flow* at which water can be collected from a *groundwater resource* or a surface source. For *groundwater*, yield is the pumping rate that can be sustained from a well or well field without lowering the water level below the pump intake. For surface water, yield is the streamflow during a given time derived from a unit area, expressed in dimensions of $L^3 \cdot T^{-1} \cdot L^{-2}$. Cf. *Average yield; Delayed yield; Firm yield; No-failure yield; Optimum yield; Safe yield; Specific yield; Steady-state flow yield; Sustained yield; Well yield*.

**Yield–depression curve** A graph of *drawdown* versus *yield* for a pumped well, which produces a curved plot that is used to determine the optimum pumping rate for a water supply well.

**Yield guarantee** The well contractor's statement that a certain rate of *flow* will be sustainable from the well drilled. Well contractors are vulnerable to lawsuits based on guarantees stipulated in well contracts if common precautions, necessitated by the unpredictability of hydrogeologic conditions, are overlooked. An example of a common yield guarantee clause in a well contract is as follows: "The contractor shall guarantee a *yield* of _ gpm after _ hours of continuous pumping."

**Yield point** The time at which the stress and the strain in a fluid are nearly constant and the *viscosity* (the fluid's resistance to a shear stress) does not change significantly when the stress is increased. The ability of the *drilling fluid* to move cuttings up the *borehole* is dependent on its viscosity and the uphole velocity. Drilling fluids commonly contain clay additives and act like a plastic material, meaning that they do not deform until the stress level has reached the yield point. The yield point is the *pressure* at which the pump begins to move the drilling fluid. The strength of the attractive forces between particles in the fluid controls the yield point. Water, a *Newtonian fluid*, deforms proportionately to an applied stress. A drilling fluid, originally similar to a plastic material, becomes Newtonian when the yield point has been reached.

**Yield strength** The tensile stress required to produce a total elongation of 0.5% of the *gauge* length as determined by an extensiometer, expressed in pounds per square inch (psi). Yield strength is considered when comparing *well casing* materials. Cf. *Column strength*.

**Young stream** A stream in the *youth* stage of watercourse development, e.g., actively eroding its channel. Cf. *Youth*.

**Youth** Also called the youthful stage. The first stage of development of a *river* or a *stream*, in which the watercourse is actively eroding a V-shaped valley with associated falls and *rapids*. A stream in the youth stage can carry a *sediment load* greater than the *load* it typically carries, and is therefore capable of eroding the landscape. A stream in youth usually has only a few short *tributaries*. Cf. *Juvenile water; Young stream*.

**Y**

**Zero-air-voids curve** The curve showing the zero-air voids unit weight as a function of water content.

**Zero-degree isotherm** The location at which a watercourse or a water body is first cooled to 0°C. When water reaches 0°C, any additional drop in temperature causes freezing. Wide, shallow *streams* or *rivers* freeze before deeper water bodies and rivers usually cool in the *downstream* direction.

**Zero discharge** The total recycling of a fluid or water, as in the delivery of pure, clean water, or that containing no substance at a concentration greater than that normally occurring in the local environment.

**Zero energy** Large bodies of water such as oceans, seas, and large *lakes* are *reservoirs* that are considered to represent a point of minimum energy from the perspective of the *hydrologic cycle*. Water particles in the hydrologic cycle have a velocity and/or position simultaneously along various travel paths and have three possible energies: potential, kinetic, and heat energy. Potential energy relates to the position or elevation of the water particle relative to the *mean sea level* (msl). The kinetic energy of a moving water particle is dependent on the mass as well as the velocity of the particle, and moving water particles encounter friction forces, causing loss of energy in terms of heat. Surface water residing at the elevation of mean sea level are at rest; therefore, the water particles do not have potential, kinetic, or heat energy. In the hydrologic cycle, they possess zero energy. These water particles acquire energy with the addition of an external driving force such as solar radiation. **Figure Z-1** shows the hydrogeologic energy cycle.

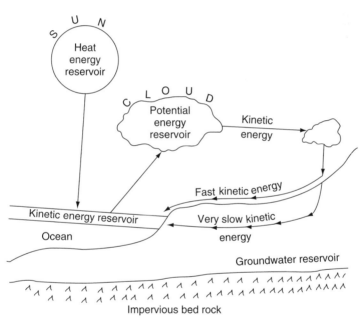

**Figure Z-1** The energy cycle incorporated into the hydrologic cycle showing areas of zero energy, i.e., oceans and other large water bodies. (Sen, 1995. Reprinted with permission from CRC Press, Inc.)

**Zero-flow record** A year of no *flow* or virtually no flow, for a given watercourse. The term "zero-flow record" is often encountered in low-flow series in *flood frequency analysis*. In arid regions, zero-flow years occur more frequently than non-zero years, indicating that *streams* were either dry or were flowing below a set recording limit. At most United States Geological Survey (USGS) *gauges*, streamflow is reported as zero flow when it falls below 0.05 ft$^3$ s$^{-1}$. Refer to **Appendix A** for unit conversions.

**Zero-flux plane** The plane of zero potential *gradient*. The measurement of *evaporation* illustrated in **Figure Z-2** uses *soil moisture* depletion supplemented with the determination of an average zero-flux plane to distinguish between the relative proportions of net water loss due to evaporation in the upward direction and *drainage* in the downward direction.

Z

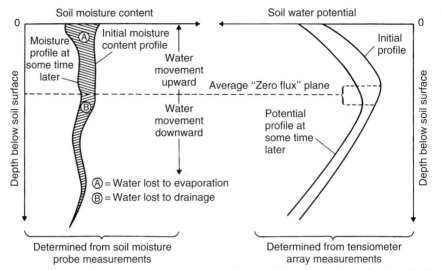

**Figure Z-2** Measurement of evaporation using soil moisture depletion and determination of zero-flux plane. (Maidmont, 1993. Reprinted with permission from McGraw-Hill, Inc.)

**Zero-order rate reaction** In some reactions, the rate is independent of the concentration of the reactant or reactants, and these are termed zero-order reactions. Photochemical reactions (in which the rate-determining factor is the light intensity rather than the concentration of the reactant) may be zero-order reactions. In such cases, the rate is expressed as:

$$-\frac{d[A]}{dt} = k_0$$

by defining $A_0 = [A]_{t=0}$ (concentration at $t = 0$; initial concentration) and $A_t = [A]_t$ (concentration at a given $t$ time). Then:

$$\int_{A_0}^{A_t} dA = -k_0 \int_0^t dt$$

$$a_t = A_0 - k_0 t$$

Consequently, there is a linear relationship between the concentration and time in zero-order reactions. The reactions can also be characterized by the half-life ($t_{1/2}$), which is the time required for one half of the starting material to disappear:

$$\frac{1}{2} A_0 = A_0 - k_0 t_{1/2} \rightarrow t_{1/2} = \frac{A_0}{2k_0}$$

**Zigzag watershed** A pattern formed as a result of the headward erosion of watercourses. This *erosion* breaks through a *drainage divide*, which retains its original location between *drainage basins*, forming a zigzag pattern in the *watershed*.

**Zone of aeration** Also called the aeration zone. The area below the ground surface where the subsurface void spaces are filled with a combination of water, moisture, and air, and where a free exchange of air and moisture occurs. Water collected within this *unsaturated zone* is called vadose water. The terms "unsaturated zone" and "*vadose zone*" have been used interchangeably with "zone of aeration." The boundary separating the *zone of saturation* and the zone of aeration is

**Z**

the groundwater table, or *water table*, where the subsurface pressure equals the atmospheric pressure. Water molecules are attracted from the water table into the overlying zone of aeration by *soil moisture* surface tension and by the molecular attraction between the liquid and solid phases, or *capillarity*. *Pressure* increases below the boundary between the zone of aeration and the zone of saturation.

**Zone of capillarity** See *capillary fringe*.

**Zone of capture** The areas that supply groundwater *recharge* to a well. Cf. *Zone of influence*.

**Zone of discharge** The part of the saturated zone that *discharges* onto or intersects the ground surface, i.e., where *groundwater* becomes surface water. Cf. *Artesian water; Zone of saturation*.

**Zone of dispersion** A zone of mixing that surrounds, in advance of and behind, an advective front of *tracers*. **Figure Z-3** illustrates a zone of dispersion at three time increments increasing as the advective front moves farther from the source. Experiments demonstrate that *dispersion* spreads some of the mass of a contaminant beyond the region in which it would exist due to *advection* alone.

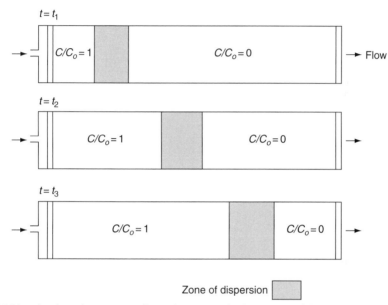

Zone of dispersion

**Figure Z-3** Dispersion shown in a porous medium at three progressive times. A zone of dispersion progressively increases between the two fluids, tracer $C/C_0 = 1$ and native fluid $C/C_0 = 0$. (Domeninco, 1990. Reprinted with permission from John Wiley & Sons, Inc.)

Cf. *Plume*.

**Zone of influence** The area experiencing change in the *water table* or *potentiometric surface* due to a well that is either pumping from or recharging the area. Cf. *Zone of capture; Cone of depression*.

**Zone of moisture** See *Belt of soil moisture*.

**Zone of saturation** Also called the saturated zone. The part of the lithosphere in which each *void space* in subsurface material is filled with water, or is saturated, under *pressure* greater than that of the atmosphere. Technically, all water below the ground surface can be called *groundwater*, and the zone of saturation is the area below the groundwater table. In the zone of saturation, the total volume of primary and/or secondary pores or openings in the rock or soil matrix are filled with water. Cf. *Phreatic water*.

**Zone of soil water** See *belt of soil moisture*.

**Zone of suspended water** See *unsaturated zone*.

**Zones of water** A subsurface-water classification system based on the location and properties of water in the subsurface media. The two fundamental zones of water that occur within the lithosphere are the saturated zone and the *unsaturated zone*. The unsaturated zone is further subdivided into the *soil moisture* zone, the *intermediate* zone, and the *capillary zone*. **Figure Z-4** shows the relative locations of the zones of water in relation to the ground surface.

**Z**

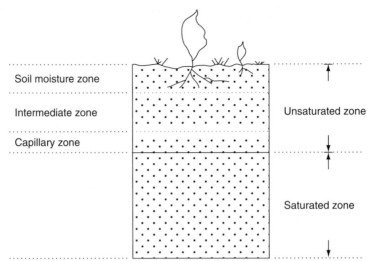

**Figure Z-4** Schematic diagram showing the water zones within the soil zones. (Sen, 1995. Reprinted with permission from CRC Press, Inc.)

Cf. *Zone of saturation; Zone of aeration.*

**Z–R relationship** The US National Weather Service has a very low ratio of rain gauges per area, making it very difficult to accurately assess rainfall amounts based on gauge data alone. To address this problem, correlation of radar reflectivity values (Z) with rainfall rates (R) has been performed. Today, the US National Weather Service uses $Z = 300R^{1.4}$ as the primary default Z–R relationship.

# Appendix A-1a: Conversions by Property

| Property | Multiply | By | To obtain |
|---|---|---|---|
| Acceleration | Centimeter per second squared (cm s$^{-2}$) | 3.28E − 02 | Foot per second squared (ft s$^{-2}$) |
| Acceleration | Centimeter per second squared (cm s$^{-2}$) | 1.00E − 04 | Kilometer per second squared (km s$^{-2}$) |
| Acceleration | Centimeter per second squared (cm s$^{-2}$) | 1.00E − 02 | Meter per second squared (m s$^{-2}$) |
| Acceleration | Foot per second squared (ft s$^{-2}$) | 3.05E + 01 | Centimeter per second squared (cm s$^{-2}$) |
| Acceleration | Foot per second squared (ft s$^{-2}$) | 3.05E − 04 | Kilometer per second squared (km s$^{-2}$) |
| Acceleration | Foot per second squared (ft s$^{-2}$) | 3.05E − 01 | Meter per second squared (m s$^{-2}$) |
| Acceleration | Kilometer per second squared (km s$^{-2}$) | 1.00E + 05 | Centimeter per second squared (cm s$^{-2}$) |
| Acceleration | Kilometer per second squared (km s$^{-2}$) | 3.28E + 03 | Foot per second squared (ft s$^{-2}$) |
| Acceleration | Kilometer per second squared (km s$^{-2}$) | 1.00E + 03 | Meter per second squared (m s$^{-2}$) |
| Acceleration | Meter per second squared (m s$^{-2}$) | 1.00E + 02 | Centimeter per second squared (cm s$^{-2}$) |
| Acceleration | Meter per second squared (m s$^{-2}$) | 3.28E + 00 | Foot per second squared (ft s$^{-2}$) |
| Acceleration | Meter per second squared (m s$^{-2}$) | 1.00E − 03 | Kilometer per second squared (km s$^{-2}$) |
| Area | Acre (A) | 4.05E + 07 | Centimeter squared (cm$^2$) |
| Area | Acre (A) | 4.36E + 04 | Foot squared (ft$^2$) |
| Area | Acre (A) | 4.05E − 01 | Hectare (ha) |
| Area | Acre (A) | 6.27E + 06 | Inch squared (in.$^2$) |
| Area | Acre (A) | 4.05E − 03 | Kilometer squared (km$^2$) |
| Area | Acre (A) | 4.05E + 03 | Meter squared (m$^2$) |
| Area | Acre (A) | 1.56E − 03 | Mile squared (mi$^2$) |
| Area | Acre (A) | 4.05E + 09 | Millimeter squared (mm$^2$) |
| Area | Acre (A) | 4.84E + 03 | Yard squared (yd$^2$) |
| Area | Centimeter squared (cm$^2$) | 2.47E − 08 | Acre (ac) |
| Area | Centimeter squared (cm$^2$) | 1.08E − 03 | Foot squared (ft$^2$) |
| Area | Centimeter squared (cm$^2$) | 1.00E − 09 | Hectare (ha) |
| Area | Centimeter squared (cm$^2$) | 1.55E − 01 | Inch squared (in.$^2$) |
| Area | Centimeter squared (cm$^2$) | 1.00E − 09 | Kilometer squared (km$^2$) |
| Area | Centimeter squared (cm$^2$) | 1.00E − 03 | Meter squared (m$^2$) |
| Area | Centimeter squared (cm$^2$) | 3.86E − 11 | Mile squared (mi$^2$) |
| Area | Centimeter squared (cm$^2$) | 1.00E + 02 | Millimeter squared (mm$^2$) |
| Area | Centimeter squared (cm$^2$) | 6.06E − 10 | Yard squared (yd$^2$) |
| Area | Foot squared (ft$^2$) | 2.30E − 05 | Acre (ac) |
| Area | Foot squared (ft$^2$) | 9.29E + 02 | Centimeter squared (cm$^2$) |
| Area | Foot squared (ft$^2$) | 9.29E − 06 | Hectare (ha) |

*(Continued)*

(Continued)

| Property | Multiply | By | To obtain |
|---|---|---|---|
| Area | Foot squared ($ft^2$) | $1.44E+02$ | Inch squared ($in.^2$) |
| Area | Foot squared ($ft^2$) | $9.29E-08$ | Kilometer squared ($km^2$) |
| Area | Foot squared ($ft^2$) | $9.29E-02$ | Meter squared ($m^2$) |
| Area | Foot squared ($ft^2$) | $3.59E-08$ | Mile squared ($mi^2$) |
| Area | Foot squared ($ft^2$) | $9.29E+04$ | Millimeter squared ($mm^2$) |
| Area | Foot squared ($ft^2$) | $1.10E-01$ | Yard squared ($yd^2$) |
| Area | Hectare (ha) | $2.47E+00$ | Acre (ac) |
| Area | Hectare (ha) | $1.00E+09$ | Centimeter squared ($cm^2$) |
| Area | Hectare (ha) | $1.08E+05$ | Foot squared ($ft^2$) |
| Area | Hectare (ha) | $1.55E+07$ | Inch squared ($in.^2$) |
| Area | Hectare (ha) | $1.00E-02$ | Kilometer squared ($km^2$) |
| Area | Hectare (ha) | $1.00E+04$ | Meter squared ($m^2$) |
| Area | Hectare (ha) | $3.86E-03$ | Mile squared ($mi^2$) |
| Area | Hectare (ha) | $1.00E+11$ | Millimeter squared ($mm^2$) |
| Area | Hectare (ha) | $1.20E+04$ | Yard squared ($yd^2$) |
| Area | Inch squared ($in.^2$) | $1.59E-07$ | Acre (ac) |
| Area | Inch squared ($in.^2$) | $6.45E+00$ | Centimeter squared ($cm^2$) |
| Area | Inch squared ($in.^2$) | $6.94E-03$ | Foot squared ($ft^2$) |
| Area | Inch squared ($in.^2$) | $6.45E-08$ | Hectare (ha) |
| Area | Inch squared ($in.^2$) | $6.45E-10$ | Kilometer squared ($km^2$) |
| Area | Inch squared ($in.^2$) | $6.45E-04$ | Meter squared ($m^2$) |
| Area | Inch squared ($in.^2$) | $2.49E-10$ | Mile squared ($mi^2$) |
| Area | Inch squared ($in.^2$) | $6.45E+02$ | Millimeter squared ($mm^2$) |
| Area | Inch squared ($in.^2$) | $7.72E-04$ | Yard squared ($yd^2$) |
| Area | Kilometer squared ($km^2$) | $2.47E+02$ | Acre (ac) |
| Area | Kilometer squared ($km^2$) | $1.00E+09$ | Centimeter squared ($cm^2$) |
| Area | Kilometer squared ($km^2$) | $1.08E+07$ | Foot squared ($ft^2$) |
| Area | Kilometer squared ($km^2$) | $1.00E+02$ | Hectare (ha) |
| Area | Kilometer squared ($km^2$) | $1.55E+09$ | Inch squared ($in.^2$) |
| Area | Kilometer squared ($km^2$) | $1.00E+06$ | Meter squared ($m^2$) |
| Area | Kilometer squared ($km^2$) | $3.86E-01$ | Mile squared ($mi^2$) |
| Area | Kilometer squared ($km^2$) | $1.00E+13$ | Millimeter squared ($mm^2$) |
| Area | Kilometer squared ($km^2$) | $1.20E+06$ | Yard squared ($yd^2$) |
| Area | Meter squared ($m^2$) | $2.47E-04$ | Acre (ac) |
| Area | Meter squared ($m^2$) | $1.00E+04$ | Centimeter squared ($cm^2$) |
| Area | Meter squared ($m^2$) | $1.08E+01$ | Foot squared ($ft^2$) |
| Area | Meter squared ($m^2$) | $1.00E-04$ | Hectare (ha) |
| Area | Meter squared ($m^2$) | $1.55E+03$ | Inch squared ($in.^2$) |
| Area | Meter squared ($m^2$) | $1.00E-06$ | Kilometer squared ($km^2$) |
| Area | Meter squared ($m^2$) | $3.86E-07$ | Mile squared ($mi^2$) |
| Area | Meter squared ($m^2$) | $1.00E+07$ | Millimeter squared ($mm^2$) |
| Area | Meter squared ($m^2$) | $3.95E-02$ | Rod squared ($rd^2$) |
| Area | Meter squared ($m^2$) | $1.20E+00$ | Yard squared ($yd^2$) |
| Area | Mile squared ($mi^2$) | $6.40E+02$ | Acre (ac) |
| Area | Mile squared ($mi^2$) | $2.59E+10$ | Centimeter squared ($cm^2$) |
| Area | Mile squared ($mi^2$) | $2.79E+07$ | Foot squared ($ft^2$) |
| Area | Mile squared ($mi^2$) | $2.59E+02$ | Hectare (ha) |
| Area | Mile squared ($mi^2$) | $4.01E+09$ | Inch squared ($in.^2$) |
| Area | Mile squared ($mi^2$) | $2.59E+00$ | Kilometer squared ($km^2$) |
| Area | Mile squared ($mi^2$) | $2.59E+06$ | Meter squared ($m^2$) |
| Area | Mile squared ($mi^2$) | $2.59E+12$ | Millimeter squared ($mm^2$) |
| Area | Mile squared ($mi^2$) | $3.10E+06$ | Yard squared ($yd^2$) |
| Area | Millimeter squared ($mm^2$) | $2.47E-10$ | Acre (ac) |
| Area | Millimeter squared ($mm^2$) | $1.00E-02$ | Centimeter squared ($cm^2$) |
| Area | Millimeter squared ($mm^2$) | $1.08E-05$ | Foot squared ($ft^2$) |
| Area | Millimeter squared ($mm^2$) | $1.00E-09$ | Hectare (ha) |
| Area | Millimeter squared ($mm^2$) | $1.50E-02$ | Inch squared ($in.^2$) |
| Area | Millimeter squared ($mm^2$) | $1.00E-11$ | Kilometer squared ($km^2$) |

(Continued)

| Property | Multiply | By | To obtain |
|---|---|---|---|
| Area | Millimeter squared (mm$^2$) | 1.00E − 05 | Meter squared (m$^2$) |
| Area | Millimeter squared (mm$^2$) | 3.86E − 13 | Mile squared (mi$^2$) |
| Area | Millimeter squared (mm$^2$) | 1.20E − 06 | Yard squared (yd$^2$) |
| Area | Rod squared (rd$^2$) | 2.72E + 02 | Foot squared (ft$^2$) |
| Area | Rod squared (rd$^2$) | 2.53E + 01 | Meter squared (m$^2$) |
| Area | Yard squared (yd$^2$) | 1.30E + 03 | Inch squared (in.$^2$) |
| Area | Yard squared (yd$^2$) | 2.07E − 04 | Acre |
| Area | Yard squared (yd$^2$) | 8.36E + 03 | Centimeter squared (cm$^2$) |
| Area | Yard squared (yd$^2$) | 9.00E + 00 | Foot squared (ft$^2$) |
| Area | Yard squared (yd$^2$) | 8.36E − 05 | Hectare (ha) |
| Area | Yard squared (yd$^2$) | 8.36E − 07 | Kilometer squared (km$^2$) |
| Area | Yard squared (yd$^2$) | 8.36E − 01 | Meter squared (m$^2$) |
| Area | Yard squared (yd$^2$) | 3.23E − 07 | Mile squared (mi$^2$) |
| Area | Yard squared (yd$^2$) | 8.36E + 05 | Millimeter squared (mm$^2$) |
| Concentration | Pound per million gal. of $H_2o$ | 1.20E − 01 | Milligram per liter (mg L$^{-1}$, ppm) |
| Density | Gram per cubic centimeter (g cm$^{-3}$) | 1.00E + 03 | Gram per liter (g L$^{-1}$) |
| Density | Gram per cubic centimeter (g cm$^{-3}$) | 1.00E + 00 | Gram per milliliter (g mL$^{-1}$) |
| Density | Gram per cubic centimeter (g cm$^{-3}$) | 1.00E + 00 | Kilogram per decimeter cubed (kg dm$^{-3}$) |
| Density | Gram per cubic centimeter (g cm$^{-3}$) | 1.00E + 03 | Kilogram per meter cubed (kg m$^{-3}$) |
| Density | Gram per cubic centimeter (g cm$^{-3}$) | 6.24E + 01 | Pound per foot cubed (lb ft$^{-3}$) |
| Density | Gram per cubic centimeter (g cm$^{-3}$) | 1.00E + 01 | Pound per gallon (Brit) (lb gal$^{-1}$) |
| Density | Gram per cubic centimeter (g cm$^{-3}$) | 8.35E + 00 | Pound per gallon (US) (lb gal$^{-1}$) |
| Density | Gram per cubic centimeter (g cm$^{-3}$) | 3.61E − 02 | Pound per inch cubed (lb in.$^{-3}$) |
| Density | Gram per cubic centimeter (g cm$^{-3}$) | 1.69E + 03 | Pound per yard cubed (lb yd$^{-3}$) |
| Density | Gram per cubic centimeter (g cm$^{-3}$) | 1.94E + 00 | Slug per foot cubed (slug ft$^{-3}$) |
| Density | Gram per liter (g L$^{-1}$) | 1.00E − 03 | Gram per centimeter cubed (g cm$^{-3}$) |
| Density | Gram per liter (g L$^{-1}$) | 1.00E − 03 | Gram per milliliter (g mL$^{-1}$) |
| Density | Gram per liter (g L$^{-1}$) | 1.00E − 03 | Kilogram per decimeter cubed (kg dm$^{-3}$) |
| Density | Gram per liter (g L$^{-1}$) | 1.00E + 00 | Kilogram per meter cubed (kg m$^{-3}$) |
| Density | Gram per liter (g L$^{-1}$) | 6.24E − 02 | Pound per foot cubed (lb ft$^{-3}$) |
| Density | Gram per liter (g L$^{-1}$) | 1.00E − 02 | Pound per gallon (Brit) (lb gal$^{-1}$) |
| Density | Gram per liter (g L$^{-1}$) | 8.35E − 03 | Pound per gallon (US) (lb gal$^{-1}$) |
| Density | Gram per liter (g L$^{-1}$) | 3.61E − 05 | Pound per inch cubed (lb in.$^{-3}$) |
| Density | Gram per liter (g L$^{-1}$) | 1.69E + 00 | Pound per yard cubed (lb yd$^{-3}$) |
| Density | Gram per liter (g L$^{-1}$) | 1.94E − 03 | Slug per foot cubed (slug ft$^{-3}$) |

(*Continued*)

(Continued)

| Property | Multiply | By | To obtain |
|---|---|---|---|
| Density | Gram per milliliter (g mL$^{-1}$) | 1.00E + 00 | Gram per centimeter cubed (g cm$^{-3}$) |
| Density | Gram per milliliter (g mL$^{-1}$) | 1.00E + 03 | Gram per liter (g L$^{-1}$) |
| Density | Gram per milliliter (g mL$^{-1}$) | 1.00E + 00 | Kilogram per decimeter cubed (kg dm$^{-3}$) |
| Density | Gram per milliliter (g mL$^{-1}$) | 1.00E + 03 | Kilogram per meter cubed (kg m$^{-3}$) |
| Density | Gram per milliliter (g mL$^{-1}$) | 6.24E + 01 | Pound per foot cubed (lb ft$^{-3}$) |
| Density | Gram per milliliter (g mL$^{-1}$) | 1.00E + 01 | Pound per gallon (Brit) (lb gal$^{-1}$) |
| Density | Gram per milliliter (g mL$^{-1}$) | 8.35E + 00 | Pound per gallon (US) (lb gal$^{-1}$) |
| Density | Gram per milliliter (g mL$^{-1}$) | 3.61E − 02 | Pound per inch cubed (lb in.$^{-3}$) |
| Density | Gram per milliliter (g mL$^{-1}$) | 1.69E + 03 | Pound per yard cubed (lb yd$^{-3}$) |
| Density | Gram per milliliter (g mL$^{-1}$) | 1.94E + 00 | Slug per foot cubed (slug ft$^{-3}$) |
| Density | Kilogram per decimeter cubed (kg dm$^{-3}$) | 1.00E + 00 | Gram per centimeter cubed (g cm$^{-3}$) |
| Density | Kilogram per decimeter cubed (kg dm$^{-3}$) | 1.00E − 03 | Gram per liter (g L$^{-1}$) |
| Density | Kilogram per decimeter cubed (kg dm$^{-3}$) | 1.00E + 00 | Gram per milliliter (g mL$^{-1}$) |
| Density | Kilogram per decimeter cubed (kg dm$^{-3}$) | 1.00E − 03 | Kilogram per meter cubed (kg m$^{-3}$) |
| Density | Kilogram per decimeter cubed (kg dm$^{-3}$) | 1.60E − 02 | Pound per foot cubed (lb ft$^{-3}$) |
| Density | Kilogram per decimeter cubed (kg dm$^{-3}$) | 9.98E − 02 | Pound per gallon (Brit) (lb gal$^{-1}$) |
| Density | Kilogram per decimeter cubed (kg dm$^{-3}$) | 1.20E − 01 | Pound per gallon (US) (lb gal$^{-1}$) |
| Density | Kilogram per decimeter cubed (kg dm$^{-3}$) | 2.77E + 01 | Pound per inch cubed (lb in.$^{-3}$) |
| Density | Kilogram per decimeter cubed (kg dm$^{-3}$) | 5.93E − 04 | Pound per yard cubed (lb yd$^{-3}$) |
| Density | Kilogram per decimeter cubed (kg dm$^{-3}$) | 5.15E − 01 | Slug per foot cubed (slug ft$^{-3}$) |
| Density | Kilogram per meter cubed (kg m$^{-3}$) | 1.00E − 03 | Gram per centimeter cubed (g cm$^{-3}$) |
| Density | Kilogram per meter cubed (kg m$^{-3}$) | 1.00E + 00 | Gram per liter (g L$^{-1}$) |
| Density | Kilogram per meter cubed (kg m$^{-3}$) | 1.00E − 03 | Gram per milliliter (g mL$^{-1}$) |
| Density | Kilogram per meter cubed (kg m$^{-3}$) | 1.00E − 03 | Kilogram per decimeter cubed (kg dm$^{-3}$) |
| Density | Kilogram per meter cubed (kg m$^{-3}$) | 6.24E − 02 | Pound per foot cubed (lb ft$^{-3}$) |
| Density | Kilogram per meter cubed (kg m$^{-3}$) | 1.00E − 02 | Pound per gallon (Brit) (lb gal$^{-1}$) |
| Density | Kilogram per meter cubed (kg m$^{-3}$) | 8.35E − 03 | Pound per gallon (US) (lb gal$^{-1}$) |
| Density | Kilogram per meter cubed (kg m$^{-3}$) | 3.61E − 05 | Pound per inch cubed (lb in.$^{-3}$) |
| Density | Kilogram per meter cubed (kg m$^{-3}$) | 1.69E + 00 | Pound per yard cubed (lb yd$^{-3}$) |

(Continued)

| Property | Multiply | By | To obtain |
|---|---|---|---|
| Density | Kilogram per meter cubed $(\text{kg m}^{-3})$ | 1.94E − 03 | Slug per foot cubed (slug $\text{ft}^{-3}$) |
| Density | Pound per foot cubed $(\text{lb ft}^{-3})$ | 1.60E − 03 | Gram per centimeter cubed $(\text{g cm}^{-3})$ |
| Density | Pound per foot cubed $(\text{lb ft}^{-3})$ | 1.60E + 01 | Gram per liter $(\text{g l}^{-1})$ |
| Density | Pound per foot cubed $(\text{lb ft}^{-3})$ | 1.60E − 02 | Gram per milliliter $(\text{g mL}^{-1})$ |
| Density | Pound per foot cubed $(\text{lb ft}^{-3})$ | 1.60E − 02 | Gram per milliliter $(\text{g mL}^{-1})$ |
| Density | Pound per foot cubed $(\text{lb ft}^{-3})$ | 1.60E − 02 | Kilogram per decimeter cubed $(\text{kg dm}^{-3})$ |
| Density | Pound per foot cubed $(\text{lb ft}^{-3})$ | 1.60E + 01 | Kilogram per meter cubed $(\text{kg m}^{-3})$ |
| Density | Pound per foot cubed $(\text{lb ft}^{-3})$ | 1.61E − 01 | Pound per gallon (Brit) $(\text{lb gal}^{-1})$ |
| Density | Pound per foot cubed $(\text{lb ft}^{-3})$ | 1.34E − 01 | Pound per gallon (US) $(\text{lb gal}^{-1})$ |
| Density | Pound per foot cubed $(\text{lb ft}^{-3})$ | 5.79E − 04 | Pound per inch cubed $(\text{lb in.}^{-3})$ |
| Density | Pound per foot cubed $(\text{lb ft}^{-3})$ | 2.70E + 01 | Pound per yard cubed $(\text{lb yd}^{-3})$ |
| Density | Pound per foot cubed $(\text{lb ft}^{-3})$ | 3.11E − 02 | Slug per foot cubed (slug $\text{ft}^{-3}$) |
| Density | Pound per gallon (Brit) $(\text{lb gal}^{-1})$ | 9.98E − 02 | Gram per centimeter cubed $(\text{g cm}^{-3})$ |
| Density | Pound per gallon (Brit) $(\text{lb gal}^{-1})$ | 9.98E + 01 | Gram per liter $(\text{g L}^{-})$ |
| Density | Pound per gallon (Brit) $(\text{lb gal}^{-1})$ | 9.98E − 02 | Gram per milliliter $(\text{g mL}^{-1})$ |
| Density | Pound per gallon (Brit) $(\text{lb gal}^{-1})$ | 9.98E − 02 | Kilogram per decimeter cubed $(\text{kg dm}^{-3})$ |
| Density | Pound per gallon (Brit) $(\text{lb gal}^{-1})$ | 9.98E + 01 | Kilogram per meter cubed $(\text{kg m}^{-3})$ |
| Density | Pound per gallon (Brit) $(\text{lb gal}^{-1})$ | 6.23E + 00 | Pound per foot cubed $(\text{lb ft}^{-3})$ |
| Density | Pound per gallon (Brit) $(\text{lb gal}^{-1})$ | 8.33E − 01 | Pound per gallon (US) $(\text{lb gal}^{-1})$ |
| Density | Pound per gallon (Brit) $(\text{lb gal}^{-1})$ | 3.60E + 00 | Pound per inch cubed $(\text{lb in.}^{-3})$ |
| Density | Pound per gallon (Brit) $(\text{lb gal}^{-1})$ | 1.68E + 02 | Pound per yard cubed $(\text{lb yd}^{-3})$ |
| Density | Pound per gallon (Brit) $(\text{lb gal}^{-1})$ | 1.94E − 01 | Slug per foot cubed (slug $\text{ft}^{-3}$) |
| Density | Pound per gallon (US) $(\text{lb gal}^{-1})$ | 1.20E − 01 | Gram per centimeter cubed $(\text{g cm}^{-3})$ |
| Density | Pound per gallon (US) $(\text{lb gal}^{-1})$ | 1.20E + 02 | Gram per liter $(\text{g L}^{-1})$ |
| Density | Pound per gallon (US) $(\text{lb gal}^{-1})$ | 1.20E − 01 | Gram per milliliter $(\text{g mL}^{-1})$ |
| Density | Pound per gallon (US) $(\text{lb gal}^{-1})$ | 1.20E − 01 | Kilogram per decimeter cubed $(\text{kg dm}^{-3})$ |
| Density | Pound per gallon (US) $(\text{lb gal}^{-1})$ | 1.20E + 02 | Kilogram per meter cubed $(\text{kg m}^{-3})$ |
| Density | Pound per gallon (US) $(\text{lb gal}^{-1})$ | 7.48E + 00 | Pound per foot cubed $(\text{lb ft}^{-3})$ |

(*Continued*)

(Continued)

| Property | Multiply | By | To obtain |
|---|---|---|---|
| Density | Pound per gallon (US) (lb gal$^{-1}$) | 1.20E + 00 | Pound per gallon (Brit) (lb gal$^{-1}$) |
| Density | Pound per gallon (US) (lb gal$^{-1}$) | 4.33E − 03 | Pound per inch cubed (lb in.$^{-3}$) |
| Density | Pound per gallon (US) (lb gal$^{-1}$) | 2.02E + 02 | Pound per yard cubed (lb yd$^{-3}$) |
| Density | Pound per gallon (US) (lb gal$^{-1}$) | 2.33E − 01 | Slug per foot cubed (slug ft$^{-3}$) |
| Density | Pound per inch cubed (lb in.$^{-3}$) | 2.77E + 01 | Gram per centimeter cubed (g cm$^{-3}$) |
| Density | Pound per inch cubed (lb in.$^{-3}$) | 2.77E + 04 | Gram per liter (g L$^{-1}$) |
| Density | Pound per inch cubed (lb in.$^{-3}$) | 2.77E + 01 | Gram per milliliter (g mL$^{-1}$) |
| Density | Pound per inch cubed (lb in.$^{-3}$) | 2.77E + 01 | Kilogram per decimeter cubed (kg dm$^{-3}$) |
| Density | Pound per inch cubed (lb in.$^{-3}$) | 2.77E + 04 | Kilogram per meter cubed (kg m$^{-3}$) |
| Density | Pound per inch cubed (lb in.$^{-3}$) | 1.73E + 03 | Pound per foot cubed (lb ft$^{-3}$) |
| Density | Pound per inch cubed (lb in.$^{-3}$) | 2.77E + 02 | Pound per gallon (Brit) (lb gal$^{-1}$) |
| Density | Pound per inch cubed (lb in.$^{-3}$) | 2.31E + 02 | Pound per gallon (US) (lb gal$^{-1}$) |
| Density | Pound per inch cubed (lb in.$^{-3}$) | 4.67E + 04 | Pound per yard cubed (lb yd$^{-3}$) |
| Density | Pound per inch cubed (lb in.$^{-3}$) | 5.37E + 01 | Slug per foot cubed (slug ft$^{-3}$) |
| Density | Pound per yard cubed (lb yd$^{-3}$) | 5.93E − 04 | Gram per centimeter cubed (g cm$^{-3}$) |
| Density | Pound per yard cubed (lb yd$^{-3}$) | 5.93E − 01 | Gram per liter (g L$^{-1}$) |
| Density | Pound per yard cubed (lb yd$^{-3}$) | 5.93E − 04 | Gram per milliliter (g mL$^{-1}$) |
| Density | Pound per yard cubed (lb yd$^{-3}$) | 5.93E − 04 | Kilogram per decimeter cubed (kg dm$^{-3}$) |
| Density | Pound per yard cubed (lb yd$^{-3}$) | 5.93E − 01 | Kilogram per meter cubed (kg m$^{-3}$) |
| Density | Pound per yard cubed (lb yd$^{-3}$) | 2.14E + 00 | Pound per cubic inch (lb in.$^{-3}$) |
| Density | Pound per yard cubed (lb yd$^{-3}$) | 3.70E − 02 | Pound per foot cubed (lb ft$^{-3}$) |
| Density | Pound per yard cubed (lb yd$^{-3}$) | 5.95E − 03 | Pound per gallon (Brit) (lb gal$^{-1}$) |
| Density | Pound per yard cubed (lb yd$^{-3}$) | 4.95E − 03 | Pound per gallon (US) (lb gal$^{-1}$) |
| Density | Pound per yard cubed (lb yd$^{-3}$) | 1.15E − 03 | Slug per foot cubed (slug ft$^{-3}$) |
| Density | Slug per foot cubed (slug ft$^{-3}$) | 5.15E − 01 | Gram per centimeter cubed (g cm$^{-3}$) |
| Density | Slug per foot cubed (slug ft$^{-3}$) | 5.15E + 02 | Gram per liter (g L$^{-1}$) |
| Density | Slug per foot cubed (slug ft$^{-3}$) | 5.15E − 01 | Gram per milliliter (g mL$^{-1}$) |
| Density | Slug per foot cubed (slug ft$^{-3}$) | 5.15E − 01 | Kilogram per decimeter cubed (kg dm$^{-3}$) |

(Continued)

| Property | Multiply | By | To obtain |
|---|---|---|---|
| Density | Slug per foot cubed (slug $ft^{-3}$) | 5.15E + 02 | Kilogram per meter cubed $(kg\,m^{-3})$ |
| Density | Slug per foot cubed (slug $ft^{-3}$) | 3.22E + 01 | Pound per foot cubed $(lb\,ft^{-3})$ |
| Density | Slug per foot cubed (slug $ft^{-3}$) | 5.17E + 00 | Pound per gallon (Brit) $(lb\,gal^{-1})$ |
| Density | Slug per foot cubed (slug $ft^{-3}$) | 4.30E + 00 | Pound per gallon (US) $(lb\,gal^{-1})$ |
| Density | Slug per foot cubed (slug $ft^{-3}$) | 1.86E − 02 | Pound per inch cubed $(lb\,in.^{-3})$ |
| Density | Slug per foot cubed (slug $ft^{-3}$) | 8.69E + 02 | Pound per yard cubed $(lb\,yd^{-3})$ |
| Discharge | Acre-feet per day (ac ft $day^{-1}$) | 5.04E − 01 | Foot cubed per second $(ft^3\,s^{-1})$ |
| Discharge | Acre-feet per day (ac ft $day^{-1}$) | 1.43E + 01 | Liter per second ($L\,s^{-1}$) |
| Discharge | Acre-feet per day (ac ft $day^{-1}$) | 1.23E + 03 | Meter cubed per day ($m^3$ $day^{-1}$) |
| Discharge | Acre-feet per day (ac ft $day^{-1}$) | 2.26E + 02 | US gallon per minute $(gal\,min^{-1}$ or gpm) |
| Discharge | Barrel per day (bbl $day^{-1}$) | | Foot cubed per second $(ft^3\,s^{-1})$ |
| Discharge | Barrel per day (bbl $day^{-1}$) | | Liter per second ($L\,s^{-1}$) |
| Discharge | Barrel per day (bbl $day^{-1}$) | | Meter cubed per day $(m^3\,day^{-1})$ |
| Discharge | Barrel per day (bbl $day^{-1}$) | 2.92E − 02 | US gallon per minute $(gal\,min^{-1}$ or gpm) |
| Discharge | Foot cubed per second $(ft^3\,s^{-1})$ | 1.98E + 00 | Acre-feet per day (ac ft $day^{-1}$) |
| Discharge | Foot cubed per second $(ft^3\,s^{-1})$ | 2.83E + 01 | Liter per second ($L\,s^{-1}$) |
| Discharge | Foot cubed per second $(ft^3\,s^{-1})$ | 2.45E + 03 | Meter cubed per day $(m^3\,day^{-1})$ |
| Discharge | Foot cubed per second $(ft^3\,s^{-1})$ | 2.83E − 02 | Meter cubed per second $(m^3\,s^{-1})$ |
| Discharge | Foot cubed per second $(ft^3\,s^{-1})$ | 6.47E + 05 | US gallon per day $(gal\,day^{-1})$ |
| Discharge | Foot cubed per second $(ft^3\,s^{-1})$ | 4.49E + 02 | US gallon per minute $(gal\,min^{-1}$ or gpm) |
| Discharge | Imperial gallons per minute (Imperial gal $min^{-1}$) | | Acre-feet per day (ac ft $day^{-1}$) |
| Discharge | Imperial gallons per minute (Imperial gal $min^{-1}$) | 7.58E − 02 | Liter per second ($L\,s^{-1}$) |
| Discharge | Imperial gallons per minute (Imperial gal $min^{-1}$) | | Meter cubed per day $(m^3\,day^{-1})$ |
| Discharge | Imperial gallons per minute (Imperial gal $min^{-1}$) | 7.58E − 05 | Meter cubed per second $(m^3\,s^{-1})$ |
| Discharge | Imperial gallons per minute (Imperial gal $min^{-1}$) | | US gallon per day (gal $day^{-1}$) |
| Discharge | Liter per second ($L\,s^{-1}$) | 3.53E − 02 | Foot cubed per second $(ft^3\,s^{-1})$ |
| Discharge | Liter per second ($L\,s^{-1}$) | 8.64E + 01 | Meter cubed per day $(m^3\,day^{-1})$ |
| Discharge | Liter per second ($L\,s^{-1}$) | 2.28E + 04 | US gallon per day (gal $day^{-1}$) |

*(Continued )*

(Continued)

| Property | Multiply | By | To obtain |
|---|---|---|---|
| Discharge | Liter per second (L s$^{-1}$) | 7.01E − 02 | US gallon per minute (gal min$^{-1}$ or gpm) |
| Discharge | Liter per second (L s$^{-1}$) | 1.59E + 01 | US gallon per minute (gal min$^{-1}$ or gpm) |
| Discharge | Liter per second (L s$^{-1}$) | 1.58E + 01 | US gallon per minute (gal min$^{-1}$ or gpm) |
| Discharge | Meter cubed per day (m$^3$ day$^{-1}$) | 6.05E + 06 | Acre-feet per day (ac ft day$^{-1}$) |
| Discharge | Meter cubed per day (m$^3$ day$^{-1}$) | 3.02E + 06 | Foot cubed per second (ft$^3$ s$^{-1}$) |
| Discharge | Meter cubed per day (m$^3$ day$^{-1}$) | 8.64E + 07 | Liter per second (L s$^{-1}$) |
| Discharge | Meter cubed per day (m$^3$ day$^{-1}$) | 1.37E + 09 | US gallon per minute (gal min$^{-1}$ or gpm) |
| Discharge | Meter cubed per second (m$^3$ s$^{-1}$) | 2.28E + 07 | US gallon per day (gal day$^{-1}$) |
| Discharge | US gallon per day (gal day$^{-1}$) | 1.55E − 06 | Foot cubed per second (ft$^3$ s$^{-1}$) |
| Discharge | US gallon per day (gal day$^{-1}$) | 4.38E − 05 | Liter per second (L s$^{-1}$) |
| Discharge | US gallon per day (gal day$^{-1}$) | 4.38E − 08 | Meter cubed per second (m$^3$ s$^{-1}$) |
| Discharge | US gallon per day (gal day$^{-1}$) | 6.94E − 04 | US gallon per minute (gal min$^{-1}$ or gpm) |
| Discharge | US gallon per minute (gal min$^{-1}$ or gpm) | 4.42E − 03 | Acre-feet per day (ac ft day$^{-1}$) |
| Discharge | US gallon per minute (gal min$^{-1}$ or gpm) | 3.43E + 01 | Barrel per day (bbl day$^{-1}$) |
| Discharge | US gallon per minute (gal min$^{-1}$ or gpm) | 2.23E − 03 | Foot cubed per sec (ft$^3$ s$^{-1}$) |
| Discharge | US gallon per minute (gal min$^{-1}$ or gpm) | 2.23E − 03 | Foot cubed per second (ft$^3$ s$^{-1}$) |
| Discharge | US gallon per minute (gal min$^{-1}$ or gpm) | 6.31E − 02 | Liter per second (L s$^{-1}$) |
| Discharge | US gallon per minute (gal min$^{-1}$ or gpm) | 6.31E − 02 | Liter per second (L s$^{-1}$) |
| Discharge | US gallon per minute (gal min$^{-1}$ or gpm) | 5.45E + 00 | Meter cubed per day (m$^3$ day$^{-1}$) |
| Discharge | US gallon per minute (gal min$^{-1}$ or gpm) | 6.31E − 05 | Meter cubed per second (m$^3$ s$^{-1}$) |
| Discharge | US gallon per minute (gal min$^{-1}$ or gpm) | 6.31E − 05 | Meter cubed per second (m$^3$ s$^{-1}$) |
| Discharge | US gallon per minute (gal min$^{-1}$ or gpm) | 1.44E + 03 | US gallon per day (gal day$^{-1}$) |
| Dry volume | Bushels | 4.00E + 00 | Peck |
| Dry volume | Centimeter cubed (cm$^3$) | 6.10E − 02 | Inch cubed (in.$^3$) |
| Dry volume | Centimeter cubed (cm$^3$) | 1.00E + 03 | Millimeter cubed (mm$^3$) |
| Dry volume | Decimeter cubed (dm3) | 1.00E + 03 | Centimeter cubed (cm$^3$) |
| Dry volume | Foot cubed (ft$^3$) | 1.73E + 03 | Inch cubed (in.$^3$) |
| Dry volume | Foot cubed (ft$^3$) | 2.83E − 02 | Meter cubed (m$^3$) |
| Dry volume | Foot cubed (ft$^3$) | 3.70E − 02 | Yard cubed (yd$^3$) |
| Dynamic viscosity | Centipoises | 1.00E − 01 | Dynes-second per square centimeter (dynes s cm$^{-2}$) |
| Dynamic viscosity | Centipoises | 1.02E − 04 | kilogram$_{force}$-second per square meter (kg$_f$ s m$^{-2}$) |

(Continued)

| Property | Multiply | By | To obtain |
|---|---|---|---|
| Dynamic viscosity | Centipoise | $1.00E-02$ | Pascal-second $(kg\,s^{-1})$ |
| Dynamic viscosity | Centipoises | $1.00E-01$ | Poises $(g\,cm^{-1}\,s^{-1})$ |
| Dynamic viscosity | Centipoises | $1.45E-07$ | Pound$_{force}$-second per square inch $(lb_f\,s\,in.^{-2})$ |
| Dynamic viscosity | Centipoises | $2.09E-05$ | Pound$_{force}$-second per square foot $(lb_f\,s\,ft^{-2})$ |
| Dynamic viscosity | Centipoises | $2.4191$ | Pound$_{mass}$ per foot-hour $(lb_m\,ft^{-1}\,h^{-1})$ |
| Dynamic viscosity | Centipoises | $6.72E-04$ | Pound$_{mass}$ per foot-second $(lb_m\,ft^{-1}\,s^{-1})$ |
| Dynamic viscosity | Dynes-second per square centimeter (dynes s cm$^{-2}$) | $1.00E+03$ | Centipoises |
| Dynamic viscosity | Dynes-second per square centimeter (dynes s cm$^{-2}$) | $1.02E-02$ | Kilogram$_{force}$-second per square meter $(kg_f\,s\,m^{-2})$ |
| Dynamic viscosity | Dynes-second per square centimeter (dynes s cm$^{-2}$) | $1.00E+00$ | Pascal-second $(kg\,m\,s^{-1})$ |
| Dynamic viscosity | Dynes-second per square centimeter (dynes s cm$^{-2}$) | $1$ | Poises $(g\,cm^{-1}\,s^{-1})$ |
| Dynamic viscosity | Dynes-second per square centimeter (dynes s cm$^{-2}$) | $1.45E-05$ | Pound$_{force}$-second per square inch $(lb_f\,s\,in.^{-2})$ |
| Dynamic viscosity | Dynes-second per square centimeter (dynes s cm$^{-2}$) | $2.09E-03$ | Pound$_{force}$-second per square foot $(lb_f\,s\,ft^{-2})$ |
| Dynamic viscosity | Dynes-second per square centimeter (dynes s cm$^{-2}$) | $2.42E+02$ | Pound$_{mass}$ per foot-hour $(lb_m\,ft^{-1}\,h^{-1})$ |
| Dynamic viscosity | Dynes-second per square centimeter (dynes s cm$^{-2}$) | $6.72E-02$ | Pound$_{mass}$ per foot-second $(lb_m\,ft^{-1}\,s^{-1})$ |
| Dynamic viscosity | Kilogram$_{force}$-second per square meter $(kg_f\,s\,m^{-2})$ | $9.81E+03$ | Centipoises |
| Dynamic viscosity | Kilogram$_{force}$-second per square meter $(kg_f\,s\,m^{-2})$ | $9.81E+01$ | Dynes-second per square centimeter (dynes s cm$^{-2}$) |
| Dynamic viscosity | Kilogram$_{force}$-second per square meter $(kg_f\,s\,m^{-2})$ | $9.8067$ | Pascal-second $(kg\,m^{-1}\,s^{-1})$ |
| Dynamic viscosity | Kilogram$_{force}$-second per square meter $(kg_f\,s\,m^{-2})$ | $9.81E+01$ | Poises $(g\,cm^{-1}\,s^{-1})$ |
| Dynamic viscosity | Kilogram$_{force}$-second per square meter $(kg_f\,s\,m^{-2})$ | $1.42E-33$ | Pound$_{force}$-second per square inch $(lb_f\,s\,in.^{-2})$ |
| Dynamic viscosity | Kilogram$_{force}$-second per square meter $(kg_f\,s\,m^{-2})$ | $2.05E-01$ | Pound$_{force}$-second per square foot $(lb_f\,s\,ft^{-2})$ |
| Dynamic viscosity | Kilogram$_{force}$-second per square meter $(kg_f\,s\,m^{-2})$ | $2.37E+04$ | Pound$_{mass}$ per foot-hour $(lb_m\,ft^{-1}\,h^{-1})$ |
| Dynamic viscosity | Kilogram$_{force}$-second per square meter $(kg_f\,s\,m^{-2})$ | $6.59$ | Pound$_{mass}$ per foot-second $(lb_m\,ft^{-1}\,s^{-1})$ |
| Dynamic viscosity | Pascal-second $(kg_f\,m^{-1}\,s^{-1})$ | $1.00E+04$ | Centipoises |

(*Continued*)

(Continued)

| Property | Multiply | By | To obtain |
|---|---|---|---|
| Dynamic viscosity | Pascal-second ($kg_f\,m^{-1}\,s^{-1}$) | 1.00E + 01 | Dynes-second per square centimeter (dynes s $cm^{-2}$) |
| Dynamic viscosity | Pascal-second ($kg_f\,m^{-1}\,s^{-1}$) | 1.02E − 01 | Kilogram$_{force}$-second per square meter ($kg_f\,s\,m^{-2}$) |
| Dynamic viscosity | Pascal-second ($kg_f\,m^{-1}\,s^{-1}$) | 1.00E + 01 | Poises ($g\,cm^{-1}\,s^{-1}$) |
| Dynamic viscosity | Pascal-second ($kg_f\,m^{-1}\,s^{-1}$) | 1.45E − 04 | Pound$_{force}$-second per square inch ($lb_f\,s\,in.^{-2}$) |
| Dynamic viscosity | Pascal-second ($kg_f\,m^{-1}\,s^{-1}$) | 2.09E − 02 | Pound$_{force}$-second per square foot ($lb_f\,s\,ft^{-2}$) |
| Dynamic viscosity | Pascal-second ($kg_f\,m\,s^{-1}$) | 2.42E + 03 | Pound$_{mass}$ per foot-hour ($lb_m\,ft^{-1}\,h^{-1}$) |
| Dynamic viscosity | Pascal-second ($kg_f\,m^{-1}\,s^{-1}$) | 6.72E − 01 | Pound$_{mass}$ per foot-second ($lb_m\,ft^{-1}\,s^{-1}$) |
| Dynamic viscosity | Poises ($g_f\,cm^{-1}\,s^{-1}$) | 1.00E + 03 | Centipoises |
| Dynamic viscosity | Poises ($g_f\,cm^{-1}\,s^{-1}$) | 1.00E + 00 | Dynes-second per square centimeter (dynes s$^{-1}$ $cm^{-2}$) |
| Dynamic viscosity | Poises ($g_f\,cm^{-1}\,s^{-1}$) | 1.02E − 02 | Kilogram$_{force}$-second per square meter ($kg_f\,s\,m^{-2}$) |
| Dynamic viscosity | Poises ($g_f\,cm^{-1}\,s^{-1}$) | 1.00E + 00 | Pascal-second ($kg\,m^{-1}\,s^{-1}$) |
| Dynamic viscosity | Poises ($g_f\,cm^{-1}\,s^{-1}$) | 1.45E − 05 | Pound$_{force}$-second per square inch ($lb_f\,s\,in.^{-2}$) |
| Dynamic viscosity | Poises ($g_f\,cm^{-1}\,s^{-1}$) | 2.42E + 02 | Pound$_{mass}$ per foot-hour ($lb_m\,ft^{-1}\,h^{-1}$) |
| Dynamic viscosity | Poises ($g_f\,cm^{-1}\,s^{-1}$) | 6.72E − 02 | Pound$_{mass}$ per foot-second ($lb_m\,ft^{-1}\,s^{-1}$) |
| Dynamic viscosity | Poises ($g_f\,cm^{-1}\,s^{-1}$) | 2.09E − 03 | Pound$_{force}$-second per square foot ($lb_f\,s\,ft^{-2}$) |
| Dynamic viscosity | Pound$_{force}$-second per square foot ($lb_f\,s\,ft^{-2}$) | 4.79E + 04 | Centipoises |
| Dynamic viscosity | Pound$_{force}$-second per square foot ($lb_f\,s\,ft^{-2}$) | 4.79E + 02 | Dynes-second per square centimeter (dynes s $cm^{-2}$) |
| Dynamic viscosity | Pound$_{force}$-second per square foot ($lb_f\,s\,ft^{-2}$) | 4.8824 | Kilogram$_{force}$-second per square meter ($kg_f\,s\,m^{-2}$) |
| Dynamic viscosity | Pound$_{force}$-second per square foot ($lb_f\,s\,ft^{-2}$) | 4.79E + 01 | Pascal-second ($kg\,m^{-1}\,s^{-1}$) |
| Dynamic viscosity | Pound$_{force}$-second per square foot ($lb_f\,s\,ft^{-2}$) | 4.79E + 02 | Poises ($g\,cm^{-1}\,s^{-1}$) |
| Dynamic viscosity | Pound$_{force}$-second per square foot ($lb_f\,s\,ft^{-2}$) | 6.95E − 03 | Pound$_{force}$-second per square inch ($lb_f\,s\,in.^{-2}$) |
| Dynamic viscosity | Pound$_{force}$-second per square foot ($lb_f\,s\,ft^{-2}$) | 1.16E + 05 | Pound$_{mass}$ per foot-hour ($lb_m\,ft^{-1}\,h^{-1}$) |
| Dynamic viscosity | Pound$_{force}$-second per square foot ($lb_f\,s\,ft^{-2}$) | 3.22E + 01 | Pound$_{mass}$ per foot-second ($lb_m\,ft^{-1}\,s^{-1}$) |
| Dynamic viscosity | Pound$_{force}$-second per square inch ($lb_f\,s\,in.^{-2}$) | 6.85E + 06 | Centipoises |
| Dynamic viscosity | Pound$_{force}$-second per square inch ($lb_f\,s\,in.^{-2}$) | 6.89E + 04 | Dynes-second per square centimeter (dynes s $cm^{-2}$) |

| Property | Multiply | By | To obtain |
|---|---|---|---|
| Dynamic viscosity | Pound$_{force}$-second per square inch (lb$_f$ s in.$^{-2}$) | 7.03E + 02 | Kilogram$_{force}$-second per square meter (kg$_f$ s m$^{-2}$) |
| Dynamic viscosity | Pound$_{force}$-second per square inch (lb$_f$ s in.$^{-2}$) | 6.89E + 03 | Pascal-second (kg m$^{-1}$ s$^{-1}$) |
| Dynamic viscosity | Pound$_{force}$-second per square inch (lb$_f$ s in.$^{-2}$) | 6.89E + 04 | Poises (g cm$^{-1}$ s$^{-1}$) |
| Dynamic viscosity | Pound$_{force}$-second per square inch (lb$_f$ s in.$^{-2}$) | 1.67E + 07 | Pound$_{force}$-second per square inch (lb$_f$ s in.$^{-2}$) |
| Dynamic viscosity | Pound$_{force}$-second per square inch (lb$_f$ s in.$^{-2}$) | 1.44E + 02 | Pound$_{force}$-second per square foot (lb$_f$ s ft$^{-2}$) |
| Dynamic viscosity | Pound$_{force}$-second per square inch (lb$_f$ s in.$^{-2}$) | 4.63E + 03 | Pound$_{mass}$ per foot-second (lb$_m$ ft$^{-1}$ s$^{-1}$) |
| Dynamic viscosity | Pound$_{mass}$ per foot-hour (lb$_m$ ft h$^{-1}$) | 4.13E − 01 | Centipoises |
| Dynamic viscosity | Pound$_{mass}$ per foot-hour (lb$_m$ ft h$^{-1}$) | 4.13E − 03 | Dynes-second per square centimeter (dynes s cm$^{-2}$) |
| Dynamic viscosity | Pound$_{mass}$ per foot-hour (lb$_m$ ft h$^{-1}$) | 4.22E − 05 | Kilogram$_{force}$-second per square meter (kg$_f$ s m$^{-2}$) |
| Dynamic viscosity | Pound$_{mass}$ per foot-hour (lb$_m$ ft h$^{-1}$) | 4.13E − 04 | Pascal-second (kg m$^{-1}$ s) |
| Dynamic viscosity | Pound$_{mass}$ per foot-hour (lb$_m$ ft h$^{-1}$) | 4.13E − 03 | Poises (g cm$^{-1}$ s$^{-1}$) |
| Dynamic viscosity | Pound$_{mass}$ per foot-hour (lb$_m$ ft h$^{-1}$) | 5.99E − 08 | Pound$_{force}$-second per square inch (lb$_f$ s in.$^{-2}$) |
| Dynamic viscosity | Pound$_{mass}$ per foot-hour (lb$_m$ ft h$^{-1}$) | 8.63E − 06 | Pound$_{force}$-second per square foot (lb$_f$ s ft$^{-2}$) |
| Dynamic viscosity | Pound$_{mass}$ per foot-hour (lb$_m$ ft h$^{-1}$) | 2.78E − 04 | Pound$_{mass}$ per foot-second (lb$_m$ ft s$^{-1}$) |
| Dynamic viscosity | Pound$_{mass}$ per foot-second (lb$_m$ ft s$^{-1}$) | 1.49E + 03 | Centipoises |
| Dynamic viscosity | Pound$_{mass}$ per foot-second (lb$_m$ ft s$^{-1}$) | 1.49E + 01 | Dynes-second per square centimeter (dynes s cm$^{-2}$) |
| Dynamic viscosity | Pound$_{mass}$ per foot-second (lb$_m$ ft s$^{-1}$) | 1.52E − 01 | Kilogram$_{force}$-second per square meter (kg$_f$ s m$^{-2}$) |
| Dynamic viscosity | Pound$_{mass}$ per foot-second (lb$_m$ ft s$^{-1}$) | 1.4882 | Pascal-second (kg m$^{-1}$ s$^{-1}$) |
| Dynamic viscosity | Pound$_{mass}$ er foot-second (lb$_m$ ft s$^{-1}$) | 1.49E + 01 | Poises (g cm$^{-1}$ s$^{-1}$) |
| Dynamic viscosity | Pound$_{mass}$ per foot-second (lb$_m$ ft s$^{-1}$) | 3.11E − 02 | Pound$_{force}$-second per square foot (lb$_f$ s ft$^{-2}$) |
| Dynamic viscosity | Pound$_{mass}$ per foot-second (lb$_m$ ft s$^{-1}$) | 2.16E − 04 | Pound$_{force}$-second per square inch (lb$_f$ s in.$^{-2}$) |
| Dynamic viscosity | Pound$_{mass}$ per foot-second (lb$_m$ ft s$^{-1}$) | 3.60E + 03 | Pound$_{mass}$ per foot-hours (lb$_m$ ft$^{-1}$ h$^{-1}$) |
| Energy and work | British thermal units (BTU) | 2.52E + 02 | Calorie, thermal (cal) |
| Energy and work | British thermal units (BTU) | 1.06E + 10 | Erg |
| Energy and work | British thermal units (BTU) | 3.68E − 01 | Foot cubed, atmosphere (ft$^3$) |
| Energy and work | British thermal units (BTU) | 7.78E + 02 | Foot-pound (ft lb), force |

(Continued)

(Continued)

| Property | Multiply | By | To obtain |
|---|---|---|---|
| Energy and work | British thermal units (BTU) | 2.50E + 04 | Foot-poundal |
| Energy and work | British thermal units (BTU) | 3.93E − 04 | Horsepower-hour |
| Energy and work | British thermal units (BTU) | 3.98E − 04 | Horsepower-hour (hp h) metric |
| Energy and work | British thermal units (BTU) | 1.06E + 03 | Joule (J) |
| Energy and work | British thermal units (BTU) | 2.52E − 01 | Kilocalorie (cal or kcal) |
| Energy and work | British thermal units (BTU) | 1.08E + 02 | Kilogram-force-meter |
| Energy and work | British thermal units (BTU) | 2.93E − 04 | Kilowatt-hour (kW h) |
| Energy and work | British thermal units (BTU) | 1.04E + 01 | Liter atmosphere |
| Energy and work | British thermal units (BTU) | 2.93E − 01 | Watt-hour |
| Energy and work | Calorie, thermal (cal) | 3.97E − 03 | British thermal units (BTU) |
| Energy and work | Calorie, thermal (cal) | 4.18E + 07 | Erg |
| Energy and work | Calorie, thermal (cal) | 1.46E − 03 | Foot cubed, atmosphere $(ft^3)$ |
| Energy and work | Calorie, thermal (cal) | 3.09E + 00 | Foot-pound (ft lb), force |
| Energy and work | Calorie, thermal (cal) | 9.93E + 01 | Foot-poundal |
| Energy and work | Calorie, thermal (cal) | 1.56E − 06 | Horsepower-hour |
| Energy and work | Calorie, thermal (cal) | 1.58E − 06 | Horsepower-hour (metric) |
| Energy and work | Calorie, thermal (cal) | 4.19E + 00 | Joule (J) |
| Energy and work | Calorie, thermal (cal) | 1.00E − 03 | Kilocalorie (cal or kcal) |
| Energy and work | Calorie, thermal (cal) | 4.27E − 01 | Kilogram-force-meter |
| Energy and work | Calorie, thermal (cal) | 1.16E − 06 | Kilowatt-hour (kW h) |
| Energy and work | Calorie, thermal (cal) | 4.13E − 02 | Liter atmosphere |
| Energy and work | Calorie, thermal (cal) | 1.16E − 03 | Watt-hour |
| Energy and work | Erg | 9.48E − 11 | British thermal units (BTU) |
| Energy and work | Erg | 2.39E − 08 | Calorie, thermal (cal) |
| Energy and work | Erg | 3.49E − 11 | Foot cubed, atmosphere $(ft^3)$ |
| Energy and work | Erg | 7.38E − 08 | Foot-pound force |
| Energy and work | Erg | 2.37E − 06 | Foot-poundal |
| Energy and work | Erg | 3.73E − 14 | Horsepower-hour |

(Continued)

| Property | Multiply | By | To obtain |
|---|---|---|---|
| Energy and work | Erg | 3.78E − 14 | Horsepower-hour (hp h) metric |
| Energy and work | Erg | 1.00E − 06 | Joule (J) |
| Energy and work | Erg | 2.39E − 11 | Kilocalorie (cal or kcal) |
| Energy and work | Erg | 1.02E − 08 | Kilogram-force-meter |
| Energy and work | Erg | 2.78E − 14 | Kilowatt-hour (kW h) |
| Energy and work | Erg | 9.87E − 10 | Liter atmosphere |
| Energy and work | Erg | 2.78E − 11 | Watt-hour |
| Energy and work | Foot cubed, atmosphere (ft$^3$) | 2.72E + 00 | British thermal units (BTU) |
| Energy and work | Foot cubed, atmosphere (ft$^3$) | 6.86E + 02 | Calorie, thermal (cal) |
| Energy and work | Foot cubed, atmosphere (ft$^3$) | 2.87E + 10 | Erg |
| Energy and work | Foot cubed, atmosphere (ft$^3$) | 2.12E + 03 | Foot-pound force |
| Energy and work | Foot cubed, atmosphere (ft$^3$) | 6.81E + 04 | Foot-poundal |
| Energy and work | Foot cubed, atmosphere (ft$^3$) | 1.07E − 03 | Horsepower-hour |
| Energy and work | Foot cubed, atmosphere (ft$^3$) | 1.08E − 03 | Horsepower-hour (hp h) metric |
| Energy and work | Foot cubed, atmosphere (ft$^3$) | 2.87E + 03 | Joule (J) |
| Energy and work | Foot cubed, atmosphere (ft$^3$) | 6.86E − 01 | Kilocalorie (cal or kcal) |
| Energy and work | Foot cubed, atmosphere (ft$^3$) | 2.93E + 02 | Kilogram-force-meter |
| Energy and work | Foot cubed, atmosphere (ft$^3$) | 7.97E − 04 | Kilowatt-hour (kW h) |
| Energy and work | Foot cubed, atmosphere (ft$^3$) | 2.83E + 01 | Liter atmosphere |
| Energy and work | Foot cubed, atmosphere (ft$^3$) | 7.97E − 01 | Watt-hour |
| Energy and work | Foot-pound (ft lb), force | 1.29E − 03 | British thermal unit (BTU) |
| Energy and work | Foot-pound (ft lb), force | 3.24E − 01 | Calorie, thermal (cal) |
| Energy and work | Foot-pound (ft lb), force | 1.36E + 07 | Erg |
| Energy and work | Foot-pound (ft lb), force | 4.73E − 04 | Foot cubed atmosphere |
| Energy and work | Foot-pound (ft lb), force | 3.22E + 01 | Foot-poundal |
| Energy and work | Foot-pound (ft lb), force | 5.05E − 02 | Horsepower-hour |
| Energy and work | Foot-pound (ft lb), force | 5.12E − 07 | Horsepower-hour (hp h) metric |
| Energy and work | Foot-pound (ft lb), force | 1.36E + 00 | Joule (J) |

(*Continued*)

(Continued)

| Property | Multiply | By | To obtain |
|---|---|---|---|
| Energy and work | Foot-pound (ft lb), force | 3.24E − 04 | Kilocalorie (cal or kcal) |
| Energy and work | Foot-pound (ft lb), force | 1.38E − 01 | Kilogram-force-meter |
| Energy and work | Foot-pound (ft lb), force | 3.77E − 07 | Kilowatt-hour (kW h) |
| Energy and work | Foot-pound (ft lb), force | 1.34E − 02 | Liter atmosphere |
| Energy and work | Foot-pound (ft lb), force | 3.77E − 04 | Watt-hour |
| Energy and work | Foot-poundal | 3.99E − 05 | British thermal units (BTU) |
| Energy and work | Foot-poundal | 1.01E − 02 | Calorie, thermal (cal) |
| Energy and work | Foot-poundal | 4.21E + 05 | Erg |
| Energy and work | Foot-poundal | 1.47E − 05 | Foot cubed, atmosphere ($ft^3$) |
| Energy and work | Foot-poundal | 3.11E − 02 | Foot-pound force |
| Energy and work | Foot-poundal | 1.57E − 08 | Horsepower-hour |
| Energy and work | Foot-poundal | 1.59E − 08 | Horsepower-hour (hp h) metric |
| Energy and work | Foot-poundal | 4.21E − 02 | Joule (J) |
| Energy and work | Foot-poundal | 1.01E − 05 | Kilocalorie (cal or kcal) |
| Energy and work | Foot-poundal | 4.30E − 03 | Kilogram-force-meter |
| Energy and work | Foot-poundal | 1.17E − 08 | Kilowatt-hour (kW h) |
| Energy and work | Foot-poundal | 4.16E − 04 | Liter atmosphere |
| Energy and work | Foot-poundal | 1.17E − 05 | Watt-hour |
| Energy and work | Horsepower-hour (hp h) | 2.54E + 03 | British thermal unit (BTU) |
| Energy and work | Horsepower-hour (hp h) | 6.42E + 05 | Calorie, thermal (cal) |
| Energy and work | Horsepower-hour (hp h) | 2.68E + 13 | Erg |
| Energy and work | Horsepower-hour (hp h) | 9.36E + 02 | Foot cubed atmosphere |
| Energy and work | Horsepower-hour (hp h) | 1.98E + 06 | Foot-pound (ft lb), force |
| Energy and work | Horsepower-hour (hp h) | 6.37E + 07 | Foot-poundal |
| Energy and work | Horsepower-hour (hp h) | 1.01E + 00 | Horsepower-hour (hp h) metric |
| Energy and work | Horsepower-hour (hp h) | 2.68E + 06 | Joule (J) |
| Energy and work | Horsepower-hour (hp h) | 6.42E + 02 | Kilocalorie (cal or kcal) |
| Energy and work | Horsepower-hour (hp h) | 2.74E + 05 | Kilogram-force-meter |

(Continued)

| Property | Multiply | By | To obtain |
|---|---|---|---|
| Energy and work | Horsepower-hour (hp h) | 7.46E − 01 | Kilowatt-hour (kW h) |
| Energy and work | Horsepower-hour (hp h) | 2.65E + 04 | Liter atmosphere |
| Energy and work | Horsepower-hour (hp h) | 7.46E + 06 | Watt-hour |
| Energy and work | Horsepower-hour (hp h) metric | 2.51E + 03 | British thermal unit (BTU) |
| Energy and work | Horsepower-hour (hp h) metric | 6.33E + 05 | Calorie, thermal (cal) |
| Energy and work | Horsepower-hour (hp h) metric | 2.65E + 13 | Erg |
| Energy and work | Horsepower-hour (hp h) metric | 9.23E + 02 | Foot cubed, atmosphere ($ft^3$) |
| Energy and work | Horsepower-hour (hp h) metric | 1.95E + 06 | Foot-pound (ft lb), force |
| Energy and work | Horsepower-hour (hp h) metric | 6.28E + 07 | Foot-poundal |
| Energy and work | Horsepower-hour (hp h) metric | 9.86E − 01 | Horsepower-hour (hp h) |
| Energy and work | Horsepower-hour (hp h) metric | 2.65E + 06 | Joule (J) |
| Energy and work | Horsepower-hour (hp h) metric | 6.33E + 02 | Kilocalorie (Cal or kcal) |
| Energy and work | Horsepower-hour (hp h) metric | 2.70E + 05 | Kilogram-force-meter |
| Energy and work | Horsepower-hour (hp h) metric | 7.36E − 01 | Kilowatt-hour (kW h) |
| Energy and work | Horsepower-hour (hp h) metric | 2.61E + 04 | Liter atmosphere |
| Energy and work | Horsepower-hour (hp h) metric | 7.36E + 06 | Watt-hour |
| Energy and work | Joules (J) | 9.48E − 04 | British thermal unit (BTU) |
| Energy and work | Joules (J) | 2.39E − 01 | Calorie, thermal (cal) |
| Energy and work | Joules (J) | 1.00E + 08 | Erg |
| Energy and work | Joules (J) | 3.49E − 04 | Foot cubed, atmosphere ($ft^3$) |
| Energy and work | Joules (J) | 7.38E − 01 | Foot-pound (ft lb), force |
| Energy and work | Joules (J) | 2.37E + 01 | Foot-poundal |
| Energy and work | Joules (J) | 3.73E − 07 | Horsepower-hour (hp h) |
| Energy and work | Joules (J) | 3.78E − 07 | Horsepower-hour (hp h) metric |
| Energy and work | Joules (J) | 2.39E − 04 | Kilocalorie (cal or kcal) |
| Energy and work | Joules (J) | 1.02E − 01 | Kilogram-force-meter |
| Energy and work | Joules (J) | 2.78E − 07 | Kilowatt-hour (kW h) |
| Energy and work | Joules (J) | 9.87E − 03 | Liter atmosphere |

(*Continued*)

| Property | Multiply | By | To obtain |
|----------|----------|-----|-----------|
| Energy and work | Joules (J) | 2.78E − 04 | Watt-hour |
| Energy and work | Kilocalorie (cal or kcal) | 3.97E + 00 | British thermal unit (BTU) |
| Energy and work | Kilocalorie (cal or kcal) | 1.00E + 03 | Calorie, thermal (cal) |
| Energy and work | Kilocalorie (cal or kcal) | 4.18E + 10 | Erg |
| Energy and work | Kilocalorie (cal or kcal) | 1.46E + 00 | Foot cubed, atmosphere ($ft^3$) |
| Energy and work | Kilocalorie (cal or kcal) | 3.09E + 03 | Foot-pound (ft lb), force |
| Energy and work | Kilocalorie (cal or kcal) | 9.93E + 04 | Foot-poundal |
| Energy and work | Kilocalorie (cal or kcal) | 1.56E − 03 | Horsepower-hour (hp h) |
| Energy and work | Kilocalorie (cal or kcal) | 1.58E − 03 | Horsepower-hour (hp h) metric |
| Energy and work | Kilocalorie (cal or kcal) | 4.18E + 03 | Joules (J) |
| Energy and work | Kilocalorie (cal or kcal) | 4.27E + 02 | Kilogram-force-meter |
| Energy and work | Kilocalorie (cal or kcal) | 1.16E − 03 | Kilowatt-hour (kW h) |
| Energy and work | Kilocalorie (cal or kcal) | 4.13E + 01 | Liter atmosphere |
| Energy and work | Kilocalorie (cal or kcal) | 1.16E + 00 | Watt-hour |
| Energy and work | Kilogram-force-meter | 2.34E + 00 | Calorie, thermal (cal) |
| Energy and work | Kilogram-force-meter | 9.81E + 07 | Erg |
| Energy and work | Kilogram-force-meter | 3.42E − 03 | Foot cubed, atmosphere ($ft^3$) |
| Energy and work | Kilogram-force-meter | 7.23E + 00 | Foot-pound (ft lb), force |
| Energy and work | Kilogram-force-meter | 2.33E + 02 | Foot-poundal |
| Energy and work | Kilogram-force-meter | 3.65E − 06 | Horsepower-hour (hp h) |
| Energy and work | Kilogram-force-meter | 3.70E − 06 | Horsepower-hour (hp h) metric |
| Energy and work | Kilogram-force-meter | 9.81E + 00 | Joules (J) |
| Energy and work | Kilogram-force-meter | 2.34E − 03 | Kilocalorie (cal or kcal) |
| Energy and work | Kilogram-force-meter | 2.72E − 06 | Kilowatt-hour (kW h) |
| Energy and work | Kilogram-force-meter | 9.68E − 02 | Liter atmosphere |
| Energy and work | Kilogram-force-meter | 2.72E − 03 | Watt-hour |
| Energy and work | Kilogram-force-meter | 9.29E − 03 | British thermal unit (BTU) |
| Energy and work | Kilowatt-hour (kW h) | 3.412.14 | British thermal unit (BTU) |

(Continued)

| Property | Multiply | By | To obtain |
|---|---|---|---|
| Energy and work | Kilowatt-hour (kW h) | 8.60E + 05 | Calorie, thermal (cal) |
| Energy and work | Kilowatt-hour (kW h) | 3.60E + 13 | Erg |
| Energy and work | Kilowatt-hour (kW h) | 1.25E + 03 | Foot cubed, atmosphere (ft$^3$) |
| Energy and work | Kilowatt-hour (kW h) | 2.66E + 06 | Foot-pound (ft lb), force |
| Energy and work | Kilowatt-hour (kW h) | 8.54E + 07 | Foot-poundal |
| Energy and work | Kilowatt-hour (kW h) | 1.34E + 00 | Horsepower-hour (hp h) |
| Energy and work | Kilowatt-hour (kW h) | 1.36E + 00 | Horsepower-hour (hp h) metric |
| Energy and work | Kilowatt-hour (kW h) | 3.60E + 06 | Joules (J) |
| Energy and work | Kilowatt-hour (kW h) | 8.60E + 02 | Kilocalorie (cal or kcal) |
| Energy and work | Kilowatt-hour (kW h) | 3.67E + 05 | Kilogram-force-meter |
| Energy and work | Kilowatt-hour (kW h) | 3.55E + 04 | Liter atmosphere |
| Energy and work | Kilowatt-hour (kW h) | 1.00E + 03 | Watt-hour |
| Energy and work | Liter atmosphere | 9.60E − 02 | British thermal unit (BTU) |
| Energy and work | Liter atmosphere | 2.42E + 01 | Calorie, thermal (cal) |
| Energy and work | Liter atmosphere | 1.01E + 09 | Erg |
| Energy and work | Liter atmosphere | 3.53E − 02 | Foot cubed, atmosphere (ft$^3$) |
| Energy and work | Liter atmosphere | 7.47E + 01 | Foot-pound (ft lb), force |
| Energy and work | Liter atmosphere | 2.40E + 03 | Foot-poundal |
| Energy and work | Liter atmosphere | 3.77E − 05 | Horsepower-hour (hp h) |
| Energy and work | Liter atmosphere | 3.83E − 05 | Horsepower-hour (hp h) metric |
| Energy and work | Liter atmosphere | 1.01E + 02 | Joules (J) |
| Energy and work | Liter atmosphere | 2.42E − 02 | Kilocalorie (cal or kcal) |
| Energy and work | Liter atmosphere | 1.03E + 01 | Kilogram-force-meter |
| Energy and work | Liter atmosphere | 2.81E − 05 | Kilowatt-hour (kW h) |
| Energy and work | Liter atmosphere | 8.15E − 05 | Watt-hour |
| Energy and work | Watt-hour | 3.41E + 00 | British thermal unit (BTU) |
| Energy and work | Watt-hour | 8.60E + 02 | Calorie, thermal (cal) |
| Energy and work | Watt-hour | 3.60E + 10 | Erg |

(Continued)

(Continued)

| Property | Multiply | By | To obtain |
|---|---|---|---|
| Energy and work | Watt-hour | 1.25E + 00 | Foot cubed, atmosphere ($ft^3$) |
| Energy and work | Watt-hour | 2.66E + 03 | Foot-pound (ft lb), force |
| Energy and work | Watt-hour | 8.54E + 04 | Foot-poundal |
| Energy and work | Watt-hour | 1.34E − 03 | Horsepower-hour (hp h) |
| Energy and work | Watt-hour | 1.36E − 03 | Horsepower-hour (hp h) metric |
| Energy and work | Watt-hour | 3.60E + 03 | Joules (J) |
| Energy and work | Watt-hour | 8.60E − 01 | Kilocalorie (cal or kcal) |
| Energy and work | Watt-hour | 3.67E + 02 | Kilogram-force-meter |
| Energy and work | Watt-hour | 1.00E − 03 | Kilowatt-hour (kW h) |
| Energy and work | Watt-hour | 3.55E + 01 | Liter atmosphere |
| Force | Dyne | 1.02E − 03 | Gram force ($g_{force}$) |
| Force | Dyne | 1.00E − 06 | Joules per centimeter ($J\,cm^{-1}$) |
| Force | Dyne | 1.02E − 06 | Kilogram force ($kg_{force}$) |
| Force | Dyne | 1.00E − 05 | Newton (N) |
| Force | Dyne | 2.25E − 06 | Pound force ($lb_f$) |
| Force | Dyne | 7.23E − 05 | Poundal |
| Force | Gram force ($g_{force}$) | 9.81E + 02 | Dyne |
| Force | Gram force ($g_{force}$) | 9.81E − 05 | Joules per centimeter ($J\,cm^{-1}$) |
| Force | Gram force ($g_{force}$) | 1.00E − 03 | Kilogram force ($kg_{force}$) |
| Force | Gram force ($g_{force}$) | 9.81E − 03 | Newton (N) |
| Force | Gram force ($g_{force}$) | 2.20E − 03 | Pound force ($lb_f$) |
| Force | Gram force ($g_{force}$) | 7.09E − 02 | Poundal |
| Force | Joules per centimeter ($J\,cm^{-1}$) | 1.00E + 08 | Dyne |
| Force | Joules per centimeter ($J\,cm^{-1}$) | 1.02E + 04 | Gram force ($g_{force}$) |
| Force | Joules per centimeter ($J\,cm^{-1}$) | 1.02E + 01 | Kilogram force ($kg_{force}$) |
| Force | Joules per centimeter ($J\,cm^{-1}$) | 1.00E + 02 | Newton (N) |
| Force | Joules per centimeter ($J\,cm^{-1}$) | 2.25E + 01 | Pound force ($lb_f$) |
| Force | Joules per centimeter ($J\,cm^{-1}$) | 7.23E + 02 | Poundal |
| Force | Kilogram force ($kg_{force}$) | 9.81E + 05 | Dyne |
| Force | Kilogram force ($kg_{force}$) | 1.00E + 03 | Gram force ($g_{force}$) |
| Force | Kilogram force ($kg_{force}$) | 9.81E − 02 | Joules per centimeter ($J\,cm^{-1}$) |
| Force | Kilogram force ($kg_{force}$) | 9.81E + 00 | Newton (N) |
| Force | Kilogram force ($kg_{force}$) | 2.20E + 00 | Pound force ($lb_f$) |
| Force | Kilogram force ($kg_{force}$) | 7.09E + 01 | Poundal |
| Force | Newtons (N) | 1.00E + 05 | Dyne |
| Force | Newtons (N) | 1.02E + 02 | Gram force ($g_{force}$) |

(Continued)

| Property | Multiply | By | To obtain |
|---|---|---|---|
| Force | Newtons (N) | 1.00E − 02 | Joules per centimeter (J cm$^{-1}$) |
| Force | Newtons (N) | 1.02E − 01 | Kilogram force (kg$_{force}$) |
| Force | Newtons (N) | 2.25E − 01 | Pound force (lb$_f$) |
| Force | Newtons (N) | 7.23E + 00 | Poundal |
| Force | Pound force (lb$_f$) | 4.45E + 04 | Dyne |
| Force | Pound force (lb$_f$) | 4.54E + 02 | Gram force (g$_{force}$) |
| Force | Pound force (lb$_f$) | 4.45E − 02 | Joules per centimeter (J cm$^{-1}$) |
| Force | Pound force (lb$_f$) | 4.54E − 01 | Kilogram force (kg$_{force}$) |
| Force | Pound force (lb$_f$) | 4.45E + 00 | Newton (N) |
| Force | Pound force (lb$_f$) | 3.22E + 01 | Poundal |
| Force | Poundal | 1.38E + 04 | Dyne |
| Force | Poundal | 1.41E + 01 | Gram force (g$_{force}$) |
| Force | Poundal | 1.38E − 03 | Joules per centimeter (J cm$^{-1}$) |
| Force | Poundal | 1.41E − 02 | Kilogram force (kg$_{force}$) |
| Force | Poundal | 1.38E − 01 | Newton (N) |
| Force | Poundal | 3.11E − 02 | Pound force (lb$_f$) |
| Hydraulic conductivity | Foot per second (ft s$^{-1}$) | 3.05E − 01 | Meter per second (m s$^{-1}$) |
| Hydraulic conductivity | US gallon per day per foot squared (gal day$^{-1}$ ft$^{-2}$) | 4.72E − 07 | Meter per second (m s$^{-1}$) |
| Length | Angstrom | 1.00E − 07 | Centimeter (cm) |
| Length | Angstrom | 5.47E − 11 | Fathom |
| Length | Angstrom | 3.28E − 10 | Foot (ft) |
| Length | Angstrom | 3.94E − 09 | Inch (in.) |
| Length | Angstrom | 1.00E − 12 | Kilometer (km) |
| Length | Angstrom | 1.00E − 09 | Meter (m) |
| Length | Angstrom | 1.00E − 03 | Micrometer |
| Length | Angstrom | 5.40E − 14 | Mile (nautical) |
| Length | Angstrom | 6.21E − 14 | Mile (statute) |
| Length | Angstrom | 1.00E − 06 | Millimeter (mm) |
| Length | Angstrom | 1.00E − 01 | Nanometer |
| Length | Angstrom | 1.09E − 10 | Yard (yd) |
| Length | Centimeter (cm) | 1.00E + 09 | Angstrom |
| Length | Centimeter (cm) | 5.47E − 03 | Fathom |
| Length | Centimeter (cm) | 3.28E − 02 | Foot (ft) |
| Length | Centimeter (cm) | 3.94E − 01 | Inch (in.) |
| Length | Centimeter (cm) | 1.00E − 04 | Kilometer (km) |
| Length | Centimeter (cm) | 1.00E − 02 | Meter (m) |
| Length | Centimeter (cm) | 1.00E + 04 | Micrometer |
| Length | Centimeter (cm) | 5.40E − 06 | Mile (nautical) |
| Length | Centimeter (cm) | 6.21E − 06 | Mile (statute) |
| Length | Centimeter (cm) | 1.00E + 01 | Millimeter (mm) |
| Length | Centimeter (cm) | 1.00E + 08 | Nanometer |
| Length | Centimeter (cm) | 1.09E − 02 | Yard (yd) |
| Length | Chain (surveyor's) | 2.01E + 01 | Meter (m) |
| Length | Chain (surveyor's) | 2.20E + 01 | Yard (yd) |
| Length | Fathom | 1.83E + 10 | Angstrom |
| Length | Fathom | 1.83E + 02 | Centimeter (cm) |
| Length | Fathom | 6.00E + 00 | Foot (ft) |
| Length | Fathom | 7.20E + 01 | Inch (in.) |

(*Continued*)

(Continued)

| Property | Multiply | By | To obtain |
|----------|----------|-----|-----------|
| Length | Fathom | 1.83E − 03 | Kilometer (km) |
| Length | Fathom | 1.83E + 00 | Meter (m) |
| Length | Fathom | 1.83E + 06 | Micrometer |
| Length | Fathom | 9.87E − 04 | Mile (nautical) |
| Length | Fathom | 1.14E − 03 | Mile (statute) |
| Length | Fathom | 1.83E + 03 | Millimeter (mm) |
| Length | Fathom | 1.83E + 09 | Nanometer |
| Length | Fathom | 2.00E + 00 | Yard (yd) |
| Length | Foot (ft) | 3.05E + 09 | Angstrom |
| Length | Foot (ft) | 3.05E + 01 | Centimeter (cm) |
| Length | Foot (ft) | 1.67E − 01 | Fathom |
| Length | Foot (ft) | 1.20E + 01 | Inch (in.) |
| Length | Foot (ft) | 3.05E − 04 | Kilometer (km) |
| Length | Foot (ft) | 3.05E − 01 | Meter (m) |
| Length | Foot (ft) | 3.05E + 05 | Micrometer |
| Length | Foot (ft) | 1.65E − 04 | Mile (nautical) |
| Length | Foot (ft) | 1.89E − 04 | Mile (statute) |
| Length | Foot (ft) | 3.05E + 02 | Millimeter (mm) |
| Length | Foot (ft) | 3.05E + 08 | Nanometer |
| Length | Foot (ft) | 3.33E − 01 | Yard (yd) |
| Length | Inch (in.) | 2.54E + 08 | Angstrom |
| Length | Inch (in.) | 2.54E + 00 | Centimeter (cm) |
| Length | Inch (in.) | 1.69E − 02 | Fathom |
| Length | Inch (in.) | 5.33E − 02 | Foot (ft) |
| Length | Inch (in.) | 2.54E − 05 | Kilometer (km) |
| Length | Inch (in.) | 2.54E − 02 | Meter (m) |
| Length | Inch (in.) | 2.54E + 04 | Micrometer |
| Length | Inch (in.) | 1.37E − 05 | Mile (nautical) |
| Length | Inch (in.) | 1.58E − 05 | Mile (statute) |
| Length | Inch (in.) | 2.54E + 01 | Millimeter (mm) |
| Length | Inch (in.) | 2.54E + 07 | Nanometer |
| Length | Inch (in.) | 2.78E − 02 | Yard (yd) |
| Length | Kilometer (km) | 1.00E + 14 | Angstrom |
| Length | Kilometer (km) | 1.00E + 05 | Centimeter (cm) |
| Length | Kilometer (km) | 5.47E + 02 | Fathom |
| Length | Kilometer (km) | 3.28E + 03 | Foot (ft) |
| Length | Kilometer (km) | 3.94E + 04 | Inch (in.) |
| Length | Kilometer (km) | 1.00E + 03 | Meter (m) |
| Length | Kilometer (km) | 1.00E + 10 | Micrometer |
| Length | Kilometer (km) | 5.40E − 01 | Mile (nautical) |
| Length | Kilometer (km) | 6.21E − 01 | Mile (statute) |
| Length | Kilometer (km) | 1.00E + 07 | Millimeter (mm) |
| Length | Kilometer (km) | 1.00E + 13 | Nanometer |
| Length | Kilometer (km) | 1.09E + 03 | Yard (yd) |
| Length | Meter (m) | 1.00E + 11 | Angstrom |
| Length | Meter (m) | 1.00E + 02 | Centimeter (cm) |
| Length | Meter (m) | 5.47E − 01 | Fathom |
| Length | Meter (m) | 3.28E + 00 | Foot (ft) |
| Length | Meter (m) | 3.94E + 01 | Inch (in.) |
| Length | Meter (m) | 1.00E − 03 | Kilometer (km) |
| Length | Meter (m) | 1.00E + 07 | Micrometer |
| Length | Meter (m) | 5.40E − 04 | Mile (nautical) |
| Length | Meter (m) | 6.21E − 04 | Mile (statute) |
| Length | Meter (m) | 1.00E + 03 | Millimeter (mm) |
| Length | Meter (m) | 1.00E + 10 | Nanometer |
| Length | Meter (m) | 1.09E + 00 | Yard (yd) |
| Length | Micrometer | 1.00E + 04 | Angstrom |
| Length | Micrometer | 1.00E − 04 | Centimeter (cm) |

(Continued)

| Property | Multiply | By | To obtain |
|----------|----------|-----|-----------|
| Length | Micrometer | 5.47E − 07 | Fathom |
| Length | Micrometer | 3.28E − 06 | Foot (ft) |
| Length | Micrometer | 3.94E − 05 | Inch (in) |
| Length | Micrometer | 1.00E − 08 | Kilometer (km) |
| Length | Micrometer | 1.00E − 05 | Meter (m) |
| Length | Micrometer | 5.40E − 10 | Mile (nautical) |
| Length | Micrometer | 6.21E − 10 | Mile (statute) |
| Length | Micrometer | 1.00E − 03 | Millimeter (mm) |
| Length | Micrometer | 1.00E + 03 | Nanometer |
| Length | Micrometer | 1.09E − 06 | Yard (yd) |
| Length | Mile (nautical) (nmi) | 1.85E + 13 | Angstrom |
| Length | Mile (nautical) (nmi) | 1.85E + 05 | Centimeter (cm) |
| Length | Mile (nautical) (nmi) | 1.01E + 03 | Fathom |
| Length | Mile (nautical) (nmi) | 6.08E + 03 | Foot (ft) |
| Length | Mile (nautical) (nmi) | 7.29E + 04 | Inch (in.) |
| Length | Mile (nautical) (nmi) | 1.85E + 00 | Kilometer (km) |
| Length | Mile (nautical) (nmi) | 1.85E + 03 | Meter (m) |
| Length | Mile (nautical) (nmi) | 1.85E + 09 | Micrometer |
| Length | Mile (nautical) (nmi) | 1.15E + 00 | Mile (statute) |
| Length | Mile (nautical) (nmi) | 1.85E + 06 | Millimeter (mm) |
| Length | Mile (nautical) (nmi) | 1.85E + 12 | Nanometer |
| Length | Mile (nautical) (nmi) | 2.03E + 03 | Yard (yd) |
| Length | Mile (statute) | 1.61E + 13 | Angstrom |
| Length | Mile (statute) | 1.61E + 05 | Centimeter (cm) |
| Length | Mile (statute) | 8.80E + 02 | Fathom |
| Length | Mile (statute) | 5.28E + 03 | Foot (ft) |
| Length | Mile (statute) | 6.34E + 04 | Inch (in.) |
| Length | Mile (statute) | 1.61E + 00 | Kilometer (km) |
| Length | Mile (statute) | 1.61E + 03 | Meter (m) |
| Length | Mile (statute) | 1.61E + 09 | Micrometer |
| Length | Mile (statute) | 8.69E − 01 | Mile (nautical) (nmi) |
| Length | Mile (statute) | 1.61E + 06 | Millimeter (mm) |
| Length | Mile (statute) | 1.61E + 12 | Nanometer |
| Length | Mile (statute) | 1.76E + 03 | Yard (yd) |
| Length | Millimeter (mm) | 1.00E + 08 | Angstrom |
| Length | Millimeter (mm) | 1.00E − 01 | Centimeter (cm) |
| Length | Millimeter (mm) | 5.47E − 04 | Fathom |
| Length | Millimeter (mm) | 3.28E − 03 | Foot (ft) |
| Length | Millimeter (mm) | 3.94E − 02 | Inch (in.) |
| Length | Millimeter (mm) | 1.00E − 05 | Kilometer (km) |
| Length | Millimeter (mm) | 1.00E − 03 | Meter (m) |
| Length | Millimeter (mm) | 1.00E + 03 | Micrometer |
| Length | Millimeter (mm) | 5.40E − 07 | Mile (nautical) (nmi) |
| Length | Millimeter (mm) | 6.21E − 07 | Mile (statute) |
| Length | Millimeter (mm) | 1.00E + 07 | Nanometer |
| Length | Millimeter (mm) | 1.09E − 03 | Yard (yd) |
| Length | Nanometer | 1.00E + 01 | Angstrom |
| Length | Nanometer | 1.00E − 06 | Centimeter (cm) |
| Length | Nanometer | 5.47E − 10 | Fathom |
| Length | Nanometer | 3.28E − 09 | Foot (ft) |
| Length | Nanometer | 3.94E − 09 | Inch (in.) |
| Length | Nanometer | 1.00E − 11 | Kilometer (km) |
| Length | Nanometer | 1.00E − 08 | Meter (m) |
| Length | Nanometer | 1.00E − 03 | Micrometer |
| Length | Nanometer | 5.40E − 13 | Mile (nautical) (nmi) |
| Length | Nanometer | 6.21E − 13 | Mile (statute) |

*(Continued)*

(Continued)

| Property | Multiply | By | To obtain |
|---|---|---|---|
| Length | Nanometer | 1.00E − 05 | Millimeter (mm) |
| Length | Nanometer | 1.09E − 09 | Yard (yd) |
| Length | Yard (yd) | 9.14E + 09 | Angstrom |
| Length | Yard (yd) | 9.14E + 01 | Centimeter (cm) |
| Length | Yard (yd) | 5.00E − 01 | Fathom |
| Length | Yard (yd) | 3.00E + 00 | Foot (ft) |
| Length | Yard (yd) | 3.60E + 01 | Inch (in.) |
| Length | Yard (yd) | 9.14E − 04 | Kilometer (km) |
| Length | Yard (yd) | 9.14E − 01 | Meter (m) |
| Length | Yard (yd) | 9.14E + 05 | Micrometer |
| Length | Yard (yd) | 4.94E − 04 | Mile (nautical) (nmi) |
| Length | Yard (yd) | 5.68E − 04 | Mile (statute) |
| Length | Yard (yd) | 9.14E + 02 | Millimeter (mm) |
| Length | Yard (yd) | 9.14E + 08 | Nanometer |
| Mass | Atomic mass unit (amu) | 9.37E − 25 | Dram (avdp.) |
| Mass | Atomic mass unit (amu) | 2.56E − 23 | Grain |
| Mass | Atomic mass unit (amu) | 1.66E − 24 | Gram (g) |
| Mass | Atomic mass unit (amu) | 3.27E − 29 | Hundred weight (long) (hundred wgt.) |
| Mass | Atomic mass unit (amu) | 3.66E − 29 | Hundred weight (short) (hundred wgt.) |
| Mass | Atomic mass unit (amu) | 1.66E − 27 | Kilogram (kg) |
| Mass | Atomic mass unit (amu) | 5.86E − 26 | Ounce (avdp.) |
| Mass | Atomic mass unit (amu) | 5.34E − 26 | Ounce (troy) |
| Mass | Atomic mass unit (amu) | 1.07E − 24 | Pennyweight |
| Mass | Atomic mass unit (amu) | 3.66E − 27 | Pound (avdp.) |
| Mass | Atomic mass unit (amu) | 1.14E − 28 | Slug |
| Mass | Atomic mass unit (amu) | 2.61E − 28 | Stone |
| Mass | Atomic mass unit (amu) | 1.63E − 30 | Ton (long) |
| Mass | Atomic mass unit (amu) | 1.83E − 30 | Ton (short) |
| Mass | Atomic mass unit (amu) | 1.66E − 30 | Tonne (metric) |
| Mass | Dram (avdp.) | 1.07E + 24 | Atomic mass unit (amu) |
| Mass | Dram (avdp.) | 2.73E + 01 | Grain |
| Mass | Dram (avdp.) | 1.77E + 00 | Gram (g) |
| Mass | Dram (avdp.) | 3.49E − 05 | Hundred weight (long) (hundred wgt.) |
| Mass | Dram (avdp.) | 3.91E − 05 | Hundred weight (short) (hundred wgt.) |
| Mass | Dram (avdp.) | 1.77E − 03 | Kilogram (kg) |
| Mass | Dram (avdp.) | 6.25E − 02 | Ounce (avdp.) |
| Mass | Dram (avdp.) | 5.70E − 02 | Ounce (troy) |
| Mass | Dram (avdp.) | 1.14E + 00 | Pennyweight |
| Mass | Dram (avdp.) | 3.91E − 03 | Pound (avdp.) |
| Mass | Dram (avdp.) | 1.21E − 07 | Slug |
| Mass | Dram (avdp.) | 2.79E − 04 | Stone |
| Mass | Dram (avdp.) | 1.74E − 06 | Ton (long) |
| Mass | Dram (avdp.) | 1.95E − 06 | Ton (short) |
| Mass | Dram (avdp.) | 1.77E − 06 | Tonne (metric) |
| Mass | Grain (gr) | 3.90E + 22 | Atomic mass unit (amu) |
| Mass | Grain (gr) | 3.66E − 02 | Dram (avdp.) |
| Mass | Grain (gr) | 6.48E − 02 | Gram (g) |
| Mass | Grain (gr) | 1.28E − 06 | Hundred weight (long) (hundred wgt.) |
| Mass | Grain (gr) | 1.43E − 06 | Hundred weight (short) (hundred wgt.) |
| Mass | Grain (gr) | 6.48E − 05 | Kilogram (kg) |

(Continued)

| Property | Multiply | By | To obtain |
|----------|----------|-----|-----------|
| Mass | Grain (gr) | $2.29E - 03$ | Ounce (avdp.) |
| Mass | Grain (gr) | $2.08E - 03$ | Ounce (troy) |
| Mass | Grain (gr) | $4.17E - 02$ | Pennyweight |
| Mass | Grain (gr) | $1.43E - 04$ | Pound (avdp.) |
| Mass | Grain (gr) | $4.44E - 06$ | Slug |
| Mass | Grain (gr) | $1.02E - 05$ | Stone |
| Mass | Grain (gr) | $6.38E - 08$ | Ton (long) |
| Mass | Grain (gr) | $7.14E - 08$ | Ton (short) |
| Mass | Grain (gr) | $6.48E - 08$ | Tonne (metric) |
| Mass | Gram (g) | $6.02E + 23$ | Atomic mass unit (amu) |
| Mass | Gram (g) | $5.64E - 01$ | Dram (avdp.) |
| Mass | Gram (g) | $1.54E + 01$ | Grain (gr) |
| Mass | Gram (g) | $1.97E - 05$ | Hundred weight (long) (hundred wgt.) |
| Mass | Gram (g) | $2.20E - 05$ | Hundred weight (short) (hundred wgt.) |
| Mass | Gram (g) | $1.00E - 03$ | Kilogram (kg) |
| Mass | Gram (g) | $3.53E - 02$ | Ounce (avdp.) |
| Mass | Gram (g) | $3.22E - 02$ | Ounce (troy) |
| Mass | Gram (g) | $6.43E - 01$ | Pennyweight |
| Mass | Gram (g) | $2.20E - 03$ | Pound (avdp.) |
| Mass | Gram (g) | $6.85E - 05$ | Slug |
| Mass | Gram (g) | $1.57E - 04$ | Stone |
| Mass | Gram (g) | $9.84E + 00$ | Ton (long) |
| Mass | Gram (g) | $1.10E - 06$ | Ton (short) |
| Mass | Gram (g) | $1.00E - 05$ | Tonne (metric) |
| Mass | Hundred weight (long) (hundred wgt.) | $3.03E + 28$ | Atomic mass unit (amu) |
| Mass | Hundred weight (long) (hundred wgt.) | $2.87E + 04$ | Dram (avdp.) |
| Mass | Hundred weight (long) (hundred wgt.) | $7.84E + 05$ | Grain (gr) |
| Mass | Hundred weight (long) (hundred wgt.) | $5.08E + 04$ | Gram (g) |
| Mass | Hundred weight (long) (hundred wgt.) | $1.12E + 00$ | Hundred weight (short) (hundred wgt.) |
| Mass | Hundred weight (long) (hundred wgt.) | $5.08E + 01$ | Kilogram (kg) |
| Mass | Hundred weight (long) (hundred wgt.) | $1.79E + 03$ | Ounce (avdp.) |
| Mass | Hundred weight (long) (hundred wgt.) | $1.63E + 03$ | Ounce (troy) |
| Mass | Hundred weight (long) (hundred wgt.) | $3.27E + 04$ | Pennyweight |
| Mass | Hundred weight (long) (hundred wgt.) | $1.12E + 02$ | Pound (avdp.) |
| Mass | Hundred weight (long) (hundred wgt.) | $3.48E + 00$ | Slug |
| Mass | Hundred weight (long) (hundred wgt.) | $8.00E + 00$ | Stone |
| Mass | Hundred weight (long) (hundred wgt.) | $5.00E - 02$ | Ton (long) |
| Mass | Hundred weight (long) (hundred wgt.) | $5.60E - 02$ | Ton (short) |
| Mass | Hundred weight (long) (hundred wgt.) | $5.08E - 02$ | Tonne (metric) |

(*Continued*)

(Continued)

| Property | Multiply | By | To obtain |
|---|---|---|---|
| Mass | Hundred weight (short) (hundred wgt.) | 2.73E + 28 | Atomic mass unit (amu) |
| Mass | Hundred weight (short) (hundred wgt.) | 2.56E + 04 | Dram (avdp.) |
| Mass | Hundred weight (short) (hundred wgt.) | 7.00E + 05 | Grain (gr) |
| Mass | Hundred weight (short) (hundred wgt.) | 4.54E + 04 | Gram (g) |
| Mass | Hundred weight (short) (hundred-wgt.) | 8.93E − 01 | Hundred weight (long) (hundred wgt.) |
| Mass | Hundred weight (short) (hundred wgt.) | 4.54E + 01 | Kilogram (kg) |
| Mass | Hundred weight (short) (hundred wgt.) | 1.60E + 03 | Ounce (avdp.) |
| Mass | Hundred weight (short) (hundred wgt.) | 1.46E + 03 | Ounce (troy) |
| Mass | Hundred weight (short) (hundred wgt.) | 2.92E + 04 | Pennyweight |
| Mass | Hundred weight (short) (hundred-wgt.) | 1.00E + 02 | Pound (avdp.) |
| Mass | Hundred weight (short) (hundred wgt.) | 3.11E + 00 | Slug |
| Mass | Hundred weight (short) (hundred wgt.) | 7.14E + 00 | Stone |
| Mass | Hundred weight (short) (hundred wgt.) | 4.46E − 02 | Ton (long) |
| Mass | Hundred weight (short) (hundred wgt.) | 5.00E − 02 | Ton (short) |
| Mass | Hundred weight (short) (hundred wgt.) | 4.54E − 02 | Tonne (metric) |
| Mass | Kilogram (kg) | 6.02E + 26 | Atomic mass unit (amu) |
| Mass | Kilogram (kg) | 5.64E + 02 | Dram (avdp.) |
| Mass | Kilogram (kg) | 1.54E + 04 | Grain (gr) |
| Mass | kilogram (kg) | 1.00E + 03 | Gram (g) |
| Mass | Kilogram (kg) | 1.97E − 02 | Hundred weight (long) (hundred wgt.) |
| Mass | Kilogram (kg) | 2.20E − 02 | Hundred weight (short) (hundred wgt.) |
| Mass | Kilogram (kg) | 3.53E + 01 | Ounce (avdp.) |
| Mass | Kilogram (kg) | 3.22E + 01 | Ounce (troy) |
| Mass | Kilogram (kg) | 6.43E + 00 | Pennyweight |
| Mass | Kilogram (kg) | 2.20E + 00 | Pound (avdp.) |
| Mass | Kilogram (kg) | 6.85E − 02 | Slug |
| Mass | Kilogram (kg) | 1.57E − 01 | Stone |
| Mass | Kilogram (kg) | 9.84E − 04 | Ton (long) |
| Mass | Kilogram (kg) | 1.10E − 03 | Ton (short) |
| Mass | Kilogram (kg) | 1.00E − 03 | Tonne (metric) |
| Mass | Ounce (avdp.) | 1.71E + 25 | Atomic mass unit (amu) |
| Mass | Ounce (avdp.) | 1.60E + 01 | Dram (avdp.) |
| Mass | Ounce (avdp.) | 4.38E + 02 | Grain (gr) |
| Mass | Ounce (avdp.) | 2.83E + 01 | Gram (g) |
| Mass | Ounce (avdp.) | 5.58E − 04 | Hundred weight (long) (hundred wgt.) |
| Mass | Ounce (avdp.) | 6.25E − 04 | Hundred weight (short) (hundred wgt.) |
| Mass | Ounce (avdp.) | 2.83E − 02 | Kilogram (kg) |

(Continued)

| Property | Multiply | By | To obtain |
|---|---|---|---|
| Mass | Ounce (avdp.) | 9.11E − 01 | Ounce (troy) |
| Mass | Ounce (avdp.) | 1.82E + 01 | Pennyweight |
| Mass | Ounce (avdp.) | 6.25E − 02 | Pound (avdp.) |
| Mass | Ounce (avdp.) | 1.64E − 03 | Slug |
| Mass | Ounce (avdp.) | 4.46E − 03 | Stone |
| Mass | Ounce (avdp.) | 2.79E − 05 | Ton (long) |
| Mass | Ounce (avdp.) | 3.13E − 05 | Ton (short) |
| Mass | Ounce (avdp.) | 2.83E + 00 | Tonne (metric) |
| Mass | Ounce (troy) | 1.87E + 25 | Atomic mass unit (amu) |
| Mass | Ounce (troy) | 1.76E + 01 | Dram (avdp.) |
| Mass | Ounce (troy) | 4.80E + 02 | Grain (gr) |
| Mass | Ounce (troy) | 3.11E + 01 | Gram (g) |
| Mass | Ounce (troy) | 6.12E − 04 | Hundred weight (long) (hundred wgt.) |
| Mass | Ounce (troy) | 6.86E − 04 | Hundred weight (short) (hundred wgt.) |
| Mass | Ounce (troy) | 3.11E − 02 | Kilogram (kg) |
| Mass | Ounce (troy) | 1.10E + 00 | Ounce (avdp.) |
| Mass | Ounce (troy) | 2.00E + 01 | Pennyweight |
| Mass | Ounce (troy) | 6.86E − 02 | Pound (avdp.) |
| Mass | Ounce (troy) | 2.13E − 03 | Slug |
| Mass | Ounce (troy) | 4.90E − 03 | Stone |
| Mass | Ounce (troy) | 3.06E − 05 | Ton (long) |
| Mass | Ounce (troy) | 3.43E + 05 | Ton (short) |
| Mass | Ounce (troy) | 3.11E − 05 | Tonne (metric) |
| Mass | Pennyweight | 9.37E + 23 | Atomic mass unit (amu) |
| Mass | Pennyweight | 8.78E − 01 | Dram (avdp.) |
| Mass | Pennyweight | 2.40E + 01 | Grain (gr) |
| Mass | Pennyweight | 1.56E + 00 | Gram (g) |
| Mass | Pennyweight | 3.06E − 05 | Hundred weight (long) (hundred wgt.) |
| Mass | Pennyweight | 3.43E − 05 | Hundred weight (short) (hundred wgt.) |
| Mass | Pennyweight | 1.56E − 03 | Kilogram (kg) |
| Mass | Pennyweight | 5.49E − 02 | Ounce (avdp.) |
| Mass | Pennyweight | 5.00E − 02 | Ounce (troy) |
| Mass | Pennyweight | 3.43E − 03 | Pound (avdp.) |
| Mass | Pennyweight | 1.07E − 04 | Slug |
| Mass | Pennyweight | 2.48E − 04 | Stone |
| Mass | Pennyweight | 1.53E − 06 | Ton (long) |
| Mass | Pennyweight | 1.71E − 06 | Ton (short) |
| Mass | Pennyweight | 1.00E + 02 | Tonne (metric) |
| Mass | Pound mass ($lb_m$) (avdp.) | 2.73E + 26 | Atomic mass unit (amu) |
| Mass | Pound mass ($lb_m$) (avdp.) | 2.56E + 02 | Dram (avdp.) |
| Mass | Pound mass ($lb_m$) (avdp.) | 7.00E + 03 | Grain (gr) |
| Mass | Pound mass ($lb_m$) (avdp.) | 4.54E + 02 | Gram (g) |
| Mass | Pound mass ($lb_m$) (avdp.) | 8.93E − 03 | Hundred weight (long) (hundred wgt.) |
| Mass | Pound mass ($lb_m$) (avdp.) | 1.00E − 02 | Hundred weight (short) (hundred wgt.) |
| Mass | Pound mass ($lb_m$) (avdp.) | 4.54E − 01 | Kilogram (kg) |
| Mass | Pound mass ($lb_m$) (avdp.) | 1.60E + 01 | Ounce (avdp.) |
| Mass | Pound mass ($lb_m$) (avdp.) | 1.46E + 01 | Ounce (troy) |
| Mass | Pound mass ($lb_m$) (avdp.) | 2.92E + 02 | Pennyweight |
| Mass | Pound mass ($lb_m$) (avdp.) | 3.11E − 02 | Slug |
| Mass | Pound mass ($lb_m$) (avdp.) | 7.14E − 02 | Stone |

(Continued)

(Continued)

| Property | Multiply | By | To obtain |
|----------|----------|-----|-----------|
| Mass | Pound mass (lb$_m$) (avdp.) | 4.46E − 04 | Ton (long) |
| Mass | Pound mass (lb$_m$) (avdp.) | 5.00E − 04 | Ton (short) |
| Mass | Pound mass (lb$_m$) (avdp.) | 4.54E − 04 | Tonne (metric) |
| Mass | Slug | 8.79E + 27 | Atomic mass unit (amu) |
| Mass | Slug | 8.24E + 03 | Dram (avdp.) |
| Mass | Slug | 2.25E + 05 | Grain (gr) |
| Mass | Slug | 1.46E + 04 | Gram (g) |
| Mass | Slug | 2.87E − 01 | Hundred weight (long) (hundred wgt.) |
| Mass | Slug | 3.22E − 01 | Hundred weight (short) (hundred wgt.) |
| Mass | Slug | 1.46E + 01 | Kilogram (kg) |
| Mass | Slug | 5.15E + 02 | Ounce (avdp.) |
| Mass | Slug | 4.69E + 02 | Ounce (troy) |
| Mass | Slug | 9.38E + 03 | Pennyweight |
| Mass | Slug | 3.22E + 01 | Pound (lb) (avdp.) |
| Mass | Slug | 2.30E + 00 | Stone |
| Mass | Slug | 1.44E − 02 | Ton (long) |
| Mass | Slug | 1.61E − 02 | Ton (short) |
| Mass | Slug | 1.46E − 02 | Tonne (metric) |
| Mass | Stone | 3.82E + 27 | Atomic mass unit (amu) |
| Mass | Stone | 3.58E + 03 | Dram (avdp.) |
| Mass | Stone | 9.80E + 04 | Grain (gr) |
| Mass | Stone | 6.35E + 03 | Gram (g) |
| Mass | Stone | 1.25E − 01 | Hundred weight (long) (hundred wgt.) |
| Mass | Stone | 1.40E − 01 | Hundred weight (short) (hundred wgt.) |
| Mass | Stone | 6.35E + 00 | Kilogram (kg) |
| Mass | Stone | 2.24E + 02 | Ounce (avdp.) |
| Mass | Stone | 2.04E + 04 | Ounce (troy) |
| Mass | Stone | 4.08E + 03 | Pennyweight |
| Mass | Stone | 1.40E + 01 | Pound (lb) (avdp.) |
| Mass | Stone | 4.35E − 01 | Slug |
| Mass | Stone | 6.25E − 03 | Ton (long) |
| Mass | Stone | 7.00E − 03 | Ton (short) |
| Mass | Stone | 6.35E − 03 | Tonne (metric) |
| Mass | Ton (long) | 6.12E + 27 | Atomic mass unit (amu) |
| Mass | Ton (long) | 5.73E + 05 | Dram (avdp.) |
| Mass | Ton (long) | 1.57E + 07 | Grain (gr) |
| Mass | Ton (long) | 1.02E + 06 | Gram (g) |
| Mass | Ton (long) | 2.00E + 01 | Hundred weight (long) (hundred wgt.) |
| Mass | Ton (long) | 2.24E + 01 | Hundred weight (short) (hundred wgt.) |
| Mass | Ton (long) | 1.02E + 03 | Kilogram (kg) |
| Mass | Ton (long) | 3.58E + 04 | Ounce (avdp.) |
| Mass | Ton (long) | 3.27E + 04 | Ounce (troy) |
| Mass | Ton (long) | 6.53E + 05 | Pennyweight |
| Mass | Ton (long) | 2.24E + 03 | Pound (lb) (avdp.) |
| Mass | Ton (long) | 6.91E + 01 | Slug |
| Mass | Ton (long) | 1.60E + 02 | Stone |
| Mass | Ton (long) | 1.12E + 00 | Ton (short) |
| Mass | Ton (long) | 1.02E + 00 | Tonne (metric) |
| Mass | Ton (short) | 5.46E + 29 | Atomic mass unit (amu) |
| Mass | Ton (short) | 5.12E + 05 | Dram (avdp.) |

(Continued)

| Property | Multiply | By | To obtain | |
|---|---|---|---|---|
| Mass | Ton (short) | 1.40E + 07 | Grain (gr) | |
| Mass | Ton (short) | 9.07E + 05 | Gram (g) | |
| Mass | Ton (short) | 1.79E + 01 | Hundred weight (long) (hundred wgt.) | |
| Mass | Ton (short) | 2.00E + 01 | Hundred weight (short) (hundred wgt.) | |
| Mass | Ton (short) | 9.07E + 02 | Kilogram (kg) | |
| Mass | Ton (short) | 3.20E + 04 | Ounce (avdp.) | |
| Mass | Ton (short) | 2.92E + 04 | Ounce (troy) | |
| Mass | Ton (short) | 5.83E + 05 | Pennyweight | |
| Mass | Ton (short) | 2.00E + 03 | Pound (lb) (avdp.) | |
| Mass | Ton (short) | 6.22E + 01 | Slug | |
| Mass | Ton (short) | 1.43E + 02 | Stone | |
| Mass | Ton (short) | 8.93E − 01 | Ton (long) | |
| Mass | Ton (short) | 9.07E − 01 | Tonne (metric) | |
| Mass | Tonne (metric) | 6.02E + 29 | Atomic mass unit (amu) | |
| Mass | Tonne (metric) | 5.64E + 05 | Dram (avdp.) | |
| Mass | Tonne (metric) | 1.54E + 07 | Grain (gr) | |
| Mass | Tonne (metric) | 1.00E + 05 | Gram (g) | |
| Mass | Tonne (metric) | 1.97E + 01 | Hundred weight (long) (hundred wgt.) | |
| Mass | Tonne (metric) | 2.20E + 01 | Hundred weight (short) (hundred wgt.) | |
| Mass | Tonne (metric) | 1.00E + 03 | Kilogram (kg) | |
| Mass | Tonne (metric) | 3.53E + 04 | Ounce (avdp.) | |
| Mass | Tonne (metric) | 3.22E + 04 | Ounce (troy) | |
| Mass | Tonne (metric) | 6.43E + 05 | Pennyweight | |
| Mass | Tonne (metric) | 2.204.62 | Pound (lb) (avdp.) | |
| Mass | Tonne (metric) | 6.85E + 01 | Slug | |
| Mass | Tonne (metric) | 1.57E + 02 | Stone | |
| Mass | Tonne (metric) | 9.84E − 01 | Ton (long) | |
| Mass | Tonne (metric) | 1.10E + 00 | Ton (short) | |
| Permeability | Centimeter per second (cm s$^{-1}$) | 1.02E − 05 | Centimeter squared (cm$^2$) | |
| Permeability | Centimeter per second (cm s$^{-1}$) | 1.03E + 03 | Darcy | |
| Permeability | Centimeter per second (cm s$^{-1}$) | 2.83E + 03 | Foot per day (ft day$^{-1}$) | |
| Permeability | Centimeter per second (cm s$^{-1}$) | 3.28E − 02 | Foot per second (ft s$^{-1}$) | |
| Permeability | Centimeter per second (cm s$^{-1}$) | 1.06E + 06 | Foot per year (ft year$^{-1}$) | |
| Permeability | Centimeter per second (cm s$^{-1}$) | 1.10E − 08 | Foot squared (ft$^2$) | |
| Permeability | Centimeter per second (cm s$^{-1}$) | 2.12E + 04 | Gallons per day (US) per foot squared (Meinzer) (gpd ft$^{-2}$) | |
| Permeability | Centimeter per second (cm s$^{-1}$) | 1.00E + 01 | Liter per second per meter squared (l s$^{-1}$ m$^{-2}$) | |
| Permeability | Centimeter per second (cm s$^{-1}$) | 1.02E − 09 | Meter squared (m$^2$) | |
| Permeability | Centimeter squared (cm$^2$) | 9.81E + 04 | Centimeter per second (cm s$^{-1}$) | Foot squared (ft$^2$) |
| Permeability | Centimeter squared (cm$^2$) | 1.01E + 08 | Darcy | Centimeter squared (cm$^2$) |

(Continued)

(Continued)

| Property | Multiply | By | To obtain | |
|---|---|---|---|---|
| Permeability | Centimeter squared (cm$^2$) | 2.78E + 08 | Foot per day (ft day$^{-1}$) | Millimeter squared (mm$^2$) |
| Permeability | Centimeter squared (cm$^2$) | 3.22E + 03 | Foot per second (ft s$^{-1}$) | |
| | | | | Hectare (ha) |
| Permeability | Centimeter squared (cm$^2$) | $1.015 \times 10^{11}$ | Foot per year (ft year$^{-1}$) | Inch squared (in.$^2$) |
| Permeability | Centimeter squared (cm$^2$) | 1.08E − 03 | Foot squared (ft$^2$) | |
| Permeability | Centimeter squared (cm$^2$) | 2.08E + 09 | Gallons per day (US) per foot squared (Meinzer) (gpd ft$^{-2}$) | Meter squared (m$^2$) |
| Permeability | Centimeter squared (cm$^2$) | 9.81E + 05 | Liter per second per meter squared (L s m$^{-2}$) | Kilometer squared (km$^2$) |
| Permeability | Centimeter squared (cm$^2$) | 3.97E + 04 | Meter squared (m$^2$) | |
| Permeability | Darcy | 9.68E − 04 | Centimeter per second (cm s$^{-1}$) | Inch cubed (in.$^3$) |
| Permeability | Darcy | 9.87E − 09 | Centimeter squared (cm$^2$) | Inch cubed (in.$^3$) |
| Permeability | Darcy | 2.74E + 00 | Foot per day (ft day$^{-1}$) | |
| Permeability | Darcy | 3.18E − 05 | Foot per second (ft s$^{-1}$) | Meter cubed (m$^3$) |
| Permeability | Darcy | 1.00E + 03 | Foot per year (ft year$^{-1}$) | Yard cubed (yd$^3$) |
| Permeability | Darcy | 1.06E − 11 | Foot squared (ft$^{-2}$) | Centimeter cubed (cm$^3$) |
| Permeability | Darcy | 2.05E + 01 | Gallons per day (US) per foot squared (Meinzer) (gpd ft$^{-2}$) | |
| Permeability | Darcy | 9.68E − 03 | Liter per second per meter squared (L s$^{-1}$ m$^{-2}$) | |
| Permeability | Darcy | 9.90E − 13 | Meter squared (m$^2$) | Millimeter cubed (mm$^3$) |
| Permeability | Foot per second (ft s$^{-1}$) | 3.05E + 01 | Centimeter per second (cm s$^{-1}$) | |
| Permeability | Foot per second (ft s$^{-1}$) | 3.11E − 04 | Centimeter squared (cm$^2$) | |
| Permeability | Foot per second (ft s$^{-1}$) | 3.15E + 04 | Darcy | |
| Permeability | Foot per second (ft s$^{-1}$) | 8.64E + 04 | Foot per day (ft day$^{-1}$) | |
| Permeability | Foot per second (ft s$^{-1}$) | 3.16E + 07 | Foot per year (ft year$^{-1}$) | |
| Permeability | Foot per second (ft s$^{-1}$) | 3.35E − 07 | Foot squared (ft$^2$) | |
| Permeability | Foot per second (ft s$^{-1}$) | 6.46E + 05 | Gallons per day (US) per foot squared (Meinzer) (gpd ft$^{-2}$) | |
| Permeability | Foot per second (ft s$^{-1}$) | 3.05E + 02 | Liter per second per meter squared (L s$^{-1}$ m$^{-2}$) | |
| Permeability | Foot per second (ft s$^{-1}$) | 3.11E − 08 | Meter squared (m$^2$) | |
| Permeability | Foot per year (ft year$^{-1}$) | 9.66E − 07 | Centimeter per second (cm s$^{-1}$) | |
| Permeability | Foot per year (ft year$^{-1}$) | 9.85E − 12 | Centimeter squared (cm$^2$) | |
| Permeability | Foot per year (ft year$^{-1}$) | 9.98E − 04 | Darcy | |
| Permeability | Foot per year (ft year$^{-1}$) | 2.74E − 03 | Foot per day (ft day$^{-1}$) | |
| Permeability | Foot per year (ft year$^{-1}$) | 3.17E − 08 | Foot per second (ft s$^{-1}$) | |
| Permeability | Foot per year (ft year$^{-1}$) | 1.06E − 14 | Foot squared (ft$^2$) | |
| Permeability | Foot per year (ft year$^{-1}$) | 2.05E − 02 | Gallons per day (US) per foot squared (Meinzer) (gpd ft$^{-2}$) | |

(Continued)

| Property | Multiply | By | To obtain | |
| --- | --- | --- | --- | --- |
| Permeability | Foot per year (ft year$^{-1}$) | 9.66E − 06 | Liter per second per meter squared (L s$^{-1}$ m$^{-2}$) | |
| Permeability | Foot per year (ft year$^{-1}$) | 9.85E − 16 | Meter squared (m$^2$) | |
| Permeability | Foot squared (ft$^2$) | 9.11E + 07 | Centimeter per second (cm s$^{-1}$) | Liter per second (L s$^{-1}$) |
| Permeability | Foot squared (ft$^2$) | 9.29E + 02 | Centimeter squared (cm$^2$) | Foot cubed per second (ft$^3$ s$^{-1}$) |
| Permeability | Foot squared (ft$^2$) | 9.41E + 10 | Darcy | Liter per second (L s$^{-1}$) |
| Permeability | Foot squared (ft$^2$) | 2.56E + 11 | Foot per day (ft day$^{-1}$) | Peck |
| Permeability | Foot squared (ft$^2$) | 2.99E + 06 | Foot per second (ft s$^{-1}$) | Meter cubed per day (m$^3$ day$^{-1}$) |
| Permeability | Foot squared (ft$^2$) | 9.43E + 13 | Foot per year (ft year$^{-1}$) | Meter cubed per second (m$^3$ s$^{-1}$) |
| Permeability | Foot squared (ft$^2$) | 1.93E + 12 | Gallons per day (US) per foot squared (Meinzer) (gpd ft$^{-2}$) | US gallon per day (gal day$^{-1}$) |
| Permeability | Foot squared (ft$^2$) | 9.11E + 08 | Liter per second per meter squared (L s$^{-1}$ m$^{-2}$) | Meter cubed per second (m$^3$ s$^{-1}$) |
| Permeability | Foot squared (ft$^2$) | 9.29E − 02 | Meter squared (m$^2$) | Foot cubed per second (ft$^3$ s$^{-1}$) |
| Permeability | Foot per day | 3.53E − 04 | Centimeter per second (cm s$^{-1}$) | |
| Permeability | Foot per day | 3.60E − 09 | Centimeter squared (cm$^2$) | |
| Permeability | Foot per day | 3.65E − 01 | Darcy | |
| Permeability | Foot per day | 1.16E − 05 | Foot per second (ft s$^{-1}$) | |
| Permeability | Foot per day | 3.65E + 02 | Foot per year (ft year$^{-1}$) | |
| Permeability | Foot per day | 3.87E − 12 | Foot squared (ft$^{-2}$) | |
| Permeability | Foot per day | 7.48E + 00 | Gallons per day (US) per foot squared (Meinzer) (gpd ft$^{-2}$) | |
| Permeability | Foot per day | 3.53E − 03 | Liter per second per meter squared (L s$^{-1}$ m$^{-2}$) | |
| Permeability | Foot per day | 3.60E − 13 | Meter squared (m$^2$) | |
| Permeability | Gallons per day (US) per foot squared (Meinzer) (gpd ft$^{-2}$) | 4.72E − 05 | Centimeter per second (cm s$^{-1}$) | |
| Permeability | Gallons per day (US) per foot squared (Meinzer) (gpd ft$^{-2}$) | 4.81E − 10 | Centimeter squared (cm$^2$) | |
| Permeability | Gallons per day (US) per foot squared (Meinzer) (gpd ft$^{-2}$) | 4.87E − 02 | Darcy | |
| Permeability | Gallons per day (US) per foot squared (Meinzer) (gpd ft$^{-2}$) | 1.55E − 06 | Foot per second (ft s$^{-1}$) | |
| Permeability | Gallons per day (US) per foot squared (Meinzer) (gpd ft$^{-2}$) | 4.99E + 01 | Foot per year (ft year$^{-1}$) | |

(*Continued*)

(Continued)

| Property | Multiply | By | To obtain | |
|---|---|---|---|---|
| Permeability | Gallons per day (US) per foot squared (Meinzer) (gpd ft$^{-2}$) | 5.17E − 13 | Foot squared (ft$^2$) | |
| Permeability | Gallons per day (US) per foot squared (Meinzer) (gpd ft$^{-2}$) | 1.34E − 01 | Foot per day | |
| Permeability | Gallons per day (US) per foot squared (Meinzer) (gpd ft$^{-2}$) | 4.72E − 04 | Liter per second per meter squared (L s$^{-1}$ m$^{-2}$) | |
| Permeability | Gallons per day (US) per foot squared (Meinzer) (gpd ft$^{-2}$) | 4.81E − 14 | Meter squared (m$^2$) | |
| Permeability | Liter per second per meter squared (L s$^{-1}$ m$^{-2}$) | 3.97E + 04 | Centimeter per second (cm s$^{-1}$) | |
| Permeability | Liter per second per meter squared (L s$^{-1}$ m$^{-2}$) | 1.02E − 06 | Centimeter squared (cm$^2$) | |
| Permeability | Liter per second per meter squared (L s$^{-1}$ m$^{-2}$) | 1.03E + 02 | Darcy | |
| Permeability | Liter per second per meter squared (L s$^{-1}$ m$^{-2}$) | 2.83E + 02 | Foot per day (ft day$^{-1}$) | |
| Permeability | Liter per second per meter squared (L s$^{-1}$ m$^{-2}$) | 3.28E − 03 | Foot per second (ft s$^{-1}$) | |
| Permeability | Liter per second per meter squared (L s$^{-1}$ m$^{-2}$) | 1.04E + 05 | Foot per year (ft year$^{-1}$) | |
| Permeability | Liter per second per meter squared (L s$^{-1}$ m$^{-2}$) | 1.10E − 09 | Foot squared (ft$^2$) | |
| Permeability | Liter per second per meter squared (L s$^{-1}$ m$^{-2}$) | 2.12E + 03 | Gallons per day (US) per foot squared (Meinzer) (gpd ft$^{-2}$) | |
| Permeability | Liter per second per meter squared (L s$^{-1}$ m$^{-2}$) | 1.02E − 10 | Meter squared (m$^2$) | |
| Permeability | Meter squared (m$^2$) | 9.81E + 08 | Centimeter per second (cm s$^{-1}$) | Foot cubed per second (ft$^3$ s$^{-1}$) |
| Permeability | Meter squared (m$^2$) | 1.04E + 02 | Centimeter squared (cm$^2$) | Yard squared (yd$^2$) |
| Permeability | Meter squared (m$^2$) | 1.01E + 12 | Darcy | |
| Permeability | Meter squared (m$^2$) | 2.78E + 12 | Foot per day (ft day$^{-1}$) | Barrel per day (bbl day$^{-1}$) |
| Permeability | Meter squared (m$^2$) | 3.22E + 07 | Foot per second (ft s$^{-1}$) | Liter per second (L s$^{-1}$) |
| Permeability | Meter squared (m$^2$) | 1.02E + 15 | Foot per year (ft year$^{-1}$) | Meter cubed per second (m$^3$ s$^{-1}$) |
| Permeability | Meter squared (m$^2$) | 1.08E + 01 | Foot squared (ft$^2$) | Acre (ac) |
| Permeability | Meter squared (m$^2$) | 2.08E + 13 | Gallons per day (US) per foot squared (Meinzer) (gpd ft$^{-2}$) | Acre-feet per day (ac ft day$^{-1}$) |
| Permeability | Meter squared (m$^2$) | $9.8067 \times 10^9$ | Liter per second per meter squared (L s$^{-1}$ m$^{-2}$) | US gallon per minute (gal min$^{-1}$ or gpm) |
| Pressure | Atmosphere (atm) | 1.01E + 00 | Bars | |
| Pressure | Atmosphere (atm) | 3.40E + 01 | Feet of water (68°F) | |
| Pressure | Atmosphere (atm) | 2.99E + 01 | Inch of mercury (32°F) | |

(Continued)

| Property | Multiply | By | To obtain |
|----------|----------|-----|-----------|
| Pressure | Atmosphere (atm) | 4.08E + 02 | Inch of water (68°F) |
| Pressure | Atmosphere (atm) | 1.01E + 02 | Kilopascal (kPa) |
| Pressure | Atmosphere (atm) | 1.03E + 00 | Kilogram per centimeter squared (kg cm$^{-2}$) |
| Pressure | Atmosphere (atm) | 1.03E + 04 | Kilogram per meter squared (kg m$^{-2}$) |
| Pressure | Atmosphere (atm) | 7.60E + 02 | Millimeter of mercury (32°F) |
| Pressure | Atmosphere (atm) | 2.12E + 03 | Pound per foot squared (lb ft$^{-2}$) |
| Pressure | Atmosphere (atm) | 1.47E + 01 | Pound per inch squared (lb in.$^{-2}$, psi) |
| Pressure | Bars | 9.87E − 01 | Atmosphere (atm) |
| Pressure | Bars | 3.35E + 01 | Feet of water (68°F) |
| Pressure | Bars | 2.95E + 01 | Inch of mercury (32°F) |
| Pressure | Bars | 4.02E + 02 | Inch of water (68°F) |
| Pressure | Bars | 1.00E + 02 | Kilopascal (kPa) |
| Pressure | Bars | 1.02E + 00 | Kilogram per centimeter squared (kg cm$^{-2}$) |
| Pressure | Bars | 1.02E + 04 | Kilogram per meter squared (kg m$^{-2}$) |
| Pressure | Bars | 7.50E + 02 | Millimeter of mercury (32°F) |
| Pressure | Bars | 2.09E + 03 | Pound per foot squared (lb ft$^{-2}$) |
| Pressure | Bars | 1.45E + 01 | Pound per inch squared (lb in.$^{-2}$) |
| Pressure | Feet of water (68°F) | 2.95E − 02 | Atmosphere (atm) |
| Pressure | Feet of water (68°F) | 2.98E − 02 | Bars |
| Pressure | Feet of water (68°F) | 1.20E + 01 | Inch of water (68°F) |
| Pressure | Feet of water (68°F) | 2.98E + 00 | Kilopascal (kPa) |
| Pressure | Feet of water (68°F) | 3.04E − 02 | Kilogram per centimeter squared (kg cm$^{-2}$) |
| Pressure | Feet of water (68°F) | 3.04E + 02 | Kilogram per meter squared (kg m$^{-2}$) |
| Pressure | Feet of water (68°F) | 2.24E + 01 | Millimeter of mercury (32°F) |
| Pressure | Feet of water (68°F) | 6.23E + 01 | Pound per foot squared (lb ft$^{-2}$) |
| Pressure | Feet of water (68°F) | 4.33E − 01 | Pound per inch squared (lb in.$^{-2}$) |
| Pressure | Feet of water (ft$_{water}$) | 8.85E − 01 | Inch of mercury Hg (in. Hg) |
| Pressure | Inch of mercury (in. Hg) | 3.39E + 03 | Pascal (Pa) |
| Pressure | Inch of mercury (in. Hg) | 1.13E + 00 | Feet of water (ft) |
| Pressure | Inch of mercury (in. Hg) | 2.53E + 01 | Millimeter of mercury (mm Hg) |
| Pressure | Inch of mercury (in. Hg) (32°F) | 3.34E − 02 | Atmosphere (atm) |
| Pressure | Inch of mercury (in. Hg) (32°F) | 3.39E − 02 | Bars |
| Pressure | Inch of mercury (in. Hg) (32°F) | 1.14E + 00 | Feet of water (68°F) |
| Pressure | Inch of mercury (in. Hg) (32°F) | 1.36E + 01 | Inch of water (68°F) |
| Pressure | Inch of mercury (in. Hg) (32°F) | 3.39E + 00 | Kilopascal (kPa) |

(*Continued*)

| Property | Multiply | By | To obtain |
|---|---|---|---|
| Pressure | Inch of mercury (in. Hg) (32°F) | 3.45E − 02 | Kilogram per centimeter squared (kg cm$^{-2}$) |
| Pressure | Inch of mercury (in. Hg) (32°F) | 3.45E + 02 | Kilogram per meter squared (kg m$^{-2}$) |
| Pressure | Inch of mercury (in. Hg) (32°F) | 2.54E + 01 | Millimeter of mercury (32°F) |
| Pressure | Inch of mercury (in. Hg) (32°F) | 7.07E + 01 | Pound per foot squared (lb ft$^{-2}$) |
| Pressure | Inch of mercury (in. Hg) (32°F) | 4.91E − 01 | Pound per inch squared (psi lb in.$^{-2}$) |
| Pressure | Inch of water (68°F) | 2.45E − 03 | Atmosphere (atm) |
| Pressure | Inch of water (68°F) | 2.49E − 03 | Bars |
| Pressure | Inch of water (68°F) | 8.33E − 02 | Feet of water (68°F) |
| Pressure | Inch of water (68°F) | 7.34E − 02 | Inch of mercury (32°F) |
| Pressure | Inch of water (68°F) | 2.49E − 01 | Kilopascal (kPa) |
| Pressure | Inch of water (68°F) | 2.53E − 03 | Kilogram per centimeter squared (kg cm$^{-2}$) |
| Pressure | Inch of water (68°F) | 2.54E + 01 | Kilogram per meter squared (kg m$^{-2}$) |
| Pressure | Inch of water (68°F) | 1.87E + 00 | Millimeter of mercury (32°F) |
| Pressure | Inch of water (68°F) | 5.20E + 00 | Pound per foot squared (lb ft$^{-2}$) |
| Pressure | Inch of water (68°F) | 3.61E − 02 | Pound per inch squared (psi lb in.$^{-2}$) |
| Pressure | Kilopascal (kPa) | 9.87E − 03 | Atmosphere (atm) |
| Pressure | Kilopascal (kPa) | 1.00E − 02 | Bars |
| Pressure | Kilopascal (kPa) | 3.35E − 01 | Feet of water (68°F) |
| Pressure | Kilopascal (kPa) | 2.95E − 01 | Inch of mercury (32°F) |
| Pressure | Kilopascal (kPa) | 4.02E + 00 | Inch of water (68°F) |
| Pressure | Kilopascal (kPa) | 1.02E − 02 | Kilogram per centimeter squared (kg cm$^{-2}$) |
| Pressure | Kilopascal (kPa) | 1.02E + 02 | Kilogram per meter squared (kg m$^{-2}$) |
| Pressure | Kilopascal (kPa) | 7.50E + 00 | Millimeter of mercury (32°F) |
| Pressure | Kilopascal (kPa) | 2.09E + 01 | Pound per foot squared (lb ft$^{-2}$) |
| Pressure | Kilopascal (kPa) | 3.61E − 02 | Pound per inch squared (psi lb in.$^{-2}$) |
| Pressure | Kilogram per centimeter squared (kg cm$^{-2}$) | 9.68E − 01 | Atmosphere (atm) |
| Pressure | Kilogram per centimeter squared (kg cm$^{-2}$) | 9.81E − 01 | Bars |
| Pressure | Kilogram per centimeter squared (kg cm$^{-2}$) | 3.29E + 01 | Feet of water (68°F) |
| Pressure | Kilogram per centimeter squared (kg cm$^{-2}$) | 2.90E + 01 | Inch of mercury (32°F) |
| Pressure | Kilogram per centimeter squared (kg cm$^{-2}$) | 3.94E + 02 | Inch of water (68°F) |
| Pressure | Kilogram per centimeter squared (kg cm$^{-2}$) | 9.81E + 01 | Kilopascal (kPa) |
| Pressure | Kilogram per centimeter squared (kg cm$^{-2}$) | 1.00E + 04 | Kilogram per meter squared (kg m$^{-2}$) |
| Pressure | Kilogram per centimeter squared (kg cm$^{-2}$) | 7.36E + 02 | Millimeter of mercury (32°F) |

(Continued)

| Property | Multiply | By | To obtain |
|---|---|---|---|
| Pressure | Kilogram per centimeter squared (kg cm$^{-2}$) | 2.05E + 03 | Pound per foot squared (lb ft$^{-2}$) |
| Pressure | Kilogram per centimeter squared (kg cm$^{-2}$) | 6.23E − 02 | Pound per inch squared (psi lb in.$^{-2}$) |
| Pressure | Kilogram per meter squared (kg m$^{-2}$) | 9.68E − 05 | Atmosphere (atm) |
| Pressure | Kilogram per meter squared (kg m$^{-2}$) | 9.81E − 05 | Bars |
| Pressure | Kilogram per meter squared (kg m$^{-2}$) | 3.29E − 03 | Feet of water (68°F) |
| Pressure | Kilogram per meter squared (kg m$^{-2}$) | 2.90E − 03 | Inch of mercury (32°F) |
| Pressure | Kilogram per meter squared (kg m$^{-2}$) | 3.94E − 02 | Inch of water (68°F) |
| Pressure | Kilogram per meter squared (kg m$^{-2}$) | 9.81E − 03 | Kilopascal (kPa) |
| Pressure | Kilogram per meter squared (kg m$^{-2}$) | 7.36E − 02 | Millimeter of mercury (32°F) |
| Pressure | Kilogram per meter squared (kg m$^{-2}$) | 2.05E − 01 | Pound per foot squared (lb ft$^{-2}$) |
| Pressure | Kilogram per meter squared (kg m$^{-2}$) | 1.42E − 03 | Pound per inch squared (psi lb in.$^{-2}$) |
| Pressure | Millibar (mb) | 1.00E + 02 | Pascal (Pa) |
| Pressure | Millimeter of mercury (Hg) (torr) | 3.95E − 02 | Inch of mercury Hg |
| Pressure | Millimeter of mercury (32°F) | 1.32E − 03 | Atmosphere (atm) |
| Pressure | Millimeter of mercury (32°F) | 1.33E − 03 | Bars |
| Pressure | Millimeter of mercury (32°F) | 4.47E − 02 | Feet of water (68°F) |
| Pressure | Millimeter of mercury (32°F) | 3.94E − 02 | Inch of mercury (32°F) |
| Pressure | Millimeter of mercury (32°F) | 5.36E − 01 | Inch of water (68°F) |
| Pressure | Millimeter of mercury (32°F) | 1.33E − 01 | Kilopascal |
| Pressure | Millimeter of mercury (32°F) | 1.36E − 03 | Kilogram per centimeter squared (kg cm$^{-2}$) |
| Pressure | Millimeter of mercury (32°F) | 1.36E + 01 | Kilogram per meter squared (kg m$^{-2}$) |
| Pressure | Millimeter of mercury (32°F) | 2.79E + 00 | Pound per feet squared (lb ft$^{-2}$) |
| Pressure | Millimeter of mercury (32°F) | 1.93E − 02 | Pound per inch squared (psi lb in.$^{-2}$) |
| Pressure | Pound per feet squared (lb ft$^{-2}$) | 4.73E − 04 | Atmosphere (atm) |
| Pressure | Pound per feet squared (lb ft$^{-2}$) | 4.79E − 04 | Bars |
| Pressure | Pound per feet squared (lb ft$^{-2}$) | 1.61E − 02 | Feet of water (68°F) |
| Pressure | Pound per feet squared (lb ft$^{-2}$) | 1.41E − 02 | Inch of mercury (32°F) |
| Pressure | Pound per feet squared (lb ft$^{-2}$) | 1.93E − 01 | Inch of water (68°F) |
| Pressure | Pound per feet squared (lb ft$^{-2}$) | 4.79E − 02 | Kilopascal (kPa) |

*(Continued)*

(Continued)

| Property | Multiply | By | To obtain |
|---|---|---|---|
| Pressure | Pound per feet squared $(\text{lb ft}^{-2})$ | 4.88E − 04 | Kilogram per centimeter squared $(\text{kg cm}^{-2})$ |
| Pressure | Pound per feet squared $(\text{lb ft}^{-2})$ | 4.88E + 00 | Kilogram per meter squared $(\text{kg m}^{-2})$ |
| Pressure | Pound per feet squared $(\text{lb ft}^{-2})$ | 3.59E − 01 | Millimeter of mercury (32°F) |
| Pressure | Pound per feet squared $(\text{lb ft}^{-2})$ | 4.79E + 01 | Pascal $(\text{Pa, N m}^{-2})$ |
| Pressure | Pound per feet squared $(\text{lb ft}^{-2})$ | 6.95E − 03 | Pound per inch squared (psi $\text{lb in.}^{-2})$ |
| Pressure | Pound per inch squared (psi) $(\text{lb in.}^{-2})$ | 6.81E − 02 | Atmosphere (atm) |
| Pressure | Pound per inch squared (psi) $(\text{lb in.}^{-2})$ | 6.80E − 02 | Atmosphere (atm) |
| Pressure | Pound per inch squared (psi) $(\text{lb in.}^{-2})$ | 6.90E − 02 | Bars |
| Pressure | Pound per inch squared (psi) $(\text{lb in.}^{-2})$ | 2.31E + 00 | Feet of water (68°F) |
| Pressure | Pound per inch squared (psi) $(\text{lb in.}^{-2})$ | 2.04E + 00 | Inch of mercury (32°F) |
| Pressure | Pound per inch squared (psi) $(\text{lb in.}^{-2})$ | 2.77E + 01 | Inch of water (68°F) |
| Pressure | Pound per inch squared (psi) $(\text{lb in.}^{-2})$ | 2.77E + 01 | Inch of water column (in.) |
| Pressure | Pound per inch squared (psi) $(\text{lb in.}^{-2})$ | 6.90E + 00 | Kilopascal (kPa) |
| Pressure | Pound per inch squared (psi) $(\text{lb in.}^{-2})$ | 7.03E − 02 | Kilogram per centimeter squared $(\text{kg cm}^{-2})$ |
| Pressure | Pound per inch squared (psi) $(\text{lb in.}^{-2})$ | 7.03E + 02 | Kilogram per meter squared $(\text{kg m}^{-2})$ |
| Pressure | Pound per inch squared (psi) $(\text{lb in.}^{-2})$ | 4.79E − 02 | Kilopascal (kpa) |
| Pressure | Pound per inch squared (psi) $(\text{lb in.}^{-2})$ | 5.17E + 01 | Millimeter of mercury (32°F) |
| Pressure | Pound per inch squared (psi) $(\text{lb in.}^{-2})$ | 6.90E + 02 | Newtons per meter squared $(\text{N m}^{-2})$ |
| Pressure | Pound per inch squared (psi) $(\text{lb in.}^{-2})$ | 6.90E + 02 | Pascal (Pa) |
| Pressure | Pound per inch squared (psi) $(\text{lb in.}^{-2})$ | 1.44E + 02 | Pound per square foot $(\text{lb ft}^{-2})$ |
| Pressure | Pound per square inch $(\text{lb ft}^{-2} \text{ or psi})$ | 2.31E + 00 | Feet of head, water (ft) |
| Pressure | Poundal per foot squared $(\text{poundal ft}^{-2})$ | 1.49E + 00 | Pascal (Pa) |
| Pressure | Atmosphere (atm) | 1.01E + 05 | Pascal (Pa) |
| Temperature | °Celsius | (°C × 1.8) + 32 | °Fahrenheit |
| Temperature | °Celsius | (°C + 273.15 × 1.8) | °Rankine |
| Temperature | °Celsius | °C + 273.15 | Kelvin |
| Temperature | °Fahrenheit | (°F − 32) / 1.8 | °Celsius |
| Temperature | °Fahrenheit | °F + 459.67 | °Rankine |
| Temperature | ° Fahrenheit | ((°F − 32) / 1.8) + 273.15 | Kelvin |
| Temperature | °Rankine | (°Rank/1.8) − 273.15 | °Celsius |
| Temperature | °Rankine | °Rank − 459.67 | °Fahrenheit |
| Temperature | °Rankine | °Rank / 1.8 | Kelvin |

| Property | Multiply | By | To obtain |
|---|---|---|---|
| Temperature | Kelvin | $K - 273.15$ | °Celsius |
| Temperature | Kelvin | $(K \times 1.8) - 459.87$ | °Fahrenheit |
| Temperature | Kelvin | $K \times 1.8$ | °Rankine |
| Thermal conductivity | British thermal unit per hour-foot-degree Fahrenheit (BTU $h^{-1} ft^{-1} °F^{-1}$) | $4.13E - 03$ | Calorie per second-degree Kelvin (cal $s^{-1} cm^{-1} °K^{-1}$) |
| Thermal conductivity | British thermal unit per hour-foot-degree Fahrenheit (BTU $h^{-1} ft^{-1} °F^{-1}$) | $1.73E + 05$ | Gram-centimeter per second cubed-degree Kelvin (g $cm^{-1} s^{-3} °K^{-1}$) |
| Thermal conductivity | British thermal unit per hour-foot-degree Fahrenheit (BTU $h^{-1} ft^{-1} °F$) | $1.7307$ | Kilogram-meter per second cubed-degree Fahrenheit (kg m $s^{-3} °F^{-1}$) |
| Thermal conductivity | British thermal unit per hour-foot-degree Fahrenheit (BTU $h^{-1} ft^{-1} °F^{-1}$) | $2.16E - 01$ | Pound$_{force}$ per second-degree Fahrenheit (lb$_f$ $s^{-1} °F^{-1}$) |
| Thermal conductivity | British thermal unit per hour-foot-degree Fahrenheit (BTU $h^{-1} ft^{-1} °F^{-1}$) | $6.9546$ | Pound$_{mass}$-foot per second cubed-degree Fahrenheit (lb$_m$ ft $s^{-3} °F^{-1}$) |
| Thermal conductivity | Calorie per second-degree Kelvin (cal $s^{-1} cm^{-1} °K^{-1}$) | $2.42E + 02$ | British thermal unit per hour-foot-degree Fahrenheit (BTU $h^{-1} ft^{-1} °F^{-1}$) |
| Thermal conductivity | Calorie per second-degree Kelvin (cal $s^{-1} cm^{-1} °K^{-1}$) | $4.18E + 07$ | Gram-centimeter per second cubed-degree Kelvin (g $cm s^{-3} °K^{-1}$) |
| Thermal conductivity | Calorie per second-degree Kelvin (cal $s^{-1} cm^{-1} °K^{-1}$) | $4.19E + 02$ | Kilogram-meter per second cubed-degree Fahrenheit (kg m $s^{-3} °F^{-1}$) |
| Thermal conductivity | Calorie per second-degree Kelvin (cal $s^{-1} cm^{-1} °K^{-1}$) | $5.23E + 01$ | Pound$_{force}$ per second-degree Fahrenheit (lb$_f$ $s^{-1} °F^{-1}$) |
| Thermal conductivity | calorie per second-degree Kelvin (cal $s^{-1} cm^{-1} °K^{-1}$) | $1.68E + 03$ | Pound$_{mass}$-foot per second cubed-degree Fahrenheit (lb$_m$ ft $s^{-3} °F^{-1}$) |
| Thermal conductivity | Gram-centimeter per second cubed-degree Kelvin (g $cm s^{-3} °K^{-1}$) | $5.79E - 06$ | British thermal unit per hour-foot-degree Fahrenheit (BTU $h^{-1} ft^{-1} °F^{-1}$) |
| Thermal conductivity | Gram-centimeter per second cubed-degree Kelvin (g $cm s^{-3} °K^{-1}$) | $2.39E - 08$ | Calorie per second-degree Kelvin (cal $s^{-1} cm^{-1} °K^{-1}$) |
| Thermal conductivity | Gram-centimeter per second cubed-degree Kelvin (g $cm s^{-3} °K^{-1}$) | $1.00E - 04$ | Kilogram-meter per second cubed-degree Fahrenheit (kg m $s^{-3} °F^{-1}$) |
| Thermal conductivity | Gram-centimeter per second cubed-degree Kelvin (g $cm s^{-3} °K^{-1}$) | $1.25E - 06$ | Pound$_{force}$ per second-degree Fahrenheit (lb$_f$ $s^{-1} °F^{-1}$) |

*(Continued)*

(Continued)

| Property | Multiply | By | To obtain |
|---|---|---|---|
| Thermal conductivity | Gram-centimeter per second cubed-degree Kelvin (g cm s$^{-3}$ °K$^{-1}$) | 4.02E − 05 | Pound$_{mass}$-foot per second cubed-degree Fahrenheit (lb$_m$ ft s$^{-3}$ °F$^{-1}$) |
| Thermal conductivity | Kilogram-meter per second cubed-degree Fahrenheit (kg m s$^{-3}$°F$^{-1}$) | 5.78E − 01 | British thermal unit per hour-foot-degree Fahrenheit (BTU h$^{-1}$ft$^{-1}$ °F$^{-1}$) |
| Thermal conductivity | Kilogram-meter per second cubed-degree Fahrenheit (kg m s$^{-3}$°F$^{-1}$) | 2.39E − 03 | Calorie per second-degree Kelvin (cal s$^{-1}$ cm$^{-1}$ °K$^{-1}$) |
| Thermal conductivity | Kilogram-meter per second cubed-degree Fahrenheit (kg m s$^{-3}$ °F$^{-1}$) | 1.00E + 06 | Gram-centimeter per second cubed-degree Kelvin (g cm s$^{-3}$ °K$^{-1}$) |
| Thermal conductivity | Kilogram-meter per second cubed-degree Fahrenheit (kg m s$^{-3}$ °F$^{-1}$) | 1.25E − 01 | Pound$_{force}$ per second-degree Fahrenheit (lb$_f$ s$^{-1}$ °F$^{-1}$) |
| Thermal conductivity | Kilogram-meter per second cubed-degree Fahrenheit (kg m s$^{-3}$ °F$^{-1}$) | 4.0183 | Pound$_{mass}$-foot per second cubed-degree Fahrenheit (lb$_m$ ft s$^{-3}$ °F$^{-1}$) |
| Thermal conductivity | Pound$_{force}$ per second-degree Fahrenheit (lb$_f$ s$^{-1}$ °F$^{-1}$) | 4.6263 | British thermal unit per hour-foot-degree Fahrenheit (BTU h$^{-1}$ft$^{-1}$ °F$^{-1}$) |
| Thermal conductivity | Pound$_{force}$ per second-degree Fahrenheit (lb$_f$ s$^{-1}$ °F$^{-1}$) | 1.91E − 02 | Calorie per second-degree Kelvin (cal s$^{-1}$ cm$^{-1}$ °K$^{-1}$) |
| Thermal conductivity | Pound$_{force}$ per second-degree Fahrenheit (lb$_f$ s$^{-1}$ °F$^{-1}$) | 8.01E + 05 | Gram-centimeter per second cubed-degree Kelvin (g cm s$^{-3}$ °K$^{-1}$) |
| Thermal conductivity | Pound$_{force}$ per second-degree Fahrenheit (lb$_f$ s$^{-1}$ °F$^{-1}$) | 8.01 | Kilogram-meter per second cubed-degree Fahrenheit (kg m s$^{-3}$ °F$^{-1}$) |
| Thermal conductivity | Pound$_{force}$ per second-degree Fahrenheit (lb$_f$ s$^{-1}$ °F$^{-1}$) | 3.22E + 01 | Pound$_{mass}$-foot per second cubed-degree Fahrenheit (lb$_m$ ft s$^{-3}$ °F$^{-1}$) |
| Thermal conductivity | Pound$_{mass}$-foot per second cubed-degree Fahrenheit (lb$_m$ ft s$^{-3}$ °F$^{-1}$) | 1.44E − 01 | British thermal unit per hour-foot-degree Fahrenheit (BTU h$^{-1}$ft$^{-1}$ °F$^{-1}$) |
| Thermal conductivity | Pound$_{mass}$-foot per second cubed-degree Fahrenheit (lb$_m$ ft s$^{-3}$ °F$^{-1}$) | 5.94E − 04 | Calorie per second-degree Kelvin (cal s$^{-1}$ cm$^{-1}$ °K$^{-1}$) |
| Thermal conductivity | Pound$_{mass}$-foot per second cubed-degree Fahrenheit (lb$_m$ ft s$^{-3}$ °F$^{-1}$) | 2.49E + 04 | Gram-centimeter per second cubed-degree Kelvin (g cm s$^{-3}$ °K$^{-1}$) |

(Continued)

| Property | Multiply | By | To obtain |
|---|---|---|---|
| Thermal conductivity | Pound$_{mass}$-foot per second cubed-degree Fahrenheit ($lb_m$ ft s$^{-3}$ $°F^{-1}$) | 2.49E − 01 | Kilogram-meter per second cubed-degree Fahrenheit (kg m s$^{-3}$ $°F^{-1}$) |
| Thermal conductivity | Pound$_{mass}$-foot per second cubed-degree Fahrenheit ($lb_m$ ft s$^{-3}$ $°F^{-1}$) | 3.11E − 02 | Pound$_{force}$ per second-degree Fahrenheit ($lb_f$ s$^{-1}$ $°F^{-1}$) |
| Time | Day | 2.40E + 01 | Hours (h) |
| Time | Day | 1.44E + 03 | Minutes (min) |
| Time | Day | 8.64E + 04 | Second |
| Time | Day | 2.74E − 03 | Standard year |
| Time | Day | 1.43E − 01 | Week |
| Time | Hours (h) | 4.17E − 02 | Day |
| Time | Hours (h) | 6.00E + 01 | Minutes (min) |
| Time | Hours (h) | 3.60E + 03 | Second (s) |
| Time | Hours (h) | 1.14E − 04 | Standard year |
| Time | Hours (h) | 5.95E − 03 | Week |
| Time | Minutes (min) | 6.94E − 04 | Day |
| Time | Minutes (min) | 1.67E − 02 | Hours (h) |
| Time | Minutes (min) | 6.00E + 01 | Second (s) |
| Time | Minutes (min) | 1.90E − 06 | Standard year |
| Time | Minutes (min) | 9.92E − 05 | Week |
| Time | Second (s) | 1.16E − 05 | Day |
| Time | Second (s) | 2.77E − 04 | Hours (h) |
| Time | Second (s) | 1.66E − 02 | Minutes (min) |
| Time | Second (s) | 3.17E − 08 | Standard year |
| Time | Second (s) | 1.65E − 06 | Week |
| Time | Standard year | 3.65E + 02 | Day |
| Time | Standard year | 8.76E + 03 | Hours (h) |
| Time | Standard year | 5.26E + 05 | Minutes (min) |
| Time | Standard year | 3.15E + 07 | Second (s) |
| Time | Standard year | 5.21E + 01 | Week |
| Time | Week | 7.00E + 00 | Day |
| Time | Week | 1.68E + 02 | Hours (h) |
| Time | Week | 1.01E + 04 | Minutes (min) |
| Time | Week | 6.05E + 05 | Second (s) |
| Time | Week | 1.92E − 02 | Standard year |
| Transmissivity | Foot squared per second (ft$^2$ s$^{-1}$) | 9.29E − 02 | Meter squared per second (m$^2$ s$^{-1}$) |
| Transmissivity | US gallon per day per foot (US gal day$^{-1}$ ft$^{-1}$) | 1.44E − 07 | Meter squared per second (m$^2$ s$^{-1}$) |
| Velocity | Centimeter per hour (cm h$^{-1}$) | 1.67E − 02 | Centimeter per min (cm min$^{-1}$) |
| Velocity | Centimeter per hour (cm h$^{-1}$) | 2.78E − 04 | Centimeter per second (cm s$^{-1}$) |
| Velocity | Centimeter per hour (cm h$^{-1}$) | 3.28E − 02 | Foot per hour (ft h$^{-1}$) |
| Velocity | Centimeter per hour (cm h$^{-1}$) | 5.47E − 04 | Foot per minute (ft min$^{-1}$) |
| Velocity | Centimeter per hour (cm h$^{-1}$) | 9.11E − 06 | Foot per second (ft s$^{-1}$) |
| Velocity | Centimeter per hour (cm h$^{-1}$) | 1.00E − 04 | Kilometer per hour (km h$^{-1}$) |

(Continued)

(Continued)

| Property | Multiply | By | To obtain |
|---|---|---|---|
| Velocity | Centimeter per hour (cm h$^{-1}$) | 1.67E − 07 | Kilometer per minute (km min$^{-1}$) |
| Velocity | Centimeter per hour (cm h$^{-1}$) | 2.78E − 09 | Kilometer per second (km s$^{-1}$) |
| Velocity | Centimeter per hour (cm h$^{-1}$) | 5.40E − 06 | Knots |
| Velocity | Centimeter per hour (cm h$^{-1}$) | 1.00E − 02 | Meter per hour (m h$^{-1}$) |
| Velocity | Centimeter per hour (cm h$^{-1}$) | 1.67E − 04 | Meter per minute (m min$^{-1}$) |
| Velocity | Centimeter per hour (cm h$^{-1}$) | 2.78E − 06 | Meter per second (m s$^{-1}$) |
| Velocity | Centimeter per hour (cm h$^{-1}$) | 6.21E − 08 | Mile per hour (mi h$^{-1}$) |
| Velocity | Centimeter per hour (cm h$^{-1}$) | 1.04E − 07 | Mile per minute (mi min$^{-1}$) |
| Velocity | Centimeter per hour (cm h$^{-1}$) | 1.73E − 09 | Mile per second (mi s$^{-1}$) |
| Velocity | Centimeter per min (cm min$^{-1}$) | 6.00E + 01 | Centimeter per hour (cm h$^{-1}$) |
| Velocity | Centimeter per min (cm min$^{-1}$) | 1.67E − 02 | Centimeter per second (cm s$^{-1}$) |
| Velocity | Centimeter per min (cm min$^{-1}$) | 1.97E + 00 | Foot per hour (ft h$^{-1}$) |
| Velocity | Centimeter per min (cm min$^{-1}$) | 3.28E − 02 | Foot per minute (ft min$^{-1}$) |
| Velocity | Centimeter per min (cm min$^{-1}$) | 5.47E − 04 | Foot per second (ft s$^{-1}$) |
| Velocity | Centimeter per min (cm min$^{-1}$) | 6.00E − 04 | Kilometer per hour (km h$^{-1}$) |
| Velocity | Centimeter per min (cm min$^{-1}$) | 1.00E − 05 | Kilometer per minute (km min$^{-1}$) |
| Velocity | Centimeter per min (cm min$^{-1}$) | 1.67E − 07 | Kilometer per second (km s$^{-1}$) |
| Velocity | Centimeter per min (cm min$^{-1}$) | 3.24E − 04 | Knots |
| Velocity | Centimeter per min (cm min$^{-1}$) | 6.00E − 01 | Meter per hour (m h$^{-1}$) |
| Velocity | Centimeter per min (cm min$^{-1}$) | 1.00E − 02 | Meter per minute (m min$^{-1}$) |
| Velocity | Centimeter per min (cm min$^{-1}$) | 1.67E − 04 | Meter per second (m s$^{-1}$) |
| Velocity | Centimeter per min (cm min$^{-1}$) | 3.73E − 04 | Mile per hour (mi h$^{-1}$) |
| Velocity | Centimeter per min (cm min$^{-1}$) | 6.21E − 06 | Mile per minute (mi min$^{-1}$) |
| Velocity | Centimeter per min (cm min$^{-1}$) | 1.04E − 07 | Mile per second (mi s$^{-1}$) |
| Velocity | Centimeter per second (cm s$^{-1}$) | 3.60E + 03 | Centimeter per hour (cm h$^{-1}$) |
| Velocity | Centimeter per second (cm s$^{-1}$) | 6.00E + 01 | Centimeter per min (cm min$^{-1}$) |
| Velocity | Centimeter per second (cm s$^{-1}$) | 1.18E + 02 | Foot per hour (ft h$^{-1}$) |
| Velocity | Centimeter per second (cm s$^{-1}$) | 1.20E + 00 | Foot per minute (ft min$^{-1}$) |

| Property | Multiply | By | To obtain |
|---|---|---|---|
| Velocity | Centimeter per second (cm s$^{-1}$) | 3.28E − 02 | Foot per second (ft s$^{-1}$) |
| Velocity | Centimeter per second (cm s$^{-1}$) | 3.60E − 02 | Kilometer per hour (km h$^{-1}$) |
| Velocity | Centimeter per second (cm s$^{-1}$) | 6.00E − 04 | Kilometer per minute (km min) |
| Velocity | Centimeter per second (cm s$^{-1}$) | 1.00E − 05 | Kilometer per second (km s$^{-1}$) |
| Velocity | Centimeter per second (cm s$^{-1}$) | 1.94E − 02 | Knots |
| Velocity | Centimeter per second (cm s$^{-1}$) | 3.60E + 01 | Meter per hour (m h$^{-1}$) |
| Velocity | Centimeter per second (cm s$^{-1}$) | 6.00E − 01 | Meter per minute (m min$^{-1}$) |
| Velocity | Centimeter per second (cm s$^{-1}$) | 1.00E − 02 | Meter per second (m s$^{-1}$) |
| Velocity | Centimeter per second (cm s$^{-1}$) | 2.24E − 02 | Mile per hour (mi h$^{-1}$) |
| Velocity | Centimeter per second (cm s$^{-1}$) | 3.73E − 04 | Mile per minute (mi min$^{-1}$) |
| Velocity | Centimeter per second (cm s$^{-1}$) | 6.21E − 06 | Mile per second (mi s$^{-1}$) |
| Velocity | Foot per day (ft day$^{-1}$) | 1.16E − 05 | Feet per second (ft s$^{-1}$) |
| Velocity | Foot per day (ft day$^{-1}$) | 1.27E − 05 | Kilometer per hour (km h$^{-1}$) |
| Velocity | Foot per day (ft day$^{-1}$) | 3.53E − 06 | Meter per second (m s$^{-1}$) |
| Velocity | Foot per day (ft day$^{-1}$) | 7.89E − 06 | Mile per hour (mi h$^{-1}$ or mph) |
| Velocity | Foot per hour (ft h$^{-1}$) | 3.05E + 01 | Centimeter per hour (cm h$^{-1}$) |
| Velocity | Foot per hour (ft h$^{-1}$) | 5.08E − 01 | Centimeter per min (cm min$^{-1}$) |
| Velocity | Foot per hour (ft h$^{-1}$) | 8.47E − 03 | Centimeter per second (cm s$^{-1}$) |
| Velocity | Foot per hour (ft h$^{-1}$) | 1.67E − 02 | Foot per minute (ft min$^{-1}$) |
| Velocity | Foot per hour (ft h$^{-1}$) | 2.78E − 04 | Foot per second (ft s$^{-1}$) |
| Velocity | Foot per hour (ft h$^{-1}$) | 3.05E − 04 | Kilometer per hour (km h$^{-1}$) |
| Velocity | Foot per hour (ft h$^{-1}$) | 5.08E − 06 | Kilometer per minute (km min$^{-1}$) |
| Velocity | Foot per hour (ft h$^{-1}$) | 8.47E − 08 | Kilometer per second (km s$^{-1}$) |
| Velocity | Foot per hour (ft h$^{-1}$) | 1.65E − 04 | Knots |
| Velocity | Foot per hour (ft h$^{-1}$) | 3.05E − 01 | Meter per hour (m h$^{-1}$r) |
| Velocity | Foot per hour (ft h$^{-1}$) | 5.08E − 03 | Meter per minute (m min$^{-1}$) |
| Velocity | Foot per hour (ft h$^{-1}$) | 8.47E − 05 | Meter per second (m s$^{-1}$) |
| Velocity | Foot per hour (ft h$^{-1}$) | 1.89E − 04 | Mile per hour (mi h$^{-1}$) |
| Velocity | Foot per hour (ft h$^{-1}$) | 3.16E − 06 | Mile per minute (mi min$^{-1}$) |
| Velocity | Foot per hour (ft h$^{-1}$) | 5.23E − 08 | Mile per second (mi s$^{-1}$) |
| Velocity | Foot per minute (ft min$^{-1}$) | 1.83E + 03 | Centimeter per hour (cm h$^{-1}$) |
| Velocity | Foot per minute (ft min$^{-1}$) | 3.05E + 01 | Centimeter per min (cm min$^{-1}$) |
| Velocity | Foot per minute (ft min$^{-1}$) | 5.08E − 01 | Centimeter per second (cm s$^{-1}$) |
| Velocity | Foot per minute (ft min$^{-1}$) | 6.00E + 01 | Foot per hour (ft h$^{-1}$) |

(Continued)

(Continued)

| Property | Multiply | By | To obtain |
|---|---|---|---|
| Velocity | Foot per minute (ft min$^{-1}$) | 1.67E − 02 | Foot per second (ft s$^{-1}$) |
| Velocity | Foot per minute (ft min$^{-1}$) | 1.83E − 02 | Kilometer per hour (km h$^{-1}$) |
| Velocity | Foot per minute (ft min$^{-1}$) | 3.05E − 04 | Kilometer per minute (km min$^{-1}$) |
| Velocity | Foot per minute (ft min$^{-1}$) | 5.08E − 06 | Kilometer per second (km s$^{-1}$) |
| Velocity | Foot per minute (ft min$^{-1}$) | 9.87E − 03 | Knots |
| Velocity | Foot per minute (ft min$^{-1}$) | 1.83E + 01 | Meter per hour (m h$^{-1}$) |
| Velocity | Foot per minute (ft min$^{-1}$) | 3.05E − 01 | Meter per minute (m min$^{-1}$) |
| Velocity | Foot per minute (ft min$^{-1}$) | 5.08E − 03 | Meter per second (m s$^{-1}$) |
| Velocity | Foot per minute (ft min$^{-1}$) | 1.14E − 02 | Mile per hour (mi h$^{-1}$) |
| Velocity | Foot per minute (ft min$^{-1}$) | 1.89E − 04 | Mile per minute (mi min$^{-1}$) |
| Velocity | Foot per minute (ft min$^{-1}$) | 3.16E − 06 | Mile per second (mi s$^{-1}$) |
| Velocity | Foot per second (ft s$^{-1}$) | 1.10E + 05 | Centimeter per hour (cm h$^{-1}$) |
| Velocity | Foot per second (ft s$^{-1}$) | 1.83E + 03 | Centimeter per min (cm min$^{-1}$) |
| Velocity | Foot per second (ft s$^{-1}$) | 3.05E + 01 | Centimeter per second (cm s$^{-1}$) |
| Velocity | Foot per second (ft s$^{-1}$) | 3.60E + 03 | Foot per hour (ft h$^{-1}$) |
| Velocity | Foot per second (ft s$^{-1}$) | 6.00E + 01 | Foot per minute (ft min$^{-1}$) |
| Velocity | Foot per second (ft s$^{-1}$) | 1.10E + 00 | Kilometer per hour (km h$^{-1}$) |
| Velocity | Foot per second (ft s$^{-1}$) | 1.83E − 02 | Kilometer per minute (km min$^{-1}$) |
| Velocity | Foot per second (ft s$^{-1}$) | 3.05E − 04 | Kilometer per second (km s$^{-1}$) |
| Velocity | Foot per second (ft s$^{-1}$) | 5.92E − 01 | Knots |
| Velocity | Foot per second (ft s$^{-1}$) | 1.10E + 03 | Meter per hour (m h$^{-1}$) |
| Velocity | Foot per second (ft s$^{-1}$) | 1.83E + 01 | Meter per minute (m min$^{-1}$) |
| Velocity | Foot per second (ft s$^{-1}$) | 3.05E − 01 | Meter per second (m s$^{-1}$) |
| Velocity | Foot per second (ft s$^{-1}$) | 6.82E − 01 | Mile per hour (mi h$^{-1}$) |
| Velocity | Foot per second (ft s$^{-1}$) | 1.14E − 02 | Mile per minute (mi min$^{-1}$) |
| Velocity | Foot per second (ft s$^{-1}$) | 1.89E − 04 | Mile per second (mi s$^{-1}$) |
| Velocity | Kilometer per hour (km h$^{-1}$) | 1.00E + 05 | Centimeter per hour (cm h$^{-1}$) |
| Velocity | Kilometer per hour (km h$^{-1}$) | 1.67E + 03 | Centimeter per min (cm min$^{-1}$) |
| Velocity | Kilometer per hour (km h$^{-1}$) | 2.78E + 01 | Centimeter per second (cm s$^{-1}$) |
| Velocity | Kilometer per hour (km h$^{-1}$) | 3.28E + 03 | Foot per hour (ft h$^{-1}$) |
| Velocity | Kilometer per hour (km h$^{-1}$) | 5.47E + 01 | Foot per minute (ft min$^{-1}$) |
| Velocity | Kilometer per hour (km h$^{-1}$) | 9.11E − 01 | Foot per second (ft s$^{-1}$) |
| Velocity | Kilometer per hour (km h$^{-1}$) | 1.67E − 02 | Kilometer per minute (km min$^{-1}$) |
| Velocity | Kilometer per hour (km h$^{-1}$) | 2.78E − 04 | Kilometer per second (km s$^{-1}$) |
| Velocity | Kilometer per hour (km h$^{-1}$) | 5.40E − 01 | Knots |
| Velocity | Kilometer per hour (km h$^{-1}$) | 1.00E + 03 | Meter per hour (m h$^{-1}$) |
| Velocity | Kilometer per hour (km h$^{-1}$) | 1.67E + 01 | Meter per minute (m min$^{-1}$) |

(Continued)

| Property | Multiply | By | To obtain |
|---|---|---|---|
| Velocity | Kilometer per hour (km h$^{-1}$) | 2.78E − 01 | Meter per second (m s$^{-1}$) |
| Velocity | Kilometer per hour (km h$^{-1}$) | 6.21E − 01 | Mile per hour (mi h$^{-1}$) |
| Velocity | Kilometer per hour (km h$^{-1}$) | 1.04E − 02 | Mile per minute (mi min$^{-1}$) |
| Velocity | Kilometer per hour (km h$^{-1}$) | 1.73E − 04 | Mile per second (mi s$^{-1}$) |
| Velocity | Kilometer per minute (km min$^{-1}$) | 6.00E + 06 | Centimeter per hour (cm h$^{-1}$) |
| Velocity | Kilometer per minute (km min$^{-1}$) | 1.00E + 05 | Centimeter per min (cm min$^{-1}$) |
| Velocity | Kilometer per minute (km min$^{-1}$) | 1.67E + 03 | Centimeter per second (cm s$^{-1}$) |
| Velocity | Kilometer per minute (km min$^{-1}$) | 1.97E + 05 | Foot per hour (ft h$^{-1}$) |
| Velocity | Kilometer per minute (km min$^{-1}$) | 3.28E + 03 | Foot per minute (ft min$^{-1}$) |
| Velocity | Kilometer per minute (km min$^{-1}$) | 5.47E + 01 | Foot per second (ft s$^{-1}$) |
| Velocity | Kilometer per minute (km min$^{-1}$) | 6.00E + 01 | Kilometer per hour (km h$^{-1}$) |
| Velocity | Kilometer per minute (km min$^{-1}$) | 1.67E − 02 | Kilometer per second (km s$^{-1}$) |
| Velocity | Kilometer per minute (km min$^{-1}$) | 3.24E + 01 | Knots |
| Velocity | Kilometer per minute (km min$^{-1}$) | 6.00E + 04 | Meter per hour (m h$^{-1}$) |
| Velocity | Kilometer per minute (km min$^{-1}$) | 1.00E + 03 | Meter per minute (m min$^{-1}$) |
| Velocity | Kilometer per minute (km min$^{-1}$) | 1.67E + 01 | Meter per second (m s$^{-1}$) |
| Velocity | Kilometer per minute (km min$^{-1}$) | 3.73E + 01 | Mile per hour (mi h$^{-1}$) |
| Velocity | Kilometer per minute (km min$^{-1}$) | 6.21E − 01 | Mile per minute (mi min$^{-1}$) |
| Velocity | Kilometer per minute (km min$^{-1}$) | 1.04E − 02 | Mile per second (mi s$^{-1}$) |
| Velocity | Kilometer per second (km s$^{-1}$) | 3.60E + 08 | Centimeter per hour (cm h$^{-1}$) |
| Velocity | Kilometer per second (km s$^{-1}$) | 6.00E + 06 | Centimeter per min (cm min$^{-1}$) |
| Velocity | Kilometer per second (km s$^{-1}$) | 1.00E + 05 | Centimeter per second (cm s$^{-1}$) |
| Velocity | Kilometer per second (km s$^{-1}$) | 1.18E + 07 | Foot per hour (ft h$^{-1}$) |
| Velocity | Kilometer per second (km s$^{-1}$) | 1.97E + 05 | Foot per minute (ft min$^{-1}$) |
| Velocity | Kilometer per second (km s$^{-1}$) | 3.28E + 03 | Foot per second (ft s$^{-1}$) |
| Velocity | Kilometer per second (km s$^{-1}$) | 3.60E + 03 | Kilometer per hour (km h$^{-1}$) |
| Velocity | Kilometer per second (km s$^{-1}$) | 6.00E + 01 | Kilometer per minute (km min$^{-1}$) |
| Velocity | Kilometer per second (km s$^{-1}$) | 1.94E + 03 | Knots (k) |

*(Continued)*

423

(Continued)

| Property | Multiply | By | To obtain |
|---|---|---|---|
| Velocity | Kilometer per second (km s$^{-1}$) | 3.60E + 06 | Meter per hour (m h$^{-1}$) |
| Velocity | Kilometer per second (km s$^{-1}$) | 6.00E + 04 | Meter per minute (m min$^{-1}$) |
| Velocity | Kilometer per second (km s$^{-1}$) | 1.00E + 03 | Meter per second (m s$^{-1}$) |
| Velocity | Kilometer per second (km s$^{-1}$) | 2.24E + 03 | Mile per hour (mi h$^{-1}$) |
| Velocity | Kilometer per second (km s$^{-1}$) | 3.73E + 01 | Mile per minute (mi min$^{-1}$) |
| Velocity | Kilometer per second (km s$^{-1}$) | 6.21E − 01 | Mile per second (mi s$^{-1}$) |
| Velocity | Knots (k) | 1.85E + 05 | Centimeter per hour (cm h$^{-1}$) |
| Velocity | Knots (k) | 3.09E + 03 | Centimeter per min (cm min$^{-1}$) |
| Velocity | Knots (k) | 5.14E + 01 | Centimeter per second (cm s$^{-1}$) |
| Velocity | Knots (k) | 6.08E + 03 | Foot per hour (ft h$^{-1}$) |
| Velocity | Knots (k) | 1.01E + 02 | Foot per minute (ft min$^{-1}$) |
| Velocity | Knots (k) | 1.69E + 00 | Foot per second (ft s$^{-1}$) |
| Velocity | Knots (k) | 3.09E − 02 | Kilometer per minute (km min$^{-1}$) |
| Velocity | Knots (k) | 1.85E + 00 | Kilometer per hour (km h$^{-1}$) |
| Velocity | Knots (k) | 5.14E − 04 | Kilometer per second (km s$^{-1}$) |
| Velocity | Knots (k) | 1.85E + 03 | Meter per hour (m h$^{-1}$) |
| Velocity | Knots (k) | 3.09E + 01 | Meter per minute (m min$^{-1}$) |
| Velocity | Knots (k) | 5.14E − 01 | Meter per second (m s$^{-1}$) |
| Velocity | Knots (k) | 1.15E + 00 | Mile per hour (mi h$^{-1}$) |
| Velocity | Knots (k) | 1.92E − 02 | Mile per minute (mi min$^{-1}$) |
| Velocity | Knots (k) | 3.20E − 04 | Mile per second (mi s$^{-1}$) |
| Velocity | Meter per hour (m h$^{-1}$) | 1.00E + 02 | Centimeter per hour (cm h$^{-1}$) |
| Velocity | Meter per hour (m h$^{-1}$) | 1.67E + 00 | Centimeter per min (cm min$^{-1}$) |
| Velocity | Meter per hour (m h$^{-1}$) | 2.78E − 02 | Centimeter per second (cm s$^{-1}$) |
| Velocity | Meter per hour (m h$^{-1}$) | 3.28E + 00 | Foot per hour (ft h$^{-1}$) |
| Velocity | Meter per hour (m h$^{-1}$) | 5.47E − 02 | Foot per minute (ft min$^{-1}$) |
| Velocity | Meter per hour (m h$^{-1}$) | 9.11E − 04 | Foot per second (ft s$^{-1}$) |
| Velocity | Meter per hour (m h$^{-1}$) | 1.00E − 03 | Kilometer per hour (km h$^{-1}$) |
| Velocity | Meter per hour (m h$^{-1}$) | 1.67E − 05 | Kilometer per minute (km min$^{-1}$) |
| Velocity | Meter per hour (m h$^{-1}$) | 2.78E − 07 | Kilometer per second (km s$^{-1}$) |
| Velocity | Meter per hour (m h$^{-1}$) | 5.40E − 04 | Knots (k) |
| Velocity | Meter per hour (m h$^{-1}$) | 1.67E − 02 | Meter per minute (m min$^{-1}$) |
| Velocity | Meter per hour (m h$^{-1}$) | 2.78E − 04 | Meter per second (m s$^{-1}$) |
| Velocity | Meter per hour (m h$^{-1}$) | 6.21E − 04 | Mile per hour (mi h$^{-1}$) |
| Velocity | Meter per hour (m h$^{-1}$) | 1.04E − 05 | Mile per minute (mi min$^{-1}$) |
| Velocity | Meter per hour (m h$^{-1}$) | 1.73E − 07 | Mile per second (mi s$^{-1}$) |

(Continued)

| Property | Multiply | By | To obtain |
|---|---|---|---|
| Velocity | Meter per minute (m min$^{-1}$) | 6.00E + 03 | Centimeter per hour (cm h$^{-1}$) |
| Velocity | Meter per minute (m min$^{-1}$) | 1.00E + 02 | Centimeter per min (cm min$^{-1}$) |
| Velocity | Meter per minute (m min$^{-1}$) | 1.67E + 00 | Centimeter per second (cm s$^{-1}$) |
| Velocity | Meter per minute (m min$^{-1}$) | 1.97E + 02 | Foot per hour (ft h$^{-1}$) |
| Velocity | Meter per minute (m min$^{-1}$) | 3.28E + 00 | Foot per minute (ft min$^{-1}$) |
| Velocity | Meter per minute (m min$^{-1}$) | 5.47E − 02 | Foot per second (ft s$^{-1}$) |
| Velocity | Meter per minute (m min$^{-1}$) | 1.00E − 03 | Kilometer per minute (km min$^{-1}$) |
| Velocity | Meter per minute (m min$^{-1}$) | 6.00E − 02 | Kilometer per hour (km h$^{-1}$) |
| Velocity | Meter per minute (m min$^{-1}$) | 1.67E − 05 | Kilometer per second (km s$^{-1}$) |
| Velocity | Meter per minute (m min$^{-1}$) | 3.24E − 02 | Knots |
| Velocity | Meter per minute (m min$^{-1}$) | 6.00E + 01 | Meter per hour (m h$^{-1}$) |
| Velocity | Meter per minute (m min$^{-1}$) | 1.67E − 02 | Meter per second (m s$^{-1}$) |
| Velocity | Meter per minute (m min$^{-1}$) | 3.73E − 02 | Mile per hour (mi h$^{-1}$) |
| Velocity | Meter per minute (m min$^{-1}$) | 6.21E − 04 | Mile per minute (mi min$^{-1}$) |
| Velocity | Meter per minute (m min$^{-1}$) | 1.04E − 05 | Mile per second (mi s$^{-1}$) |
| Velocity | Meter per second (m s$^{-1}$) | 3.60E + 05 | Centimeter per hour (cm h$^{-1}$) |
| Velocity | Meter per second (m s$^{-1}$) | 6.00E + 03 | Centimeter per min (cm min$^{-1}$) |
| Velocity | Meter per second (m s$^{-1}$) | 1.00E + 02 | Centimeter per second (cm s$^{-1}$) |
| Velocity | Meter per second (m s$^{-1}$) | 1.18E + 04 | Foot per hour (ft h$^{-1}$) |
| Velocity | Meter per second (m s$^{-1}$) | 1.97E + 02 | Foot per minute (ft min$^{-1}$) |
| Velocity | Meter per second (m s$^{-1}$) | 3.28E + 00 | Foot per second (ft s$^{-1}$) |
| Velocity | Meter per second (m s$^{-1}$) | 6.00E − 02 | Kilometer per minute (km min$^{-1}$) |
| Velocity | Meter per second (m s$^{-1}$) | 3.60E + 00 | Kilometer per hour (km h$^{-1}$) |
| Velocity | Meter per second (m s$^{-1}$) | 1.00E − 03 | Kilometer per second (km s$^{-1}$) |
| Velocity | Meter per second (m s$^{-1}$) | 1.95E + 00 | Knots |
| Velocity | Meter per second (m s$^{-1}$) | 3.60E + 03 | Meter per hour (m h$^{-1}$) |
| Velocity | Meter per second (m s$^{-1}$) | 6.00E + 01 | Meter per minute (m min$^{-1}$) |
| Velocity | Meter per second (m s$^{-1}$) | 2.24E + 00 | Mile per hour (mi h$^{-1}$) |
| Velocity | Meter per second (m s$^{-1}$) | 3.73E − 02 | Mile per minute (mi min$^{-1}$) |
| Velocity | Meter per second (m s$^{-1}$) | 6.21E − 04 | Mile per second (mi s$^{-1}$) |
| Velocity | Mile per minute (mi min$^{-1}$) | 9.66E + 06 | Centimeter per hour (cm h$^{-1}$) |
| Velocity | Mile per minute (mi min$^{-1}$) | 1.61E + 05 | Centimeter per min (cm min$^{-1}$) |
| Velocity | Mile per minute (mi min$^{-1}$) | 2.68E + 03 | Centimeter per second (cm s$^{-1}$) |
| Velocity | Mile per minute (mi min$^{-1}$) | 3.17E + 05 | Foot per hour (ft h$^{-1}$) |
| Velocity | Mile per minute (mi min$^{-1}$) | 5.28E + 03 | Foot per minute (ft min$^{-1}$) |
| Velocity | Mile per minute (mi min$^{-1}$) | 8.80E + 01 | Foot per second (ft s$^{-1}$) |
| Velocity | Mile per minute (mi min$^{-1}$) | 1.61E + 00 | Kilometer per minute (km min$^{-1}$) |
| Velocity | Mile per minute (mi min$^{-1}$) | 9.66E + 01 | Kilometer per hour (km h$^{-1}$) |

*(Continued)*

(Continued)

| Property | Multiply | By | To obtain |
|---|---|---|---|
| Velocity | Mile per minute (mi min$^{-1}$) | 2.68E − 02 | Kilometer per second (km s$^{-1}$) |
| Velocity | Mile per minute (mi min$^{-1}$) | 5.21E + 01 | Knots |
| Velocity | Mile per minute (mi min$^{-1}$) | 9.66E + 04 | Meter per hour (m h$^{-1}$) |
| Velocity | Mile per minute (mi min$^{-1}$) | 1.61E + 03 | Meter per minute (m min$^{-1}$) |
| Velocity | Mile per minute (mi min$^{-1}$) | 2.68E + 01 | Meter per second (m s$^{-1}$) |
| Velocity | Mile per minute (mi min$^{-1}$) | 1.67E − 02 | Mile per second (mi s$^{-1}$) |
| Velocity | Mile per minute (mi min$^{-1}$) | 6.00E + 01 | Mile per hour (mi h$^{-1}$ or mph) |
| Velocity | Mile per second (mi s$^{-1}$) | 5.79E + 08 | Centimeter per hour (cm h$^{-1}$) |
| Velocity | Mile per second (mi s$^{-1}$) | 9.66E + 06 | Centimeter per min (cm min$^{-1}$) |
| Velocity | Mile per second (mi s$^{-1}$) | 1.61E + 05 | Centimeter per second (cm s$^{-1}$) |
| Velocity | Mile per second (mi s$^{-1}$) | 1.90E + 07 | Foot per hour (ft h$^{-1}$) |
| Velocity | Mile per second (mi s$^{-1}$) | 3.17E + 05 | Foot per minute (ft min$^{-1}$) |
| Velocity | Mile per second (mi s$^{-1}$) | 5.28E + 03 | Foot per second (ft s$^{-1}$) |
| Velocity | Mile per second (mi s$^{-1}$) | 9.66E + 01 | Kilometer per minute (km min$^{-1}$) |
| Velocity | Mile per second (mi s$^{-1}$) | 5.79E + 03 | Kilometer per hour (km h$^{-1}$) |
| Velocity | Mile per second (mi s$^{-1}$) | 1.61E + 00 | Kilometer per second (km s$^{-1}$) |
| Velocity | Mile per second (mi s$^{-1}$) | 3.13E + 03 | Knots |
| Velocity | Mile per second (mi s$^{-1}$) | 5.79E + 06 | Meter per hour (m h$^{-1}$) |
| Velocity | Mile per second (mi s$^{-1}$) | 9.66E + 04 | Meter per minute (m min$^{-1}$) |
| Velocity | Mile per second (mi s$^{-1}$) | 1.61E + 03 | Meter per second (m s$^{-1}$) |
| Velocity | Mile per second (mi s$^{-1}$) | 6.00E + 01 | Mile per minute (mi min$^{-1}$) |
| Velocity | Mile per second (mi s$^{-1}$) | 3.60E + 03 | Mile per hour (mi h$^{-1}$ or mph) |
| Velocity | Mile per hour (mi h$^{-1}$ or mph) | 1.61E + 05 | Centimeter per hour (cm h$^{-1}$) |
| Velocity | Mile per hour (mi h$^{-1}$ or mph) | 2.68E + 03 | Centimeter per min (cm min$^{-1}$) |
| Velocity | Mile per hour (mi h$^{-1}$ or mph) | 4.47E + 01 | Centimeter per second (cm s$^{-1}$) |
| Velocity | Mile per hour (mi h$^{-1}$ or mph) | 5.28E + 03 | Foot per hour (ft h$^{-1}$) |
| Velocity | Mile per hour (mi h$^{-1}$ or mph) | 8.80E + 01 | Foot per minute (ft min$^{-1}$) |
| Velocity | Mile per hour (mi h$^{-1}$ or mph) | 1.47E + 00 | Foot per second (ft s$^{-1}$) |
| Velocity | Mile per hour (mi h$^{-1}$ or mph) | 2.68E − 02 | Kilometer per minute (km min$^{-1}$) |
| Velocity | Mile per hour (mi h$^{-1}$ or mph) | 1.61E + 00 | Kilometer per hour (km h$^{-1}$) |
| Velocity | Mile per hour (mi h$^{-1}$ or mph) | 4.47E − 04 | Kilometer per second (km s$^{-1}$) |
| Velocity | Mile per hour (mi h$^{-1}$ or mph) | 8.69E − 01 | Knots |
| Velocity | Mile per hour (mi h$^{-1}$ or mph) | 1.61E + 03 | Meter per hour (m h$^{-1}$) |
| Velocity | Mile per hour (mi h$^{-1}$ or mph) | 2.68E + 01 | Meter per minute (m min$^{-1}$) |

426

(Continued)

| Property | Multiply | By | To obtain |
|---|---|---|---|
| Velocity | Mile per hour (mi h$^{-1}$ or mph) | 4.47E − 01 | Meter per second (m s$^{-1}$) |
| Velocity | Mile per hour (mi h$^{-1}$ or mph) | 1.67E − 02 | Mile per minute (mi min$^{-1}$) |
| Velocity | Mile per hour (mi h$^{-1}$ or mph) | 2.78E − 04 | Mile per second (mi s$^{-1}$) |
| Volume | Acre-inch | 2.72E + 04 | US gallon (gal) |
| Volume | Acre-feet | 4.36E + 04 | Cubic feet (ft$^3$) |
| Volume | Acre-feet | 7.53E + 07 | Cubic inch (in$^3$) |
| Volume | Acre-feet | 1.61E + 03 | Cubic yard (yd$^3$) |
| Volume | Acre-feet | 1.23E − 01 | Hectare-meter (ha m) |
| Volume | Acre-feet | 1.23E + 06 | Liter (L) |
| Volume | Acre-feet | 1.23E + 06 | Liters (L) |
| Volume | Acre-feet | 1.23E + 03 | Meter cubed (m$^3$) |
| Volume | Acre-feet | 3.26E + 05 | US gallon (gal) |
| Volume | Acre-feet | 3.26E + 05 | US gallon (gal) |
| Volume | Barrel | 1.23E + 03 | US gallon (gal) |
| Volume | Centiliter (cL) | 3.38E − 01 | Fluid ounce (fl oz) |
| Volume | Cup | 8.00E + 00 | Fluid ounce (fl oz) |
| Volume | Cup | 2.40E + 02 | Milliliter (mL) |
| Volume | Cup | 1.60E + 01 | Tablespoon (tbls) |
| Volume | Cup | 4.80E + 01 | Teaspoon (tsp) |
| Volume | Dram (dr) | 6.25E − 02 | Ounce (oz) |
| Volume | Feet cubed of water (ft$^3_{water}$) | 6.23E + 01 | lb of water (lb) |
| Volume | Fluid ounce (fl oz) | 3.00E + 01 | Milliliter (mL) |
| Volume | Fluid ounce (fl oz) | 2.00E + 00 | Tablespoon (tbls) |
| Volume | Fluid ounce (fl oz) | 6.00E + 00 | Teaspoon (tsp) |
| Volume | Foot cubed (ft$^3$) | 2.83E + 01 | Liter (L) |
| Volume | Foot cubed (ft$^3$) | 7.48E + 00 | US gallon (gal) |
| Volume | Gallon (gal) | 3.07E − 06 | Acre-feet (ac ft) |
| Volume | Gallon (gal) | 3.68E − 05 | Acre-inch (ac in.) |
| Volume | Gallon (gal) | 2.38E − 02 | Barrel (bbl) |
| Volume | Gallon (gal) | 1.60E + 01 | Cup |
| Volume | Gallon (gal) | 1.28E + 02 | Fluid ounce (fl oz) |
| Volume | Gallon (gal) | 1.34E − 01 | Foot cubed (ft$^3$) |
| Volume | Gallon (gal) | 3.79E − 07 | Hectare-meter (ha m) |
| Volume | Gallon (gal) | 2.31E + 02 | Inch cubed (in.$^3$) |
| Volume | Gallon (gal) | 3.79E + 00 | Liter (L) |
| Volume | Gallon (gal) | 3.79E − 03 | Meter cubed (m$^3$) |
| Volume | Gallon (gal) | 3.79E + 03 | Milliliter (mL) |
| Volume | Gallon (gal) | 8.00E + 00 | Pint (pt) |
| Volume | Gallon (gal) | 4.00E + 00 | Quart (qt) |
| Volume | Gallon (gal) | 4.95E − 03 | Yard cubed (yd$^3$) |
| Volume | Hectare-meter | 8.10E + 00 | Acre-feet (ac ft) |
| Volume | Hectare-meter | 2.64E + 06 | Gallon (gal) |
| Volume | Imperial gallon | 4.55E + 00 | Liter (L) |
| Volume | Inch cubed (in.$^3$) | 1.33E − 08 | Acre-feet (ac ft) |
| Volume | Inch cubed (in.$^3$) | 1.64E + 01 | Centimeter cubed (cm$^3$) |
| Volume | Inch cubed (in.$^3$) | 5.79E − 04 | Foot cubed (ft$^3$) |
| Volume | Inch cubed (in.$^3$) | 1.64E − 02 | Liter (L) |
| Volume | Inch cubed (in.$^3$) | 1.64E − 05 | Meter cubed (m$^3$) |
| Volume | Inch cubed (in.$^3$) | 4.33E − 03 | US gallon (gal) |
| Volume | Inch cubed (in.$^3$) | 2.14E − 05 | Yard cubed (yd$^3$) |
| Volume | Liter (L) | 8.13E − 07 | Acre-feet (ac ft) |

(Continued)

| Property | Multiply | By | To obtain |
|---|---|---|---|
| Volume | Liter (L) | 8.11E − 07 | Acre-feet (ac ft) |
| Volume | Liter (L) | 6.10E + 01 | Cubic inch (in.$^3$) |
| Volume | Liter (L) | 1.31E − 03 | Cubic yard (yd$^3$) |
| Volume | Liter (L) | 2.11E + 00 | Fluid pint (fl pt) |
| Volume | Liter (L) | 1.06E + 00 | Fluid quart (fl qt) |
| Volume | Liter (L) | 3.53E − 02 | Foot cubed (ft$^3$) |
| Volume | Liter (L) | 1.00E − 03 | Meter cubed (m$^3$) |
| Volume | Liter (L) | 1.00E − 03 | Meter cubed (m$^3$) |
| Volume | Liter (L) | 3.38E + 01 | Ounce (oz) |
| Volume | Liter (L) | 2.64E − 01 | US gallon (gal) |
| Volume | Liter (L) | 1.31E − 03 | Yard cubed (yd$^3$) |
| Volume | Liter of water | 1.00E + 03 | Gram of water (g) |
| Volume | Meter cubed (m$^3$) | 8.13E − 04 | Acre-feet (ac ft) |
| Volume | Meter cubed (m$^3$) | 8.11E − 04 | Acre-feet (ac ft) |
| Volume | Meter cubed (m$^3$) | 3.53E + 01 | Cubic foot (ft$^3$) |
| Volume | Meter cubed (m$^3$) | 6.10E + 04 | Cubic inch (in.$^3$) |
| Volume | Meter cubed (m$^3$) | 1.31E + 00 | Cubic yard (yd$^3$) |
| Volume | Meter cubed (m$^3$) | 1.00E + 03 | Decimeter cubed (dm$^3$) |
| Volume | Meter cubed (m$^3$) | 2.27E + 02 | Dry gallon (dry gal) |
| Volume | Meter cubed (m$^3$) | 1.00E + 03 | Liter (L) |
| Volume | Meter cubed (m$^3$) | 1.00E + 03 | Liter (L) |
| Volume | Meter cubed (m$^3$) | 2.64E + 02 | US gallon (gal) |
| Volume | Meter cubed (m$^3$) | 2.64E + 02 | US gallon (gal) |
| Volume | Milliliter (mL) | 3.40E − 02 | Fluid ounce (fl oz) |
| Volume | Peck | 2.50E − 01 | Bushel (bu) |
| Volume | Peck | 8.00E + 00 | Quart (qt) |
| Volume | Pennyweight (dwt) | 5.49E − 02 | Ounce (oz) |
| Volume | Pica (p) | 1.20E + 01 | Point (pt) |
| Volume | Pint | 2.00E + 00 | Cups |
| Volume | Pint | 1.60E + 01 | Fluid ounce (fl oz) |
| Volume | Pint | 4.70E + 02 | Milliliter (mL) |
| Volume | Pint | 5.00E − 01 | Quarts (qt) |
| Volume | Pound of water | 1.60E − 02 | Cubic feet of water (ft$^3$) |
| Volume | Quart | 1.25E − 01 | Peck |
| Volume | Quart | 2.00E + 00 | Pint |
| Volume | Quart (qt) | 4.00E + 00 | Cup |
| Volume | Quart (qt) | 3.20E + 01 | Fluid ounce (fl oz) |
| Volume | Quart (qt) | 9.50E + 02 | Milliliter (mL) |
| Volume | Tablespoon (tbls) | 1.50E + 01 | Milliliter (mL) |
| Volume | Tablespoon (tbls) | 3.00E + 00 | Teaspoon (tsp) |
| Volume | Teaspoon (tsp) | 5.00E + 00 | Milliliter (mL) |
| Volume | US gallon (gal) | 3.79E + 00 | Liter (L) |
| Volume | Yard cubed (yd$^3$) | 6.20E − 04 | Acre-feet (ac ft) |
| Volume | Yard cubed (yd$^3$) | 2.70E + 01 | Foot cubed (ft$^3$) |
| Volume | Yard cubed (yd$^3$) | 4.67E + 04 | Inch cubed (in.$^3$) |
| Volume | Yard cubed (yd$^3$) | 7.65E + 02 | Liter (L) |
| Volume | Yard cubed (yd$^3$) | 7.65E − 01 | Meter cubed (m$^3$) |
| Volume | Yard cubed (yd$^3$) | 2.02E + 02 | US gallon (gal) |
| Weight | Ounce (oz) | 2.83E + 01 | Gram (gm) |
| Weight | Ounce (oz) | 2.84E − 02 | Kilogram (kg) |
| Weight | Ounce (oz) | 2.96E − 02 | Liter (L) |
| Weight | Ounce (oz) | 2.79E − 05 | Long ton |
| Weight | Ounce (oz) | 2.89E − 03 | Metric slug |
| Weight | Ounce (oz) | 2.84E − 05 | Metric ton (tonne) |
| Weight | Ounce (oz) | 2.96E + 01 | Milliliter (mL) |
| Weight | Ounce (oz) | 6.25E − 02 | Pound (lb) |

| Property | Multiply | By | To obtain |
|---|---|---|---|
| Weight | Ounce (oz) | 3.13E − 05 | Short ton |
| Weight | Ounce (oz) | 1.94E − 03 | Slug |
| Weight | Pound (lb) | 4.46E − 04 | Long ton |
| Weight | Pound (lb) | 1.60E + 01 | Ounce (oz) |
| Weight | Stone (st) | 1.40E + 01 | Pound (lb) |
| Weight | Ton (t) | 2.00E + 03 | Pound (lb) |
| Weight | Tonne (t) | 2.20E + 03 | Pound (lb) |
| Weight | Tonne (t) | 1.10E + 00 | Tons |

## Appendix A-1b: Conversions by Unit

| Multiply | By | To obtain | Property |
|---|---|---|---|
| Centimeter per second squared (cm s$^{-2}$) | 3.28E − 02 | Foot per second squared (ft s$^{-2}$) | Acceleration |
| Centimeter per second squared (cm s$^{-2}$) | 1.00E − 04 | Kilometer per second squared (km s$^{-2}$) | Acceleration |
| Centimeter per second squared (cm s$^{-2}$) | 1.00E − 02 | Meter per second squared (m s$^{-2}$) | Acceleration |
| Foot per second squared (ft s$^{-2}$) | 3.05E + 01 | Centimeter per second squared (cm s$^{-2}$) | Acceleration |
| Foot per second squared (ft s$^{-2}$) | 3.05E − 04 | Kilometer per second squared (km s$^{-2}$) | Acceleration |
| Foot per second squared (ft s$^{-2}$) | 3.05E − 01 | Meter per second squared (m s$^{-2}$) | Acceleration |
| Kilometer per second squared (km s$^{-2}$) | 1.00E + 05 | Centimeter per second squared (cm s$^{-2}$) | Acceleration |
| Kilometer per second squared (km s$^{-2}$) | 3.28E + 03 | Foot per second squared (ft s$^{-2}$) | Acceleration |
| Kilometer per second squared (km s$^{-2}$) | 1.00E + 03 | Meter per second squared (m s$^{-2}$) | Acceleration |
| Meter per second squared (m s$^{-2}$) | 1.00E + 02 | Centimeter per second squared (cm s$^{-2}$) | Acceleration |
| Meter per second squared (m s$^{-2}$) | 3.28E + 00 | Foot Per second squared (ft s$^{-2}$) | Acceleration |
| Meter per second squared (m s$^{-2}$) | 1.00E − 03 | Kilometer per second squared (km s$^{-2}$) | Acceleration |
| Acre (A) | 4.05E + 07 | Centimeter squared (cm$^2$) | Area |
| Acre (A) | 4.36E + 04 | Foot squared (ft$^2$) | Area |
| Acre (A) | 4.05E − 01 | Hectare (ha) | Area |
| Acre (A) | 6.27E + 06 | Inch squared (in.$^2$) | Area |
| Acre (A) | 4.05E − 03 | Kilometer squared (km$^2$) | Area |
| Acre (A) | 4.05E + 03 | Meter squared (m$^2$) | Area |
| Acre (A) | 1.56E − 03 | Mile squared (mi$^2$) | Area |
| Acre (A) | 4.05E + 09 | Millimeter squared (mm$^2$) | Area |
| Acre (A) | 4.84E + 03 | Yard squared (yd$^2$) | Area |
| Centimeter squared (cm$^2$) | 2.47E − 08 | Acre (ac) | Area |
| Centimeter squared (cm$^2$) | 1.08E − 03 | Foot squared (ft$^2$) | Area |

(Continued)

(Continued)

| Multiply | By | To obtain | Property |
|---|---|---|---|
| Centimeter squared (cm$^2$) | 1.00E − 09 | Hectare (ha) | Area |
| Centimeter squared (cm$^2$) | 1.55E − 01 | Inch squared (in.$^2$) | Area |
| Centimeter squared (cm$^2$) | 1.00E − 09 | Kilometer squared (km$^2$) | Area |
| Centimeter squared (cm$^2$) | 1.00E − 03 | Meter squared (m$^2$) | Area |
| Centimeter squared (cm$^2$) | 3.86E − 11 | Mile squared (mi$^2$) | Area |
| Centimeter squared (cm$^2$) | 1.00E + 02 | Millimeter squared (mm$^2$) | Area |
| Centimeter squared (cm$^2$) | 6.06E − 10 | Yard squared (yd$^2$) | Area |
| Foot squared (ft$^2$) | 2.30E − 05 | Acre (ac) | Area |
| Foot squared (ft$^2$) | 9.29E + 02 | Centimeter squared (cm$^2$) | Area |
| Foot squared (ft$^2$) | 9.29E − 06 | Hectare (ha) | Area |
| Foot squared (ft$^2$) | 1.44E + 02 | Inch squared (in.$^2$) | Area |
| Foot squared (ft$^2$) | 9.29E − 08 | Kilometer squared (km$^2$) | Area |
| Foot squared (ft$^2$) | 9.29E − 02 | Meter squared (m$^2$) | Area |
| Foot squared (ft$^2$) | 3.59E − 08 | Mile squared (mi$^2$) | Area |
| Foot squared (ft$^2$) | 9.29E + 04 | Millimeter squared (mm$^2$) | Area |
| Foot squared (ft$^2$) | 1.10E − 01 | Yard squared (yd$^2$) | Area |
| Hectare (ha) | 2.47E + 00 | Acre (ac) | Area |
| Hectare (ha) | 1.00E + 09 | Centimeter squared (cm$^2$) | Area |
| Hectare (ha) | 1.08E + 05 | Foot squared (ft$^2$) | Area |
| Hectare (ha) | 1.55E + 07 | Inch squared (in.$^2$) | Area |
| Hectare (ha) | 1.00E − 02 | Kilometer squared (km$^2$) | Area |
| Hectare (ha) | 1.00E + 04 | Meter squared (m$^2$) | Area |
| Hectare (ha) | 3.86E − 03 | Mile squared (mi$^2$) | Area |
| Hectare (ha) | 1.00E + 11 | Millimeter squared (mm$^2$) | Area |
| Hectare (ha) | 1.20E + 04 | Yard squared (yd$^2$) | Area |
| Inch squared (in.$^2$) | 1.59E − 07 | Acre (ac) | Area |
| Inch squared (in.$^2$) | 6.45E + 00 | Centimeter squared (cm$^2$) | Area |
| Inch squared (in.$^2$) | 6.94E − 03 | Foot squared (ft$^2$) | Area |
| Inch squared (in.$^2$) | 6.45E − 08 | Hectare (ha) | Area |
| Inch squared (in.$^2$) | 6.45E − 10 | Kilometer squared (km$^2$) | Area |
| Inch squared (in.$^2$) | 6.45E − 04 | Meter squared (m$^2$) | Area |
| Inch squared (in.$^2$) | 2.49E − 10 | Mile squared (mi$^2$) | Area |
| Inch squared (in.$^2$) | 6.45E + 02 | Millimeter squared (mm$^2$) | Area |
| Inch squared (in.$^2$) | 7.72E − 04 | Yard squared (yd$^2$) | Area |
| Kilometer squared (km$^2$) | 2.47E + 02 | Acre (ac) | Area |
| Kilometer squared (km$^2$) | 1.00E + 09 | Centimeter squared (cm$^2$) | Area |
| Kilometer squared (km$^2$) | 1.08E + 07 | Foot squared (ft$^2$) | Area |
| Kilometer squared (km$^2$) | 1.00E + 02 | Hectare (ha) | Area |
| Kilometer squared (km$^2$) | 1.55E + 09 | Inch squared (in$^2$) | Area |
| Kilometer squared (km$^2$) | 1.00E + 06 | Meter squared (m$^2$) | Area |
| Kilometer squared (km$^2$) | 3.86E − 01 | Mile squared (mi$^2$) | Area |
| Kilometer squared (km$^2$) | 1.00E + 13 | Millimeter squared (mm$^2$) | Area |
| Kilometer squared (km$^2$) | 1.20E + 06 | Yard squared (yd$^2$) | Area |
| Meter squared (m$^2$) | 2.47E − 04 | Acre (ac) | Area |
| Meter squared (m$^2$) | 1.00E + 04 | Centimeter squared (cm$^2$) | Area |
| Meter squared (m$^2$) | 1.08E + 01 | Foot squared (ft$^2$) | Area |
| Meter squared (m$^2$) | 1.00E − 04 | Hectare (ha) | Area |
| Meter squared (m$^2$) | 1.55E + 03 | Inch squared (in.$^2$) | Area |
| Meter squared (m$^2$) | 1.00E − 06 | Kilometer squared (km$^2$) | Area |
| Meter squared (m$^2$) | 3.86E − 07 | Mile squared (mi$^2$) | Area |
| Meter squared (m$^2$) | 1.00E + 07 | Millimeter squared (mm$^2$) | Area |
| Meter squared (m$^2$) | 3.95E − 02 | Rod squared (rd$^2$) | Area |
| Meter squared (m$^2$) | 1.20E + 00 | Yard squared (yd$^2$) | Area |
| Mile squared (mi$^2$) | 6.40E + 02 | Acre (ac) | Area |
| Mile squared (mi$^2$) | 2.59E + 10 | Centimeter squared (cm$^2$) | Area |
| Mile squared (mi$^2$) | 2.79E + 07 | Foot squared (ft$^2$) | Area |

(Continued)

| Multiply | By | To obtain | Property |
|---|---|---|---|
| Mile squared (mi$^2$) | 2.59E + 02 | Hectare (ha) | Area |
| Mile squared (mi$^2$) | 4.01E + 09 | Inch squared (in.$^2$) | Area |
| Mile squared (mi$^2$) | 2.59E + 00 | Kilometer squared (km$^2$) | Area |
| Mile squared (mi$^2$) | 2.59E + 06 | Meter squared (m$^2$) | Area |
| Mile squared (mi$^2$) | 2.59E + 12 | Millimeter squared (mm$^2$) | Area |
| Mile squared (mi$^2$) | 3.10E + 06 | Yard squared (yd$^2$) | Area |
| Millimeter squared (mm$^2$) | 2.47E − 10 | Acre (ac) | Area |
| Millimeter squared (mm$^2$) | 1.00E − 02 | Centimeter squared (cm$^2$) | Area |
| Millimeter squared (mm$^2$) | 1.08E − 05 | Foot squared (ft$^2$) | Area |
| Millimeter squared (mm$^2$) | 1.00E − 09 | Hectare (ha) | Area |
| Millimeter squared (mm$^2$) | 1.50E − 02 | Inch squared (in.$^2$) | Area |
| Millimeter squared (mm$^2$) | 1.00E − 11 | Kilometer squared (km$^2$) | Area |
| Millimeter squared (mm$^2$) | 1.00E − 05 | Meter squared (m$^2$) | Area |
| Millimeter squared (mm$^2$) | 3.86E − 13 | Mile squared (mi$^2$) | Area |
| Millimeter squared (mm$^2$) | 1.20E − 06 | Yard squared (yd$^2$) | Area |
| Rod squared (rd$^2$) | 2.72E + 02 | Foot squared (ft$^2$) | Area |
| Rod squared (rd$^2$) | 2.53E + 01 | Meter squared (m$^2$) | Area |
| Yard squared (yd$^2$) | 1.30E + 03 | Inch squared (in.$^2$) | Area |
| Yard squared (yd$^2$) | 2.07E − 04 | Acre | Area |
| Yard squared (yd$^2$) | 8.36E + 03 | Centimeter squared (cm$^2$) | Area |
| Yard squared (yd$^2$) | 9.00E + 00 | Foot squared (ft$^2$) | Area |
| Yard squared (yd$^2$) | 8.36E − 05 | Hectare (ha) | Area |
| Yard squared (yd$^2$) | 8.36E − 07 | Kilometer squared (km$^2$) | Area |
| Yard squared (yd$^2$) | 8.36E − 01 | Meter squared (m$^2$) | Area |
| Yard squared (yd$^2$) | 3.23E − 07 | Mile squared (mi$^2$) | Area |
| Yard squared (yd$^2$) | 8.36E + 05 | Millimeter squared (mm$^2$) | Area |
| Pound per million gal. of H$_2$O | 1.20E − 01 | Milligram per liter (mg L$^{-1}$, ppm) | Concentration |
| Gram per cubic centimeter (g cm$^{-3}$) | 1.00E + 03 | Gram per liter (g L$^{-1}$) | Density |
| Gram per cubic centimeter (g cm$^{-3}$) | 1.00E + 00 | Gram per milliliter (g mL$^{-1}$) | Density |
| Gram per cubic centimeter (g cm$^{-3}$) | 1.00E + 00 | Kilogram per decimeter cubed (kg dm$^{-3}$) | Density |
| Gram per cubic centimeter (g cm$^{-3}$) | 1.00E + 03 | Kilogram per meter cubed (kg m$^{-3}$) | Density |
| Gram per cubic centimeter (g cm$^{-3}$) | 6.24E + 01 | Pound per foot cubed (lb ft$^{-3}$) | Density |
| Gram per cubic centimeter (g cm$^{-3}$) | 1.00E + 01 | Pound per gallon (Brit) (lb gal$^{-1}$) | Density |
| Gram per cubic centimeter (g cm$^{-3}$) | 8.35E + 00 | Pound per gallon (US) (lb gal$^{-1}$) | Density |
| Gram per cubic centimeter (g cm$^{-3}$) | 3.61E − 02 | Pound per inch cubed (lb in.$^{-3}$) | Density |
| Gram per cubic centimeter (g cm$^{-3}$) | 1.69E + 03 | Pound per yard cubed (lb yd$^{-3}$) | Density |
| Gram per cubic centimeter (g cm$^{-3}$) | 1.94E + 00 | Slug per foot cubed (slug ft$^{-3}$) | Density |
| Gram per liter (g L$^{-1}$) | 1.00E − 03 | Gram per centimeter cubed (g cm$^{-3}$) | Density |
| Gram per liter (g L$^{-1}$) | 1.00E − 03 | Gram per milliliter (g mL$^{-1}$) | Density |
| Gram per liter (g L$^{-1}$) | 1.00E − 03 | Kilogram per decimeter cubed (kg dm$^{-3}$) | Density |
| Gram per liter (g L$^{-1}$) | 1.00E + 00 | Kilogram per meter cubed (kg m$^{-3}$) | Density |
| Gram per liter (g L$^{-1}$) | 6.24E − 02 | Pound per foot cubed (lb ft$^{-3}$) | Density |

(Continued)

(Continued)

| Multiply | By | To obtain | Property |
|---|---|---|---|
| Gram per liter (g L$^{-1}$) | 1.00E − 02 | Pound per gallon (Brit) (lb gal$^{-1}$) | Density |
| Gram per liter (g L$^{-1}$) | 8.35E − 03 | Pound per gallon (US) (lb gal$^{-1}$) | Density |
| Gram per liter (g L$^{-1}$) | 3.61E − 05 | Pound per inch cubed (lb in.$^{-3}$) | Density |
| Gram per liter (g L$^{-1}$) | 1.69E + 00 | Pound per yard cubed (lb yd$^{-3}$) | Density |
| Gram per liter (g L$^{-1}$) | 1.94E − 03 | Slug per foot cubed (slug ft$^{-3}$) | Density |
| Gram per milliliter (g mL$^{-1}$) | 1.00E + 00 | Gram per centimeter cubed (g cm$^{-3}$) | Density |
| Gram per milliliter (g mL$^{-1}$) | 1.00E + 03 | Gram per liter (g L$^{-1}$) | Density |
| Gram per milliliter (g mL$^{-1}$) | 1.00E + 00 | Kilogram per decimeter cubed (kg dm$^{-3}$) | Density |
| Gram per milliliter (g mL$^{-1}$) | 1.00E + 03 | Kilogram per meter cubed (kg m$^{-3}$) | Density |
| Gram per milliliter (g mL$^{-1}$) | 6.24E + 01 | Pound per foot cubed (lb ft$^{-3}$) | Density |
| Gram per milliliter (g mL$^{-1}$) | 1.00E + 01 | Pound per gallon (Brit) (lb gal$^{-1}$) | Density |
| Gram per milliliter (g mL$^{-1}$) | 8.35E + 00 | Pound per gallon (US) (lb gal$^{-1}$) | Density |
| Gram per milliliter (g mL$^{-1}$) | 3.61E − 02 | Pound per inch cubed (lb in.$^{-3}$) | Density |
| Gram per milliliter (g mL$^{-1}$) | 1.69E + 03 | Pound per yard cubed (lb yd$^{-3}$) | Density |
| Gram per milliliter (g mL$^{-1}$) | 1.94E + 00 | Slug per foot cubed (slug ft$^{-3}$) | Density |
| Kilogram per decimeter cubed (kg dm$^{-3}$) | 1.00E + 00 | Gram per centimeter cubed (g cm$^{-3}$) | Density |
| Kilogram per decimeter cubed (kg dm$^{-3}$) | 1.00E − 03 | Gram per liter (g L$^{-1}$) | Density |
| Kilogram per decimeter cubed (kg dm$^{-3}$) | 1.00E + 00 | Gram per milliliter (g mL$^{-1}$) | Density |
| Kilogram per decimeter cubed (kg dm$^{-3}$) | 1.00E − 03 | Kilogram per meter cubed (kg m$^{-3}$) | Density |
| Kilogram per decimeter cubed (kg dm$^{-3}$) | 1.60E − 02 | Pound per foot cubed (lb ft$^{-3}$) | Density |
| Kilogram per decimeter cubed (kg dm$^{-3}$) | 9.98E − 02 | Pound per gallon (Brit) (lb gal$^{-1}$) | Density |
| Kilogram per decimeter cubed (kg dm$^{-3}$) | 1.20E − 01 | Pound per gallon (US) (lb gal$^{-1}$) | Density |
| Kilogram per decimeter cubed (kg dm$^{-3}$) | 2.77E + 01 | Pound per inch cubed (lb in.$^{-3}$) | Density |
| Kilogram per decimeter cubed (kg dm$^{-3}$) | 5.93E − 04 | Pound per yard cubed (lb yd$^{-3}$) | Density |
| Kilogram per decimeter cubed (kg dm$^{-3}$) | 5.15E − 01 | Slug per foot cubed (slug ft$^{-3}$) | Density |
| Kilogram per meter cubed (kg m$^{-3}$) | 1.00E − 03 | Gram per centimeter cubed (g cm$^{-3}$) | Density |
| Kilogram per meter cubed (kg m$^{-3}$) | 1.00E + 00 | Gram per liter (g L$^{-1}$) | Density |
| Kilogram per meter cubed (kg m$^{-3}$) | 1.00E − 03 | Gram per milliliter (g mL$^{-1}$) | Density |
| Kilogram per meter cubed (kg m$^{-3}$) | 1.00E − 03 | Kilogram per decimeter cubed (kg dm$^{-3}$) | Density |
| Kilogram per meter cubed (kg m$^{-3}$) | 6.24E − 02 | Pound per foot cubed (lb ft$^{-3}$) | Density |
| Kilogram per meter cubed (kg m$^{-3}$) | 1.00E − 02 | Pound per gallon (Brit) (lb gal$^{-1}$) | Density |
| Kilogram per meter cubed (kg m$^{-3}$) | 8.35E − 03 | Pound per gallon (US) (lb gal$^{-1}$) | Density |
| Kilogram per meter cubed (kg m$^{-3}$) | 3.61E − 05 | Pound per inch cubed (lb in.$^{-3}$) | Density |
| Kilogram per meter cubed (kg m$^{-3}$) | 1.69E + 00 | Pound per yard cubed (lb yd$^{-3}$) | Density |

| Multiply | By | To obtain | Property |
|---|---|---|---|
| Kilogram per meter cubed (kg m$^{-3}$) | 1.94E − 03 | Slug per foot cubed (slug ft$^{-3}$) | Density |
| Pound per foot cubed (lb ft$^{-3}$) | 1.60E − 03 | Gram per centimeter cubed (g cm$^{-3}$) | Density |
| Pound per foot cubed (lb ft$^{-3}$) | 1.60E + 01 | Gram per liter (g L$^{-1}$) | Density |
| Pound per foot cubed (lb ft$^{-3}$) | 1.60E − 02 | Gram per milliliter (g mL$^{-1}$) | Density |
| Pound per foot cubed (lb ft$^{-3}$) | 1.60E − 02 | Gram per milliliter (g mL$^{-1}$) | Density |
| Pound per foot cubed (lb ft$^{-3}$) | 1.60E − 02 | Kilogram per decimeter cubed (kg dm$^{-3}$) | Density |
| Pound per foot cubed (lb ft$^{-3}$) | 1.60E + 01 | Kilogram per meter cubed (kg m$^{-3}$) | Density |
| Pound per foot cubed (lb ft$^{-3}$) | 1.61E − 01 | Pound per gallon (Brit) (lb gal$^{-1}$) | Density |
| Pound per foot cubed (lb ft$^{-3}$) | 1.34E − 01 | Pound per gallon (US) (lb gal$^{-1}$) | Density |
| Pound per foot cubed (lb ft$^{-3}$) | 5.79E − 04 | Pound per inch cubed (lb in.$^{-3}$) | Density |
| Pound per foot cubed (lb ft$^{-3}$) | 2.70E + 01 | Pound per yard cubed (lb yd$^{-3}$) | Density |
| Pound per foot cubed (lb ft$^{-3}$) | 3.11E − 02 | Slug per foot cubed (slug ft$^{-3}$) | Density |
| Pound per gallon (Brit) (lb gal$^{-1}$) | 9.98E − 02 | Gram per centimeter cubed (g cm$^{-3}$) | Density |
| Pound per gallon (Brit) (lb gal$^{-1}$) | 9.98E + 01 | Gram per liter (g L$^{-1}$) | Density |
| Pound per gallon (Brit) (lb gal$^{-1}$) | 9.98E − 02 | Gram per milliliter (g mL$^{-1}$) | Density |
| Pound per gallon (Brit) (lb gal$^{-1}$) | 9.98E − 02 | Kilogram per decimeter cubed (kg dm$^{-3}$) | Density |
| Pound per gallon (Brit) (lb gal$^{-1}$) | 9.98E + 01 | Kilogram per meter cubed (kg m$^{-3}$) | Density |
| Pound per gallon (Brit) (lb gal$^{-1}$) | 6.23E + 00 | Pound per foot cubed (lb ft$^{-3}$) | Density |
| Pound per gallon (Brit) (lb gal$^{-1}$) | 8.33E − 01 | Pound per gallon (US) (lb gal$^{-1}$) | Density |
| Pound per gallon (Brit) (lb gal$^{-1}$) | 3.60E + 00 | Pound per inch cubed (lb in.$^{-3}$) | Density |
| Pound per gallon (Brit) (lb gal$^{-1}$) | 1.68E + 02 | Pound per yard cubed (lb yd$^{-3}$) | Density |
| Pound per gallon (Brit) (lb gal$^{-1}$) | 1.94E − 01 | Slug per foot cubed (slug ft$^{-3}$) | Density |
| Pound per gallon (US) (lb gal$^{-1}$) | 1.20E − 01 | Gram per centimeter cubed (g cm$^{-3}$) | Density |
| Pound per gallon (US) (lb gal$^{-1}$) | 1.20E + 02 | Gram per liter (g L$^{-1}$) | Density |
| Pound per gallon (US) (lb gal$^{-1}$) | 1.20E − 01 | Gram per milliliter (g mL$^{-1}$) | Density |
| Pound per gallon (US) (lb gal$^{-1}$) | 1.20E − 01 | Kilogram per decimeter cubed (kg dm$^{-3}$) | Density |
| Pound per gallon (US) (lb gal$^{-1}$) | 1.20E + 02 | Kilogram per meter cubed (kg m$^{-3}$) | Density |
| Pound per gallon (US) (lb gal$^{-1}$) | 7.48E + 00 | Pound per foot cubed (lb ft$^{-3}$) | Density |
| Pound per gallon (US) (lb gal$^{-1}$) | 1.20E + 00 | Pound per gallon (Brit) (lb gal$^{-1}$) | Density |
| Pound per gallon (US) (lb gal$^{-1}$) | 4.33E − 03 | Pound per inch cubed (lb in.$^{-3}$) | Density |
| Pound per gallon (US) (lb gal$^{-1}$) | 2.02E + 02 | Pound per yard cubed (lb yd$^{-3}$) | Density |
| Pound per gallon (US) (lb gal$^{-1}$) | 2.33E − 01 | Slug per foot cubed (slug ft$^{-3}$) | Density |
| Pound per inch cubed (lb in.$^{-3}$) | 2.77E + 01 | Gram per centimeter cubed (g cm$^{-3}$) | Density |
| Pound per inch cubed (lb in.$^{-3}$) | 2.77E + 04 | Gram per liter (g L$^{-1}$) | Density |
| Pound per inch cubed (lb in.$^{-3}$) | 2.77E + 01 | Gram per milliliter (g mL$^{-1}$) | Density |
| Pound per inch cubed (lb in.$^{-3}$) | 2.77E + 01 | Kilogram per decimeter cubed (kg dm$^{-3}$) | Density |
| Pound per inch cubed (lb in.$^{-3}$) | 2.77E + 04 | Kilogram per meter cubed (kg m$^{-3}$) | Density |
| Pound per inch cubed (lb in.$^{-3}$) | 1.73E + 03 | Pound per foot cubed (lb ft$^{-3}$) | Density |
| Pound per inch cubed (lb in.$^{-3}$) | 2.77E + 02 | Pound per gallon (Brit) (lb gal$^{-1}$) | Density |
| Pound per inch cubed (lb in.$^{-3}$) | 2.31E + 02 | Pound per gallon (US) (lb gal$^{-1}$) | Density |
| Pound per inch cubed (lb in.$^{-3}$) | 4.67E + 04 | Pound per yard cubed (lb yd$^{-3}$) | Density |
| Pound per inch cubed (lb in.$^{-3}$) | 5.37E + 01 | Slug per foot cubed (slug ft$^{3}$) | Density |
| Pound per yard cubed (lb yd$^{-3}$) | 5.93E − 04 | Gram per centimeter cubed (g cm$^{-3}$) | Density |

(Continued)

(Continued)

| Multiply | By | To obtain | Property |
|---|---|---|---|
| Pound per yard cubed (lb yd$^{-3}$) | 5.93E − 01 | Gram per liter (g L$^{-1}$) | Density |
| Pound per yard cubed (lb yd$^{-3}$) | 5.93E − 04 | Gram per milliliter (g mL$^{-1}$) | Density |
| Pound per yard cubed (lb yd$^{-3}$) | 5.93E − 04 | Kilogram per decimeter cubed (kg dm$^{-3}$) | Density |
| Pound per yard cubed (lb yd$^{-3}$) | 5.93E − 01 | Kilogram per meter cubed (kg m$^{-3}$) | Density |
| Pound per yard cubed (lb yd$^{-3}$) | 2.14E + 00 | Pound per cubic inch (lb in.$^{-3}$) | Density |
| Pound per yard cubed (lb yd$^{-3}$) | 3.70E − 02 | Pound per foot cubed (lb ft$^{-3}$) | Density |
| Pound per yard cubed (lb yd$^{-3}$) | 5.95E − 03 | Pound per gallon (Brit) (lb gal$^{-1}$) | Density |
| Pound per yard cubed (lb yd$^{-3}$) | 4.95E − 03 | Pound per gallon (US) (lb gal$^{-1}$) | Density |
| Pound per yard cubed (lb yd$^{-3}$) | 1.15E − 03 | Slug per foot cubed (slug ft$^{-3}$) | Density |
| Slug per foot cubed (slug ft$^{-3}$) | 5.15E − 01 | Gram per centimeter cubed (g cm$^{-3}$) | Density |
| Slug per foot cubed (slug ft$^{-3}$) | 5.15E + 02 | Gram per liter (g L$^{-1}$) | Density |
| Slug per foot cubed (slug ft$^{-3}$) | 5.15E − 01 | Gram per milliliter (g mL$^{-1}$) | Density |
| Slug per foot cubed (slug ft$^{-3}$) | 5.15E − 01 | Kilogram per decimeter cubed (kg dm$^{-3}$) | Density |
| Slug per foot cubed (slug ft$^{-3}$) | 5.15E + 02 | Kilogram per meter cubed (kg m$^{-3}$) | Density |
| Slug per foot cubed (slug ft$^{-3}$) | 3.22E + 01 | Pound per foot cubed (lb ft$^{-3}$) | Density |
| Slug per foot cubed (slug ft$^{-3}$) | 5.17E + 00 | Pound per gallon (Brit) (lb gal$^{-1}$) | Density |
| Slug per foot cubed (slug ft$^{-3}$) | 4.30E + 00 | Pound per gallon (US) (lb gal$^{-1}$) | Density |
| Slug per foot cubed (slug ft$^{-3}$) | 1.86E − 02 | Pound per inch cubed (lb in.$^{-3}$) | Density |
| Slug per foot cubed (slug ft$^{-3}$) | 8.69E + 02 | Pound per yard cubed (lb yd$^{-3}$) | Density |
| Acre-feet per day (ac ft day$^{-1}$) | 5.04E − 01 | Foot cubed per second (ft$^3$ s$^{-1}$) | Discharge |
| Acre-feet per day (ac ft day$^{-1}$) | 1.43E + 01 | Liter per second (L s$^{-1}$) | Discharge |
| Acre-feet per day (ac ft day$^{-1}$) | 1.23E + 03 | Meter cubed per day (m$^3$ day$^{-1}$) | Discharge |
| Acre-feet per day (ac ft day$^{-1}$) | 2.26E + 02 | US gallon per minute (gal min$^{-1}$ or gpm) | Discharge |
| Barrel per day (bbl day$^{-1}$) | | Foot cubed per second (ft$^3$ s$^{-1}$) | Discharge |
| Barrel per day (bbl day$^{-1}$) | | Liter per second (L s$^{-1}$) | Discharge |
| Barrel per day (bbl day$^{-1}$) | | Meter cubed per day (m$^3$ day$^{-1}$) | Discharge |
| Barrel per day (bbl day$^{-1}$) | 2.92E − 02 | US gallon per minute (gal min$^{-1}$ or gpm) | Discharge |
| Foot cubed per second (ft$^3$ s$^{-1}$) | 1.98E + 00 | Acre-feet per day (ac ft day$^{-1}$) | Discharge |
| Foot cubed per second (ft$^3$ s$^{-1}$) | 2.83E + 01 | Liter per second (L s$^{-1}$) | Discharge |
| Foot cubed per second (ft$^3$ s$^{-1}$) | 2.45E + 03 | Meter cubed per day (m$^3$ day$^{-1}$) | Discharge |
| Foot cubed per second (ft$^3$ s$^{-1}$) | 2.83E − 02 | Meter cubed per second (m$^3$ s$^{-1}$) | Discharge |
| Foot cubed per second (ft$^3$ s$^{-1}$) | 6.47E + 05 | US gallon per day (gal day$^{-1}$) | Discharge |
| Foot cubed per second (ft$^3$ s$^{-1}$) | 4.49E + 02 | US gallon per minute (gal min$^{-1}$ or gpm) | Discharge |
| Imperial gallons per minute (Imperial gal min$^{-1}$) | | Acre-feet per day (ac ft day$^{-1}$) | Discharge |
| Imperial gallons per minute (Imperial gal min$^{-1}$) | 7.58E − 02 | Liter per second (L s$^{-1}$) | Discharge |
| Imperial gallons per minute (Imperial gal min$^{-1}$) | | Meter cubed per day (m$^3$ day$^{-1}$) | Discharge |
| Imperial gallons per minute (Imperial gal min$^{-1}$) | 7.58E − 05 | Meter cubed per second (m$^3$ s$^{-1}$) | Discharge |
| Imperial gallons per minute (Imperial gal min$^{-1}$) | | US gallon per day (gal day$^{-1}$) | Discharge |
| Liter per second (L s$^{-1}$) | 3.53E − 02 | Foot cubed per second (ft$^3$ s$^{-1}$) | Discharge |
| Liter per second (L s$^{-1}$) | 8.64E + 01 | Meter cubed per day (m$^3$ day$^{-1}$) | Discharge |
| Liter per second (L s$^{-1}$) | 2.28E + 04 | US gallon per day (gal day$^{-1}$) | Discharge |
| Liter per second (L s$^{-1}$) | 7.01E − 02 | US gallon per minute (gal min$^{-1}$ or gpm) | Discharge |

(Continued)

| Multiply | By | To obtain | Property |
|---|---|---|---|
| Liter per second (L s$^{-1}$) | 1.59E + 01 | US gallon per minute (gal min$^{-1}$ or gpm) | Discharge |
| Liter per second (L s$^{-1}$) | 1.58E + 01 | US gallon per minute (gal min$^{-1}$ or gpm) | Discharge |
| Meter cubed per day (m$^3$ day$^{-1}$) | 6.05E + 06 | Acre-feet per day (ac ft day$^{-1}$) | Discharge |
| Meter cubed per day (m$^3$ day$^{-1}$) | 3.02E + 06 | Foot cubed per second (ft$^3$ s$^{-1}$) | Discharge |
| Meter cubed per day (m$^3$ day$^{-1}$) | 8.64E + 07 | Liter per second (L s$^{-1}$) | Discharge |
| Meter cubed per day (m$^3$ day$^{-1}$) | 1.37E + 09 | US gallon per minute (gal min$^{-1}$ or gpm) | Discharge |
| Meter cubed per second (m$^3$ s$^{-1}$) | 2.28E + 07 | US gallon per day (gal day$^{-1}$) | Discharge |
| US gallon per day (gal day$^{-1}$) | 1.55E − 06 | Foot cubed per second (ft$^3$ s$^{-1}$) | Discharge |
| US gallon per day (gal day$^{-1}$) | 4.38E − 05 | Liter per second (L s$^{-1}$) | Discharge |
| US gallon per day (gal day$^{-1}$) | 4.38E − 08 | Meter cubed per second (m$^3$ s$^{-1}$) | Discharge |
| US gallon per day (gal day$^{-1}$) | 6.94E − 04 | US gallon per minute (gal min$^{-1}$ or gpm) | Discharge |
| US gallon per minute (gal min$^{-1}$ or gpm) | 4.42E − 03 | Acre-feet per day (ac ft day$^{-1}$) | Discharge |
| US gallon per minute (gal min$^{-1}$ or gpm) | 3.43E + 01 | Barrel per day (bbl d$^{-1}$) | Discharge |
| US gallon per minute (gal min$^{-1}$ or gpm) | 2.23E − 03 | Foot cubed per sec (ft$^3$ s$^{-1}$) | Discharge |
| US gallon per minute (gal min$^{-1}$ or gpm) | 2.23E − 03 | Foot cubed per second (ft$^3$ s$^{-1}$) | Discharge |
| US gallon per minute (gal min$^{-1}$ or gpm) | 6.31E − 02 | Liter per second (L s$^{-1}$) | Discharge |
| US gallon per minute (gal min$^{-1}$ or gpm) | 6.31E − 02 | Liter per second (L s$^{-1}$) | Discharge |
| US gallon per minute (gal min$^{-1}$ or gpm) | 5.45E + 00 | Meter cubed per day (m$^3$ day$^{-1}$) | Discharge |
| US gallon per minute (gal min$^{-1}$ or gpm) | 6.31E − 05 | Meter cubed per second (m$^3$ s$^{-1}$) | Discharge |
| US gallon per minute (gal min$^{-1}$ or gpm) | 6.31E − 05 | Meter cubed per second (m$^3$ s$^{-1}$) | Discharge |
| US gallon per minute (gal min$^{-1}$ or gpm) | 1.44E + 03 | US gallon per day (gal day$^{-1}$) | Discharge |
| Bushels | 4.00E + 00 | Peck | Dry volume |
| Centimeter cubed (cm$^3$) | 6.10E − 02 | Inch cubed (in.$^3$) | Dry volume |
| Centimeter cubed (cm$^3$) | 1.00E + 03 | Millimeter cubed (mm$^3$) | Dry volume |
| Decimeter cubed (dm$^3$) | 1.00E + 03 | Centimeter cubed (cm$^3$) | Dry volume |
| Foot cubed (ft$^3$) | 1.73E + 03 | Inch cubed (in.$^3$) | Dry volume |
| Foot cubed (ft$^3$) | 2.83E − 02 | Meter cubed (m$^3$) | Dry volume |
| Foot cubed (ft$^3$) | 3.70E − 02 | Yard cubed (yd$^3$) | Dry volume |
| Centipoises | 1.00E − 01 | Dynes-second per square centimeter (dynes s cm$^{-2}$) | Dynamic viscosity |
| Centipoises | 1.02E − 04 | Kilogram$_{force}$-second per square meter (kg$_f$ s m$^{-2}$) | Dynamic viscosity |
| Centipoises | 1.00E − 02 | Pascal-second (kg s$^{-1}$) | Dynamic viscosity |
| Centipoises | 1.00E − 01 | Poises (g cm$^{-1}$ s$^{-1}$) | Dynamic viscosity |
| Centipoises | 1.45E − 07 | Pound$_{force}$-second per square inch (lb$_f$ s in.$^{-2}$) | Dynamic viscosity |
| Centipoises | 2.09E − 05 | Pound$_{force}$-second per square foot (lb$_f$ s ft$^{-2}$) | Dynamic viscosity |
| Centipoises | 2.4191 | Pound$_{mass}$ per foot-hour (lb$_m$ ft$^{-1}$ h$^{-1}$) | Dynamic viscosity |

*(Continued)*

(Continued)

| Multiply | By | To obtain | Property |
|---|---|---|---|
| Centipoises | 6.72E − 04 | Pound$_{mass}$ per foot-second (lb$_m$ ft$^{-1}$ s$^{-1}$) | Dynamic viscosity |
| Dynes-second per square centimeter (dynes s cm$^{-2}$) | 1.00E + 03 | Centipoises | Dynamic viscosity |
| Dynes-second per square centimeter (dynes s cm$^{-2}$) | 1.02E − 02 | Kilogram$_{force}$-second per square meter (kg$_f$ s m$^{-2}$) | Dynamic viscosity |
| Dynes-second per square centimeter (dynes s cm$^{-2}$) | 1.00E + 00 | Pascal-second (kg m$^{-1}$ s$^{-1}$) | Dynamic viscosity |
| Dynes-second per square centimeter (dynes s cm$^{-2}$) | 1 | Poises (g cm$^{-1}$ s$^{-1}$) | Dynamic viscosity |
| Dynes-second per square centimeter (dynes s cm$^{-2}$) | 1.45E − 05 | Pound$_{force}$-second per square inch (lb$_f$ s in.$^{-2}$) | Dynamic viscosity |
| Dynes-second per square centimeter (dynes s cm$^{-2}$) | 2.09E − 03 | Pound$_{force}$-second per square foot (lb$_f$ s ft$^{-2}$) | Dynamic viscosity |
| Dynes-second per square centimeter (dynes s cm$^{-2}$) | 2.42E + 02 | Pound$_{mass}$ per foot-hour (lb$_m$ ft$^{-1}$ h$^{-1}$) | Dynamic viscosity |
| Dynes-second per square centimeter (dynes s cm$^{-2}$) | 6.72E − 02 | Pound$_{mass}$ per foot-second (lb$_m$ ft$^{-1}$ s$^{-1}$) | Dynamic viscosity |
| Kilogram$_{force}$-second per square meter (kg$_f$ s m$^{-2}$) | 9.81E + 03 | Centipoises | Dynamic viscosity |
| Kilogram$_{force}$-second per square meter (kg$_f$ s m$^{-2}$) | 9.81E + 01 | Dynes-second per square centimeter (dynes s cm$^{-2}$) | Dynamic viscosity |
| Kilogram$_{force}$-second per square meter (kg$_f$ s m$^{-2}$) | 9.8067 | Pascal-second (kg m$^{-1}$ s$^{-1}$) | Dynamic viscosity |
| Kilogram$_{force}$-second per square meter (kg$_f$ s m$^{-2}$) | 9.81E + 01 | Poises (g cm$^-$ s$^{-1}$) | Dynamic viscosity |
| Kilogram$_{force}$-second per square meter (kg$_f$ s m$^{-2}$) | 1.42E − 33 | Pound$_{force}$-second per square inch (lb$_f$ s in.$^{-2}$) | Dynamic viscosity |
| Kilogram$_{force}$-second per square meter (kg$_f$ s m$^{-2}$) | 2.05E − 01 | Pound$_{force}$-second per square foot (lb$_f$ s ft$^{-2}$) | Dynamic viscosity |
| Kilogram$_{force}$-second per square meter (kg$_f$ s m$^{-2}$) | 2.37E + 04 | Pound$_{mass}$ per foot-hour (lb$_m$ ft$^{-1}$ h$^{-1}$) | Dynamic viscosity |
| Kilogram$_{force}$-second per square meter (kg$_f$ s m$^{-2}$) | 6.59 | Pound$_{mass}$ per foot-second (lb$_m$ ft$^{-1}$ s$^{-1}$) | Dynamic viscosity |
| Pascal-second (kg$_f$ m$^{-1}$ s$^{-1}$) | 1.00E + 04 | Centipoises | Dynamic viscosity |
| Pascal-second (kg$_f$ m$^{-1}$ s$^{-1}$) | 1.00E + 01 | Dynes-second per square centimeter (dynes s cm$^{-2}$) | Dynamic viscosity |
| Pascal-second (kg$_f$ m$^{-1}$ s$^{-1}$) | 1.02E − 01 | Kilogram$_{force}$-second per square meter (kg$_f$ s m$^{-2}$) | Dynamic viscosity |
| Pascal-second (kg$_f$ m$^{-1}$ s$^{-1}$) | 1.00E + 01 | Poises (g cm s$^{-1}$) | Dynamic viscosity |
| Pascal-second (kg$_f$ m$^{-1}$ s$^{-1}$) | 1.45E − 04 | Pound$_{force}$-second per square inch (lb$_f$ s in.$^{-2}$) | Dynamic viscosity |
| Pascal-second (kg$_f$ m$^{-1}$ s$^{-1}$) | 2.09E − 02 | Pound$_{force}$-second per square foot (lb$_f$ s ft$^{-2}$) | Dynamic viscosity |
| Pascal-second (kg$_f$ m$^{-1}$ s$^{-1}$) | 2.42E + 03 | Pound$_{mass}$ per foot-hour (lb$_m$ ft$^{-1}$ h$^{-1}$) | Dynamic viscosity |
| Pascal-second (kg$_f$ m$^{-1}$ s$^{-1}$) | 6.72E − 01 | Pound$_{mass}$ per foot-second (lb$_m$ ft$^{-1}$ s$^{-1}$) | Dynamic viscosity |
| Poises (g$_f$ cm$^{-1}$ s$^{-1}$) | 1.00E + 03 | Centipoises | Dynamic viscosity |
| Poises (g$_f$ cm$^{-1}$ s$^{-1}$) | 1.00E + 00 | Dynes-second per square centimeter (dynes s cm$^{-2}$) | Dynamic viscosity |
| Poises (g$_f$ cm$^{-1}$ s$^{-1}$) | 1.02E − 02 | Kilogram$_{force}$-second per square meter (kg$_f$ s m$^{-2}$) | Dynamic viscosity |
| Poises (g$_f$ cm$^{-1}$ s$^{-1}$) | 1.00E + 00 | Pascal-second (kg m$^{-1}$ s) | Dynamic viscosity |
| Poises (g$_f$ cm$^{-1}$ s$^{-1}$) | 1.45E − 05 | Pound$_{force}$-second per square inch (lb$_f$ s in.$^{-2}$) | Dynamic viscosity |

(Continued)

| Multiply | By | To obtain | Property |
|---|---|---|---|
| Poises ($g_f$ cm$^{-1}$ s$^{-1}$) | 2.42E + 02 | Pound$_{mass}$ per foot-hour ($lb_m$ ft$^{-1}$ h$^{-1}$) | Dynamic viscosity |
| Poises ($g_f$ cm$^{-1}$ s$^{-1}$) | 6.72E − 02 | Pound$_{mass}$ per foot-second ($lb_m$ ft$^{-1}$ s$^{-1}$) | Dynamic viscosity |
| Poises ($g_f$ cm$^{-1}$ s$^{-1}$) | 2.09E − 03 | Pound$_{force}$-second per square foot ($lb_f$ s ft$^{-2)}$ | Dynamic viscosity |
| Pound$_{force}$-second per square foot ($lb_f$ s ft$^{-2}$) | 4.79E + 04 | Centipoises | Dynamic viscosity |
| Pound$_{force}$-second per square foot ($lb_f$ s ft$^{-2}$) | 4.79E + 02 | Dynes-second per square centimeter (dynes s cm$^{-2}$) | Dynamic viscosity |
| Pound$_{force}$-second per square foot ($lb_f$ s ft$^{-2}$) | 4.8824 | Kilogram$_{force}$-second per square meter ($kg_f$ s m$^{-2}$) | Dynamic viscosity |
| Pound$_{force}$-second per square foot ($lb_f$ s ft$^{-2}$) | 4.79E + 01 | Pascal-second (kg m s$^{-1}$) | Dynamic viscosity |
| Pound$_{force}$-second per square foot ($lb_f$ s ft$^{-2}$) | 4.79E + 02 | Poises (g cm s$^{-1}$) | Dynamic viscosity |
| Pound$_{force}$-second per square foot ($lb_f$ s ft$^{-2}$) | 6.95E − 03 | Pound$_{force}$-second per square inch ($lb_f$ s in.$^{-2}$) | Dynamic viscosity |
| Pound$_{force}$-second per square foot ($lb_f$ s ft$^{-2}$) | 1.16E + 05 | Pound$_{mass}$ per foot-hour ($lb_m$ ft$^{-1}$ h$^{-1}$) | Dynamic viscosity |
| Pound$_{force}$-second per square foot ($lb_f$ s ft$^{-2}$) | 3.22E + 01 | Pound$_{mass}$ per foot-second ($lb_m$ ft$^{-1}$ s$^{-1}$) | Dynamic viscosity |
| Pound$_{force}$-second per square inch ($lb_f$ s in.$^{-2}$) | 6.85E + 06 | Centipoises | Dynamic viscosity |
| Pound$_{force}$-second per square inch ($lb_f$ s in.$^{-2}$) | 6.89E + 04 | Dynes-second per square centimeter (dynes s$^{-1}$cm$^{-2}$) | Dynamic viscosity |
| Pound$_{force}$-second per square inch ($lb_f$ s in.$^{-2}$) | 7.03E + 02 | Kilogram$_{force}$-second per square meter ($kg_f$ s m$^{-2}$) | Dynamic viscosity |
| Pound$_{force}$-second per square inch ($lb_f$ s in.$^{-2}$) | 6.89E + 03 | Pascal-second (kg m$^{-1}$ s$^{-1}$) | Dynamic viscosity |
| Pound$_{force}$-second per square inch ($lb_f$ s in.$^{-2}$) | 6.89E + 04 | Poises (g cm$^{-1}$ s$^{-1}$) | Dynamic viscosity |
| Pound$_{force}$-second per square inch ($lb_f$ s in.$^{-2}$) | 1.67E + 07 | Pound$_{force}$-second per square inch ($lb_f$ s in.$^{-2}$) | Dynamic viscosity |
| Pound$_{force}$-second per square inch ($lb_f$ s in.$^{-2}$) | 1.44E + 02 | Pound$_{force}$-second per square foot ($lb_f$ s ft$^{-2}$) | Dynamic viscosity |
| Pound$_{force}$-second per square inch ($lb_f$ s in.$^{-2}$) | 4.63E + 03 | Pound$_{mass}$ per foot-second ($lb_m$ ft$^{-1}$ s$^{-1}$) | Dynamic viscosity |
| Pound$_{mass}$ per foot-hour ($lb_m$ ft h$^{-1}$) | 4.13E − 01 | Centipoises | Dynamic viscosity |
| Pound$_{mass}$ per foot-hour ($lb_m$ ft h$^{-1}$) | 4.13E − 03 | Dynes-second per square centimeter (dynes s cm$^{-2}$) | Dynamic viscosity |
| Pound$_{mass}$ per foot-hour ($lb_m$ ft h$^{-1}$) | 4.22E − 05 | Kilogram$_{force}$-second per square meter ($kg_f$ s m$^{-2}$) | Dynamic viscosity |
| Pound$_{mass}$ per foot-hour ($lb_m$ ft h$^{-1}$) | 4.13E − 04 | Pascal-second (kg m$^{-1}$ s$^{-1}$) | Dynamic viscosity |
| Pound$_{mass}$ per foot-hour ($lb_m$ ft h$^{-1}$) | 4.13E − 03 | Poises (g cm$^{-1}$ s$^{-1}$) | Dynamic viscosity |
| Pound$_{mass}$ per foot-hour ($lb_m$ ft h$^{-1}$) | 5.99E − 08 | Pound$_{force}$-second per square inch ($lb_f$ s in.$^{-2}$) | Dynamic viscosity |
| Pound$_{mass}$ per foot-hour ($lb_m$ ft h$^{-1}$) | 8.63E − 06 | Pound$_{force}$-second per square foot ($lb_f$ s ft$^{-2}$) | Dynamic viscosity |
| Pound$_{mass}$ per foot-hour ($lb_m$ ft h$^{-1}$) | 2.78E − 04 | Pound$_{mass}$ per foot-second ($lb_m$ ft$^{-1}$ s$^{-1}$) | Dynamic viscosity |
| Pound$_{mass}$ per foot-second ($lb_m$ ft s$^{-1}$) | 1.49E + 03 | Centipoises | Dynamic viscosity |

(Continued)

| Multiply | By | To obtain | Property |
|---|---|---|---|
| Pound$_{mass}$ per foot-second (lb$_m$ ft s$^{-1}$) | 1.49E + 01 | Dynes-second per square centimeter (dynes s cm$^{-2}$) | Dynamic viscosity |
| Pound$_{mass}$ per foot-second (lb$_m$ ft s$^{-1}$) | 1.52E − 01 | Kilogram$_{force}$-second per square meter (kg$_f$ s m$^{-2}$) | Dynamic viscosity |
| Pound$_{mass}$ per foot-second (lb$_m$ ft s$^{-1}$) | 1.4882 | Pascal-second (kg m$^{-1}$ s$^{-1}$) | Dynamic viscosity |
| Pound$_{mass}$ per foot-second (lb$_m$ ft s$^{-1}$) | 1.49E + 01 | Poises (g cm$^{-1}$ s$^{-1}$) | Dynamic viscosity |
| Pound$_{mass}$ per foot-second (lb$_m$ ft s$^{-1}$) | 3.11E − 02 | Pound$_{force}$-second per square foot (lb$_f$ s ft$^{-2}$) | Dynamic viscosity |
| Pound$_{mass}$ per foot-second (lb$_m$ ft s$^{-1}$) | 2.16E − 04 | Pound$_{force}$-second per square inch (lb$_f$ s in.$^{-2}$) | Dynamic viscosity |
| Pound$_{mass}$ per foot-second (lb$_m$ ft s$^{-1}$) | 3.60E + 03 | Pound$_{mass}$ per foot-hour (lb$_m$ ft h$^{-1}$) | Dynamic viscosity |
| British thermal units (BTU) | 2.52E + 02 | Calorie, thermal (cal) | Energy and work |
| British thermal units (BTU) | 1.06E + 10 | Erg | Energy and work |
| British thermal units (BTU) | 3.68E − 01 | Foot cubed, atmosphere (ft$^3$) | Energy and work |
| British thermal units (BTU) | 7.78E + 02 | Foot-pound (ft lb), force | Energy and work |
| British thermal units (BTU) | 2.50E + 04 | Foot-poundal | Energy and work |
| British thermal units (BTU) | 3.93E − 04 | Horsepower-hour | Energy and work |
| British thermal units (BTU) | 3.98E − 04 | Horsepower-hour (hp h) metric | Energy and work |
| British thermal units (BTU) | 1.06E + 03 | Joule (J) | Energy and work |
| British thermal units (BTU) | 2.52E − 01 | Kilocalorie (cal or kcal) | Energy and work |
| British thermal units (BTU) | 1.08E + 02 | Kilogram-force-meter | Energy and work |
| British thermal units (BTU) | 2.93E − 04 | Kilowatt-hour (kW h) | Energy and work |
| British thermal units (BTU) | 1.04E + 01 | Liter atmosphere | Energy and work |
| British thermal units (BTU) | 2.93E − 01 | Watt-hour | Energy and work |
| Calorie, thermal (cal) | 3.97E − 03 | British thermal units (BTU) | Energy and work |
| Calorie, thermal (cal) | 4.18E + 07 | Erg | Energy and work |
| Calorie, thermal (cal) | 1.46E − 03 | Foot cubed, atmosphere (ft$^3$) | Energy and work |
| Calorie, thermal (cal) | 3.09E + 00 | Foot-pound (ft lb), force | Energy and work |
| Calorie, thermal (cal) | 9.93E + 01 | Foot-poundal | Energy and work |
| Calorie, thermal (cal) | 1.56E − 06 | Horsepower-hour | Energy and work |
| Calorie, thermal (cal) | 1.58E − 06 | Horsepower-hour (metric) | Energy and work |
| Calorie, thermal (cal) | 4.19E + 00 | Joule (J) | Energy and work |
| Calorie, thermal (cal) | 1.00E − 03 | Kilocalorie (cal or kcal) | Energy and work |
| Calorie, thermal (cal) | 4.27E − 01 | Kilogram-force-meter | Energy and work |
| Calorie, thermal (cal) | 1.16E − 06 | Kilowatt-hour (kW h) | Energy and work |
| Calorie, thermal (cal) | 4.13E − 02 | Liter atmosphere | Energy and work |
| Calorie, thermal (cal) | 1.16E − 03 | Watt-hour | Energy and work |
| Erg | 9.48E − 11 | British thermal units (BTU) | Energy and work |
| Erg | 2.39E − 08 | Calorie, thermal (cal) | Energy and work |
| Erg | 3.49E − 11 | Foot cubed, atmosphere (ft$^3$) | Energy and work |
| Erg | 7.38E − 08 | Foot-pound force | Energy and work |
| Erg | 2.37E − 06 | Foot-poundal | Energy and work |
| Erg | 3.73E − 14 | Horsepower-hour | Energy and work |
| Erg | 3.78E − 14 | Horsepower-hour (hp h) metric | Energy and work |
| Erg | 1.00E − 06 | Joule (J) | Energy and work |
| Erg | 2.39E − 11 | Kilocalorie (cal or kcal) | Energy and work |
| Erg | 1.02E − 08 | Kilogram-force-meter | Energy and work |
| Erg | 2.78E − 14 | Kilowatt-hour (kW h) | Energy and work |
| Erg | 9.87E − 10 | Liter atmosphere | Energy and work |
| Erg | 2.78E − 11 | Watt-hour | Energy and work |
| Foot cubed, atmosphere (ft$^3$) | 2.72E + 00 | British thermal units (BTU) | Energy and work |
| Foot cubed, atmosphere (ft$^3$) | 6.86E + 02 | Calorie, thermal (cal) | Energy and work |

| Multiply | By | To obtain | Property |
|---|---|---|---|
| Foot cubed, atmosphere (ft$^3$) | 2.87E + 10 | Erg | Energy and work |
| Foot cubed, atmosphere (ft$^3$) | 2.12E + 03 | Foot-pound force | Energy and work |
| Foot cubed, atmosphere (ft$^3$) | 6.81E + 04 | Foot-poundal | Energy and work |
| Foot cubed, atmosphere (ft$^3$) | 1.07E − 03 | Horsepower-hour | Energy and work |
| Foot cubed, atmosphere (ft$^3$) | 1.08E − 03 | Horsepower-hour (hp h) metric | Energy and work |
| Foot cubed, atmosphere (ft$^3$) | 2.87E + 03 | Joule (J) | Energy and work |
| Foot cubed, atmosphere (ft$^3$) | 6.86E − 01 | Kilocalorie (cal or kcal) | Energy and work |
| Foot cubed, atmosphere (ft$^3$) | 2.93E + 02 | Kilogram-force-meter | Energy and work |
| Foot cubed, atmosphere (ft$^3$) | 7.97E − 04 | Kilowatt-hour (kW h) | Energy and work |
| Foot cubed, atmosphere (ft$^3$) | 2.83E + 01 | Liter atmosphere | Energy and work |
| Foot cubed, atmosphere (ft$^3$) | 7.97E − 01 | Watt-hour | Energy and work |
| Foot-pound (ft lb), force | 1.29E − 03 | British thermal unit (BTU) | Energy and work |
| Foot-pound (ft lb), force | 3.24E − 01 | Calorie, thermal (cal) | Energy and work |
| Foot-pound (ft lb), force | 1.36E + 07 | Erg | Energy and work |
| Foot-pound (ft lb), force | 4.73E − 04 | Foot cubed atmosphere | Energy and work |
| Foot-pound (ft lb), force | 3.22E + 01 | Foot-poundal | Energy and work |
| Foot-pound (ft lb), force | 5.05E − 02 | Horsepower-hour | Energy and work |
| Foot-pound (ft lb), force | 5.12E − 07 | Horsepower-hour (hp h) metric | Energy and work |
| Foot-pound (ft lb), force | 1.36E + 00 | Joule (J) | Energy and work |
| Foot-pound (ft lb), force | 3.24E − 04 | Kilocalorie (cal or kcal) | Energy and work |
| Foot-pound (ft lb), force | 1.38E − 01 | Kilogram-force-meter | Energy and work |
| Foot-pound (ft lb), force | 3.77E − 07 | Kilowatt-hour (kW h ) | Energy and work |
| Foot-pound (ft lb), force | 1.34E − 02 | Liter atmosphere | Energy and work |
| Foot-pound (ft lb), force | 3.77E − 04 | Watt-hour | Energy and work |
| Foot-poundal | 3.99E − 05 | British thermal units (BTU) | Energy and work |
| Foot-poundal | 1.01E − 02 | Calorie, thermal (cal) | Energy and work |
| Foot-poundal | 4.21E + 05 | Erg | Energy and work |
| Foot-poundal | 1.47E − 05 | Foot cubed, atmosphere (ft$^3$) | Energy and work |
| Foot-poundal | 3.11E − 02 | Foot-pound force | Energy and work |
| Foot-poundal | 1.57E − 08 | Horsepower-hour | Energy and work |
| Foot-poundal | 1.59E − 08 | Horsepower-hour (hp h) metric | Energy and work |
| Foot-poundal | 4.21E − 02 | Joule (J) | Energy and work |
| Foot-poundal | 1.01E − 05 | Kilocalorie (cal or kcal) | Energy and work |
| Foot-poundal | 4.30E − 03 | Kilogram-force-meter | Energy and work |
| Foot-poundal | 1.17E − 08 | Kilowatt-hour (kW h) | Energy and work |
| Foot-poundal | 4.16E − 04 | Liter atmosphere | Energy and work |
| Foot-poundal | 1.17E − 05 | Watt-hour | Energy and work |
| Horsepower-hour (hp h) | 2.54E + 03 | British thermal unit (BTU) | Energy and work |
| Horsepower-hour (hp h) | 6.42E + 05 | Calorie, thermal (cal) | Energy and work |
| Horsepower-hour (hp h) | 2.68E + 13 | Erg | Energy and work |
| Horsepower-hour (hp h) | 9.36E + 02 | Foot cubed atmosphere | Energy and work |
| Horsepower-hour (hp h) | 1.98E + 06 | Foot-pound (ft lb), force | Energy and work |
| Horsepower-hour (hp h) | 6.37E + 07 | Foot-poundal | Energy and work |
| Horsepower-hour (hp h) | 1.01E + 00 | Horsepower-hour (hp h) metric | Energy and work |
| Horsepower-hour (hp h) | 2.68E + 06 | Joule (J) | Energy and work |
| Horsepower-hour (hp h) | 6.42E + 02 | Kilocalorie (cal or kcal) | Energy and work |
| Horsepower-hour (hp h ) | 2.74E + 05 | Kilogram-force-meter | Energy and work |
| Horsepower-hour (hp h) | 7.46E − 01 | Kilowatt-hour (kW h) | Energy and work |
| Horsepower-hour (hp h) | 2.65E + 04 | Liter atmosphere | Energy and work |
| Horsepower-hour (hp h) | 7.46E + 06 | Watt-hour | Energy and work |
| Horsepower-hour (hp h) metric | 2.51E + 03 | British thermal unit (BTU) | Energy and work |
| Horsepower-hour (hp h) metric | 6.33E + 05 | Calorie, thermal (cal) | Energy and work |
| Horsepower-hour (hp h) metric | 2.65E + 13 | Erg | Energy and work |
| Horsepower-hour (hp h) metric | 9.23E + 02 | Foot cubed, atmosphere (ft$^3$) | Energy and work |
| Horsepower-hour (hp h) metric | 1.95E + 06 | Foot-pound (ft lb), force | Energy and work |
| Horsepower-hour (hp h) metric | 6.28E + 07 | Foot-poundal | Energy and work |
| Horsepower-hour (hp h) metric | 9.86E − 01 | Horsepower-hour (hp h) | Energy and work |

(Continued)

| Multiply | By | To obtain | Property |
|----------|-----|-----------|----------|
| Horsepower-hour (hp h) metric | 2.65E + 06 | Joule (J) | Energy and work |
| Horsepower-hour (hp h) metric | 6.33E + 02 | Kilocalorie (Cal or kcal) | Energy and work |
| Horsepower-hour (hp h) metric | 2.70E + 05 | Kilogram-force-meter | Energy and work |
| Horsepower-hour (hp h) metric | 7.36E − 01 | Kilowatt-hour (kW h) | Energy and work |
| Horsepower-hour (hp h) metric | 2.61E + 04 | Liter atmosphere | Energy and work |
| Horsepower-hour (hp h) metric | 7.36E + 06 | Watt-hour | Energy and work |
| Joules (J) | 9.48E − 04 | British thermal unit (BTU) | Energy and work |
| Joules (J) | 2.39E − 01 | Calorie, thermal (cal) | Energy and work |
| Joules (J) | 1.00E + 08 | erg | Energy and work |
| Joules (J) | 3.49E − 04 | Foot cubed, atmosphere (ft$^3$) | Energy and work |
| Joules (J) | 7.38E − 01 | Foot-pound (ft lb), force | Energy and work |
| Joules (J) | 2.37E + 01 | Foot-poundal | Energy and work |
| Joules (J) | 3.73E − 07 | Horsepower-hour (hp h) | Energy and work |
| Joules (J) | 3.78E − 07 | Horsepower-hour (hp h) metric | Energy and work |
| Joules (J) | 2.39E − 04 | Kilocalorie (cal or kcal) | Energy and work |
| Joules (J) | 1.02E − 01 | Kilogram-force-meter | Energy and work |
| Joules (J) | 2.78E − 07 | Kilowatt-hour (kW h) | Energy and work |
| Joules (J) | 9.87E − 03 | Liter atmosphere | Energy and work |
| Joules (J) | 2.78E − 04 | Watt-hour | Energy and work |
| Kilocalorie (cal or kcal) | 3.97E + 00 | British thermal unit (BTU) | Energy and work |
| Kilocalorie (cal or kcal) | 1.00E + 03 | Calorie, thermal (cal) | Energy and work |
| Kilocalorie (cal or kcal) | 4.18E + 10 | Erg | Energy and work |
| Kilocalorie (cal or kcal) | 1.46E + 00 | Foot cubed, atmosphere (ft$^3$) | Energy and work |
| Kilocalorie (cal or kcal) | 3.09E + 03 | Foot-pound (ft lb), force | Energy and work |
| Kilocalorie (cal or kcal) | 9.93E + 04 | Foot-poundal | Energy and work |
| Kilocalorie (cal or kcal) | 1.56E − 03 | Horsepower-hour (hp h) | Energy and work |
| Kilocalorie (cal or kcal) | 1.58E − 03 | Horsepower-hour (hp h) metric | Energy and work |
| Kilocalorie (cal or kcal) | 4.18E + 03 | Joules (J) | Energy and work |
| Kilocalorie (cal or kcal) | 4.27E + 02 | Kilogram-force-meter | Energy and work |
| Kilocalorie (cal or kcal) | 1.16E − 03 | Kilowatt-hour (kW h) | Energy and work |
| Kilocalorie (cal or kcal) | 4.13E + 01 | Liter atmosphere | Energy and work |
| Kilocalorie (cal or kcal) | 1.16E + 00 | Watt-hour | Energy and work |
| Kilogram-force-meter | 2.34E + 00 | Calorie, thermal (cal) | Energy and work |
| Kilogram-force-meter | 9.81E + 07 | Erg | Energy and work |
| Kilogram-force-meter | 3.42E − 03 | Foot cubed, atmosphere (ft$^3$) | Energy and work |
| Kilogram-force-meter | 7.23E + 00 | Foot-pound (ft lb), force | Energy and work |
| Kilogram-force-meter | 2.33E + 02 | Foot-poundal | Energy and work |
| Kilogram-force-meter | 3.65E − 06 | Horsepower-hour (hp h) | Energy and work |
| Kilogram-force-meter | 3.70E − 06 | Horsepower-hour (hp h) metric | Energy and work |
| Kilogram-force-meter | 9.81E + 00 | Joules (J) | Energy and work |
| Kilogram-force-meter | 2.34E − 03 | Kilocalorie (cal or Kcal) | Energy and work |
| Kilogram-force-meter | 2.72E − 06 | Kilowatt-hour (kW h) | Energy and work |
| Kilogram-force-meter | 9.68E − 02 | Liter atmosphere | Energy and work |
| Kilogram-force-meter | 2.72E − 03 | Watt-hour | Energy and work |
| Kilogram-force-meter | 9.29E − 03 | British thermal unit (BTU) | Energy and work |
| Kilowatt-hour (kW h) | 3.412.14 | British thermal unit (BTU) | Energy and work |
| Kilowatt-hour (kW h) | 8.60E + 05 | calorie, thermal (cal) | Energy and work |
| Kilowatt-hour (kW h) | 3.60E + 13 | Erg | Energy and work |
| Kilowatt-hour (kW h) | 1.25E + 03 | Foot cubed, atmosphere (ft$^3$) | Energy and work |
| Kilowatt-hour (kW h) | 2.66E + 06 | Foot-pound (ft lb), force | Energy and work |
| Kilowatt-hour (kW h) | 8.54E + 07 | Foot-poundal | Energy and work |
| Kilowatt-hour (kW h) | 1.34E + 00 | Horsepower-hour (hp h) | Energy and work |
| Kilowatt-hour (kW h) | 1.36E + 00 | Horsepower-hour (hp h) metric | Energy and work |
| Kilowatt-hour (kW h) | 3.60E + 06 | Joules (J) | Energy and work |
| Kilowatt-hour (kW h) | 8.60E + 02 | Kilocalorie (cal or kcal) | Energy and work |
| Kilowatt-hour (kW h) | 3.67E + 05 | Kilogram-force-meter | Energy and work |

| Multiply | By | To obtain | Property |
|---|---|---|---|
| Kilowatt-hour (kW h) | 3.55E + 04 | Liter atmosphere | Energy and work |
| Kilowatt-hour (kW h) | 1.00E + 03 | Watt-hour | Energy and work |
| Liter atmosphere | 9.60E − 02 | British thermal unit (BTU) | Energy and work |
| Liter atmosphere | 2.42E + 01 | Calorie, thermal (cal) | Energy and work |
| Liter atmosphere | 1.01E + 09 | Erg | Energy and work |
| Liter atmosphere | 3.53E − 02 | Foot cubed, atmosphere ($ft^3$) | Energy and work |
| Liter atmosphere | 7.47E + 01 | Foot-pound (ft lb), force | Energy and work |
| Liter atmosphere | 2.40E + 03 | Foot-poundal | Energy and work |
| Liter atmosphere | 3.77E − 05 | Horsepower-hour (hp h) | Energy and work |
| Liter atmosphere | 3.83E − 05 | Horsepower-hour (hp h) metric | Energy and work |
| Liter atmosphere | 1.01E + 02 | Joules (J) | Energy and work |
| Liter atmosphere | 2.42E − 02 | Kilocalorie (cal or kcal) | Energy and work |
| Liter atmosphere | 1.03E + 01 | Kilogram-force-meter | Energy and work |
| Liter atmosphere | 2.81E − 05 | Kilowatt-hour (kW h) | Energy and work |
| Liter atmosphere | 8.15E − 05 | Watt-hour | Energy and work |
| Watt-hour | 3.41E + 00 | British thermal unit (BTU) | Energy and work |
| Watt-hour | 8.60E + 02 | Calorie, thermal (cal) | Energy and work |
| Watt-hour | 3.60E + 10 | Erg | Energy and work |
| Watt-hour | 1.25E + 00 | Foot cubed, atmosphere ($ft^3$) | Energy and work |
| Watt-hour | 2.66E + 03 | Foot-pound (ft lb), force | Energy and work |
| Watt-hour | 8.54E + 04 | Foot-poundal | Energy and work |
| Watt-hour | 1.34E − 03 | Horsepower-hour (hp h) | Energy and work |
| Watt-hour | 1.36E − 03 | Horsepower-hour (hp h) metric | Energy and work |
| Watt-hour | 3.60E + 03 | Joules (J) | Energy and work |
| Watt-hour | 8.60E − 01 | Kilocalorie (cal or kcal) | Energy and work |
| Watt-hour | 3.67E + 02 | Kilogram-force-meter | Energy and work |
| Watt-hour | 1.00E − 03 | Kilowatt-hour (kW h) | Energy and work |
| Watt-hour | 3.55E + 01 | Liter atmosphere | Energy and work |
| Dyne | 1.02E − 03 | Gram force ($g_{force}$) | Force |
| Dyne | 1.00E − 06 | Joules per centimeter ($J\,cm^{-1}$) | Force |
| Dyne | 1.02E − 06 | Kilogram force ($kg_{force}$) | Force |
| Dyne | 1.00E − 05 | Newton (N) | Force |
| Dyne | 2.25E − 06 | Pound force ($lb_f$) | Force |
| Dyne | 7.23E − 05 | Poundal | Force |
| Gram force ($g_{force}$) | 9.81E + 02 | Dyne | Force |
| Gram force ($g_{force}$) | 9.81E − 05 | Joules per centimeter ($J\,cm^{-1}$) | Force |
| Gram force ($g_{force}$) | 1.00E − 03 | Kilogram force ($kg_{force}$) | Force |
| Gram force ($g_{force}$) | 9.81E − 03 | Newton (N) | Force |
| Gram force ($g_{force}$) | 2.20E − 03 | Pound force ($lb_f$) | Force |
| Gram force ($g_{force}$) | 7.09E − 02 | Poundal | Force |
| Joules per centimeter ($J\,cm^{-1}$) | 1.00E + 08 | Dyne | Force |
| Joules per centimeter ($J\,cm^{-1}$) | 1.02E + 04 | Gram force ($g_{force}$) | Force |
| Joules per centimeter ($J\,cm^{-1}$) | 1.02E + 01 | Kilogram force ($kg_{force}$) | Force |
| Joules per centimeter ($J\,cm^{-1}$) | 1.00E + 02 | Newton (N) | Force |
| Joules per centimeter ($J\,cm^{-1}$) | 2.25E + 01 | Pound force ($lb_f$) | Force |
| Joules per centimeter ($J\,cm^{-1}$) | 7.23E + 02 | Poundal | Force |
| Kilogram force ($kg_{force}$) | 9.81E + 05 | Dyne | Force |
| Kilogram force ($kg_{force}$) | 1.00E + 03 | Gram force ($g_{force}$) | Force |
| Kilogram force ($kg_{force}$) | 9.81E − 02 | Joules per centimeter ($J\,cm^{-1}$) | Force |
| Kilogram force ($kg_{force}$) | 9.81E + 00 | Newton (N) | Force |
| Kilogram force ($kg_{force}$) | 2.20E + 00 | Pound force ($lb_f$) | Force |
| Kilogram force ($kg_{force}$) | 7.09E + 01 | Poundal | Force |
| Newtons (N) | 1.00E + 05 | Dyne | Force |
| Newtons (N) | 1.02E + 02 | Gram force ($g_{force}$) | Force |
| Newtons (N) | 1.00E − 02 | Joules per centimeter ($J\,cm^{-1}$) | Force |

(Continued)

(Continued)

| Multiply | By | To obtain | Property |
|---|---|---|---|
| Newtons (N) | 1.02E − 01 | Kilogram force (kg$_{force}$) | Force |
| Newtons (N) | 2.25E − 01 | Pound force (lb$_f$) | Force |
| Newtons (N) | 7.23E + 00 | Poundal | Force |
| Pound force (lb$_f$) | 4.45E + 04 | Dyne | Force |
| Pound force (lb$_f$) | 4.54E + 02 | Gram force (g$_{force}$) | Force |
| Pound force (lb$_f$) | 4.45E − 02 | Joules per centimeter (J cm$^{-1}$) | Force |
| Pound force (lb$_f$) | 4.54E − 01 | Kilogram force (kg$_{force}$) | Force |
| Pound force (lb$_f$) | 4.45E + 00 | Newton (N) | Force |
| Pound force (lb$_f$) | 3.22E + 01 | Poundal | Force |
| Poundal | 1.38E + 04 | Dyne | Force |
| Poundal | 1.41E + 01 | Gram force (g$_{force}$) | Force |
| Poundal | 1.38E − 03 | Joules per centimeter (J cm$^{-1}$) | Force |
| Poundal | 1.41E − 02 | Kilogram force (kg$_{force}$) | Force |
| Poundal | 1.38E − 01 | Newton (N) | Force |
| Poundal | 3.11E − 02 | Pound force (lb$_f$) | Force |
| Foot per second (ft s$^{-1}$) | 3.05E − 01 | Meter per second (m s$^{-1}$) | Hydraulic conductivity |
| US gallon per day per foot squared (gal day$^{-1}$ ft$^{-2}$) | 4.72E − 07 | Meter per second (m s$^{-1}$) | Hydraulic conductivity |
| Angstrom | 1.00E − 07 | Centimeter (cm) | Length |
| Angstrom | 5.47E − 11 | Fathom | Length |
| Angstrom | 3.28E − 10 | Foot (ft) | Length |
| Angstrom | 3.94E − 09 | Inch (in.) | Length |
| Angstrom | 1.00E − 12 | Kilometer (km) | Length |
| Angstrom | 1.00E − 09 | Meter (m) | Length |
| Angstrom | 1.00E − 03 | Micrometer | Length |
| Angstrom | 5.40E − 14 | Mile (nautical) | Length |
| Angstrom | 6.21E − 14 | Mile (statute) | Length |
| Angstrom | 1.00E − 06 | Millimeter (mm) | Length |
| Angstrom | 1.00E − 01 | Nanometer | Length |
| Angstrom | 1.09E − 10 | Yard (yd) | Length |
| Centimeter (cm) | 1.00E + 09 | Angstrom | Length |
| Centimeter (cm) | 5.47E − 03 | Fathom | Length |
| Centimeter (cm) | 3.28E − 02 | Foot (ft) | Length |
| Centimeter (cm) | 3.94E − 01 | Inch (in.) | Length |
| Centimeter (cm) | 1.00E − 04 | Kilometer (km) | Length |
| Centimeter (cm) | 1.00E − 02 | Meter (m) | Length |
| Centimeter (cm) | 1.00E + 04 | Micrometer | Length |
| Centimeter (cm) | 5.40E − 06 | Mile (nautical) | Length |
| Centimeter (cm) | 6.21E − 06 | Mile (statute) | Length |
| Centimeter (cm) | 1.00E + 01 | Millimeter (mm) | Length |
| Centimeter (cm) | 1.00E + 08 | Nanometer | Length |
| Centimeter (cm) | 1.09E − 02 | Yard (yd) | Length |
| Chain (surveyor's) | 2.01E + 01 | Meter (m) | Length |
| Chain (surveyor's) | 2.20E + 01 | Yard (yd) | Length |
| Fathom | 1.83E + 10 | Angstrom | Length |
| Fathom | 1.83E + 02 | Centimeter (cm) | Length |
| Fathom | 6.00E + 00 | Foot (ft) | Length |
| Fathom | 7.20E + 01 | Inch (in.) | Length |
| Fathom | 1.83E − 03 | Kilometer (km) | Length |
| Fathom | 1.83E + 00 | Meter (m) | Length |
| Fathom | 1.83E + 06 | Micrometer | Length |
| Fathom | 9.87E − 04 | Mile (nautical) | Length |
| Fathom | 1.14E − 03 | Mile (statute) | Length |
| Fathom | 1.83E + 03 | Millimeter (mm) | Length |

(Continued)

| Multiply | By | To obtain | Property |
|---|---|---|---|
| Fathom | 1.83E + 09 | Nanometer | Length |
| Fathom | 2.00E + 00 | Yard (yd) | Length |
| Foot (ft) | 3.05E + 09 | Angstrom | Length |
| Foot (ft) | 3.05E + 01 | Centimeter (cm) | Length |
| Foot (ft) | 1.67E − 01 | Fathom | Length |
| Foot (ft) | 1.20E + 01 | Inch (in.) | Length |
| Foot (ft) | 3.05E − 04 | Kilometer (km) | Length |
| Foot (ft) | 3.05E − 01 | Meter (m) | Length |
| Foot (ft) | 3.05E + 05 | Micrometer | Length |
| Foot (ft) | 1.65E − 04 | Mile (nautical) | Length |
| Foot (ft) | 1.89E − 04 | Mile (statute) | Length |
| Foot (ft) | 3.05E + 02 | Millimeter (mm) | Length |
| Foot (ft) | 3.05E + 08 | Nanometer | Length |
| Foot (ft) | 3.33E − 01 | Yard (yd) | Length |
| Inch (in.) | 2.54E + 08 | Angstrom | Length |
| Inch (in.) | 2.54E + 00 | Centimeter (cm) | Length |
| Inch (in.) | 1.69E − 02 | Fathom | Length |
| Inch (in.) | 5.33E − 02 | Foot (ft) | Length |
| Inch (in.) | 2.54E − 05 | Kilometer (km) | Length |
| Inch (in.) | 2.54E − 02 | Meter (m) | Length |
| Inch (in.) | 2.54E + 04 | Micrometer | Length |
| Inch (in.) | 1.37E − 05 | Mile (nautical) | Length |
| Inch (in.) | 1.58E − 05 | Mile (statute) | Length |
| Inch (in.) | 2.54E + 01 | Millimeter (mm) | Length |
| Inch (in.) | 2.54E + 07 | Nanometer | Length |
| Inch (in.) | 2.78E − 02 | Yard (yd) | Length |
| Kilometer (km) | 1.00E + 14 | Angstrom | Length |
| Kilometer (km) | 1.00E + 05 | Centimeter (cm) | Length |
| Kilometer (km) | 5.47E + 02 | Fathom | Length |
| Kilometer (km) | 3.28E + 03 | Foot (ft) | Length |
| Kilometer (km) | 3.94E + 04 | Inch (in.) | Length |
| Kilometer (km) | 1.00E + 03 | Meter (m) | Length |
| Kilometer (km) | 1.00E + 10 | Micrometer | Length |
| Kilometer (km) | 5.40E − 01 | Mile (nautical) | Length |
| Kilometer (km) | 6.21E − 01 | Mile (statute) | Length |
| Kilometer (km) | 1.00E + 07 | Millimeter (mm) | Length |
| Kilometer (km) | 1.00E + 13 | Nanometer | Length |
| Kilometer (km) | 1.09E + 03 | Yard (yd) | Length |
| Meter (m) | 1.00E + 11 | Angstrom | Length |
| Meter (m) | 1.00E + 02 | Centimeter (cm) | Length |
| Meter (m) | 5.47E − 01 | Fathom | Length |
| Meter (m) | 3.28E + 00 | Foot (ft) | Length |
| Meter (m) | 3.94E + 01 | Inch (in.) | Length |
| Meter (m) | 1.00E − 03 | Kilometer (km) | Length |
| Meter (m) | 1.00E + 07 | Micrometer | Length |
| Meter (m) | 5.40E − 04 | Mile (nautical) | Length |
| Meter (m) | 6.21E − 04 | Mile (statute) | Length |
| Meter (m) | 1.00E + 03 | Millimeter (mm) | Length |
| Meter (m) | 1.00E + 10 | Nanometer | Length |
| Meter (m) | 1.09E + 00 | Yard (yd) | Length |
| Micrometer | 1.00E + 04 | Angstrom | Length |
| Micrometer | 1.00E − 04 | Centimeter (cm) | Length |
| Micrometer | 5.47E − 07 | Fathom | Length |
| Micrometer | 3.28E − 06 | Foot (ft) | Length |
| Micrometer | 3.94E − 05 | Inch (in) | Length |
| Micrometer | 1.00E − 08 | Kilometer (km) | Length |
| Micrometer | 1.00E − 05 | Meter (m) | Length |

*(Continued)*

(Continued)

| Multiply | By | To obtain | Property |
|---|---|---|---|
| Micrometer | 5.40E − 10 | Mile (nautical) | Length |
| Micrometer | 6.21E − 10 | Mile (statute) | Length |
| Micrometer | 1.00E − 03 | Millimeter (mm) | Length |
| Micrometer | 1.00E + 03 | Nanometer | Length |
| Micrometer | 1.09E − 06 | Yard (yd) | Length |
| Mile (nautical) (nmi) | 1.85E + 13 | Angstrom | Length |
| Mile (nautical) (nmi) | 1.85E + 05 | Centimeter (cm) | Length |
| Mile (nautical) (nmi) | 1.01E + 03 | Fathom | Length |
| Mile (nautical) (nmi) | 6.08E + 03 | Foot (ft) | Length |
| Mile (nautical) (nmi) | 7.29E + 04 | Inch (in.) | Length |
| Mile (nautical) (nmi) | 1.85E + 00 | Kilometer (km) | Length |
| Mile (nautical) (nmi) | 1.85E + 03 | Meter (m) | Length |
| Mile (nautical) (nmi) | 1.85E + 09 | Micrometer | Length |
| Mile (nautical) (nmi) | 1.15E + 00 | Mile (statute) | Length |
| Mile (nautical) (nmi) | 1.85E + 06 | Millimeter (mm) | Length |
| Mile (nautical) (nmi) | 1.85E + 12 | Nanometer | Length |
| Mile (nautical) (nmi) | 2.03E + 03 | Yard (yd) | Length |
| Mile (statute) | 1.61E + 13 | Angstrom | Length |
| Mile (statute) | 1.61E + 05 | Centimeter (cm) | Length |
| Mile (statute) | 8.80E + 02 | Fathom | Length |
| Mile (statute) | 5.28E + 03 | Foot (ft) | Length |
| Mile (statute) | 6.34E + 04 | Inch (in.) | Length |
| Mile (statute) | 1.61E + 00 | Kilometer (km) | Length |
| Mile (statute) | 1.61E + 03 | Meter (m) | Length |
| Mile (statute) | 1.61E + 09 | Micrometer | Length |
| Mile (statute) | 8.69E − 01 | Mile (nautical) (nmi) | Length |
| Mile (statute) | 1.61E + 06 | Millimeter (mm) | Length |
| Mile (statute) | 1.61E + 12 | Nanometer | Length |
| Mile (statute) | 1.76E + 03 | Yard (yd) | Length |
| Millimeter (mm) | 1.00E + 08 | Angstrom | Length |
| Millimeter (mm) | 1.00E − 01 | Centimeter (cm) | Length |
| Millimeter (mm) | 5.47E − 04 | Fathom | Length |
| Millimeter (mm) | 3.28E − 03 | Foot (ft) | Length |
| Millimeter (mm) | 3.94E − 02 | Inch (in.) | Length |
| Millimeter (mm) | 1.00E − 05 | Kilometer (km) | Length |
| Millimeter (mm) | 1.00E − 03 | Meter (m) | Length |
| Millimeter (mm) | 1.00E + 03 | Micrometer | Length |
| Millimeter (mm) | 5.40E − 07 | Mile (nautical) (nmi) | Length |
| Millimeter (mm) | 6.21E − 07 | Mile (statute) | Length |
| Millimeter (mm) | 1.00E + 07 | Nanometer | Length |
| Millimeter (mm) | 1.09E − 03 | Yard (yd) | Length |
| Nanometer | 1.00E + 01 | Angstrom | Length |
| Nanometer | 1.00E − 06 | Centimeter (cm) | Length |
| Nanometer | 5.47E − 10 | Fathom | Length |
| Nanometer | 3.28E − 09 | Foot (ft) | Length |
| Nanometer | 3.94E − 09 | Inch (in.) | Length |
| Nanometer | 1.00E − 11 | Kilometer (km) | Length |
| Nanometer | 1.00E − 08 | Meter (m) | Length |
| Nanometer | 1.00E − 03 | Micrometer | Length |
| Nanometer | 5.40E − 13 | Mile (nautical) (nmi) | Length |
| Nanometer | 6.21E − 13 | Mile (statute) | Length |
| Nanometer | 1.00E − 05 | Millimeter (mm) | Length |
| Nanometer | 1.09E − 09 | Yard (yd) | Length |
| Yard (yd) | 9.14E + 09 | Angstrom | Length |
| Yard (yd) | 9.14E + 01 | Centimeter (cm) | Length |
| Yard (yd) | 5.00E − 01 | Fathom | Length |
| Yard (yd) | 3.00E + 00 | Foot (ft) | Length |

| Multiply | By | To obtain | Property |
|---|---|---|---|
| Yard (yd) | 3.60E + 01 | Inch (in.) | Length |
| Yard (yd) | 9.14E − 04 | Kilometer (km) | Length |
| Yard (yd) | 9.14E − 01 | Meter (m) | Length |
| Yard (yd) | 9.14E + 05 | Micrometer | Length |
| Yard (yd) | 4.94E − 04 | Mile (nautical) (nmi) | Length |
| Yard (yd) | 5.68E − 04 | Mile (statute) | Length |
| Yard (yd) | 9.14E + 02 | Millimeter (mm) | Length |
| Yard (yd) | 9.14E + 08 | Nanometer | Length |
| Atomic mass unit (amu) | 9.37E − 25 | Dram (avdp.) | Mass |
| Atomic mass unit (amu) | 2.56E − 23 | Grain | Mass |
| Atomic mass unit (amu) | 1.66E − 24 | Gram (g) | Mass |
| Atomic mass unit (amu) | 3.27E − 29 | Hundred weight (long) (hundred wgt.) | Mass |
| Atomic mass unit (amu) | 3.66E − 29 | Hundred weight (short) (hundred wgt.) | Mass |
| Atomic mass unit (amu) | 1.66E − 27 | Kilogram (kg) | Mass |
| Atomic mass unit (amu) | 5.86E − 26 | Ounce (avdp.) | Mass |
| Atomic mass unit (amu) | 5.34E − 26 | Ounce (troy) | Mass |
| Atomic mass unit (amu) | 1.07E − 24 | Pennyweight | Mass |
| Atomic mass unit (amu) | 3.66E − 27 | Pound (avdp.) | Mass |
| Atomic mass unit (amu) | 1.14E − 28 | Slug | Mass |
| Atomic mass unit (amu) | 2.61E − 28 | Stone | Mass |
| Atomic mass unit (amu) | 1.63E − 30 | Ton (long) | Mass |
| Atomic mass unit (amu) | 1.83E − 30 | Ton (short) | Mass |
| Atomic mass unit (amu) | 1.66E − 30 | Tonne (metric) | Mass |
| Dram (avdp.) | 1.07E + 24 | Atomic mass unit (amu) | Mass |
| Dram (avdp.) | 2.73E + 01 | Grain | Mass |
| Dram (avdp.) | 1.77E + 00 | Gram (g) | Mass |
| Dram (avdp.) | 3.49E − 05 | Hundred weight (long) (hundred wgt.) | Mass |
| Dram (avdp.) | 3.91E − 05 | Hundred weight (short) (hundred wgt.) | Mass |
| Dram (avdp.) | 1.77E − 03 | Kilogram (kg) | Mass |
| Dram (avdp.) | 6.25E − 02 | Ounce (avdp.) | Mass |
| Dram (avdp.) | 5.70E − 02 | Ounce (troy) | Mass |
| Dram (avdp.) | 1.14E + 00 | Pennyweight | Mass |
| Dram (avdp.) | 3.91E − 03 | Pound (avdp.) | Mass |
| Dram (avdp.) | 1.21E − 07 | Slug | Mass |
| Dram (avdp.) | 2.79E − 04 | Stone | Mass |
| Dram (avdp.) | 1.74E − 06 | Ton (long) | Mass |
| Dram (avdp.) | 1.95E − 06 | Ton (short) | Mass |
| Dram (avdp.) | 1.77E − 06 | Tonne (metric) | Mass |
| Grain (gr) | 3.90E + 22 | Atomic mass unit (amu) | Mass |
| Grain (gr) | 3.66E − 02 | Dram (avdp.) | Mass |
| Grain (gr) | 6.48E − 02 | Gram (g) | Mass |
| Grain (gr) | 1.28E − 06 | Hundred weight (long) (hundred wgt.) | Mass |
| Grain (gr) | 1.43E − 06 | Hundred weight (short) (hundred wgt.) | Mass |
| Grain (gr) | 6.48E − 05 | Kilogram (kg) | Mass |
| Grain (gr) | 2.29E − 03 | Ounce (avdp.) | Mass |
| Grain (gr) | 2.08E − 03 | Ounce (troy) | Mass |
| Grain (gr) | 4.17E − 02 | Pennyweight | Mass |
| Grain (gr) | 1.43E − 04 | Pound (avdp.) | Mass |
| Grain (gr) | 4.44E − 06 | Slug | Mass |
| Grain (gr) | 1.02E − 05 | Stone | Mass |

*(Continued)*

(Continued)

| Multiply | By | To obtain | Property |
|---|---|---|---|
| Grain (gr) | 6.38E − 08 | Ton (long) | Mass |
| Grain (gr) | 7.14E − 08 | Ton (short) | Mass |
| Grain (gr) | 6.48E − 08 | Tonne (metric) | Mass |
| Gram (g) | 6.02E + 23 | Atomic mass unit (amu) | Mass |
| Gram (g) | 5.64E − 01 | Dram (avdp.) | Mass |
| Gram (g) | 1.54E + 01 | Grain (gr) | Mass |
| Gram (g) | 1.97E − 05 | Hundred weight (long) (hundred wgt.) | Mass |
| Gram (g) | 2.20E − 05 | Hundred weight (short) (hundred wgt.) | Mass |
| Gram (g) | 1.00E − 03 | Kilogram (kg) | Mass |
| Gram (g) | 3.53E − 02 | Ounce (avdp.) | Mass |
| Gram (g) | 3.22E − 02 | Ounce (troy) | Mass |
| Gram (g) | 6.43E − 01 | Pennyweight | Mass |
| Gram (g) | 2.20E − 03 | Pound (avdp.) | Mass |
| Gram (g) | 6.85E − 05 | Slug | Mass |
| Gram (g) | 1.57E − 04 | Stone | Mass |
| Gram (g) | 9.84E + 00 | Ton (long) | Mass |
| Gram (g) | 1.10E − 06 | Ton (short) | Mass |
| Gram (g) | 1.00E − 05 | Tonne (metric) | Mass |
| Hundred weight (long) (hundred wgt.) | 3.03E + 28 | Atomic mass unit (amu) | Mass |
| Hundred weight (long) (hundred wgt.) | 2.87E + 04 | Dram (avdp.) | Mass |
| Hundred weight (long) (hundred wgt.) | 7.84E + 05 | Grain (gr) | Mass |
| Hundred weight (long) (hundred wgt.) | 5.08E + 04 | Gram (g) | Mass |
| Hundred weight (long) (hundred wgt.) | 1.12E + 00 | Hundred weight (short) (hundred wgt.) | Mass |
| Hundred weight (long) (hundred wgt.) | 5.08E + 01 | Kilogram (kg) | Mass |
| Hundred weight (long) (hundred wgt.) | 1.79E + 03 | Ounce (avdp.) | Mass |
| Hundred weight (long) (hundred wgt.) | 1.63E + 03 | Ounce (troy) | Mass |
| Hundred weight (long) (hundred wgt.) | 3.27E + 04 | Pennyweight | Mass |
| Hundred weight (long) (hundred wgt.) | 1.12E + 02 | Pound (avdp.) | Mass |
| Hundred weight (long) (hundred wgt.) | 3.48E + 00 | Slug | Mass |
| Hundred weight (long) (hundred wgt.) | 8.00E + 00 | Stone | Mass |
| Hundred weight (long) (hundred wgt.) | 5.00E − 02 | Ton (long) | Mass |
| Hundred weight (long) (hundred wgt.) | 5.60E − 02 | Ton (short) | Mass |
| Hundred weight (long) (hundred wgt.) | 5.08E − 02 | Tonne (metric) | Mass |
| Hundred weight (short) (hundred wgt.) | 2.73E + 28 | Atomic mass unit (amu) | Mass |
| Hundred weight (short) (hundred wgt.) | 2.56E + 04 | Dram (avdp.) | Mass |
| Hundred weight (short) (hundred wgt.) | 7.00E + 05 | Grain (gr) | Mass |

(Continued)

| Multiply | By | To obtain | Property |
|---|---|---|---|
| Hundred weight (short) (hundred wgt.) | 4.54E + 04 | Gram (g) | Mass |
| Hundred weight (short) (hundred wgt.) | 8.93E − 01 | Hundred weight (long) (hundred wgt.) | Mass |
| Hundred weight (short) (hundred wgt.) | 4.54E + 01 | Kilogram (kg) | Mass |
| Hundred weight (short) (hundred wgt.) | 1.60E + 03 | Ounce (avdp.) | Mass |
| Hundred weight (short) (hundred wgt.) | 1.46E + 03 | Ounce (troy) | Mass |
| Hundred weight (short) (hundred wgt.) | 2.92E + 04 | Pennyweight | Mass |
| Hundred weight (short) (hundred wgt.) | 1.00E + 02 | Pound (avdp.) | Mass |
| Hundred weight (short) (hundred wgt.) | 3.11E + 00 | Slug | Mass |
| Hundred weight (short) (hundred wgt.) | 7.14E + 00 | Stone | Mass |
| Hundred weight (short) (hundred wgt.) | 4.46E − 02 | Ton (long) | Mass |
| Hundred weight (short) (hundred wgt.) | 5.00E − 02 | Ton (short) | Mass |
| Hundred weight (short) (hundred wgt.) | 4.54E − 02 | Tonne (metric) | Mass |
| Kilogram (kg) | 6.02E + 26 | Atomic mass unit (amu) | Mass |
| Kilogram (kg) | 5.64E + 02 | Dram (avdp.) | Mass |
| Kilogram (kg) | 1.54E + 04 | Grain (gr) | Mass |
| Kilogram (kg) | 1.00E + 03 | Gram (g) | Mass |
| Kilogram (kg) | 1.97E − 02 | Hundred weight (long) (hundred wgt.) | Mass |
| Kilogram (kg) | 2.20E − 02 | Hundred weight (short) (hundred wgt.) | Mass |
| Kilogram (kg) | 3.53E + 01 | Ounce (avdp.) | Mass |
| Kilogram (kg) | 3.22E + 01 | Ounce (troy) | Mass |
| Kilogram (kg) | 6.43E + 00 | Pennyweight | Mass |
| Kilogram (kg) | 2.20E + 00 | Pound (avdp.) | Mass |
| Kilogram (kg) | 6.85E − 02 | Slug | Mass |
| Kilogram (kg) | 1.57E − 01 | Stone | Mass |
| Kilogram (kg) | 9.84E − 04 | Ton (long) | Mass |
| Kilogram (kg) | 1.10E − 03 | Ton (short) | Mass |
| Kilogram (kg) | 1.00E − 03 | Tonne (metric) | Mass |
| Ounce (avdp.) | 1.71E + 25 | Atomic mass unit (amu) | Mass |
| Ounce (avdp.) | 1.60E + 01 | Dram (avdp.) | Mass |
| Ounce (avdp.) | 4.38E + 02 | Grain (gr) | Mass |
| Ounce (avdp.) | 2.83E + 01 | Gram (g) | Mass |
| Ounce (avdp.) | 5.58E − 04 | Hundred weight (long) (hundred wgt.) | Mass |
| Ounce (avdp.) | 6.25E − 04 | Hundred weight (short) (hundred wgt.) | Mass |
| Ounce (avdp.) | 2.83E − 02 | Kilogram (kg) | Mass |
| Ounce (avdp.) | 9.11E − 01 | Ounce (troy) | Mass |
| Ounce (avdp.) | 1.82E + 01 | Pennyweight | Mass |
| Ounce (avdp.) | 6.25E − 02 | Pound (avdp.) | Mass |
| Ounce (avdp.) | 1.64E − 03 | Slug | Mass |
| Ounce (avdp.) | 4.46E − 03 | Stone | Mass |
| Ounce (avdp.) | 2.79E − 05 | Ton (long) | Mass |

(Continued)

(Continued)

| Multiply | By | To obtain | Property |
|----------|-----|-----------|----------|
| Ounce (avdp.) | 3.13E − 05 | Ton (short) | Mass |
| Ounce (avdp.) | 2.83E + 00 | Tonne (metric) | Mass |
| Ounce (troy) | 1.87E + 25 | Atomic mass unit (amu) | Mass |
| Ounce (troy) | 1.76E + 01 | Dram (avdp.) | Mass |
| Ounce (troy) | 4.80E + 02 | Grain (gr) | Mass |
| Ounce (troy) | 3.11E + 01 | Gram (g) | Mass |
| Ounce (troy) | 6.12E − 04 | Hundred weight (long) (hundred wgt.) | Mass |
| Ounce (troy) | 6.86E − 04 | Hundred weight (short) (hundred wgt.) | Mass |
| Ounce (troy) | 3.11E − 02 | Kilogram (kg) | Mass |
| Ounce (troy) | 1.10E + 00 | Ounce (avdp.) | Mass |
| Ounce (troy) | 2.00E + 01 | Pennyweight | Mass |
| Ounce (troy) | 6.86E − 02 | Pound (avdp.) | Mass |
| Ounce (troy) | 2.13E − 03 | Slug | Mass |
| Ounce (troy) | 4.90E − 03 | Stone | Mass |
| Ounce (troy) | 3.06E − 05 | Ton (long) | Mass |
| Ounce (troy) | 3.43E + 05 | Ton (short) | Mass |
| Ounce (troy) | 3.11E − 05 | Tonne (metric) | Mass |
| Pennyweight | 9.37E + 23 | Atomic mass unit (amu) | Mass |
| Pennyweight | 8.78E − 01 | Dram (avdp.) | Mass |
| Pennyweight | 2.40E + 01 | Grain (gr) | Mass |
| Pennyweight | 1.56E + 00 | Gram (g) | Mass |
| Pennyweight | 3.06E − 05 | Hundred weight (long) (hundred wgt.) | Mass |
| Pennyweight | 3.43E − 05 | Hundred weight (short) (hundred wgt.) | Mass |
| Pennyweight | 1.56E − 03 | Kilogram (kg) | Mass |
| Pennyweight | 5.49E − 02 | Ounce (avdp.) | Mass |
| Pennyweight | 5.00E − 02 | Ounce (troy) | Mass |
| Pennyweight | 3.43E − 03 | Pound (avdp.) | Mass |
| Pennyweight | 1.07E − 04 | Slug | Mass |
| Pennyweight | 2.48E − 04 | Stone | Mass |
| Pennyweight | 1.53E − 06 | Ton (long) | Mass |
| Pennyweight | 1.71E − 06 | Ton (short) | Mass |
| Pennyweight | 1.00E + 02 | Tonne (metric) | Mass |
| Pound mass ($lb_m$) (avdp.) | 2.73E + 26 | Atomic mass unit (amu) | Mass |
| Pound mass ($lb_m$) (avdp.) | 2.56E + 02 | Dram (avdp.) | Mass |
| Pound mass ($lb_m$) (avdp.) | 7.00E + 03 | Grain (gr) | Mass |
| Pound mass ($lb_m$) (avdp.) | 4.54E + 02 | Gram (g) | Mass |
| Pound mass ($lb_m$) (avdp.) | 8.93E − 03 | Hundred weight (long) (hundred wgt.) | Mass |
| Pound mass ($lb_m$) (avdp.) | 1.00E − 02 | Hundred weight (short) (hundred wgt.) | Mass |
| Pound mass ($lb_m$) (avdp.) | 4.54E − 01 | Kilogram (kg) | Mass |
| Pound mass ($lb_m$) (avdp.) | 1.60E + 01 | Ounce (avdp.) | Mass |
| Pound mass ($lb_m$) (avdp.) | 1.46E + 01 | Ounce (troy) | Mass |
| Pound mass ($lb_m$) (avdp.) | 2.92E + 02 | Pennyweight | Mass |
| Pound mass ($lb_m$) (avdp.) | 3.11E − 02 | Slug | Mass |
| Pound mass ($lb_m$) (avdp.) | 7.14E − 02 | Stone | Mass |
| Pound mass ($lb_m$) (avdp.) | 4.46E − 04 | Ton (long) | Mass |
| Pound mass ($lb_m$) (avdp.) | 5.00E − 04 | Ton (short) | Mass |
| Pound mass ($lb_m$) (avdp.) | 4.54E − 04 | Tonne (metric) | Mass |
| Slug | 8.79E + 27 | Atomic mass unit (amu) | Mass |
| Slug | 8.24E + 03 | Dram (avdp.) | Mass |
| Slug | 2.25E + 05 | Grain (gr) | Mass |
| Slug | 1.46E + 04 | Gram (g) | Mass |

(Continued)

| Multiply | By | To obtain | Property |
|---|---|---|---|
| Slug | 2.87E − 01 | Hundred weight (long) (hundred wgt.) | Mass |
| Slug | 3.22E − 01 | Hundred weight (short) (hundred wgt.) | Mass |
| Slug | 1.46E + 01 | Kilogram (kg) | Mass |
| Slug | 5.15E + 02 | Ounce (avdp.) | Mass |
| Slug | 4.69E + 02 | Ounce (troy) | Mass |
| Slug | 9.38E + 03 | Pennyweight | Mass |
| Slug | 3.22E + 01 | Pound (lb) (avdp.) | Mass |
| Slug | 2.30E + 00 | Stone | Mass |
| Slug | 1.44E − 02 | Ton (long) | Mass |
| Slug | 1.61E − 02 | Ton (short) | Mass |
| Slug | 1.46E − 02 | Tonne (metric) | Mass |
| Stone | 3.82E + 27 | Atomic mass unit (amu) | Mass |
| Stone | 3.58E + 03 | Dram (avdp.) | Mass |
| Stone | 9.80E + 04 | Grain (gr) | Mass |
| Stone | 6.35E + 03 | Gram (g) | Mass |
| Stone | 1.25E − 01 | Hundred weight (long) (hundred wgt.) | Mass |
| Stone | 1.40E − 01 | Hundred weight (short) (hundred wgt.) | Mass |
| Stone | 6.35E + 00 | Kilogram (kg) | Mass |
| Stone | 2.24E + 02 | Ounce (avdp.) | Mass |
| Stone | 2.04E + 04 | Ounce (troy) | Mass |
| Stone | 4.08E + 03 | Pennyweight | Mass |
| Stone | 1.40E + 01 | Pound (lb) (avdp.) | Mass |
| Stone | 4.35E − 01 | Slug | Mass |
| Stone | 6.25E − 03 | Ton (long) | Mass |
| Stone | 7.00E − 03 | Ton (short) | Mass |
| Stone | 6.35E − 03 | Tonne (metric) | Mass |
| Ton (long) | 6.12E + 27 | Atomic mass unit (amu) | Mass |
| Ton (long) | 5.73E + 05 | Dram (avdp.) | Mass |
| Ton (long) | 1.57E + 07 | Grain (gr) | Mass |
| Ton (long) | 1.02E + 06 | Gram (g) | Mass |
| Ton (long) | 2.00E + 01 | Hundred weight (long) (hundred wgt.) | Mass |
| Ton (long) | 2.24E + 01 | Hundred weight (short) (hundred wgt.) | Mass |
| Ton (long) | 1.02E + 03 | Kilogram (kg) | Mass |
| Ton (long) | 3.58E + 04 | Ounce (avdp.) | Mass |
| Ton (long) | 3.27E + 04 | Ounce (troy) | Mass |
| Ton (long) | 6.53E + 05 | Pennyweight | Mass |
| Ton (long) | 2.24E + 03 | Pound (lb) (avdp.) | Mass |
| Ton (long) | 6.91E + 01 | Slug | Mass |
| Ton (long) | 1.60E + 02 | Stone | Mass |
| Ton (long) | 1.12E + 00 | Ton (short) | Mass |
| Ton (long) | 1.02E + 00 | Tonne (metric) | Mass |
| Ton (short) | 5.46E + 29 | Atomic mass unit (amu) | Mass |
| Ton (short) | 5.12E + 05 | Dram (avdp.) | Mass |
| Ton (short) | 1.40E + 07 | Grain (gr) | Mass |
| Ton (short) | 9.07E + 05 | Gram (g) | Mass |
| Ton (short) | 1.79E + 01 | Hundred weight (long) (hundred wgt.) | Mass |
| Ton (short) | 2.00E + 01 | Hundred weight (short) (hundred wgt.) | Mass |
| Ton (short) | 9.07E + 02 | Kilogram (kg) | Mass |

(Continued)

(Continued)

| Multiply | By | To obtain | Property |
|---|---|---|---|
| Ton (short) | 3.20E + 04 | Ounce (avdp.) | Mass |
| Ton (short) | 2.92E + 04 | Ounce (troy) | Mass |
| Ton (short) | 5.83E + 05 | Pennyweight | Mass |
| Ton (short) | 2.00E + 03 | Pound (lb) (avdp.) | Mass |
| Ton (short) | 6.22E + 01 | Slug | Mass |
| Ton (short) | 1.43E + 02 | Stone | Mass |
| Ton (short) | 8.93E − 01 | Ton (long) | Mass |
| Ton (short) | 9.07E − 01 | Tonne (metric) | Mass |
| Tonne (metric) | 6.02E + 29 | Atomic mass unit (amu) | Mass |
| Tonne (metric) | 5.64E + 05 | Dram (avdp.) | Mass |
| Tonne (metric) | 1.54E + 07 | Grain (gr) | Mass |
| Tonne (metric) | 1.00E + 05 | Gram (g) | Mass |
| Tonne (metric) | 1.97E + 01 | Hundred weight (long) (hundred wgt.) | Mass |
| Tonne (metric) | 2.20E + 01 | Hundred weight (short) (hundred wgt.) | Mass |
| Tonne (metric) | 1.00E + 03 | Kilogram (kg) | Mass |
| Tonne (metric) | 3.53E + 04 | Ounce (avdp.) | Mass |
| Tonne (metric) | 3.22E + 04 | Ounce (troy) | Mass |
| Tonne (metric) | 6.43E + 05 | Pennyweight | Mass |
| Tonne (metric) | 2.204.62 | Pound (lb) (avdp.) | Mass |
| Tonne (metric) | 6.85E + 01 | Slug | Mass |
| Tonne (metric) | 1.57E + 02 | Stone | Mass |
| Tonne (metric) | 9.84E − 01 | Ton (long) | Mass |
| Tonne (metric) | 1.10E + 00 | Ton (short) | Mass |
| Centimeter per second (cm s$^{-1}$) | 1.02E − 05 | Centimeter squared (cm$^2$) | Permeability |
| Centimeter per second (cm s$^{-1}$) | 1.03E + 03 | Darcy | Permeability |
| Centimeter per second (cm s$^{-1}$) | 2.83E + 03 | Foot per day (ft day$^{-1}$) | Permeability |
| Centimeter per second (cm s$^{-1}$) | 3.28E − 02 | Foot per second (ft s$^{-1}$) | Permeability |
| Centimeter per second (cm s$^{-1}$) | 1.06E + 06 | Foot per year (ft year$^{-1}$) | Permeability |
| Centimeter per second (cm s$^{-1}$) | 1.10E − 08 | Foot squared (ft$^2$) | Permeability |
| Centimeter per second (cm s$^{-1}$) | 2.12E + 04 | Gallons per day (US) per foot squared (Meinzer) (gpd ft$^{-2}$) | Permeability |
| Centimeter per second (cm s$^{-1}$) | 1.00E + 01 | Liter per second per meter squared (L s$^{-1}$ m$^{-2}$) | Permeability |
| Centimeter per second (cm s$^{-1}$) | 1.02E − 09 | Meter squared (m$^2$) | Permeability |
| Centimeter squared (cm$^2$) | 9.81E + 04 | Centimeter per second (cm s$^{-1}$) | Permeability |
| Centimeter squared (cm$^2$) | 1.01E + 08 | Darcy | Permeability |
| Centimeter squared (cm$^2$) | 2.78E + 08 | Foot per day (ft day$^{-1}$) | Permeability |
| Centimeter squared (cm$^2$) | 3.22E + 03 | Foot per second (ft s$^{-1}$) | Permeability |
| Centimeter squared (cm$^2$) | 1.015 10$^{11}$ | Foot per year (ft year$^{-1}$) | Permeability |
| Centimeter squared (cm$^2$) | 1.08E − 03 | Foot squared (ft$^2$) | Permeability |
| Centimeter squared (cm$^2$) | 2.08E + 09 | Gallons per day (US) per foot squared (Meinzer) (gpd ft$^{-2}$) | Permeability |
| Centimeter squared (cm$^2$) | 9.81E + 05 | Liter per second per meter squared (L s$^{-1}$ m$^{-2}$) | Permeability |
| Centimeter squared (cm$^2$) | 3.97E + 04 | Meter squared (m$^2$) | Permeability |
| Darcy | 9.68E − 04 | Centimeter per second (cm s$^{-1}$) | Permeability |
| Darcy | 9.87E − 09 | Centimeter squared (cm$^2$) | Permeability |
| Darcy | 2.74E + 00 | Foot per day (ft day$^{-1}$) | Permeability |
| Darcy | 3.18E − 05 | Foot per second (ft s$^{-1}$) | Permeability |
| Darcy | 1.00E + 03 | Foot per year (ft year$^{-1}$) | Permeability |
| Darcy | 1.06E − 11 | Foot squared (ft$^2$) | Permeability |
| Darcy | 2.05E + 01 | Gallons per day (US) per foot squared (Meinzer) (gpd ft$^{-2}$) | Permeability |

| Multiply | By | To obtain | Property |
|---|---|---|---|
| Darcy | 9.68E − 03 | Liter per second per meter squared ($L\,s\,m^{-2}$) | Permeability |
| Darcy | 9.90E − 13 | Meter squared ($m^2$) | Permeability |
| Foot per second (ft s$^{-1}$) | 3.05E + 01 | Centimeter per second (cm s$^{-1}$) | Permeability |
| Foot per second (ft s$^{-1}$) | 3.11E − 04 | Centimeter squared (cm$^2$) | Permeability |
| Foot per second (ft s$^{-1}$) | 3.15E + 04 | Darcy | Permeability |
| Foot per second (ft s$^{-1}$) | 8.64E + 04 | Foot per day (ft day$^{-1}$) | Permeability |
| Foot per second (ft s$^{-1}$) | 3.16E + 07 | Foot per year (ft year$^{-1}$) | Permeability |
| Foot per second (ft s$^{-1}$) | 3.35E − 07 | Foot squared (ft$^2$) | Permeability |
| Foot per second (ft s$^{-1}$) | 6.46E + 05 | Gallons per day (US) per foot squared (Meinzer) (gpd ft$^{-2}$) | Permeability |
| Foot per second (ft s$^{-1}$) | 3.05E + 02 | Liter per second per meter squared ($L\,s\,m^{-2}$) | Permeability |
| Foot per second (ft s$^{-1}$) | 3.11E − 08 | Meter squared ($m^2$) | Permeability |
| Foot per year (ft year$^{-1}$) | 9.66E − 07 | Centimeter per second (cm s$^{-1}$) | Permeability |
| Foot per year (ft year$^{-1}$) | 9.85E − 12 | Centimeter squared (cm$^2$) | Permeability |
| Foot per year (ft year$^{-1}$) | 9.98E − 04 | Darcy | Permeability |
| Foot per year (ft year$^{-1}$) | 2.74E − 03 | Foot per day (ft day$^{-1}$) | Permeability |
| Foot per year (ft year$^{-1}$) | 3.17E − 08 | Foot per second (ft s$^{-1}$) | Permeability |
| Foot per year (ft year$^{-1}$) | 1.06E − 14 | Foot squared (ft$^2$) | Permeability |
| Foot per year (ft year$^{-1}$) | 2.05E − 02 | Gallons per day (US) per foot squared (Meinzer) (gpd ft$^{-2}$) | Permeability |
| Foot per year (ft year$^{-1}$) | 9.66E − 06 | Liter per second per meter squared ($L\,s^{-1}\,m^{-2}$) | Permeability |
| Foot per year (ft year$^{-1}$) | 9.85E − 16 | Meter squared ($m^2$) | Permeability |
| Foot squared (ft$^2$) | 9.11E + 07 | Centimeter per second (cm s$^{-1}$) | Permeability |
| Foot squared (ft$^2$) | 9.29E + 02 | Centimeter squared (cm$^2$) | Permeability |
| Foot squared (ft$^2$) | 9.41E + 10 | Darcy | Permeability |
| Foot squared (ft$^2$) | 2.56E + 11 | Foot per day (ft day$^{-1}$) | Permeability |
| Foot squared (ft$^2$) | 2.99E + 06 | Foot per second (ft s$^{-1}$) | Permeability |
| Foot squared (ft$^2$) | 9.43E + 13 | Foot per year (ft year$^{-1}$) | Permeability |
| Foot squared (ft$^2$) | 1.93E + 12 | Gallons per day (US) per foot squared (Meinzer) (gpd ft$^{-2}$) | Permeability |
| Foot squared (ft$^2$) | 9.11E + 08 | Liter per second per meter squared ($L\,s^{-1}\,m^{-2}$) | Permeability |
| Foot squared (ft$^2$) | 9.29E − 02 | Meter squared ($m^2$) | Permeability |
| Foot per day | 3.53E − 04 | Centimeter per second (cm s$^{-1}$) | Permeability |
| Foot per day | 3.60E − 09 | Centimeter squared (cm$^2$) | Permeability |
| Foot per day | 3.65E − 01 | Darcy | Permeability |
| Foot per day | 1.16E − 05 | Foot per second (ft s$^{-1}$) | Permeability |
| Foot per day | 3.65E + 02 | Foot per year (ft year$^{-1}$) | Permeability |
| Foot per day | 3.87E − 12 | Foot squared (ft$^2$) | Permeability |
| Foot per day | 7.48E + 00 | Gallons per day (US) per foot squared (Meinzer) (gpd ft$^{-2}$) | Permeability |
| Foot per day | 3.53E − 03 | Liter per second per meter squared ($L\,s\,m^{-2}$) | Permeability |
| Foot per day | 3.60E − 13 | Meter squared ($m^2$) | Permeability |
| Gallons per day (US) per foot squared (Meinzer) (gpd ft$^{-2}$) | 4.72E − 05 | Centimeter per second (cm s$^{-1}$) | Permeability |
| Gallons per day (US) per foot squared (Meinzer) (gpd ft$^{-2}$) | 4.81E − 10 | Centimeter squared (cm$^2$) | Permeability |
| Gallons per day (US) per foot squared (Meinzer) (gpd ft$^{-2}$) | 4.87E − 02 | Darcy | Permeability |
| Gallons per day (US) per foot squared (Meinzer) (gpd ft$^{-2}$) | 1.55E − 06 | Foot per second (ft s$^{-1}$) | Permeability |
| Gallons per day (US) per foot squared (Meinzer) (gpd ft$^{-2}$) | 4.99E + 01 | Foot per year (ft year$^{-1}$) | Permeability |

| Multiply | By | To obtain | Property |
|---|---|---|---|
| Gallons per day (US) per foot squared (Meinzer) (gpd ft$^{-2}$) | 5.17E − 13 | Foot squared (ft$^2$) | Permeability |
| Gallons per day (US) per foot squared (Meinzer) (gpd ft$^{-2}$) | 1.34E − 01 | Foot per day | Permeability |
| Gallons per day (US) per foot squared (Meinzer) (gpd ft$^{-2}$) | 4.72E − 04 | Liter per second per meter squared (L s$^{-1}$ m$^{-2}$) | Permeability |
| Gallons per day (US) per foot squared (Meinzer) (gpd ft$^{-2}$) | 4.81E − 14 | Meter squared (m$^2$) | Permeability |
| Liter per second per meter squared (L s$^{-1}$ m$^{-2}$) | 3.97E + 04 | Centimeter per second (cm s$^{-1}$) | Permeability |
| Liter per second per meter squared (L s$^{-1}$ m$^{-2}$) | 1.02E − 06 | Centimeter squared (cm$^2$) | Permeability |
| Liter per second per meter squared (L s$^{-1}$ m$^{-2}$) | 1.03E + 02 | Darcy | Permeability |
| Liter per second per meter squared (L s$^{-1}$ m$^{-2}$) | 2.83E + 02 | Foot per day (ft day$^{-1}$) | Permeability |
| Liter per second per meter squared (L s$^{-1}$ m$^{-2}$) | 3.28E − 03 | Foot per second (ft s$^{-1}$) | Permeability |
| Liter per second per meter squared (L s$^{-1}$ m$^{-2}$) | 1.04E + 05 | Foot per year (ft year$^{-1}$) | Permeability |
| Liter per second per meter squared (L s$^{-1}$ m$^{-2}$) | 1.10E − 09 | Foot squared (ft$^2$) | Permeability |
| Liter per second per meter squared (L s$^{-1}$ m$^{-2}$) | 2.12E + 03 | Gallons per day (US) per foot squared (Meinzer) (gpd ft$^{-2}$) | Permeability |
| Liter per second per meter squared (L s$^{-1}$ m$^{-2}$) | 1.02E − 10 | Meter squared (m$^2$) | Permeability |
| Meter squared (m$^2$) | 9.81E + 08 | Centimeter per second (cm s$^{-1}$) | Permeability |
| Meter squared (m$^2$) | 1.04E + 02 | Centimeter squared (cm$^2$) | Permeability |
| Meter squared (m$^2$) | 1.01E + 12 | Darcy | Permeability |
| Meter squared (m$^2$) | 2.78E + 12 | Foot per day (ft day$^{-1}$) | Permeability |
| Meter squared (m$^2$) | 3.22E + 07 | Foot per second (ft s$^{-1}$) | Permeability |
| Meter squared (m$^2$) | 1.02E + 15 | Foot per year (ft year$^{-1}$) | Permeability |
| Meter squared (m$^2$) | 1.08E + 01 | Foot squared (ft$^2$) | Permeability |
| Meter squared (m$^2$) | 2.08E + 13 | Gallons per day (US) per foot squared (Meinzer) (gpd ft$^{-2}$) | Permeability |
| Meter squared (m$^2$) | $9.8067 \times 10^9$ | Liter per second per meter squared (L s$^{-1}$ m$^{-2}$) | Permeability |
| Atmosphere (atm) | 1.01E + 00 | Bars | Pressure |
| Atmosphere (atm) | 3.40E + 01 | Feet of water (68°F) | Pressure |
| Atmosphere (atm) | 2.99E + 01 | Inch of mercury (32°F) | Pressure |
| Atmosphere (atm) | 4.08E + 02 | Inch of water (68°F) | Pressure |
| Atmosphere (atm) | 1.01E + 02 | Kilopascal (kPa) | Pressure |
| Atmosphere (atm) | 1.03E + 00 | Kilogram per centimeter squared (kg cm$^{-2}$) | Pressure |
| Atmosphere (atm) | 1.03E + 04 | Kilogram per meter squared (kg m$^{-2}$) | Pressure |
| Atmosphere (atm) | 7.60E + 02 | Millimeter of mercury (32°F) | Pressure |
| Atmosphere (atm) | 2.12E + 03 | Pound per foot squared (lb ft$^{-2}$) | Pressure |
| Atmosphere (atm) | 1.47E + 01 | Pound per inch squared (lb in.$^{-2}$, psi) | Pressure |
| Bars | 9.87E − 01 | Atmosphere (atm) | Pressure |
| Bars | 3.35E + 01 | Feet of water (68°F) | Pressure |
| Bars | 2.95E + 01 | Inch of mercury (32°F) | Pressure |
| Bars | 4.02E + 02 | Inch of water (68°F) | Pressure |
| Bars | 1.00E + 02 | Kilopascal (kPa) | Pressure |

(Continued)

| Multiply | By | To obtain | Property |
|---|---|---|---|
| Bars | 1.02E + 00 | Kilogram per centimeter squared (kg cm$^{-2}$) | Pressure |
| Bars | 1.02E + 04 | Kilogram per meter squared (kg m$^{-2}$) | Pressure |
| Bars | 7.50E + 02 | Millimeter of mercury (32°F) | Pressure |
| Bars | 2.09E + 03 | Pound per foot squared (lb ft$^{-2}$) | Pressure |
| Bars | 1.45E + 01 | Pound per inch squared (lb in.$^{-2}$) | Pressure |
| Feet of water (68°F) | 2.95E − 02 | Atmosphere (atm) | Pressure |
| Feet of water (68°F) | 2.98E − 02 | Bars | Pressure |
| Feet of water (68°F) | 1.20E + 01 | Inch of water (68°F) | Pressure |
| Feet of water (68°F) | 2.98E + 00 | Kilopascal (kPa) | Pressure |
| Feet of water (68°F) | 3.04E − 02 | Kilogram per centimeter squared (kg cm$^{-2}$) | Pressure |
| Feet of water (68°F) | 3.04E + 02 | Kilogram per meter squared (kg m$^{-2}$) | Pressure |
| Feet of water (68°F) | 2.24E + 01 | Millimeter of mercury (32°F) | Pressure |
| Feet of water (68°F) | 6.23E + 01 | Pound per foot squared (lb ft$^{-2}$) | Pressure |
| Feet of water (68°F) | 4.33E − 01 | Pound per inch squared (lb in.$^{-2}$) | Pressure |
| Feet of water (ft$_{water}$) | 8.85E − 01 | Inch of mercury (in. Hg) | Pressure |
| Inch of mercury (in Hg) | 3.39E + 03 | Pascal (Pa) | Pressure |
| Inch of mercury (in Hg) | 1.13E + 00 | Feet of water (ft) | Pressure |
| Inch of mercury (in Hg) | 2.53E + 01 | Millimeter of mercury (mm Hg) | Pressure |
| Inch of mercury (in. Hg) (32°F) | 3.34E − 02 | Atmosphere (atm) | Pressure |
| Inch of mercury (in. Hg) (32°F) | 3.39E − 02 | Bars | Pressure |
| Inch of mercury (in. Hg) (32°F) | 1.14E + 00 | Feet of water (68°F) | Pressure |
| Inch of mercury (in. Hg) (32°F) | 1.36E + 01 | Inch of water (68°F) | Pressure |
| Inch of mercury (in. Hg) (32°F) | 3.39E + 00 | Kilopascal (kPa) | Pressure |
| Inch of mercury (in. Hg) (32°F) | 3.45E − 02 | Kilogram per centimeter squared (kg cm$^{-2}$) | Pressure |
| Inch of mercury (in. Hg) (32°F) | 3.45E + 02 | Kilogram per meter squared (kg m$^{-2}$) | Pressure |
| Inch of mercury (in. Hg) (32°F) | 2.54E + 01 | Millimeter of mercury (32°F) | Pressure |
| Inch of mercury (in. Hg) (32°F) | 7.07E + 01 | Pound per foot squared (lb ft$^{-2}$) | Pressure |
| Inch of mercury (in. Hg) (32°F) | 4.91E − 01 | Pound per inch squared (psi lb in.$^{-2}$) | Pressure |
| Inch of water (68°F) | 2.45E − 03 | Atmosphere (atm) | Pressure |
| Inch of water (68°F) | 2.49E − 03 | Bars | Pressure |
| Inch of water (68°F) | 8.33E − 02 | Feet of water (68°F) | Pressure |
| Inch of water (68°F) | 7.34E − 02 | Inches of mercury (32°F) | Pressure |
| Inch of water (68°F) | 2.49E − 01 | Kilopascal (kPa) | Pressure |
| Inch of water (68°F) | 2.53E − 03 | Kilogram per centimeter squared (kg cm$^{-2}$) | Pressure |
| Inch of water (68°F) | 2.54E + 01 | Kilogram per meter squared (kg m$^{-2}$) | Pressure |
| Inch of water (68°F) | 1.87E + 00 | Millimeter of mercury (32°F) | Pressure |
| Inch of water (68°F) | 5.20E + 00 | Pound per foot squared (lb ft$^{-2}$) | Pressure |
| Inch of water (68°F) | 3.61E − 02 | Pound per inch squared (psi lb in.$^{-2}$) | Pressure |
| Kilopascal (kPa) | 9.87E − 03 | Atmosphere (atm) | Pressure |
| Kilopascal (kPa) | 1.00E − 02 | Bars | Pressure |
| Kilopascal (kPa) | 3.35E − 01 | Feet of water (68°F) | Pressure |
| Kilopascal (kPa) | 2.95E − 01 | Inch of mercury (32°F) | Pressure |
| Kilopascal (kPa) | 4.02E + 00 | Inch of water (68°F) | Pressure |
| Kilopascal (kPa) | 1.02E − 02 | Kilogram per centimeter squared (kg cm$^{-2}$) | Pressure |

(Continued)

(Continued)

| Multiply | By | To obtain | Property |
|---|---|---|---|
| Kilopascal (kPa) | 1.02E + 02 | Kilogram per meter squared (kg m$^{-2}$) | Pressure |
| Kilopascal (kPa) | 7.50E + 00 | Millimeter of mercury (32°F) | Pressure |
| Kilopascal (kPa) | 2.09E + 01 | Pound per foot squared (lb ft$^{-2}$) | Pressure |
| Kilopascal (kPa) | 3.61E − 02 | Pound per inch squared (psi lb in.$^{-2}$) | Pressure |
| Kilogram per centimeter squared (kg cm$^{-2}$) | 9.68E − 01 | Atmosphere (atm) | Pressure |
| Kilogram per centimeter squared (kg cm$^{-2}$) | 9.81E − 01 | Bars | Pressure |
| Kilogram per centimeter squared (kg cm$^{-2}$) | 3.29E + 01 | Feet of water (68°F) | Pressure |
| Kilogram per centimeter squared (kg cm$^{-2}$) | 2.90E + 01 | Inch of mercury (32°F) | Pressure |
| Kilogram per centimeter squared (kg cm$^{-2}$) | 3.94E + 02 | Inch of water (68°F) | Pressure |
| Kilogram per centimeter squared (kg cm$^{-2}$) | 9.81E + 01 | Kilopascal (kPa) | Pressure |
| Kilogram per centimeter squared (kg cm$^{-2}$) | 1.00E + 04 | Kilogram per meter squared (kg m$^{-2}$) | Pressure |
| Kilogram per centimeter squared (kg cm$^{-2}$) | 7.36E + 02 | Millimeter of mercury (32°F) | Pressure |
| Kilogram per centimeter squared (kg cm$^{-2}$) | 2.05E + 03 | Pound per foot squared (lb ft$^{-2}$) | Pressure |
| Kilogram per centimeter squared (kg cm$^{-2}$) | 6.23E − 02 | Pound per inch squared (psi lb in.$^{-2}$) | Pressure |
| Kilogram per meter squared (kg m$^{-2}$) | 9.68E − 05 | Atmosphere (atm) | Pressure |
| Kilogram per meter squared (kg m$^{-2}$) | 9.81E − 05 | Bars | Pressure |
| Kilogram per meter squared (kg m$^{-2}$) | 3.29E − 03 | Feet of water (68°F) | Pressure |
| Kilogram per meter squared (kg m$^{-2}$) | 2.90E − 03 | Inch of mercury (32°F) | Pressure |
| Kilogram per meter squared (kg m$^{-2}$) | 3.94E − 02 | Inch of water (68°F) | Pressure |
| Kilogram per meter squared (kg m$^{-2}$) | 9.81E − 03 | Kilopascal (kPa) | Pressure |
| Kilogram per meter squared (kg m$^{-2}$) | 7.36E − 02 | Millimeter of mercury (32°F) | Pressure |
| Kilogram per meter squared (kg m$^{-2}$) | 2.05E − 01 | Pound per foot squared (lb ft$^{-2}$) | Pressure |
| Kilogram per meter squared (kg m$^{-2}$) | 1.42E − 03 | Pound per inch squared (psi lb in.$^{-2}$) | Pressure |
| Millibar (mb) | 1.00E + 02 | Pascal (Pa) | Pressure |
| Millimeter of mercury (Hg) (torr) | 3.95E − 02 | Inch of mercury | Pressure |
| Millimeter of mercury (32°F) | 1.32E − 03 | Atmosphere (atm) | Pressure |
| Millimeter of mercury (32°F) | 1.33E − 03 | Bars | Pressure |
| Millimeter of mercury (32°F) | 4.47E − 02 | Feet of water (68°F) | Pressure |
| Millimeter of mercury (32°F) | 3.94E − 02 | Inch of mercury (32°F) | Pressure |
| Millimeter of mercury (32°F) | 5.36E − 01 | Inch of water (68°F) | Pressure |
| Millimeter of mercury (32°F) | 1.33E − 01 | Kilopascal | Pressure |
| Millimeter of mercury (32°F) | 1.36E − 03 | Kilogram per centimeter squared (kg cm$^{-2}$) | Pressure |
| Millimeter of mercury (32°F) | 1.36E + 01 | Kilogram per meter squared (kg m$^{-2}$) | Pressure |

(Continued)

| Multiply | By | To obtain | Property |
|---|---|---|---|
| Millimeter of mercury (32°F) | 2.79E + 00 | Pound per feet squared (lb ft$^{-2}$) | Pressure |
| Millimeter of mercury (32°F) | 1.93E − 02 | Pound per inch squared (psi lb in.$^{-2}$) | Pressure |
| Pound per feet squared (lb ft$^{-2}$) | 4.73E − 04 | Atmosphere (atm) | Pressure |
| Pound per feet squared (lb ft$^{-2}$) | 4.79E − 04 | Bars | Pressure |
| Pound per feet squared (lb ft$^{-2}$) | 1.61E − 02 | Feet of water (68°F) | Pressure |
| Pound per feet squared (lb ft$^{-2}$) | 1.41E − 02 | Inch of mercury (32°F) | Pressure |
| Pound per feet squared (lb ft$^{-2}$) | 1.93E − 01 | Inch of water (68°F) | Pressure |
| Pound per feet squared (lb ft$^{-2}$) | 4.79E − 02 | Kilopascal (kPa) | Pressure |
| Pound per feet squared (lb ft$^{-2}$) | 4.88E − 04 | Kilogram per centimeter squared (kg cm$^{-2}$) | Pressure |
| Pound per feet squared (lb ft$^{-2}$) | 4.88E + 00 | Kilogram per meter squared (kg m$^{-2}$) | Pressure |
| Pound per feet squared (lb ft$^{-2}$) | 3.59E − 01 | Millimeter of mercury (32°F) | Pressure |
| Pound per feet squared (lb ft$^{-2}$) | 4.79E + 01 | Pascal ( Pa, N m$^{-2}$) | Pressure |
| Pound per feet squared (lb ft$^{-2}$) | 6.95E − 03 | Pound per inch squared (psi lb in.$^{-2}$) | Pressure |
| Pound per inch squared (psi) (lb in.$^{-2}$) | 6.81E − 02 | Atmosphere (atm) | Pressure |
| Pound per inch squared (psi) (lb in.$^{-2}$) | 6.80E − 02 | Atmosphere (atm) | Pressure |
| Pound per inch squared (psi) (lb in.$^{-2}$) | 6.90E − 02 | Bars | Pressure |
| Pound per inch squared (psi) (lb in.$^{-2}$) | 2.31E + 00 | Feet of water (68°F) | Pressure |
| Pound per inch squared (psi) (lb in.$^{-2}$) | 2.04E + 00 | Inch of mercury (32°F) | Pressure |
| Pound per inch squared (psi) (lb in.$^{-2}$) | 2.77E + 01 | Inch of water (68°F) | Pressure |
| Pound per inch squared (psi) (lb in.$^{-2}$) | 2.77E + 01 | Inch of water column (in.) | Pressure |
| Pound per inch squared (psi) (lb in.$^{-2}$) | 6.90E + 00 | Kilopascal (kPa) | Pressure |
| Pound per inch squared (psi) (lb in.$^{-2}$) | 7.03E − 02 | Kilogram per centimeter squared (kg cm$^{-2}$) | Pressure |
| Pound per inch squared (psi) (lb in.$^{-2}$) | 7.03E + 02 | Kilogram per meter squared (kg m$^{-2}$) | Pressure |
| Pound per inch squared (psi) (lb in.$^{-2}$) | 4.79E − 02 | Kilopascal (kPa) | Pressure |
| Pound per inch squared (psi) (lb in.$^{-2}$) | 5.17E + 01 | Millimeter of mercury (32°F) | Pressure |
| Pound per inch squared (psi) (lb in.$^{-2}$) | 6.90E + 02 | Newtons per meter squared (N m$^{-2}$) | Pressure |
| Pound per inch squared (psi) (lb in.$^{-2}$) | 6.90E + 02 | Pascal (Pa) | Pressure |
| Pound per inch squared (psi) (lb in.$^{-2}$) | 1.44E + 02 | Pound per square foot (lb ft$^{-2}$) | Pressure |
| Pound per square inch (lb ft$^{-2}$ or psi) | 2.31E + 00 | Feet of head, water (ft) | Pressure |
| Poundal per foot squared (poundal ft$^{-2}$) | 1.49E + 00 | Pascal (Pa) | Pressure |
| Atmosphere (atm) | 1.01E + 05 | Pascal (Pa) | Pressure |
| °Celsius | (°C × 1.8) + 32 | °Fahrenheit | Temperature |
| °Celsius | (°C + 273.15) × 1.8 | °Rankine | Temperature |
| °Celsius | °C + 273.15 | Kelvin | Temperature |
| °Fahrenheit | (°F − 32) / 1.8 | °Celsius | Temperature |

*(Continued )*

(Continued)

| Multiply | By | To obtain | Property |
|---|---|---|---|
| °Fahrenheit | °F + 459.67 | °Rankine | Temperature |
| °Fahrenheit | ((°F − 32) / 1.8) + 273.15 | Kelvin | Temperature |
| °Rankine | (°Rank / 1.8) − 273.15 | °Celsius | Temperature |
| °Rankine | °Rank − 459.67 | °Fahrenheit | Temperature |
| °Rankine | °Rank / 1.8 | Kelvin | Temperature |
| Kelvin | K − 273.15 | °Celsius | Temperature |
| Kelvin | (K × 1.8) − 459.87 | °Fahrenheit | Temperature |
| Kelvin | K × 1.8 | °Rankine | Temperature |
| British thermal unit per hour-foot-degree Fahrenheit ($BTU\,h^{-1}\,ft^{-1}\,°F^{-1}$) | 4.13E − 03 | Calorie per second-degree Kelvin ($cal\,s^{-1}\,cm^{-1}\,°K^{-1}$) | Thermal conductivity |
| British thermal unit per hour-foot-degree Fahrenheit ($BTU\,h^{-1}\,ft^{-1}\,°F^{-1}$) | 1.73E + 05 | Gram-centimeter per second cubed-degree Kelvin ($g\,cm\,s^{-3}\,°K^{-1}$) | Thermal conductivity |
| British thermal unit per hour-foot-degree Fahrenheit ($BTU\,h^{-1}\,ft^{-1}\,°F^{-1}$) | 1.7307 | Kilogram-meter per second cubed-degree Fahrenheit ($kg\,m\,s^{-3}\,°F^{-1}$) | Thermal conductivity |
| British thermal unit per hour-foot-degree Fahrenheit ($BTU\,h^{-1}\,ft^{-1}\,°F^{-1}$) | 2.16E − 01 | Pound$_{force}$ per second-degree Fahrenheit ($lb_f\,s^{-1}\,°F^{-1}$) | Thermal conductivity |
| British thermal unit per hour-foot-degree Fahrenheit ($BTU\,h^{-1}\,ft^{-1}\,°F^{-1}$) | 6.9546 | Pound$_{mass}$-foot per second cubed-degree Fahrenheit ($lb_m\,ft\,s^{-3}\,°F^{-1}$) | Thermal conductivity |
| Calorie per second-degree Kelvin ($cal\,s^{-1}\,cm^{-1}\,°K^{-1}$) | 2.42E + 02 | British thermal unit per hour-foot-degree Fahrenheit ($BTU\,h^{-1}\,ft^{-1}\,°F^{-1}$) | Thermal conductivity |
| Calorie per second-degree Kelvin ($cal\,s^{-1}\,cm^{-1}\,°K^{-1}$) | 4.18E + 07 | Gram-centimeter per second cubed-degree Kelvin ($g\,cm\,s^{-3}\,°K^{-1}$) | Thermal conductivity |
| Calorie per second-degree Kelvin ($cal\,s^{-1}\,cm^{-1}\,°K^{-1}$) | 4.19E + 02 | Kilogram-meter per second cubed-degree Fahrenheit ($kg\,m\,s^{-3}\,°F^{-1}$) | Thermal conductivity |
| Calorie per second-degree Kelvin ($cal\,s^{-1}\,cm^{-1}\,°K^{-1}$) | 5.23E + 01 | Pound$_{force}$ per second-degree Fahrenheit ($lb_f\,s^{-1}\,°F^{-1}$) | Thermal conductivity |
| Calorie per second-degree Kelvin ($cal\,s^{-1}\,cm^{-1}\,°K^{-1}$) | 1.68E + 03 | Pound$_{mass}$-foot per second cubed-degree Fahrenheit ($lb_m\,ft\,s^{-3}\,°F^{-1}$) | Thermal conductivity |
| Gram-centimeter per second cubed-degree Kelvin ($g\,cm\,s^{-3}\,°K^{-1}$) | 5.79E − 06 | British thermal unit per hour-foot-degree Fahrenheit ($BTU\,h^{-1}\,ft^{-1}\,°F^{-1}$) | Thermal conductivity |
| Gram-centimeter per second cubed-degree Kelvin ($g\,cm\,s^{-3}\,°K^{-1}$) | 2.39E − 08 | Calorie per second-degree Kelvin ($cal\,s^{-1}\,cm^{-1}\,°K^{-1}$) | Thermal conductivity |
| Gram-centimeter per second cubed-degree Kelvin ($g\,cm\,s^{-3}\,°K^{-1}$) | 1.00E − 04 | Kilogram-meter per second cubed-degree Fahrenheit ($kg\,m\,s^{-3}\,°F^{-1}$) | Thermal conductivity |
| Gram-centimeter per second cubed-degree Kelvin ($g\,cm\,s^{-3}\,°K^{-1}$) | 1.25E − 06 | Pound$_{force}$ per second-degree Fahrenheit ($lb_f\,s^{-1}\,°F^{-1}$) | Thermal conductivity |
| Gram-centimeter per second cubed-degree Kelvin ($g\,cm\,s^{-3}\,°K^{-1}$) | 4.02E − 05 | Pound$_{mass}$-foot per second cubed-degree Fahrenheit ($lb_m\,ft\,s^{-3}\,°F^{-1}$) | Thermal conductivity |
| Kilogram-meter per second cubed-degree Fahrenheit ($kg\,m\,s^{-3}\,°F^{-1}$) | 5.78E − 01 | British thermal unit per hour-foot-degree Fahrenheit ($BTU\,h^{-1}\,ft^{-1}\,°F^{-1}$) | Thermal conductivity |

(Continued)

| Multiply | By | To obtain | Property |
|---|---|---|---|
| Kilogram-meter per second cubed-degree Fahrenheit ($kg\,m\,s^{-3}\,°F^{-1}$) | 2.39E − 03 | Calorie per second-degree Kelvin ($cal\,s^{-1}\,cm^{-1}\,°K^{-1}$) | Thermal conductivity |
| Kilogram-meter per second cubed-degree Fahrenheit ($kg\,m\,s^{-3}\,°F^{-1}$) | 1.00E + 06 | Gram-centimeter per second cubed-degree Kelvin ($g\,cm\,s^{-3}\,°K^{-1}$) | Thermal conductivity |
| Kilogram-meter per second cubed-degree Fahrenheit ($kg\,m\,s^{-3}\,°F^{-1}$) | 1.25E − 01 | Pound$_{force}$ per second-degree Fahrenheit ($lb_f\,s^{-1}\,°F^{-1}$) | Thermal conductivity |
| Kilogram-meter per second cubed-degree Fahrenheit ($kg\,m\,s^{-3}\,°F^{-1}$) | 4.0183 | Pound$_{mass}$-foot per second cubed-degree Fahrenheit ($lb_m\,ft\,s^{-3}\,°F^{-1}$) | Thermal conductivity |
| Pound$_{force}$ per second-degree Fahrenheit ($lb_f\,s^{-1}\,°F^{-1}$) | 4.6263 | British thermal unit per hour-foot-degree Fahrenheit ($BTU\,h^{-1}\,ft^{-1}\,°F^{-1}$) | Thermal conductivity |
| Pound$_{force}$ per second-degree Fahrenheit ($lb_f\,s^{-1}\,°F^{-1}$) | 1.91E − 02 | Calorie per second-degree Kelvin ($cal\,s^{-1}\,cm^{-1}\,°K^{-1}$) | Thermal conductivity |
| Pound$_{force}$ per second-degree Fahrenheit ($lb_f\,s^{-1}\,°F^{-1}$) | 8.01E + 05 | Gram-centimeter per second cubed-degree Kelvin ($g\,cm\,s^{-3}\,°K^-$) | Thermal conductivity |
| Pound$_{force}$ per second-degree Fahrenheit ($lb_f\,s^{-1}\,°F^{-1}$) | 8.01 | Kilogram-meter per second cubed-degree Fahrenheit ($kg\,m\,s^{-3}\,°F^{-1}$) | Thermal conductivity |
| Pound$_{force}$ per second-degree Fahrenheit ($lb_f\,s^{-1}\,°F^{-1}$) | 3.22E + 01 | Pound$_{mass}$-foot per second cubed-degree Fahrenheit ($lb_m\,ft\,s^{-3}\,°F^{-1}$) | Thermal conductivity |
| Pound$_{force}$ per second-degree Fahrenheit ($lb_f\,s^{-1}\,°F^{-1}$) | 1.44E − 01 | British thermal unit per hour-foot-degree Fahrenheit ($BTU\,h^{-1}\,ft^{-1}\,°F^{-1}$) | Thermal conductivity |
| Pound$_{force}$ per second-degree Fahrenheit ($lb_f\,s^{-1}\,°F^{-1}$) | 5.94E − 04 | Calorie per second-degree Kelvin ($cal\,s^{-1}\,cm^{-1}\,°K^{-1}$) | Thermal conductivity |
| Pound$_{force}$ per second-degree Fahrenheit ($lb_f\,s^{-1}\,°F^{-1}$) | 2.49E + 04 | Gram-centimeter per second cubed-degree Kelvin ($g\,cm\,s^{-3}\,°K^{-1}$) | Thermal conductivity |
| Pound$_{force}$ per second-degree Fahrenheit ($lb_f\,s^{-1}\,°F^{-1}$) | 2.49E − 01 | Kilogram-meter per second cubed-degree Fahrenheit ($kg\,m\,s^{-3}\,°F^{-1}$) | Thermal conductivity |
| Pound$_{force}$ per second-degree Fahrenheit ($lb_f\,s^{-1}\,°F^{-1}$) | 3.11E − 02 | Pound$_{force}$ per second-degree Fahrenheit ($lb_f\,s^{-1}\,°F^{-1}$) | Thermal conductivity |
| Day | 2.40E + 01 | Hours (h) | Time |
| Day | 1.44E + 03 | Minutes (min) | Time |
| Day | 8.64E + 04 | Second | Time |
| Day | 2.74E − 03 | Standard year | Time |
| Day | 1.43E − 01 | Week | Time |
| Hours (h) | 4.17E − 02 | Day | Time |
| Hours (h) | 6.00E + 01 | Minutes (min) | Time |
| Hours (h) | 3.60E + 03 | Second (s) | Time |
| Hours (h) | 1.14E − 04 | Standard year | Time |
| Hours (h) | 5.95E − 03 | Week | Time |
| Minutes (min) | 6.94E − 04 | Day | Time |
| Minutes (min) | 1.67E − 02 | Hours (h) | Time |
| Minutes (min) | 6.00E + 01 | Second (s) | Time |
| Minutes (min) | 1.90E − 06 | Standard year | Time |
| Minutes (min) | 9.92E − 05 | Week | Time |
| Second (s) | 1.16E − 05 | Day | Time |

*(Continued)*

(Continued)

| Multiply | By | To obtain | Property |
|---|---|---|---|
| Second (s) | 2.77E − 04 | Hours (h) | Time |
| Second (s) | 1.66E − 02 | Minutes (min) | Time |
| Second (s) | 3.17E − 08 | Standard year | Time |
| Second (s) | 1.65E − 06 | Week | Time |
| Standard year | 3.65E + 02 | Day | Time |
| Standard year | 8.76E + 03 | Hours (h) | Time |
| Standard year | 5.26E + 05 | Minutes (min) | Time |
| Standard year | 3.15E + 07 | Second (s) | Time |
| Standard year | 5.21E + 01 | Week | Time |
| Week | 7.00E + 00 | Day | Time |
| Week | 1.68E + 02 | Hours (h) | Time |
| Week | 1.01E + 04 | Minutes (min) | Time |
| Week | 6.05E + 05 | Second (s) | Time |
| Week | 1.92E − 02 | Standard year | Time |
| Foot squared per second (ft$^2$ s$^{-1}$) | 9.29E − 02 | Meter squared per second (m$^2$ s$^{-1}$) | Transmissivity |
| US gallon per day per foot (US gal day$^{-1}$ ft$^{-1}$) | 1.44E − 07 | Meter squared per second (m$^2$ s$^{-1}$) | Transmissivity |
| Centimeter per hour (cm h$^{-1}$) | 1.67E − 02 | Centimeter per min (cm min$^{-1}$) | Velocity |
| Centimeter per hour (cm h$^{-1}$) | 2.78E − 04 | Centimeter per second (cm s$^{-1}$) | Velocity |
| Centimeter per hour (cm h$^{-1}$) | 3.28E − 02 | Foot per hour (ft h$^{-1}$) | Velocity |
| Centimeter per hour (cm h$^{-1}$) | 5.47E − 04 | Foot per minute (ft min$^{-1}$) | Velocity |
| Centimeter per hour (cm h$^{-1}$) | 9.11E − 06 | Foot per second (ft s$^{-1}$) | Velocity |
| Centimeter per hour (cm h$^{-1}$) | 1.00E − 04 | Kilometer per hour (km h$^{-1}$) | Velocity |
| Centimeter per hour (cm h$^{-1}$) | 1.67E − 07 | Kilometer per minute (km min$^{-1}$) | Velocity |
| Centimeter per hour (cm h$^{-1}$) | 2.78E − 09 | Kilometer per second (km s$^{-1}$) | Velocity |
| Centimeter per hour (cm h$^{-1}$) | 5.40E − 06 | Knots | Velocity |
| Centimeter per hour (cm h$^{-1}$) | 1.00E − 02 | Meter per hour (m h$^{-1}$) | Velocity |
| Centimeter per hour (cm h$^{-1}$) | 1.67E − 04 | Meter per minute (m min$^{-1}$) | Velocity |
| Centimeter per hour (cm h$^{-1}$) | 2.78E − 06 | Meter per second (m s$^{-1}$) | Velocity |
| Centimeter per hour (cm h$^{-1}$) | 6.21E − 08 | Mile per hour (mi h$^{-1}$) | Velocity |
| Centimeter per hour (cm h$^{-1}$) | 1.04E − 07 | Mile per minute (mi min$^{-1}$) | Velocity |
| Centimeter per hour (cm h$^{-1}$) | 1.73E − 09 | Mile per second (mi s$^{-1}$) | Velocity |
| Centimeter per min (cm min$^{-1}$) | 6.00E + 01 | Centimeter per hour (cm h$^{-1}$) | Velocity |
| Centimeter per min (cm min$^{-1}$) | 1.67E − 02 | Centimeter per second (cm s$^{-1}$) | Velocity |
| Centimeter per min (cm min$^{-1}$) | 1.97E + 00 | Foot per hour (ft h$^{-1}$) | Velocity |
| Centimeter per min (cm min$^{-1}$) | 3.28E − 02 | Foot per minute (ft min$^{-1}$) | Velocity |
| Centimeter per min (cm min$^{-1}$) | 5.47E − 04 | Foot per second (ft s$^{-1}$) | Velocity |
| Centimeter per min (cm min$^{-1}$) | 6.00E − 04 | Kilometer per hour (km h$^{-1}$) | Velocity |
| Centimeter per min (cm min$^{-1}$) | 1.00E − 05 | Kilometer per minute (km min$^{-1}$) | Velocity |
| Centimeter per min (cm min$^{-1}$) | 1.67E − 07 | Kilometer per second (km s$^{-1}$) | Velocity |
| Centimeter per min (cm min$^{-1}$) | 3.24E − 04 | Knots | Velocity |
| Centimeter per min (cm min$^{-1}$) | 6.00E − 01 | Meter per hour (m h$^{-1}$) | Velocity |
| Centimeter per min (cm min$^{-1}$) | 1.00E − 02 | Meter per minute (m min$^{-1}$) | Velocity |
| Centimeter per min (cm min$^{-1}$) | 1.67E − 04 | Meter per second (m s$^{-1}$) | Velocity |
| Centimeter per min (cm min$^{-1}$) | 3.73E − 04 | Mile per hour (mi h$^{-1}$) | Velocity |
| Centimeter per min (cm min$^{-1}$) | 6.21E − 06 | Mile per minute (mi min$^{-1}$) | Velocity |
| Centimeter per min (cm min$^{-1}$) | 1.04E − 07 | Mile per second (mi s$^{-1}$) | Velocity |
| Centimeter per second (cm s$^{-1}$) | 3.60E + 03 | Centimeter per hour (cm h$^{-1}$) | Velocity |
| Centimeter per second (cm s$^{-1}$) | 6.00E + 01 | Centimeter per min (cm min$^{-1}$) | Velocity |
| Centimeter per second (cm s$^{-1}$) | 1.18E + 02 | Foot per hour (ft h$^{-1}$) | Velocity |
| Centimeter per second (cm s$^{-1}$) | 1.20E + 00 | Foot per minute (ft min$^{-1}$) | Velocity |
| Centimeter per second (cm s$^{-1}$) | 3.28E − 02 | Foot per second (ft s$^{-1}$) | Velocity |
| Centimeter per second (cm s$^{-1}$) | 3.60E − 02 | Kilometer per hour (km h$^{-1}$) | Velocity |

| Multiply | By | To obtain | Property |
|---|---|---|---|
| Centimeter per second (cm s$^{-1}$) | 6.00E − 04 | Kilometer per minute (km min$^{-1}$) | Velocity |
| Centimeter per second (cm s$^{-1}$) | 1.00E − 05 | Kilometer per second (km s$^{-1}$) | Velocity |
| Centimeter per second (cm s$^{-1}$) | 1.94E − 02 | Knots | Velocity |
| Centimeter per second (cm s$^{-1}$) | 3.60E + 01 | Meter per hour (m h$^{-1}$) | Velocity |
| Centimeter per second (cm s$^{-1}$) | 6.00E − 01 | Meter per minute (m min$^{-1}$) | Velocity |
| Centimeter per second (cm s$^{-1}$) | 1.00E − 02 | Meter per second (m s$^{-1}$) | Velocity |
| Centimeter per second (cm s$^{-1}$) | 2.24E − 02 | Mile per hour (mi h$^{-1}$) | Velocity |
| Centimeter per second (cm s$^{-1}$) | 3.73E − 04 | Mile per minute (mi min$^{-1}$) | Velocity |
| Centimeter per second (cm s$^{-1}$) | 6.21E − 06 | Mile per second (mi s$^{-1}$) | Velocity |
| Foot per day (ft day$^{-1}$) | 1.16E − 05 | Feet per second (ft s$^{-1}$) | Velocity |
| Foot per day (ft day$^{-1}$) | 1.27E − 05 | Kilometer per hour (km h$^{-1}$) | Velocity |
| Foot per day (ft day$^{-1}$) | 3.53E − 06 | Meter per second (m s$^{-1}$) | Velocity |
| Foot per day (ft day$^{-1}$) | 7.89E − 06 | Mile per hour (mi h$^{-1}$ or mph) | Velocity |
| Foot per hour (ft h$^{-1}$) | 3.05E + 01 | Centimeter per hour (cm h$^{-1}$) | Velocity |
| Foot per hour (ft h$^{-1}$) | 5.08E − 01 | Centimeter per min (cm min$^{-1}$) | Velocity |
| Foot per hour (ft h$^{-1}$) | 8.47E − 03 | Centimeter per second (cm s$^{-1}$) | Velocity |
| Foot per hour (ft h$^{-1}$) | 1.67E − 02 | Foot per minute (ft min$^{-1}$) | Velocity |
| Foot per hour (ft h$^{-1}$) | 2.78E − 04 | Foot per second (ft s$^{-1}$) | Velocity |
| Foot per hour (ft h$^{-1}$) | 3.05E − 04 | Kilometer per hour (km h$^{-1}$) | Velocity |
| Foot per hour (ft h$^{-1}$) | 5.08E − 06 | Kilometer per minute (km min$^{-1}$) | Velocity |
| Foot per hour (ft h$^{-1}$) | 8.47E − 08 | Kilometer per second (km s$^{-1}$) | Velocity |
| Foot per hour (ft h$^{-1}$) | 1.65E − 04 | Knots | Velocity |
| Foot per hour (ft h$^{-1}$) | 3.05E − 01 | Meter per hour (m h$^{-1}$) | Velocity |
| Foot per hour (ft h$^{-1}$) | 5.08E − 03 | Meter per minute (m min$^{-1}$) | Velocity |
| Foot per hour (ft h$^{-1}$) | 8.47E − 05 | Meter per second (m s$^{-1}$) | Velocity |
| Foot per hour (ft h$^{-1}$) | 1.89E − 04 | Mile per hour (mi h$^{-1}$) | Velocity |
| Foot per hour (ft h$^{-1}$) | 3.16E − 06 | Mile per minute (mi min$^{-1}$) | Velocity |
| Foot per hour (ft h$^{-1}$) | 5.23E − 08 | Mile per second (mi s$^{-1}$) | Velocity |
| Foot per minute (ft min$^{-1}$) | 1.83E + 03 | Centimeter per hour (cm h$^{-1}$) | Velocity |
| Foot per minute (ft min$^{-1}$) | 3.05E + 01 | Centimeter per min (cm min$^{-1}$) | Velocity |
| Foot per minute (ft min$^{-1}$) | 5.08E − 01 | Centimeter per second (cm s$^{-1}$) | Velocity |
| Foot per minute (ft min$^{-1}$) | 6.00E + 01 | Foot per hour (ft h$^{-1}$) | Velocity |
| Foot per minute (ft min$^{-1}$) | 1.67E − 02 | Foot per second (ft s$^{-1}$) | Velocity |
| Foot per minute (ft min$^{-1}$) | 1.83E − 02 | Kilometer per hour (km h$^{-1}$) | Velocity |
| Foot per minute (ft min$^{-1}$) | 3.05E − 04 | Kilometer per minute (km min$^{-1}$) | Velocity |
| Foot per minute (ft min$^{-1}$) | 5.08E − 06 | Kilometer per second (km s$^{-1}$) | Velocity |
| Foot per minute (ft min$^{-1}$) | 9.87E − 03 | Knots | Velocity |
| Foot per minute (ft min$^{-1}$) | 1.83E + 01 | Meter per hour (m h$^{-1}$) | Velocity |
| Foot per minute (ft min$^{-1}$) | 3.05E − 01 | Meter per minute (m min$^{-1}$) | Velocity |
| Foot per minute (ft min$^{-1}$) | 5.08E − 03 | Meter per second (m s$^{-1}$) | Velocity |
| Foot per minute (ft min$^{-1}$) | 1.14E − 02 | Mile per hour (mi h$^{-1}$) | Velocity |
| Foot per minute (ft min$^{-1}$) | 1.89E − 04 | Mile per minute (mi min$^{-1}$) | Velocity |
| Foot per minute (ft min$^{-1}$) | 3.16E − 06 | Mile per second (mi s$^{-1}$) | Velocity |
| Foot per second (ft s$^{-1}$) | 1.10E + 05 | Centimeter per hour (cm h$^{-1}$) | Velocity |
| Foot per second (ft s$^{-1}$) | 1.83E + 03 | Centimeter per min (cm min$^{-1}$) | Velocity |
| Foot per second (ft s$^{-1}$) | 3.05E + 01 | Centimeter per second (cm s$^{-1}$) | Velocity |
| Foot per second (ft s$^{-1}$) | 3.60E + 03 | Foot per hour (ft h$^{-1}$) | Velocity |
| Foot per second (ft s$^{-1}$) | 6.00E + 01 | Foot per minute (ft min$^{-1}$) | Velocity |
| Foot per second (ft s$^{-1}$) | 1.10E + 00 | Kilometer per hour (km h$^{-1}$) | Velocity |
| Foot per second (ft s$^{-1}$) | 1.83E − 02 | Kilometer per minute (km min$^{-1}$) | Velocity |
| Foot per second (ft s$^{-1}$) | 3.05E − 04 | Kilometer per second (km s$^{-1}$) | Velocity |
| Foot per second (ft s$^{-1}$) | 5.92E − 01 | Knots | Velocity |
| Foot per second (ft s$^{-1}$) | 1.10E + 03 | Meter per hour (m h$^{-1}$) | Velocity |
| Foot per second (ft s$^{-1}$) | 1.83E + 01 | Meter per minute (m min$^{-1}$) | Velocity |

(Continued)

(Continued)

| Multiply | By | To obtain | Property |
|---|---|---|---|
| Foot per second (ft s$^{-1}$) | 3.05E − 01 | Meter per second (m s$^{-1}$) | Velocity |
| Foot per second (ft s$^{-1}$) | 6.82E − 01 | Mile per hour (mi h$^{-1}$) | Velocity |
| Foot per second (ft s$^{-1}$) | 1.14E − 02 | Mile per minute (mi min$^{-1}$) | Velocity |
| Foot per second (ft s$^{-1}$) | 1.89E − 04 | Mile per second (mi s$^{-1}$) | Velocity |
| Kilometer per hour (km h$^{-1}$) | 1.00E + 05 | Centimeter per hour (cm h$^{-1}$) | Velocity |
| Kilometer per hour (km h$^{-1}$) | 1.67E + 03 | Centimeter per min (cm min$^{-1}$) | Velocity |
| Kilometer per hour (km h$^{-1}$) | 2.78E + 01 | Centimeter per second (cm s$^{-1}$) | Velocity |
| Kilometer per hour (km h$^{-1}$) | 3.28E + 03 | Foot per hour (ft h$^{-1}$) | Velocity |
| Kilometer per hour (km h$^{-1}$) | 5.47E + 01 | Foot per minute (ft min$^{-1}$) | Velocity |
| Kilometer per hour (km h$^{-1}$) | 9.11E − 01 | Foot per second (ft s$^{-1}$) | Velocity |
| Kilometer per hour (km h$^{-1}$) | 1.67E − 02 | Kilometer per minute (km min$^{-1}$) | Velocity |
| Kilometer per hour (km h$^{-1}$) | 2.78E − 04 | Kilometer per second (km s$^{-1}$) | Velocity |
| Kilometer per hour (km h$^{-1}$) | 5.40E − 01 | Knots | Velocity |
| Kilometer per hour (km h$^{-1}$) | 1.00E + 03 | Meter per hour (m h$^{-1}$) | Velocity |
| Kilometer per hour (km h$^{-1}$) | 1.67E + 01 | Meter per minute (m min$^{-1}$) | Velocity |
| Kilometer per hour (km h$^{-1}$) | 2.78E − 01 | Meter per second (m s$^{-1}$) | Velocity |
| Kilometer per hour (km h$^{-1}$) | 6.21E − 01 | Mile per hour (mi h$^{-1}$) | Velocity |
| Kilometer per hour (km h$^{-1}$) | 1.04E − 02 | Mile per minute (mi min$^{-1}$) | Velocity |
| Kilometer per hour (km h$^{-1}$) | 1.73E − 04 | Mile per second (mi s$^{-1}$) | Velocity |
| Kilometer per minute (km min$^{-1}$) | 6.00E + 06 | Centimeter per hour (cm h$^{-1}$) | Velocity |
| Kilometer per minute (km min$^{-1}$) | 1.00E + 05 | Centimeter per min (cm min$^{-1}$) | Velocity |
| Kilometer per minute (km min$^{-1}$) | 1.67E + 03 | Centimeter per second (cm s$^{-1}$) | Velocity |
| Kilometer per minute (km min$^{-1}$) | 1.97E + 05 | Foot per hour (ft h$^{-1}$) | Velocity |
| Kilometer per minute (km min$^{-1}$) | 3.28E + 03 | Foot per minute (ft min$^{-1}$) | Velocity |
| Kilometer per minute (km min$^{-1}$) | 5.47E + 01 | Foot per second (ft s$^{-1}$) | Velocity |
| Kilometer per minute (km min$^{-1}$) | 6.00E + 01 | Kilometer per hour (km h$^{-1}$) | Velocity |
| Kilometer per minute (km min$^{-1}$) | 1.67E − 02 | Kilometer per second (km s$^{-1}$) | Velocity |
| Kilometer per minute (km min$^{-1}$) | 3.24E + 01 | Knots | Velocity |
| Kilometer per minute (km min$^{-1}$) | 6.00E + 04 | Meter per hour (m h$^{-1}$) | Velocity |
| Kilometer per minute (km min$^{-1}$) | 1.00E + 03 | Meter per minute (m min$^{-1}$) | Velocity |
| Kilometer per minute (km min$^{-1}$) | 1.67E + 01 | Meter per second (m s$^{-1}$) | Velocity |
| Kilometer per minute (km min$^{-1}$) | 3.73E + 01 | Mile per hour (mi h$^{-1}$) | Velocity |
| Kilometer per minute (km min$^{-1}$) | 6.21E − 01 | Mile per minute (mi min$^{-1}$) | Velocity |
| Kilometer per minute (km min$^{-1}$) | 1.04E − 02 | Mile per second (mi s$^{-1}$) | Velocity |
| Kilometer per second (km s$^{-1}$) | 3.60E + 08 | Centimeter per hour (cm h$^{-1}$) | Velocity |
| Kilometer per second (km s$^{-1}$) | 6.00E + 06 | Centimeter per min (cm min$^{-1}$) | Velocity |
| Kilometer per second (km s$^{-1}$) | 1.00E + 05 | Centimeter per second (cm s$^{-1}$) | Velocity |
| Kilometer per second (km s$^{-1}$) | 1.18E + 07 | Foot per hour (ft h$^{-1}$) | Velocity |
| Kilometer per second (km s$^{-1}$) | 1.97E + 05 | Foot per minute (ft min$^{-1}$) | Velocity |
| Kilometer per second (km s$^{-1}$) | 3.28E + 03 | Foot per second (ft s$^{-1}$) | Velocity |
| Kilometer per second (km s$^{-1}$) | 3.60E + 03 | Kilometer per hour (km h$^{-1}$) | Velocity |
| Kilometer per second (km s$^{-1}$) | 6.00E + 01 | Kilometer per minute (km min$^{-1}$) | Velocity |
| Kilometer per second (km s$^{-1}$) | 1.94E + 03 | Knots (k) | Velocity |
| Kilometer per second (km s$^{-1}$) | 3.60E + 06 | Meter per hour (m h$^{-1}$) | Velocity |
| Kilometer per second (km s$^{-1}$) | 6.00E + 04 | Meter per minute (m min$^{-1}$) | Velocity |
| Kilometer per second (km s$^{-1}$) | 1.00E + 03 | Meter per second (m s$^{-1}$) | Velocity |
| Kilometer per second (km s$^{-1}$) | 2.24E + 03 | Mile per hour (mi h$^{-1}$) | Velocity |
| Kilometer per second (km s$^{-1}$) | 3.73E + 01 | Mile per minute (mi min$^{-1}$) | Velocity |
| Kilometer per second (km s$^{-1}$) | 6.21E − 01 | Mile per second (mi s$^{-1}$) | Velocity |
| Knots (k) | 1.85E + 05 | Centimeter per hour (cm h$^{-1}$) | Velocity |
| Knots (k) | 3.09E + 03 | Centimeter per min (cm min$^{-1}$) | Velocity |
| Knots (k) | 5.14E + 01 | Centimeter per second (cm s$^{-1}$) | Velocity |
| Knots (k) | 6.08E + 03 | Foot per hour (ft h$^{-1}$) | Velocity |
| Knots (k) | 1.01E + 02 | Foot per minute (ft min$^{-1}$) | Velocity |
| Knots (k) | 1.69E + 00 | Foot per second (ft s$^{-1}$) | Velocity |

| Multiply | By | To obtain | Property |
|---|---|---|---|
| Knots (k) | 3.09E − 02 | Kilometer per minute (km min$^{-1}$) | Velocity |
| Knots (k) | 1.85E + 00 | Kilometer per hour (km h$^{-1}$) | Velocity |
| Knots (k) | 5.14E − 04 | Kilometer per second (km s$^{-1}$) | Velocity |
| Knots (k) | 1.85E + 03 | Meter per hour (m h$^{-1}$) | Velocity |
| Knots (k) | 3.09E + 01 | Meter per minute (m min$^{-1}$) | Velocity |
| Knots (k) | 5.14E − 01 | Meter per second (m s$^{-1}$) | Velocity |
| Knots (k) | 1.15E + 00 | Mile per hour (mi h$^{-1}$) | Velocity |
| Knots (k) | 1.92E − 02 | Mile per minute (mi min$^{-1}$) | Velocity |
| Knots (k) | 3.20E − 04 | Mile per second (mi s$^{-1}$) | Velocity |
| Meter per hour (m h$^{-1}$) | 1.00E + 02 | Centimeter per hour (cm h$^{-1}$) | Velocity |
| Meter per hour (m h$^{-1}$) | 1.67E + 00 | Centimeter per min (cm min$^{-1}$) | Velocity |
| Meter per hour (m h$^{-1}$) | 2.78E − 02 | Centimeter per second (cm s$^{-1}$) | Velocity |
| Meter per hour (m h$^{-1}$) | 3.28E + 00 | Foot per hour (ft h$^{-1}$) | Velocity |
| Meter per hour (m h$^{-1}$) | 5.47E − 02 | Foot per minute (ft min$^{-1}$) | Velocity |
| Meter per hour (m h$^{-1}$) | 9.11E − 04 | Foot per second (ft s$^{-1}$) | Velocity |
| Meter per hour (m h$^{-1}$) | 1.00E − 03 | Kilometer per hour (km h$^{-1}$) | Velocity |
| Meter per hour (m h$^{-1}$) | 1.67E − 05 | Kilometer per minute (km min$^{-1}$) | Velocity |
| Meter per hour (m h$^{-1}$) | 2.78E − 07 | Kilometer per second (km s$^{-1}$) | Velocity |
| Meter per hour (m h$^{-1}$) | 5.40E − 04 | Knots (k) | Velocity |
| Meter per hour (m h$^{-1}$) | 1.67E − 02 | Meter per minute (m min$^{-1}$) | Velocity |
| Meter per hour (m h$^{-1}$) | 2.78E − 04 | Meter per second (m s$^{-1}$) | Velocity |
| Meter per hour (m h$^{-1}$) | 6.21E − 04 | Mile per hour (mi h$^{-1}$) | Velocity |
| Meter per hour (m h$^{-1}$) | 1.04E − 05 | Mile per minute (mi min$^{-1}$) | Velocity |
| Meter per hour (m h$^{-1}$) | 1.73E − 07 | Mile per second (mi s$^{-1}$) | Velocity |
| Meter per minute (m min$^{-1}$) | 6.00E + 03 | Centimeter per hour (cm h$^{-1}$) | Velocity |
| Meter per minute (m min$^{-1}$) | 1.00E + 02 | Centimeter per min (cm min$^{-1}$) | Velocity |
| Meter per minute (m min$^{-1}$) | 1.67E + 00 | Centimeter per second (cm s$^{-1}$) | Velocity |
| Meter per minute (m min$^{-1}$) | 1.97E + 02 | Foot per hour (ft h$^{-1}$) | Velocity |
| Meter per minute (m min$^{-1}$) | 3.28E + 00 | Foot per minute (ft min$^{-1}$) | Velocity |
| Meter per minute (m min$^{-1}$) | 5.47E − 02 | Foot per second (ft s$^{-1}$) | Velocity |
| Meter per minute (m min$^{-1}$) | 1.00E − 03 | Kilometer per minute (km min$^{-1}$) | Velocity |
| Meter per minute (m min$^{-1}$) | 6.00E − 02 | Kilometer per hour (km h$^{-1}$) | Velocity |
| Meter per minute (m min$^{-1}$) | 1.67E − 05 | Kilometer per second (km s$^{-1}$) | Velocity |
| Meter per minute (m min$^{-1}$) | 3.24E − 02 | Knots | Velocity |
| Meter per minute (m min$^{-1}$) | 6.00E + 01 | Meter per hour (m h$^{-1}$) | Velocity |
| Meter per minute (m min$^{-1}$) | 1.67E − 02 | Meter per second (m s$^{-1}$) | Velocity |
| Meter per minute (m min$^{-1}$) | 3.73E − 02 | Mile per hour (mi h$^{-1}$) | Velocity |
| Meter per minute (m min$^{-1}$) | 6.21E − 04 | Mile per minute (mi min$^{-1}$) | Velocity |
| Meter per minute (m min$^{-1}$) | 1.04E − 05 | Mile per second (mi s$^{-1}$) | Velocity |
| Meter per second (m s$^{-1}$) | 3.60E + 05 | Centimeter per hour (cm h$^{-1}$) | Velocity |
| Meter per second (m s$^{-1}$) | 6.00E + 03 | Centimeter per min (cm min$^{-1}$) | Velocity |
| Meter per second (m s$^{-1}$) | 1.00E + 02 | Centimeter per second (cm s$^{-1}$) | Velocity |
| Meter per second (m s$^{-1}$) | 1.18E + 04 | Foot per hour (ft h$^{-1}$) | Velocity |
| Meter per second (m s$^{-1}$) | 1.97E + 02 | Foot per minute (ft min$^{-1}$) | Velocity |
| Meter per second (m s$^{-1}$) | 3.28E + 00 | Foot per second (ft s$^{-1}$) | Velocity |
| Meter per second (m s$^{-1}$) | 6.00E − 02 | Kilometer per minute (km min$^{-1}$) | Velocity |
| Meter per second (m s$^{-1}$) | 3.60E + 00 | Kilometer per hour (km h$^{-1}$) | Velocity |
| Meter per second (m s$^{-1}$) | 1.00E − 03 | Kilometer per second (km s$^{-1}$) | Velocity |
| Meter per second (m s$^{-1}$) | 1.95E + 00 | Knots | Velocity |
| Meter per second (m s$^{-1}$) | 3.60E + 03 | Meter per hour (m h$^{-1}$) | Velocity |
| Meter per second (m s$^{-1}$) | 6.00E + 01 | Meter per minute (m min$^{-1}$) | Velocity |
| Meter per second (m s$^{-1}$) | 2.24E + 00 | Mile per hour (mi h$^{-1}$) | Velocity |
| Meter per second (m s$^{-1}$) | 3.73E − 02 | Mile per minute (mi min$^{-1}$) | Velocity |
| Meter per second (m s$^{-1}$) | 6.21E − 04 | Mile per second (mi s$^{-1}$) | Velocity |
| Mile per minute (mi min$^{-1}$) | 9.66E + 06 | Centimeter per hour (cm h$^{-1}$) | Velocity |

(Continued)

| Multiply | By | To obtain | Property |
|---|---|---|---|
| Mile per minute (mi min$^{-1}$) | 1.61E + 05 | Centimeter per min (cm min$^{-1}$) | Velocity |
| Mile per minute (mi min$^{-1}$) | 2.68E + 03 | Centimeter per second (cm s$^{-1}$) | Velocity |
| Mile per minute (mi min$^{-1}$) | 3.17E + 05 | Foot per hour (ft h$^{-1}$) | Velocity |
| Mile per minute (mi min$^{-1}$) | 5.28E + 03 | Foot per minute (ft min$^{-1}$) | Velocity |
| Mile per minute (mi min$^{-1}$) | 8.80E + 01 | Foot per second (ft s$^{-1}$) | Velocity |
| Mile per minute (mi min$^{-1}$) | 1.61E + 00 | Kilometer per minute (km min$^{-1}$) | Velocity |
| Mile per minute (mi min$^{-1}$) | 9.66E + 01 | Kilometer per hour (km h$^{-1}$) | Velocity |
| Mile per minute (mi min$^{-1}$) | 2.68E − 02 | Kilometer per second (km s$^{-1}$) | Velocity |
| Mile per minute (mi min$^{-1}$) | 5.21E + 01 | Knots | Velocity |
| Mile per minute (mi min$^{-1}$) | 9.66E + 04 | Meter per hour (m h$^{-1}$) | Velocity |
| Mile per minute (mi min$^{-1}$) | 1.61E + 03 | Meter per minute (m min$^{-1}$) | Velocity |
| Mile per minute (mi min$^{-1}$) | 2.68E + 01 | Meter per second (m s$^{-1}$) | Velocity |
| Mile per minute (mi min$^{-1}$) | 1.67E − 02 | Mile per second (mi s$^{-1}$) | Velocity |
| Mile per minute (mi min$^{-1}$) | 6.00E + 01 | Mile per hour (mi h$^{-1}$ or mph) | Velocity |
| Mile per second (mi s$^{-1}$) | 5.79E + 08 | Centimeter per hour (cm h$^{-1}$) | Velocity |
| Mile per second (mi s$^{-1}$) | 9.66E + 06 | Centimeter per min (cm min$^{-1}$) | Velocity |
| Mile per second (mi s$^{-1}$) | 1.61E + 05 | Centimeter per second (cm s$^{-1}$) | Velocity |
| Mile per second (mi s$^{-1}$) | 1.90E + 07 | Foot per hour (ft h$^{-1}$) | Velocity |
| Mile per second (mi s$^{-1}$) | 3.17E + 05 | Foot per minute (ft min$^{-1}$) | Velocity |
| Mile per second (mi s$^{-1}$) | 5.28E + 03 | Foot per second (ft s$^{-1}$) | Velocity |
| Mile per second (mi s$^{-1}$) | 9.66E + 01 | Kilometer per minute (km min$^{-1}$) | Velocity |
| Mile per second (mi s$^{-1}$) | 5.79E + 03 | Kilometer per hour (km h$^{-1}$) | Velocity |
| Mile per second (mi s$^{-1}$) | 1.61E + 00 | Kilometer per second (km s$^{-1}$) | Velocity |
| Mile per second (mi s$^{-1}$) | 3.13E + 03 | Knots | Velocity |
| Mile per second (mi s$^{-1}$) | 5.79E + 06 | Meter per hour (m h$^{-1}$) | Velocity |
| Mile per second (mi s$^{-1}$) | 9.66E + 04 | Meter per minute (m min$^{-1}$) | Velocity |
| Mile per second (mi s$^{-1}$) | 1.61E + 03 | Meter per second (m s$^{-1}$) | Velocity |
| Mile per second (mi s$^{-1}$) | 6.00E + 01 | Mile per minute (mi min$^{-1}$) | Velocity |
| Mile per second (mi s$^{-1}$) | 3.60E + 03 | Mile per hour (mi h$^{-1}$ or mph) | Velocity |
| Mile per hour (mi h$^{-1}$ or mph) | 1.61E + 05 | Centimeter per hour (cm h$^{-1}$) | Velocity |
| Mile per hour (mi h$^{-1}$ or mph) | 2.68E + 03 | Centimeter per min (cm min$^{-1}$) | Velocity |
| Mile per hour (mi h$^{-1}$ or mph) | 4.47E + 01 | Centimeter per second (cm s$^{-1}$) | Velocity |
| Mile per hour (mi h$^{-1}$ or mph) | 5.28E + 03 | Foot per hour (ft h$^{-1}$) | Velocity |
| Mile per hour (mi h$^{-1}$ or mph) | 8.80E + 01 | Foot per minute (ft min$^{-1}$) | Velocity |
| Mile per hour (mi h$^{-1}$ or mph) | 1.47E + 00 | Foot per second (ft s$^{-1}$) | Velocity |
| Mile per hour (mi h$^{-1}$ or mph) | 2.68E − 02 | Kilometer per minute (km min$^{-1}$) | Velocity |
| Mile per hour (mi h$^{-1}$ or mph) | 1.61E + 00 | Kilometer per hour (km h$^{-1}$) | Velocity |
| Mile per hour (mi h$^{-1}$ or mph) | 4.47E − 04 | Kilometer per second (km s$^{-1}$) | Velocity |
| Mile per hour (mi h$^{-1}$ or mph) | 8.69E − 01 | Knots | Velocity |
| Mile per hour (mi h$^{-1}$ or mph) | 1.61E + 03 | Meter per hour (m h$^{-1}$) | Velocity |
| Mile per hour (mi h$^{-1}$ or mph) | 2.68E + 01 | Meter per minute (m min$^{-1}$) | Velocity |
| Mile per hour (mi h$^{-1}$ or mph) | 4.47E − 01 | Meter per second (m s$^{-1}$) | Velocity |
| Mile per hour (mi h$^{-1}$ or mph) | 1.67E − 02 | Mile per minute (mi min$^{-1}$) | Velocity |
| Mile per hour (mi h$^{-1}$ or mph) | 2.78E − 04 | Mile per second (mi s$^{-1}$) | Velocity |
| Acre-inch | 2.72E + 04 | US gallon (gal) | Volume |
| Acre-feet | 4.36E + 04 | Cubic feet (ft$^3$) | Volume |
| Acre-feet | 7.53E + 07 | Cubic inch (in.$^3$) | Volume |
| Acre-feet | 1.61E + 03 | Cubic yard (yd$^3$) | Volume |
| Acre-feet | 1.23E − 01 | Hectare-meter (ha m) | Volume |
| Acre-feet | 1.23E + 06 | Liter (L) | Volume |
| Acre-feet | 1.23E + 06 | Liter (L) | Volume |
| Acre-feet | 1.23E + 03 | Meter cubed (m$^3$) | Volume |
| Acre-feet | 3.26E + 05 | US gallon (gal) | Volume |

(Continued)

| Multiply | By | To obtain | Property |
|---|---|---|---|
| Acre-feet | 3.26E + 05 | US gallon (gal) | Volume |
| Barrel | 1.23E + 03 | US gallon (gal) | Volume |
| Centiliter (cL) | 3.38E − 01 | Fluid ounce (fl oz) | Volume |
| Cup | 8.00E + 00 | Fluid ounce (fl oz) | Volume |
| Cup | 2.40E + 02 | Milliliter (mL) | Volume |
| Cup | 1.60E + 01 | Tablespoon (tbls) | Volume |
| Cup | 4.80E + 01 | Teaspoon (tsp) | Volume |
| Dram (dr) | 6.25E − 02 | Ounce (oz) | Volume |
| Feet cubed of water ($ft^3_{water}$) | 6.23E + 01 | lb of water (lb) | Volume |
| Fluid ounce (fl oz) | 3.00E + 01 | Milliliter (mL) | Volume |
| Fluid ounce (fl oz) | 2.00E + 00 | Tablespoon (tbls) | Volume |
| Fluid ounce (fl oz) | 6.00E + 00 | Teaspoon (tsp) | Volume |
| Foot cubed ($ft^3$) | 2.83E + 01 | Liter (L) | Volume |
| Foot cubed ($ft^3$) | 7.48E + 00 | US gallon (gal) | Volume |
| Gallon (gal) | 3.07E − 06 | Acre-feet (ac ft) | Volume |
| Gallon (gal) | 3.68E − 05 | Acre-inch (ac in.) | Volume |
| Gallon (gal) | 2.38E − 02 | Barrel (bbl) | Volume |
| Gallon (gal) | 1.60E + 01 | Cup | Volume |
| Gallon (gal) | 1.28E + 02 | Fluid ounce (fl oz) | Volume |
| Gallon (gal) | 1.34E − 01 | Foot cubed ($ft^3$) | Volume |
| Gallon (gal) | 3.79E − 07 | Hectare-meter (ha m) | Volume |
| Gallon (gal) | 2.31E + 02 | Inch cubed ($in.^3$) | Volume |
| Gallon (gal) | 3.79E + 00 | Liter (L) | Volume |
| Gallon (gal) | 3.79E − 03 | Meter cubed ($m^3$) | Volume |
| Gallon (gal) | 3.79E + 03 | Milliliter (mL) | Volume |
| Gallon (gal) | 8.00E + 00 | Pint (pt) | Volume |
| Gallon (gal) | 4.00E + 00 | Quart (qt) | Volume |
| Gallon (gal) | 4.95E − 03 | Yard cubed ($yd^3$) | Volume |
| Hectare-meter | 8.10E + 00 | Acre-feet (ac ft) | Volume |
| Hectare-meter | 2.64E + 06 | Gallon (gal) | Volume |
| Imperial gallon | 4.55E + 00 | Liter (L) | Volume |
| Inch cubed ($in.^3$) | 1.33E − 08 | Acre-feet (ac ft) | Volume |
| Inch cubed ($in.^3$) | 1.64E + 01 | Centimeter cubed ($cm^3$) | Volume |
| Inch cubed ($in.^3$) | 5.79E − 04 | Foot cubed ($ft^3$) | Volume |
| Inch cubed ($in.^3$) | 1.64E − 02 | Liter (L) | Volume |
| Inch cubed ($in.^3$) | 1.64E − 05 | Meter cubed ($m^3$) | Volume |
| Inch cubed ($in.^3$) | 4.33E − 03 | US gallon (gal) | Volume |
| Inch cubed ($in.^3$) | 2.14E − 05 | Yard cubed ($yd^3$) | Volume |
| Liter (L) | 8.13E − 07 | Acre-feet (ac ft) | Volume |
| Liter (L) | 8.11E − 07 | Acre-feet (ac ft) | Volume |
| Liter (L) | 6.10E + 01 | Cubic inch ($in.^3$) | Volume |
| Liter (L) | 1.31E − 03 | Cubic yard ($yd^3$) | Volume |
| Liter (L) | 2.11E + 00 | Fluid pint (fl pt) | Volume |
| Liter (L) | 1.06E + 00 | Fluid quart (fl qt) | Volume |
| Liter (L) | 3.53E − 02 | Foot cubed ($ft^3$) | Volume |
| Liter (L) | 1.00E − 03 | Meter cubed ($m^3$) | Volume |
| Liter (L) | 1.00E − 03 | Meter cubed ($m^3$) | Volume |
| Liter (L) | 3.38E + 01 | Ounce (oz) | Volume |
| Liter (L) | 2.64E − 01 | US gallon (gal) | Volume |
| Liter (L) | 1.31E − 03 | Yard cubed ($yd^3$) | Volume |
| Liter of water | 1.00E + 03 | Gram of water (g) | Volume |
| Meter cubed ($m^3$) | 8.13E − 04 | Acre-feet (ac ft) | Volume |
| Meter cubed ($m^3$) | 8.11E − 04 | Acre-feet (ac ft) | Volume |
| Meter cubed ($m^3$) | 3.53E + 01 | Cubic foot ($ft^3$) | Volume |
| Meter cubed ($m^3$) | 6.10E + 04 | Cubic inch ($in.^3$) | Volume |

(Continued)

(Continued)

| Multiply | By | To obtain | Property |
|---|---|---|---|
| Meter cubed (m$^3$) | 1.31E + 00 | Cubic yard (yd$^3$) | Volume |
| Meter cubed (m$^3$) | 1.00E + 03 | Decimeter cubed (dm$^3$) | Volume |
| Meter cubed (m$^3$) | 2.27E + 02 | Dry gallon (dry gal) | Volume |
| Meter cubed (m$^3$) | 1.00E + 03 | Liter (L) | Volume |
| Meter cubed (m$^3$) | 1.00E + 03 | Liter (L) | Volume |
| Meter cubed (m$^3$) | 2.64E + 02 | US gallon (gal) | Volume |
| Meter cubed (m$^3$) | 2.64E + 02 | US gallon (gal) | Volume |
| Milliliter (mL) | 3.40E − 02 | Fluid ounce (fl oz) | Volume |
| Peck | 2.50E − 01 | Bushel (bu) | Volume |
| Peck | 8.00E + 00 | Quart (qt) | Volume |
| Pennyweight (dwt) | 5.49E − 02 | Ounce (oz) | Volume |
| Pica (p) | 1.20E + 01 | Point (pt) | Volume |
| Pint | 2.00E + 00 | Cups | Volume |
| Pint | 1.60E + 01 | Fluid ounce (fl oz) | Volume |
| Pint | 4.70E + 02 | Milliliter (mL) | Volume |
| Pint | 5.00E − 01 | Quarts (qt) | Volume |
| Pound of water | 1.60E − 02 | Cubic feet of water (ft$^3$) | Volume |
| Quart | 1.25E − 01 | Peck | Volume |
| Quart | 2.00E + 00 | Pint | Volume |
| Quart (qt) | 4.00E + 00 | Cup | Volume |
| Quart (qt) | 3.20E + 01 | Fluid ounce (fl oz) | Volume |
| Quart (qt) | 9.50E + 02 | Milliliter (mL) | Volume |
| Tablespoon (tbls) | 1.50E + 01 | Milliliter (mL) | Volume |
| Tablespoon (tbls) | 3.00E + 00 | Teaspoon (tsp) | Volume |
| Teaspoon (tsp) | 5.00E + 00 | Milliliter (mL) | Volume |
| US gallon (gal) | 3.79E + 00 | Liter (L) | Volume |
| Yard cubed (yd$^3$) | 6.20E − 04 | Acre-feet (ac ft) | Volume |
| Yard cubed (yd$^3$) | 2.70E + 01 | Foot cubed (ft$^3$) | Volume |
| Yard cubed (yd$^3$) | 4.67E + 04 | Inch cubed (in.$^3$) | Volume |
| Yard cubed (yd$^3$) | 7.65E + 02 | Liter (L) | Volume |
| Yard cubed (yd$^3$) | 7.65E − 01 | Meter cubed (m$^3$) | Volume |
| Yard cubed (yd$^3$) | 2.02E + 02 | US gallon (gal) | Volume |
| Ounce (oz) | 2.83E + 01 | Gram (gm) | Weight |
| Ounce (oz) | 2.84E − 02 | Kilogram (kg) | Weight |
| Ounce (oz) | 2.96E − 02 | Liter (L) | Weight |
| Ounce (oz) | 2.79E − 05 | Long ton | Weight |
| Ounce (oz) | 2.89E − 03 | Metric slug | Weight |
| Ounce (oz) | 2.84E − 05 | Metric ton (tonne) | Weight |
| Ounce (oz) | 2.96E + 01 | Milliliter (mL) | Weight |
| Ounce (oz) | 6.25E − 02 | Pound (lb) | Weight |
| Ounce (oz) | 3.13E − 05 | Short ton | Weight |
| Ounce (oz) | 1.94E − 03 | Slug | Weight |
| Pound (lb) | 4.46E − 04 | Long ton | Weight |
| Pound (lb) | 1.60E + 01 | Ounce (oz) | Weight |
| Stone (st) | 1.40E + 01 | Pound (lb) | Weight |
| Ton (t) | 2.00E + 03 | Pound (lb) | Weight |
| Tonne (t) | 2.20E + 03 | Pound (lb) | Weight |
| Tonne (t) | 1.10E + 00 | Tons | Weight |

# Appendix A-2: Abbreviations

Although many of the abbreviations listed in this appendix are specific to the United States of America, you will find that many other countries have either adopted the same terms or have been required to understand their use. Therefore, if a term is found in most world wide literatures, we have included them to the extent possible so that they may be interpreted. Because of the extent of definitions that have been incorporated into the text, we have often excluded abbreviations in the following list. The meaning of the included abbreviations may be found by using the Internet.

| Abbreviation | Term |
|---|---|
|  | Elastic silty, sandy silt, clayey silt, or gravelly silt |
| °C | Degrees Celsius |
| °F | Degrees Fahrenheit |
| °K | Degrees Kelvin |
| °R | Degrees Rankin |
| μs | Microsecond |
| AAQ | Ambient air quality |
| AAQS | Ambient air quality standard |
| AAS | Atomic absorption spectroscopy |
| ABS | Acrylonitrile butadiene styrene |
| ADI | Acceptable daily intake |
| AE | Actual evapotranspiration |
| AMC | Antecedent moisture conditions |
| ANSI | American National Standards Institute |
| APCD | Air pollution control district |
| API | American Petroleum Institute |
| API | Antecedent precipitation index |
| APR | Air purifying respirator |
| AQMD | Air quality management district |
| ARCs | Antecedent runoff conditions |
| ASTM | American society for Testing and Materials |
| AWS |  |
| AWWA | American Water Well Association |
| BAT | Best available technology (water) |
| BCT | Best control technology for conventional pollutants (CWA) |
| BEI | Biological exposure indices |
| Bhp | Brake horsepower |
| BLEVE | Boiling-Liquid-Expanding-Vapor Explosion |
| bls | Below land surface |
| BMP | Best management practices |
| BOD | Biochemical oxygen demand |
| BOD | Biological oxygen demand |
| BPT | Best practical technology |
| BTU | British thermal unit |
| C | Runoff coefficient (dimensionless) |
| CEQA | California Environmental Quality Act |
| CERCLA | Comprehensive Environmental Response, Compensation and Liability Act (USA) |
| cfm | Cubic feet per minute |
| CH | Fat clay, sandy clay, silty clay, or gravelly clay |
| CL | Lean clay, sandy clay, or gravelly silt |
| cm$^2$ | Centimeters squared |
| CN | Runoff curve number |
| COC | Constituent of concern |
| COD | Chemical oxygen demand |

(*Continued*)

(Continued)

| Abbreviation | Term |
|---|---|
| cpm | Counts per minute |
| CFR | Code of Federal Regulation (USA) |
| CWA | Clean Water Act |
| CWRCB | California Water Resources Control Act (USA) |
| DHS | Department of Health Services (CA, USA) |
| DNAPL | Dense nonaqueous phase liquid |
| DO | Dissolved oxygen |
| DRI | Direct reading instrument |
| EIR | Environmental Impact Report |
| EPA | Environmental protection agency |
| FEMA | Federal emergency management agency |
| FIRM | Flood insurance rate map |
| fps | Feet per second |
| FSP | Field sampling plan |
| ft | Feet |
| ft min$^{-1}$ | Feet per minute |
| ft s$^{-1}$ | Feet per second |
| ft$^2$ | Square feet or feet squared |
| ft$^3$ | Cubic feet or feet cubed |
| ft lb | Foot pound |
| g | Gram |
| GAC | Granulated activated carbon |
| gal | Gallons |
| GC | Gas chromatograph (analyzer) |
| GC | Glayey gravel and clayey gravel with sand |
| GC/MS | Gas chromatograph/mass spectrometer (analyzer) |
| GM | Silty gravel and silty gravel with sand |
| GP | Poorly graded gravel and poorly graded gravel with sand |
| gpd | Gallons per day |
| gpd ft$^{-1}$ | Gallons per day per foot |
| gpg | Grains per gallon |
| gpm | Gallons per minute |
| GW | Well grade gravel and well graded gravel with sand |
| hr | Hour |
| HASP | Health and safety plan |
| hp | Horsepower |
| HRS | Hazard ranking system |
| HSWA | Hazard and Solid Waste Act (1984 amendments to RCRA) |
| Hz | Hertz |
| I.D. | Inside diameter |
| IDLH | Immediately dangerous to life or health |
| in. | Inch |
| in.$^2$ | Inches squared or square inches |
| in.$^3$ | Inches cubed or cubic inches |
| IP | Ionization potential |
| J | Joule |
| kg | Kilogram |
| km | Kilometer |
| km$^2$ | Kilometer squared or square kilometers |
| km$^3$ | Kilometer cubed or cubic kilometers |
| kPa | Kilopascals |
| L | Liter |
| lb | Pound |
| LEL | Lower explosive limit of a chemical |
| LFL | Lower flammable liquid |
| LNAPL | Light nonaqueous phase liquid |

(Continued)

| Abbreviation | Term |
|---|---|
| LUST | Leaking underground storage tank |
| LVF | Liquid-volume fraction |
| m | Meter |
| $m\,day^{-1}$ | Meter per day |
| $m\,min^{-1}$ | Meter per minute |
| $m\,s^{-1}$ | meter per second |
| $m^2$ | Meter squared or square meters |
| $m^2\,day^{-1}$ | Meter squared per day or square meters per day |
| $m^3$ | Meter cubed or cubic meters |
| $m^3\,day^{-1}$ | Meter cubed per day or cubic meters per day |
| $m^3\,s^{-1}$ | Meter cubed per second or cubic meters per second |
| mb | Millibars |
| MCL | Maximum contaminant level (water) |
| MCLG | Maximum contaminant level goal (water) |
| MDL | Method detection limit |
| $mg\,L^{-1}$ | Milligram per liter |
| MHz | Megahertz |
| mi | Mile |
| $mi^2$ | Mile squared or square mile |
| $mi^3$ | Mile cubed or cubic mile |
| min | Minute |
| mL | Milliliter |
| ML | Silty, sandy silty, clayey silt, or graveley silt |
| mm | Millimeter |
| $mm^2$ | Millimeter squared or square millimeters |
| $mm^3$ | Millimeter cubed or cubic millimeters |
| MSDS | Material safety data sheet |
| msl | Mean sea level |
| NIOSH | National Institute of Occupational Safety and Health (USA) |
| NPDES | National pollutant discharge elimination series |
| NPL | National priorities list (USA) |
| NPSH | Net positive suction head |
| NWWA | National Water Well Association |
| O.D. | Outside diameter |
| OH | Organic clay, silt, sandy clay, silty clay, clayey silt, sandy silt, or gravelly silts and clays |
| OL | Organic silt, clay, sandy silt, sandy clay, gravelly clay, or gravelly silt |
| OSHA | Occupational Safety and Health Administration (USA) |
| OVA | Organic vapor analyzer |
| oz | Ounce |
| Pa | Pascal |
| PAC | Powered activated carbon |
| PAH | Polynuclear aromatic hydrocarbons (also polycarbons) |
| PCB | Polychlorinated biphenyl |
| $pCi\,L^{-1}$ | Picocurie per liter |
| PDS | Partial-duration series |
| PE | Potential evapotranspiration |
| PEL | Permissible Exposure Limit (used by OSHA – USA) |
| pHs | Saturation pH |
| PID | Photo ionization detector |
| POT | Peaks over threshold |
| PPB | Parts per billion (concentration) |
| PPE | Personal protective equipment |
| PPM | Parts per million (concentration) |
| PRP | Potential responsible party |
| psi | Pounds per square inch |

(Continued)

(Continued)

| Abbreviation | Term |
|---|---|
| psig | Pounds per square inch at gauge |
| PT | Peat or other highly organic soil |
| PVC | Polyvinyl chloride |
| PWL | Pumping water level |
| Q | Discharge |
| QA/QC | Quality assurance/quality control |
| qt | Quart |
| R | Bowen's ratio |
| RCRA | Resource Conservation and Recovery Act (1976 Amendment to Solid Waste Disposal Act – USA) |
| REL | Recommended exposure limits |
| RI | Remedial investigation |
| RI/FS | Remedial investigation/feasibility study |
| RMCL | Recommended maximum contaminant level |
| RO | Reverse osmosis |
| ROD | Record of decision (US EPA) |
| RP | Responsible party |
| rpm | Revolutions per minute |
| rps | Revolutions per second |
| RQ | Reportable quantity |
| RWQCB | Regional Water Control Board (California) |
| s | Second |
| SAR | Sodium adsorbtion rate |
| SAR | Sodium adsorption ratio |
| SARA | Superfund Amendments and Reauthorization Act |
| SC | Clayey sand and clayey sand with gravel |
| SCBA | Self-contained breathing apparatus |
| SCS | Soil Conservation Service |
| SDR | Standard dimension ratio |
| SDWA | Safe Drinking Water Act (USA) |
| SM | Silty sand and silty sand with gravel |
| SMEWW | Standard methods for the examination of water and waste water |
| SOC | Synthetic organic chemical |
| SOP | Standard operating procedures |
| SOW | Statement of Work |
| SP | Poorly graded sand and poorly graded sand with gravel |
| SP | Spontaneous potential |
| sp. gr. | Specific gravity |
| SR | Styrene rubber |
| SS | Suspended soils |
| STEL | Short-term exposure level (15-min average) |
| STLC | Soluble threshold limit concentration (mg $L^{-1}$, CA, USA) |
| SW | Well-graded sand and well-graded sand with gravel |
| SWL | Static water level |
| SWRBC | State Water Resources Control Board California, USA |
| t | Time |
| t | Ton |
| T | Tonne |
| TDS | Total dissolved solids |
| THM | Trihalomethane |
| TLV | Threshold limit value |
| TOC | Total organic carbon |
| TSD | Treatment, storage, or disposal |
| TSDF | Treatment, storage, or disposal facility |
| TTLC | Total threshold limit concentration (mg $kg^{-1}$, CA, USA) |
| USEPA | United States Environmental Protection Agency |

| Abbreviation | Term |
|---|---|
| UEL | Upper explosive limit of a chemical |
| USGS | United States Geological Society |
| UST | Underground storage tank |
| VOC | Volatile organic chemicals |
| WET | Waste extraction test (CA, USA) |
| WHO | World Health Organization |
| Whp | Water horsepower |
| WOGA | Western Oil and Gas Association (USA) |
| Year | Year |
| $Z_{pc}$ | Point of zero change |
| $\mu g\,L^{-1}$ | Microgram per liter |

# Appendix A-3: Symbols

Symbols often have several interpretations and several terms have different symbols depending on their use and/or the discipline in which they are used. We have attempted to be as complete as possible and included as many symbols and meanings that are common to hydrogeology.

| Symbol | Represents |
|---|---|
| $u$ | $1.87\ r^2 S/Tt$ |
| $B$ | $264/T \log (0.03\ Tt/r2S)$ |
| $T$ | Absolute temperature ($K = 273.16 + {}^\circ C$) |
| $g$ | Acceleration of gravity |
| $\gamma_i$ | Activity coefficient |
| $H\sim$ | Adjusted penetration depth for each column |
| $q_a$ | Air flow velocity |
| $\theta a$ | Air-filled porosity (volume of air/total volume) |
| $G_a$ | Air-space ratio |
| $b$ | Aquifer spacing |
| $B$ | Aquifer thickness |
| $A$ | Area |
| $a$ | Array spacing |
| $\nu$ | Average linear velocity |
| $\theta$ | Average volumetric water content |
| $R_a$ | Basin area ratio |
| $L_b$ | Basin length |
| $P$ | Basin perimeter |
| $H$ | Basin relief |
| $R_c$ | Basin-circularity ratio |
| $R$ | Bowen's ratio |
| $m$ | Bulk contraction |
| $\rho_b$ | Bulk density of the soil |
| $C$ | Coefficient of discharge |
| $S$ | Coefficient of storage |
| $D$ | Combined dispersion coefficient |
| $\theta$ | Combined retardation factor |
| $C_w$ | Concentration of a contaminant in liquid (water) phase |

(Continued)

| Symbol | Represents |
|---|---|
| $C_s$ | Concentration of a contaminant in solid phase |
| $C_a$ | Concentration of a contaminant in vapor (air) phase |
| $C_{aq}$ | Concentration of horizontal groundwater influx |
| $C$ | Constant |
| $i_c$ | Critical gradient |
| $A_{aq}$ | Cross-sectional aquifer area perpendicular to the groundwater flow direction |
| $A_{soil}$ | Cross-sectional area perpendicular to the vertical infiltration in the soil column |
| $I$ | Current |
| $q_{aq}$ | Darcy velocity in the aquifer |
| $\gamma$ | Decay rate of the solute source due to either degradation or flushing by the infiltration |
| $\rho$ | Density |
| $D$ | Depth |
| $d$ | Depth to water |
| $d$ | Diameter |
| $D_{air}$ | Diffusion coefficient of the contaminant in the free air |
| $K_H$ | Dimensional form of Henry's law constant |
| $Q$ | Discharge |
| $i$ | Discretized soil column cell (column index in finite difference grid) |
| $D_w$ | Dispersion coefficient for the liquid phase contaminant in the pore water |
| $r$ | Distance from center of a pumped well to a point where the drawdown is measured |
| $\Delta x$ | Distance of lattice point in horizontal space |
| $\Delta y$ | Distance of lattice point in vertical space |
| $K_d$ | Distribution coefficient between the solid phase and liquid phase |
| $s$ | Drawdown |
| $t_O$ | Duration of solute release |
| $\mu$ | Dynamic viscosity |
| $\sqcap_e$ | Effective velocity |
| $E$ | Efficiency of well |
| $z$ | Elevation above a certain datum |
| $z$ | Elevation head |
| $e$ | Exponential function |
| $\mu_a$ | First-order decay rate of a contaminant in gaseous phase |
| $\mu_s$ | First-order decay rate of a contaminant in solid phase |
| $\mu_w$ | First-order decay rate of a contaminant in water phase |
| $\mu$ | First-order decay rate of a contaminant |
| $V$ | Flow function |
| $q$ | Flow through each foot of aquifer width |
| $\rho$ | Fluid pressure |
| $F_s$ | Force exerted by the soil |
| $f_{oc}$ | Fraction organic carbon of the soil |
| $D_a$ | Gaseous phase diffusion coefficient in the pore air |
| $dh/dl$ | Gradient |
| $CQ^2$ | Head loss attributable to turbulent flow |
| $h_L$ | Head loss |
| $BQ$ | Head loss attributable to laminar flow |
| $h$ | Head potential |
| $N$ | Head to a reasonable number of points |
| $h$ | Head, hydraulic head |
| $h$ | Height |

| Symbol | Represents |
|--------|-----------|
| $K_H$ | Dimensional form of Henry's law constant |
| $H$ | Henry's partition coefficient between the air phase and the water phase |
| $L$ | Horizontal length dimension of the waste |
| $K$ | Hydraulic conductivity |
| $I$ | Hydraulic gradient |
| $R$ | Hydraulic radius |
| $R_h$ | Hydraulic resistance |
| $C_s(z,0)$ | Initial solid-phase concentration |
| $k$ | Intrinsic permeability |
| $m$ | Iteration index |
| $L$ | Length or distance |
| $x$ | Length or distance |
| $C_O$ | Liquid-phase solute concentration in the infiltration water |
| $D$ | Longitudinal dispersion coefficient |
| $a_L$ | Longitudinal dispersivity of the vadose zone |
| $C_{mx(i)}$ | Mixed concentration in groundwater |
| $\omega$ | Net groundwater recharge rate |
| $N_f$ | Number of flow channels in the flow net |
| $N_c$ | Number of potential drops along each channel |
| $K_{oc}$ | Organic carbon–water partition coefficient |
| $H_d$ | Penetration depth |
| $\pi$ | Pi (3.14) |
| $Z_{pc}$ | Point of zero charge |
| $\eta$ | Porosity |
| $P$ | Pressure |
| $p$ | Pressure |
| $\psi$ | Pressure head |
| $r$ | Radius |
| $R$ | Radius |
| $R_o$ | Radius of influence |
| $r$ | Radius of influence |
| $c$ | Residual between two successive Gauss–Sedial iterations |
| $ś$ | Residual drawdown |
| $R$ | Resistance |
| $\rho$ | Resistivity |
| $j$ | Row index in finite difference grid |
| $v$ | Seepage velocity |
| $C$ | Solute concentration |
| $C_a$ | Solute concentration in air |
| $C_w$ | Solute concentration in water |
| $Q/s$ | Specific capacity |
| $q_x$ | Specific flow rate in x direction |
| $q_y$ | specific flow rate in y direction |
| $N_s$ | Specific speed |
| $\gamma$ | Specific weight |
| $S_y$ | Specific yield |
| $S$ | Storage coefficient |
| $T$ | Temperature |
| $z_{hl}$ | Thickness of impermeable materials placed landward of structure |
| $b$ | Thickness of the saturated part of an aquifer |
| $t$ | Time |
| $\Delta t$ | Time increment |
| $k$ | Time level |
| $M$ | Total contaminant mass per unit volume of the soil |

(*Continued*)

| Symbol | Represents |
|---|---|
| $H$ | Total head |
| $n$ | Total porosity of the soil: equals the sum of water filled porosity and air filled porosity |
| $T$ | Transmissivity |
| $\alpha_v$ | Transverse (vertical) dispersivity of the aquifer |
| $x$ | Unit length |
| $R$ | Universal gas constant |
| $F_w$ | Upward force on a soil column caused by reservoir pressure |
| $h_{ij}$ | Value of the head at the lattice point |
| $V$ | Velocity |
| $z$ | Vertical coordinate with positive being downward |
| $L_u$ | Vertical distance from the source to the observation point (i.e., the water table) |
| $V$ | Voltage |
| $v$ | Volume |
| $\theta_a$ | Volumetric air content (volume of air/total volume) |
| $\theta$ | Volumetric content |
| $\theta_w$ | Volumetric water content (volume of water/total volume) |
| $q_w$ | Water flow velocity (recharge rate) |
| $W(u)$ | well function of u, represents an exponential integral |
| $W$ | Width |
| $w$ | Width |

# Appendix B: Tables of Properties of Water

## Chemical characteristics of common organic chemicals

| CAS No. | Chemical | Molecular weight (g mole⁻¹) | Specific gravity (g mL⁻¹l) | Solubility in water (S) (mg L⁻¹) | Diffusivity in air ($D_i$) (cm² s⁻¹) | Diffusivity in water ($D_w$) (cm² s⁻¹) | Dimensionless Henry's law constant ($H'$) (25°C) | Organic carbon partition coefficient ($K_{oc}$) (L kg⁻¹) | First-order degradation constant ($\lambda$) (d⁻¹) |
|---|---|---|---|---|---|---|---|---|---|
| **Neutral organics** | | | | | | | | | |
| 83-32-9 | Acenaphthene | 154.21 | 1.024 | 4.24 | 0.0421 | 7.69E-6 | 0.00636 | 7080 | 0.0034 |
| 67-64-1 | Acetone | 58.08 | 0.79 | 1,000,000 | 0.124 | 1.14E-5 | 0.00159 | 0.575 | 0.0495 |
| 79-06-1 | Acrylamide | 71.08 | 1.0511 | 2,100,000 | | | | 0.17 | No data |
| 79-10-7 | Acrylic acid | 72.06 | 1.05 | 1,000,000 | | | | 1.45 | No data |
| 15972-60-8 | Alachlor | 269.5 | | 242 | 0.0198 | 5.69E-6 | 0.00000132 | 394 | No data |
| 116-06-3 | Aldicarb | 190.27 | 1.195 | 6 000 | 0.0305 | 7.19E-6 | 0.0000000574 | 12 | 0.00109 |
| 309-00-2 | Aldrin | 364.8 | | 0.18 | 0.0132 | 4.86E-6 | 0.00697 | 2,450,000 | 0.00059 |
| 120-12-7 | Anthracene | 178 | 1.283 | 0.0434 | 0.0324 | 7.74E-6 | 0.00267 | 29,500 | 0.00075 |
| 1912-24-9 | Atrazine | 215.5 | 1.187 | 70 | 0.0258 | 6.69E-6 | 0.00000005 | 451 | No data |
| 71-43-2 | Benzene | 78.11 | 0.87865 | 1,750 | 0.088 | 9.80E-6 | 0.228 | 58.9 | 0.0009 |
| 56-55-3 | Benzo(a)anthracene | 228.29 | | 0.0094 | 0.0510 | 9.00E-6 | 0.000137 | 398,000 | 0.00051 |
| 205-99-2 | Benzo(b)fluoranthene | 252.32 | | 0.0015 | 0.0226 | 5.56E-6 | 0.00455 | 1,230,000 | 0.00057 |
| 207-08-9 | Benzo(k)fluoranthene | 252.32 | | 0.0008 | 0.0226 | 5.56E-6 | 0.000034 | 1,230,000 | 0.00016 |
| 65-85-0 | Benzoic acid | 122.12 | 1.2659 | 3,500 | 0.0536 | 7.97E-6 | 0.00000631 | 0.600 | No data |
| 50-32-8 | Benzo(a)pyrene | 252.3 | 1.24 | 0.00162 | 0.043 | 9.00E-6 | 0.0000463 | 1,020,000 | 0.00065 |
| 111-44-4 | Bis(2-chloroethyl)ether | 142.9 | 1.220 | 17,200 | 0.0692 | 7.53E-6 | 0.000738 | 15.5 | 0.0019 |
| 117-81-7 | Bis(2-ethylhexyl) phthalate | 390.56 | 0.981 | 0.34 | 0.0351 | 3.66E-6 | 0.00000418 | 15,100,000 | 0.0018 |
| 75-27-4 | Bromodichloromethane | 163.83 | 1.98 | 6,740 | 0.0298 | 1.06E-5 | 0.0656 | 55.0 | No data |
| 75-25-2 | Bromoform | 252.8 | 2.890 | 3100 | 0.0149 | 1.03E-5 | 0.0219 | 87.1 | 0.0019 |
| 71-36-3 | Butanol | | | 74,000 | 0.0800 | 9.30E-6 | 0.000361 | 6.92 | 0.01283 |
| 85-68-7 | Butyl benzyl phthalate | 312 | 1.1 | 2.69 | 0.0174 | 4.83E-6 | 0.0000517 | 57,500 | 0.00385 |
| 86-74-8 | Carbazole | 201 | 1.1 | 7.48 | 0.0390 | 7.03E-6 | 0.000000626 | 3390 | No data |
| 1563-66-2 | Carbofuran | 257 | | 320 | 0.0249 | 6.63E-6 | 0.00377 | 37 | No data |
| 75-15-0 | Carbon disulfide | 76.1 | 1.263 | 1190 | 0.104 | 1.00E-5 | 1.24 | 45.7 | No data |
| 56-23-5 | Carbon tetrachloride | 153.8 | 1.594 | 793 | 0.0780 | 8.80E-6 | 1.25 | 174 | 0.0019 |
| 57-74-9 | Chlordane | 409.6 | 1.59–1.63 | 0.056 | 0.0118 | 4.37E-6 | 0.00199 | 120,000 | 0.00025 |
| 106-47-8 | p-Chloroaniline | 127.45 | 1.169 | 5 300 | 0.0483 | 1.01E-5 | 0.0000136 | 66.1 | No data |
| 108-09-7 | Chlorobenzene | 112.6 | 1.11 | 472 | 0.0730 | 8.70E-6 | 0.152 | 219 | 0.0023 |
| 124-48-1 | Chlorodibromomethane | 208.28 | 2.451 | 2600 | 0.0196 | 1.05E-5 | 0.0321 | 63.1 | 0.00385 |
| 67-66-3 | Chloroform | 119.4 | 1.483 | 7920 | 0.104 | 1.00E-5 | 0.15 | 39.8 | 0.00039 |
| 95-57-8 | 2-Chlorophenol | 128.6 | 1.263 | 22,000 | 0.0501 | 9.46E-6 | 0.016 | 388 | No data |

| | | | | | | | | | |
|---|---|---|---|---|---|---|---|---|---|
| 218-01-9 | Chrysene | 228.28 | 1.274 | 0.0016 | 0.0248 | 6.21E-6 | 0.00388 | 398.000 | 0.00035 |
| 94-75-7 | 2,4-D | 221 | | 680 | 0.0231 | 7.31E-6 | 0.00000041 | 451 | 0.00385 |
| 72-54-8 | 4,4'-DDD | 320.04 | 1.385 | 0.09 | 0.0169 | 4.76E-6 | 0.000164 | 1.000.000 | 0.000062 |
| 72-55-9 | 4,4'-DDE | 318.03 | | 0.12 | 0.0144 | 5.87E-6 | 0.000861 | 4.470.000 | 0.000062 |
| 50-29-3 | 4,4'-DDT | 354.5 | 0.98–0.99 | 0.025 | 0.0137 | 4.95E-6 | 0.000332 | 2.630.000 | 0.000062 |
| 75-99-0 | Dalapon | 142.97 | 1.4014 | 900.000 | 0.0414 | 9.46E-6 | 0.00000264 | 5.8 | 0.005775 |
| 53-70-3 | Dibenzo(a,h)anthracene | 278.4 | 1.28 | 0.00249 | 0.0202 | 5.18E-6 | 0.00000000603 | 3.800.000 | 0.00037 |
| 96-12-8 | 1,2-Dibromo-3-chloropropane | 236.32 | 2.050 | 1.200 | 0.0212 | 7.02E-6 | 0.00615 | 182 | 0.001925 |
| 106-93-4 | 1,2-Dibromoethane | 187.86 | 2.17 | 4.200 | 0.0287 | 8.06E-6 | 0.0303 | 93 | 0.005775 |
| 84-74-2 | Di-n-butyl phthalate | 278.34 | 1.047 | 11.2 | 0.0438 | 7.86E-6 | 0.0000000385 | 33.900 | 0.03013 |
| 95-50-1 | 1,2-Dichlorobenzene | 147 | 1.305 | 156 | 0.0690 | 7.90E-6 | 0.0779 | 617 | 0.0019 |
| 106-46-7 | 1,4-Dichlorobenzene | 147 | 1.25 | 73.8 | 0.0690 | 7.90E-6 | 0.0996 | 617 | 0.0019 |
| 91-94-1 | 3,3-Dichlorobenzidine | 253.13 | | 3.11 | 0.0194 | 6.74E-6 | 0.000000164 | 724 | 0.0019 |
| 75-34-3 | 1,1-Dichloroethane | 98.96 | 1.176 | 5.060 | 0.0742 | 1.05E-5 | 0.23 | 31.6 | 0.0019 |
| 107-06-2 | 1,2-Dichloroethane | 98.96 | 1.235 | 8.520 | 0.104 | 9.90E-6 | 0.0401 | 17.4 | 0.0019 |
| 75-35-4 | 1,1-Dichloroethylene | 96.94 | 1.218 | 2.250 | 0.0900 | 1.04E-5 | 1.07 | 58.9 | 0.0053 |
| 156-59-2 | Cis-1,2-dichloroethylene | 96.94 | 1.257 | 3.500 | 0.0736 | 1.13E-5 | 0.167 | 35.5 | 0.00024 |
| 156-60-5 | Trans-1,2-dichloroethylene | 96.94 | 1.257 | 6.300 | 0.0707 | 1.19E-5 | 0.385 | 52.5 | 0.00024 |
| 120-83-2 | 2,4-Dichlorophenol | 162.9 | 1.383 | 4.500 | 0.0346 | 8.77E-6 | 0.00013 | 147 | 0.00027 |
| 78-87-5 | 1,2-Dichloropropane | 113 | 1.56 | 2.800 | 0.0782 | 8.73E-6 | 0.115 | 43.7 | 0.00027 |
| 542-75-6 | 1,3-Dichloropropylene (cis + trans) | 110.9 | 1.224, 1.182 | 2.800 | 0.0626 | 1.00E-5 | 0.726 | 45.7 | 0.061 |
| 60-57-1 | Dieldrin | 380.7 | 1.75 | 0.195 | 0.0125 | 4.74E-6 | 0.000619 | 21.400 | 0.00032 |
| 84-66-2 | Diethyl phthalate | 222 | 1.118 | 1.080 | 0.0256 | 6.35E-6 | 0.0000185 | 288 | 0.00619 |
| 105-67-9 | 2,4-Dimethylphenol | 244 | 0.9650 | 7.870 | 0.0584 | 8.69E-6 | 0.0000082 | 209 | 0.0495 |
| 51-28-5 | 2,4-Dinitrophenol | 184 | 1.683 | 2.790 | 0.0273 | 9.06E-6 | 0.0000182 | 0.01 | 0.00132 |
| 121-14-2 | 2,4-Dinitrotoluene | 182.1 | 1.32 | 270 | 0.203 | 7.06E-6 | 0.0000038 | 95.5 | 0.00192 |
| 606-20-2 | 2,6-Dinitrotoluene | 182.1 | | 182 | 0.0327 | 7.26E-6 | 0.0000306 | 69.2 | 0.00192 |
| 88-85-7 | Dinoseb | 240 | 1.26 | 52 | 0.0215 | 6.62E-6 | 0.0000189 | 1120 | 0.002817 |
| 117-84-0 | Di-n-octyl phthalate | 390.56 | 0.978 | 0.02 | 0.0151 | 3.58E-6 | 0.00274 | 83.200.000 | 0.0019 |
| 115-29-7 | Endosulfan | 406.8 | 1.745 | 0.51 | 0.0115 | 4.55E-6 | 0.000459 | 2140 | 0.07629 |
| 145-73-3 | Endothall | 186.18 | 1.431 | 21.000 | 0.0291 | 8.07E-6 | 0.0000000107 | 0.29 | No data |
| 72-20-8 | Endrin | 380.93 | 1.64 | 0.25 | 0.0125 | 4.74E-6 | 0.000308 | 12.300 | 0.00032 |
| 100-41-4 | Ethylbenzene | 106.2 | 0.867 | 169 | 0.0750 | 7.80E-6 | 0.323 | 363 | 0.003 |
| 206-44-0 | Fluoranthene | 202 | 1.252 | 0.206 | 0.0302 | 6.35E-6 | 0.00066 | 107.000 | 0.00019 |
| 86-73-7 | Fluorene | 166.22 | 1.202 | 1.98 | 0.0363 | 7.88E-6 | 0.00261 | 13.800 | 0.000691 |
| 76-44-8 | Heptachlor | 173.7 | 1.58 | 0.18 | 0.0112 | 5.69E-6 | 60.7 | 1.410.000 | 0.13 |

*(Continued)*

**Chemical characteristics of common organic chemicals (Continued)**

| CAS No. | Chemical | Molecular weight (g mole⁻¹) | Specific gravity (g mL⁻¹) | Solubility in water (S) (mg L⁻¹) | Diffusivity in air (Di) (cm² s⁻¹) | Diffusivity in water (Dw) (cm² s⁻¹) | Dimensionless Henry's law constant (H') (25°C) | Organic carbon partition coefficient (Koc) (L kg⁻¹) | First-order degradation constant (λ) (d⁻¹) |
|---|---|---|---|---|---|---|---|---|---|
| 1024-57-3 | Heptachlor epoxide | 389.3 | | 0.2 | 0.0132 | 4.23E-6 | 0.00039 | 83,200 | 0.00063 |
| 118-74-1 | Hexachlorobenzene | 284.8 | 2.04 | 6.2 | 0.0542 | 5.91E-6 | 0.0541 | 55,000 | 0.00017 |
| 319-84-6 | Alpha-HCH (alpha-BHC) | 290.85 | 1.87 | 2.0 | 0.0142 | 7.34E-6 | 0.000435 | 1230 | 0.0025 |
| 58-89-9 | Gamma-HCH (lindane) | 290.828 | | 6.8 | 0.0142 | 7.34E-6 | 0.000574 | 1070 | 0.0029 |
| 77-47-4 | Hexachlorocyclopentadiene | 272.7 | 1.702 | 1.8 | 0.0161 | 7.21E-6 | 1.11 | 200,000 | 0.012 |
| 67-72-1 | Hexachloroethane | 236.7 | 2.09 | 50 | 0.0025 | 6.80E-6 | 0.159 | 1780 | 0.00192 |
| 193-39-5 | Indeno(1,2,3-c,d)pyrene | 276.3 | | 0.000022 | 0.0190 | 5.66E-6 | 0.0000656 | 3,470,000 | 0.00047 |
| 78-59-1 | Isophorone | 138.21 | 0.92 | 12,000 | 0.0623 | 6.76E-6 | 0.000272 | 46.8 | 0.01238 |
| 7439-97-6 | Mercury | 200.59 | 13.534 | — | 0.0307 | 6.30E-6 | 0.467 | — | No data |
| 72-43-5 | Methoxychlor | 345.5 | | 0.045 | 0.0156 | 4.46E-6 | 0.000648 | 97,700 | 0.0019 |
| 74-83-9 | Methyl bromide | 94.94 | 1.73(liq; 0°C), 3.974 (gas; 20°C) | 15,200 | 0.0728 | 1.21E-5 | 0.256 | 10.5 | 0.01824 |
| 1634-04-4 | Methyl tertiary-butyl ether | 88.15 | 0.7404 | 51,000 | 0.102 | 1.10E-5 | 0.0241 | 11.5 | No data |
| 75-09-2 | Methylene chloride | 84.93 | 1.327 | 13,000 | 0.101 | 1.17E-5 | 0.0898 | 11.7 | 0.012 |
| 95-48-7 | 2-Methylphenol | 108.14 | 2.5 | 26,000 | 0.0740 | 8.30E-6 | 0.0000492 | 91.2 | 0.0495 |
| 91-20-3 | Naphthalene | 128.17 | 4.42 | 31.0 | 0.0590 | 7.50E-6 | 0.0198 | 2000 | 0.0027 |
| 98-95-3 | Nitrobenzene | 123.1 | 1.199 | 2,090 | 0.0760 | 8.60E-6 | 0.000984 | 64.6 | 0.00176 |
| 86-30-6 | N-Nitroso diphenylamine | 198.23 | 1.23 | 35.1 | 0.0312 | 6.35E-6 | 0.000205 | 1290 | 0.01 |
| 621-64-7 | N-Nitrosodi-n-propylamine | 130.22 | 0.9160 | 9890 | 0.0545 | 8.17E-6 | 0.0000923 | 24.0 | 0.0019 |
| 87-86-5 | Pentachlorophenol | 266.3 | 1.978 | 1950 | 0.0560 | 6.10E-6 | 0.000001 | 592 | 0.00045 |
| 108-95-2 | Phenol | 94.1 | 1.07 | 82,800 | 0.0820 | 9.10E-6 | 0.0000163 | 28.8 | 0.099 |
| 1918-02-1 | Picloram | 241.46 | | 430 | 0.0255 | 5.28E-6 | 0.00000000000166 | 1.98 | No data |
| 1336-36-3 | Polychlorinated biphenyls (PCBs) | | | 0.7 | —[a] | —[a] | —[a] | 309,000 | No data |

| | | | | | | | | | |
|---|---|---|---|---|---|---|---|---|---|
| 129-00-0 | Pyrene | 202.25 | 1.271 | 0.135 | 0.0272 | 7.24E-6 | 0.000451 | 105,000 | 0.00018 |
| 122-34-9 | Simazine | 201.657 | 1.3 | 5 | 0.027 | 7.36E-6 | 0.0000000133 | 133 | No data |
| 100-42-5 | Styrene | 104.15 | 0.909 | 310 | 0.0710 | 8.00E-6 | 0.113 | 776 | 0.0033 |
| 93-72-1 | 2,4,5-TP (silvex) | 269.51 | 1.2085 | 31 | 0.0194 | 5.83E-6 | 0.0000000000032 | 5440 | No data |
| 127-18-4 | Tetrachloroethylene | 165.8 | 1.623 | 200 | 0.0720 | 8.20E-6 | 0.754 | 155 | 0.00096 |
| 108-88-3 | Toluene | 92.1 | 0.8669 | 526 | 0.0870 | 8.60E-6 | 0.272 | 182 | 0.011 |
| 8001-35-2 | Toxaphene | 414 | 1.65 | 0.74 | 0.0116 | 4.34E-6 | 0.000246 | 257,000 | No data |
| 120-82-1 | 1,2,4-Trichlorobenzene | 314.8 | 1.454 | 300 | 0.0300 | 8.23E-6 | 0.0582 | 1780 | 0.0019 |
| 71-55-6 | 1,1,1-Trichloroethane | 133.4 | 1.339 | 1330 | 0.0780 | 8.80E-6 | 0.705 | 110 | 0.0013 |
| 79-00-5 | 1,1,2-Trichloroethane | 133.4 | 1.440 | 4420 | 0.0780 | 8.80E-6 | 0.0374 | 50.1 | 0.00095 |
| 79-01-6 | Trichloroethylene | 131.4 | 1.464 | 1100 | 0.0790 | 9.10E-6 | 0.422 | 166 | 0.00042 |
| 95-95-4 | 2,4,5-Trichlorophenol | 197.45 | 1.68 | 1200 | 0.0291 | 7.03E-6 | 0.000178 | 1600 | 0.00038 |
| 88-06-2 | 2,4,6-Trichlorophenol | 197.45 | 1.675 | 800 | 0.0318 | 6.25E-6 | 0.000319 | 381 | 0.00038 |
| 108-05-4 | Vinyl acetate | 86.09 | 0.934 | 20,000 | 0.0850 | 9.20E-6 | 0.021 | 5.25 | No data |
| 57-01-4 | Vinyl chloride | 62.5 | 0.91 | 2760 | 0.106 | 1.23E-6 | 1.11 | 18.6 | 0.00024 |
| 108-38-3 | m-Xylene | 106.2 | 0.86 | 161 | 0.070 | 7.80E-6 | 0.301 | 407 | 0.0019 |
| 95-47-6 | o-Xylene | 106.2 | 0.86 | 178 | 0.087 | 1.00E-5 | 0.213 | 363 | 0.0019 |
| 106-42-3 | p-Xylene | 106.2 | 0.86 | 185 | 0.0769 | 8.44E-6 | 0.314 | 389 | 0.0019 |
| 1330-20-7 | Xylenes (total) | 106.2 | 0.86 | 186 | 0.0720 | 9.34E-6 | 0.25 | 260 | 0.0019 |

Chemical Abstracts Service CAS, registry number. This number in the format xxx-xx-x is unique for each chemical and allows efficient searching on computerized data bases.

Sources: 26 Ill. Reg. 2683, effective February 5, 2002, Missouri Department of Natural Resources, Cleanup Levels for Missouri (CALM); Cohen, R. M., Mercer, J. W. and Matthew, J. 1993. *DNAPL Site Evaluation*, CRC Press, Boca Raton. FL; Spectrum Laboratories; InChem; Wikipedia.

## Soil saturation limits ($C_{sat}$) for chemicals whose melting point is less than 30°C

| CAS No. | Chemical name | $C_{sat}$ (mg kg$^{-1}$) |
|---|---|---|
| 67-64-1 | Acetone | 100,000 |
| 71-43-2 | Benzene | 870 |
| 111-44-4 | Bis(2-chloroethyl)ether | 3 300 |
| 117-81-7 | Bis(2-ethylhexyl)phthalate | 31,000 |
| 75-27-4 | Bromodichloromethane (dichlorobromomethane) | 3 000 |
| 75-25-2 | Bromoform | 1 900 |
| 71-36-3 | Butanol | 10,000 |
| 85-68-7 | Butyl benzyl phthalate | 930 |
| 75-15-0 | Carbon disulfide | 720 |
| 56-23-5 | Carbon tetrachloride | 1 100 |
| 108-90-7 | Chlorobenzene (monochlorobenzene) | 680 |
| 124-48-1 | Chlorodibromomethane (dibromochloromethane) | 1 300 |
| 67-66-3 | Chloroform | 2 900 |
| 96-12-8 | 1,2-Dibromo-3-chloropropane | 1 400 |
| 106-93-4 | 1,2-Dibromoethane (ethylene dibromide) | 2 800 |
| 84-74-2 | Di-*n*-butyl phthalate | 2 300 |
| 95-50-1 | 1,2-Dichlorobenzene (*o*-dichlorobenzene) | 560 |
| 75-34-3 | 1,1-Dichloroethane | 1 700 |
| 107-06-2 | 1,2-Dichloroethane (ethylene dichloride) | 1 800 |
| 75-35-4 | 1,1-Dichloroethylene | 1 500 |
| 156-59-2 | *cis*-1,2-dichloroethylene | 1 200 |
| 156-60-5 | *trans*-1,2-dichloroethylene | 3 100 |
| 78-87-5 | 1,2-Dichloropropane | 1 100 |
| 542-75-6 | 1,3-Dichloropropene (1,3-Dichloropropylene, *cis* + *trans*) | 1 400 |
| 84-66-2 | Diethyl phthalate | 2 000 |
| 117-84-0 | Di-*n*-octyl phthalate | 10,000 |
| 100-41-4 | Ethylbenzene | 400 |
| 77-47-4 | Hexachlorocyclopentadiene | 2 200 |
| 78-59-1 | Isophorone | 4 600 |
| 74-83-9 | Methyl bromide (bromomethane) | 3 200 |
| 1634-04-4 | Methyl tertiary-butyl ether | 8 800 |
| 75-09-2 | Methylene chloride (dichloromethane) | 2 400 |
| 98-95-3 | Nitrobenzene | 1 000 |
| 100-42-5 | Styrene | 1 500 |
| 127-18-4 | Tetrachloroethylene (perchloroethylene) | 240 |
| 108-88-3 | Toluene | 650 |
| 120-82-1 | 1,2,4-Trichlorobenzene | 3 200 |
| 71-55-6 | 1,1,1-Trichloroethane | 1 200 |
| 79-00-5 | 1,1,2-Trichloroethane | 1 800 |
| 79-01-6 | Trichloroethylene | 1 300 |
| 108-05-4 | Vinyl acetate | 2 700 |
| 75-01-4 | Vinyl chloride | 1 200 |
| 108-38-3 | *m*-Xylene | 420 |
| 95-47-6 | *o*-Xylene | 410 |
| 106-42-3 | *p*-Xylene | 460 |
| 1330-20-7 | Xylenes (total) | 320 |
| **Ionizable organics** | | |
| 95-57-8 | 2-Chlorophenol | 53,000 |

*Source:* Amended at 26 Ill. Reg. 2683, effective February 5, 2002.

**Chemicals used for treatment by public water-supply systems in the United States and Canada (Based on data from 430 of the largest US utilities and 24 of the 75 largest Canadian utilities)**

| Chemical | Total use (tons) |
|---|---|
| Quick lime | 330,988 |
| Aluminum sulfate | 188,986 |
| Chlorine | 79,034 |
| Hydrated lime | 44,679 |
| Caustic soda | 39,030 |
| Carbon dioxide | 18,111 |
| Soda ash | 13,750 |
| Ferrous sulfate | 10,590 |
| Powdered activated carbon | 9,016 |
| Ferric sulfate | 7,956 |
| Sodium silicofluoride | 7,903 |
| Polyelectrolytes | 5,915 |
| Ammonia | 5,232 |
| Phosphate | 3,970 |
| Copper sulfate | 2,825 |
| Granular activated carbon | 2,587 |
| Potassium permanganate | 1,231 |
| Sodium aluminate | 1,129 |
| Hypochlorites | 1,112 |
| Sodium chloride | 828 |
| Clays | 133 |

*Source:* American Water Works Association, 1998. Gring, N.S., 1984 Water Utility Operating Data, Summary Report. Reprinted with permission.

Values of the functions $W(u, r/B)$ for various values of $u$

| $u$ \ $r/B$ | 0.002 | 0.004 | 0.006 | 0.008 | 0.01 | 0.02 | 0.04 | 0.06 | 0.08 | 0.1 | 0.2 | 0.4 | 0.6 | 0.8 | 1 | 2 | 4 | 6 | 8 |
|---|---|---|---|---|---|---|---|---|---|---|---|---|---|---|---|---|---|---|---|
| 0 | 12.7 | 11.3 | 10.5 | 9.89 | 9.44 | 8.06 | 6.67 | 5.87 | 5.29 | 4.85 | 3.51 | 2.23 | 1.55 | 1.13 | 0.842 | 0.228 | 0.0223 | 0.0025 | 0.0003 |
| 0.000002 | 12.1 | 11.2 | 10.5 | 9.89 | 9.44 | | | | | | | | | | | | | | |
| 0.000004 | 11.6 | 11.1 | 10.4 | 9.88 | 9.44 | | | | | | | | | | | | | | |
| 0.000006 | 11.3 | 10.9 | 10.4 | 9.87 | 9.44 | | | | | | | | | | | | | | |
| 0.000008 | 11.0 | 10.7 | 10.3 | 9.84 | 9.43 | | | | | | | | | | | | | | |
| 0.00001 | 10.8 | 10.6 | 10.2 | 9.80 | 9.42 | 8.06 | | | | | | | | | | | | | |
| 0.00002 | 10.2 | 10.1 | 9.84 | 9.58 | 9.30 | 8.06 | | | | | | | | | | | | | |
| 0.00004 | 9.52 | 9.45 | 9.34 | 9.19 | 9.01 | 8.03 | | | | | | | | | | | | | |
| 0.00006 | 9.13 | 9.08 | 9.00 | 8.89 | 8.77 | 7.98 | 6.67 | | | | | | | | | | | | |
| 0.00008 | 8.84 | 8.81 | 8.75 | 8.67 | 8.57 | 7.91 | | | | | | | | | | | | | |
| 0.0001 | 8.62 | 8.59 | 8.55 | 8.48 | 8.40 | 7.84 | 6.67 | 5.87 | 5.29 | | | | | | | | | | |
| 0.0002 | 7.94 | 7.92 | 7.90 | 7.86 | 7.82 | 7.50 | 6.62 | 5.86 | 5.29 | 4.85 | | | | | | | | | |
| 0.0004 | 7.24 | 7.24 | 7.22 | 7.21 | 7.19 | 7.01 | 6.45 | 5.83 | 5.29 | 4.85 | | | | | | | | | |
| 0.0006 | 6.84 | 6.84 | 6.83 | 6.82 | 6.80 | 6.68 | 6.27 | 5.77 | 5.27 | 4.85 | | | | | | | | | |
| 0.0008 | 6.55 | 6.55 | 6.54 | 6.53 | 6.52 | 6.43 | 6.11 | 5.69 | 5.25 | 4.84 | | | | | | | | | |
| 0.001 | 6.33 | 6.33 | 6.32 | 6.32 | 6.31 | 6.23 | 5.97 | 5.61 | 5.21 | 4.83 | 3.51 | | | | | | | | |
| 0.002 | 5.64 | 5.64 | 5.63 | 5.63 | 5.63 | 5.59 | 5.45 | 5.24 | 4.98 | 4.71 | 3.50 | | | | | | | | |
| 0.004 | 4.95 | 4.95 | 4.95 | 4.94 | 4.94 | 4.92 | 4.85 | 4.74 | 4.59 | 4.42 | 3.48 | 2.23 | | | | | | | |
| 0.006 | 4.54 | | | | 4.54 | 4.53 | 4.48 | 4.41 | 4.30 | 4.18 | 3.43 | 2.23 | | | | | | | |
| 0.008 | 4.26 | | | | 4.26 | 4.25 | 4.21 | 4.15 | 4.08 | 3.98 | 3.36 | 2.23 | | | | | | | |
| 0.01 | 4.04 | | | | 4.04 | 4.03 | 4.00 | 3.95 | 3.89 | 3.81 | 3.29 | 2.23 | 1.55 | 1.13 | | | | | |
| 0.02 | 3.35 | | | | 3.35 | 3.35 | 3.34 | 3.31 | 3.28 | 3.24 | 2.95 | 2.18 | 1.55 | 1.13 | | | | | |
| 0.04 | 2.68 | | | | 2.68 | 2.68 | 2.67 | 2.66 | 2.65 | 2.63 | 2.48 | 2.02 | 1.52 | 1.13 | 0.842 | | | | |
| 0.06 | 2.30 | | | | 2.30 | 2.29 | 2.29 | 2.28 | 2.27 | 2.26 | 2.17 | 1.85 | 1.46 | 1.11 | 0.839 | | | | |
| 0.08 | 2.03 | | | | | 2.03 | 2.02 | 2.02 | 2.01 | 2.00 | 1.94 | 1.69 | 1.39 | 1.08 | 0.832 | | | | |
| 0.1 | 1.82 | | | | | | 1.82 | 1.82 | 1.81 | 1.80 | 1.75 | 1.56 | 1.31 | 1.05 | 0.819 | 0.228 | | | |
| 0.2 | 1.22 | | | | | | 1.22 | 1.22 | 1.22 | 1.22 | 1.19 | 1.11 | 0.996 | 0.857 | 0.715 | 0.227 | | | |
| 0.4 | 0.702 | | | | | | 0.702 | 0.702 | 0.701 | 0.700 | 0.693 | 0.665 | 0.621 | 0.565 | 0.502 | 0.210 | | | |
| 0.6 | 0.454 | | | | | | 0.454 | 0.454 | 0.454 | 0.453 | 0.450 | 0.436 | 0.415 | 0.387 | 0.354 | 0.177 | 0.0222 | | |
| 0.8 | 0.311 | | | | | | 0.311 | 0.310 | 0.310 | 0.310 | 0.308 | 0.301 | 0.289 | 0.273 | 0.254 | 0.144 | 0.0218 | | |
| 1 | 0.219 | | | | | | | | | 0.219 | 0.218 | 0.213 | 0.206 | 0.197 | 0.185 | 0.114 | 0.0207 | 0.0025 | |
| 2 | 0.049 | | | | | | | | | | 0.049 | 0.048 | 0.047 | 0.046 | 0.044 | 0.034 | 0.011 | 0.0021 | 0.0003 |
| 4 | 0.0038 | | | | | | | | | | | 0.0038 | 0.0037 | 0.0037 | 0.0036 | 0.0031 | 0.0016 | 0.0006 | 0.0002 |
| 6 | 0.0004 | | | | | | | | | | | | | | 0.0004 | 0.0003 | 0.0002 | 0.0001 | |
| 8 | 0 | | | | | | | | | | | | | | | | | | 0 |

Source: After M. S. Hantush, "Analysis of Data from Pumping Test in Leaky Aquifers," Transactions, American Geophysical Union, 37 (1956):702–714.

## Values of the functions $K_0(X)$ and exp $(X)K_0(X)$

| $X$ | $K_0(X)$ | exp $(X)K_0(X)$ | $X$ | $K_0(X)$ | exp $(X)K_0(X)$ |
|---|---|---|---|---|---|
| 0.001 | 7.02 | 7.03 | 0.25 | 1.54 | 1.98 |
| 0.005 | 5.41 | 5.44 | 0.30 | 1.37 | 1.85 |
| 0.01 | 4.72 | 4.77 | 0.35 | 1.23 | 1.75 |
| 0.015 | 4.32 | 4.38 | 0.40 | 1.11 | 1.66 |
| 0.02 | 4.03 | 4.11 | 0.45 | 1.01 | 1.59 |
| 0.025 | 3.81 | 3.91 | 0.50 | 0.92 | 1.52 |
| 0.03 | 3.62 | 3.73 | 0.55 | 0.85 | 1.47 |
| 0.035 | 3.47 | 3.59 | 0.60 | 0.78 | 1.42 |
| 0.04 | 3.34 | 3.47 | 0.65 | 0.72 | 1.37 |
| 0.045 | 3.22 | 3.37 | 0.70 | 0.66 | 1.33 |
| 0.05 | 3.11 | 3.27 | 0.75 | 0.61 | 1.29 |
| 0.055 | 3.02 | 3.19 | 0.80 | 0.57 | 1.26 |
| 0.06 | 2.93 | 3.11 | 0.85 | 0.52 | 1.23 |
| 0.065 | 2.85 | 3.05 | 0.90 | 0.49 | 1.20 |
| 0.07 | 2.78 | 2.98 | 0.95 | 0.45 | 1.17 |
| 0.075 | 2.71 | 2.92 | 1.0 | 0.42 | 1.14 |
| 0.08 | 2.65 | 2.87 | 1.5 | 0.21 | 0.96 |
| 0.085 | 2.59 | 2.82 | 2.0 | 0.11 | 0.84 |
| 0.09 | 2.53 | 2.77 | 2.5 | 0.062 | 0.760 |
| 0.095 | 2.48 | 2.72 | 3.0 | 0.035 | 0.698 |
| 0.10 | 2.43 | 2.68 | 3.5 | 0.020 | 0.649 |
| 0.15 | 2.03 | 2.36 | 4.0 | 0.011 | 0.609 |
| 0.20 | 1.75 | 2.14 | 4.5 | 0.006 | 0.576 |
|  |  |  | 5.0 | 0.004 | 0.548 |

*Source*: Adapted from M. S. Hantush. "Analysis of Data from Pumping Testis in Leaky Aquifers," *Transactions. American Geophysical Union*, 37 (1956): 702–714.

## Physical properties of water

| Temperature | | Density | | Specific weight | | Absolute viscosity | | Kinematic viscosity | | Surface tension | | Thermal capacity, | Enthalpy | Heat of vaporization | | Saturation vapor pressure | | Saturation vapor pressure head | |
|---|---|---|---|---|---|---|---|---|---|---|---|---|---|---|---|---|---|---|---|
| °F | °C | $kg \cdot m^{-3}$ | $slugs\ ft^{-3}$ | $N\ m^{-3}$ | $lb_f\ ft^{-3}$ | $Kg\ m^{-1}\ s^{-1a}$ | $slugs\ ft^{-1}\ s^{-1b}$ | $m^2\ s^{-1c}$ | $ft^2\ s^{-1b}$ | $N\ m^{-1}$ | $lb_f\ ft^{-1}$ | $J\ g^{-1}\ °C^{-1}$ | $J\ g^{-1}$ | $J\ kg^{-1}$ | $But\ lb^{-1}$ | $Pa$ | $psia$ | $m$ | $ft$ |
| 32.0 | 0 | 999.87 | 1.940 | 9805.4 | 62.419 | 0.001787 | 3.732E−05 | 1.787E−06 | 1.924E−05 | 0.076 | 0.00518 | 4.2177 | 0.1026 | 2.501E+06 | 1075.1 | 611 | 0.0886 | 0.062 | 0.204 |
| 33.8 | 1 | 999.93 | 1.940 | 9805.9 | 62.423 | 0.001728 | 3.608E−05 | 1.728E−06 | 1.860E−05 | | | 4.2141 | 4.3184 | 2.499E+06 | 1074.2 | 657 | 0.0952 | 0.067 | 0.220 |
| 35.6 | 2 | 999.97 | 1.940 | 9806.3 | 62.426 | 0.001671 | 3.491E−05 | 1.671E−06 | 1.799E−05 | | | 4.2107 | 8.5308 | 2.496E+06 | 1073.2 | 705 | 0.1023 | 0.072 | 0.236 |
| 37.4 | 3 | 999.99 | 1.940 | 9806.6 | 62.427 | 0.001618 | 3.379E−05 | 1.618E−06 | 1.741E−05 | | | 4.2077 | 12.7400 | 2.494E+06 | 1072.1 | 758 | 0.1099 | 0.077 | 0.253 |
| 39.2 | 4 | 1000.00 | 1.940 | 9806.6 | 62.428 | 0.001567 | 3.272E−05 | 1.567E−06 | 1.687E−05 | | | 4.2048 | 16.9462 | 2.492E+06 | 1071.1 | 813 | 0.1179 | 0.083 | 0.272 |
| 41.0 | 5 | 999.99 | 1.940 | 9806.6 | 62.427 | 0.001518 | 3.171E−05 | 1.518E−06 | 1.634E−05 | 0.075 | 0.00513 | 4.2022 | 21.1408 | 2.489E+06 | 1070.1 | 872 | 0.1265 | 0.089 | 0.292 |
| 42.8 | 6 | 999.97 | 1.940 | 9806.3 | 62.426 | 0.001472 | 3.074E−05 | 1.472E−06 | 1.585E−05 | | | 4.1999 | 25.5496 | 2.487E+06 | 1069.1 | 935 | 0.1356 | 0.095 | 0.313 |
| 44.6 | 7 | 999.93 | 1.940 | 9806.0 | 62.423 | 0.001428 | 2.982E−05 | 1.428E−06 | 1.537E−05 | | | 4.1977 | 29.5496 | 2.484E+06 | 1068.1 | 1001 | 0.1452 | 0.102 | 0.335 |
| 46.4 | 8 | 999.88 | 1.940 | 9805.4 | 62.420 | 0.001386 | 2.894E−05 | 1.386E−06 | 1.492E−05 | | | 4.1957 | 33.7463 | 2.482E+06 | 1067.1 | 1072 | 0.1555 | 0.109 | 0.359 |
| 48.2 | 9 | 999.81 | 1.940 | 9804.8 | 62.416 | 0.001346 | 2.810E−05 | 1.346E−06 | 1.449E−05 | | | 4.1939 | 37.9410 | 2.480E+06 | 1066.1 | 1147 | 0.1664 | 0.117 | 0.384 |
| 50.0 | 10 | 999.73 | 1.940 | 9804.0 | 62.411 | 0.001307 | 2.730E−05 | 1.308E−06 | 1.407E−05 | 0.074 | 0.00509 | 4.1922 | 42.1341 | 2.477E+06 | 1065.0 | 1227 | 0.1780 | 0.125 | 0.411 |
| 51.8 | 11 | 999.63 | 1.940 | 9803.1 | 62.405 | 0.001270 | 2.653E−05 | 1.271E−06 | 1.368E−05 | | | 4.1907 | 46.3255 | 2.475E+06 | 1064.0 | 1312 | 0.1903 | 0.134 | 0.439 |
| 53.6 | 12 | 999.53 | 1.939 | 9802.0 | 62.398 | 0.001235 | 2.580E−05 | 1.236E−06 | 1.330E−05 | | | 4.1893 | 50.7041 | 2.473E+06 | 1063.0 | 1402 | 0.2033 | 0.143 | 0.469 |
| 55.4 | 13 | 999.41 | 1.939 | 9800.8 | 62.391 | 0.001202 | 2.510E−05 | 1.202E−06 | 1.294E−05 | | | 4.1880 | 54.7041 | 2.470E+06 | 1062.0 | 1497 | 0.2171 | 0.153 | 0.501 |
| 57.2 | 14 | 999.27 | 1.939 | 9799.5 | 62.382 | 0.001169 | 2.442E−05 | 1.170E−06 | 1.260E−05 | | | 4.1869 | 58.8916 | 2.468E+06 | 1061.0 | 1598 | 0.2317 | 0.163 | 0.535 |
| 59.0 | 15 | 999.13 | 1.939 | 9798.1 | 62.373 | 0.001139 | 2.378E−05 | 1.140E−06 | 1.227E−05 | 0.073 | 0.00504 | 4.1858 | 63.0779 | 2.466E+06 | 1060.0 | 1704 | 0.2472 | 0.174 | 0.571 |
| 60.8 | 16 | 998.97 | 1.938 | 9796.6 | 62.363 | 0.001109 | 2.316E−05 | 1.110E−06 | 1.195E−05 | | | 4.1849 | 67.2632 | 2.463E+06 | 1058.9 | 1817 | 0.2636 | 0.186 | 0.609 |
| 62.6 | 17 | 998.80 | 1.938 | 9794.9 | 62.353 | 0.001081 | 2.257E−05 | 1.082E−06 | 1.164E−05 | | | 4.1840 | 71.4476 | 2.461E+06 | 1057.9 | 1937 | 0.2809 | 0.198 | 0.649 |
| 64.4 | 18 | 998.62 | 1.938 | 9793.2 | 62.342 | 0.001053 | 2.200E−05 | 1.055E−06 | 1.135E−05 | | | 4.1832 | 75.6312 | 2.459E+06 | 1056.9 | 2063 | 0.2992 | 0.211 | 0.691 |
| 66.2 | 19 | 998.43 | 1.937 | 9791.3 | 62.330 | 0.001027 | 2.145E−05 | 1.029E−06 | 1.107E−05 | | | 4.1825 | 79.8141 | 2.456E+06 | 1055.9 | 2196 | 0.3186 | 0.224 | 0.736 |
| 68.0 | 20 | 998.23 | 1.937 | 9789.3 | 62.317 | 0.001002 | 2.093E−05 | 1.004E−06 | 1.080E−05 | 0.073 | 0.00498 | 4.1819 | 83.9963 | 2.454E+06 | 1054.9 | 2337 | 0.3390 | 0.239 | 0.783 |
| 69.8 | 21 | 998.02 | 1.936 | 9787.3 | 62.304 | 0.0009780 | 2.043E−05 | 9.799E−07 | 1.055E−05 | | | 4.1813 | 88.1778 | 2.451E+06 | 1053.9 | 2486 | 0.3606 | 0.254 | 0.833 |
| 71.6 | 22 | 997.80 | 1.936 | 9785.1 | 62.290 | 0.0009548 | 1.994E−05 | 9.570E−07 | 1.030E−05 | | | 4.1808 | 92.3589 | 2.449E+06 | 1052.9 | 2643 | 0.3833 | 0.270 | 0.886 |
| 73.4 | 23 | 997.57 | 1.936 | 9782.8 | 62.276 | 0.0009326 | 1.948E−05 | 9.349E−07 | 1.006E−05 | | | 4.1804 | 96.5395 | 2.447E+06 | 1051.8 | 2809 | 0.4074 | 0.287 | 0.942 |
| 75.2 | 24 | 997.33 | 1.935 | 9780.4 | 62.261 | 0.0009111 | 1.903E−05 | 9.136E−07 | 9.834E−06 | | | 4.1800 | 100.7196 | 2.444E+06 | 1050.8 | 2983 | 0.4327 | 0.305 | 1.001 |
| 77.0 | 25 | 997.08 | 1.935 | 9778.0 | 62.245 | 0.0008905 | 1.860E−05 | 8.931E−07 | 9.613E−06 | 0.072 | 0.00493 | 4.1796 | 104.8994 | 2.442E+06 | 1049.8 | 3167 | 0.4593 | 0.324 | 1.063 |
| 78.8 | 26 | 996.81 | 1.934 | 9775.4 | 62.229 | 0.0008705 | 1.818E−05 | 8.733E−07 | 9.400E−06 | | | 4.1793 | 109.0788 | 2.440E+06 | 1048.8 | 3361 | 0.4874 | 0.344 | 1.128 |

| | | | | | | | | | | | | | | | | | | |
|---|---|---|---|---|---|---|---|---|---|---|---|---|---|---|---|---|---|---|
| 80.6 | 27 | 996.54 | 1.934 | 9772.8 | 62.212 | 0.0008513 | 1.778E−05 | 8.543E−07 | | 9.195E−06 | | 4.1790 | 113.2580 | 2.437E+06 | 1047.8 | 3565 | 0.5170 | 0.365 | 1.197 |
| 82.4 | 28 | 996.26 | 1.933 | 9770.0 | 62.194 | 0.0008328 | 1.739E−05 | 8.359E−07 | | 8.998E−06 | | 4.1788 | 117.4369 | 2.435E+06 | 1046.8 | 3780 | 0.5482 | 0.387 | 1.269 |
| 84.2 | 29 | 995.98 | 1.933 | 9767.2 | 62.176 | 0.0008149 | 1.702E−05 | 8.182E−07 | | 8.807E−06 | | 4.1786 | 121.6157 | 2.433E+06 | 1045.8 | 4006 | 0.5809 | 0.410 | 1.345 |
| 86.0 | 30 | 995.68 | 1.932 | 9764.3 | 62.158 | 0.0007976 | 1.666E−05 | 8.011E−07 | 0.071 | 8.622E−06 | | 4.1785 | 125.7943 | 2.430E+06 | 1044.7 | 4243 | 0.6154 | 0.435 | 1.426 |
| 87.8 | 31 | 995.37 | 1.931 | 9761.3 | 62.139 | 0.0007809 | 1.631E−05 | 7.845E−07 | | 8.444E−06 | 0.00488 | 4.1784 | 129.9727 | 2.428E+06 | 1043.7 | 4493 | 0.6516 | 0.460 | 1.510 |
| 89.6 | 32 | 995.06 | 1.931 | 9758.2 | 62.119 | 0.0007647 | 1.597E−05 | 7.686E−07 | | 8.273E−06 | | 4.1783 | 134.1510 | 2.425E+06 | 1042.7 | 4755 | 0.6897 | 0.487 | 1.599 |
| 91.4 | 33 | 994.73 | 1.930 | 9755.0 | 62.099 | 0.0007491 | 1.565E−05 | 7.531E−07 | | 8.106E−06 | | 4.1783 | 138.3293 | 2.423E+06 | 1041.7 | 5031 | 0.7296 | 0.516 | 1.692 |
| 93.2 | 34 | 994.40 | 1.929 | 9751.8 | 62.078 | 0.0007340 | 1.533E−05 | 7.823E−07 | | 7.945E−06 | | 4.1782 | 142.5078 | 2.421E+06 | 1040.7 | 5320 | 0.7716 | 0.546 | 1.790 |
| 95.0 | 35 | 994.06 | 1.929 | 9748.4 | 62.057 | 0.0007194 | 1.503E−05 | 7.237E−07 | | 7.790E−06 | | 4.1782 | 146.6858 | 2.418E+06 | 1039.7 | 5624 | 0.8156 | 0.577 | 1.893 |

*Notes:* 1. Density computed from the Thiesen–Scheel–Diesselhorst equation.

2. Specific weight = density × 9.806650 m s⁻².

3. Dynamic or absolute viscosity from US National Bureau of Standards (Hardy and Cottington,[26] Swidells, unpublished data, and Weast[78]),

4. Kinematic viscosity = dynamic viscosity divided by density.

5. Surface tension values from Weast.[78]

6. Thermal capacity at 1 atm in absolute J from Weast.[78] Note that 1 J = 0.23885 cal = 9.4782 × 10⁻⁴ Btu.

7. Enthalpy values from Weast.[78]

8. Heat of vaporization values from $H_t = 2.501 \times 10^6 - 2361\,T$.

9. Vapor pressure from the Goff–Gratch formula used to compute values in the Smithsonian Meteorological Tables.[38] See Chap. 1 for conversion factors to other typical pressure units. The Goff–Gratch formula is different from the Smithsonian tables by 0.5 percent.

ᵃ 1 kg m⁻¹ s⁻¹ = 10 poise = 1000 centipoise = 10 g cm⁻¹ s⁻¹ = 10 dynes s cm⁻² = 10 g cm⁻¹ s⁻¹.

ᵇ 1 slug ft⁻¹ s⁻¹ = 1 lb_f s ft⁻².

ᶜ 1 m² s⁻¹ = 10,000 stokes = 10,000 cm² s⁻¹ = 100,000 centistokes.

## Properties of water in English units

| Temp. (°F) | Specific gravity | Specific weight, lb ft$^{-3}$ | Heat of vaporization, Btu lb$^{-1}$ | Viscosity | | Vapor pressure | | |
|---|---|---|---|---|---|---|---|---|
| | | | | Dynamic lb·s ft$^{-2}$ | Kinematic, ft$^2$ s$^{-1}$ | In Hg | Millibar | lb/in$^{-2}$ |
| 32 | 0.99986 | 62.418 | 1075.5 | $3.746 \times 10^{-5}$ | $1.931 \times 10^{-5}$ | 0.180 | 6.11 | 0.089 |
| 40 | 0.99998 | 62.426[†] | 1071.0 | 3.229 | 1.664 | 0.248 | 8.39 | 0.122 |
| 50 | 0.99971 | 62.409 | 1065.3 | 2.735 | 1.410 | 0.362 | 12.27 | 0.178 |
| 60 | 0.99902 | 62.366 | 1059.7 | 2.359 | 1.217 | 0.522 | 17.66 | 0.256 |
| 70 | 0.99798 | 62.301 | 1054.0 | 2.050 | 1.058 | 0.739 | 25.03 | 0.363 |
| 80 | 0.99662 | 62.216 | 1048.4 | 1.799 | 0.930 | 1.032 | 34.96 | 0.507 |
| 90 | 0.99497 | 62.113 | 1042.7 | 1.595 | 0.826 | 1.422 | 48.15 | 0.698 |
| 100 | 0.99306 | 61.994 | 1037.1 | 1.424 | 0.739 | 1.933 | 65.47 | 0.950 |
| 120 | 0.98856 | 61.713 | 1025.6 | 1.168 | 0.609 | 3.448 | 116.75 | 1.693 |
| 140 | 0.98321 | 61.379 | 1014.0 | 0.981 | 0.514 | 5.884 | 199.26 | 2.890 |
| 160 | 0.97714 | 61.000 | 1002.2 | 0.838 | 0.442 | 9.656 | 326.98 | 4.742 |
| 180 | 0.97041 | 60.580 | 990.2 | 0.726 | 0.386 | 15.295 | 517.95 | 7.512 |
| 200 | 0.96306 | 60.121 | 977.9 | 0.637 | 0.341 | 23.468 | 794.72 | 11.526 |
| 212 | 0.95837 | 59.828 | 970.3 | 0.593 | 0.319 | 29.921 | 1013.25 | 14.696 |

[†] Maximum specific weight is 62.427 lb ft$^{-3}$ at 39.2°F.

## Properties of water in metric units

| Temp. (°C) | Specific gravity | Density (g cm$^{-3}$) | Heat of vaporization (cal g$^{-1}$) | Viscosity | | Vapor pressure | | |
|---|---|---|---|---|---|---|---|---|
| | | | | Dynamic (centipoise)[b] | Kinematic (centistokes)[c] | mm Hg | Millibar | g cm$^{-2}$ |
| 0 | 0.99987 | 0.99984 | 597.3 | 1.79 | 1.79 | 4.58 | 6.11 | 6.23 |
| 5 | 0.99999 | 0.99996[a] | 594.5 | 1.52 | 1.52 | 6.54 | 8.72 | 8.89 |
| 10 | 0.99973 | 0.99970 | 591.7 | 1.31 | 1.31 | 9.20 | 12.27 | 12.51 |
| 15 | 0.99913 | 0.99910 | 588.9 | 1.14 | 1.14 | 12.78 | 17.04 | 17.38 |
| 20 | 0.99824 | 0.99821 | 586.0 | 1.00 | 1.00 | 17.53 | 23.37 | 23.83 |
| 25 | 0.99708 | 0.99705 | 583.2 | 0.890 | 0.893 | 23.76 | 31.67 | 32.30 |
| 30 | 0.99568 | 0.99565 | 580.4 | 0.798 | 0.801 | 31.83 | 42.43 | 43.27 |
| 35 | 0.99407 | 0.99404 | 577.6 | 0.719 | 0.723 | 42.18 | 56.24 | 57.34 |
| 40 | 0.99225 | 0.99222 | 574.7 | 0.653 | 0.658 | 55.34 | 73.78 | 75.23 |
| 50 | 0.98807 | 0.98804 | 569.0 | 0.547 | 0.554 | 92.56 | 123.40 | 125.83 |
| 60 | 0.98323 | 0.98320 | 563.2 | 0.466 | 0.474 | 149.46 | 199.26 | 203.19 |
| 70 | 0.97780 | 0.97777 | 557.4 | 0.404 | 0.413 | 233.79 | 311.69 | 317.84 |
| 80 | 0.97182 | 0.97179 | 551.4 | 0.355 | 0.365 | 355.28 | 473.67 | 483.01 |
| 90 | 0.96534 | 0.96531 | 545.3 | 0.315 | 0.326 | 525.89 | 701.13 | 714.95 |
| 100 | 0.95839 | 0.95836 | 539.1 | 0.282 | 0.294 | 760.00 | 1013.25 | 1033.23 |

[a] Maximum density is 0.999973 g cm$^{-3}$ at 3.98°C.
[b] Centipoise = (g cm$^{-1}$ · s) × 10$^2$ = (Pa · s) × 10$^3$
[c] Centistokes = (cm$^2$ s$^{-1}$) × 10$^2$ = (m$^2$ s$^{-1}$) × 10$^6$

| $r/B$ | $e^{(r/B)}$ | $K_0(r/B)$ | $r/B$ | $e^{(r/B)}$ | $K_0(r/B)$ | $r/B$ | $e^{(r/B)}$ | $K_0(r/B)$ |
|---|---|---|---|---|---|---|---|---|
| 0.010 | 1.0101 | 4.7212 | 0.10 | 1.1052 | 2.4271 | 1.0 | 2.7183 | 0.4210 |
| 0.011 | 1.0111 | 4.6260 | 0.11 | 1.1163 | 2.3333 | 1.1 | 3.0042 | 0.3656 |
| 0.012 | 1.0121 | 4.5390 | 0.12 | 1.1275 | 2.2479 | 1.2 | 3.3201 | 0.3185 |
| 0.013 | 1.0131 | 4.4590 | 0.13 | 1.1388 | 2.1695 | 1.3 | 3.6693 | 0.2782 |
| 0.014 | 1.0141 | 4.3849 | 0.14 | 1.1503 | 2.0972 | 1.4 | 4.0552 | 0.2437 |
| 0.015 | 1.0151 | 4.3159 | 0.15 | 1.1618 | 2.0300 | 1.5 | 4.4817 | 0.2138 |
| 0.016 | 1.0161 | 4.2514 | 0.16 | 1.1735 | 1.9674 | 1.6 | 4.9530 | 0.1880 |
| 0.017 | 1.0171 | 4.1908 | 0.17 | 1.1853 | 1.9088 | 1.7 | 5.4739 | 0.1655 |
| 0.018 | 1.0182 | 4.1337 | 0.18 | 1.1972 | 1.8537 | 1.8 | 6.0496 | 0.1459 |
| 0.019 | 1.0192 | 4.0797 | 0.19 | 1.2092 | 1.8018 | 1.9 | 6.6859 | 0.1288 |
| 0.020 | 1.0202 | 4.0285 | 0.20 | 1.2214 | 1.7527 | 2.0 | 7.3891 | 0.1139 |
| 0.021 | 1.0212 | 3.9797 | 0.21 | 1.2337 | 1.7062 | 2.1 | 8.1662 | 0.1008 |
| 0.022 | 1.0222 | 3.9332 | 0.22 | 1.2461 | 1.6620 | 2.2 | 9.0250 | 0.0893 |
| 0.023 | 1.0233 | 3.8888 | 0.23 | 1.2586 | 1.6199 | 2.3 | 9.9742 | 0.0791 |
| 0.024 | 1.0243 | 3.8463 | 0.24 | 1.2712 | 1.5798 | 2.4 | 11.0232 | 0.0702 |
| 0.025 | 1.0253 | 3.8056 | 0.25 | 1.2840 | 1.5415 | 2.5 | 12.1825 | 0.0623 |
| 0.026 | 1.0263 | 3.7664 | 0.26 | 1.2969 | 1.5048 | 2.6 | 13.4637 | 0.0554 |
| 0.027 | 1.0274 | 3.7287 | 0.27 | 1.3100 | 1.4697 | 2.7 | 14.8797 | 0.0493 |
| 0.028 | 1.0284 | 3.6924 | 0.28 | 1.3231 | 1.4360 | 2.8 | 16.4446 | 0.0438 |
| 0.029 | 1.0294 | 3.6574 | 0.29 | 1.3364 | 1.4036 | 2.9 | 18.1741 | 0.0390 |
| 0.030 | 1.0305 | 3.6235 | 0.30 | 1.3499 | 1.3725 | 3.0 | 20.0855 | 0.0347 |
| 0.031 | 1.0315 | 3.5908 | 0.31 | 1.3634 | 1.3425 | 3.1 | 22.1980 | 0.0310 |
| 0.032 | 1.0325 | 3.5591 | 0.32 | 1.3771 | 1.3136 | 3.2 | 24.5325 | 0.0276 |
| 0.033 | 1.0336 | 3.5284 | 0.33 | 1.3910 | 1.2857 | 3.3 | 27.1126 | 0.0246 |
| 0.034 | 1.0346 | 3.4986 | 0.34 | 1.4049 | 1.2587 | 3.4 | 29.9641 | 0.0220 |
| 0.035 | 1.0356 | 3.4697 | 0.35 | 1.4191 | 1.2327 | 3.5 | 33.1155 | 0.0196 |
| 0.036 | 1.0367 | 3.4416 | 0.36 | 1.4333 | 1.2075 | 3.6 | 36.5982 | 0.0175 |
| 0.037 | 1.0377 | 3.4143 | 0.37 | 1.4477 | 1.1832 | 3.7 | 40.4473 | 0.0156 |
| 0.038 | 1.0387 | 3.3877 | 0.38 | 1.4623 | 1.1596 | 3.8 | 44.7012 | 0.0140 |
| 0.039 | 1.0398 | 3.3618 | 0.39 | 1.4770 | 1.1367 | 3.9 | 49.4024 | 0.0125 |
| 0.040 | 1.0408 | 3.3365 | 0.40 | 1.4918 | 1.1145 | 4.0 | 54.5982 | 0.0112 |
| 0.041 | 1.0419 | 3.3119 | 0.41 | 1.5068 | 1.0930 | 4.1 | 60.3403 | 0.0100 |
| 0.042 | 1.0429 | 3.2879 | 0.42 | 1.5220 | 1.0721 | 4.2 | 66.6863 | 0.0089 |
| 0.043 | 1.0439 | 3.2645 | 0.43 | 1.5373 | 1.0518 | 4.3 | 73.6998 | 0.0080 |
| 0.044 | 1.0450 | 3.2415 | 0.44 | 1.5527 | 1.0321 | 4.4 | 81.4509 | 0.0071 |
| 0.045 | 1.0460 | 3.2192 | 0.45 | 1.5683 | 1.0129 | 4.5 | 90.0171 | 0.0064 |
| 0.046 | 1.0471 | 3.1973 | 0.46 | 1.5841 | 0.9943 | 4.6 | 99.4843 | 0.0057 |
| 0.047 | 1.0481 | 3.1758 | 0.47 | 1.6000 | 0.9761 | 4.7 | 109.9472 | 0.0051 |
| 0.048 | 1.0492 | 3.1549 | 0.48 | 1.6161 | 0.9584 | 4.8 | 121.5104 | 0.0046 |
| 0.049 | 1.0502 | 3.1343 | 0.49 | 1.6323 | 0.9412 | 4.9 | 134.2898 | 0.0041 |
| 0.050 | 1.0513 | 3.1142 | 0.50 | 1.6487 | 0.9244 | 5.0 | 148.4132 | 0.0037 |
| 0.051 | 1.0523 | 3.0945 | 0.51 | 1.6653 | 0.9081 | 5.1 | 164.0219 | 0.0033 |
| 0.052 | 1.0534 | 3.0752 | 0.52 | 1.6820 | 0.8921 | 5.2 | 181.2722 | 0.0030 |
| 0.053 | 1.0544 | 3.0562 | 0.53 | 1.6989 | 0.8766 | 5.3 | 200.3368 | 0.0027 |
| 0.054 | 1.0555 | 3.0376 | 0.54 | 1.7160 | 0.8614 | 5.4 | 221.4064 | 0.0024 |
| 0.055 | 1.0565 | 3.0194 | 0.55 | 1.7333 | 0.8466 | 5.5 | 244.6919 | 0.0021 |
| 0.056 | 1.0576 | 3.0015 | 0.56 | 1.7507 | 0.8321 | 5.6 | 270.4264 | 0.0019 |
| 0.057 | 1.0587 | 2.9839 | 0.57 | 1.7683 | 0.8180 | 5.7 | 298.8674 | 0.0017 |
| 0.058 | 1.0597 | 2.9666 | 0.58 | 1.7860 | 0.8042 | 5.8 | 330.2996 | 0.0015 |
| 0.059 | 1.0608 | 2.9496 | 0.59 | 1.8040 | 0.7907 | 5.9 | 365.0375 | 0.0014 |
| 0.060 | 1.0618 | 2.9329 | 0.60 | 1.8221 | 0.7775 | 6.0 | 403.4288 | 0.0012 |
| 0.061 | 1.0629 | 2.9165 | 0.61 | 1.8404 | 0.7646 | 6.1 | 445.8578 | 0.0011 |
| 0.062 | 1.0640 | 2.9003 | 0.62 | 1.8589 | 0.7520 | 6.2 | 492.7490 | 0.0010 |
| 0.063 | 1.0650 | 2.8844 | 0.63 | 1.8776 | 0.7397 | 6.3 | 544.5719 | 0.0009 |
| 0.064 | 1.0661 | 2.8688 | 0.64 | 1.8965 | 0.7277 | 6.4 | 601.8450 | 0.0008 |
| 0.065 | 1.0672 | 2.8534 | 0.65 | 1.9155 | 0.7159 | 6.5 | 665.1416 | 0.0007 |
| 0.066 | 1.0682 | 2.8382 | 0.66 | 1.9348 | 0.7043 | 6.6 | 735.0952 | 0.0007 |

(*Continued*)

| r/B | e^(r/B) | $K_0(r/B)$ | r/B | e^(r/B) | $K_0(r/B)$ | r/B | e^(r/B) | $K_0(r/B)$ |
|---|---|---|---|---|---|---|---|---|
| 0.067 | 1.0693 | 2.8233 | 0.67 | 1.9542 | 0.6930 | 6.7 | 812.4058 | 0.0006 |
| 0.068 | 1.0704 | 2.8086 | 0.68 | 1.9739 | 0.6820 | 6.8 | 897.8473 | 0.0005 |
| 0.069 | 1.0714 | 2.7941 | 0.69 | 1.9937 | 0.6711 | 6.9 | 992.2747 | 0.0005 |
| 0.070 | 1.0725 | 2.7798 | 0.70 | 2.0138 | 0.6605 | 7.0 | 1096.6332 | 0.0004 |
| 0.071 | 1.0736 | 2.7657 | 0.71 | 2.0340 | 0.6501 | 7.1 | 1211.9671 | 0.0004 |
| 0.072 | 1.0747 | 2.7519 | 0.72 | 2.0544 | 0.6399 | 7.2 | 1339.4308 | 0.0003 |
| 0.073 | 1.0757 | 2.7382 | 0.73 | 2.0751 | 0.6300 | 7.3 | 1480.2999 | 0.0003 |
| 0.074 | 1.0768 | 2.7247 | 0.74 | 2.0959 | 0.6202 | 7.4 | 1635.9844 | 0.0003 |
| 0.075 | 1.0779 | 2.7114 | 0.75 | 2.1170 | 0.6106 | 7.5 | 1808.0424 | 0.0002 |
| 0.076 | 1.0790 | 2.6983 | 0.76 | 2.1383 | 0.6012 | 7.6 | 1998.1959 | 0.0002 |
| 0.077 | 1.0800 | 2.6853 | 0.77 | 2.1598 | 0.5920 | 7.7 | 2208.3480 | 0.0002 |
| 0.078 | 1.0811 | 2.6726 | 0.78 | 2.1815 | 0.5829 | 7.8 | 2440.6020 | 0.0002 |
| 0.079 | 1.0822 | 2.6599 | 0.79 | 2.2034 | 0.5740 | 7.9 | 2697.2823 | 0.0002 |
| 0.080 | 1.0833 | 2.6475 | 0.80 | 2.2255 | 0.5653 | 8.0 | 2980.9580 | 0.0001 |
| 0.081 | 1.0844 | 2.6352 | 0.81 | 2.2479 | 0.5568 | 8.1 | 3294.4681 | 0.0001 |
| 0.082 | 1.0855 | 2.6231 | 0.82 | 2.2705 | 0.5484 | 8.2 | 3640.9503 | 0.0001 |
| 0.083 | 1.0865 | 2.6111 | 0.83 | 2.2933 | 0.5402 | 8.3 | 4023.8724 | 0.0001 |
| 0.084 | 1.0876 | 2.5992 | 0.84 | 2.3164 | 0.5321 | 8.4 | 4447.0667 | 0.0001 |
| 0.085 | 1.0887 | 2.5875 | 0.85 | 2.3396 | 0.5242 | 8.5 | 4914.7688 | 0.0001 |
| 0.086 | 1.0898 | 2.5759 | 0.86 | 2.3632 | 0.5165 | 8.6 | 5431.6596 | 0.0001 |
| 0.087 | 1.0909 | 2.5645 | 0.87 | 2.3869 | 0.5088 | 8.7 | 6002.9122 | 0.0001 |
| 0.088 | 1.0920 | 2.5532 | 0.88 | 2.4109 | 0.5013 | 8.8 | 6634.2440 | 0.0001 |
| 0.089 | 1.0931 | 2.5421 | 0.89 | 2.4351 | 0.4940 | 8.9 | 7331.9735 | 0.0001 |
| 0.090 | 1.0942 | 2.5310 | 0.90 | 2.4596 | 0.4867 | 9.0 | 8103.0839 | 0.0001 |
| 0.091 | 1.0953 | 2.5201 | 0.91 | 2.4843 | 0.4796 | 9.1 | 8955.2927 | 0.0000 |
| 0.092 | 1.0964 | 2.5093 | 0.92 | 2.5093 | 0.4727 | 9.2 | 9897.1291 | 0.0000 |
| 0.093 | 1.0975 | 2.4986 | 0.93 | 2.5345 | 0.4658 | 9.3 | 10938.0192 | 0.0000 |
| 0.094 | 1.0986 | 2.4881 | 0.94 | 2.5600 | 0.4591 | 9.4 | 12088.3807 | 0.0000 |
| 0.095 | 1.0997 | 2.4776 | 0.95 | 2.5857 | 0.4524 | 9.5 | 13359.7268 | 0.0000 |
| 0.096 | 1.1008 | 2.4673 | 0.96 | 2.6117 | 0.4459 | 9.6 | 14764.7816 | 0.0000 |
| 0.097 | 1.1019 | 2.4571 | 0.97 | 2.6379 | 0.4396 | 9.7 | 16317.6072 | 0.0000 |
| 0.098 | 1.1030 | 2.4470 | 0.98 | 2.6645 | 0.4333 | 9.8 | 18033.7449 | 0.0000 |
| 0.099 | 1.1041 | 2.4370 | 0.99 | 2.6912 | 0.4271 | 9.9 | 19930.3704 | 0.0000 |

**Representative open areas of screens (Driscoll, 1986. Reprinted with permission from Johnson Screens/a Weatherford Company)**

| Screen diameter | Slot size | Continuous slot | | Louvered (minimum open area) | | Louvered (maximum open area) | | Bridge slot | | Mill slotted* (vertical) | | Slotted pipe (horizontal) | | | Plastic continuous slot | | Slotted plastic† | | Concrete | | Fiberglass reinforced plastic continuous slot‡ | |
|---|---|---|---|---|---|---|---|---|---|---|---|---|---|---|---|---|---|---|---|---|---|---|
| | | in.$^2$/ft | % | in.$^2$/ft | % | in.$^2$/ft | % | in.$^2$/ft | % | in.$^2$/ft | % | Slots/ft | in.$^2$/ft | % | in.$^2$/ft | % | in.$^2$/ft | % | in.$^2$/ft | % | in.$^2$/ft | % |
| 4″ ID | 20 | 44 | 25 | | | | | | | | | | | | 22 | 13 | | | | | 25 | 12 |
| | 30 | 58 | 33 | | | | | | | | | | | | 31 | 18 | | | | | 37 | 17 |
| | 40 | 72 | 41 | | | | | 13 | 8 | 13 | 8 | | | | 40 | 23 | 13 | 8 | | | 48 | 23 |
| | 50 | 78 | 45 | | | | | | | | | | | | 47 | 27 | | | | | | |
| | 60 | 90 | 52 | | | | | 19 | 12 | 8 | 5 | 90 | 16 | 3 | 52 | 30 | 18 | 11 | | | | |
| | | | | | | | | | | | | 130 | 23 | 5 | | | | | | | | |
| | | | | | | | | | | | | 240 | 43 | 9 | | | | | | | | |
| | 80 | 102 | 59 | | | | | | | | | | | | | | | | | | | |
| | 90 | 105 | 60 | | | | | 29 | 17 | 12 | 7 | 90 | 22 | 5 | | | 26 | 16 | | | | |
| | | | | | | | | | | | | 130 | 31 | 7 | | | | | | | | |
| | | | | | | | | | | | | 240 | 58 | 12 | | | | | | | | |
| | 95 | 106 | 61 | | | | | | | | | | | | | | | | | | | |
| | 100 | 112 | 64 | | | | | | | | | | | | | | 29 | 17 | | | | |
| | 120 | 99 | 57 | | | | | | | | | | | | | | | | | | | |
| | 125 | 100 | 58 | | | | | | | | | | | | | | | | | | | |
| 6″ ID | 20 | 45 | 18 | | | | | 41 | 24 | 16 | 10 | | | | 25 | 10 | 12 | 5 | | | 14 | 5 |
| | 30 | 61 | 25 | | | | | | | | | | | | 36 | 14 | 18 | 7 | | | | |
| | 40 | 77 | 31 | | | | | 14 | 6 | 9 | 4 | | | | 45 | 18 | 23 | 9 | | | 29 | 9 |
| | 50 | 88 | 35 | | | | | | | 11 | 5 | | | | 54 | 22 | 28 | 11 | | | | |
| | 60 | 100 | 40 | | | | | 21 | 8 | | | | | | 62 | 25 | 32 | 13 | | | 41 | 14 |
| | 90 | 124 | 50 | 2 | 1 | 7 | 3 | 31 | 12 | 17 | 7 | 90 | 27 | 6 | | | | | | | 59 | 20 |
| | | | | | | | | | | | | 130 | 39 | 8 | | | | | | | | |
| | | | | | | | | | | | | 240 | 72 | 15 | | | | | | | | |
| | 95 | 127 | 51 | | | | | | | | | | | | | | | | | | | |
| | 100 | 131 | 53 | 4 | 2 | 11 | 5 | | | | | | | | | | | | | | 65 | 22 |
| | 120 | 141 | 57 | 5 | 2 | 15 | 7 | | | | | | | | | | | | | | | |
| | 125 | 127 | 51 | | | | | | | | | | | | | | | | | | | |
| 8″ ID | 20 | 58 | 18 | | | | | 43 | 17 | 23 | 9 | | | | 41 | 12 | | | | | | |
| | 30 | 80 | 25 | | | | | | | | | | | | 57 | 18 | | | | | | |
| | 40 | 98 | 30 | | | | | 8 | 3 | | | | | | 72 | 22 | | | | | | |
| | 50 | 114 | 35 | | | | | | | | | | | | 86 | 26 | | | | | | |
| | 60 | 135 | 41 | 4 | 1 | 10 | 3 | 17 | 6 | 23 | 7 | | | | 93 | 29 | 26 | 8 | | | | |
| | 95 | 165 | 51 | | | | | | | | | | | | | | | | | | | |
| | 100 | 169 | 52 | 6 | 2 | 15 | 5 | 38 | 13 | 31 | 10 | | | | | | 47 | 14 | | | | |
| | 125 | 166 | 51 | 7 | 2 | 20 | 6 | | | | | | | | | | 67 | 21 | | | | |

*(Continued)*

**Representative open areas of screens (Driscoll, 1986. Reprinted with permission from Johnson Screens/a Weatherford Company) (Continued)**

| Screen diameter | Slot size | Continuous slot in.²/ft | Continuous slot % | Louvered (min) in.²/ft | Louvered (min) % | Louvered (max) in.²/ft | Louvered (max) % | Bridge slot in.²/ft | Bridge slot % | Mill slotted* (vert.) in.²/ft | Mill slotted % | Slotted pipe (horiz.) Slots/ft | Slotted pipe in.²/ft | Slotted pipe % | Plastic cont. slot in.²/ft | Plastic cont. slot % | Slotted plastic† in.²/ft | Slotted plastic† % | Concrete in.²/ft | Concrete % | Fiberglass reinf. plastic cont. slot‡ in.²/ft | Fiberglass % |
|---|---|---|---|---|---|---|---|---|---|---|---|---|---|---|---|---|---|---|---|---|---|---|
| 10" ID | 20 | 72 | 18 | | | | | 10 | 2 | | | | | | | | 18 | 4 | | | 32 | 6 |
| | 30 | 100 | 25 | | | | | | | | | | | | | | 26 | 6 | | | 64 | 12 |
| | 40 | 122 | 30 | | | | | | | | | | | | | | 33 | 8 | | | 94 | 18 |
| | 50 | 143 | 35 | 4 | 1 | 16 | 44 | | | 15 | 4 | | | | | | 39 | 10 | | | | |
| | 60 | 135 | 33 | 7 | 2 | 24 | 6 | 22 | 5 | 19 | 5 | | | | | | 45 | 11 | | | | |
| | 90 | 174 | 43 | | | | | | | | | | | | | | | | | | | |
| | 95 | 179 | 44 | | | | | | | 28 | 7 | | | | | | 65 | 16 | | | | |
| | 100 | 186 | 46 | | | | | | | | | | | | | | 72 | 18 | | | | |
| | 120 | 203 | 50 | 9 | 2 | 32 | 8 | 48 | 12 | 38 | 9 | | | | | | | | | | | |
| | 125 | 207 | 51 | | | | | | | | | | | | | | | | | | | |
| 12" ID | 20 | 69 | 14 | | | | | 12 | 3 | | | | | | | | | | | | 32 | 6 |
| | 30 | 77 | 16 | | | | | 22 | 5 | | | | | | | | | | | | 64 | 12 |
| | 40 | 99 | 21 | | | | | | | | | | | | | | 38 | 8 | | | 94 | 18 |
| | 50 | 117 | 24 | | | | | | | | | | | | | | | | | | | |
| | 60 | 135 | 28 | 6 | 1 | 20 | 4 | 27 | 6 | 17 | 4 | | | | | | | | | | | |
| | 90 | 176 | 37 | | | | | 49 | 10 | 21 | 5 | | | | | | 52 | 11 | | | | |
| | 95 | 182 | 38 | 9 | 2 | | | | | | | | | | | | | | | | | |
| | 100 | 189 | 39 | | | | | | | 32 | 7 | | | | | | | | | | | |
| | 120 | 209 | 44 | | | | | | | | | | | | | | | | | | | |
| | 125 | 214 | 45 | 12 | 3 | 39 | 9 | 59 | 13 | 43 | 9 | | | | | | | | | | | |
| 16" OD | 20 | 68 | 11 | | | | | 16 | 3 | | | | | | | | | | | | | |
| | 30 | 97 | 16 | | | | | | | | | | | | | | | | | | | |
| | 40 | 124 | 21 | | | | | | | | | | | | | | | | | | | |
| | 50 | 146 | 24 | 7 | 1 | 24 | 4 | | | 22 | 4 | | | | | | | | | | | |
| | 60 | 169 | 28 | | | | | 35 | 6 | 27 | 5 | 104 | 19 | 3 | | | | | | | | |
| | | | | | | | | | | | | 141 | 29 | 5 | | | | | | | | |
| | | | | | | | | | | | | 192 | 35 | 6 | | | | | | | | |
| | 80 | 206 | 34 | | | | | | | | | 104 | 25 | 4 | | | 138 | 23 | | | | |
| | | | | | | | | | | | | 141 | 34 | 6 | | | | | | | | |
| | | | | | | | | | | | | 192 | 46 | 8 | | | | | | | | |
| | 90 | 221 | 37 | | | 35 | 6 | | | 41 | 7 | | | | | | | | | | | |
| | 95 | 228 | 38 | 11 | 2 | | | | | | | 104 | 31 | 5 | | | | | | | | |
| | 100 | 238 | 40 | | | | | | | | | 141 | 42 | 7 | | | | | | | | |
| | | | | | | | | | | | | 192 | 58 | 10 | | | | | | | | |
| | 125 | 268 | 45 | 15 | 2 | 47 | 8 | | | | | | | | | | | | | | | |

| Casing size | | | | | | | | | | | | | | | |
|---|---|---|---|---|---|---|---|---|---|---|---|---|---|---|---|
| 17"OD (13"ID) | 190 | | | | | | | | | | | | | 25 | 4 |
| 18"OD | 20 | 76 | 11 | | | | | | | | 12 | 22 | 3 | | |
| | 30 | 109 | 16 | | | | | | | | 199 | 36 | 5 | | |
| | 40 | 137 | 20 | | | | | | | | 336 | 61 | 9 | | |
| | 60 | 187 | 28 | 7 | 1 | 28 | 4 | 18 | 40 | 3 | 6 | 121 | 29 | 4 | |
| | 80 | 228 | 34 | | | | | | | | 199 | 48 | 7 | | |
| | 95 | 255 | 38 | 11 | 2 | 41 | 6 | | | | 336 | 81 | 12 | | |
| | 100 | 263 | 39 | | | | | | | | 121 | 36 | 5 | | |
| | 125 | 236 | 35 | 16 | 2 | 55 | 8 | 88 | 13 | | 199 | 60 | 9 | | |
| 21"OD (16"ID) | 190 | | 30 | | | | | | | | 336 | 101 | 15 | | |
| 23"OD (18"ID) | 190 | | 30 | | | | | | | | | | | 4 | 4 |

This chart is to give the reader a representative guide to typical open areas for various types of screens. The actual open area may vary somewhat above or below these values depending on the materials used in the screen construction, the collapse and column strength requirements, and the manufacturing techniques. Screen manufacturers should be contacted for specific open area data.

\* Diameter of 4" screen is 4.5" OD.

† Screen diameters are 4.5" OD, 6.5" OD, and 12.5" OD

‡ Screen diameters are 4.5" OD, 6.5" OD, 12" OD, and 16" OD.

## Volume of a well (modified from Driscoll, 1986)

| Diameter of casing or hole (in.) | US gallons per foot of depth | Imp. gallons per foot of depth | Cubic feet per foot of depth | Liters per meter of depth | Cubin meters per meter of depth |
|---|---|---|---|---|---|
| 1 | 0.041 | 0.034 | 0.0055 | 0.509 | $0.509 \times 10^{-3}$ |
| 1 ½ | 0.092 | 0.077 | 0.0123 | 1.142 | $1.142 \times 10^{-3}$ |
| 2 | 0.163 | 0.135 | 0.0218 | 2.024 | $2.024 \times 10^{-3}$ |
| 2 ½ | 0.255 | 0.212 | 0.0341 | 3.167 | $3.167 \times 10^{-3}$ |
| 3 | 0.367 | 0.306 | 0.0491 | 4.558 | $4.558 \times 10^{-3}$ |
| 3 ½ | 0.500 | 0.416 | 0.0668 | 6.209 | $6.209 \times 10^{-3}$ |
| 4 | 0.653 | 0.544 | 0.0873 | 8.110 | $8.110 \times 10^{-3}$ |
| 4 ½ | 0.826 | 0.689 | 0.1104 | 10.26 | $10.26 \times 10^{-3}$ |
| 5 | 1.020 | 0.849 | 0.1364 | 12.67 | $12.67 \times 10^{-3}$ |
| 5 ½ | 1.234 | 1.027 | 0.1650 | 15.33 | $15.33 \times 10^{-3}$ |
| 6 | 1.469 | 1.223 | 0.1963 | 18.24 | $18.24 \times 10^{-3}$ |
| 7 | 2.000 | 1.665 | 0.2673 | 24.84 | $24.84 \times 10^{-3}$ |
| 8 | 2.611 | 2.174 | 0.3491 | 32.43 | $32.43 \times 10^{-3}$ |
| 9 | 3.305 | 2.752 | 0.4418 | 41.04 | $41.04 \times 10^{-3}$ |
| 10 | 4.080 | 3.397 | 0.5454 | 50.67 | $50.67 \times 10^{-3}$ |
| 11 | 4.937 | 4.111 | 0.6600 | 61.31 | $61.31 \times 10^{-3}$ |
| 12 | 5.875 | 4.892 | 0.7854 | 72.96 | $72.96 \times 10^{-3}$ |
| 14 | 8.000 | 6.662 | 1.069 | 99.35 | $99.35 \times 10^{-3}$ |
| 16 | 10.44 | 8.693 | 1.396 | 129.65 | $129.65 \times 10^{-3}$ |
| 18 | 13.22 | 11.01 | 1.767 | 164.18 | $164.18 \times 10^{-3}$ |
| 20 | 16.32 | 13.59 | 2.182 | 202.68 | $202.68 \times 10^{-3}$ |
| 22 | 19.75 | 16.45 | 2.640 | 245.28 | $245.28 \times 10^{-3}$ |
| 24 | 23.50 | 19.57 | 3.142 | 291.85 | $291.85 \times 10^{-3}$ |
| 26 | 27.58 | 22.97 | 3.687 | 342.52 | $342.52 \times 10^{-3}$ |
| 28 | 32.00 | 26.65 | 4.276 | 397.41 | $397.41 \times 10^{-3}$ |
| 30 | 36.72 | 30.58 | 4.909 | 456.02 | $456.02 \times 10^{-3}$ |
| 32 | 41.78 | 34.79 | 5.585 | 518.87 | $518.87 \times 10^{-3}$ |
| 34 | 47.16 | 39.27 | 6.305 | 585.68 | $585.68 \times 10^{-3}$ |
| 36 | 52.88 | 44.03 | 7.069 | 656.72 | $656.72 \times 10^{-3}$ |

1 US gallon = 3.785 L.
1 US gallon = 0.833 imperial gallons.
1 imperial gallon = 4.55 L.
1 US gallon water weighs 8.33 lbs = 3.785 K.
1 L water weighs 1 k = 2.205 lbs.
1 gallon per foot of depth = 1.2419 liters per foot of depth.
1 gallon per meter of depth = $12.419 \times 10^{-3}$ cubic meters per meter depth of water.

**Typical water quality values in rivers and streams. (Maidment, 1993. Reprinted with permission from McGraw-Hill, Inc.)**

| Water quality parameter | Typical value | Range of values observed[a] | Units | Alternative units[b] |
|---|---|---|---|---|
| Temperature | Variable | 0–30 | °C | °Fahrenheit<br>Kelvin[c]<br>°Rankine |
| pH | 4.5–8.5 | 1–9 | pH units<br>$-\log[H^+]$ | |
| Dissolved oxygen ($O_2$) | 3–9 | 0–19 | mg/L | % saturation[d] |
| Total nitrogen (N) | 0.1–10 | 0.004–> 100 | mg/L | millimoles/L |
| Organic nitrogen | 0.1–9 | <0.2–20 | mg/L | |
| Ammonia ($NH_3$-N) | 0.01–10 | <0.01–45 | mg N/L | mg $NH_3$/L[e]<br>moles/L |
| Nitrite ($NO_2$-N) | 0.01–0.5 | <0.002–10 | mg N/L | mg $NO_2$/L[f] |
| Nitrate ($NO_3$-N) | 0.23 | 0.01–250 | mg N/L | mg $NO_3$/L[g]<br>moles/L |
| Nitrogen gas ($N_2$) | 0–18.4 | | mL/L | |
| Total phosphorus (P) | 0.02–6 | 0.01–30 | mg P/L | moles/L |
| Orthophosphate ($PO_4$) | 0.01–0.5 | <0.01–14 | mg P/L | mg $PO_4$/L<br>moles/L |
| Total organic carbon (C) | 1–10 | 0.01–40 | mg C/L | moles/L |
| Dissolved organic carbon (C) | 1–6 | 0.3–32 | mg C/L | moles/L |
| Volatile organic carbon | 0.05 | | mg C/L | |
| Total organic matter | 2–20 | 0.02–80 | mg/L | |
| Inorganic carbon | 50 | 5–250 | mg $CaCO_3$/L | mg $HCO_3$/L<br>moles/L<br>meq/L |
| Carbon dioxide ($CO_2$) | 0–5 | 0–50 | mg $CO_2$/L | |
| Alkalinity (as $CaCO_3$) | 150 | 5–250 | mg $CaCO_3$/L | mg $HCO_3$/L<br>moles/L<br>meq/L |
| Acidity (as $CaCO_3$) | | 2.8–23.3 | meq/L | |
| Biochemical oxygen demand ($BOD_5$) | 2–15 | <2–65 | mg/L | |
| Chemical oxygen demand (COD) | | <2–100 | mg/L | |
| Hardness (as $CaCO_3$) | 47–54 | 1–1,000 | mg $CaCO_3$/L | |
| Color | 1–10 | 0–500 | Color units | |
| Turbidity | | 0–3 | NTU[h] | |
| Specific conductance | 70 | 40–1,500[i] | µS/cm at 25°C | mS/m at 25°C[j] |
| Dissolved solids (total) | 73–89 | 5–317 | mg/L | |
| Suspended solids | 10–110 | 0.3–50,000 | mg/L | |
| Total solids | | 20–1,000 | mg/L | |
| Cyanide | 1–4 | | µg/L | |
| Phenol | <1 | <1–6 | µg/L | |
| Chloride ($Cl^-$) | 8 | ≈0–158,000 | mg/L | moles/L |
| Chlorine (Cl) | | 0.5–2 | mg/L | moles/L |
| Iron (Fe) | 0.04 | 0.01–2000 | mg/L | |
| Manganese (Mn) | 8.2 | 0.01–2200 | µg/L | |
| Fluoride (F) | 0.1–0.3 | 0–5 | mg/L | |
| Bicarbonate ($HCO_3^-$) (as $CaCO_3$) | 58.4 | ≈0–4467 | mg/L | |
| Carbonate ($CO_3^{2-}$) | ≈0 | | mg/L | |
| Sulfate ($SO_4$) | 8.3–11.2 | 0.13 to 3930 | mg/L | |
| Hydrogen sulfide ($H_2S$) | | <0.25 | µg/L | |
| Calcium (Ca) | 13–15 | ≈0–954 | mg/L | |
| Magnesium (Mg) | 4 | 0–379 | mg/L | |

(Continued)

**Typical water quality values in rivers and streams. (Maidment, 1993. Reprinted with permission from McGraw-Hill, Inc.) (Continued)**

| Water quality parameter | Typical value | Range of values observed[a] | Units | Alternative units[b] |
|---|---|---|---|---|
| Sodium (Na) | 5.1–6.3 | 0.7–1220 | mg/L | |
| Potassium (K) | 1.3–2.3 | 0.02–189 | mg/L | |
| Silicate ($SiO_2$) | 10–14 | 0.15–101 | mg/L | |
| Aluminum (Al) | 50 | 7–4400 | μg/L | |
| Boron (B) | 18 | 0.7–840 | μg/L | |
| Lithium (Li) | 12 | 0.01–400 | μg/L | |
| Rubidium (Rb) | 1.5 | 0.3–7.4 | μg/L | |
| Cesium (Cs) | 0.035 | 0.004–0.2 | μg/L | |
| Beryllium (Be) | 0.013 | 0.01–1 | μg/L | |
| Strontium (Sr) | 60 | 6.3–<1500 | μg/L | |
| Barium (Ba) | 60 | 18–152 | μg/L | |
| Titanium (Ti) | 10 | <0.01–107 | μg/L | |
| Vanadium (V) | 1 | ≈0–171 | μg/L | |
| Chromium (Cr) | 1 | <0.01–84 | μg/L | |
| Molybdenum (Mo) | 0.5 | ≈0–145 | μg/L | |
| Cobalt (Co) | 0.2 | <0.001–15 | μg/L | |
| Nickel (Ni) | 2.2 | 0.001–530 | μg/L | |
| Copper (Cu) | 10 | 0.05–>100 | μg/L | |
| Silver (Ag) | 0.30 | 0.03–2 | μg/L | |
| Zinc (Zn) | 30 | ≈0–<5000 | μg/L | |
| Cadmium (Cd) | ≈0–5 | 0.09–130 | μg/L | |
| Mercury (Hg) | $1^k$ | <0.1–$5^k$ | μg/L | |
| Lead (Pb) | 1 | <0.01–55 | μg/L | |
| Gold (Au) | 0.002 | <0.001–1 | μg/L | |
| Tin (Sn) | ≈0–2.1 | <100 | μg/L | |
| Bismuth (Bi) | <10 | | μg/L | |
| Thallium (Tl) | <10 | | μg/L | |
| Platinum (Pt) | | | μg/L | |
| Arsenic (As) | 2 | <0.10–1100 | μg/L | |
| Antimony (Sb) | 1 | 0.26–5.1 | μg/L | |
| Selenium (Se) | 0.20 | 0.11–2680 | μg/L | |
| Bromine (Br) | 20 | 0.5–4400 | μg/L | |
| Iodine (I) | 2 to 7 | 0.2–100 | μg/L | |
| Uranium (U) | 0.04 | 0.016–47 | μg/L | |
| Thorium (Th) | 0.1 | 0.044–10 | μg/L | |
| Radon (Rn) | 1 | | picocurie/L | |
| Radium (Ra) | $0.3 \times 10^{-7}$ to $4 \times 10^{-7}$ | $0.02 \times 10^{-7}$ to $34 \times 10^{-7}$ | μg/L | |
| Zirconium (Zr) | | 0.05–22.5 | μg/L | |
| Scandium (Sc) | 0.004 | 0.001–0.011 | μg/L | |
| Gallium (Ga) | 0.09 | 0.089–1 | μg/L | |
| Tungsten (W) | 0.03 | | μg/L | |
| Yttrium (Y) | 0.7 | <9.0 | μg/L | |
| Lanthanum (La) | 0.05 | 0.001–1.76 | μg/L | |
| Cerium (Ce) | 0.08 | | μg/L | |
| Praseodymium (Pr) | 0.007 | | μg/L | |
| Neodymium (Nd) | 0.04 | | μg/L | |
| Samarium (Sm) | 0.008 | | μg/L | |
| Europium (Eu) | 0.001 | | μg/L | |
| Gadolinium (Gd) | 0.008 | | μg/L | |
| Terbium (Tb) | 0.001 | | μg/L | |
| Dysprosium (Dy) | 0.05 | | μg/L | |
| Holmium (Ho) | 0.001 | | μg/L | |

**Typical water quality values in rivers and streams. (Maidment, 1993. Reprinted with permission from McGraw-Hill, Inc.) (Continued)**

| Water quality parameter | Typical value | Range of values observed[a] | Units | Alternative units[b] |
|---|---|---|---|---|
| Thulium (Tm) | 0.001 | | μg/L | |
| Ytterbium (Yb) | 0.05 | <0.9 | μg/L | |
| Lutetium (Lu) | 0.001 | | μg/L | |
| Erbium (Er) | 0.004 | | μg/L | |
| Germanium (Ge) | | n.d. to 3.7 | | |

Note: From McCutcheon.[45] Many typical values taken from 1963 estimates by Livingstone[39] that may not fully incorporate more recent pollution effects, including effects of acid rain on increased leaching of geochemicals and from estimates by Martin and Meybeck[43] that excluded polluted rivers in the eastern United States and western Europe. Both investigators concentrated on the loads in large rivers draining directly into the oceans.

[a] High concentrations are observed at the mouth of rivers where mixing with seawater occurs. Concentrations approach maximum values found in seawater. Typical seawater concentrations representing the highest concentrations of many elements in streams are given in Table 11.1.5.

[b] See Table 11.1.2 for some conversions.

[c] SI unit.

[e] % saturation = $(DO\ mg\ L^{-1}/DO_{141}mg\ L^{-1}) \times 100$.

[e] Conversion is 1 mg N/L = 1.2159 mg $NH_3$/L.

[f] Conversion is 1 mg N/L = 3.2845 mg $NO_2$/L.

[g] Conversion is 1 mg N/L = 4.4268 mg $NO_3$/L.

[h] NTU = nephelometric turbidity units. A measure of light scatter in a hydrazine sulfate (concentration of 1.25 mg/L at 1 NTU) and hexamethy-lenetetramine (concentration of 12.5 mg/L at 1 NTU) solution that forms a formazine suspension. Before the development of modern light scattering measuring devices (nephelometers), turbidity was measured with the Jackson candle turbidimeter in Jackson turbidity units (JTU) which are not exactly convertible to NTU but are approximately the same, i.e., 40 NTU = 40 JTU.

[i] Extreme values have been observed in streams dominated by acid mine drainage (Spring Creek, Calif.; 350 μS/cm). Some industrial wastes may contain in excess of 10,000 μS/cm. Seawater has a conductivity of approximately 50,000 μS/cm. Some brines associated with halite may have conductivities of approximately 500,000 μS/cm. Precipitation may have as little as 2 μS/cm.

[j] SI units are millisiemens/m at 25C (abbreviated mS/m). 1 millisiemen/m = 10 microsiemen/cm (abbreviated μS/cm). 1 μS/cm = 1 micromho/cm at 2C (the tpical English unit).

[k] It is suspected that many measurements of Hg are incorrect because of ubiquitous laboratory contamination.

# Primary and secondary drinking water criteria[a]

## Primary Standards

### Inorganic Compounds

| Parameter | Current MCL | Parameter | Current MCL |
|---|---|---|---|
| | All values in mg L$^{-1}$ (except where noted) | | |
| Arsenic | 0.05 | Mercury | 0.02 |
| Asbestos | 7 MFL[b] | Nitrogen, Nitrate | 10 |
| Barium | 2.0 | Nitrogen Nitrite | 1.0 |
| Cadmium | 0.005 | Nitrogen, Nitrate–Nitrite | 10 |
| Chromium | 0.1 | Selenium | 0.05 |
| Fluoride | 4.0 | Silver | Note 2 |
| Lead | Note 1 | Turbidity | 1 – 5 NTU |

**Note 1** – (Lead and Copper):

**Lead Action Level** is exceeded if the concentration of lead in more than 10% of the water samples collected during any monitoring period is greater than 0.015 mg/l.

**Copper Action Level** is exceeded if the concentration of copper in more than 10% of the water samples collected in any monitoring period is greater than 1.3 mg/l.

Exceeding action levels requires the institution of corrosion control techniques and testing for additional corrosion control parameters as defined within the regulations.

**Note 2** – Silver is now a secondary parameter.

### Pesticides, Herbicides and PCBs

| Parameter | Current MCL | Parameter | Current MCL |
|---|---|---|---|
| | All values in mg L$^{-1}$ (except where noted) | | |
| Alachlor | 0.002 | Ethylene dibromide (EDB) | 0.00005 |
| Aldicarb | 0.003 | Heptachlor | 0.0004 |
| Aldicarb Sulfoxide | 0.003 | Heptachlor Epoxide | 0.0002 |
| Aldicarb Sulfone | 0.003 | Lindane | 0.0002 |
| Atrazine | 0.003 | Methoxychlor | 0.04 |
| Carbofuran | 0.04 | PCBs | 0.0005 |
| Chlordane | 0.002 | Pentachlorophenol | 0.001 |
| Dibromochloropropane (DBCP) | 0.0002 | Toxaphene | 0.003 |
| Endrin | 0.0002 | 2,4,5-TP (Silvex) | 0.05 |
| | | 2,4-D | 0.07 |

### Microbiological

| Parameter | | Current MCL |
|---|---|---|
| Coliform | - | Note 1 |
| Giardia lambia | - | Treatment Technique |
| Viruses | - | Treatment Technique |
| HPC | - | Treatment Technique |
| Legioella | - | Treatment Technique |

**Note 1** – compliance based on system collecting 40 or more samples per month, no more than 5% can be total coliform positive. A system collecting less than 40 samples per month, no more than one sample can be total coliform positive.

(Continued)

## Primary and secondary drinking water criteria[a] (Continued)

### Volatile Organic Compounds

| Parameter | Current MCL | Parameter | Current MCL |
|---|---|---|---|
| | All values in mg $L^{-1}$ (exceed where noted) | | |
| Benzene | 0.005 | Ethylbenzene | 0.700 |
| Vinyl chloride | 0.002 | Chlorobenzene | 0.100 |
| Carbon tetrachloride | 0.005 | o-Dichlorobenzene | 0.600 |
| 1,2 Dichloroethane | 0.005 | Styrene | 0.100 |
| Trichloroethylene | 0.005 | Tetrachloroethylene | 0.005 |
| p-Dichlorobenzene | 0.075 | Toluene | 1.0 |
| 1,1-Dichloroethylene | 0.007 | Trans-1,2-Dichloroethylene | 0.100 |
| 1,1,1-Trichloroethane | 0.200 | Xylenes, total | 10.0 |
| Cis-1,2-Dichloroethylene | 0.070 | Trihalomethanes, total | 0.100 |
| 1,2-Dichloropropane | 0.005 | | |

### Radioactivity

| Parameter | Current MCL | Parameter | Current MCL |
|---|---|---|---|
| | All values in pico-Curies per liter (pCi $L^{-1}$) | | |
| Radium 226 & 228 | 5 | Gross Beta | 50 |
| Gross Alpha | 15 | | |

### Secondary Maximum Contaminant Levels

| Parameter | Current MCL | Units | Parameter | Current MCL | Units |
|---|---|---|---|---|---|
| Aluminum | 0.05-0.2 | mg $L^{-1}$ | Manganese | 0.05 | mg $L^{-1}$ |
| Chloride | 250 | mg $L^{-1}$ | Odor | 3 | T.O.N. |
| Color | 15 | APHA[c] | pH | 6.5-8.5 | units |
| Copper | 1.0 | mg $L^{-1}$ | Silver | 0.10 | mg $L^{-1}$ |
| Corrosivity | Non-corrosive | | Sulfate | 250 | mg $L^{-1}$ |
| Fluoride | 2.0 | mg $L^{-1}$ | Total dissolved solids | 500 | mg $L^{-1}$ |
| Foaming Agents | 0.5 | mg $L^{-1}$ | Zinc | 5.0 | mg $L^{-1}$ |
| Iron | 0.3 | mg $L^{-1}$ | | | |

[a] U.S. EPA.
[b] MFL = million fibers/liter.
[c] American Public Health Association.

**Range of water quality parameters in typical streams and rivers. (Maidment, 1993. Reprinted with permission from McGraw-Hill.)**

| Water quality parameter | Typical value | Range of values observed[a] | Units | Alternative units[b] |
|---|---|---|---|---|
| Temperature | Variable | 0–30 | °C | °Fahrenheit Kelvin[c] °Rankine |
| pH | 4.5–8.5 | 1–9 | pH units $-\log[H^+]$ | |
| Dissolved oxygen ($O_2$) | 3–9 | 0–19 | mg/L | % saturation[d] |
| Total nitrogen (N) | 0.1–10 | 0.004–>100 | mg/L | millimoles/L |
| Organic nitrogen | 0.1–9 | < 0.2–20 | mg/L | |
| Ammonia ($NH_3$-N) | 0.01–10 | < 0.01–45 | mg N/L | mg $NH_3$/L[e] moles/L |
| Nitrite ($NO_2$-N) | 0.01–0.5 | < 0.002–10 | mg N/L | mg $NO_2$/l[f] |
| Nitrate ($NO_3$-N) | 0.23 | 0.01–250 | mg N/L | mg $NO_3$/l[g] moles/L |
| Nitrogen gas ($N_2$) | 0–18.4 | | mL/L | |
| Total phosphorus (P) | 0.02–6 | 0.01–30 | mg P/L | moles/L |
| Orthophosphate ($PO_4$) | 0.01–0.5 | < 0.01–14 | mg P/L | mg $PO_4$/L moles/L |
| Total organic carbon (C) | 1–10 | 0.01–40 | mg C/L | moles/L |
| Dissolved organic carbon (C) | 1–6 | 0.3–32 | mg C/L | moles/L |
| Volatile organic carbon | 0.05 | | mg C/L | |
| Total organic matter | 2–20 | 0.02–80 | mg/L | |
| Inorganic carbon | 50 | 5–250 | mg $CaCO_3$/L | mg $HCO_3$/L moles/L meq/L |
| Carbon dioxide ($CO_2$) | 0–5 | 0–50 | mg $CO_2$/L | |
| Alkalinity (as $CaCO_3$) | 150 | 5–250 | mg $CaCO_3$/L | mg $HCO_3$/L moles/L meq/L |
| Acidity (as $CaCO_3$) | | 2.8–23.3 | meq/L | |
| Biochemical oxygen demand ($BOD_5$) | 2–15 | < 2–65 | mg/L | |
| Chemical oxygen demand (COD) | | < 2–100 | mg/L | |
| Hardness (as $CaCO_3$) | 47–54 | 1–1,000 | mg $CaCO_3$/L | |
| Color | 1–10 | 0–500 | Color units | |
| Turbidity | | 0–3 | NTU[h] | |
| Specific conductance | 70 | 40–1,500 | μS/cm at 25°C | mS/m at 25°C[i] |
| Dissolved solids (total) | 73–89 | 5–317 | mg/L | |
| Suspended solids | 10–110 | 0.3–50,000 | mg/L | |
| Total solids | | 20–1,000 | mg/L | |
| Cyanide | 1–4 | < 1–6 | μg/L | |
| Phenol | < 1 | ≈0–158,000 | μg/L | |
| Chloride ($Cl^-$) | 8 | 0.5–2 | mg/L | moles/L |
| Chlorine (Cl) | | 0.01–2000 | mg/L | moles/L |
| Iron (Fe) | 0.04 | 0.01–2200 | mg/L | |
| Manganese (Mn) | 8.2 | 0–5 | μg/L | |
| Fluoride (F) | 0.1–0.3 | ≈0–4467 | mg/L | |
| Bicarbonate ($HCO_3^-$) (as $CaCO_3$) | 58.4 | | mg/L | |
| Carbonate ($CO_3^{2-}$) | ≈0 | 0.13 to 3930 | mg/L | |
| Sulfate ($SO_4$) | 8.3–11.2 | | mg/L | |

| Constituent | | | Units |
|---|---|---|---|
| Hydrogen sulfide (H$_2$S) | | <0.25 | μg/L |
| Calcium (Ca) | 13–15 | ≈0–954 | mg/L |
| Magnesium (Mg) | 4 | 0–379 | mg/L |
| Sodium (Na) | 5.1–6.3 | 0.7–1220 | mg/L |
| Potassium (K) | 1.3–2.3 | 0.02–189 | mg/L |
| Silicate (SiO$_2$) | 10–14 | 0.15–101 | mg/L |
| Aluminum (Al) | 50 | 7–4400 | μg/L |
| Boron (B) | 18 | 0.7–840 | μg/L |
| Lithium (Li) | 12 | 0.01–400 | μg/L |
| Rubidium (Rb) | 1.5 | 0.3–7.4 | μg/L |
| Cesium (Cs) | 0.035 | 0.004–0.2 | μg/L |
| Beryllium (Be) | 0.013 | 0.01–1 | μg/L |
| Strontium (Sr) | 60 | 6.3–<1500 | μg/L |
| Barium (Ba) | 60 | 18–152 | μg/L |
| Titanium (Ti) | 10 | <0.01–107 | μg/L |
| Vanadium (V) | 1 | ≈0–171 | μg/L |
| Chromium (Cr) | 1 | <0.01–84 | μg/L |
| Molybdenum (Mo) | 0.5 | ≈0–145 | μg/L |
| Cobalt (Co) | 0.2 | <0.001–15 | μg/L |
| Nickel (Ni) | 2.2 | 0.001–530 | μg/L |
| Copper (Cu) | 10 | 0.05–>100 | μg/L |
| Silver (Ag) | 0.30 | 0.03–2 | μg/L |
| Zinc (Zn) | 30 | ≈0–<5000 | μg/L |
| Cadmium (Cd) | ≈0–5 | 0.09–130 | μg/L |
| Mercury (Hg) | 1[k] | <0.1–5[k] | μg/L |
| Lead (Pb) | 1 | <0.01–55 | μg/L |
| Gold (Au) | 0.002 | <0.001–1 | μg/L |
| Tin (Sn) | ≈0–2.1 | <100 | μg/L |
| Bismuth (Bi) | <10 | | μg/L |
| Thallium (Tl) | <10 | | μg/L |
| Platinum (Pt) | | | μg/L |
| Arsenic (As) | 2 | <0.10–1100 | μg/L |
| Antimony (Sb) | 1 | 0.26–5.1 | μg/L |
| Selenium (Se) | 0.20 | 0.11–2680 | μg/L |
| Bromine (Br) | 20 | 0.5–4400 | μg/L |
| Iodine (I) | 2 to 7 | 0.2–100 | μg/L |
| Uranium (U) | 0.04 | 0.016–47 | μg/L |
| Thorium (Th) | 0.1 | 0.044–10 | μg/L |
| Radon (Rn) | | | picocurie/L |
| Radium (Ra) | $0.3 \times 10^{-7}$ to $4 \times 10^{-7}$ | $0.02 \times 10^{-7}$ to $34 \times 10^{-7}$ | μg/L |

(Continued)

**Range of water quality parameters in typical streams and rivers. (Maidment, 1993. Reprinted with permission from McGraw-Hill.) (Continued)**

| Water quality parameter | Typical value | Range of values observed[a] | Units | Alternative units[b] |
|---|---|---|---|---|
| Zirconium (Zr) | 0.004 | 0.05–22.5 | µg/L | |
| Scandium (Sc) | 0.09 | 0.001–0.011 | µg/L | |
| Gallium (Ga) | 0.03 | 0.089–1 | µg/L | |
| Tungsten (W) | 0.7 | | µg/L | |
| Yttrium (Y) | 0.05 | <9.0 | µg/L | |
| Lanthanum (La) | 0.08 | 0.001–1.76 | µg/L | |
| Cerium (Ce) | 0.007 | | µg/L | |
| Praseodymium (Pr) | 0.04 | | µg/L | |
| Neodymium (Nd) | 0.008 | | µg/L | |
| Samarium (Sm) | 0.001 | | µg/L | |
| Europium (Eu) | 0.008 | | µg/L | |
| Gadolinium (Gd) | 0.001 | | µg/L | |
| Terbium (Tb) | 0.05 | | µg/L | |
| Dysprosium (Dy) | 0.001 | | µg/L | |
| Holmium (Ho) | 0.001 | | µg/L | |
| Thulium (Tm) | 0.05 | <0.9 | µg/L | |
| Ytterbium (Yb) | 0.001 | | µg/L | |
| Lutetium (Lu) | 0.004 | | µg/L | |
| Erbium (Er) | | n.d. to 3.7 | | |
| Germanium (Ge) | | | | |

Note: From McCutcheon.[45] Many typical values taken from 1963 estimates by Livingstone[39] that may not fully incorporate more recent pollution effects, including effects of acid rain on increased leaching of geochemicals and from estimates by Martin and Meybeck[43] that excluded polluted rivers in the eastern United States and western Europe. Both investigators concentrated on the loads in large rivers draining directly into the oceans.

[a] High concentrations are observed at the mouth of rivers where mixing with seawater occurs. Concentrations approach maximum values found in seawater. Typical seawater concentrations representing the highest concentrations of many elements in streams are given in Table 11.1.5.

[b] See Table 11.1.2 for some conversions.

[c] SI unit.

[d] % saturation = (DO mg $L^{-1}$/DO$_{sat}$ mg $L^{-1}$) × 100.

[e] Conversion is 1 mg N/L = 1.2159 mg $NH_3$/L.

[f] Conversion is 1 mg N/L = 3.2845 mg $NO_2$/L.

[g] Conversion is 1 mg N/L = 4.4268 mg $NO_3$/L.

[h] NTU = nephelometric turbidity units. A measure of light scatter in a hydrazine sulfate (concentration of 1.25 mg/L at 1 NTU) and hexamethylenetetramine (concentration of 12.5 mg/L at 1 NTU) solution that forms a formazine suspension. Before the development of modern light scattering measuring devices (nephelometers), turbidity was measured with the Jackson candle turbidimeter in Jackson turbidity units (JTU) which are not exactly convertible to NTU but are approximately the same, i.e., ~40 NTU = 40 JTU.

[i] Extreme values have been observed in streams dominated by acid mine drainage (Spring Creek, Calif.: 350 µS/cm). Some industrial wastes may contain in excess of 10,000 µS/cm. Seawater has a conductivity of approximately 50,000 µS/cm. Some brines associated with halite may have conductivities of approximately 500,000 µS/cm. Precipitation may have as little as 2 µS/cm.) SI units are millisiemens/m at 25°C (abbreviated mS/m). 1 millisiemen/m = 10 microsiemen/cm (abbreviated µS/cm). 1 µS/cm = 1 micromho/cm at 25°C (the typical English unit).

[k] It is suspected that many measurements of Hg are incorrect because of ubiquitous laboratory contamination.

**Values of W(u,r/B) (from Walton, 1984. Reprinted with permission from National Water Well Association)**

| U | | | | | | r/B | | | | | | |
|---|---|---|---|---|---|---|---|---|---|---|---|---|
| | 0.01 | 0.03 | 0.05 | 0.1 | 0.2 | 0.4 | 0.6 | 0.8 | 1.0 | 1.5 | 2.0 | 2.5 |
| $5.0 \times 10^{-6}$ | $9.44 \times 10^{0}$ | | | | | | | | | | | |
| $1.0 \times 10^{-5}$ | $9.42 \times 10^{0}$ | | | | | | | | | | | |
| $5.0 \times 10^{-5}$ | $8.88 \times 10^{0}$ | $7.25 \times 10^{0}$ | | | | | | | | | | |
| $1.0 \times 10^{-4}$ | $8.40 \times 10^{0}$ | $7.21 \times 10^{0}$ | $6.23 \times 10^{0}$ | | | | | | | | | |
| $5.0 \times 10^{-4}$ | $6.98 \times 10^{0}$ | $6.62 \times 10^{0}$ | $6.08 \times 10^{0}$ | $4.85 \times 10^{0}$ | | | | | | | | |
| $1.0 \times 10^{-3}$ | $6.31 \times 10^{0}$ | $6.12 \times 10^{0}$ | $5.80 \times 10^{0}$ | $4.83 \times 10^{0}$ | $3.51 \times 10^{0}$ | | | | | | | |
| $5.0 \times 10^{-3}$ | $4.72 \times 10^{0}$ | $4.69 \times 10^{0}$ | $4.61 \times 10^{0}$ | $4.30 \times 10^{0}$ | $3.46 \times 10^{0}$ | $2.23 \times 10^{0}$ | | | | | | |
| $1.0 \times 10^{-3}$ | $4.04 \times 10^{0}$ | $4.02 \times 10^{0}$ | $3.98 \times 10^{0}$ | $3.82 \times 10^{0}$ | $3.20 \times 10^{0}$ | $2.23 \times 10^{0}$ | $1.56 \times 10^{0}$ | $1.13 \times 10^{0}$ | | | | |
| $5.0 \times 10^{-2}$ | $2.47 \times 10^{0}$ | $2.46 \times 10^{0}$ | $2.46 \times 10^{0}$ | $2.43 \times 10^{0}$ | $2.31 \times 10^{0}$ | $1.71 \times 10^{0}$ | $1.30 \times 10^{0}$ | $1.12 \times 10^{0}$ | $8.41 \times 10^{-1}$ | | | |
| $1.0 \times 10^{-2}$ | $1.82 \times 10^{0}$ | $1.82 \times 10^{0}$ | $1.82 \times 10^{0}$ | $1.81 \times 10^{0}$ | $1.75 \times 10^{0}$ | $1.56 \times 10^{0}$ | $1.31 \times 10^{0}$ | $1.05 \times 10^{0}$ | $8.19 \times 10^{-1}$ | $4.27 \times 10^{-1}$ | $2.28 \times 10^{-1}$ | |
| $5.0 \times 10^{-1}$ | $5.60 \times 10^{-1}$ | $5.60 \times 10^{-1}$ | $5.60 \times 10^{-1}$ | $5.58 \times 10^{-1}$ | $5.53 \times 10^{-1}$ | $5.34 \times 10^{-1}$ | $5.04 \times 10^{-1}$ | $4.66 \times 10^{-1}$ | $4.21 \times 10^{-1}$ | $3.01 \times 10^{-1}$ | $1.94 \times 10^{-1}$ | $1.17 \times 10^{-1}$ |
| $1.0 \times 10^{-3}$ | $2.19 \times 10^{-1}$ | $2.19 \times 10^{-1}$ | $2.19 \times 10^{-1}$ | $2.19 \times 10^{-1}$ | $2.18 \times 10^{-1}$ | $2.14 \times 10^{-1}$ | $2.07 \times 10^{-1}$ | $2.02 \times 10^{-1}$ | $1.86 \times 10^{-1}$ | $1.51 \times 10^{-1}$ | $1.14 \times 10^{-1}$ | $8.03 \times 10^{-2}$ |
| $5.0 \times 10^{0}$ | $1.10 \times 10^{-3}$ | $1.10 \times 10^{-3}$ | $1.10 \times 10^{-3}$ | $1.10 \times 10^{-3}$ | $1.10 \times 10^{-3}$ | $1.10 \times 10^{-3}$ | $1.10 \times 10^{-3}$ | $1.10 \times 10^{-3}$ | $1.10 \times 10^{-3}$ | $1.00 \times 10^{-3}$ | $1.00 \times 10^{-3}$ | $9.00 \times 10^{-4}$ |

**Atmospheric pressures at various altitudes. (Driscoll, 1986. Reprinted by permission of Johnson Screens/a Weatherford Company.)**

| Altitude | | Barometer reading | | | Atmospheric pressure | | |
|---|---|---|---|---|---|---|---|
| ft | m | in Hg | mm Hg | psia | kPa | ft of water | m of water |
| −1,000 | −305 | 31.10 | 787 | 15.2 | 105 | 34.2 | 10.7 |
| 0 | 0 | 29.9 | 759 | 14.7 | 101 | 33.9 | 10.3 |
| 1,000 | 305 | 28.9 | 734 | 14.2 | 97.9 | 32.8 | 10.0 |
| 2,000 | 610 | 27.8 | 706 | 13.7 | 94.5 | 31.5 | 9.6 |
| 3,000 | 914 | 26.8 | 681 | 13.2 | 91.0 | 30.4 | 9.3 |
| 4,000 | 1,220 | 25.9 | 655 | 12.7 | 87.6 | 29.2 | 8.9 |
| 5,000 | 1,520 | 24.9 | 632 | 12.2 | 84.1 | 28.2 | 8.6 |
| 6,000 | 1,830 | 24.0 | 610 | 11.8 | 81.4 | 27.2 | 8.3 |
| 7,000 | 2,130 | 23.1 | 587 | 11.3 | 77.9 | 26.2 | 8.0 |
| 8,000 | 2,440 | 22.2 | 564 | 10.9 | 75.2 | 25.2 | 7.7 |
| 9,000 | 2,740 | 21.4 | 544 | 10.5 | 72.4 | 24.3 | 7.4 |
| 10,000 | 3,050 | 20.6 | 523 | 10.1 | 69.6 | 13.2 | 7.1 |

**Vapor pressures of water. (Driscoll, 1986. Reprinted by permission of Johnson Screens/a Weatherford Company.)**

| Temperature | | Absolute vapor pressure | | | |
|---|---|---|---|---|---|
| °F | °C | psia | kPa | ft of water | m of water |
| 32 | 0 | 0.09 | 0.62 | 0.2 | 0.06 |
| 40 | 4.4 | 0.12 | 0.83 | 0.28 | 0.09 |
| 50 | 10.0 | 0.18 | 1.24 | 0.41 | 0.13 |
| 60 | 15.6 | 0.26 | 1.79 | 0.59 | 0.18 |
| 70 | 21.1 | 0.36 | 2.48 | 0.89 | 0.27 |
| 80 | 26.7 | 0.51 | 3.52 | 1.2 | 0.37 |
| 90 | 32.2 | 0.70 | 4.83 | 1.6 | 0.49 |
| 100 | 37.8 | 0.95 | 6.55 | 2.2 | 0.67 |
| 110 | 43.3 | 1.28 | 8.83 | 3.0 | 0.91 |
| 120 | 48.9 | 1.69 | 11.7 | 3.9 | 1.19 |
| 130 | 54.4 | 2.22 | 15.3 | 5.0 | 1.52 |
| 140 | 60.0 | 2.89 | 19.9 | 6.8 | 2.07 |
| 150 | 65.6 | 3.72 | 25.6 | 8.8 | 2.68 |
| 160 | 71.1 | 4.74 | 32.7 | 11.2 | 3.41 |
| 170 | 76.7 | 5.99 | 41.3 | 1.2 | 4.33 |
| 180 | 82.2 | 7.51 | 51.8 | 17.8 | 5.43 |
| 190 | 87.8 | 9.34 | 64.4 | 22.3 | 6.80 |
| 200 | 93.3 | 11.5 | 79.3 | 27.6 | 8.41 |
| 210 | 98.9 | 14.1 | 97.2 | 33.9 | 10.30 |

## Values of $W(u)$ for various values of $u$

| u | 1.0 | 2.0 | 3.0 | 4.0 | 5.0 | 6.0 | 7.0 | 8.0 | 9.0 |
|---|---|---|---|---|---|---|---|---|---|
| $\times 1$ | 0.219 | 0.049 | 0.013 | 0.0038 | 0.0011 | 0.00036 | 0.00012 | 0.000038 | 0.000012 |
| $\times 10^{-1}$ | 1.82 | 1.22 | 0.91 | 0.70 | 0.56 | 0.45 | 0.37 | 0.31 | 0.26 |
| $\times 10^{-2}$ | 4.04 | 3.35 | 29.6 | 2.68 | 2.47 | 2.30 | 2.15 | 2.03 | 1.92 |
| $\times 10^{-3}$ | 6.33 | 5.64 | 5.23 | 4.95 | 4.73 | 4.54 | 4.39 | 4.26 | 4.14 |
| $\times 10^{-4}$ | 8.63 | 7.94 | 7.53 | 7.25 | 7.02 | 6.84 | 6.69 | 6.55 | 6.44 |
| $\times 10^{-5}$ | 10.94 | 10.24 | 9.84 | 9.55 | 9.33 | 9.14 | 8.99 | 8.86 | 8.74 |
| $\times 10^{-6}$ | 13.24 | 12.55 | 12.14 | 11.85 | 11.63 | 11.45 | 11.29 | 11.16 | 11.04 |
| $\times 10^{-7}$ | 15.54 | 14.85 | 14.44 | 14.15 | 13.93 | 13.75 | 13.60 | 13.46 | 13.34 |
| $\times 10^{-8}$ | 17.84 | 17.15 | 16.74 | 16.46 | 16.23 | 16.05 | 15.90 | 15.76 | 15.65 |
| $\times 10^{-9}$ | 20.15 | 19.45 | 19.05 | 18.76 | 18.54 | 18.35 | 18.20 | 18.07 | 17.95 |
| $\times 10^{-10}$ | 22.45 | 21.76 | 21.35 | 21.06 | 20.84 | 20.66 | 20.50 | 20.37 | 20.25 |
| $\times 10^{-11}$ | 24.75 | 24.06 | 23.65 | 23.36 | 23.14 | 22.96 | 22.81 | 22.67 | 22.55 |
| $\times 10^{-12}$ | 27.05 | 26.36 | 25.96 | 25.67 | 24.44 | 25.26 | 25.11 | 24.97 | 24.86 |
| $\times 10^{-13}$ | 29.36 | 28.66 | 28.26 | 27.97 | 27.75 | 27.56 | 27.41 | 27.28 | 27.16 |
| $\times 10^{-14}$ | 31.66 | 30.97 | 30.56 | 30.27 | 30.05 | 29.876 | 29.71 | 29.58 | 29.46 |
| $\times 10^{-15}$ | 33.96 | 33.27 | 32.86 | 32.58 | 32.35 | 32.17 | 32.02 | 31.88 | 31.76 |

# Appendix C: Drilling Methods

## Introduction

The authors recognize that drilling methods formed an integral part of hydrogeological investigations and as a result, this book would not be complete without providing some information on drilling methods. In the authors' professional experience, there is no substitute for working with an experienced driller in performing field investigations. Further, as noted in the text, well design is intrinsically related to the drilling equipment available to install the well and knowledge of the site geology, both of which involve working with an experienced driller.

Wells have been used for obtaining water for thousands of years. Wells are mentioned in the Bible. In Europe, sinking of small-diameter wells was first employed in Artois,[1] in the North of France. Within the gardens of a former Dominican convent at Lillerse, in Artois, a deep well has flowed continuously since the year 1126.

In consolidated formations, the drill bit may be constructed of hardened steel, industrial diamonds, or other hard materials to cut the rock. *Diamond drilling* refers to the use of a bit constructed of industrial diamonds to cut a rock core to geologically log the borehole. The rock core is removed from the drill stem assembly using a core barrel attached to a wireline to bring the core to the surface for examination and testing.

Roller bits or tricone bits grind and abrade the rock, which is then washed to the surface. Other drill bits such as drag bits have hardened steel teeth to break up the rock, while carbide-tipped bits are used in very hard formations. **Table App-C1** (modified from Driscoll, 1986) presents a guide for well-drilling techniques.

Drilling fluids are circulated in the hole to cool the bit and lift drill cuttings out of the hole. Typical drill fluids include bentonite mud, water, air, foam, and bipolymer slurries. Conventional circulation of drilling fluids in the borehole is through the drill stem and then up through the annular space between the drill rod and the borehole walls carrying the drill cuttings to the surface. *Reverse-circulation* methods involve circulating the drilling fluid from the annular space and up through the drill stem. The benefits of reverse-circulation methods are the following:

- Large-diameter holes can be drilled.
- Penetration rates are high in unconsolidated deposits.
- Less fluid additives are required to lift the drill cuttings.
- Well development time is reduced over conventional circulation methods.

**Table App-C1** Well-drilling selection guide

| Well drilling selection guide | | | | | | | | |
|---|---|---|---|---|---|---|---|---|
| **Type of formation** | | | | | | | | |
| **Geologic origin** | Igneous and metamorphic | | Sedimentary | | | | | |
| **Examples** | Granite<br>Basalt | Quartzite<br>Gneiss    Schist | Limestone | Sandstone | Shale | Clay | Sand | Gravel |
| **Hardness** | Very hard to hard | | Hard to soft | | | Unconsolidated | | |
| **Drilling methods** | | Air or foam rotary | Air or foam rotary | | | Mud rotary | | |
| **Bit type** | Carbide insert bit | Carbide tooth bits | Steel tooth bits | | | Steel tooth bits | | |
| **Diameter** | Small (4–8 in.) | | Small to Medium (6–12 in.) | | | | | |
| **Depth** | Shallow (50–200 ft) | | Shallow to Deep (50–1000 ft) | | | | | |

---

[1] The word artesian is believed to have been derived from Artois.

The reason well development times are reduced is that the fluids in contact with the borehole walls have lower suspended solids compared to conventional circulation methods. Reverse-circulation methods have increased drill stem handling times over conventional methods, increasing the time and cost to drill the borehole. *Dual-wall reverse-circulation methods* involve a drill stem with an annular space for circulation of drilling fluids. Dual-wall methods result in reduced amounts of drilling fluids and filter cake build up on the borehole wall.

Drilling methods have evolved and are continuing to evolve over time. Presented below is a synopsis of drilling techniques.

## Air Hammer Drilling

This represents a modern version of the centuries-old cable tool method (see below) using an air hammer versus gravity drop of a weighted tool. There are three main air hammer techniques: casing hammer, downhole underreaming hammer, and percussion hammer. Each air hammer method allows the hole to be drilled and cased in one operation.

*Casing hammer* permits drilling and casing to be controlled independently, allowing the operator to coordinate drilling and casing varied and difficult formations. The casing drivers are equipped to drive upward for well abandonment or to reuse the casing.

*Downhole underreaming hammer* is a method well-suited for difficult overburden drilling or bedrock formations where lost circulation may occur. An underreaming bit allows for simultaneous drilling and casing, preventing the hole from collapsing above and behind the drill bit.

*Percussion hammer* is well-suited for formations with large boulders or drilling through landfills where subsurface obstacles are typically encountered. This method advances a dual-wall casing as drilling progresses. Soil, rocks, and cobbles are recovered intact. Reverse-circulation dual-tube percussion drilling has been in use since the 1950s, but has found more use since 1985 for the installation of monitoring wells in environmental-sensitive areas. This method is suited for landfills and other difficult geologic environments where water and drilling muds are not allowed. It enables the recovery of soil, rocks, and cobbles intact. The method uses an above-ground pile hammer, where the drill pipe is driven into the ground rather than being rotated. Compressed air is forced through the annular space between the tubes, returning through the middle tube with the cuttings as illustrated in **Figure App-C1**. Boreholes drilled using this method can be inclined up to an angle of 30° from vertical.

## Bucket Auger/Rotary Bucket Drilling

This method is suited for drilling large-diameter wells (up to 60 in. in diameter) for recovery wells and is also used for construction of caissons. This method has found application for leachate and methane collection wells in landfills.

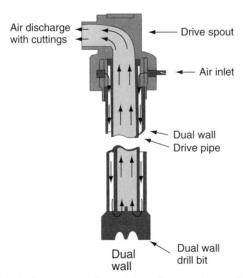

**Figure App-C1** Schematic of dual-tube percussion drilling technique. (Courtesy of Layne, 1900 Shawnee Mission Parkway, Mission Woods, Kansas, USA 66205).

# Cable Tool Drilling

This method of percussion drilling operates by repeatedly lifting and dropping a heavy string of tools in a borehole. Its origins go back approximately 4000 years to ancient China. The reciprocating action of the tools mixes the crushed or loosened particles with water to form a slurry, which is removed using a *dart valve bailer* at intervals as drilling proceeds. The casing is advanced as drilling proceeds. Large-diameter wells can be effectively installed using this method. **Figure App-C2** shows a cable tool drilling rig that would have been in use in the late nineteenth and early twentieth centuries.

# Direct Rotary Drilling

The *direct rotary method* was developed to increase the rate of drilling and to reach greater depths compared to bale tool techniques, such as bucket auger or rotary bucket auger drilling. Direct rotary methods are the most commonly used techniques today. In a most general sense, rotary methods involve rotating a drill stem on to which a bit is attached to cut into the geologic formation. In unconsolidated formations, the rotation speed is typically less than that employed in consolidated formations. In unconsolidated formations, the bit is cooled by moisture and groundwater in the soil. The bit is advanced by the pull-down force that can be applied and the feed rate. The soil augers (*hollow stem* or flight augers; **Table App-C2**) also serve to advance the drill stem into the soil. Reversing the rotation of the drill stem serves to remove the drill cuttings from the hole without removal of the drill stem.

Drilling in rock using rotary methods involves either coring of the rock formation or a grinding action using a tricone bit or an abrasion bit. Drilling fluids include air, water, drilling mud, and synthetic materials. The purpose of the drilling fluid is to cool the bit and return drill cuttings to the surface to keep the hole clear. If the hole is being cored, the hole can be directly logged by examining the rock core. The rock core is retrieved from the core barrel via a wireline. That is, a retrieving tool is sent from the surface that latches onto the core barrel and the drill core is brought to the surface via the wireline. The core barrel is emptied and the empty core barrel is lowered down the hole for the next core run.

## Cable tool drilling

**Figure App-C2** Schematic of cable tool drilling method. (Courtesy of Wisconsin Department of Natural Resources).

**Table App-C2**   CME Auger sizes

**Hollow Stem Augers**

| Inside diameter | Model | Flight O.D. | Auger head size (hole size) | Center rod system |
|---|---|---|---|---|
| $2\,^1/_4''$ (5.7 cm) | Std. | $5\,^5/_8''$ (14.3 cm) | $6\,^1/_4''$ (15.9 cm) | Drill rod |
| $2\,^3/_4''$ (7.0 cm) | Std. | $6\,^1/_8''$ (15.6 cm) | $6\,^3/_4''$ (17.1 cm) | Drill rod |
| $3\,^1/_4''$ (8.3 cm) | Std. | $6\,^5/_8''$ (16.8 cm) | $7\,^1/_4''$ (18.4 cm) | Drill rod or hex rod |
| $3\,^1/_4''$ (8.3 cm) | H.D. | $6\,^5/_8''$ (16.8 cm) | $7\,^1/_4''$ (18.4 cm) | Drill rod or hex rod |
| $3\,^3/_4''$ (9.5 cm) | Std. | $7\,^1/_8''$ (18.1 cm) | $7\,^3/_4''$ (19.7 cm) | Drill rod or hex rod |
| $3\,^3/_4''$ (9.5 cm) | H.D. | $7\,^1/_8''$ (18.1 cm) | $7\,^3/_4''$ (19.7 cm) | Drill rod or hex rod |
| $4\,^1/_4''$ (10.8 cm) | Std. | $7\,^5/_8''$ (19.4 cm) | $8\,^1/_4''$ (21.0 cm) | Drill rod or hex rod |
| $4\,^1/_4''$ (10.8 cm) | H.D. | $8\,^1/_8''$ (20.6 cm) | $9''$ (22.9 cm) | Drill rod or hex rod |
| $4\,^1/_4''$ (10.8 cm) | X.H.D. | $8\,^1/_4''$ (21.0 cm) | $9''$ (22.9 cm) | Drill rod or hex rod |
| $5\,^1/_4''$ (13.3 cm) | H.D. | $9\,^1/_8''$ (23.2 cm) | $10''$ (25.4 cm) | Drill rod or hex rod |
| $6\,^1/_4''$ (15.9 cm) | Std. | $9\,^5/_8''$ (24.4 cm) | $10\,^1/_2''$ (26.7 cm) | Drill rod or hex rod |
| $6\,^1/_4''$ (15.9 cm) | H.D. | $10\,^1/_4''$ (26.0 cm) | $11''$ (27.9 cm) | Drill rod or hex rod |
| $6\,^1/_4''$ (15.9 cm) | X.H.D. | $10\,^1/_4''$ (26.0 cm) | $11''$ (27.9 cm) | Drill rod or hex rod |
| $7\,^1/_4''$ (18.4 cm) | X.H.D. | $11\,^1/_4''$ (28.6 cm) | $12''$ (30.5 cm) | Hex rod |
| $8\,^1/_4''$ (21.0 cm) | H.D. | $12\,^1/_4''$ (31.1 cm) | $13''$ (33.0 cm) | Drill rod or hex rod |
| $10\,^1/_4''$ (26.0 cm) | H.D. | $14''$ (35.6 cm) | $14\,^3/_4''$ (37.5 cm) | Hex rod |
| $12\,^1/_4''$ (31.1 cm) | H.D. | $17\,^1/_4''$ (43.8 cm) | $18\,^1/_2''$ (47.0 cm) | Hex rod |

**1500 Series continuous flight augers**

| | | | | | |
|---|---|---|---|---|---|
| **Hole Size** | $3''$ (7.6 cm) | $3\,^1/_8''$ (7.9 cm) | $4\,^1/_2''$ (11.4 cm) | $5''$ (12.7 cm) | $6''$ (15.2 cm) |
| **Auger O.D.** | $2\,^1/_2''$ (6.4 cm) | $3''$ (7.6 cm) | $4''$ (10.1 cm) | $4\,^1/_2''$ (11.4 cm) | $5\,^1/_2''$ (14 cm) |

**2000 Series continuous flight augers**

| | | | | | |
|---|---|---|---|---|---|
| **Hole Size** | $4\,^1/_2''$ (11.4 cm) | $5''$ (12.7 cm) | $6''$ (15.2 cm) | $6\,^3/_4''$ (17.1 cm) | $8\,^1/_4''$ (21.0 cm) |
| **Auger O.D.** | $4''$ (10.2 cm) | $4\,^1/_2''$ (11.4 cm) | $5\,^1/_2''$ (14.0 cm) | $6''$ (15.2 cm) | $7''$ (17.8 cm) |

**2875 Series continuous flight augers**

| | | | |
|---|---|---|---|
| **Hole Size** | $9''$ (22.9 cm) | $10''$ (25.4 cm) | $12''$ (30.5 cm) |
| **Auger O.D.** | $7\,^7/_8''$ (20.0 cm) | $8\,^7/_8''$ (22.5 cm) | $11''$ (27.9 cm) |

   For holes that are being drilled using roller bits, tricone bits, etc., the hole is logged by monitoring the rock chips being brought up to the surface with the drilling fluid. Typically, where holes are not cored, the hole is subsequently logged using downhole geophysical techniques.

   *Dual-tube reverse-rotary drilling* is a method where a continuous formation sample is recovered through the center section of a double-wall drill pipe. The dual-tube method obtains quick and accurate profiles of subsurface soils and enables the collection of water samples at select intervals, and can also be used to install wells. This method is illustrated in **Figure App-C3**. This method uses a dual wall pipe, top drive rotation, and a side inlet for injecting the drilling fluid. Air, water, or both are forced under pressure down the annulus of the dual-wall pipe the drill bit, where it is directed to the center of the pipe. The drilling fluid, which is under pressure, returns drill cuttings up the center of the pipe [at more than 1200 m s$^{-1}$ (4000 ft s$^{-1}$)]. High rates of drilling are possible [up to 24 m s$^{-1}$ (80 ft s$^{-1}$)]. Further, this method eliminates lost circulation problems and minimizes the need for double casing or cementing the casing in the borehole to maintain circulation or to prevent cross contamination.

## Jet Drilling

This represents a relatively inexpensive method for installing small-diameter wells. A chisel bit or a drive point fixed on a well screen (*well point*) is driven into the ground using a high-velocity stream of water through the bit or the well point. This method is generally limited to installing wells to a depth of 60 m (200 ft). This method works very well for sand and bouldery till deposits.

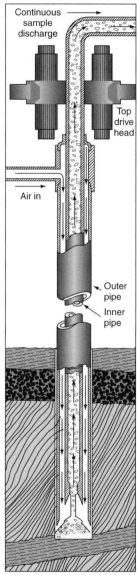

**Figure App-C3** Schematic of dual-tube rotary drilling method. (Courtesy of Layne, 1992 Shawnee Mission Parkway, Mission Woods, Kansas, USA 66205).

## Direct Push or Drive Methods

In unconsolidated formations, boreholes can be advanced using a standard penetration test pneumatic hammer into the ground. A commercial application of this method is known as GeoProbe®. Small-diameter wells and piezometers have been installed using these methods. Since these are small-diameter wells installed in small-diameter boreholes, such installations are considered temporary since well seals cannot be practically installed. Typically this is done for one-time sampling and water level measurements to provide input to design a more permanent monitoring network.

The following are examples of rotary drill rigs manufactured by Central Mine Equipment (CME) of St. Louis, Missouri. This information is provided from specifications provided by CME at the time of preparing this book. Central Mine Equipment reserves the right to substitute or modify the specifications at any time. The capability of any drill rig is dependent upon a number of factors:

- The geologic formation being drilled;
- The condition of the equipment (both mechanical and downhole);
- Modifications made to the drill rig (both as options and after market);
- The diameter of the hole to be drilled; and
- The skill and experience of the driller.

Local drillers should be contacted for all drilling needs to determine rig availability and capability as part of any hydrogeologic investigation or well design project.

# CME - 45C Truck Mounted Drill Rig

### Power
59 horsepower Deutz 187 cubic inch air-cooled 3 cylinder diesel engine

### Rotary Drive
Standard:
    Rotary torque ................................ 3,337 ft. lbs. (4,525 Nm)
    Rotary speed .................................................. 698 rpm max
High torque:
    Rotary torque ................................ 4,730 ft. lbs. (6,414 Nm)
    Rotary speed .................................................. 492 rpm max
High speed:
    Rotary torque ................................ 3,063 ft. lbs. (4,166 Nm)
    Rotary speed .................................................. 760 rpm max
Clutch, heavy duty ............................................ 12 in. (30.5 cm)
Transmission .................................... 4 speed fwd., 1 speed rev.

### Hydraulic Feed System
Retract force ............................................ 19,600 lbs. (8,891 Kg)
Pulldown force ......................................... 13,650 lbs. (6,192 Kg)
Retract rate (max) ......................................... 55 ft./min. (17 m/min)
Feed rate (max) ........................................... 80 ft./min. (24 m/min)
Stroke ...................................................................... 68 in. (173 cm)

Typical truck mounted configuration with optional deck platform and 18' mast.
Dimensions will vary, depending on truck wheel base and all-wheel drive applications.

**Specifications**: Courtesy of Central Mine Equipment, St. Louis, Missouri.

# CME - 45C Track Mounted Drill Rig

## Power
Deutz, 187 cubic inch (3.1 L) 59 horsepower air-cooled 3 cylinder diesel engine

## Carrier
Track width (standard) . . . . . . . . . . . . . . . . . . . .12.6 in. (32 cm)
Average ground bearing pressure . . . . . .5.2 psi (.366 kgf/cm²)
Track width (optional) . . . . . . . . . . . . . . . . . . . . .17.7 in. (45 cm)
Average ground bearing pressure . . . . . .3.7 psi (.260 kgf/cm²)
Suspension . . . . . . . . . . . . . . . . . . . . . . . .triple walking beams
Turning radius . . . . . . . . . . . . . . . . . . . . . . . . . . . . . . .in place
Drive . . . . . . . . . . . . . . .hydraulic motor/planetary wheel drives
Steering . . . . . . . . . . . . . . . . . . .remote radio controlled guidance
Hydraulic front winch . . . . . . . . . . . . . . . .12,000 lb. (5,443 kg)
Auger & rod racks . . . . . . . . . . . . . . . . . . . . . . . . . . . .standard
Tool boxes . . . . . . . . . . . . . . . . . . . . . . . . . . . . . . . . . .standard

## Gradeability
Straight-ahead climb . . . . . . . . . . . . . . . . . . . . . . . .50% grade
Side-hill traverse . . . . . . . . . . . . . . . . . . . . . . . . . . . .36% grade

## Rotary Drive
Standard:
   Rotary torque . . . . . . . . . . . . . . 3,335 foot pounds ( 4,522 Nm)
   Rotary speed . . . . . . . . . . . . . . . . . . . . . . . up to 700 rpm max
High torque:
   Rotary torque . . . . . . . . . . . . . . 4,730 foot pounds (6,413 Nm)
   Rotary speed . . . . . . . . . . . . . . . . . . . . . . . up to 490 rpm max
High speed:
   Rotary torque . . . . . . . . . . . . . . 3,060 foot pounds (4,149 Nm)
   Rotary speed . . . . . . . . . . . . . . . . . . . . . . . up to 760 rpm max
Clutch, heavy duty . . . . . . . . . . . . . . . . . . . . . 12 inch (30.5 cm)
Transmission . . . . . . . . . . . . 4 speed forward, 1 speed reverse

## Hydraulic Feed System
Retract force. . . . . . . . . . . . . . . . . . . 19,600 pounds (8,891 Kg)
Pulldown force . . . . . . . . . . . . . . . . . 13,650 pounds (6,192 Kg)
Retract rate (max) . . . . . . . . . . . . . . 55 feet (17 m) per minute
Feed rate (max). . . . . . . . . . . . . . . . . 80 feet (24 m) per minute
Stroke . . . . . . . . . . . . . . . . . . . . . . . . . . . . . . . 68 inch (173 cm)

## Leveling System
Four jacks, inverted design with chrome-plated piston rods enclosed at all times
Stroke . . . . . . . . . . . . . . . . . . . . . . . . . . . . . . . . . .36 in. (91 cm)

## Weight (Approximate)
Without tools . . . .12,000 lbs. (5,443 kg) - 13,200 lbs. (5,988 kg)

17' 4"

7' 1"

7' 3"

13' 2"

14' 11"

**Specifications:** Courtesy of Central Mine Equipment, St. Louis, Missouri.

# CME - 75 Truck Mounted Drill Rig

**Power**
Cummins 6BT 5.9 turbo charged 6 cylinder diesel engine

**Rotary Drive**
Clutch, heavy duty.................................................13 in. (33 cm)
Transmission .....................................5 speed fwd., 1 speed rev.
Rotary torque ....................................10,000 ft. lbs. (13,560 Nm)
Rotary torque (optional)....................13,000 ft. lbs. (17,628 Nm)
Rotary speed......................................................725 rpm max
Rotary speed (optional) .........................................911 rpm max
Hollow spindle I.D........2 3/4 in. (7 cm ) {3 3/4 in.(9.5 cm) avail.}

**Hydraulic Feed System**
Retract force ..........................................30,000 lbs. (13,608 Kg)
Pulldown force .......................................20,000 lbs. (9,072 Kg)
Hoist rate (max)........................................95 ft./min. (29 m/min)
Feed rate (max)........................................58 ft./min. (18 m/min)
Stroke....................................................................72 in. (1.8 m)

**Typical single rear axle truck configuration with 26' mast and optional deck platform.**
Dimensions will vary, depending on truck wheel base and all-wheel drive or tandem rear axle applications.

**Specifications:** Courtesy of Central Mine Equipment, St. Louis, Missouri.

## CME - 85 Truck Mounted Drill Rig

### Power

Cummins 6BT 359 cu. in. 6 cylinder turbocharged diesel engine

### Rotary Drive

Clutch, heavy-duty ............................ 13 in. (33 cm)
Transmission ....................... 5 speed fwd., 1 speed rev.
Rotary torque, (standard) ......... 20,000 ft. lbs. max. (27120 Nm)
Rotary speed, (standard) ........................ 465 rpm max.
Hollow spindle I.D. ........................ 3 3/4 in. (9.5 cm)

### Hydraulic Feed System

Retract force ......................... 48,000 lbs. (21,773 Kg)
Pulldown force ....................... 28,000 lbs. (12,700 Kg)
Hoist rate (max.) ...................... 95 ft./min. (29 m/min)
Feed rate (max.) ...................... 84 ft./min. (26 m/min)
Stroke ..................................... 75 in. (1.9 m)

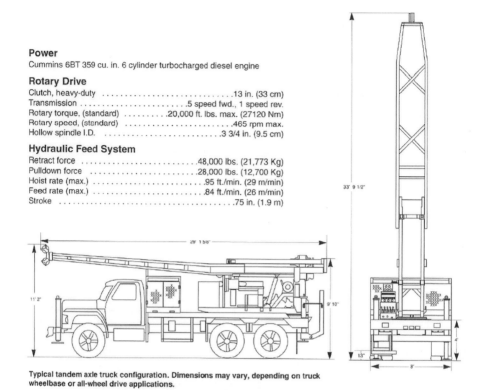

Typical tandem axle truck configuration. Dimensions may vary, depending on truck wheelbase or all-wheel drive applications.

**Specifications:** Courtesy of Central Mine Equipment, St. Louis, Missouri.

510

# CME - 95 Truck Mounted Drill Rig

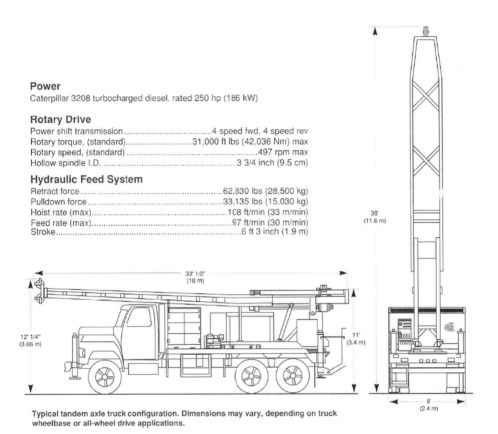

## Power
Caterpillar 3208 turbocharged diesel, rated 250 hp (186 kW)

## Rotary Drive
Power shift transmission .....................................4 speed fwd, 4 speed rev
Rotary torque, (standard)............................31,000 ft lbs (42,036 Nm) max
Rotary speed, (standard) ........................................................497 rpm max
Hollow spindle I.D. ........................................................3 3/4 inch (9.5 cm)

## Hydraulic Feed System
Retract force.............................................................62,830 lbs (28,500 kg)
Pulldown force..........................................................33,135 lbs (15,030 kg)
Hoist rate (max)......................................................108 ft/min (33 m/min)
Feed rate (max).........................................................97 ft/min (30 m/min)
Stroke..............................................................................6 ft 3 inch (1.9 m)

38'
(11.6 m)

33' 1/2"
(10 m)

12' 1/4"
(3.66 m)

11'
(3.4 m)

8'
(2.4 m)

Typical tandem axle truck configuration. Dimensions may vary, depending on truck
wheelbase or all-wheel drive applications.

**Specifications:** Courtesy of Central Mine Equipment, St. Louis, Missouri.

# CME - 50DD Track Mounted Directional Drill Rig

## Engine

John Deere 6068T turbocharged diesel engine rated at 185 hp

## Rack and pinion feed and retract system

Dual range thrust and pullback

Low range mode:
    Force.............................. 50,250 pounds (max)
    Speed ............................ 55 feet per minute (max)
High range mode:
    Force.............................. 25,125 pounds (max)
    Speed ............................ 110 feet per minute (max)
Drilling angle ................................... 0% to 46%

## Rotary drive

Dual range rotation system:
Low range mode ......... 9,100 foot pounds torque, 75 rpm (max)
High range mode ........ 4,550 foot pounds torque, 150 rpm (max)

## Hydraulic break-out system

6" opposing cylinders, open top with quick change jaws

## Drill pipe capacities

Length........................................... 10 feet
Outside diameter ........................... up to 3 1/2 inch

## On-board drilling fluid system

FMC Triplex pump, dual speed with infinite control
    max fluid pressure ........................... 1,500 psi
    max fluid flow rate............................... 75 gpm
60 gallon on-board fresh water storage

## Optional mud motor package

FMC Triplex pump, dual speed with infinite control
    max fluid pressure ........................... 1,500 psi
    max fluid flow rate............................... 137 gpm
Spindle brake to prevent unwanted rotation when running mud motor

## Anchoring system

Primary.......... hydraulic front anchor system with eight 2' augers
Secondary... 80 lb. hydraulic hammer with four drive-in style anchors

## Radio remote controlled tracked carrier

Rubber tracks, steel core reinforced with embedded metal grousers
Individual hydraulic planetary drives
Walking beam suspension with heavy duty, maintenance free, sealed
    idler wheels
Automatic track tensioning system

## On-board crane

400 - 3,200 lb. lifting capacity (depending on boom extension)
Double extension boom
Cable pendant control with manual override
Pressure sensitive lockout based on boom position

## Protective equipment

Zap alert system
Zap alert tester
4' by 10' grounding mat with cable
Emergency shut down switches at operator's console, stake
    down control station and crane control station

## Additional standard equipment

Rod clamp for crane, rod wiper, umbrella with stand,
pressure washer wand with hose, spare jaw teeth for vise,
spare saver sub

## Weight

20,200 lbs (approx.)

## 12 month/1500 hour limited warranty

*Parts and components are warranted to be free from defects in material and workmanship
under normal use and service for a period of 12 months/1500 hours, whichever occurs first.
Labor for replacement of parts and components will be covered for the first 6 months or 750
hours of service, whichever occurs first. Contact CME for details and exclusions.

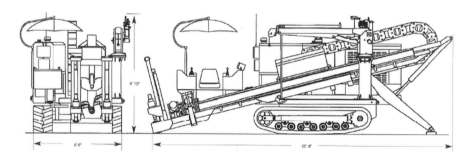

**Specifications:** Courtesy of Central Mine Equipment, St. Louis, Missouri.

# CME - 550X All Terrain Vehicle Mounted Drill Rig

**Power**
*Cummins 4BT, 239 cubic inch (3.9 L) 4 cylinder turbocharged diesel engine*

**Carrier**
*Tire size, single front, dual rear ....44 inch (112 cm) diameter x 18.4 inch (47 cm) x 16.1 (41 cm) x 6-ply*
*Ground bearing pressure (without tools).........................6.4 psi (.45 kgf/cm²)*
*Transmission...........................................5 speed forward, 1 speed reverse*
*Axles (front and rear)................................planetary with no-spin differentials*
*Front axle disconnect..........................................................................standard*
*Steering ...........................................................................hydraulic power*
*Hydraulic front winch ...............................................12,000 pound (5,443 kg)*

**Rotary Drive**
Standard :
*   Rotary torque.......................................8,515 foot pounds (11,545 Nm) max*
*   Rotary speed...........................................................up to 693 rpm max*
High torque:
*   Rotary torque ................................ 10,640 foot pounds (14,426 Nm) max*
*   Rotary speed...........................................................up to 554 rpm max*
High speed:
*   Rotary torque ...................................6,385 foot pounds (8,657 Nm) max*
*   Rotary speed...........................................................up to 924 rpm max*
*Clutch, heavy duty ................................................................13 inch (33 cm)*
*Transmission ...........................................5 speed forward, 1 speed reverse*
*Hollow spindle I.D. ....................................2 3/4 inch (3 3/4 inch available)*

**Hydraulic Feed System**
*Retract force .......................................................28,275 pounds (12,826 Kg)*
*Pulldown force ....................................................18,650 pounds (8,460 Kg)*
*Retract rate (max) ..............................................35 feet (10.7 m) per minute*
*Feed rate (max) ..................................................53 feet (16.2 m) per minute*
*Stroke ................................................................................72 inch (183 cm)*

**Leveling System**
*Three jacks, inverted design with chrome-plated piston rods enclosed at all times*
*Stroke. . . . . . . . . . . . . . . . . . . . . . . . . . . . . . . . . . . . . . 36 inch (91.4 cm)*

**Specifications:** Courtesy of Central Mine Equipment, St. Louis, Missouri.

## CME - 850 Track Mounted Drill Rig

### Power
Cummins 4 BT 3.9 turbocharged 4 cylinder diesel engine
Rated horsepower . . . . . . . . . . . . . . . . . .110 hp @ 2500 rpm

### Carrier
Track width . . . . . . . . . . . . . . . . . . . . . . . .33 in. (84 cm)
Tractive force . . . . . . . . . . . . . . .43,500 lbs. (19,732 kg)
Ground bearing pressure . . . . . . . .2.5 psi (.175 kgf/cm²)
Inside turning radius . . . . . . . . . . . . . . . . . . .11 ft. (3.4 m)
Transmission . . . . . . . . . . . . . .3 speed fwd., 3 speed rev.
Hydraulic front winch . . . . . . . . . . .15,000 lb. (6,804 kg)

### Gradeability
Straight-ahead climb . . . . . . . . . . . . . . . . . . .60% grade
Side-hill traverse . . . . . . . . . . . . . . . . . . . . . .40% grade

### Rotary Drive
Transmission . . . . . . . . . . . . . .5 speed fwd., 1 speed rev.
Rotary torque . . . . . . . . .14,255 ft. lbs. (19,330 Nm) max
Rotary speed . . . . . . . . . . . . . . . . . . . . . . .728 rpm max
Hollow spindle I.D 2 3/4 in. (7 cm) [3 3/4 in. {10 cm} avail.]

### Hydraulic Feed System
Retract force . . . . . . . . . . . . . . . .28,275 lbs. (12,826 kg)
Pulldown force . . . . . . . . . . . . . . .18,650 lbs. (8,460 kg)
Hoist rate (max) . . . . . . . . . . . . . . . . . .90 ft. (27 m)/min.
Feed rate (max) . . . . . . . . . . . . . . . . .58 ft. (18 m)/min.
Stroke . . . . . . . . . . . . . . . . . . . . . . . . .72 in. (183 cm)

### Leveling System
Three jacks, inverted design with chrome-plated piston
rods enclosed at all times
Stroke . . . . . . . . . . . . . . . . . . . . . . . . . . .36 in. (91 cm)

### Dimensions
Length (bumper to bumper) . . . . . . . . .18 ft. 4 in. (5.6 m)
        (including mast) . . . . . . . . . . . .23 ft.6 in. (7.2 m)
Width . . . . . . . . . . . . . . . . . . . . . . . . . . . . . .8 ft. (2.4m)
Height (mast down) . . . . . . . . . . . . . . . . .95 in. (241 cm)
Height (mast up, sheaves to ground) . . . . . . .26 ft. (8 m)
Height of deck . . . . . . . . . . . . . . . . . . . . .42 in. (107 cm)
Ground clearance . . . . . . . . . . . . . . . . .13.5 in. (34 cm)
Approx. Weight
(less drilling tools) . .18,000 - 22,000 lbs. (8,165 - 9,980 kg)

## We make our own tracks

**Specifications:** Courtesy of Central Mine Equipment, St. Louis, Missouri.

# CME - 1050 All Terrain Vehicle Mounted Drill Rig

### Power
Cummins 6BT (5.9 L) 6 cylinder turbocharged diesel engine

### Carrier
Tire sizes: Front..50 inch (127 cm) diameter x 25 inch (63 cm) wide x 10-ply
Rear..50 inch (127 cm) diameter x 31 inch (79 cm) wide x 12-ply
Ground bearing pressure ....................................................8 psi (.56 kg/cm²)
Power-shift transmission ...........................4 speed forward,4 speed reverse
Axles (front and rear) ................................planetary with no-spin differentials
Front axle disconnect...................................................................standard
Steering ...........................................................................hydraulic power
Wheel brakes...................................................................hydraulic power
Parking brake...................................................................................standard
Hydraulic front winch ...........................................15,000 pound (6,804 kg)

### Rotary Drive
Power shift transmission ...........................4 speed forward,4 speed reverse
Rotary torque.....................................24,000 foot pounds (32,544 Nm) max
Rotary speed.................................................................up to 489 rpm max

### Hydraulic Feed System
Retract force ...............................................48,000 pounds (21,773 kg)
Pulldown force ............................................28,000 pounds (12,701 kg)
Retract rate (max)...................................95 feet (29 m) per minute
Feed rate (max) ......................................84 feet (26 m) per minute
Stroke...........................................................................75 inch (190 cm)

### Leveling System
Three jacks, inverted design with chrome-plated piston rods enclosed at all times
Stroke..........................................................................36 inch (91 cm)

### Hydraulic slide bases (standard)
In/out travel...............................................................20 inches (51 cm)
Sideways travel ................................................up to 8 inches (20 cm)

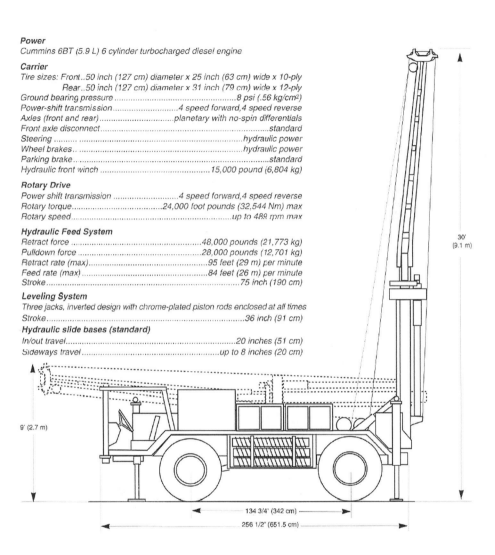

30'
(9.1 m)

9' (2.7 m)

134 3/4' (342 cm)

256 1/2" (651.5 cm)

**Specifications:** Courtesy of Central Mine Equipment, St. Louis, Missouri.

## Bibliography

Appelo, C.A.J. and Postma, D., 1993. *Geochemistry, Groundwater and Pollution*, Balkema, Rotterdam, Netherlands.

Air Force Center for Environmental Excellence, 1995. Technology Transfer Division, Technical Protocol for Implementing Intrinsic Remediation with Long-Term Monitoring for Natural Attenuation of Fuel Contamination Dissolved in Groundwater, Vol. 1. p. B3-2.

American Society of Civil Engineers and Water Pollution Control Federation, Design and Construction of Sanitary Storm Sewers, ASCE Manuals and Reports on Engineering Practice No. 37 and WPCF Manual of Practice No. 9, 1969.

Attewell, P.B. and Farmer, I.W., 1976. *Principles of Engineering Geology*, Chapman and Hall, London. CRC Press/ Taylor & Francis.

Bear, J., 1977. *Dynamics of Fluids on Porous Media*, Elsevier, New York.

Black, P.E., 1996. *Watershed Hydrology*, 2nd ed., Ann Arbor Press, Inc., Chelsea, Michigan.

Bouwer, H., 1989. The Bouwer and Rice slug test – An update. *Ground Water*, Vol. 27, No. 3:304–309.

Bruington, A. E. "Control of sea-water intrusion in a groundwater aquifer" Ground Water, Vol. 7, No. 3, May – June 1969.

Chorley, D.W., Schwartz, F.W. and Crowe, A.S., 1982. Inventory and potential Applications of Groundwater Flow and Chemistry Models. Research Management Division, Alberta Environment, John Wiley & Sons, Inc., New York.

Clark, I.D. and Fritz, P., 1997. *Environmental Isotopes in Hydrogeology*. CRC Press LLC, Boca Raton, Florida.

Cohen, R.M., Mercer, J.W., and Matthews, J., 1993. *DNAPL Site Evaluation*, CRC Press, Boca Raton, Florida.

Constantz, J., 1982. Temperature dependence of unsaturated hydraulic conductivity of two soils. *Soil Science Society of America Journal* 46, No. 3:466–470.

County of Los Angeles, 1989. *Hydrology Manual*, Los Angeles County Department of Public Works, Alhambra, Calif.

Daniel, D.E. and Trautwein, S.J., 1994. *Hydraulic Conductivity and Waste Contaminant Transport in Soil*, American Society for Testing and Materials (ASTM) STP 1142, Philadelphia, PA.

Davis, S.N. and DeWiest, R.J.M., 1966. *Hydrogeology*, John Wiley & Sons, Inc., New York, NY.

Deutsch, W.J., 1997. *Groundwater Geochemistry: Fundamentals and Applications to Contamination*, Lewis Publishers, Boca Raton, Fl.

Domeninco, P.A. and Schwaartz, F.W., 1998. *Physical and Chemical Hydrogeology*, John Wiley & Sons, Inc., New York, NY.

Drever, J.I., 1997. *The Geochemistry of Natural Waters: The Surface and Groundwater Environments*, 3rd ed., Prentice-Hall, New York.

Driscoll, F.G., 1986. *Groundwater and Wells*, 2nd ed., Johnson Filtration Systems, Inc., St. Paul, Minnesota.

Emiliani, C., 1988. *The Scientific Companion – Exploring the Physical World with Facts, Figures, and Formulas*, John Wiley & Sons, Inc., New York, NY.

Erdelyi, M. and Galfi, J., 1988. *Surface and Subsurface Mapping in Hydrogeology*, John Wiley & Sons, Inc., New York, NY.

FEMA, 1984. *Conditions and Criteria for Appeals of Proposed Base Flood Evaluation Determination*, Federal Emergency Management Agency, Washington, D.C.

Ferris, J.G., Knowles, D.B., Brown, R.H., and Stallman, R.W., *Theory of Aquifer Tests*, U.S. Geological Survey Water-Supply Paper 1536-E, 1962.

Fetter, C.W., 1988. *Applied Hydrogeology*, 2nd ed., Merrill Publishing, Company, Columbus, Ohio.

Fetter, C.W., 1994. *Applied Hydrogeology*, 3rd ed., Macmillan College Publishing Comp., New York.

Fetter, C.W., 1999. *Contaminant Hydrogeology*, 2nd ed., Prentice Hall, New Jersey.

Fetter, C.W. Attenuation of waste water elutriated through glacial outwash. *Ground Water* 15, No. 5:365–371.

Fitts, C.R., 2002. *Groundwater Science*, Academic Press, an imprint of Elsevier Science, Ltd., San Diego, Calif.

Fletcher, F.W., 1997. *Basic Hydrogeologic Methods – A field and Laboratory Manual with Microcomputer Applications*, Technomic Publishing Comp., Lancaster PA.

Freeze, R.A. and Cherry, J.A., 1979. *Groundwater*. Prentice-Hall, Inc., New Jersey.

Geological Society of America 'Abstracts with Programs', Springs: Keys to Understanding Geochemical Processes in Aquifers, Vol. 37, No. 7, 2005.

Heath, R.C., 1983. *Basic Ground-Water Hydrology*, U.S. Geological Survey Water-Supply Paper 2220.

Hem, J.D., 1985. *Study and Interpretation of the Chemical Characteristics of Natural Water*, 3rd ed., U.S. Geological Survey Water-Supply Paper 2254.

Hough, B.K. 1969. *Basic Soils Engineering*, 2nd ed., The Ronald Press Company, New York, NY.

Hounslow, A.W., 1995. *Water Quality Data – Analysis and Interpretation*, CRC Press, LLC, Boca Raton, Fl.

Hubbert, M. King, 1953, Entrapment of petroleum under hydrodynamic conditions. *AAPG Bulletin*, Vol. 37, No. 8, pp. 1954–2026.

Javandel, I., and Tsang, C. F. "Capture-Zone Type Curves: A Tool for Aquifer Cleanup", Ground Water, Vol, 24, No. 5, pp. 616–625, 1986.

Jerseyrauskopf, K.B. and Bird, D.K., 1994. *Introduction to Geochemistry*, 3rd ed., McGraw Hill, Inc., New York, NY.

Johansson, T., Adestam, L., 2004. Slug tests in groundwater monitoring wells in soil in the Simpevarp area, Oskarshamn site investigation WSP Sweden Air Base, p. 24.

Keys, W.S., 1989. *Borehole Geophysics Applied to GroundWater Investigations*, National Water Well Association, Dublin, Ohio.

Keys, W.S. and MacCary, L.M., 1971. Application of Borehole Geophysics to Water Resources Investigation. *Technical Techniques of Water-Resources Investigations of the U.S. Geological Survey, Book 2*, U.S. Geological Survey.

Kruseman, G.P. and deRidder, N.A., 1991. *Analysis and Evaluation of Pumping Test Data, 2nd ed.*, International Institute for Land Reclamation and Improvement/ILRI, The Netherlands.

Krynine, D.P. and Judd, W.R., 1957. *Principles of Engineering Geology and Geotechnics*, McGraw-Hill Inc., New York, NY.

Langmuir, D., 1996. *Aqueous Environmental Geochemistry*, Prentice-Hall, New Jersey.

Leopold, L.B., Wolman, M.G., and Miller, J.P., 1992. *Fluvial Processes in Geomorphology*, Dover Publication, Inc., New York.

Linsley, R.K. Jr., Kohler, M.A., and Paulhus, J.L.H., 1982. *Hydrology for Engineers*, McGraw-Hill, Inc., New York, NY.

Lohman, S.W., 1972. *Well Hydraulics*, U.S. Geological Survey Professional Paper 708.

Maidment, D.R., 1993. *Handbook of Hydrology*, McGraw-Hill, Inc., New York, NY.

Mazor, E., 1991. *Chemical and Isotopic Groundwater Hydrology-The applied Approach*, 2nd ed., Halsted Press, a division of John Wiley & Sons, Inc., New York, NY.

McGeary, D., Plummer, C.C., Carlson, D.H., 2001. *Earth Revealed*, 4th edition, McGraw-Hill Comp. Inc. New York, N.Y.

Montana State University, Bozeman. Stream and Riparian Area Management. Streams and Watersheds, p. 5.

Narasimhan, T.N., 1982. *Recent Trends in Hydrogeology*. Special Paper 189. The Geological Society of America, Boulder, CO.

North, F.K., 1985. *Petroleum Geology*. Allen & Unwin, Boston, MA, p. 607.

Nyer, E.K., 1985. *Groundwater Treatment Technology*, Van Nostrand Reinhold Company, New York, NY.

Parker, S.P., 1984. *McGraw-Hill Dictionary of Scientific and Technical Terms*, 3rd ed., McGraw Hill, Inc., New York, NY.

Palmer, C.M., 1996. *Principles of Contaminant Hydrogeology*, 2nd ed., Lewis Publishers an Imprint of CRC Press, Boca Raton, Fl.

Peterson, B.R. *Practical Use and Application of Geophysical Logs for Hydrological and Environmental Projects*, Century Geophysical Corp., Tulsa, Oklahoma.

Plummer, C.C., McGeary, D., and Carlson, D.H., 2001. *Physical Geology*, 8th edition, McGraw-Hill Comp. Inc. New York, N.Y.

Por, F.D., 1972. *Hydrobiological notes on the high-salinity waters of the Sinai Peninsula*. Mar. Biol., 14(2): 111–119.

Powers, J.P., 1981. *Construction Dewatering – A Guide to Theory and Practice*, John Wiley & Sons, New York, N.Y.

Roscoe Moss Company, 1990. *Handbook of Ground Water Development*, John Wiley & Sons, Inc. New York, NY.

Sanders, L.L., 1998. *A Manual of Field Hydrogeology*, Prentice Hall, Inc., New Jersey.

Schmidt, A. R., Yen, B. C., 2001. Proceedings of the 2001 International Symposium on Environmental Hydraulics-Stage-Discharge Relationship in Open Channels.

Schumm, S.A., 1977. *The Fluvial System*, John Wiley & Sons, Inc., New York, NY.

Schwarzenbach, R.P., Gschend, P.M., and Imboden, D.M., 1993. *Environmental Organic Chemistry*, John Wiley & Sons, Inc., New York, NY.

Sen, Z., 1995. *Applied Hydrology*, CRC Press, Inc., Boca Raton, Fl.

Shaw, E.M., 1983. *Hydrology in Practice*, Van Nostrand Reinhold (UK) Co. Ltd

Singh, V.P. (Ed.), 1995. *Environmental Hydrology Series: Water Science and Technology Library*, Vol. 15, Kluwer Academic Publishers. Netherlands.

Stevens, D.B., 1996. *Vadose Zone Hydrology*, Lewis Publishers, Boca Raton, Fl.

Strack, O.D.L., 1989. *Groundwater Mechanics*, Prentice Hall, Englewood Cliffs, NJ

Streeter, V.L. and Wylie, E.B., 1975. *Fluid Mechanics*, 6th ed., McGraw-Hill, Inc., New York, NY.

Stumm, W. and Morgan, J.J., 1981. *Aquatic Chemistry*, 3rd ed., John Wiley & Sons, Inc., New York.

Tank, R.W., 1983. *Environmental Geology*, Oxford University Press, New York.

Tank, R.W., 1983. *Legal Aspects of Geology*, Plenum Press, New York.

Telford, W.M., Geldart, L.P., Sheriff, R.E., and Keys, X.X., 1976. *Applied Geophysics*, Cambridge University Press, London.

Thornthwaite, C.W. Instructions and tables for computing potential evapotranspiration and the water balance. Publications in Climatology, John Hopkins University, Vol. 10, No. 3.

Toth, J. A. 1962. A theory of ground-water motion in small drainage basins in central Alberta, Canada. Journal of Geophysical Research 67, No. 11, 4375–4387.

U.S. Standard Atmosphere, 1962. *National Aeronautics and Space Administration*, U.S. Airforce, and U.S. Weather Bureau.

Walesh, S.G., 1989. *Urban Surface Water Management*, John Wiley & Sons, Inc., New York, NY.

Walton, W.C., 1970. *Groundwater Resource Evaluation*, McGraw-Hill, Inc., New York, NY.

Walton, W.C., 1984. *Practical Aspects of Groundwater Modeling*. National Water Well Association, Dublin, OH.

Weight, W.D. and Sonderegger, J.L., 2001. *Manual of Applied Field Hydrogeology*, McGraw Hill, Inc., New York, NY.

Welenco, 1996. *Water and Environmental Geophysical Well Logs*, 8th ed., Vol. 1, Technical Information and Data, Welenco Inc., Bakersfield, Calif.

Wilson, L.G., Everett, L.G., and Cullen, S.J., 1995. *Handbook of Vadose Zone Characterization and Monitoring*, Lewis Publishers, Boca Raton, Fl.

Wilson, E.W. and Moore, J.E., 1998. *Glossary of Hydrology*. American Geological Institute, Alexandria, VA.

Winter, T. C., Judson, W. H., Franke, O. L., Alley W. M., 1998. Groundwater and surface water a single resource. Circular 1139, US Geological Survey.